Springer Collected Works in Mathematics

More information about this series at http://www.springer.com/series/11104

Karl Menger

Selecta Mathematica II

Edited by
Bert Schweizer, Abe Sklar, Karl Sigmund, Peter Gruber, Edmund Hlawka,
Ludwig Reich, Leopold Schmetterer

Reprint of the 2003 Edition

 Springer

Author
Karl Menger (1902 – 1985)
Illinois Institute of Technology
Chicago
USA

Editors
Bert Schweizer
University of Massachusetts
Amherst, MA
USA

Abe Sklar
Department of Mathematics
Illinois Institute of Technology
Chicago, IL
USA

Karl Sigmund
Institute of Mathematics
University of Vienna
Vienna
Austria

Peter Gruber
Technical University
Vienna
Austria

Edmund Hlawka
Technical University
Vienna
Austria

Ludwig Reich
Institute of Mathematics
University of Graz
Graz
Austria

Leopold Schmetterer
Medical University of Vienna
Vienna
Austria

ISSN 2194-9875
Springer Collected Works in Mathematics
ISBN 978-3-7091-4863-1 (Softcover)

Library of Congress Control Number: 2012954381

Mathematical Subject Classification (2010): 01A75, 60.0X, 27.2X, 30.0X, 09.0

Printed on acid-free paper

This Springer imprint is published by Springer Nature
The registered company is Springer-Verlag GmbH Austria
The registered company address is: Prinz-Eugen-Strasse 8-10, 1040 Wien, Austria

Karl Menger
Selecta Mathematica
Volume 2

Bert Schweizer, Abe Sklar,
Karl Sigmund, Peter Gruber,
Edmund Hlawka, Ludwig Reich,
Leopold Schmetterer (eds.)

Springer-Verlag Wien GmbH

Prof. Dr. Bert Schweizer
Department of Mathematics and Statistics
University of Massachusetts, Lederle Graduate Research Center
Amherst, MA, USA

Prof. Dr. Abe Sklar
Department of Mathematics, Illinois Institute of Technology
Chicago, IL, USA

O. Univ.-Prof. Dr. Karl Sigmund
em. O. Univ.-Prof. Dr. Leopold Schmetterer
Institute of Mathematics, University of Vienna, Austria

O. Univ.-Prof. Dr. DDr. h.c. Peter Gruber
em. O. Univ.-Prof. Dr. DDr. h.c. Edmund Hlawka
Institute of Analysis and Technical Mathematics
Technical University, Vienna, Austria

O. Univ.-Prof. Dr. Ludwig Reich
Institute of Mathematics, University of Graz, Austria

© 2003 Springer-Verlag Wien

Product Liability: The publisher can give no guarantee
for all the information contained in this book

Printed on acid-free and chlorine-free bleached paper
SPIN: 10777146

With 11 Figures

CIP data applied for

ISBN 978-3-211-83734-4 Springer-Verlag Wien New York

Preface

Karl Menger was born in Vienna 100 years ago, the son of the famous Austrian economist, Carl Menger. He was precocious and by the time he was 22 years old had established himself as one of the founders of dimension theory. He went on to an outstanding career, making important contributions to geometry, lattice theory, logic and foundations, mathematical economics and other fields. In 1964 he listed those of his papers which should be included in an eventual collection of his selected works. In the two volumes of "Karl Menger – Selecta Mathematica", we have essentially followed his lead. We have, however, excluded those papers which Menger himself published in his "Selected Papers in Logic and Foundations, Didactics, Economics" (Vienna Circle Collection, Volume 10, Kluwer, Amsterdam, 1979) and those which are included in the recent reprint of the "Ergebnisse eines Mathematischen Kolloquiums" (edited by E. Dierker and Karl Sigmund, Springer-Verlag, Vienna, 1998). Thus, with the publication of these Selecta, virtually all of Karl Menger's major papers will be available in print – as a tribute to his wide-ranging achievements and as a service to the mathematical community.

We wish to thank Dr. Hans Ploss, Karin Picek and Martina Obermaier for their help in preparing these Selecta. We also wish to express our gratitude to the Austrian Academy of Science and to the Department of Mathematics of the Illinois Institute of Technology for their generous financial support.

The Editors

Table of Contents

Menger's Work on Logic and Foundations 1

Van Dalen, D.: Commentary on Menger and Intuitionism 3

Bemerkungen zu Grundlagenfragen I: Über
Verzweigungsmengen, Jahresbericht der Deutschen
Mathematiker-Vereinigung 37 (1928), 213–226 9

Selected Papers on Analysis 23

Reich, L.: Commentary on Karl Menger's
Contributions to Analysis 25

On Cauchy's Integral Theorem in the Real Plane,
Proc. Nat. Acad. Sci. 25 (1939), 621–625 35

On Green's Formula, Proc. Nat. Acad. Sci.
26 (1940), 660–664 41

The Behavior of a Complex Function at Infinity,
Proc. Nat. Acad. Sci. 41 (1955), 512–513 47

A Characterization of Weierstrass Analytic Functions,
Proc. Nat. Acad. Sci. 54 (1965), 1025–1026 49

Analytische Funktionen, Wissenschaftliche
Abhandlungen der Arbeitsgemeinschaft für
Forschung des Landes Nordrhein-Westfalen,
Vol. 33, Köln (1965), 609–612 51

Une caractérisation des fonctions analytiques,
Comptes Rendus Acad. Paris 261 (1965), 4968–4969 55

Weierstrass Analytic Functions, Mathematische
Annalen 167 (1966), 177–194 57

Selected Papers on Length 75

Senechal, L.: Commentary on Menger's
Path Length Papers 77

On Shortest Polygonal Approximations to a Curve,
Reports of Math. Colloq. Notre Dame, Indiana
2 (1940), 33–38 81

Définition intrinsèque de la notion de chemin,
Comptes Rendus Acad. Paris 221 (1945), 739–741 87

Stieltjes Integrals Considered as Lengths, Annales
Société Polonaise Math. 21 (1948), 173–175 91

What Paths Have Length?, Fundamenta Mathematicae
36 (1949), 109–118 95

A Topological Characterization of the Length of Paths,
Proc. Nat. Acad. Sci. 38 (1952), 66–69 105

Selected Papers on Algebra of Analysis 109

Sklar, A.: Commentary on the Algebra of Analysis
and Algebra of Functions 111

Kaiser, H.: The Influence of Menger's Ideas
on the Work of Nöbauer and His School 127

Algebra of Analysis, Report Notre Dame Mathematical
Lectures, Vol. 3, 1944 (pp. I–II, 1–44) 135

Tri-Operational Algebra, Reports of Math. Colloq.
Notre Dame, Indiana 5/6 (1944), 3–10 181

General Algebra of Analysis, Reports of Math. Colloq.
Notre Dame, Indiana 7 (1946), 46–60 189

An Axiomatic Theory of Functions and Fluents, in:
The Axiomatic Method (eds. L. Henkin, P. Suppes
and A. Tarski) North Holland, 1959, 454–473 205

The Algebra of Functions: Past, Present, Future,
Rendiconti di Matematica 20 (1961), 409–430 225

On Substitutive Algebra and Its Syntax, Zeitschrift
für mathematische Logik und Grundlagen
der Mathematik 10 (1964), 81–104 247

Superassociative Systems and Logical Functors,
Mathematische Annalen 157 (1964), 278–295 271

Two Theorems on the Generation of Systems of Functions
(with H. I. Whitlock), Fundamenta Mathematicae
58 (1966), 229–240 289

Selected Papers on Didactics, Variables, and Fluents 301

Schweizer, B., Sklar, A.: Commentary on Didactics,
Variables, Fluents 303

Senechal, L., Schweizer, B.: A Mengerian Tour Along
Caratheodory's Royal Road 317

Calculus – A Modern Approach, Introduction and Chapter 1,
Ginn and Co., Boston, 1955, xi–xvii and 1–18 325

The Mathematics of Elementary Thermodynamics,
American Journal of Physics 18 (1950), 89–103 351

Random Variables from the Point of View of a General
Theory of Variables, Proceedings of the Third Berkeley
Symposium on Mathematical Statistics and Probability
(1954–1955), eds. L. M. Le Cam and O. Neyman, University
of California Press, Berkeley, 1956, Vol. 2, 215–229 367

What are x and y? The Mathematical Gazette 40 (1956),
246–255 383

Rates of Change and Derivatives, Fundamenta Mathematicae
46 (1958), 89–102 393

Selected Papers on Probabilistic Geometry 407

Schweizer, B.: Commentary on Probabilistic Geometry 409

Statistical Metrics, Proc. Nat. Acad. Sci. 28 (1942), 535–537 433

Probabilistic Theories of Relations, Proc. Nat. Acad. Sci.
37 (1951), 178–180 437

Probabilistic Geometry, Proc. Nat. Acad. Sci. 37 (1951),
226–229 441

Ensembles flous et fonctions aléatoires, Comptes Rendus
Acad. Paris 232 (1951), 2001–2003 445

Selected Papers on Group Theory and Algebra 449

Lausch, H.: Commentary on Menger's Work on Algebra 451

Beiträge zur Gruppentheorie I. Über eine Gruppenmetrik,
Mathematische Zeitschrift 33 (1931), 396–418 463

Une théorie axiomatique générale des déterminants,
Comptes Rendus Acad. Paris 234 (1952), 1941–1943 487

Selected Papers on Sociology 491

Leonard, R.: Commentary on Menger's Work on Sociology 493

An Exact Theory of Social Groups and Relations, American
Journal of Sociology 43 (1938), 790–798 501

On Social Groups and Relations, Mathematical Social
Sciences 6 (1983), 13–25 511

Menger's Work on Economics 525

Sigmund, K.: Commentary on Menger's 'Austrian
Marginalism and Mathematical Economics' 527

Austrian Marginalism and Mathematical Economics, in:
Carl Menger and the Austrian School of Economics, eds.
R. O. Hicks and W. Weber, Oxford, 1973, 38–60 531

Miscellaneous 555

Editors Comments 557

Otto Schreier, Nachruf, Monatshefte für Mathematik
37 (1930), 1–6 559

Hans Hahn †, Fundamenta Mathematicae 24 (1935),
317–320 565

Memories of Moritz Schlick, in: Rationality and Science,
ed. by E. T. Gadol, Springer, Wien-New York,
1982, 83–103 569

Introduction to: Ernst Mach: The Science of Mechanics,
6th American Edition, Open Court, La Salle, Illinois
1960, v–xxi 591

Introduction to: Hans Hahn: Empiricism, Logic and
Mathematics, Vienna Circle Collection Vol. 13, Reidel,
Dordrecht, 1980, ix–xviii 609

You Will Like Geometry, A Guide Book for the Illinois
Institute of Technology Geometry Exhibition at the
Museum of Science and Industry, Chicago, 1952, 1–36 619

Table of Contents – Vol. I 655

Table of Contents of Karl Menger: "Selected Papers
in Logic and Foundations, Didactics, Economics" 659

List of Publications – Karl Menger 663

List of Authors

Prof. Dr. *Dirk van Dalen*, Department of Philosophy, Universiteit Utrecht, Heidelberglaan 8, NL-3584 CS Utrecht, The Netherlands

Prof. Dr. *Hans Kaiser*, Institut für Algebra und Computermathematik, Technische Universität Wien, Wiedner Hauptstraße 8–10/118, A-1040 Wien, Austria (h.kaiser@tuwien.ac.at)

Prof. Dr. *Hans Lausch*, School of Mathematical Sciences, Monash University, Victoria 3800, Australia (h.lausch@sci.monash.edu.au)

Prof. Dr. *Robert Leonard*, Département des Sciences Économiques, Université du Québec à Montréal, Succursale Centre-Ville, Montréal, Québec H3C 3P8, Canada

Prof. Dr. *Ludwig Reich*, Institut für Mathematik, Universität Graz, Heinrichstraße 36, A-8010 Graz, Austria (Ludwig.reich@kfunigraz.ac.at)

Prof. Dr. *Bert Schweizer*, Department of Mathematics and Statistics, University of Massachusetts, Amherst, MA 01003-9305, USA (bert@math.umass.edu)

Prof. Dr. *Lester Senechal*, Mount Holyoke, South Hadley, MA 01075, USA (lsenecha@mtholyoke.edu)

Prof. Dr. *Karl Sigmund*, Institut für Mathematik, Universität Wien, Strudlhofgasse 4, A-1090 Wien, Austria (karl.sigmund@univie.ac.at)

Prof. Dr. *Abe Sklar*, 5044 Marine Drive, Chicago, IL 60640, USA

Menger's Work on Logic and Foundations

Commentary on Menger and Intuitionism

Dirk van Dalen

1. Ramification Sets and Spreads

Of all the enigmas with which intuitionism confronted the world, that of "spread" was the most perplexing. Brouwer's first definition [3], 13 lines long, was so exotic that hardly anyone could grasp it. When he returned to the definition in [4, 5], he apologetically remarked, "The circumstances that the definition of spread is long-winded can unfortunately not be helped". The notion indeed suffered from a number of pedagogical deficiencies. The worst one seems to be the choice of terminology; in his German presentation Brouwer used the term "Menge", thereby almost inviting confusion. In order to understand Brouwer's terminology, one has to bear in mind that ever since his dissertation (1907), he sought to characterise the "possible sets" of the constructive universe [1, 2].

The proposals of his dissertation eventually could not meet the requirements of constructiveness that he envisaged. So, only after he saw how to incorporate the "choice" aspect of mathematics did he arrive at the notion that was to become the fundamental one of intuitionistic mathematics. Basically, the prime objects of mathematics were infinite sequences of (previously obtained) mathematical entities subject to a condition that the choices were to be made in (lawlike) subtrees of the universal tree of all finite sequences of natural numbers [16, Ch. 4]. It was this definition that baffled the foundational experts. Fraenkel, for example, in his famous "Zehn Vorlesungen über die Grundlegung der Mengenlehre" [8], mentioned that Brouwer's intuitionism is based on the "principle of constructive set definition", but cautiously avoided any explication. For Brouwer spreads (*Mengen*) were the sets that a constructivist could accept as things that he could generate. The traditional Cantorian sets (i.e., collections of previously obtained objects, specified by a property) were called *species* by Brouwer.

Menger too was puzzled by this notion of spread; but in 1925, during his stay in Amsterdam, he suddenly saw the close parallel between Brouwer's spreads and the analytic sets of Suslin [15, p. 246].

Menger's paper [11] (see [15, pp. 79–87] for a somewhat edited translation) is a report of this similarity. It is a lucid presentation of the two notions "spread" and "analytic set". Menger introduced the auxiliary notion of *ramification set*[1] which contains the notion of analytic set, as given by Suslin, as a special case. The correspondence "Spread – Analytic Set" is convincingly demonstrated. As Menger himself pointed out, he was one of the few who could see this correspondence, as most topologists were not in a position to see the merits of analytic sets, basically because of "their lifelong aversion to set theory", Weyl being a case in point [15, p. 246].

In the Brouwer Archive there is a note referring to Menger's paper, dated June 1927. There are three specific comments on the manuscript. It seems likely that the comments were discussed with Menger; indeed, the points raised by Brouwer were apparently addressed in the published version. The last sentence of Footnote 3 takes into account the first comment of Brouwer. The second comment must have indicated a mistake in a first version of the Brouwerian counterexample on pages 217, 218. The published version is correct.

The content of the third comment is somewhat mysterious. Possibly Menger credited Hurewicz with some specific observation concerning the correspondence. Brouwer wrote "Delete remark about Hurewicz because this observation has been made by many, namely so far by everyone who got to know my "Punktmengen" (spreads) and also the Alexandrov, or Souslin [Brouwer's spelling] sets."

This may perhaps explain what Menger saw as Brouwer's dislike [15, p. 246]. Brouwer, as his comment shows, considered the correspondence obvious.

The problems raised by Menger, if properly pursued, would have shown a weakness in Menger's "analytic-spread" analogy. Brouwer, in his 1925 presentation of intuitionistic mathematics [4], had widened the notion of spread by allowing higher-order choices. This makes the correspondence rather problematic. A modern formulation of Menger's result would mention that the correspondence holds for the holistic universe of choice sequences (in the style of [10]). In particular, Menger's claim (p. 226) that specific notions

---

[1] Verzweigungsmenge; it is interesting to note that Brouwer used a somewhat similar term before he reached his final formulation of spread. Brouwer's term was „vertakkingsagglomeraat", branching agglomerate.

do not enter into the distinction between intuitionistic and formalistic mathematics has to be abandoned.

Menger's correspondence paper was in a way far ahead of its time; it provided a correspondence between two foundational systems. The drawback of being ahead of one's time was that there were no formal systems around to make the correspondence fully precise. Nonetheless, we may note that Menger was most careful to keep track of the applications of the principle of the excluded middle (PEM), a precision that nowadays is routine, but certainly was not at that time. The paper belongs to the happy informal tradition of the twenties; only after the introduction of formal systems could the high standards of precision of Gödel, Gentzen and others be attained.

2. Constructivity

There is another point in the correspondence paper [11] where Menger was ahead of his time. He inquired explicitly into the notion of "constructive". Footnote 12 deals with problems of this sort. Menger considers a number of different issues, e.g., that of the degree of constructiveness of classes of sets more general than the already familiar classes of Borel and analytic sets.

Another interesting question of his concerns a typical constructive issue: how to compare familiar subsets of the Baire space of all infinite sequences of natural numbers relativized to choice sequences or to lawlike sequences. He specifically mentioned the "lawlike" kernel of this Baire space.

Menger continued his lively interest in the notion of "constructive" in two short notes to the Austrian Academy [12, 13] and in Section 8 of his paper [14], which is reprinted in Volume I of these Selecta (pp. 535–538). The main point here was a surprisingly modern, not to say premature, question "What is a constructive number-theoretic function?" As in Footnote 12 of [11], he pointed out that there are many notions that might compete for the predicate "constructive". The central topic of the three notes is constructiveness of number-theoretic functions in an arithmetical setting, that is, without choice sequences. The discussion is in an informal, mathematical tone and no logical notation is used. Since the notes roughly coincide in time with the first papers of Gödel, the historian may wonder whether the topic was discussed between the two. Alas, no information exists on that point.

The first note [12] discusses the issue of uniformity: there are functions f such that for each n the value $f(n)$ is computed by a law and, on the other hand, there are those for which a law that computes $f(n)$ for all n is given in advance. In order to maintain this distinction one has to adopt a classical point of view. Take, for example, the characteristic function of a subset A of N.

Each value is computable, but there need not be an algorithm for the computation of $f(n)$ from n. Note that each total number-theoretic function is computable, at least in the first sense, so the notion is not vacuous, though in this case it is trivial. So much is clear. In a constructive setting, however, the situation is not so simple. In the absence of the principle of the excluded middle it is hard to see how one gets functions of the first kind (locally constructive) which are not of the second kind (globally computable). Even stronger, it is consistent with intuitionistic arithmetic that all number-theoretic functions are recursive (and hence constructive). Even without such meta-mathematical results, one may conclude on the basis of the (Brouwer – Heyting) proof interpretation that if there are local algorithms for f, there is also a global algorithm for f.

This shows that Menger raised a point that could not be answered at the time. One had to wait for recursion theory and for the publication of the proof interpretation [9].

In the second note [13] and in Section 8 of [14] Menger presents a number of rules which derive functions from given functions in a constructive way. The rules are clearly didactic examples; there is no attempt to maximize the set of constructive functions, as in the papers of Gödel, Church, Turing and others. Menger's rules are intended to characterize a notion like "constructive in"; in particular, there are no starting functions.

Section 8 of [14] more or less subsumes the two preceding notes. In addition to the topics sketched above, Menger criticizes Brouwer's claim that equality for reals is undecidable: in logical terminology, $(x = 0)$ or $(x \neq 0)$ is in general not true. Menger's argument runs as follows: consider a Brouwerian number r for which it is unknown whether it coincides with 0 or not. Now r is built by an infinite process of approximation, whereas the equality $r = 0$ has to be decided in a finite time (number of steps). Hence – according to Menger – there are two distinct conceptual levels involved in the argument. On this point Brouwer and Menger would agree; but they would disagree on the foundational conclusion. Nowadays we would side with Brouwer and recognize that there is no basic law in mathematics that tells us that mathematical objects must be "measured" by methods at a single conceptual level. Indeed, the later theory of recursive reals showed that even for a real, defined by a finite number of recursion equations (or a finite number of applications of the rules of recursion theory), the equality $r = 0$ is undecidable. We must conclude that Menger was again forming an opinion on the basis of insufficient empirical evidence. Menger (at that time) evidently could not share Brouwer's foundational views, but he failed to see that – assuming Brouwer's principles – the resulting facts were totally coherent. One might add that Brouwer had done better than what appears from Menger's papers, since in the mid-twenties he had already shown that it is not the case that every real is rational or irrational [6] (see also [7, p. 50]), that is to say, he had established a strong mathematical fact about the undecidability of equality rather than a weak Brouwerian counterexample of the sort "at present we have no right to assert . . .". One should keep in mind, however, that Menger explicitly stuck to arithmetical functions, which means that choice sequences could not play a role; in particular, no continuity arguments would be available.

This brief series of notes can best be considered as interesting suggestions for future research. No new insights emerge from them. This should not surprise us, as Menger ventured into what nowadays is called "recursive analysis", and he did not possess even the most basic tools needed to make progress in this area. Indeed, his guesses mostly turned out to be wrong, e.g., his remark on the existence of discontinuous functions. For example, the traditional step function, given by $f(x) = 0$ for $x \leq 0$ and $f(x) = 1$ for $x > 0$, is, constructively speaking, not a total function since it is not defined at points for which it is undecided whether they are to the left, to the right or "on" 0. There is a strong argument hidden in Brouwer's theorem "all real functions are continuous" [16, Sec. 6.3]; the theorem is also valid for the recursive universe, where it goes by the name "Kreisel-Lacombe-Shoenfield-Tcheitin theorem". For a completely elementary proof, see [17].

Yet, presciently, Menger was right in his main point, i.e., the assertion that there are more varieties of constructiveness than just the intuitionistic one.

References

1. L. E. J. Brouwer: Over de grondslagen der wiskunde, Ph. D. Thesis, Amsterdam (1907). English translation. In: L. E. J. Brouwer, Collected Works, Vol. 1, ed. by A. Heyting, North-Holland, Amsterdam, 1975.
2. L. E. J. Brouwer: Die möglichen Mächtigkeiten, Atti. IV Congr. Intern. Mat. Roma, Vol. 3 (1908) 569–571.
3. L. E. J. Brouwer: Begründung der Mengenlehre unabhängig vom logischen Satz vom ausgeschlossenen Dritten, Erster Teil: Allgemeine Mengenlehre, Koninklijke Akad. van Wetenschappen, Verhandelingen, Sect. 1, No. 12 (1918) 3–43.
4. L. E. J. Brouwer: Zur Begründung der intuitionistischen Mathematik. I. Mathematische Annalen 93 (1925) 244–257.
5. L. E. J. Brouwer: Zur Begründung der intuitionistischen Mathematik. II. Mathematische Annalen 95 (1926) 453–472.
6. L. E. J. Brouwer: Intuitionistische Betrachtungen über den Formalismus, Koninklijke Akad. van Wetenschappen, Proceedings 31 (1928) 374–379.

7. L. E. J. Brouwer: Intuitionismus (Berlin Lectures, 1927), ed. by D. van Dalen, Bibliographisches Institut, Wissenschaftsverlag, Mannheim, 1992.

8. A. Fraenkel: Zehn Vorlesungen über die Grundlegung der Mengenlehre, Teubner, Leipzig, 1927.

9. A. Heyting: Mathematische Grundlagenforschung, Intuitionismus, Beweistheorie, Springer, Berlin, 1934.

10. S. C. Kleene, R. E. Vesley: The Foundations of Intuitionistic Mathematics especially in relation to Recursive Functions, North-Holland, Amsterdam, 1965.

11. K. Menger: Bemerkungen zu Grundlagenfragen. I: Über Verzweigungsmengen, Jahresbericht der Deutschen Mathematiker-Vereinigung 37 (1928) 213–226.

12. K. Menger: Über die sogenannte Konstruktivität bei arithmetischen Definitionen, Akademie der Wissenschaften zu Wien, Anzeiger 67 (1930) 257–258.

13. K. Menger: Über den Konstruktivitätsbegriff. Zweite Mitteilung, Akademie der Wissenschaften zu Wien, Anzeiger 68 (1931) 7–9.

14. K. Menger: Some Applications of Point Set Methods, Annals of Mathematics, II. Ser. 32 (1931) 739–760.

15. K. Menger: Selected Papers in Logic and Foundations, Didactics, Economics, Vienna Circle Collection Vol. 10, D. Reidel Publishing Co., Dordrecht-Boston-London, 1979.

16. A. S. Troelstra, D. van Dalen: Constructivism in Mathematics, I, II. North-Holland, Amsterdam, 1998.

17. M. van Atten, D. van Dalen: Arguments for the continuity principle, Bulletin of Symbolic Logic 8 (2002) 329–347.

Bemerkungen zu Grundlagenfragen.

Von KARL MENGER in Wien.

Eine Reihe kurzer Noten sei einigen einfachen Bemerkungen über Grundlagenfragen gewidmet. Die vier ersten Noten enthalten ein kurzes intuitionistisch-formalistisches Wörterbuch der Mengenlehre, eine neue Aufklärung der mengentheoretischen Paradoxien, einige Bemerkungen über Potenzmengen und eine Axiomatik der endlichen Mengen und der projektiven Verknüpfungsbeziehungen.

I. Über Verzweigungsmengen.

Im folgenden soll auf eine (in der Literatur verwunderlicherweise nirgends erwähnte) enge Beziehung zwischen dem der intuitionistischen Mengenlehre zugrunde liegenden *Brouwerschen Mengenbegriff*[1]) und dem für die formalistische Mengenlehre bedeutungsvollen Begriff der *analytischen Menge*[2]) hingewiesen werden. Als Ausgangspunkt dient dabei der folgende:

1. Begriff der Verzweigungsmenge. Es sei $\mathfrak{D}$ ein System von irgendwelchen Dingen, für welche eine Verknüpfung V definiert ist von der Art, daß jeder geordneten endlichen oder abzählbar unendlichen Folge von Dingen aus $\mathfrak{D}$ ein (jedoch nicht notwendig zu $\mathfrak{D}$ gehöriges) Ding zugeordnet ist. Das der Folge $D^{(1)}, D^{(2)}, \ldots, D^{(k)}, \ldots$ von Dingen aus $\mathfrak{D}$ zugeordnete Ding bezeichnen wir mit $V(D^{(1)}, D^{(2)}, \ldots, D^{(k)}, \ldots)$.

Es sei nun jedem endlichen Komplex von natürlichen Zahlen ein Ding des Systems $\mathfrak{D}$ zugeordnet; das dem Komplex der Zahlen $(n_1, n_2, \ldots, n_k)$ zugeordnete Ding heiße $D_{n_1 n_2 \ldots n_k}$. Ist $\nu = (n_1, n_2, \ldots, n_k, \ldots)$ eine vor-

---

1) Über den Brouwerschen Mengenbegriff vgl. Brouwer, Jahresber. d. D. M.-V. **23**, S. 79 (1914); **28**, S. 204 sq. (1920); Amsterdamer Verslagen **25**, S. 1419 sq. (1917); **29**, S. 798 sq. (1920); Nieuw Archief voor Wiskunde (2) **12**, S. 440 sq. (1918); Amsterdamer Verhandelingen (1) **12**, Nr. 5, S. 3 (1918); Math. Annalen **93**, S. 244 sq. (1924).

2) Über den Begriff der analytischen Menge (bisweilen auch kurz (*A*)-Menge oder Suslinsche Menge genannt) vgl. Suslin, Comptes Rendus **164**, S. 88 (1917), ferner die zusammenfassende Darstellung bei Lusin, Fund. Math. **10**, S. 1 (1926), wo sich auch Angaben über die zahlreichen Abhandlungen über analytische Mengen finden, sowie die zweite Auflage von Hausdorffs Mengenlehre. Aus diesem Buche, in welchem von den mengentheoretischen Errungenschaften des letzten Jahrzehntes gerade die analytischen Mengen (Suslinschen Mengen) eingehende Berücksichtigung finden, stammt übrigens der Begriff der *abstrakten* analytischen Menge, während sonst bloß analytische Punktmengen untersucht wurden.

gegebene unendliche Folge von natürlichen Zahlen, dann betrachten wir die Folge der Dinge $D_{n_1}, D_{n_1 n_2}, \ldots, D_{n_1 n_2 \ldots n_k}, \ldots$ und bezeichnen mit D_ν das Ding $V(D_{n_1}, D_{n_1 n_2}, \ldots, D_{n_1 \ldots n_k}, \ldots)$. Von jedem der Dinge $D_{n_1}, D_{n_1 n_2}, \ldots, D_{n_1 n_2 \ldots n_k}, \ldots$ sagen wir, daß es an der Bestimmung des Dinges D_ν *mitwirkt*. Die Menge aller Dinge D_ν, wo ν alle unendlichen Folgen natürlicher Zahlen durchläuft, bezeichnen wir als eine *Verzweigungsmenge* und, wofern es sich um ihre nähere Bestimmung handelt, als die *durch das System der Dinge* $\{D_{n_1 n_2 \ldots n_k}\}$ $(n_1, n_2, \ldots, n_k, k = 1, 2 \ldots \text{ad inf.})$ *hinsichtlich der Verknüpfung V erzeugte Verzweigungsmenge*. Das System der Dinge $\{D_{n_1 n_2 \ldots n_k}\}$ nennen wir auch das die Verzweigungsmenge D *vermöge der Verknüpfung V erzeugende System*.

2. Analytische Mengen. Sind die Dinge des Systems $\mathfrak{D}$ Mengen und ist V die Durchschnittsbildung, dann ist die Verzweigungsmenge D die Menge aller Durchschnitte $D_\nu = D_{n_1}, D_{n_1 n_2}, \ldots, D_{n_1 n_2 \ldots n_k}, \ldots$, wo $\nu = (n_1, n_2, \ldots, n_k, \ldots)$ die unendlichen Folgen natürlicher Zahlen durchläuft. Die Summe aller Mengen D_ν (also die Summenmenge der Verzweigungsmenge D) ist identisch mit dem, was als die durch das System der Mengen $\{D_{n_1 n_2 \ldots nk}\}$ erzeugte *analytische Menge* bezeichnet wird.

Offenbar kann jede analytische Menge durch *verschiedene* Mengensysteme erzeugt werden. Wählen wir beispielsweise als D_1' eine D_1 als echten Teil enthaltende Menge derart, daß die Menge $D_1' - D_1$ zu allen Mengen $D_{1 n_2 \ldots n_k}$ $(n_2, \ldots, n_k, k = 1, 2, \ldots$ ad inf.) fremd ist, und setzen wir für jede natürliche Zahl $n_1 \neq 1$, $D_n' = D_{n_1}$ und für jeden Zahlenkomplex $n_1, n_2, \ldots, n_k$ $(k \geq 1)$ $D_{n_1 n_2 \ldots n_k}' = D_{n_1 n_2 \ldots n_k}$, so wird durch die Systeme $\{D_{n_1 n_2 \ldots n_k}\}$ und $\{D_{n_1 n_2 \ldots n_k}'\}$ offenbar dieselbe Menge erzeugt, obwohl die beiden Systeme, wegen $D_1 \neq D_1'$, nicht identisch sind.

Oder setzt man

$$D_{2 n_1, n_2, \ldots, n_k}' = D_{n_1, n_2, \ldots, n_k} \atop D_{2 n_1 - 1, n_2, \ldots, n_k}' = D_{1, n_1, n_2, \ldots, n_k} \right\} \quad (n_1, n_2, \ldots, n_k, k = 1, 2, \ldots \text{ad inf.}),$$

so wird durch die beiden Systeme $\{D_{n_1 n_2 \ldots n_k}'\}$ und $\{D_{n_1 n_2 \ldots n_k}\}$, von denen bloß das zweite die Menge D_1 enthält, wie man leicht bestätigt, dieselbe analytische Menge erzeugt.

Setzt man

$$D_{n_1}' = D_{n_1} {\scriptstyle (n_1 = 1, 2, \ldots)} \quad \text{und} \quad D_{n_1 n_2 \ldots n_k}' = D_{n_1} D_{n_1 n_2}, \ldots, D_{n_1 \ldots n_k},$$

so wird durch die beiden (nicht notwendig voneinander verschiedenen) Systeme $\{D_{n_1 n_2 \ldots n_k}\}$ und $\{D_{n_1 n_2 \ldots n_k}'\}$ dieselbe analytische Menge erzeugt. Das System der Mengen $D_{n_1 n_2 \ldots n_k}'$ erfüllt für jeden Zahlenkomplex $n_1, n_2, \ldots, n_k$ die Beziehung, daß $D_{n_1, n_2, \ldots, n_{k-1}, n_k}'$ Teilmenge von $D_{n_1 n_2 \ldots n_{k-1}}$ ist. Ein solches erzeugendes System wollen wir ein *mono-*

tones System nennen, und wir sehen also: *Zu jeder analytischen Menge existiert ein sie erzeugendes monotones System.*

3. Analytische Punktmengen. Wir betrachten nun den Spezialfall, daß die Mengen $D_{n_1 \ldots n_k}$ *abgeschlossene Intervalle* des R_n (des n-dimensionalen euklidischen Raumes) sind, wobei die *leere Menge* zu den abgeschlossenen Intervallen gezählt wird. Eine durch ein System von abgeschlossenen Intervallen des R_n erzeugte analytische Menge heißt *analytische Punktmenge des R_n*. Es möge hier eine Bemerkung eingeschaltet werden, auf die wir (in § 6) zurückkommen werden.

Man kann in etwas weitläufiger, aber prinzipiell einfacher Weise[3]) *ohne Verwendung des Satzes vom ausgeschlossenen Dritten* einsehen, daß zu jeder analytischen Teilmenge des R_n ein sie erzeugendes System $\{ W_{n_1 n_2 \cdots n_k} \}$ existiert von folgenden Eigenschaften:

1. *Jede Menge* $W_{n_1 n_2 \cdots n_k}$ *ist ein n-dimensionaler Würfel mit rationalen Eckpunkten. (Unter diese Würfel wird auch die leere Menge gezählt.)*

2. *Für jeden Zahlenkomplex* $n_1, n_2, \ldots, n_{k-1}, n_k$ *ist der Würfel* $W_{n_1 n_2 \cdots n_{k-1} n_k}$ *ganz im Innern des Würfels* $W_{n_1 n_2 \cdots n_{k-1}}$ *enthalten.*

3. *Für jede unendliche Zahlenfolge* $\nu = (n_1, n_2, \ldots, n_k, \ldots)$ *ist die Menge* $W_\nu = W_{n_1} \cdot W_{n_1 n_2} \ldots W_{n_1 n_2 \cdots n_k} \ldots$ *entweder leer oder sie enthält genau einen Punkt.*

Die durch das Würfelsystem erzeugte Verzweigungsmenge ist die Menge aller dieser höchstens je einen Punkt enthaltenden Mengen W_ν, wo ν alle Folgen natürlicher Zahlen durchläuft. Die durch das Würfelsystem erzeugte analytische Menge ist die Menge aller Punkte p, zu denen eine unendliche Folge natürlicher Zahlen $(n_1, n_2, \ldots, n_k, \ldots)$ existiert, so daß der Punkt p den Durchschnitt der Würfel W_{n_1}, $W_{n_1 n_2}$, $\ldots$, $W_{n_1 n_2 \cdots n_k}$, $\ldots$ bildet.

4. Bemerkungen über Verzweigungsmengen. Wir wenden uns nun wieder den allgemeinen Verzweigungsmengen zu und geben zunächst ein Beispiel von Verzweigungsmengen, welche zu *nicht analytischen* Teilmengen des R_n führen. Es möge das erzeugende System $\{ D_{n_1 n_2 \cdots n_k} \}$ ebenso wie bei den analytischen Punktmengen aus Intervallen des R_n bestehen; es sei aber die Verknüpfung V nicht die Durchschnitts-, sondern die Summenbildung. Die so erzeugte Verzweigungsmenge ist dann nach Definition identisch mit der Menge aller Mengen

---

3) Vgl. analoge Überlegungen z. B. in Hausdorffs Mengenlehre, 2. Aufl., S. 92. Die Bedingung 3) kann in einfachster Weise realisiert werden, indem man alle Würfel des erzeugenden Systems mit k Indizes von einer Kantenlänge $\dfrac{1}{k}$ wählt.

Vgl. über die Durchführung dieser Umformung des erzeugenden Systems Brouwer, Verh. Akad. Amsterdam (1) 12, Nr. 7, S. 5 sq.

$S_\nu = D_{n_1} + D_{n_1 n_2} + \cdots + D_{n_1 n_2 \cdots n_k} + \cdots$, wo $(\nu = n_1, n_2, \ldots, n_k, \ldots)$
alle unendlichen Folgen natürlicher Zahlen durchläuft. Der Durchschnitt
aller dieser Mengen S_ν, also die zur erzeugten Verzweigungsmenge ge-
hörige Durchschnittsmenge, ist dann eine zu einer analytischen Punktmenge komplementäre [also nach bekannten Sätzen[4]) im allgemeinen
keine analytische] Punktmenge.

Sind die Dinge des erzeugenden Systems $\{D_{n_1 n_2 \cdots n_k}\}$ Mengen und
ist die Verknüpfung V die Durchschnittsbildung, dann sind die Elemente der erzeugten Verzweigungsmenge D Mengen D_ν, welche Durchschnitte unendlicher Folgen von Mengen des erzeugenden Systems sind.
Für manche Zwecke (beispielsweise wenn man von der erzeugten Verzweigungsmenge zu ihrer Summenmenge, d. i. zur erzeugten analytischen
Menge, übergeht) sind jene Mengen D_ν, welche leer sind, ohne Interesse, und es empfiehlt sich daher bisweilen, diese leeren Elemente der
Verzweigungsmenge einfach wegzulassen und von der Folge ν natürlicher Zahlen, für welche die entsprechende Menge D_ν leer ist, zu sagen,
sie liefere keinen Beitrag zur erzeugten Verzweigungsmenge D. Wenn
für einen gewissen Zahlenkomplex $n_1, n_2, \ldots, n_k$ die Menge $D_{n_1 n_2 \cdots n_k}$
gleich der *leeren* Menge ist, dann liefert jede unendliche Zahlenfolge,
welche mit den Zahlen $n_1, n_2, \ldots, n_k$ beginnt, im erwähnten Sinn keinen
Beitrag zur erzeugten Verzweigungsmenge.

Auch bei allgemeinen Verzweigungsmengen, deren erzeugendes
System aus Dingen eines Dingbereiches $\mathfrak{D}$ besteht, ist es bisweilen
zweckmäßig, gewisse Elemente zu vernachlässigen und von den diesen
Elementen entsprechenden Zahlenfolgen zu sagen, sie liefern keinen Beitrag zur erzeugten Verzweigungsmenge. Es kann dann sein, daß im
Bereich $\mathfrak{D}$ ein Ding N existiert von folgender Eigenschaft: Ist für einen
Zahlenkomplex $n_1, n_2, \ldots, n_k$ das Ding $D_{n_1 n_2 \cdots n_k}$ gleich dem Ding N, dann
liefert jede mit den Zahlen $n_1, n_2, \ldots, n_k$ beginnende Zahlenfolge keinen Beitrag zur erzeugten Verzweigungsmenge. Falls ein Ding N von dieser Art
existiert, so wollen wir es als *Nullding* im erzeugenden System bezeichnen.

Besteht das erzeugende System aus abgeschlossenen Intervallen des
R_n, ist das System im oben (§ 2) definierten Sinn monoton und ist die
Verknüpfung V die Durchschnittsbildung, dann ist die *leere Menge* ein
Nullding des Systems und, wegen der Monotonie des Systems mit Rück-
sicht auf den Cantorschen Durchschnittssatz, offenbar auch *das einzige
Nullding* des Systems.

---

4) *Das Komplement einer analytischen Menge A ist dann und nur dann eine
analytische Menge, wenn A eine Borelsche Menge ist. Es existieren analytische
Mengen, welche nicht Borelsche Mengen sind.* Vgl. über diese beiden Sätze z. B.
Hausdorffs Mengenlehre, 2. Aufl., S. 181—193.

Unter Verwendung des Satzes vom ausgeschlossenen Dritten kann man nun zeigen: *Aus jedem eine Verzweigungsmenge D erzeugenden System S kann (bloß dadurch, daß eventuell gewisse Elemente von S durch ein gewisses Nullding ersetzt werden) ein dieselbe Verzweigungsmenge D erzeugendes System S' hergeleitet werden, welches die Eigenschaft besitzt, daß jedes von dem erwähnten Nullding verschiedene Element von S' mitwirkt an der Bestimmung von einem zur Verzweigungsmenge D einen Beitrag liefernden Element.*

Sei nämlich zunächst $\{D_{n_1 n_2 \cdots n_k}\}$ irgendein die Verzweigungsmenge D erzeugendes System von Dingen. Für jeden Zahlenkomplex $n_1, n_2, \ldots, n_k$ steht nach dem Satz vom ausgeschlossenen Dritten fest, daß das Ding $D_{n_1 n_2 \cdots n_k}$ entweder an der Bestimmung von einem zur Menge D einen Beitrag liefernden Element mitwirkt oder daß es an der Bestimmung von keinem zur Menge D einen Beitrag liefernden Element mitwirkt. Im ersten Falle setzen wir $D'_{n_1 n_2 \cdots n_k} = D_{n_1 n_2 \cdots n_k}$, im zweiten Falle setzen wir $D'_{n_1 n_2 \cdots n_k}$ gleich dem Nullding N. Das System der Dinge $\{D'_{n_1 n_2 \cdots n_k}\}$ ist dann offenbar ein die Menge D erzeugendes System von der geforderten Art, d. h. so, daß jedes vom Nullding N verschiedene Ding des Systems an der Bestimmung von einem zur Menge D einen Beitrag liefernden Element mitwirkt.

Insbesondere kann also zu jeder *analytischen Punktmenge* unter Verwendung des Satzes vom ausgeschlossenen Dritten die Existenz eines erzeugenden Systems bewiesen werden, welches die oben (§ 3) erwähnten Eigenschaften 1., 2., 3. besitzt und *überdies* die Eigenschaft 4. *Jeder nicht-leere Würfel* $W_{n_1 n_2 \cdots n_k}$ *enthält mindestens einen Punkt der analytischen Menge.* Man hat, um dies einzusehen, bloß auf ein System mit den Eigenschaften 1., 2., 3. das eben beschriebene Verfahren anzuwenden.

Man kann jedoch nach der bekannten Methode der Bezugnahme auf ein gegenwärtig ungelöstes Problem[4a]) analytische Mengen konstruieren, für welche die *Angabe* eines erzeugenden Systems mit der erwähnten Eigenschaft 4. erst durch die Lösung irgendeines Problems möglich wird, d. h. es existieren analytische Mengen, für welche die *Angabe* eines erzeugenden Systems mit der Eigenschaft 4. gegenwärtig unmöglich ist und vielleicht niemals möglich sein wird.

Beispielsweise wird eine solche analytische Punktmenge A erzeugt durch ein System $\{D_{n_1 n_2 \cdots n_k}\}$, welches den folgenden Bedingungen genügt: Jede Menge $D_{n_1 n_2 \cdots n_k}$, für welche einer der Indizes $n_i \neq 1$ ist, ist leer. Jene Menge des Systems, deren Indizes k Einsen sind, bezeichnen'

4a) Vgl. Brouwer, Jahresber. d. D. M. V. **33**, S. 242; Crelles Journal **154**, S. 3, sowie Sierpiński, Fund. Math. **2**, S. 114 f.

wir mit D_{1_k} und treffen nun über diese Menge die Festsetzungen: Die
Menge D_{1_k} soll gleich dem abgeschlossenen Intervall aller reellen Zahlen
zwischen $-\frac{1}{2^k}$ und $+\frac{1}{2^k}$ sein, wofern an der $k$ten Stelle der Dezimalbruchentwicklung der Zahl π keine Sequenz von zehn Siebnern beginnt;
die Menge D_{1_k} soll hingegen leer sein, wofern an der $k$ten Stelle der
Dezimalbruchentwicklung von π eine Siebnerdekade beginnt.

Die so erzeugte analytische Punktmenge A ist leer, falls in der
Dezimalbruchentwicklung von π eine Siebnerdekade existiert; sie enthält
den Nullpunkt des R_1, falls keine Siebnerdekade in der Dezimalbruchentwicklung von π existiert.[5])

Die Angabe eines erzeugenden Systems $\{D'_{n_1 n_2 \cdots n_k}\}$ dieser Menge A,
welches die Eigenschaft 4. besitzt, ist (zum mindesten gegenwärtig) unmöglich. Schon die Mengen D'_n mit *einem* Index von einem derartigen
System können ja nicht angegeben werden. Denn sie müssen durchweg
leer sein, falls A leer ist, d. h. falls in der Dezimalbruchentwicklung
von π eine Siebnerdekade existiert; es muß unter ihnen dagegen ein
nicht-leeres Intervall auftreten, falls die Menge A nicht leer ist, d. h.
falls in der Dezimalbruchentwicklung von π keine Siebnerdekade auf-
tritt. Die Angabe der Mengen D'_n kann also erst nach Beantwortung
der Frage nach der Existenz einer Siebnerdekade in der Dezimalbruchentwicklung von π erfolgen.

Es kann sich in einer Verzweigungsmenge auch ereignen, daß für
eine gewisse Zahlenfolge $\nu = (n_1, n_2, \ldots, n_k, \ldots)$ von einem gewissen k
angefangen alle Dinge $V(D_{n_1}, D_{n_1 n_2}, \ldots, D_{n_1 n_2 \cdots n_k}, D_{n_1 n_2 \cdots n_k \cdots n_{k+l}})$
$(l = 1, 2, \ldots$ ad inf.) und auch das Ding D_ν mit dem Ding $D_{n_1 n_2 \cdots n_k}$ identisch
sind. In diesem Fall können wir sagen, daß das Element D_ν der Verzweigungsmenge bereits durch die endlich vielen Zahlen $n_1, n_2, \ldots, n_k$
bestimmt ist. Sind beispielsweise die Dinge des erzeugenden Systems
Mengen, ist die Verknüpfung V die Durchschnittsbildung, und setzen
wir voraus, daß das vorliegende System im oben definierten Sinn mono-
ton ist, dann ist, falls für eine Zahlenfolge $\nu = (n_1, n_2, \ldots, n_k, \ldots)$ alle
Dinge $D_{n_1}, D_{n_1 n_2}, \ldots, D_{n_1 n_2 \cdots n_k} \cdots$ von einem bestimmten, etwa dem
$k$ten, angefangen, miteinander identisch sind, die Menge $D_\nu = D_{n_1 n_2 \cdots n_k}$;
also ist die Menge D_ν bereits durch die Zahlen $n_1, n_2, \ldots, n_k$ bestimmt.

---

5) Der oben mit Hilfe des Satzes vom ausgeschlossenen Dritten durchgeführte
Beweis für die *Existenz* eines erzeugenden Systems mit der Eigenschaft 4. zeigt in
den Fällen, wo die *Angabe* eines derartigen Systems (zum mindesten gegenwärtig)
unmöglich ist, *daß* (unter Voraussetzung der Widerspruchsfreiheit des Satzes vom
ausgeschlossenen Dritten in der formalistischen Mathematik) *die Annahme der Existenz eines derartigen erzeugenden Systems nicht zu einem Widerspruch führen kann.*

Sind überhaupt alle Mengen, deren Indizes mit den Zahlen $n_1, n_2, \ldots, n_k$ beginnen, mit der Menge $D_{n_1 n_2 \cdots n_k}$ identisch, dann wird für jede mit den Zahlen $n_1, n_2, \ldots, n_k$ beginnende Zahlenfolge ν die Menge D_ν bereits durch die Zahlen $n_1, n_2, \ldots, n_k$ bestimmt.

5. Der Brouwersche Mengenbegriff. Es liege eine unendliche Folge von Zeichenreihen vor (von Zeichenreihen, welche aus endlich vielen Zeichen kombiniert sind, so wie die im dekadischen System geschriebenen Zahlenzeichen aus den zehn Zeichen der Ziffern kombiniert sind). Es liege ferner [6] „ein Gesetz vor, auf Grund dessen, wenn immer wieder eine willkürliche Nummer gewählt wird, jede dieser Wahlen entweder eine bestimmte Zeichenreihe mit oder ohne Beendigung des Prozesses erzeugt oder aber die Hemmung des Prozesses mitsamt der definitiven Vernichtung seines Resultates herbeiführt, wobei für jedes $n > 1$ nach jeder unbeendigten und ungehemmten Folge von $n - 1$ Wahlen wenigstens eine Nummer angegeben werden kann, die, wenn sie als $n$te Nummer gewählt wird, nicht die Hemmung des Prozesses herbeiführt". Jede in dieser Weise von einer unbegrenzten Wahlfolge erzeugte Folge von Zeichenreihen heiße ein „*Element* der Menge". Das vorliegende Gesetz und ebenso die gemeinsame Entstehungsart der Elemente einer Menge nennt Brouwer eine *Menge.*[6]

Es liege also m. a. W. ein Gesetz vor, welches jedem Zahlenkomplex eine gewisse Zeichenreihe zuordnet; die dem Zahlenkomplex $n_1, n_2, \ldots, n_k$ zugeordnete Zeichenreihe heiße $D_{n_1 n_2 \cdots n_k}$. Element der durch dieses Gesetz erzeugten Brouwerschen Menge ist jede in der beschriebenen Weise erzeugte unendliche Folge von Zeichenreihen; d. h. wenn $\nu = (n_1, n_2, \ldots, n_k, \ldots)$ irgendeine unendliche Folge natürlicher Zahlen ist, so ist die Folge von Zeichenreihen $D_\nu = (D_{n_1}, D_{n_1 n_2}, \ldots, D_{n_1 n_2 \cdots n_k}, \cdots)$ Element der Menge. Es ist also m. a. W. ein Ding dann und nur dann Element der durch das angeführte Gesetz gegebenen Brouwerschen Menge, wenn es in unserer obigen Terminologie Element ist von der durch das System der Dinge (Zeichenreihen $\{ D_{n_1 n_2 \cdots n_k} \}$) erzeugten *Verzweigungsmenge*, wobei die Verknüpfung V der Dinge des erzeugenden Systems *im einfachen Aneinanderschreiben zu einer Folge von Dingen* (Folge von Zeichenreihen) besteht. Die „Hemmung des Prozesses mitsamt der definitiven Vernichtung seines Resultates" für den Zahlenkomplex $n_1, n_2, \ldots, n_k$ entspricht dem Auftreten des Nulldinges des erzeugenden Systems an der Stelle von $D_{n_1 n_2 \cdots n_k}$ in unserer obigen Terminologie. „Beendigung des Prozesses" für einen Zahlenkomplex $n_1, n_2, \ldots, n_k$ entspricht dem oben erwähnten Fall, daß für alle mit den Zahlen $n_1, n_2, \ldots, n_k$ be-

---

6) l. c. 1).

ginnenden Zahlenfolgen die Folge von Zeichenreihen ν bereits durch Zahlen $n_1, n_2, \ldots, n_k$ bestimmt ist.

Nun sehen wir aus der angeführten Definition, daß Brouwer an ein Mengengesetz noch eine Forderung stellt: Ist die dem Zahlenkomplex $n_1, n_2, \ldots, n_{k-1}$ entsprechende Zeichenreihe $D_{n_1 n_2 \cdots n_{k-1}}$ des erzeugenden Systems nicht das Nullding des erzeugenden Systems, dann soll das Gesetz mindestens eine Zahl n_k angeben, so daß auch die Zeichenreihe $D_{n_1 n_2 \cdots n_{k-1} n_k}$ nicht das Nullding des erzeugenden Systems ist. M. a. W. Brouwer fordert von einem Mengengesetz, daß es die Sicherheit gewähre, daß jede vom Nullding des erzeugenden Systems verschiedene Zeichenreihe mitwirkt an der Bestimmung von mindestens einem Element der Menge, welches einen Beitrag zur Menge liefert. Es soll nicht möglich sein, daß etwa, wenn D_1 eine vom Nullding verschiedene Zeichenreihe ist, an Stelle aller Zeichenreihen mit zwei Indizes, von denen der erste 1 ist, durchwegs Nulldinge stehen, so daß keine mit 1 beginnende Zahlenfolge ν existiert, welche einen Beitrag zur erzeugten Menge liefert.

So viel zur Erläuterung von Brouwers Definition. Zugleich ergibt sich aus dem Gesagten ihre Beziehung zum Begriff der analytischen Mengen. Indem wir von einigen terminologischen Unterschieden absehen, können wir nämlich offenbar sagen: *Die Brouwerschen Mengen sind identisch mit Verzweigungsmengen, in deren erzeugendem System jedes vom Nullding verschiedene Ding an der Bestimmung von mindestens einem Element mitwirkt.*

Wir haben gesehen, daß unter wesentlicher Verwendung des Satzes vom ausgeschlossenen Dritten bewiesen werden kann, daß zu jeder Verzweigungsmenge ein sie erzeugendes System existiert, in welchem jedes vom Nullding verschiedene Ding mitwirkt an der Bestimmung eines zur Menge einen Beitrag liefernden Elementes. Indem wir wieder von terminologischen Unterschieden absehen, können wir also sagen: *Unter wesentlicher Verwendung des Satzes vom ausgeschlossenen Dritten kann bewiesen werden, daß jede Verzweigungsmenge eine Brouwersche Menge ist.*

Die erwähnten Unterschiede, von welchen man absehen muß, um die beiden letzten Behauptungen richtig zu verstehen, entspringen insbesondere dem Umstand, daß in der Brouwerschen Definition zunächst der Begriff „Element einer Menge" definiert wird, und zwar unter Berufung auf ein Gesetz der geschilderten Art. Da Brouwer den Mengenbegriff nicht voraussetzt, sondern erst durch elementerzeugende Gesetze *einführen will* (den Mengenbegriff mit dem Begriff eines elementerzeugenden Gesetzes identifiziert), kann Brouwer naturgemäß nicht von der „Menge aller durch ein Gesetz der geschilderten Art erzeugten Elemente" sprechen. Dem Wortlaut der Brouwerschen Definition entsprechend müßte

man also sagen: Ein Gesetz, welches mit einer Brouwerschen Menge
identisch ist, ist zugleich ein Gesetz, welches eine Verzweigungsmenge
liefert, nämlich eine Zuordnung von Dingen (Zeichenreihen) zu den
Zahlenkomplexen, durch welche eine Verzweigungsmenge erzeugt wird,
und zwar so erzeugt wird, daß jedes vom Nullding verschiedene Ding
des erzeugenden Systems an der Bestimmung von mindestens einem
Element mitwirkt, das zur Menge einen Beitrag liefert. Und umgekehrt:
Jede Zuordnung von Dingen zu den Zahlenkomplexen, welche der erwähnten Zusatzbedingung genügt, d. h. jede der Zusatzbedingung genügende Erzeugungsvorschrift für eine Verzweigungsmenge ist ein Gesetz,
welches mit einer Brouwerschen Menge identisch ist. Hierbei läßt sich
unter Verwendung des Satzes vom ausgeschlossenen Dritten und nur
unter Verwendung dieses Satzes aus jedem eine Verzweigungsmenge erzeugenden Gesetz ein dieselbe Verzweigungsmenge erzeugendes Gesetz
herleiten, welches der erwähnten Zusatzbedingung genügt. Wie gesagt,
scheinen mir die Unterschiede der letzterwähnten Behauptungen von
den kürzeren oben angegebenen bloß terminologischer Natur.

Noch klarer wird das Letztgesagte, wenn man bedenkt, daß auch
Brouwer sich einer Bezeichnungsweise für die Totalität der Elemente
einer Menge in seinem Sinn, d. h. für die Totalität der durch ein Verzweigungssystem definierten Elemente bedient, für dieselbe allerdings,
da er das Wort Menge für gewisse Gesetze reserviert, einen anderen
Ausdruck wählen muß und so von der *Spezies* aller Elemente einer
Menge spricht.

6. Die Brouwerschen Punktmengen. Wir betrachten nun speziell
Punktmengen eines R_n. Eine *Brouwersche Punktmenge*[7]) *des* R_n ist eine
Verzweigungsmenge, deren erzeugendes System die oben angeführten
Eigenschaften 1., 2., 3., 4. besitzt, d. h. aus abgeschlossenen Würfeln
mit rationalen Eckpunkten besteht, wobei die leere Menge (als „Hemmung des Prozesses mitsamt der definitiven Vernichtung seines Resultates“) zu den Würfeln hinzugerechnet wird, wobei ferner die Würfel die
Monotoniebedingungen 2. und 3. erfüllen und wobei endlich 4. jeder
nicht-leere Würfel mindesten einen Punkt der Menge enthält. (Der Fall,
daß ein Element einer Brouwerschen Punktmenge schon durch *endlich*
viele Würfel des erzeugenden Systems bestimmt wird, kann wegen der
Bedingung 3. nicht eintreten.) *Eine Brouwersche Punktmenge ist also
eine analytische Punktmenge des* R_n. Umgekehrt ist jede durch ein
System mit den Eigenschaften 1., 2., 3., 4. erzeugte analytische Menge

---

7) Vgl. Brouwer, Verh. Akad. Amsterdam (1) **12**, Nr. 7, sowie die demnächst
in den Mathem. Annalen erscheinende Abhandlung: „Zur Begründung der intuitionistischen Mathematik. IV.“

eine Brouwersche Punktmenge. Nun haben wir oben (§ 3) erwähnt, daß
zu jeder beliebigen analytischen Punktmenge des R_n ein die betreffende
Menge erzeugendes System, welches die Eigenschaften 1., 2., 3. besitzt,
ohne Verwendung des Satzes vom ausgeschlossenen Dritten hergeleitet
werden kann. Wir sahen ferner, daß unter wesentlicher Verwendung des
Satzes vom ausgeschlossenen Dritten überdies zu jeder analytischen Menge
die Existenz eines erzeugenden Systems, welches auch die Eigenschaft 4.
besitzt, bewiesen werden kann. *Unter wesentlicher Verwendung des Satzes
vom ausgeschlossenen Dritten ergibt sich also auch, daß jede analytische
Punktmenge des R_n eine Brouwersche Punktmenge ist.*

Wir betrachten nun zwei spezielle Klassen von Brouwerschen
Mengen, und zwar für den Fall der Punktmengen.

Brouwer nennt jene seiner Punktmengen *finit*[8]), für welche das
erzeugende Würfelsystem nur endlich viele Würfel mit einem Index, nur
endlich viele Würfel mit zwei Indizes und allgemein für jede natürliche
Zahl k nur endlich viele Würfel mit k Indizes enthält. *Unter wesentlicher Verwendung des Satzes vom ausgeschlossenen Dritten kann man
nun zeigen: Jede finite Brouwersche Punktmenge ist eine beschränkte abgeschlossene Punktmenge des R_n, und jede beschränkte abgeschlossene Punktmenge des R_n ist eine finite Brouwersche Punktmenge*

Sei nämlich *erstens* eine beschränkte abgeschlossene Punktmenge A
im R_n gegeben. Wir können jedem Punkt von A für jede natürliche
Zahl k einen Würfel mit rationalen Eckpunkten und einer Kantenlänge
$\frac{1}{2^k}$ zuordnen. Nach dem Borelschen Theorem [zu dessen Beweis in der
hier verwendeten Allgemeinheit der Satz vom ausgeschlossenen Dritten
wesentlich verwendet wird[9])] ist für jedes k die Menge A bereits in
der Summe von endlich vielen unter den zugeordneten Würfeln von
einer Kantenlänge $\frac{1}{2^k}$ enthalten, etwa in der Summe von W_1^k, W_2^k, ..., $W_{n_k}^k$.
Die Gesamtheit dieser sukzessive für $k = 1, 2, \ldots$ ad inf. bestimmten
Systeme von je endlich vielen Würfeln bildet (nach einer einfachen
Modifikation, damit der Bedingung 3. genügt werde) offenbar das erzeugende System einer finiten Brouwerschen Menge, und zwar ist die erzeugte finite Menge offenbar identisch mit der Menge A. — Es sei
zweitens eine finite Brouwersche Menge A durch das Würfelsystem
$S = \{ W_{n_1 n_2 \ldots n_k} \}$ gegeben. Die Menge A ist offenbar beschränkt. Um

---

8) Vgl. Brouwer, Mathem. Annalen **93**, S. 245.

9) Vgl. Brouwer, Journal f. d. r. u. a. Mathem. **154**, S. 4 sq. Über eine intuitionistische Form des Borelschen Theorems vgl. Brouwer, Proc. Ac. Amsterdam
29, S. 866.

zu zeigen, daß sie auch abgeschlossen ist, geben wir irgendeinen Häufungspunkt p von A vor und zeigen, daß p auch Punkt von A ist. Nun liegt ja jeder Punkt von A in einem Würfel aus S mit einem Index. Wegen der Finitheit von A existieren nur endlich viele Würfel von S mit einem Index. Jede unendliche Teilmenge von A enthält also unendlich viele Punkte, welche in einem und demselben Würfel aus S mit einem Index liegen. Der Punkt p ist nach Voraussetzung Häufungspunkt von A; es liegen also in jeder Umgebung von p unendlich viele Punkte von A und daher nach dem eben Bemerkten unendlich viele Punkte, welche in einem bestimmten Würfel aus S mit einem Index, etwa in W_{n_1}, liegen. Folglich ist p selbst Punkt von W_{n_1}. In derselben Weise zeigt man, daß p Punkt von einem Würfel $W_{n_1 n_2}$ aus S mit zwei Indizes und allgemein, daß p für jede natürliche Zahl k Punkt von einem Würfel $W_{n_1 n_2 \cdots n_k}$ aus S mit k Indizes ist. Daraus folgt aber, daß p Punkt von A ist, wie behauptet.

Brouwer nennt jene seiner Punktmengen *individualisiert*[10]), welche die Eigenschaft haben, daß für je zwei nicht identische Folgen natürlicher Zahlen $v = (n_1, n_2, \ldots, n_k, \ldots)$ und $v' = (n_1', n_2', \ldots, n_k', \ldots)$, die beiden Punkte $p_v = W_{n_1} \cdot W_{n_1 n_2} \cdots W_{n_1 n_2 \cdots n_k} \cdots$ und $p_{v'} = W_{n_1'} \cdot W_{n_1' n_2'} \cdots W_{n_1' n_2' \cdots n_k'} \cdots$ verschieden sind. Nun sind unter den analytischen Mengen durch die Existenz eines erzeugenden Systems von der erwähnten Eigenschaft die Borelschen Mengen charakterisiert[11]), das sind jene Mengen, welche durch iterierte Summen- und Durchschnittsbildungen aus den abgeschlossenen und den offenen Mengen hervorgehen oder, wie sie auch definiert werden können, jene analytischen Mengen, deren Komplemente analytische Mengen sind. Unter Verwendung des Satzes vom ausgeschlossenen Dritten ergibt sich also *die Identität der individualisierten und der Borelschen Mengen*.

Es sei nun M eine beliebige Teilmenge eines euklidischen Raumes, $\overline{M}$ die abgeschlossene Hülle von M. Wir können ein die Menge $\overline{M}$ erzeugendes Verzweigungssystem $\{M_{n_1 n_2 \cdots n_k}\}$ bilden, etwa ein solches, welches aus Intervallen des euklidischen Raumes besteht. Jeder Punkt von $\overline{M}$ ist für eine gewisse Folge $v = (n_1, n_2, \ldots, n_k, \ldots)$ von natürlichen Zahlen Durchschnitt der Mengenfolge $M_{n_1}, M_{n_1 n_2}, \ldots, M_{n_1 n_2 \cdots n_k}, \ldots$, und umgekehrt ist für jede Zahlenfolge v der Durchschnitt der entsprechenden Mengenfolge ein Punkt von $\overline{M}$. Betrachten wir das Verhältnis dieses ($\overline{M}$ erzeugenden) Verzweigungssystems zur Menge M, so sehen wir: Jeder Punkt von M ist für eine gewisse Folge v von natürlichen Zahlen Durchschnitt der entsprechenden Folge von Mengen des Verzweigungssystems, und für jede Folge von natürlichen Zahlen ist

---

10) l. c. 8), S. 245.   11) l. c. 4).

der Durchschnitt der entsprechenden Folge von Mengen des Verzweigungssystems entweder Punkt oder Häufungspunkt von M. Ein Mengensystem dieser Art wollen wir ein Verzweigungs*schema* der Menge M nennen. Damit zur Menge M nicht nur Verzweigungsschemata, sondern ein aus abgeschlossenen Mengen bestehendes Verzweigungs*system* existiere, d. h. ein System von abgeschlossenen Mengen $A_{n_1 n_2 \cdots n_k}$ derart, daß jeder Punkt von M für eine gewisse Folge von natürlichen Zahlen Durchschnitt der entsprechenden Folge von Mengen des Verzweigungssystems sei und daß für jede Folge natürlicher Zahlen der Durchschnitt der entsprechenden Folge von Mengen des Verzweigungssystems (nicht nur Häufungspunkt, sondern) Punkt von M sei, ist notwendig und hinreichend, daß M eine *analytische* Menge sei. Ist M nicht-analytisch, so gibt es zu jedem Verzweigungsschema von M Zahlenfolgen, so daß der Durchschnitt der ihnen entsprechenden Folgen von Mengen des Verzweigungsschemas *nicht* Punkt der Menge M ist. Auf Grund des Cantorschen Durchschnittssatzes gilt offenbar die Aussage: Zu einem Verzweigungsschema $\{A_{n_1 n_2 \cdots n_k}\}$ der Menge M, welches kein Verzweigungssystem von M ist, existieren Zahlenfolgen $(n_1, n_2, \ldots, n_k, \ldots)$, so daß alle entsprechenden Mengen $A_{n_1}, A_{n_1 n_2}, \ldots, A_{n_1 n_2 \cdots n_k}, \ldots$ nicht-leer sind, der Durchschnitt dieser Mengen aber (der nach dem Cantorschen Durchschnittssatz nicht-leer ist) nicht zur Menge M gehört. In unserer obigen Terminologie können wir dies auch so ausdrücken: Die Verzweigungssysteme sind unter den Verzweigungsschemata dadurch charakterisiert, daß in ihnen die leere Menge ein Nullding bildet. Die analytischen Mengen sind unter den Teilmengen euklidischer Räume dadurch charakterisiert, daß unter ihren Verzweigungsschemata Verzweigungssysteme existieren.

Bekanntlich existiert nach dem verallgemeinerten Borelschen Theorem in jedem separabeln metrischen Raum ein System von abgeschlossenen Teilmengen $\{A_{n_1 n_2 \cdots n_k}\}$, so daß jeder Punkt des Raumes für eine gewisse Folge von natürlichen Zahlen Durchschnitt der entsprechenden Folge von Mengen des Systems ist und daß für jede Folge natürlicher Zahlen der Durchschnitt der entsprechenden Folge von Mengen des Systems höchstens einen Punkt enthält (d. h. entweder aus einem Punkt besteht oder leer ist). Wir wollen ein derartiges System ein *Verzweigungsschema* und einen separabeln Raum *analytisch* nennen, wenn in ihm ein *Verzweigungssystem* existiert, d. h. ein Verzweigungsschema mit der Eigenschaft, daß für jede Folge natürlicher Zahlen der Durchschnitt der entsprechenden Folge von Mengen des Verzweigungsschemas, wofern diese durchweg nicht-leer sind, wirklich einen Punkt des Raumes enthält. Auf diese Weise werden die analytischen Räume zwischen die kompakten und die separabeln Räume eingeschaltet (zwischen die kompakten und analytischen Räume könnte man noch Borelsche Räume einschalten).

Es entsprechen einander also, wie wir in Wörterbuchform zusammenfassen können, hinsichtlich ihrer Verzweigungseigenschaften folgende

Teilmengen euklidischer Räume:	Mengenklassen in der Brouwerschen Terminologie:	Räume:
die beschränkten abgeschlossenen Mengen	die finiten Mengen	die kompakten Räume
die Borelschen Mengen	die individualisierten Mengen	
die analytischen Mengen	die Mengen	die analytischen Räume
die beliebigen Mengen	die Spezies	die separabeln Räume

7. Methodische Konsequenzen. Die erwähnte Beziehung zwischen den Brouwerschen und den analytischen Mengen scheint mir von einer gewissen methodischen Bedeutung. Die analytischen Mengen werden nämlich in der letzten Zeit nicht nur eingehend untersucht, sondern vielfach von formalistischen Mengentheoretikern als die in gewissem Sinn einzigen wirklich definierbaren Mengen aufgefaßt[12]), so daß hier also eine Annäherung des formalistischen Standpunktes an den Brouwer-

---

12) Zur Präzisierung dieser Auffassung schiene mir sehr interessant, wenn es gelänge, die Definition der analytischen Mengen aus gewissen Konstruktivitätsforderungen an den Mengenbegriff zwangsläufig herzuleiten oder wenigstens gewisse Konstruktivitätseigenschaften anzugeben, durch welche die analytischen Mengen gekennzeichnet sind. Eine solche Charakterisierbarkeit der analytischen Mengen halte ich für ziemlich wahrscheinlich.

Dabei möchte ich betonen, daß ich das Wort „Konstruktivität" für ein wenn überhaupt, so vermutlich *auf verschiedene Arten* und *in verschiedenen Abstufungen* präzisierbares (bisher noch nicht präzisiertes) Wort halte. Die separabeln Mengen und die zu analytischen Mengen komplementären Mengen sind weniger konstruktiv als die analytischen Mengen, aber doch konstruktiver als irgendwelche beliebige Mengen zu nennen, da sie immerhin gewisse Ansatzpunkte für konstruktive Behandlung bieten. Andererseits könnte man den endlichen Mengen, den abzählbaren Folgen von Elementen und den insichkompakten Mengen vielleicht in gewissem Sinn eine höhere Konstruktivität als den analytischen Mengen zuschreiben. Oder, um ein anderes Beispiel zu erwähnen, welches sich freilich formalistisch gar nicht recht präzisieren läßt: Auch jene (vorläufig durchwegs abzählbaren) Mengen reeller Zahlen, in denen jedes einzelne Element, so wie Borel dies zu wiederholten Malen forderte, durch ein Gesetz sich definieren läßt, sind vielleicht in gewissem Sinn konstruktiver als die analytischen Mengen reeller Zahlen, welche ja definiert sind als Mengen aller Elemente, die durch irgendwelche unendlich viele Auswahlakte bestimmter Art entstehen. Ebenso kann vielleicht auch für die analytische Menge aller Folgen von natürlichen Zahlen als ein konstruktiverer Kern die Menge aller gesetzmäßig definierten Folgen natürlicher Zahlen bezeichnet werden.

Jahresbericht d. Deutschen Mathem.-Vereinigung. XXXVII. 1. Abt. Heft 6/10. 15

−21−

schen vorliegt. Zugleich ergibt sich, daß für einige der wichtigsten Teile der Mengenlehre (nämlich für die Lehre von den Teilmengen kompakter Mengen und für die Lehre von den Mengen reeller Zahlen) die Unterschiede zwischen intuitionistischer und formalistischer Behandlungsweise *nicht die spezifischen Begriffsbildungen,* sondern so wie in Arithmetik und Zahlentheorie bloß die *Bearbeitung der Begriffe mit dem Satz vom ausgeschlossenen Dritten* betreffen.

(Eingegangen am 15. 6. 1927.)

Selected Papers on Analysis

Commentary on Karl Menger's Contributions to Analysis

Ludwig Reich

1. Complete Weierstrass Analytic Functions

Let $P(z) = w_0 + \sum_1^n a_n(z - z_0)^n$ be a power series with complex coefficients and with a radius of convergence different from 0. Then K. Weierstrass introduced the notion of "analytisches Gebilde" (complete analytic function) defined by P as the set of all power series $w_1 + \sum_1^n a_n^{(1)}(z - z_1)^n$ obtained from P by direct and indirect analytic (i.e., holomorphic) continuation. In his article A.7, which is supplemented by his papers A.4, A.5 and A.6, K. Menger emphasizes that for many decades this was the only exact alternative to introducing functions as "laws or rules associating or pairing numbers with numbers" and multifunctions as rules of pairing numbers with sets of numbers. It is well-known today that Weierstrass' notion of a complete analytic function leads in a natural way to the concept of "analytisches Gebilde" as given by H. Weyl in his famous book [W], §2, §3. This again is, after introducing an appropriate natural topology, a nontrivial example of a Riemann surface, and it includes, in contrast to Weierstrass' complete analytic functions, also poles and algebraic ramification points (see also C. L. Siegel's lectures [S], Chapter 1, 3, Chapter 1, 4). It is also well-known to mathematicians today that the definition of Riemann surfaces as a class of two-dimensional manifolds satisfying a certain regularity condition involves the use of a class of changes of the local parameters (coordinates), namely exactly those which are given by locally biholomorphic functions. It is obvious that Menger disliked this way of defining manifolds of a certain regularity class by referring to local changes of coordinates of this regularity class. Even H. Weyl makes a short remark in [W], p. 33, against this way of defining, say, differentiable manifolds, although the famous book [W] is entirely built upon this idea, which goes back to Hilbert: It is much more in the spirit of pure geometry to try to give a characterization of the (original) Weierstrass complete analytic function by using as many notions from topology as necessary and as few from analysis as possible, in particular to avoid changes of local parameters of a certain regularity class. Menger really achieves this goal in A.7. As we already said, the complete Weierstrass function F defined by the power series $P(z) = w_0 + \sum_1^\infty a_n(z - z_0)^n$ is the set

$$F := \left\{ Q(z) = w_1 + \sum_{n=1}^\infty a_n^{(1)}(z - z_1)^n \,\middle|\, Q \text{ is a direct}\right.$$

$$\left. \text{or indirect analytic continuation of } P \right\}.$$

F may clearly be identified with a subset $\hat{F}$ of $\mathbb{C}^2$, namely

$$\hat{F} := \{(z_1, w_1) \mid w_1 = Q(z_1) \text{ for } Q \in F\} \subseteq \mathbb{C}^2 \cong \mathbb{R}^4,$$

which Menger calls the graph of F. In fact, Menger gives a characterization of the sets $\hat{F}$ in the spirit of pure geometry. Menger's characterization theorems are based on purely topological notions such as local arcwise connectnedness, and, in particular, topological dimension ≤ 2 and some basic facts of dimension theory. This is not surprising if one remembers that among Menger's most influential ideas are his contributions to dimension theory. From set theory he takes the notion of a non-vertical set which is here a subset S of $\mathbb{C} \times \mathbb{C}$ such that the projection of S on the first factor $\mathbb{C}$ contains more than one point. But it is also clear that these very "geometric" notions cannot suffice to characterize complete Weierstrass functions which are so closely related to holomorphic functions. In Menger's treatment in A.7 analyticity comes in via the following definition: A subset S of $\mathbb{C}^2$ is smooth at $(z_0, w_0) \in S$ if for every pair of sequences (z'_n, w'_n) and (z''_n, w''_n), $n \in \mathbb{N}$, in S converging to (z_0, w_0) and such that $(z'_n \neq z''_n)$ the sequence $\left(\frac{w''_n - w'_n}{z''_n - z'_n}\right)_{n \in \mathbb{N}}$ has a (finite) limit as $n \to \infty$. [A modification of this concept is also useful in metric geometry, see A.6].

In order to present his characterization of complete Weierstrass analytic functions in a very concise way, Menger introduces "patches", which are by definition the non-vertical sets $S \subset \mathbb{C}^2$ that are connected, locally arcwise connected, locally compact, more than 1-dimensional and uniform in the sense that to every pair of points $(p, p') \in S \times S$ there exist neighbourhoods of p, resp. p', in S which are homeomorphic, with p being mapped onto p'. A patch is smooth if it is everywhere smooth. A set S is called smoothly connected if it has more than one point and any two of its points are in a patch contained in S. We now give Menger's main results (in almost his own words from A.7):

1. The complete Weierstrass analytic functions (or equivalently, their graphs) are exactly the maximal nonvertical smoothly connected subsets of $\mathbb{C}^2$ which are unions of countably many smooth patches.
2. A smoothly connected subset of $\mathbb{C}^2$ is a subset of a complete Weierstrass function if and only if it is a union of countably many smooth patches.

The proofs of these results can be understood by everyone having a standard knowledge of topology, but nevertheless show Menger's high command of the topic. The various examples he gives to illustrate the role of the concepts mentioned above in the characterization of complete Weierstrass analytic functions are very interesting.

It seems that the community of mathematicians working in complex analysis did not really follow Menger's idea to rebuild an essential part of their field in terms of "pure geometry". We may see the reason for this in the fact that, already in the 19th century, the theory of Riemann surfaces and the theory of functions, differentials and integrals on them led to results which are considered to be one of the most beautiful and important achievements of mathematics and which were put on a totally sound basis allowing further very successful research by H. Weyl and others.

However, we feel that Menger's ideas deserve to be reexamined and worked out in more general and hence more demanding situations. Such problems are:

1. The characterization of "analytisches Gebilde" in the sense of H. Weyl, including poles and algebraic ramification points.
2. A characterization of Riemann surfaces and, in the context of the theory of holomorphic functions of several complex variables, a characterization of Stein spaces (see [K], Ch. 3, §32, Ch. 5, §51) as the most general manifolds on which analytic functions live, in the spirit of pure geometry. Clearly, one would have to generalize the notion "S is smooth at a point" and, very likely, also the topological parts will be much more complicated than in the case treated by Menger.

In his Comptes Rendus note A.6, Menger indicated a generalization of the notion "$S \subset \mathbb{C}^2$ is smooth at a point (z_0, w_0)" and the concept of a patch to the product $A \times B$ of two metric spaces. According to Menger this can be done by defining a subset E of $A \times B$ to be B/A-bounded at (a_0, b_0) if there exist a neighbourhood V_0 of (a_0, b_0) and an $r_0 > 0$ such that for all points $p' = (a', b')$, $p'' = (a'', b'')$ in $V_0 \cap E$ one has $d_B(b', b'')/d_A(a', a'') < r_0$. Here d_B (d_A) means the metric on B (A). This concept can be used to define the generalization of the notion of a patch (patch over A) when A is in $\mathbb{R}^n$ and separable and locally compact. Menger announces the result that each (generalized) patch is an n-dimensional manifold and says that his characterization of the complete Weierstrass analytic functions can be embedded in this framework using a theorem of H. Bohr and specializing A and B to $\mathbb{C}$.

In the articles A.4 and A.5 he gives a very readable survey of his results without presenting proofs.

2. The Behaviour of a Complex Function at Infinity

The trend of thought that led Menger to characterize complete Weierstrass analytic functions in the spirit of pure geometry was probably the motivation that had previously led him to propose, in A.3, a new definition of the behaviour of a complex function f at $z = \infty$ and, in particular, a definition of the derivative of f at infinity which is different from the usual one, at least in some cases, and which can also be applied to define a derivative at some points on the boundary of the domain of f (considered as a subset of the Riemann sphere R) which are not covered by the usual procedure. Menger points out that in the formulation of several basic theorems of complex function theory one has to distinguish the cases $z = \infty$ and $z \in \mathbb{C}$, which is not necessary if one adopts Menger's definition. This definition of the derivative follows Bouligand's notion of the "paratingent" and is also very similar to the concept "smooth at a point (z_0, w_0)" discussed in 1. Let G be an open subset of the Riemann sphere R, let f be a function $f: G \to R$, and assume that a is a point in G or on the boundary of G. Let $(z'_n)_{n \in \mathbb{N}}$, $(z''_n)_{n \in \mathbb{N}}$ be two sequences in G such that $z'_n \neq a$, $z''_n \neq a$, $z'_n \neq \infty$, $z''_n \neq \infty$; $z'_n \neq z''_n$; and $\lim_{n \to \infty} z'_n = \lim z''_n = a$. If then, in R,

$$\lim_{n \to \infty} \frac{f(z'_n) - f(z''_n)}{z'_n - z''_n} = b = (Ef)(a)$$

for all pairs of such sequences, where $b \in R$, then b will be called the exterior derivative of f at a. Similarly, the exterior derivatives of (higher) order k, $E^k f$, can be defined by induction. If $a \in G$, $a \neq \infty$, and if f is holomorphic at $a = \infty$, then the exterior derivative of f is the same as the usual one. Menger also proposes to define a as a point of extensibility of $f: G \to R$ if a belongs to the closure of G and if $\lim_{\substack{z \to a \\ z \in G}} f(z)$ exists; f is called steep at a point a of extensibility if $(E^k f)(a) = \infty$ for almost all k; it is called flat if $(E^k f)(a) = 0$ for almost all k, $E^k f$ not being constant. If at a point of extensibility of f all $(E^k f)(a)$ exist, but f is neither steep nor flat at a, then a will be said to be ordinary for f. Using this terminology, Menger shows that in the formulation of Cauchy's integral theorem for functions $f \colon G \to R$ one need not distinguish points $z \in \mathbb{C}$ and $z = \infty$. The same is true for the definition of the residue of an analytic function at a point $a \in R$. These are two examples of applications of the (geometric) definition of a derivative proposed by Menger.

In spite of Menger's proposal, the usual definition of the behaviour of a function at ∞ (see e.g. [A], 7.4.3.) is still used in complex analysis because it is very easy to handle in calculations, e.g. in the theory of differential equations in the complex domain, in particular in Fuchsian theory (see for instance E. Hille [H], Ch. 9, Ch. 10 or L. Bieberbach [Bie], Ch. 6, Ch. 7). Also, Menger does not mention that $\zeta = \frac{1}{z}, |z| > M$ is only one special local coordinate in the neighbourhood of $a = \infty$ on R. Using another local coordinate at $a = \infty$ will produce a "derivative" of f at ∞ which is in general different from the one produced by taking the coordinate $\zeta = \frac{1}{z}$, but leaves invariant the differential, the integral and the basic classifications like "holomorphic", "meromorphic" etc.

It would be an interesting topic for future research to test the role of Menger's exterior derivations in the investigation of boundary values of functions $f \colon G \to R$. Menger included some examples in this direction in his paper and the geometric nature of his definition seems to be an advantage for this purpose. Also, in the theory of differential equations, when studying an equation and its solutions in a neighbourhood of infinity, Menger's ideas may be useful.

3. Cauchy's Integral Theorem in the Plane. Green's Formula

1. Closely related to the theory of analytic functions, but also of importance in the theory of real functions and its applications (e.g., to differential equations) is the question of how the path integral $J(C) = \int_C p(x,y)dx + q(x,y)dy$ behaves as a function of the path C in a given plane region R. While this problem is usually treated under some strong regularity (i.e., differentiability) assumptions on the functions p and q in the integrand, in his paper A.1, Menger investigates this question under the hypothesis that p and q be continuous functions only and that R is the interior of a rectangle whose sides are parallel to the coordinate axes. Using geometric concepts and notions from the calculus of variations, he gives necessary and sufficient

conditions in order that $J(C) = \int_C p\,dx + q\,dy$ has (for fixed functions p and q) the same value for all rectifiable curves C in R joining fixed (but arbitrary) initial and end points. This is a natural and interesting question in the theory of path integrals which was not discussed in detail before Menger and seems to share the fate of Menger's other contributions to Analysis i.e., not to be taken up by other researchers. Menger's tools are rectangular nets Z and "dotted" rectangular nets Z' inscribed into R and constructed from Z by interpolating additional points to Z in an appropriate way, but too technical to be explained here in detail. Furthermore, he uses the "variational distance" of two points $z_0 = (x_0, y_0)$ and $z_1 = (x_1, y_1)$ in R associated to (p, q), defined by $\delta_1(z_0, z_1) = p(x_0, y_0)(x_1 - x_0) + q(x_0, y_0)(y_1 - y_0)$, the variational length of a polygon $[z_1, z_2, \ldots, z_n]$ defined by $\lambda_1(Z) = \sum_{i=1}^{n-1} \delta_1(z_i, z_{i+1})$, as well as the Euclidean distance and the Euclidean length of the polygon. Then he introduces the meshes M_{ij} of a dotted net, formed by neighbouring points z_{ij}, $z_{i+1,j}$, $z_{i,j+1}$, $z_{i+1,j+1}$ and bounded by the polygons X_{ij}, Y_{ij}, $X_{i,j+1}$, $Y_{i+1,j+1}$; he denotes by v_{ij} the maximum length of the sides of M_{ij} and by μ_{ij} the area of the rectangle defined by M_{ij}. The maximum v of the v_{ij}'s is called the norm of the dotted net Z'. The variational quantities

$$\rho_{ij} = \frac{1}{\mu_{ij}}[\lambda_1(X_{ij}) + \lambda_1(Y_{i+1,j}) - \lambda_1(Y_{ij}) - \lambda_1(X_{i,j+1})]$$

for the mesh M_{ij} are also introduced and their maximum (taken over the dotted net) is called the quotient of the net.

Menger's main result is:

In order that the integral $J(C) = \int_C p\,dx + q\,dy$ be independent of the path C in the rectangle R it is necessary and sufficient that for each $\varepsilon > 0$ there exists a dotted rectangular net in R for which both norm and quotient are less than ε.

One should be aware of the fact that these conditions involve, by definition, the functions p and q, but in a rather implicit way. On the other hand, the condition is indeed very natural in its geometrical formulation, and requires the existence of only one (appropriate) dotted net for each $\varepsilon > 0$.

As Menger points out, from his result it is not difficult to deduce the original method that Goursat used (the "Goursat-Heffter" – condition, as he says) in the proof of Cauchy's integral theorem. (Goursat's method of using rectangles with sides parallel to the coordinate axes in this context can be found, e.g., in [A], Ch. 4, 1.4.)

Menger also indicates that from his result one can easily deduce a sufficient condition which only involves certain right partial derivatives of p and q on a set A that is dense in R. Also, this sufficient condition implies that the difference of difference quotients (introduced above in a slightly different form),

$$\rho_{ij} = \frac{1}{x_{i+1} - x_i} (q(x_{i+1}, y_j) - q(x_i, y_j)) - \frac{1}{y_{j+1} - y_j} (p(x_i, y_{j+1}) - p(x_i, y_j))$$

tends to 0 if the norm of the dotted nets goes to 0. Is this sufficient? The most striking examples given by Menger are integrals $J(C)$ where the integrand is nowhere differentiable. Namely, let $f: I \to \mathbb{R}$ be a continuous nowhere differentiable function defined on an interval such that $x + y \in I$ for $(x, y) \in R$. Then, as a consequence of Menger's main theorem, $\int_C f(x + y)(dx + dy)$ is independent of C.

We believe that Menger's ideas deserve to be developed in several directions:

a) As he says himself, the nets need not be rectangular in the strict sense, but can be deformations of rectangular nets. However, they have to be of "quadrangular" structure. It would be interesting to see whether "triangular" nets would also work, in a way similar to the replacement of Goursat's rectangles by triangles in the proof of Cauchy' integral theorem given by A. Pringsheim (see [R], 7.1.3 and 7.1.4.)

b) Examples such as $\int_C f(x + y)(dx + dy)$ with a continuous nowhere differentiable function f could lead to a wide class of examples of integrals $J(C)$ "independent of C". A first step would be to try to replace the inner function $(x, y) \mapsto x + y$ (in f) by a more general law $(x, y) \mapsto G(x, y)$ of composition.

c) How to generalize Menger's results to integrals $\int_e p\, dx + q\, dy$ where even (strongly) weaker assumptions than the continuity of p and q are made, provided the integral makes sense is also an open problem.

In A.2 Menger gives a proof of Green's formula based upon an algebraic identity and avoiding iterated integration which eventually leads to allowing weaker conditions on the integrand than in the usual proofs. This idea makes the proof very suitable for introductory lectures, although Menger also needs several rather technical steps to get from the original situation of a rectangle with sides parallel to the coordinate axes to more general domains and integration paths.

The basic identity is

$$\sum_{ij} \frac{u(x_i, y_{j+1}) - u(x_i, y_j)}{y_{j+1} - y_j} (x_{i+1} - x_i)(y_{j+1} - y_j)$$
$$= \sum_{i} (u(x_i, B) - u(x_i, b))(x_{i+1} - x_i), \tag{I}$$

where u is a real-valued function defined in a rectangle $R = \{(x, y) \mid a \leq x \leq A, b \leq y \leq B\}$. If $\partial u/\partial y$ exists in R, then from this identity one obtains, via the mean value theorem, the relation $(y_j \leq \eta_{ij} < y_{j+1})$

$$\sum_{i,j} \frac{\partial u}{\partial y}(x_i, \eta_{ij})(x_{i+1} - x_i)(y_{j+1} - y_i) = \sum_{i} (u(x_i, B) - u(x_i, b))(x_{i+1} - x_i), \tag{II}$$

and this easily yields

$$\iint_R \frac{\partial u}{\partial y}(x, y) dx dy = \int_a^A (u(x, B) - u(x, b)) dx, \tag{G'}$$

if both Riemann integrals exist. If $\int_a^A u(x, B) dx$ also exists, then

$$\iint_R \frac{\partial u}{\partial y}(x, y) dx dy = -\int_{\partial R} u(x, y) dx. \tag{G}$$

A similar relation, valid under the above conditions, namely

$$\sum_{i,j} \frac{\partial u}{\partial y}(\xi_i, \eta_{ij})(x_{i+1} - x_i)(y_{j+1} - y_i) = \sum_{i} (u(\xi_i, B) - u(\xi_i, b))(x_{i+1} - x_i)$$

is used by Menger to point out that the existence of the Riemann integral

$$\iint_R \frac{\partial u}{\partial y}(x, y) \, dx \, dy$$

implies the existence of the Riemann integral $\int_a^A (u(x, B) - u(x, b)) \, dx$, and hence, if $\int_a^A u(x, B) \, dx$ also exists (as a Riemann integral), the validity of

Green's formula for the rectangle R. But he shows more: If the left hand side of (I) has a limit for a sequence of decompositions tending to "zero" in the usual sense (Menger calls this limit a Weierstrass integral and gives [Bo], §31 as a reference for this concept), and if the Riemann integral $\int_a^A (u(B, x) - u(b, x))dx$ exists, then we get the version (G') of Green's formula for the Weierstrass integral. The latter R-integral exists, if $u(B, x) - u(b, x)$ is almost everywhere continuous, and hence no further assumptions on u in the interior of the rectangle are needed. The Weierstrass double integral exists for a wider class of integrands than the double Riemann integral, and is equal to the Riemann integral, if the latter exists.

The remaining sections of A.2 extend Green's formula to much more general domains and paths of integration which are needed for the usual applications of Green's formula. Typical of Menger's style of exposition seems to be his way of describing the necessary technical details by using as few calculations as possible. However, in these parts the considerations are not too far from those used in other expositions of the topic.

As Menger says at the end of A.2, the other fundamental theorems of vector analysis (at least in the plane) can also be proved on the basis of the ideas that he used to deduce Green's formula. It is an open problem to apply generalizations of his identity (I) and relations like (II) in the higher dimensional case, at least for "rectangular" domains. [The generalization to more general domains and their boundaries clearly would then require overcoming the well-known geometric and topological difficulties which naturally arise in higher dimensions or, even more so, in the analysis on manifolds.]

References

I. Selected Papers of K. Menger on Analysis

A.1. On Cauchy's Integral Theorem in the Plane, Proc. Nat. Acad. Sci. USA 25 (1939) 621–625.

A.2. On Green's Formula, Proc. Nat. Acad. Sci. USA 26 (1940) 660–664.

A.3. The Behavior of a Complex Function at Infinity, Proc. Nat. Acad. Sci. USA 41 (1955) 512–513.

A.4. A Characterization of Weierstrass Analytic Functions, Proc. Nat. Acad. Sci. USA 54 (1965) 1025–1026.

A.5. Analytische Funktionen, Wiss. Abh. Forschungsgem. Nordrhein-Westfalen 33 (1965) 609–612. (= Festschrift Gedächtnisfeier K. Weierstrass, Westdeutscher Verlag, Köln, 1966).

A.6. Une charactérisation des fonctions analytiques, CRP 261 (1965) 4968–4969.

A.7. Weierstrass Analytic Functions, Math. Ann. 167 (1966) 177–194.

II. Further References

[A]   Ahlfors, L.: Complex Analysis, Second Edition. McGraw-Hill Book Company. New York 1966.

[Bie] Bieberbach, L.: Theorie der gewöhnlichen Differentialgleichungen auf funktionentheoretischer Grundlage dargestellt. Grundlehren der mathematischen Wissenschaften Band LXVI, Springer Verlag, Berlin, 1953.

[Bo]   Bolza, O.: Lectures on the Calculus of Variations, Chicago, 1904.

[H]   Hille, E.: Ordinary Differential Equations in the Complex Domain, John Wiley, New York, 1976.

[K]   Kaup, L., Kaup, B.: Holomorphic Functions in Several Variables, Walter de Gruyter, Berlin, 1983.

[R]   Remmert, R.: Funktionentheorie, Dritte Auflage. Springer Verlag, Berlin, 1992.

[S]   Siegel, C. L.: Topics in Complex Function Theory, Vol. 1. Elliptic Functions and Uniformization Theory. Wiley-Interscience, New York, 1969.

[W]   Weyl, H.: Die Idee der Riemannschen Fläche, 5. Auflage, B.G. Teubner, Stuttgart, 1974.

ON CAUCHY'S INTEGRAL THEOREM IN THE REAL PLANE

BY KARL MENGER

DEPARTMENT OF MATHEMATICS, UNIVERSITY OF NOTRE DAME

Communicated November 8, 1939

1. *Introduction.*—In a rectangle R let $p(x, y)$ and $q(x, y)$ be two continuous functions, and associate with each rectifiable curve C the number $J(C) = \int_C p\,dx + q\,dy$. Under which conditions is J the same for any two coterminal curves in R?

The ordinary conditions are based on the existence of partial derivatives of p and q, or at least of $p_2' = \dfrac{\partial p}{\partial y}$ and $q_1' = \dfrac{\partial q}{\partial x}$, everywhere or at least almost everywhere, and on the equality of p_2' and q_1' everywhere or almost everywhere. However, as the following example shows, no assumptions whatever about differentiability are necessary in order that the integral be independent of the path. Let $f(z)$ be any continuous but nowhere differentiable function of one real variable z, and set $p(x, y) = q(x, y) = f(x + y)$. Although neither p nor q has any partial derivative, $\int p\,dx + q\,dy$ is independent of the path.

We shall formulate a geometric condition that is both necessary and sufficient. The two classical sufficient conditions (viz., that p and q have differentials at each point of R, though their partial derivatives need not be continuous, or that p_2' and q_1' are continuous) directly imply our general condition; the second even after being generalized to such an extent that existence, continuity and equality of p_2' and q_1' are merely required at denumerably many points (thus almost nowhere). Besides also nowhere differentiable cases subsume under our general theorem.

Just as in the metrical treatment of line integrals in the calculus of variations[1]) it is useful to associate with any two points $z_0 = (x_0, y_0)$ and $z_1 = (x_1, y_1)$ beside their Euclidean distance $\delta(z_0, z_1)$ another ("variational") distance, viz.,

$$\delta_1(z_0, z_1) = p(x_0, y_0) \cdot (x_1 - x_0) + q(x_0, y_0) \cdot (y_1 - y_0),$$

and with each polygon $Z = [z_1, z_2, \ldots, z_n]$ beside its Euclidean length $\lambda(Z) = \Sigma\, \delta(z_i, z_{i+1})$ a variational length, viz., $\lambda_1(Z) = \Sigma\, \delta_1(z_i, z_{i+1})$. Since $\lambda_1(Z)$ for a polygon Z inscribed in the curve C is a Riemann sum of $\int_C p\,dx + q\,dy$, the number $J(C)$ is nothing but the length of C derived from the δ_1-distance.

2. *Nets and Dotted Nets.*—Let R be the rectangle $a \leq x \leq b$, $c \leq y \leq d$. We call *rectangular net of points* in R a matrix of points $z_{ij} = (x_i, y_j)$ [$x_0 = a$, $x_{m+1} = b$, $y_0 = c$, $y_{n+1} = d$, $x_i < x_{i+1}$, $y_j < y_{j+1}$, $0 \leq i \leq m + 1$, $0 \leq j \leq n + 1$].

Let $Z = z_{ij}$ be a rectangular net. For i, j with $0 \leq i \leq m, 0 \leq j \leq n + 1$ we form X_{ij}, a finite ordered set of points between z_{ij} and z_{i+1j} and including these points; for i, j with $0 \leq i \leq m + 1, 0 \leq j \leq n$ we form Y_{ij}, a finite ordered set of points between z_{ij} and z_{ij+1} and including these points. The set $Z' = \Sigma X_{ij} + \Sigma Y_{ij}$ will be called a *dotted rectangular net* derived from Z. In the following we shall apply this name merely to dotted rectangular nets with the property that for each i $(0 \leq i \leq m)$ the $n + 2$ sets X_{ij} $(0 \leq j \leq n + 1)$ have the same horizontal projection, and for each j the $m + 2$ sets Y_{ij} have the same vertical projection. The points of the dotted net then may be denoted by

$$z_{ik,j} = (x_{ik}, y_j) \ (0 \leq i \leq m, 0 \leq k \leq k_i, 0 \leq j \leq n + 1)$$

and $\quad z_{i,jl} = (x_i, y_{jl}) \ (0 \leq i \leq m + 1, 0 \leq j \leq n, 0 \leq l \leq l_j)$

where $\quad x_{00} = a, x_{mk_m} = b, x_{ik} < x_{ik+1} \leq x_{ik_i} = x_{i+1} = x_{i+10}$

$$y_{00} = c, y_{n_n} = d, y_{jl} < y_{jl+1} \leq y_{jl_j} = y_{j+1} = y_{j+10}.$$

Z' is undotted ($= Z$) if $k_i = 1, l_j = 1$ for $0 \leq i \leq m, 0 \leq j \leq n$. The segments $x = x_i, c \leq y \leq d, 1 \leq i \leq m$ and $y = y_j, a \leq x \leq b, 1 \leq j \leq n$ will be called the *threads* of Z'.

By the mesh M_{ij} of Z' we mean the quadrangle $z_{ij}, z_{i+1j}, z_{ij+1}, z_{i+1j+1}$ bounded by the polygons $X_{ij}, Y_{ij}, X_{ij+1}, Y_{i+1j}$. By ν_{ij} we denote the Euclidean length of the longest of these four polygons (that is, in our rectangular case, the greater of the two numbers $x_{i+1} - x_i$ and $y_{j+1} - y_j$), by μ_{ij} the area enclosed by the four polygons (in our rectangular case, the area of the rectangle). The largest of the numbers ν_{ij} will be called the *norm* of the dotted net Z' and denoted by $\nu(Z')$.

3. *A Necessary and Sufficient Condition.*—For the mesh M_{ij} we set

$$\rho_{ij} = \frac{[\lambda_1(X_{ij}) + \lambda_1(Y_{i+1j})] - [\lambda_1(Y_{ij}) + \lambda_1[X_{ij+1})]}{\mu_{ij}},$$

λ_1 denoting the variational length (see introduction). The largest of the numbers ρ_{ij} will be called the *quotient* of Z' and denoted by $\rho(Z')$.

In this terminology we can prove: *In order that in the rectangle R the integral $J(C) = \int_C p dx + q dy$ be independent of the path it is necessary and sufficient that for each $\epsilon > 0$ there exists a dotted rectangular net in R for which both norm and quotient are $< \epsilon$.* Since, as is well known, the functional $J(C)$ is continuous, there is no difficulty in extending the theorem to non-rectangular domains. Moreover it is clear that the rectangular character of the nets is inessential. The dotted rectangular nets in our general condition may be replaced by dotted *quadrangular* nets of points that can be obtained from dotted rectangular nets by deforming their threads provided that ν_{ij} denotes the Euclidean length of the longest of the four polygons bounding the mesh M_{ij}, and μ_{ij} the area bounded by these four polygons.

Our condition implies that *in order that $J(C)$ be independent of the path it is sufficient that for each $\epsilon > 0$ there exists a rectangular (or quadrangular) undotted net of points whose norm and quotient are $< \epsilon$.* If M_{ij} is a mesh of an undotted rectangular net, then each of the four polygons that form the boundary of M_{ij} consists of two corners of a rectangle, and we have $\lambda_1(X_{ij})$ $= \delta_1(z_{ij}, z_{i+1j}) = p(x_i, y_j).(x_{i+1} - x_i), \lambda_1(Y_{i+1j}) = q(x_{i+1}, y_j).(y_{j+1} - y_j),$ $\lambda_1(Y_{ij}) = q(x_i, y_j).(y_{j+1} - y_j), \lambda_1(X_{ij+1}) = p(x_i, y_{j+1}).(x_{i+1} - x_i),$ and hence

$$\rho_{ij} = \frac{q(x_{i+1} y_j) - q(x_i, y_j)}{x_{i+1} - x_i} - \frac{p(x_i, y_{j+1}) - p(x_i, y_j)}{y_{j+1} - y_j}.$$

Thus our sufficient condition implies that the difference of the *difference quotients* of q with respect to x, and of p with respect to y tend toward 0 without assuming that these difference quotients themselves tend toward any limit, i.e., without assuming the existence of the partial *derivatives* q_1' and p_2'. The question of the necessity of our last condition is open.

4. *The Goursat-Heffter Condition.*—For a given p and q and a given $\epsilon > 0$, a rectangle R' of area $|R'|$ will be called an ϵ-figure if $\int_{R'} p dx + q dy \leq \epsilon |R'|$. If both p and q have total differentials at each point $z = (x, y)$ of R, then for each $\epsilon > 0$ and each point z there exists a circle $C_\epsilon(z)$ with z as center and the following property: each rectangle R' with sides parallel to the axes and contained in $C_\epsilon(z)$ is an ϵ-figure. Now, if R^* is a rectangle in the ϵ-neighborhood of R interior to R, we can associate with each point z of R^* a rectangle $R(z)$ that is an ϵ-figure contained in $C_\epsilon(z)$ and in R, and containing z in its interior. A finite number of these rectangles $R(z)$ covers R^*. Extending the sides of these rectangles to the boundary of R we get the threads of a rectangular net of points from which we can derive a dotted net of points whose quotient is $< \epsilon$. Interpolating other vertical and horizontal threads, if necessary, we get a dotted rectangular net for which both norm and quotient are $< \epsilon$. Thus R satisfies our general condition. It is clear that we even might admit points at which p and q do not have differentials provided that the set of these points can be covered by a finite number of rectangles whose total area is arbitrarily small.

5. *A Sufficient Condition Generalizing the Continuity of p_2' and q_1.*— Let $\epsilon > 0$ be given and assume the existence of 1) a set A_ϵ dense in R such that for each point (x, y) of A_ϵ we have $|p_2'(x, y + 0) - q_1'(x + 0, y)| < \epsilon$ where these numbers denote the right partial derivatives of p and q at (x, y), and 2) a number $\delta_\epsilon > 0$ such that for each point (x, y) of A_ϵ and each positive $\delta \leq \delta_\epsilon$ we have

$$|p_2'(x, y + \delta) - p_2'(x, y + 0)| < \epsilon \text{ and } |q_1'(x + \delta, y) - q_1'(x + 0, y)| < \epsilon.$$

Then R contains even undotted rectangular nets of arbitrarily small norms whose quotients are $< 3\epsilon$. Thus *if the conditions 1) and 2) are satisfied for each $\epsilon > 0$, then the sufficient condition of section 3 holds.*

The hypothesis 2) may be weakened. It is sufficient to assume that along the short vertical and horizontal segments through the points (x, y) of A_ϵ the upper and lower derivatives $\bar{p}_2'$, $\underline{p}_2'$, $\bar{q}_1'$, $\underline{q}_1'$ do not differ too much from $p_2'(x, y + 0)$ and $q_1'(x + 0, y)$, respectively. We need not assume the existence of the partial derivatives p_2' and q_1' along these segments. In the hypothesis 1) it is sufficient to assume that the sets A_ϵ are denumerable or even finite provided that they are sufficiently dense in R. At any rate, by considering the set $A_1 + A_{1/2} + A_{1/4} + \ldots$ we see that it is only at the points of a denumerable set that we require the existence and approximative equality of partial derivatives p_2' and q_1', or more precisely, of right derivatives only.

6. *Nowhere Differentiable p and q.*—Let f be a continuous function of one real variable, and consider the integral $\int f(x + y)(dx + dy)$. If M_{ij} is any mesh of any rectangular net of points, we have

$$[\lambda_1(X_{ij}) + \lambda_1(Y_{i\,+1j})] - [\lambda_1(Y_{ij}) + \lambda_1(X_{ij\,+1})] =$$
$$[p(x_i, y_j)\cdot(x_{i\,+1} - x_i) + q(x_{i\,+1}, y_j)\cdot(y_{j\,+1} - y_j)] -$$
$$[q(x_i, y_j)\cdot(x_{i\,+1} - x_i) + p(x_i, y_{j\,+1})\cdot(y_{j\,+1} - y_j)] =$$
$$[f(x_i + y_j)(x_{i\,+1} - x_i) + f(x_{i\,+1} + y_j)(y_{j\,+1} - y_j)] -$$
$$[f(x_i + y_j)(y_{j\,+1} - y_j) + f(x_i + y_{j\,+1})(x_{i\,+1} - x_i)].$$

If M_{ij} is a square then this expression is 0, and hence $\rho_{ij} = 0$. Consequently, there exist undotted square nets of points (i.e., nets consisting of the corners of squares) whose norms are arbitrarily small and whose quotients are exactly 0. In particular, this holds for the case mentioned in the introduction in which f is nowhere differentiable. In a similar way $\int f(ax + by)(adx + bdy)$ admits undotted nets of rectangles of a certain shape whose norms are arbitrarily small and whose quotients are exactly 0.

A sufficient condition for the independence of $\int p dx + q dy$ of the path is the existence of *one sequence* of undotted rectangular nets of points whose norms and quotients tend toward 0. As a further remark Dr. A. Wald pointed out in a conversation that if *all* rectangular nets of points whose norms are sufficiently small have arbitrarily small quotients, then the partial derivatives p_2' and q_1' necessarily exist. This remark shows clearly in which respect the classical arguments are redundant, viz., that they implicitly make assumptions for *all* nets which, according to our theorem, it is sufficient to make for *one sequence* of nets, this latter assumption being at the same time necessary.

Clearly our method is also applicable to higher dimensional integrability conditions and to the theory of complex functions.

In concluding I wish to express my thanks to Dr. A. N. Milgram and Dr. A. Wald for several suggestions simplifying the formulation and the proofs of the theorems presented in this paper.

[1] *Proc. Nat. Acad. Sci.*, **23**, 246–248 (1937) and **25**, 474–478 (1939), and Ergebnisse eines mathematischen Kolloquiums, **8**, Vienna (1937).

ON GREEN'S FORMULA

BY KARL MENGER

UNIVERSITY OF NOTRE DAME

Communicated October 11, 1940

Green's formula can be derived in the following way: For rectangular domains as the limit of an identity, for more complicated domains by virtue of elementary continuity properties of the integrals.

1. *Rectangular Domains.*—Let $u(x, y)$ be a function defined in the rectangle $R(a \leq x \leq A, b \leq y \leq B)$. If $a = x_0 < x_1 < \ldots < x_m = A$ and $b = y_0 < y_1 < \ldots < y_n = B$, then the following identity may be verified at a glance

$$\sum_{ij} \frac{u(x_i, y_{j+1}) - u(x_i, y_j)}{y_{j+1} - y_j} (x_{i+1} - x_i)(y_{j+1} - y_j) =$$
$$\sum_i [u(x_i, B) - u(x_i, b)](x_{i+1} - x_i). \quad \text{(I)}$$

If $u_y = \dfrac{\partial u}{\partial y}$ exists in R, then by the mean value theorem

$$\frac{u(x_i, y_{j+1}) - u(x_i, y_j)}{y_{j+1} - y_j} = u_y(x_i, \eta_{ij})$$

where $y_j < \eta_{ij} < y_{j+1}$. Thus from (I) we get

$$\sum_{ij} u_y(x_i, \eta_{ij})(x_{i+1} - x_i)(y_{j+1} - y_j) =$$
$$\sum_i [u(x_i, B) - u(x_i, b)](x_{i+1} - x_i). \quad \text{(II)}$$

If we assume that the Riemann integrals $\int\int_R u_y \, dx \, dy$ and $\int_a^A [u(x, B) - u(x, b)] dx$ exist, then in (II) the left side is a Riemann sum for the former, the right side one for the latter integral. By applying (II) to a sequence of decomposition of R whose norms approach 0, we thus get from (II)

$$\int\int_R u_y \, dx \, dy = \int_a^A [u(x, B) - u(x, b)] dx. \quad \text{(III)}$$

If we assume that $\int_a^A u(x, B) dx$ exists, then this integral is $=$

$- \int_A^a u(x, B) dx$, and from (III) Green's formula for R follows:

$$\int\int_R u_y \, dx \, dy = - \int_R u(x, y) dx \quad \text{(IV)}$$

the integral on the right side being taken on the contour of R in the counter-clockwise sense.

2. *Remarks and Corollaries.*—Traditionally, Green's formula for simple domains is proved by iterated integration. But in order to apply the formula $\int\int_R f(x, y)dx\,dy = \int_a^A dx \int_b^B f(x, y)dy$ to $f(x, y) = u_y(x, y)$ one has to assume[1] that the function $F(x) = \int_b^B u_y(x, y)dy$ is integrable in $[a, A]$ while our proof of (III) does not require such an assumption. By means of Volterra's bounded function which is a derivative without admitting a Riemann integral it is easy to construct a function $u(x, y)$ for which $\int\int_R u_y\,dx\,dy$ and $\int_R u\,dx$ exist (thus our proof works) and yet $\int_b^B u_y(x, y)dy$, for some values of x, does not exist (thus the classical proof does not work).

In the same way as (II) we derive the identity (II')

$$\sum_{ij} u_y(\xi_i, \eta_{ij})(x_{i+1} - x_i)(y_{j+1} - y_i) = \sum_i [u(\xi_i, B) - u(\xi_i, b)](x_{i+1} - x_i)$$

for any choice of ξ_i between x_i and x_{i+1}. If S is any Riemann sum for $\int_a^A [u(x, B) - u(x, b)]dx$ of a norm $< \nu$, then S is the right side of (II') for a proper choice of the x_i and ξ_i. By (II'), S is equal to a Riemann sum for $\int\int_R u_y\,dx\,dy$ whose norm is $< \nu$. Thus from the existence of $\int\int_R u_y$ $dx\,dy$ we conclude: $\int_a^A [u(x, B) - u(x, b)]dx$ and, in general, $\int_a^A [u(x, y')$ $- u(x, y'')]dx$ for each y' and y'', exists. If $\int_a^A u(x, B)dx$ exists, then (IV) holds. If $\int_a^A u(x, y)dx$ exists for *one* value of y, it exists for *each* value of y. But $\int\int_R u_y\,dx\,dy$ may exist without $\int_a^A u(x, y)dx$ existing for any y. Example: $u(x, y) = f(x)$, for each y, where f is not Riemann integrable in $[a, A]$.

If we *define* the integral $\int\int_R u_y\,dx\,dy$ as the limit of the sums on the left side of (I) which may be called its Weierstrass sums,[2] then this integral exists and is equal to the Riemann integral $\int\int_R u_y\,dx\,dy$ whenever the latter exists. But the Weierstrass integral exists also in many other cases. Moreover, the Weierstrass integral automatically satisfies Green's formula whenever $\int_a^A [u(x, B) - u(x, b)]dx$ exists, i.e., whenever $u(x, B) - u(x, b)$ is almost everywhere continuous, without any assumption about the behavior of u in the interior of R.

3. *Simple Closed Curves.*—In a closed domain D let $u(p) = u(x, y)$ be continuous, $|u(p)| \le U$, and $|u(p) - u(p')| < \sigma$ whenever the distance between $p = (x, y)$ and $p' = (x', y')$ is $< \delta(\sigma)$. We set $\lambda(p, p') = u(p)$ $(x' - x)$ and $\lambda(\pi) = \sum \lambda(p_i, p_{i+1})$ if π is the ordered set $\{p_0, p_1, \ldots, p_n\}$. We remark that $|\lambda(\pi) - \lambda(p_0, p_n)| < \sigma l(\pi)$ if π is contained in a circle of diameter $\delta(\sigma)$, and $l(\pi)$ is the length of π. If C is a rectifiable curve, then we denote by $\lambda(C)$ the $\lim \lambda(\pi)$ for the finite subsets of C, properly ordered, as they get indefinitely dense in C; that is to say, $\lambda(C) = \int_C u\,dx$. We have (1) $\lambda(-S) = -\lambda(S)$ if S and $-S$ are opposite segments; (2) $\lambda(V) =$

0 if V is a vertical segment; (3) $|\lambda(C) - \lambda(S)| < 2\sigma l(C)$ if the curve C, of length $l(C)$, from p to p' is contained in a circle of diameter $\delta(\sigma)$, and S is the segment pp'. For from the previous remark we have $|\lambda(C) - \lambda(p, p')|$ $< \sigma l(C)$ and $|\lambda(S) - \lambda(p, p')| < \sigma l(S) \leq \sigma l(C)$. If in D the function u_y exists, is bounded, say, $|u_y| \leq U_y$, and admits a Riemann integral κ, then (4) $|\kappa(A)| \leq U_y a(A)$ for each domain A of area $a(A)$ for which $\kappa(A) = \int\int_A u_y \, dx \, dy$ exists.

If $T = \{p_0, p', p''\}$ is a triangle whose side $p'p''$ is vertical, then, by vertical segments, T may be decomposed into a triangle T_0 with the vertex p_0 and trapezoids $T_1, \ldots, T_n$ such that each T_i is the sum of a rectangle R_i and two triangles T_i' and T_i'' (or one if T has a horizontal side), all oriented in the same sense as T. In view of (2) and (3) we have $|\sum\lambda(T_i) - \sum\lambda(R_i)|$ $< \sigma l(T)$ if the slices are so thin that each non-vertical side of the T_i is $< \delta(\sigma)$. Hence $|\lambda(T) - \sum\lambda(R_i)| < \sigma l(T) + Ul(T_0)$ which can be made as small as we please by choosing σ and T_0 sufficiently small. If, moreover, the slices are so thin that the area of $T_0 + \sum(T_i' + T_i'')$ is sufficiently small, then, in view of (4), $|\kappa(T) - \sum\kappa(R_i)|$ is arbitrarily small. Since Green's formula holds for each R_i it thus holds for T.

If P is a simple closed polygon, P can be decomposed into a finite number of triangles without common interior points, and each of them into at most two triangles with a common vertical side. Since Green's formula holds for each of these triangles, in view of (1) it holds for P.

If C: $p(t)$ $(0 \leq t \leq 1)$, $p(0) = p(1)$ is a closed rectifiable curve, we divide C into parts C_i between $p(t_i) = p_i$ and p_{i+1} $(0 = t_0 < t_1 < \ldots < t_n = 1)$ so small that each C_i is contained in a circle of diameter $\delta(\sigma)$. By adding the inequalities $|\lambda(C_i) - \lambda(S_i)| < 2\sigma l(C_i)$ resulting from (3), we get $|\lambda(C) - \lambda(P)| < 2\sigma l(C)$ where P is the polygonal line $\{p_0, p_1, \ldots, p_n\}$. Being rectifiable C can be covered with a finite number of squares whose sum, Q, has an arbitrarily small area. By choosing P so dense in C that it is contained in a Q whose area is sufficiently small, we make the area between C and P, and hence, by (4), $|\kappa(C) - \kappa(P)|$ as small as we please.[3] In view of Green's formula for P we conclude: If the simple closed rectifiable curve C admits simple closed inscribed polygons, arbitrarily dense in C, which together with their interior domains are contained in D, then Green's formula holds for C. In particular, the assumption is satisfied if D contains all points common to (1) the smallest convex set containing C, and (2) some open set containing C and its interior domain: e.g., if C and the interior domain of C are contained in the interior of D; or if C is convex and D is the domain bounded by C.

If D is a domain of the particular form $[a \leq x \leq A, \varphi(x) \leq y \leq \psi(x)]$ where φ and ψ are continuous in $[a, A]$, then without assuming any continuity properties of u we can prove by the method of sections 1 and 2 (i.e., by taking the limit of an identity): If in D the function u_y has a Riemann double integral, then

$$\int_a^A [u(x, \psi(x)) - u(x, \varphi(x))]dx$$

exists and is equal to the double integral. If $\int_a^A u(x, \omega(x))dx$ exists for *one* curve $y = \omega(x)$ $(a \leq x \leq A)$ contained in D, then the latter integral exists for *each* such curve. The integrability of u_y thus implies that the function u (if bounded) is almost everywhere continuous along each curve if it is so along one.

4. *Closed Curves.*—Let ρ be a point not on the closed curve C. We denote by $\alpha(\rho, t)$ the angle which the vector from ρ to $p(t)$ includes with the positive x-axis, and by $\mu_C(\rho)$ the integer $[\alpha(\rho, 1) - \alpha(\rho, 0)]/2\pi$. By a complementary domain of C we mean a component of the complementary set of C. For any two points of the same complementary domain U of C, μ_C has the same value which we call the multiplicity of U rel. C. The unbounded complementary domain of C has the multiplicity 0 rel. C.

If P is a closed polygon with the complementary domains $U_1, \ldots, U_n$ whose multiplicities rel. P are $\mu_1, \ldots, \mu_n$, respectively, then $\lambda(P) = \sum \mu_i \kappa (U_i)$, where it is sufficient to extend the sum over the complementary domains whose multiplicities rel. P are $\neq 0$. If P is a simple closed polygon, the statement is Green's formula. In the general case let P_i be the perimeter of U_i traversed in the counterclockwise sense. Since each P_i is a simple closed polygon, $\sum \mu_i \kappa(U_i) = \sum \mu_i \lambda(P_i)$. On the other hand, $\sum \mu_i \lambda (P_i) = \lambda(P)$. For let S be a segment of P common to P_i and P_j, ρ_i and ρ_j points of U_i and U_j close to the center of S. Then on $0 \leq t \leq 1$ the function $\alpha(\rho_i, t) - \alpha(\rho_j, t)$ increases by $2\pi(-2\pi)$ each time when S is traversed with U_i to the left (right), and only then. Thus in traversing P the algebraic number of times S is traversed with U_i to the left is $\mu_i - \mu_j$. But this is also the number of times S occurs in $\sum \mu_i \lambda(P_i)$. Segments bounding one domain only are easily seen to be traversed 0 times, all in all.

If C is a rectifiable curve with a finite number of complementary domains admitting inscribed polygons, arbitrarily dense in C, which together with their complementary domains of multiplicities $\neq 0$ lie in D, then

$$\int \int_C \mu(x, y)u_y(x, y)dx\, dy = \int_C u(x, y)dx.$$

Since for an inscribed polygon P the complementary domains approach those of C as P gets indefinitely dense in C, and $\lim \mu_P(\rho) = \mu_C(\rho)$ for each point ρ not on C, we get the theorem from that about closed polygons by the method used at the end of section 3.

5. *Other Formulae of Vector Analysis.*—The method of proving the formula for a rectangular domain as the limit of an identity, and for more

complicated domains by virtue of the elementary continuity properties of the integrals can be applied to many formulae of vector analysis. In this way E. Hemmingsen[4] proved $\int \int (u_x v_x + u v_{xx}) dx\, dy = \int u v_x\, dy$. The Weierstrass sums of the left side contain second difference quotients. The identity underlying the rectangular case can be proved by an Abel transformation of the sum.

[1] Cf. de la Vallée Poussin, *Cours d'Analyse*, pp. 333–342.

[2] Bolza, *Lectures on the Calculus of Variations*, §31. A survey of the results obtained by considering *single* integrals as limits of their Weierstrass sums is contained in my paper, "Metric Geometry and Analysis," *Rice Pamphlets*, January, 1940.

[3] Cf. H. E. Bray, *Ann. Math.*, 26, 278 (1925). We do not have to leave the domain bounded by C if we use (1) the geometric fact that there exists a simple closed polygon in the interior of C as close as we please to C, and such that the area of $C - P$ is arbitrarily small; (2) the continuity of $\lambda(C)$, for which an elementary proof can be derived from the author's more general theory (*Ergebn. e. math. Koll.*, 8, 14–19 (1937)).

[4] In his Master's thesis, University of Notre Dame, 1940.

THE BEHAVIOR OF A COMPLEX FUNCTION AT INFINITY*

By Karl Menger

ILLINOIS INSTITUTE OF TECHNOLOGY

Communicated by Lars V. Ahlfors, May 11, 1955

Traditionally, the behavior at ∞ of a complex function f is defined as the behavior at 0 of the function obtained by substituting the -1st power into f. This definition adequately describes the class of values $f(z)$ for large $|z|$. For instance, the range near ∞ of the identity function j (whose value for any z is $j(z) = z$) coincides with the range near 0 of the function j^{-1}. But that definition does not in any way describe the structural behavior of f near ∞, reflected in properties of the class of pairs $(z, f(z))$ for large $|z|$. In fact, the association of the value $f(z)$ with z may, by the substitution of j^{-1}, completely change its character. For instance, the derivative of j near ∞ is the constant[1] function 1, whereas that of j^{-1} near 0 goes even faster to ∞ than does j^{-1}.

As a result of their emphasis on the mere functional range near ∞, analysts had to assign to ∞ a role differing from that of other points on the Riemann sphere, R, much more profoundly than the arithmetical peculiarity of ∞ seems to warrant. In what follows, the function-theoretical discrepancies between ∞ and the other points on R will be reduced to a minimum. Inevitable are only differences which result from the fact that quotients of the form $(f(z) - f(\infty))/(z - \infty)$ remain undefined, which, in turn, is a direct consequence of the arithmetical peculiarity of ∞, expressed in the formula $a + \infty = \infty$ for any $a \neq \infty$.

Let f be a function whose domain is an open subset of R and whose values belong to R. For any point, a, in Dom f or on its boundary ($a \neq \infty$ or $= \infty$), I will define the *exterior derivative* of f at a as the element b of R (if there is such an element) which has the property

$$\lim [(f(z_n') - f(z_n))/(z_n' - z_n)] = b$$

for any two sequences $\{z_n\}$ and $\{z_n'\}$ of elements $\neq a$ and $\neq \infty$ for which $\lim z_n = a$ and $\lim (z_n' - z_n) = 0$. The number b, which is a complex analogue of the slope of a "paratingent" in the sense of Bouligand,[2] will be denoted by $Ef(a)$. More generally,

$$E^k f(a) = \lim [(E^{k-1}f(z_n') - E^{k-1}f(z_n))/(z_n' - z_n)],$$

where $E^0 f = f$.

If a is in Dom f and $\neq \infty$, then $E^k f(a)$ exists and is equal to the classical k-th-order derivative $D^k f(a)$ if and only if f is regular at a (*i.e.*, differentiable at a and in a neighborhood of a). At ∞, one finds, for instance, $Ej(\infty) = 1$, $Ej^2(\infty) = \infty$, $E^n j^n(\infty) = n!$ for any positive integer n, and $E^k j^n(\infty) = 0$ for any negative integer n. If $f(z) = (az + b)/(cz + d)$, where $ad - bc \neq 0$, then $Ef(\infty) = 0$ or a/d, according to whether $c \neq 0$ or $= 0$. Except for their ranges, these functions "behave" at ∞ as they "behave" at 0.

If a point a (finite or ∞) belongs to Dom f or to its boundary, then it will be called a *point of extensibility* of f if $\lim_a f$ exists on R. I will say that, at such a point, f is *steep* if $E^k f(a) = \infty$ for almost all $k > 0$ and *flat* if $E^k f(a) = 0$ for almost all

512

k > 0 without $\mathbf{E}^k f$ being constant. Points of extensibility at which f is neither steep nor flat may be called *ordinary;* and f will be said to be ordinary in an open set if and only if the latter consists of ordinary points of f.

If the point a belongs to Dom f or is an isolated boundary point, then f is steep at a if and only if a $\neq \infty$ and a is a pole of f, whereas f is flat at a if and only if a $= \infty$ and f is not a polynomial in the neighborhood of ∞. If Dom f is a wedge containing the real axis and possessing a sufficiently pointed cusp at 0, and if f assumes the value e^{-1/z^2} for any z in that wedge, then 0 is a nonisolated boundary point at which f is flat. If Dom f is the region between the real axis and a curve having this axis as an asymptote of sufficiently high order, and if f assumes the value $sin \sqrt{z}$ for any z in that region, then $\mathbf{E}f(\infty) = 0$, even though ∞ is a point of inextensibility of f.

In many celebrated theorems, the traditional distinction between ∞ and finite points on R, as well as between the interior and the exterior of simple closed curves, becomes unnecessary. For instance, Cauchy's integral theorem can be formulated as follows: If f is ordinary in either complementary region of a rectifiable, simple closed curve Γ and satisfies proper conditions on Γ, then $\int_\Gamma f = 0$. (For, if f is ordinary in the complementary region containing ∞, then f is a polynomial in that region.) Similarly, Cauchy's integral formula follows from

$$\int_\Gamma \frac{f(z) - f(a)}{z - a} \, dz = 0; \qquad \text{briefly,} \qquad \int_\Gamma \frac{f - fa}{j - ja} = 0,$$

for any point a not on Γ and $\neq \infty$.

Suppose that a is an isolated point on the boundary of Dom f, and let Γ be a curve such that the complementary region of Γ containing a consists, with the possible exception of a, of ordinary points of f. Define the residue of f at a as $(1/2\pi i)\int_\Gamma f$, where Γ is traversed once in such a way that the domain containing a is on the left side of Γ. Then, regardless of whether a $\neq \infty$ or $= \infty$, one finds $\text{Res}_a f = 2\pi i c_{-1}$, where c_{-1} is the coefficient of z^{-1} in that Laurent expansion of $f(z)$ about a which is "closest" to a. An expansion about an isolated singularity at ∞ that is closest to ∞ is, of course, a series $\sum_{-\infty}^{\infty} c_n(z - a)^n$ that converges outside some circle (punctured at ∞) about some center a.

If $|f|$ is bounded on the sphere R, punctured anywhere, then f is constant. If, in a punctured circle about a, $|f|$ has either an upper bound $<\infty$ or a lower bound >0 and no upper bound, then a is a point of extensibility of f, and the value at a of the extension is $\neq \infty$ in the first case and $= \infty$ in the second.

The ideas here developed can readily be extended to multifunctions.[3]

---

* Part of the work on this paper was done as part of Project TB 2-0001(888), sponsored by the Office of Ordnance Research. In writing this paper, the author has profited from several suggestions made by Professor R. Struble.

[1] In this paper, numbers (0, 1, i; only exception π) are denoted in roman type; functions (e.g., the constant function *1* and *j*, assuming the values *1*(z) = 1 and *j*(z) = z for any z), in italics. Correspondingly, roman letters (z, a, . . .) are used as numerical variables and *f* as a function variable. (Cf. K. Menger, *Calculus: A Modern Approach* [Boston, 1955].)

[2] Bouligand, *Introduction a la géométrie infinitésimale directe* (Paris, 1932).

[3] K. Menger, "A Simple Definition of Analytic Functions and General Multifunctions," these PROCEEDINGS, **40**, 819–821, 1954.

A CHARACTERIZATION OF WEIERSTRASS ANALYTIC FUNCTIONS

By Karl Menger

ILLINOIS INSTITUTE OF TECHNOLOGY

Communicated by Anton Zygmund, July 30, 1965

The sum theorem and other results of dimension theory make it possible to characterize the graphs of Weierstrass analytic functions among the subsets of C^2 (the set of all ordered pairs of complex numbers) *without assuming local complex parametrizations and in fact without any reference to plane topology*. A complete Weierstrass function F is a set consisting of a power series and all its analytic continuations. Its graph, $\varphi(F)$, is the set of all points (z_0, w_0) such that F includes at least one power series $w_0 + a_1(z - z_0) + \ldots + a_n(z - z_0)^n + \ldots$ of positive radius of convergence.

A subset S of C^2 will be called *smooth* at the point (z_0, w_0) if, for any two sequences of points, (z_n', w_n') and (z_n'', w_n''), contained in S, converging to (z_0, w_0), and such that $z_n' \neq z_n''$, the quotients $(w_n'' - w_n')/(z_n'' - z_n')$ have a finite limit. A set that is *vertical*, i.e., whose points have the same first coordinate, is (vacuously) smooth at each point. Sets that do not include distinct points with the same first coordinate will be called *Z-simple*; they are the geometric counterpart of (single-valued) complex functions.

As a special case of theorems on Cartesian products of metric spaces, one obtains the following basic lemma. A nonvertical, locally arcwise connected, smooth subset S of C^2 is the union of denumerably many sets that are open in S and have Z-simple closures contained in S. If S is also locally compact, then those sets can be so chosen that their closures are homeomorphic to subsets of the plane and hence at most 2-dimensional. According to the sum theorem of dimension theory, S is at most 2-dimensional. If S contains a point p at which S is more than 1-dimensional, then p can be shown to be an accumulation point of a subset of S which is (a) open in S, (b) a *quasicell*, i.e., homoeomorphic to a circular region, and (c) *regular*, i.e., the graph of a power series.

By a *patch*, I mean a connected, locally arcwise connected, locally compact set P of a dimension >1, any two points of which lie in two disjoint homeomorphic sets that are open in P. A nonvertical smooth patch P is the union of denumerably many regular quasicells, each of which is open in P; and P is a subset of the graph of exactly one complete Weierstrass function. The graph of a Weierstrass function need not be smooth, as is demonstrated by the set $\varphi(F^*)$ of all points $(z,w) \neq (0,0)$ such that $w^2 - z(z - 1)^2 = 0$, which is not smooth at $(1,0)$. But the graph of every complete Weierstrass function F is *smoothly connected* in the sense that $\varphi(F)$ contains more than one point and that any two points of $\varphi(F)$ lie in a smooth patch contained in $\varphi(F)$. Moreover, $\varphi(F)$ is the union of denumerably many smooth patches. One further proves that, if a nonvertical smoothly connected set T contains a smooth patch P and a point not lying in the graph of the complete Weierstrass function containing P, then T is not the union of denumerably many smooth patches. From here, it is one step to the following

CHARACTERIZATION THEOREM. *The graphs of complete Weierstrass functions are, besides maximal vertical sets, the only subsets of C^2 which are (1) smoothly connected, (2) unions of denumerably many smooth patches, and (3) maximal or saturated with regard to properties (1) and (2).*

Riemann surfaces of complete Weierstrass functions can be obtained without any reference to plane topology by considering *punctured smooth patches* contained in subsets S of C^2 that satisfy the conditions of the Characterization Theorem. (P,q) is obtained by deleting the point q from the smooth patch P; and (P',q') and (P'',q'') are called *equivalent* if and only if $q' = q''$ and q' is an accumulation point of the intersection of P' and P''. The set of all maximal classes[1] of mutually equivalent punctured patches in S is a Riemann surface of the function with the graph S provided that a sequence (P_n,q_n) $(n = 1,2,\ldots)$ is said to converge to (P_0,q_0) if and only if S contains, for each n, a smooth patch Q_n containing a neighborhood of q_n in P_n and a neighborhood of q_0 in P_0 and such that the diameters[2] of the Q_n converge to 0. (That $\lim q_n = q$ in C^2 is necessary but not sufficient.)

Incomplete analytic functions can be introduced on the basis of the following theorem. *In order that a smoothly connected subset S of C^2 be a subset of the graph of exactly one Weierstrass function it is necessary and sufficient that S be nonvertical and the union of denumerably many smooth patches.*[3]

Note added in proof: Local arcwise connectedness may be replaced by connectedness in the basic lemma, and can be shown to be a consequence of the other properties of a smooth nonvertical patch. In a later publication, the ideas will be applied to higher dimensional manifolds. Smoothness will be weakened to boundedness of the ratios $d(w_n'',w_n')/d(z_n'',z_n')$.

[1] Such a maximal class may also be characterized by a punctured maximal regular quasicell. The previously mentioned set $\varphi(F^*)$ contains two maximal classes punctured at $(1,0)$.

[2] The distance of two points (z', w') and (z'', w'') in C^2 may be defined as $|z'' - z'| + |w'' - w'|$.

[3] The theory can be made symmetric in the coordinates by postulating that the quotients $(w_n'' - w_n')/(z_n'' - z_n')$ have finite limits $\neq 0$.

Analytische Funktionen

Von *Karl Menger*

Weierstraß' Einführung der analytischen Funktionen als gewisse Mengen von Potenzreihen ist eine der bekanntesten Schöpfungen dieses kollossalen Mathematikers. Ein besonderer Aspekt seiner Definition aber ist vielleicht nicht genügend betont worden: daß sie nämlich viele Jahrzehnte hindurch die einzige rein mengentheoretische Definition von Funktionen war. Sie spielte diese Rolle, bis in den zwanziger und dreißiger Jahren dieses Jahrhunderts die Definition von Funktionen als gewisse Mengen von Zahlenpaaren in der Analysis aufkam; denn Dirichlet's und verwandte ältere Definitionen können bei aller ihrer enormen Fruchtbarkeit nicht als rein mengentheoretisch bezeichnet werden. Inhaltlich sind Weierstraß' Mengen von Potenzreihen einerseits viel spezieller als die modernen Funktionen, da er sich auf *analytische* Funktionen beschränkt, anderseits aber auch allgemeiner, da er z. B. die komplexe Quadratwurzel einschließt, die nicht unter die modernen Funktionen fällt, sondern eine mehrdeutige binäre *Relation* darstellt.

Was ist das genaue Verhältnis zwischen den beiden mengentheoretischen Definitionen – der Weierstraßschen und der modernen? In einer Arbeit in den Mathematischen Annalen[1] charakterisiere ich unter den Mengen von Paaren komplexer Zahlen die Graphen Weierstraßscher Mengen von Potenzreihen. Dabei gilt als Graph einer *vollständigen* Weierstraßschen Menge F (bestehend aus einer Potenzreihe mit positivem Konvergenzradius und *allen* ihren analytischen Fortsetzungen) die Menge aller derjenigen Paare (z_0, w_0), für die F eine Reihe der Form $w_0 + a_1(z - z_0) + \cdots + a_n(z - z_0)^n + \ldots$ enthält. Die zur Charakterisierung der Graphen vollständiger Weierstraßscher Mengen verwendete Methode führt auch zu einer Klasse besonderer echter Teilmengen solcher Graphen, die man naturgemäß als Graphen *unvollständiger* analytischer Funktionen bezeichnen kann.

[1] Weierstraß' Analytic Functions and Riemann Surfaces, erscheint in den Mathematischen Annalen.

Die entsprechenden Mengen von Potenzreihen können dann zwanglos als unvollständige Weierstraßsche Mengen angesprochen werden, während bisher in der Definition solcher Mengen eine gewisse Willkür herrschte[2]. Die bereits ziemlich einfache Charakterisierung loc.cit.[1] habe ich seither nicht nur auf höhere Dimensionen ausgedehnt, sondern, wie im Folgenden ausgeführt werden soll, noch weiter vereinfacht.

Zugrunde liege der 4-dimensionale reelle Raum, welcher als das Cartesische Produkt $Z \times W$ zweier Argand-Ebenen aufgefaßt werden möge. In diesem Raum heiße eine Menge ein 2-dimensionaler Fleck – und im folgenden, wo von höherdimensionalen Flecken nicht die Rede sein wird, kurz ein *Fleck*, wenn sie die folgenden fünf Eigenschaften hat:

1. M ist *nicht vertikal*, d. h. die Projektion von M auf Z enthält mehr als einen Punkt.

2. M erfüllt eine Lipschitz-Bedingung, die ich als W/Z-*Beschränktheit* bezeichnen will: Jeder Punkt $m_0 = (z_0, w_0)$ von M hat eine Umgebung V_0 und eine ihm zugeordnete Zahl k_0 derart, daß je zwei Punkte (z', w') und (z'', w'') von M, die in V_0 liegen, der Ungleichung

$$|w'' - w'| \, / \, |z'' - z'| < k_0$$

genügen.

3. M ist *zusammenhängend* im Sinne von Lennes und Hausdorff, d. h., M ist nicht die Vereinigung zweier fremder, nichtleerer Mengen, die beide in M abgeschlossen sind.

4. M ist *mindestens 2-dimensional* im Sinne der Dimensionstheorie, d. h., M enthält mindestens einen Punkt m* mit der Eigenschaft, daß die Begrenzungen aller hinlänglich kleinen Umgebungen von m* mit M mindestens 1-dimensionale Durchschnitte haben (wobei eine Menge L mindestens 1-dimensional heißt, wenn sie mindestens einen Punkt l* enthält, für den die Begrenzungen aller hinlänglich kleinen Umgebungen mit L nichtleere Durchschnitte haben).

5. M ist *gleichförmig* in folgendem Sinn: Je zwei Punkte von M liegen in zwei zu einander fremden, homöomorphen, in M offenen Mengen.

Als Anwendung dimensionstheoretischer Sätze ergibt sich dann folgendes Theorem: *Jeder Fleck ist eine unberandete 2-dimensionale Mannigfaltigkeit.*

---

[2] Vgl. *Ahlfors, L. V.*, Complex Analysis, New York 1953, p. 210.

Eine Menge M soll *glatt* genannt werden, wenn sie in jedem ihrer Punkte $m_0 = (z_0, w_0)$ die folgende Eigenschaft hat:

6. Für jede gegen m_0 konvergente Folge von zu M gehörigen Punkten (z_n, w_n) mit $z_n \neq z_0 (n = 1, 2, \ldots)$ haben die Quotienten $(w_n - w_0)/(z_n - z_0)$ einen endlichen Grenzwert.

Es stellt sich dann heraus: *Jeder glatte Fleck ist der Graph einer Menge von Potenzreihen, die alle zu genau einer vollständigen Weierstraßschen Menge gehören.* Umgekehrt ist der Graph jeder eindeutigen vollständigen Weierstraßschen Menge ein glatter Fleck. Glatte Flecken umfassen aber auch die Graphen vieler mehrdeutiger Weierstraßscher Mengen. Zum Beispiel ist die Menge der Paare (w^2, w) für alle Zahlen $w \neq 0$, d. i. der Graph der vollständigen komplexen Quadratwurzel, ein glatter Fleck. Dennoch sind die Eigenschaften 1–6 nur hinreichend und nicht auch notwendig für die Graphen Weierstraßscher Mengen. Dies sieht man etwa am Beispiel der Menge aller Paare $(z, w) \neq (0, 0)$, für die $w^2 - z(z - 1)^2 = 0$ ist. Sie ist der Graph einer vollständigen Weierstraßschen Menge von Potenzreihen, aber kein Fleck, da sie für $(1, 0)$ die Bedingung 2 verletzt.

Es lassen sich jedoch aus 1–6 Bedingungen herleiten, die sowohl hinreichend als auch notwendig sind. Eine Menge M möge *fleckenverknüpft* bzw. *glatt-verknüpft* heißen, wenn je zwei Punkte von M in einem in M enthaltenen Flecken bzw. glatten Flecken liegen. Die Charakterisierung der Graphen Weierstraßscher Mengen läßt sich dann folgendermaßen formulieren: Damit eine glatt-verknüpfte Teilmenge des 4-dimensionalen Raumes der Graph einer Teilmenge einer (und natürlich nur einer) vollständigen Weierstraßschen Menge sei, ist notwendig und hinreichend, *daß M die Vereinigung abzählbar vieler Flecken sei.* Die Mengen von Potenzreihen, deren Graphen glatte Vereinigungen abzählbar vieler (eo ipso glatter) Flecken sind, können daher zwanglos als *analytische Funktionen im Weierstraßschen Sinne* bezeichnet werden, wobei diese Begriffsbildung eindeutige sowie mehrdeutige und vollständige sowie unvollständige Funktionen erfaßt. Der Graph M einer *vollständigen* Weierstraßschen Menge ist unter den Teilmengen des 4-dimensionalen Raumes dadurch charakterisiert, daß M *maximal ist hinsichtlich der Eigenschaften glatt-verknüpft und Vereinigung abzählbar vieler Flecken zu sein;* mit anderen Worten, M ist der Graph einer vollständigen Weierstraßschen Menge dann und nur dann, wenn im Raume keine Menge mit jenen zwei Eigenschaften existiert, die M als echte Teilmenge enthielte.

Wenn man bedenkt, wie wesentlich in die traditionelle Definition von differenzierbaren Mannigfaltigkeiten und in Weyl's Definition Riemannscher

Flächen lokale Parametrisierungen eingehen[3] und daß überdies Annahmen bezüglich der Verknüpfungen sich überschneidender Parametrisierungen unentbehrlich sind, so wird man vielleicht den Umstand würdigen, daß die obige Charakterisierung von solchen Voraussetzungen völlig frei ist. Keine der Eigenschaften 1–6 von M, insbesondere weder 4 noch 5, setzt Parametrisierungen irgendwelcher Art voraus oder enthält irgendwelche sonstigen Hinweise auf die Topologie der Ebene. Noch auch brauchte Bogenverknüpfbarkeit vorausgesetzt werden – Zusammenhang im Sinne von Lennes und Hausdorff ist ja viel schwächer. Zusammenhang im Kleinen wurde nicht postuliert, ja nicht einmal lokale Kompaktheit!

---

[3] In seinem Nachruf auf Hilbert schrieb *Weyl, H.* (Bull. Amer. Math. Soc. *50*, 1944, 638f.): "Hilbert defines a two-dimensional manifold by means of neighborhoods, and requires that a class of 'admissible' one-to-one mappings of a neighborhood upon Jordan domains in an x, y-plane be designated, any two of which are connected by continuous transformations ... When it comes to explaining what a *differentiable* manifold is, we are to this day bound to Hilbert's roundabout way."

THÉORIE DES FONCTIONS. — *Une caractérisation des fonctions analytiques.*
Note (*) de M. **Karl Menger**, présentée par M. Georges Bouligand.

En 1902 Hilbert ([1]) introduit des variétés à deux dimensions au
moyen de voisinages dont chacun est en correspondance biunivoque et
continue avec un domaine de Jordan contenu dans un plan. Il admettait
encore l'hypothèse que toute correspondance biunivoque entre les images
planes de deux voisinages, due à des points communs de ces derniers,
soit continue. Dans son éloge de Hilbert, H. Weyl ([2]) soulignait en 1944
qu'en définissant des variétés différentiables on était encore obligé de
suivre ce chemin indirect (« round-about way ») de Hilbert. Dès lors,
malgré la critique subtile de Weyl, cette méthode a pénétré la littérature
géométrique et analytique.

Dans ce qui suit nous introduisons d'une façon directe et intrinsèque
(c'est-à-dire sans supposer des correspondances entre les voisinages et
des domaines euclidiens et, *a fortiori*, sans hypothèse concernant des
paramétrisations chevauchantes) une classe de variétés contenues dans
un espace cartésien. Nous définirons même des objets plus généraux et
importants en géométrie différentielle, qui comprennent en particulier
les graphiques des fonctions analytiques. Ces dernières se prêtent donc
à une caractérisation directe et intrinsèque parmi les sous-ensembles de
l'espace euclidien à quatre dimensions.

Soient donnés deux espaces métriques A et B au sens de M. Fréchet.
Nous dirons qu'un sous-ensemble E du produit cartésien $A \times B$ est B/A-*borné
au point* $p_0 = (a_0, b_0)$ si E satisfait à la condition suivante de Lipschitz :
il existe un voisinage V_0 de p_0 et un nombre $r_0 > 0$ tel qu'on ait
$B(b', b'')/A(a', a'') < r_0$ pour tout couple de points $p' = (a', b')$ et $p'' = (a'', b'')$
de E en V_0, où $A(a', a'')$ et $B(b', b'')$ sont les distances dans les espaces A
et B. Un ensemble E sera appelé B/A-*borné* s'il est B/A-borné en tout
point de E et contient deux points p' et p'' tels que $a' \neq a''$.

Supposons maintenant que A soit l'espace cartésien à n dimensions et
que B soit séparable et localement compact. Par *morceau au-dessus de* A
nous entendrons un sous-ensemble M de $A \times B$ jouissant des quatre
propriétés suivantes : 1. M est B/A-borné; 2. M est connexe au sens de
Lennes et Hausdorff; 3. M est de dimension $> n - 1$ au sens de la théorie
de dimension; 4. M est *uniforme* au sens suivant : deux points quelconques
de M sont contenus dans deux voisinages disjoints et homéomorphes.
On remarquera que la condition d'uniformité est très faible ([3]). D'ailleurs,
en définissant des morceaux on ne suppose ni qu'ils soient localement
connexes, ni qu'ils contiennent des arcs, ni même qu'ils soient localement
compacts. Néanmoins, en vertu de résultats de la théorie de dimension,

on démontre le théorème suivant : *Tout morceau au-dessus de* A *est une variété à n dimensions.*

Nous appellerons M un *morceau entre* A *et* B si la condition 1 est remplacée par la suivante : 1*. M est B/A- et A/B-borné. *Si* (a_0, b_0) *est un point quelconque d'un morceau entre* A *et* B, M *détermine une homéomorphie entre un voisinage de* a_0 *et un voisinage de* b_0. Si A et B sont des plans euclidiens, un morceau entre A et B est appelé *lisse* si, pour tout point (a_0, b_0) de M et pour toute suite de points (a_n, b_n) de M $(a_n \neq a_0)$ qui converge vers (a_0, b_0), les quotients $B(b_n, b_0)/A(a_n, a_0)$ ont une limite. D'un théorème fondamental de H. Bohr [4] on déduit : *Si* M *est lisse et si* A *et* B *sont des plans d'Argand, l'homéomorphie sus-dite détermine une fonction analytique ou la conjuguée d'une telle fonction.* La fonction est analytique si M est *diagonal*, c'est-à-dire s'il contient deux arcs ayant le même point initial et des projections en A et B formant des angles du même sens.

On obtient des généralisations importantes en considérant les ensembles E *connexes par morceaux*, c'est-à-dire tels que tout couple de points de E soit situé dans un morceau entre A et B contenu en E. La portée de cette notion apparaît au cas où A et B sont deux plans euclidiens. D'après Weierstrass, une fonction analytique complète est un ensemble F qui consiste en une série de puissances d'un rayon de convergence positif et en tous ses prolongements analytiques. Par le *graphique* de F nous entendons l'ensemble E_F de tous les points (z_0, w_0) du produit de deux plans d'Argand $Z \times W$ tels que F contienne au moins une série de la forme

$$w_0 + a_1(z - z_0) + \ldots + a_n(z - z_0)^n + \ldots .$$

Pour qu'un sous-ensemble E *de* $Z \times W$ *soit le graphique d'une fonction analytique complète il est nécessaire et suffisant que* E *satisfasse aux conditions suivantes : 1.* E *est connexe par morceaux lisses et diagonaux; 2.* E *est l'union d'une famille dénombrable de morceaux au-dessus de* A; *3.* E *est maximal par rapport à ces propriétés,* c'est-à-dire E n'est pas contenu dans un sous-ensemble de $Z \times W$ plus tendu et jouissant des propriétés 1 et 2. Ces dernières (sans la condition de maximalité) donnent une définition satisfaisante des fonctions analytiques, complètes ou incomplètes [5].

[*] Séance du 15 novembre 1965.

[1] *Göttinger Nachrichten*, 1902.

[2] *Bull. Amer. Math. Soc.*, 50, 1944, p. 638 sq.

[3] Elle est satisfaite, par exemple, par la courbe triangulaire de M. Sierpinski (*Comptes rendus*, 173, 1915, p. 302) qui contient des points dont l'ordre de ramification est 3 ou 4 et trois sommets d'ordre 2.

[4] *Mathem. Z.*, 1, 1918, p. 403-420.

[5] Une caractérisation semblable, mais basée sur une notion plus restreinte de morceau, se trouve dans un Mémoire de l'auteur qui est sous presse au *Mathematische Annalen*.

(*Illinois Institute of Technology,*
Chicago, Illinois, 60616, U. S. A.)

MENGER, K.
Math. Annalen 167, 177—194 (1966)

Weierstrass Analytic Functions

KARL MENGER*

1. Introduction

WEIERSTRASS introduced a complete analytic function as a set consisting of a power series and all its analytic continuations. This definition — a fundamental expansion of LAGRANGE's introduction of a function as a power series — was for many decades the only alternative to the classical introduction of functions as laws or rules associating or pairing numbers with numbers or — à la DIRICHLET — as those associations themselves. These traditional introductions are not entirely transparent; for, while there exist definite procedures for operations with sets (e.g., with sets of power series or of numbers or of pairs of numbers) traditional mathematics lacks definitions of operations with laws or rules or associationes [1]. WEIERSTRASS' alternative was confined to the realm of analytic functions but it was perfectly clear.

A second strictly mathematical definition evolved only in the 1920's and 1930's, focusing on the *result* of the association or pairing of numbers with numbers, that is, on the set of pairs of numbers obtained [2]. But only the sets corresponding to single-valued functions have been characterized within the realm of the sets of pairs of numbers (called *binary numerical relations*), namely, by the obvious property that they do not include pairs with equal first, and unequal second, members [3]. Other sets that warrant the name function have not in the traditional literature been extricated from the enormous variety of binary relations.

---

* The author wishes to thank Prof. A. SKLAR and Mr. H. I. WHITLOCK for numerous valuable suggestions and Mr. RONALD DETRY for careful reading of the proofsheets.

[1] Hence, when operating with functions, one rarely can replace the term *function* by its supposed definiens — the words *a law of a certain kind* or the like. — Functions have also frequently been introduced by saying that a function is defined if numbers are associated with numbers in a certain way. But this is, strictly speaking, a definition of definitions of functions rather than of functions.

[2] In the middle of this century a third method (confined to a narrow class of functions) was added by recursive definitions. A fourth method — the introduction of a system of undefined elements (called functions) by algebraic postulates which imply that the system is isomorphic with a set of explicitly defined functions — is still in the process of development [cf. K. MENGER, Math. Ann. **157**, 278—295 (1964) for the history and bibliography of this idea].

[3] Even for a single-valued function (e.g., a constant function or the exponential function), Weierstrass' definition as a set of power series is of course essentially different from that as a set of pairs of numbers.

If F is any analytic function according to WEIERSTRASS, one can of course consider the set $\varphi(F)$ of all pairs (z_0, w_0) such that F includes an element (i.e., a power series) of the form

(1)
$$w_0 + \sum_1^\infty a_n(z - z_0)^n .$$

I shall call $\varphi(F)$ the *graph* of F. But only for single-valued complex functions with open domains have these graphs been characterized, namely, on the basis of Cauchy's theorem that such a function is locally a power series if and only if it is everywhere differentiable. There does not seem to exist in the literature a characterization of the graphs of multivalued analytic functions.

The main result of this paper is such a characterization. One might think of the analogue for sets of the differentiability of functions. A set S of pairs of numbers may be called *almost smooth* at (z_0, w_0) if, for each sequence of pairs (z_n, w_n) in S converging to (z_0, w_0), the quotients $(w_n - w_0)/(z_n - z_0)$ have a (finite) limit. That this property is not sufficient is demonstrated by the simple

Example 1. The set, A_1, of all pairs (z, w) such that $(w - z)^2 - z^3 = 0$ is almost smooth at each point, in particular at $(0, 0)$, where the difference quotients have a finite limit $\neq 0$, namely 1. Yet A_1 is not the graph of a Weierstrass function; for $(0, 0)$ is a branch point, wherefore the function cannot be developed in a power series with $z_0 = w_0 = 0$.

Important for the characterization problem is the following more stringent concept. I shall call a set S of pairs of complex numbers *smooth* at (z_0, w_0) if the quotients $(w_n'' - w_n')/(z_n'' - z_n')$ have a finite limit for every pair of sequences (z_n', w_n') and (z_n'', w_n'') belonging to S, converging to (z_0, w_0), and such that $z_n'' \neq z_n'$. The set A_1 in Example 1, while almost smooth, is not smooth at $(0, 0)$. Indeed, there is no finite limit of the quotients associated, e.g., with the pairs (z_n', w_n') and (z_n'', w_n'') such that

$z_n' = (2n)^{-1}, w_n' = (2n)^{-1} + (2n)^{-3/2}$ and
$z_n'' = (2n+1)^{-1}, w_n'' = (2n+1)^{-1} - (2n+1)^{-3/2}$.

But while almost smoothness of a set is not sufficient for its being the graph of a Weierstrass function, smoothness is too stringent a property to be necessary, as demonstrated by

Example 2. The complete Weierstrass function F_2 whose graph $\varphi(F_2)$ consists of all pairs (z, w) such that $w^2 - z(z - 1)^2 = 0$ except $(0, 0)$ is not smooth at $(1, 0)$, as is seen by setting

$$z_n' = (2n+1)/2n, w_n' = \sqrt{z_n'} \quad \text{and} \quad z_n'' = 2(n+1)/(2n+1), w_n'' = -\sqrt{z_n''} .$$

F_2 includes two power series with $z_0 = 1$, $w_0 = 0$, but of course none with the center $(0, 0)$. (For examples of a different nature, see Section 5.)

It will be proved, however, that every graph $\varphi(F)$ is what I shall call *smoothly connected*; that is to say, that any two pairs belonging to $\varphi(F)$ lie in a smooth patch contained in $\varphi(F)$, a *patch* being defined by simple topological properties (connectedness, local arcwise connectedness, local compactness, dimension > 1, and a very weak homogeneity) but without any reference to the complex plane of to plane topology. Secondly, each $\varphi(F)$ is the union of denumerably many smoothly connected patches. Thirdly and lastly, each $\varphi(F)$ is *saturated* or *maximal* with regard to these two properties in the sense of not being a proper subset of any set with the two properties. But these three properties can be shown to be shared by the sets $\varphi(F)$ with only one other type of sets, which I shall call *complete vertical sets* each of which, for some number z_0, consists of all pairs (z_0, w). In other words, one arrives in this way at a characterization of the graphs of complete Weierstrass functions.

In possession of this characterization one may *introduce* complete analytic functions and certain types and generalizations of analytic functions as sets of pairs of numbers with certain properties: multivalued as well as single-valued functions and, notably, complete as well as incomplete functions[4].

This introduction may also be compared with that by RIEMANN, who, in contrast to WEIERSTRASS, considered analytic functions more along the lines of DIRICHLET. He paired complex numbers with elements of a set, but not of complex numbers. As domains of analytic functions he introduced what are today called Riemann surfaces. WEYL made this idea precise by an abstract definition in terms of a topological space with local complex parametrization, thus with essential reference to plane topology. The elegance of Weyl's definition is, moreover, somewhat impaired by inevitable assumptions concerning analytic interrelations of overlapping local parametrizations. It might also be pointed out that, for many purposes, it is not sufficient to consider sets of pairs whose first members are elements of a Riemann surface while their second members are complex numbers. One is forced to consider sets of ordered pairs both of whose members are elements of (generally different) Riemann surfaces. Complex numbers, which are so essential to the idea of analytic functions, enter only through the back door.

From the point of view taken in this paper, the function, being a set of pairs of complex numbers, may also be considered as a "curve" in the complex "plane", consisting of all such pairs, or as a 2-dimensional surface in a real 4-dimensional space. It is customary to distinguish these graphs from, or even to contrast them with, the functions themselves, defined as laws or associations. But since the mathematically tangible content of the latter definitions — the products of those laws and the results of those associations — are the very same sets of pairs of numbers that constitute the graphs, it is clear that functions and their graphs in pure geometry[5] are identical. The use of geometric and

---

[4] Objections have been raised against Weierstrass' way of introducing incomplete analytic functions (see Section 10 of this paper). — The idea of characterizing analytic functions among the sets of pairs of complex numbers as well as the core of the concept of smoothness has been anticipated in two notes [K. MENGER, *A simple definition of analytic functions and general multifunctions*, Proc. Nat. Acad. Sci. U.S. **40**, 819—821 (1954) and *The behavior of a complex function at infinity.* ibid. **41**, 512—513 (1955)], which do not seem to have attracted much attention. In absence of the concept of smooth connectedness, the characterization given loc. cit., on the one hand, transcended analytic function, on the other hand, was rather cumbersome.

[5] *Pure geometry*, as opposed to *postulational* and to *physical* geometry.

13*

function-theoretical synonyms is indeed very convenient[6]; but, in principle, either the geometric or the analytic terminology might be dispensed with.

From the point of view of this paper, a complete analytic function is itself its own concrete Riemann surface of a sort. Its domain, whose specification in the Argand plane or on a Riemann surface is (unless one follows WEIERSTRASS) the traditional starting point in defining the function, will herein be introduced only secondarily, namely, as the set of the first members of the pairs belonging to the function, just as its range is the set of the second members. In geometric terminology, domain and range are projections of the function-surface.

2. Weierstrass sets

In the literature, the term *power series* is used in two senses: for a *function of a special kind* as well as for an *expression* of the form (1), often referred to as a *formal power series*. In the present paper this equivocation will be avoided by adopting the following

Convention I. A formal power series will be considered simply as a *sequence of complex numbers*[7]. Only sequences satisfying the Cauchy-Hadamard condition for a positive radius of convergence will play a role. They will be referred to as Cauchy-Hadamard (briefly, *CH-sequences*). Such a sequence, more specifically, the sequence corresponding to the expression (1), will be denoted by

$$(2) \qquad \alpha = (z_0, w_0, a_1, a_2, ..., a_n, ...), \quad \text{where} \quad r(\alpha) = \liminf |a_n|^{-1/n} > 0.$$

The pair (z_0, w_0) will be called the *center* of the sequence α. The term *power series* will be reserved for a *function* in the modern sense. More specifically, the set of the pairs (z, w), where w stands for the expression (1), for all z such that $|z - z_0| < r(\alpha)$, will be called the power series associated with the CH-sequence (2) and will be denoted by p_α.

In order to keep the definition of Weierstrass' concept free of any reference to functions other than sequences of numbers (which of course are functions whose domain is the set of all natural numbers), we shall consider Weierstrass

---

[6] Also the set of the pairs (x, x^2) for all real numbers x has two names. It is called a *parabola* in pure Cartesian geometry, and a *function* (namely, the square function or, if one uses the symbol j for the identity function, the function j^2) in real analysis.

[7] In possession of a symbol, say j, for the identity function one can, of course, also dispense with the letter z in the definition of the *function p*; one can define p as $w_0 + \sum a_n (j - z_0)^n$, where z_0 and w_0 are the constant function of the values z_0 and w_0, respectively, and the a_n are scalar multipliers.

It will be noted that *functions* in the modern sense are herein denoted by lower case letters in *italic* type while references to (complex and real) numbers are in roman type; as in the sentence, the function p assumes the value $p(z)$ for z. The method of typographical distinctions between symbols referring to unlike categories — in analysis, between symbols for numbers, functions, and operators — offers too great advantages in clarity and readability to be ignored. The systematic (conceptual as well as symbolic) distinction between a function f and its value $f(x)$ is only delayed by the widespread reluctance against the introduction of a symbol, such as j, for the identity function.

functions as sets of CH-sequences rather then as sets of power series. Any confusion that might arise from the equivocal use of the term (complex) *function* for sets of pairs of (complex) numbers as well as for sets of CH-sequences will be forestalled by the following

Convention II. Sets of CH-sequences will be referred to as *Weierstrass sets* and denoted by capital italics while functions will be denoted by lower case italics. Thus a complete Weierstrass set is a set consisting of all CH-sequences that are continuations of one of them.

The CH-sequence $\alpha' = (z_0', w_0', a_1', \ldots, a_n', \ldots)$ is said to be a *direct continuation* of α if and only if $|z_0'| < r(\alpha)$ and, for every number z such that $|z - z_0| < < r(\alpha)$ and $|z - z_0'| < r(\alpha)$,

$$w_0 + \sum_1^\infty a_n(z - z_0)^n = w_0' + \sum_1^\infty a_n'(z - z_0')^n .$$

The CH-sequence β is a *continuation* of α if there exists a finite chain $\alpha = \alpha_0, \alpha_1, \ldots, \alpha_{m-1}, \alpha_m = \beta$ of CH-sequences such that α_k is a direct continuation of α_{k-1} for $k = 1, \ldots, m$.

For example, a *constant* complete Weierstrass set is, for some number c, the set of the CH-sequences $(z_0, c, 0, \ldots, 0, \ldots)$ for all numbers z_0. The *reciprocal* complete Weierstrass set or the complete (-1)st power set the so-called "function $\dfrac{1}{z}$" is the set of the sequences $(z_0, z_0^{-1}, -z_0^{-2}, z_0^{-3}, \ldots, (-1)^{n-1}z_0^{-n}, \ldots)$ for all numbers $z_0 \neq 0$. Even the counterpart of the power series p_α can be introduced in this way. By the Weierstrass set P_α having the graph p_α or, as one might say, by the *arithmon* of p_α, will be meant the set consisting of α and all direct continuations $\alpha' = (z_0', \ldots)$ of α. Clearly, $\varphi(P_\alpha) = p_\alpha$. In order that P_α be a complete Weierstrass set it is necessary and sufficient that, for every direct continuation $\alpha' = (z_0', \ldots)$ of α, the equality $r(\alpha') = r(\alpha) - |z - z_0|$ hold. In this case, the graph p_α is said to have the circle of convergence as its natural boundary.

The use of the term *multivalued function* for something that is not a *function* is awkward and will be avoided by the following

Convention III. The term *function* will hereinafter consistently be used in the sense of *single-valued* function. The set of the first (the second) members of all pairs belonging to a function f will be called the *domain* of f (the *range* of f) — briefly, dom$f$ (and ran$f$). Some sets of pairs of numbers which are not functions will be called *multifunctions*, but by no means all such sets.

In fact, it is the purpose of this paper to characterize analytic multifunctions. It will also be convenient to adopt

Convention IV. Finite sets and even the empty set will herein be regarded as *denumerable*.

3. The Three-point Lemma and its consequences

Let M be the Cartesian product, $A \times B$, of two metric spaces, A and B. The distances between two elements a_1 and a_2, and b_1 and b_2, are denoted by

$A(a_1, a_2)$ and $B(b_1, b_2)$, respectively. The elements of M will be referred to as *points*. A topology in M will be defined by considering, for each point $p_0 = (a_0, b_0)$ of M and any two positive real numbers e_1 and e_2, the set of all points (a, b) such that $A(a, a_0) < e_1$ and $B(b, b_0) < e_2$ as a basic neighborhood of p_0. A set will be called *open*, and a *neighborhood* of each of its points, if and only if it is the union of a family of basic neighborhoods.

If S is a subset of M, then by the *first* projection or A-*projection* of S will be meant the set of all elements of A such that S includes a point (a, b) for at least one element b of B. If the A-projection of S consists of a single element, then S will be said to be *vertical*. If, on the contrary, no two points of S have the same A-projection — in other words, if S does not contain a vertical pair of points — then S will be said to be A-*simple*. If the intersection of S with a neighborhood of the point p is vertical or A-simple, then S will be said to be *vertical at* p or A-*simple at* p, respectively.

Lemma 1 *(Three-point Lemma). If a subset S of M contains a vertical pair of points,* $p' = (a, b')$ *and* $p'' = (a, b'')$*, and an arc T joining* p' *to a point* q *whose A-projection is* $\neq a$*, then S contains, for each natural number n, two points* $p_n = (a, b_n'')$ *and* $q_n = (a_n, b_n)$ *such that* $a_n \neq a$ *and*

(2)
$$\frac{B(b_n'', b_n)}{A(a, a_n)} > n .$$

If no segment of T beginning at p' *is vertical, then one may choose* $p_n = p''$ *for each n. If T does contain an initial vertical segment, then* p_n *and* q_n *can be so chosen that they converge to a point r of T.*

Let $r = (a, \bar{b})$ be the first point met in traversing T from p' to q that is an accumulation point of points whose A-projection is $\neq a$. If no initial segment of T is vertical then $r = p'$. Then T contains, for each n, a point $q_n = (a_n, b_n)$ with $a_n \neq a$ and so close to p' that (2) holds for p'' and q_n. If $r \neq p'$ then, for each n, let U_n be the neighborhood of r consisting of the points (x, y) for which $|x - a| < 1/n$ and $|y - \bar{b}| < 1/n$. Since the segment of T from p' to r is vertical, S contains 1) a point $p_n = (a, b_n'')$ with $b_n'' \neq \bar{b}$, and 2) a point $q_n = (a_n, b_n)$ with $a_n \neq a$ and so close to r that (2) holds. Clearly, p_n and q_n converge to r.

A set S will be said to be *metrically smooth at the point* $q = (a, b)$, more specifically, *metrically s-smooth at* q, where s is a finite (and obviously nonnegative real) number if and only if s is the limit of the quotients $B(b_n'', b_n')/A(a_n'', a_n')$ for any two sequences of points (a_n', b_n') and (a_n'', b_n'') of S converging to (a, b) and such that $a_n' \neq a_n''$. A set will be called *metrically smooth* if it is metrically smooth at each of its points. If S is vertical at q, then no two such sequences converging to q exist. Hence the condition for smoothness at q is (vacuously) satisfied (for each s_0).

The Three-Point Lemma has a consequence for a set S that is metrically smooth at a point q at which S is *locally arcwise connected*; this means that each neighborhood U of q contains a neighborhood V of q with the following property: each point of V is connected with q (and, therefore, with any other point in V and S) by an arc contained in U and S.

Lemma 2. *If* S *is locally arcwise connected at the point* p *of* S *and metrically smooth at* p, *then* S *is either* A-*simple at* p *or vertical at* p. *An arcwise connected, metrically smooth set that is vertical at one of its points is vertical and homeomorphic to a subset of* B. *A compact* A-*simple set is homeomorphic to a subset of* A; *more specifically, the mapping of each point of* S *on its* A-*projection is topological.*

In order to prove the first part, let U be any neighborhood of p. There exists a neighborhood V of p such that each point of the intersection, W, of S and V is joined to p by an arc contained in S and U. Suppose that S is, say, s-smooth at p. Assume further that S is not A-simple at p. Then S will be shown to be vertical at p. For W contains a vertical pair of points, p' and p", and, if S were not vertical at p, then W would contain a point with an A-projection different from that of p'. Let n be an integer $>s$. By the Three-Point Lemma, W would contain two points p_n and q_n satisfying (2). The existence of such points in each neighborhood of p would contradict the metric s-smoothness of S at p. — If S is arcwise connected and vertical at the point p' of S without being vertical, then S is not smooth at the point r mentioned in the Three-Point Lemma. — If the element b of B is associated with each point (a_0, b) of a vertical set the mapping obviously is topological. If S is A-simple and with each point (a, b) of S its A-projection, a, is associated the mapping is one-to-one and continuous; and, if S is compact, it is topological.

4. Applications to C^2

In what follows, C will denote the set of all complex numbers, and C^2, the set of all ordered pairs of complex numbers (also referred to as *points*), endowed with the ordinary limit definition. The first projection of a subset of C^2 will also be called its Z-*projection*. Clearly, the term Z-*simple subset of* C^2 is a geometric synonym (see Introduction) of the term *complex function*, the Z-*projection of the set* being the equivalent of the *domain of the function*.

If for the set $A \times B$ considered in Section 3, $A = B = C$ and $C(c', c'') = |c'' - c'|$, then metric smoothness can be sharpened to smoothness, as defined in the Introduction. A subset S of C^2 will be called *smooth* at (z_0, w_0) if not only the absolute values of the quotients $(w_n'' - w_n')/(z_n'' - z_n')$ but these quotients themselves have a finite limit for any pair of sequences converging to (z_0, w_0) provided $z_n'' \neq z_n'$. Every set S that is vertical at (z_0, w_0) is smooth at this point. While each smooth subset of C^2 is metrically smooth the converse statement is of course false. If f is any function that is analytic in a region R, and L is any straight line in C, then, for each z in R, let $f^*(z)$ denote the reflection of $f(z)$ in L. The set of the points $(z, f^*(z))$ for all z in R is metrically smooth without being smooth.

Let S be a locally arcwise connected, nonvertical, smooth (and, therefore metrically smooth) subset of C^2. According to Lemma 2, S is not vertical at any of its points; and each point of S lies in an open subset of C^2 whose intersection with S is Z-simple, i.e., a complex function. Obviously these open sets may be chosen in such a way that even the closures of their intersections with

S are Z-simple. Since C^2 is separable, denumerably many such open sets cover S. One thus has the first part of

Lemma 3. *A locally arcwise connected, nonvertical, smooth subset S of C^2 is the union of denumerably many sets that are open in S and whose closures are Z-simple. If S is also locally compact, then S is the union of denumerably many compact Z-simple sets each of which is the closure of a relatively open set and is homeomorphic to its Z-projection; and S is at most 2-dimensional in the sense of dimension theory.*

Here, by *locally compact* is meant that each point p of S lies in a set N_p that is open in S and whose closure is a compact subset of S. The intersection of N_p with a set containing p that is open in S and whose closure is Z-simple is a neighborhood of p in S whose closure is a compact and Z-simple subset of S. Denumerably many such neighborhoods cover S. According to Lemma 2, each of these compact sets is homeomorphic to its Z-projection which, being a subset of the plane, is at most 2-dimensional. Since S is the union of denumerably many compact sets that are at most 2-dimensional it follows from the Sum Theorem of dimension theory that S is at most 2-dimensional.

Of course, S need not be 2-dimensional. The set of the points (z, z) for all z such that $|z| = 1$ is compact and a continuous function but 1-dimensional. Nor is S necessarily compact or a continuous function, as is demonstrated by

Example 3. Let D be the subset of C consisting of the number 0 and the numbers $x + yi$ belonging to the union D' of the three segments

$$D_1 : 0 < x \leqq 1, y = 0 ; \quad D_2 : x = 0, 0 < y \leqq 1 ; \quad D_3 : x + y = 1, x \geqq 0, y \geqq 0.$$

Let A_3 be the subset of C^2 consisting of the point (0, 0) and the points $(z, e^{-1/z^2})$ for all z in D'. Clearly, the set A_3 is homeomorphic to a halfopen interval in the straight line. A_3 is smooth at each point. For any point of A_3 whose Z-projection belongs to D', including $(1, e^{-1})$ and (i, e), this follows (see Section 5) from the analyticity of the function f assuming for each $z \neq 0$ the value $f(z) = e^{-1/z^2}$. At (0, 0), the complete function f is not smooth; but S is, since the only points of S converging to (0, 0) have Z-projections on D_1. Yet the function S is not even continuous, let alone differentiable, at 0, since the points $(yi, e^{1/y^2})$ have no limit as y approaches 0. While Example 1 has demonstrated that an almost smooth set need not be a function, Example 3 shows that even a smooth function need not be continuous.

Lemma 4. *A compact Z-simple subset of C^2 is a continuous function. If, at the number z_0 of its domain, a function f is continuous and almost smooth (or, which is more, smooth), then f has a finite derivative $f'(z_0)$, at z_0.*

If S is a compact Z-simple set then the mapping of S on its Z-projection, being topological, has a continuous inverse. Hence S is a continuous complex function whose domain is that Z-projection (of course, without necessarily being a one-to-one function; in fact, S may be even a constant function). Suppose now that, at the number z_0 in dom$f$, a function f is continuous and almost smooth. If z_n are numbers in dom$f$ converging to z_0, then, since f is continuous at z_0, the points $(z_n, f(z_n))$ vonverge to $(z_0, f(z_0))$. Since f is almost smooth at z_0, the quotients $(f(z_n) - f(z_0))/(z_n - z_0)$ have a finite limit. This completes the proof of Lemma 4.

In conjunction with local arcwise connectedness and local compactness, the smoothness of a nonvertical set S thus implies that S is the union of denumerably many differentiable functions each of which is open in S. But all these assumptions fail to guarantee that any of those differentiable functions admits local developments in power series (since Cauchy's theorem is confined to differentiable functions whose domains are open in C). This is demonstrated by

Example 4. Let g be a real-valued function whose domain is the real interval $[0, 2\pi]$ satisfying the following conditions: the first and second derivatives g' and g'' exist and are continuous in $[0, 2\pi]$; they as well as g assume equal values at 0 and 2π; but g''' does not exist anywhere. Now consider the set A_4 consisting of the points $(e^{ti}, g(t))$ for all t in $[0, 2\pi]$. This set is an arc which, in view of the existence of g'', is easily seen to be smooth (and, in fact, is a differentiable function) but, because of the nonexistence of g''', does not contain any relatively open subset in which it would be equal to a power series.

A step toward power series is taken by assuming that S has a dimension > 1. First, the main concepts needed here and in the sequel will be summarized. By a *quasidisc* (a disc) is meant a circle without (with) its boundary. A topological image of a quasidisc (a disc) will be called a *quasicell* (a *cell*), the word *2-dimensional* being omitted for the sake of brevity. A power series p_α, as defined in Section 2, is a quasicell homeomorphic with its Z-projection. If $\alpha = (z_0, w_0, ...)$, then p_α will also be referred to as a *maximal* power series, in contrast to its *concentric restrictions* consisting, for some positive $r < r(\alpha)$, of the points $(z, p_\alpha(z))$ for all z such that $|z - z_0| < r$. Geometric synonyms will be convenient. Maximal as well as concentrically restricted power series will also be referred to as *regular quasicells* (the maximal power series, as maximal regular quasicells). By a *regular cell* with the center (z_0, w_0) will be meant, for some $\alpha = (z_0, w_0, ...)$ and some positive r less than (not $\leqq$) $r(\alpha)$, the set of the points $(z, p_\alpha(z))$ for all z such that $|z - z_0| \leqq r(\alpha)$.

Lemma 5. *If the locally compact, locally arcwise connected, nonvertical, smooth set S contains at least one point at which S is more than 1-dimensional, then S contains a regular quasicell that is open in S.*

By Lemma 3, S is the union of denumerably many compact Z-simple sets each of which is homeomorphic to a subset of the plane. At least one of these compact sets, say N, is more than 1-dimensional by the Sum Theorem of dimension theory, since S is more than 1-dimensional. The Z-projection of N, being homeomorphic to N, is a 2-dimensional subset of the plane and, therefore, contains the interior of a circle. This quasidisc belongs to the domain of the function N which, according to Lemma 4, is continuous and differentiable, since N is compact. Let M be the set that is open in N and whose closure is N. By Cauchy's theorem, N contains a regular quasicell, R, that is open in M and, therefore, in S.

R (and, in general, an even larger subset of S) is contained in the graph of a complete Weierstrass set. But this cannot always be said about S itself, even if S is 2-dimensional at each of its points, as is demonstrated by

Example 5. Let A_5 be the set consisting of the point $(0, 0)$, the points (z, z) for all $z = re^{si}$ such that $|z^2 - 1| \leq 1$ and $\frac{3}{4}\pi < s < \frac{5}{4}\pi$, and the points $(z, z^2 + z)$ for all numbers $z = re^{si}$ such that $|z^2 - 1| < 1$ and $-\frac{1}{4}\pi < s < \frac{1}{4}\pi$. The set A_5 is a smooth locally arcwise connected continuum that is 2-dimensional at each point. It is a function and everywhere differentiable. A proof is required only for $(0, 0)$. Let (z'_n, w'_n) and (z''_n, w''_n) be two sequences with $z'_n \neq z''_n$ converging to $(0, 0)$. If z'_n and z''_n lie inside of the same loop of the lemniscate, then the quotients $(w''_n - w'_n)/(z''_n - z'_n)$ clearly converge to 1. So they do also if z'_n and z''_n lie in opposite loops since

$$|(z''^2_n + z''_n - z'_n)/(z''_n - z'_n) - 1| \leq |z''^2_n/(z''_n - z'_n) - 1| \leq |z''_n|.$$

The set A_5 is the union of two smooth cells. If B_5 consists of $(0, 0)$ and those points (z, w) of A_5 for which $|z^2 - 1| < 1$, then every point of B_5, except $(0, 0)$, lies in a smooth quasicell contained in B_5; and B_5 is also the union of denumerably many smooth cells (at least one of which contains $(0, 0)$ on its rim). Clearly, neither A_5 nor B_5 is a subset of the graph of a complete Weierstrass set.

5. Smooth patches

A further step toward the graphs of complete Weierstrass sets will now be taken by considering smooth patches. By a *patch*, I mean a connected, locally arcwise connected, locally compact, more than 1-dimensional set that is *uniform* in the following sense. Any two points, p and p', of S lie in two relative neighborhoods which are homeomorphic, p and p' being one another's image. It should be noted that this definition of patches is free of any reference to C or to plane topology. (The union of the set A_3 and the set of the points $(x, e^{-1/x^2})$ for all real numbers $x < 0$ is uniform and has all properties of a smooth patch except that it is 1-dimensional.)

Being connected and locally arcwise connected, a patch P is *arcwise connected*. (The set of all points connected with a point p of P by an arc contained in P is both open and closed in P.) Being more than 1-dimensional and uniform, P has the same dimension >1 at each of its points. According to Lemma 3, P is at most 2-dimensional. Hence P is *2-dimensional at each point* of P.

Examples of smooth patches include any (vertical) set consisting of the points (z_0, w) for all w belonging to some region in C.

Remark 1. *Each regular quasicell is an* (of course nonvertical) *smooth patch.* All that requires proof is that P is uniform and smooth. In order to show the first, one merely has to map the Z-projection of P (which is a quasidisc) on itself topologically in such a way that the Z-projections of two given points are interchanged. In order to prove that P is smooth, at (z_0, w_0), let (z_n, w_n) and (z_n', w_n') $(z_n' \neq z_n)$ $(n = 1, 2, \ldots)$ be two sequences of points of P converging to (z_0, w_0). A simple computation shows that the quotients $(p_\alpha(z_n') - p_\alpha(z_n))/(z_n' - z_n)$ have a (finite) limit, namely, $p_\alpha'(z_0)$. (This fact was used in Example 3.)

Lemma 6. *A nonvertical smooth patch* P *is the union of denumerably many regular quasicells each of which is open in* P *; and* P *is a subset of the graph of a complete Weierstrass set.*

Since P is more than 1-dimensional Lemma 5 implies that at least one point of P lies in a neighborhood in P which is homeomorphic to a quasidisc. Since P is uniform each point of P does. Hence, by Cauchy's theorem each point of P lies in a regular quasicell that is open in P.

The second half of Lemma 6 follows from elementary results of function theory. If the intersection of two nonmaximal regular cells or of two regular quasicells is infinite it is indenumerable. If two (nonmaximal or maximal) regular quasicells, R and R', have an indenumerable intersection L, then each point of L lies in a regular quasicell contained in L; and R and R' are said to *overlap*. L will be called a *lense*. R and R' overlap if and only if the intersection of their Z-projections is the Z-projection of their intersection and is nonempty. From this definition, in view of Remark 1, one readily proves

Remark 2. *The union of two overlapping regular quasicells is a smooth patch.*

It is convenient to have at one's disposal geometric synonyms of the terms used in Weierstrass' theory of analytic continuation. A finite sequence $(R_1, \ldots, R_n)$ of regular quasicells such that any two consecutive members overlap will be called a *regular chain* between R and R'. If the center of each R_{i+1} lies in R_i the chain will be said to *continue* R_1 to R_n.

If α and α' are two CH-sequences corresponding to the maximal regular quasicells R and R', then α' is a direct continuation of α, as defined in Section 2, if and only if R and R' overlap and R contains the center of R'; and α' is a continuation of α if and only if there exists a regular chain of maximal quasicells continuing R to R'. If $(Q_1, \ldots, Q_n)$ is a regular chain of (not necessarily maximal) quasicells continuing Q_1 to Q_n, then the chain of the corresponding maximal quasicells continues its first member to its last. By a simple construction one proves that if R and R' are overlapping regular quasicells, then their union contains a chain continuing R to R'. Applying this remark to each pair of consecutive members of a regular chain one obtains

Remark 3. *The union of a regular chain between* R *and* R' *contains a chain continuing* R *to* R'.

In order to prove the second half of Lemma 6, let p be any point of a nonvertical, smooth patch, P. According to the first half of the lemma, p lies in a regular quasicell, R, contained in P. Let q be any point $\neq$ p of P. Being arcwise connected, P contains an arc, A, joining p and q. Each point of A lies in a regular quasicell, R, that is open in P. A finite set of such quasicells covers the compact set A. From this set, one can extract a regular chain between R and a quasicell R' containing q. According to Remark 3, the union of this chain (and hence P) contains a chain continuing R to R'. Let F be the complete Weierstrass set such that $\varphi(F)$ contains R. Clearly, $\varphi(F)$ also contains R'. Consequently, each point of P belongs to $\varphi(F)$. This completes the proof of Lemma 6.

Some smooth patches are the graphs of complete Weierstrass sets, including all those that are functions and some that are multifunctions, e.g. the square root, i.e., the set of the points (z^2, z) for all $z \neq 0$. But some smooth patches, e.g., the concentric restrictions of power series, are only proper subsets of such graphs. On the other hand, the graphs of some complete Weierstrass sets are not smooth patches.

If $\varphi(F)$ contains a point p_0 each neighborhood of which includes pairs of vertical points of $\varphi(F)$, then $\varphi(F)$ is not smooth at p_0; nor does $\varphi(F)$ contain a quasicell containing p_0 that is open in $\varphi(F)$.

Vertical pairs of points of $\varphi(F)$ may be contained in each neighborhood of a point p_0 for two reasons. The point may constitute the intersection of two relatively open sets that are *almost disjoint*, i.e., have only one point in common. In this case, p_0 may be called a *point of selfcontact* of $\varphi(F)$. Such a point exists in the set $\varphi(F_2)$ in Example 2. Secondly, p_0 may be a point where $\varphi(F)$ is touched by its rim, which is illustrated by the following

Example 6. Let R be the Riemann surface of the square root, that is, the set of all equivalence classes in the set of all pairs (r, a) of real numbers with $r > 0$, produced by the relation

$$(r, a) = (r', a') \quad \text{if and only if} \quad r = r' \quad \text{and} \quad a \equiv a' \,(\text{mod } 4\pi).$$

Let A be the relatively open subset of R consisting of the pairs (r, a) such that $1 < r < 3$ and $0 < a < 3\pi$. Let $k(r, a)$ be the complex number in C whose absolute value is r, and whose amplitude is $\equiv a(\text{mod } 2\pi)$. Moreover, associate with each (r, a) in A a complex number $g(r, a)$ in such a way that g continuously maps the closure of A on the unit disc D in C consisting of all numbers z with $|z| \leq 1$, and that A is conformally mapped onto the interior of D. Now let f be a power series that is continuous in D, analytic in its interior, and has the circumference of D as its natural boundary. There clearly exists a complete Weierstrass set F_6 such that $\varphi(F_6)$ is the set of the pairs $\big(k(r, a), f(g(r, a))\big)$ for all (r, a) in A. The point $p_0 = (z_0, w_0)$ of $\varphi(F_6)$ which is the image of $(2, \pi)$ in A does not lie in two almost disjoint quasicells contained in $\varphi(F_6)$. But if (z_0, w_1) is the image of the point $(2, 3\pi)$ on the boundary of A, and if $w_1 = w_0$, then each neighborhood of p_0 clearly includes vertical pairs of points of $\varphi(F_6)$. If $w_1 \neq w_0$, then set $h(r, a) = \left[f(g(r, a)) - \frac{1}{2}(w_0 + w_1) \right]^2$. The set of the points $(k(r, a), h(r, a))$ for all (r, a) in A is the graph of a complete Weierstrass set, G_6, such that $\varphi(G_6)$ contains vertical pairs of points in every neighborhood of the point $p_0' = \left(z_0, \frac{1}{4}(w_1 - w_0)^2\right)$, which $\varphi(G_6)$ has in common with its rim. Neither $\varphi(G_6)$ nor $\varphi(F_6)$ contains a point of selfcontact. But these sets as well as $\varphi(F_2)$ are neither smooth nor patches, since they obviously are not uniform.

In the sequel, reference will also be made to the following set which is neither a patch nor smooth nor the graph of a complete Weierstrass set.

Example 7. Let A_7 be the set of the points (z, z) and the points $(z, z^2 + z)$ for all numbers z. This set is connected, locally arcwise connected and the union of two quasicells. Each point of A_7 lies in a regular quasicell contained in A_7. At each point, A_7 is almost smooth (see Introduction); it is smooth, except at $(0, 0)$, which lies in two almost disjoint regular quasicells neither of which is open in A_7.

The graph of a complete Weierstrass set may contain a relatively open subset that is homeomorphic to a neighborhood of $(0, 0)$ in A_7. This is demonstrated by the set $\varphi(F_2)$ in Example 2.

6. Smoothly connected sets

In order to characterize the graphs of complete Weierstrass sets, one needs a concept partly stronger, partly weaker than smoothness — stronger, in order to exclude not only A_7 but also A_5; weaker, in order to include $\varphi(F_2)$, $\varphi(F_6)$, and $\varphi(G_6)$. Such a concept is smooth connectedness, to be defined without any reference to C or to plane topology.

A set S will be called *smoothly connected* if S contains more than one point and any two of its points lie in a smooth patch contained in S. Each smooth patch is of course smoothly connected. The sets $\varphi(F_2)$, $\varphi(F_6)$, and $\varphi(G_6)$, which are not uniform, are not smooth but are smoothly connected — this will follow from Lemma 7. The set A_5 is smooth but not smoothly connected. The set A_1, after the deletion of $(0, 0)$ is a smooth patch. (A_1 itself is not smooth but is a patch and, in fact, a quasicell, as can be seen by setting $z = t^2$ and $w = t^3 + t^2$ for any t in C.) The set A_7 is not smoothly connected. Of importance for what follows is

Example 8. The set C^2 is smoothly connected. So is its proper subset consisting of the points (z, w) for all w and all z such that $|z| < 1$.

The most important examples of smoothly connected sets are supplied by

Lemma 7. *If F is a complete Weierstrass set, then $\varphi(F)$ is smoothly connected.*

Let p and p' be two points of $\varphi(F)$. It must be shown that $\varphi(F)$ contains a smooth patch containing p and p'. Certainly, p and p' lie in regular quasicells, R and R' respectively, contained in $\varphi(F)$; and $\varphi(F)$ contains a regular chain **R** between R and R' (in fact, a chain continuing R to R'). The cells in **R** may (and will) be assumed to be nonmaximal so that any two nonoverlapping quasicells in **R** have a finite intersection. The union, S, of the quasicells in **R** (if there are more than two of them) need not be either smooth or a patch. But S will be shown to be smoothly connected and, in fact, to contain a *quasicell* containing p and p'. For this purpose, one first extracts from **R** a (minimal) chain **R'** $= (R = R_1, R_2, ..., R_{n-1}, R_n = R')$ such that R_i and R_j do, or do not, overlap according as $|i - j| = 1$ or > 1. If R_i and R_j do not overlap their intersection is finite. Let F be the (finite) set of all points belonging to two nonconsecutive member of **R'**. Let q be a point of F. For each of the quasicells R_i containing q, choose a quasicell K_i^q according to the following conditions: a) the closure of K_i^q is contained in R_i but does not contain p or p'; b) if R_i and R_j contain q, then K_i^q and K_j^q have only q in common; c) if q and r belong to F, then any K_i^q and any K_j^r are disjoint. If the union of the quasicells K_i^q for all q in F and all i is deleted from the union, S', of $\mathbf{R}'$, then the complement contains p and p' but no point of F. There is in S' a regular chain $\mathbf{Q}$ between quasicells containing p and p'. From $\mathbf{Q}$, one can obviously select a (minimal) sequence $\mathbf{Q}'$ of regular quasicells $(Q_1, ..., Q_t)$ such that Q_1 and Q_t contain p and p', respectively, and that Q_i and Q_j do, or do not, overlap according as $|i-j| = 1$ or > 1. Any two nonoverlapping quasicells Q_i and Q_j are disjoint. The union T_t of $\mathbf{Q}'$ is smooth. For if q is any point of T_t then 1) q lies in at most two of the t regular quasicells, and their union is smooth by Remark 2, and 2) q is not an accumulation point of the union of the other quasicells in $\mathbf{Q}'$.

In order to show that S is smoothly connected it will be proved that T_t is a quasicell, that is, homeomorphic to a quasidisc or, which is tantamount, to a hexagon. (In the sequel, all hexagons are without boundary.) Q_1, being a regular quasicell, clearly is homeomorphic to a regular hexagon,

$$H_1 = (abcz_1 z_1' z_1'' a).$$

The quasicell Q_2 has a lense L_1 in common with Q_1. By a topological mapping of H_1 on itself one can achieve that the image of L_1 is the rhombus $P_1 = (z_1 z_1' z_1'' z_1^* z_1)$. One now can map Q_2 on a hexagon

$$H_2 = (z_1'' z_1^* z_1 z_2 z_2' z_2'' z_1'')$$

such that 1) c, z_1, and z_2 as well as a, z_1'', and z_2'' is a collinear triple; 2) the sides $z_1'' z_1^*$ and $z_2 z_2'$ are parallel to the side ab; and $z_1^* z_1$ and $z_2' z_2''$ are parallel to bc; 3) the sides $z_1 z_2$ and $z_1'' z_2'$, which are equally long, are longer than the other four sides of H_2 of H_2; 4) the mapping of L_1, as a subset of Q_2, is identical with the mapping (on P_1) of L_1, as a subset of Q_1. Because of precautions 1) and 2), the union of H_1 and H_2 is a hexagon,

$$J_2 = (abcz_2 z_2' z_2'' a).$$

Because of 4), the union of Q_1 and Q_2 is topologically mapped on J_2. Because of 3), the rhombus $P_2 = (z_2 z_2' z_2'' z_2^* z_2)$ is disjoint from P_1. Hence by a topological mapping of H_2 on itself, which leaves L_1 fixed, one can achieve that the image of L_2, which is disjoint from L_1 is P_2. Continuing in this way, that is, observing with each step the precautions 1)—4), one can topologically map, for each k, the union of $Q_1, ..., Q_k$ on a hexagon

$$J_k = (abcz_k z_k' z_k'' a)$$

and, finally, T_t on a hexagon J_t.

7. A decomposition property of the sets $\varphi(F)$

Besides smooth connectedness, the sets $\varphi(F)$ have the following second fundamental property. *Each $\varphi(F)$ is the union of denumerably many smooth patches.* This formulation is free of any reference to C or plane topology. But since, according to Lemma 6, each smooth patch is the union of denumerably many regular quasicells, the second property may also be expressed in the following form

Lemma 8. *If F is a complete Weierstrass set, then $\varphi(F)$ is the union of denumerably many regular quasicells. If G is a complete Weierstrass set $\neq F$, then $\varphi(F)$ and $\varphi(G)$ have a denumerable intersection. The intersection of $\varphi(F)$ with each smooth patch not contained in $\varphi(F)$ is denumerable.*

Each point of $\varphi(F)$ lies in a regular quasicell. But since such a quasicell need not be open in $\varphi(F)$ one cannot make global use of covering theorems. If R is a maximal regular quasicell, let $\kappa(R)$ denote the union of R and all direct continuations of R, i.e. all maximal regular quasicells that overlap R and have their centers in R. If S is any such direct continuation of R, then S as well as R itself is open in $\kappa(R)$. (Indeed, any two maximal regular quasicells in $\kappa(R)$ either overlap or are disjoint, whence the Z-projections of $\kappa(R) - S$ and S are disjoint.) Hence the covering theorem is applicable and implies that $\kappa(R)$ is the union of denumerably many regular quasicells, say $R_1, R_2, ..., R_n, ...$. Let $\kappa^2(R)$ be the union of $\kappa(R)$ and all sets $\kappa(R_n)$ $(n = 1, 2, ...)$. For each n_1, the set $\kappa(R_{n_1})$ is the union of denumerably many maximal regular quasicells, say $R_{n_1 1}, R_{n_1 2}, ..., R_{n_1 n_2}, ...$. Hence $\kappa^2(R)$ is the union of denumerably many maximal regular quasicells. Continuing in this way, one can define $\kappa^{m+1}(R)$ as the union of $\kappa^m(R)$ and all sets $\kappa(R_{n_1 n_2 ... n_m})$. Let U be the union of these sets $\kappa^m(R)$ for all m. Clearly, U is the union of denumerably many regular quasicells. From what has been said in introducing the geometric synonyms of the terms about analytic continuation, it is clear that $U = \varphi(F)$, which proves the first part of the Lemma.

If $\varphi(F)$ and $\varphi(G)$ have an indenumerable intersection then, since either graph is the union of denumerably many regular quasicells, a regular quasicell in $\varphi(F)$ and one in $\varphi(G)$ must have an indenumerable intersection and, therefore, overlap. Hence $\varphi(F)$ and $\varphi(G)$ have a regular quasicell in common and must be identical.

If a smooth patch P has an indenumerable intersection with $\varphi(F)$, then so does the graph of the complete Weierstrass set whose subset is P. Consequently, this graph must be identical with $\varphi(F)$.

8. A set-theoretical lemma and its application

Lemma 9. *Let G be a family of indenumerable sets any two of which have a denumerable intersection; and let H be a family of subsets of sets belonging to G. For each indenumerable set H in H, let G(H) be the (only) set belonging to G that contains the subset H. Let T be a set containing 1) an indenumerable subset R belonging to H, 2) an element t not belonging to G(R), 3) an inde-*

numerable subset H_r *(for each element* r *of* R*) such that* H_r *belongs to* **H** *and contains* r *and* t. *Then* T *is not a subset of the union of denumerably many sets belonging to* **G**.

Each H_r includes t. Hence no H_r is a subset of **G**(R). But since each H_r belongs to **H**, the intersection of H_r with the indenumerable set R is denumerable. The family **K** of all distinct sets H_r (for the various elements r of R) consequently is indenumerable. Assume now that T is a subset of the union of denumerably many sets G_1, G_2, ... belonging to **G**. Each G_i either contains one of the sets in **K** as a subset or has a denumerable intersection with each of them. Being indenumerable, the family **K** must include a set H_{r*} whose intersection with each G_i, and hence with T, is denumerable. But this contradicts the hypothesis that H_{r*} is an indenumerable subset of T.

Now let **G** be the family of the (indenumerable) graphs of all complete Weierstrass sets. By Lemma 8, any two sets in **G** have a denumerable intersection. Let **H** be the family of all nonvertical smooth patches. Each element of the family **H** is indenumerable and, by Lemma 6, subset of at least (and hence of exactly) one element of **G**. From Lemma 9 we thus infer

Lemma 10. *If* T *is a nonvertical, smoothly connected set containing* 1) *a smooth patch,* P, *and* 2) *a point not lying in the graph of a complete Weierstrass set that contains* P, *then* T *is not the union of denumerably many smooth patches.*

P is an indenumerable subset of a set belonging to the family **H**. Being smoothly connected, T contains 3) for each point r of R, a smooth patch containing r and t, which of course is indenumerable. According to Lemma 9, T is not a subset of the union of denumerably many graphs of complete Weierstrass sets and, a fortiori, not contained in the union of denumerably many smooth patches.

9. The main result

Characterization theorem. *The graphs of complete Weierstrass sets are, besides maximal vertical sets, the only subsets of* C^2 *which are* 1) *smoothly connected,* 2) *unions of denumerably many smooth patches, and* 3) *maximal or saturated with regard to properties* 1) *and* 2).

First, let F be a complete Weierstrass set. The set $\varphi(F)$ is nonvertical and, by Lemmas 7 and 8, smoothly connected and the union of denumerably many smooth patches. $\varphi(F)$ is a proper subset of some smoothly connected sets, certainly of C^2. [If $\varphi(F)$ is a power series having the unit circle as its natural boundary, then $\varphi(F)$ is also a proper subset of A_8.] But, according to Lemma 10, no smoothly connected set containing $\varphi(F)$ as a proper subset is the union of denumerably many smooth patches. It follows that $\varphi(F)$ is maximal with regard to properties 1) and 2).

Conversely, let S be a subset of C^2 with the properties 1) and 2), and maximal with regard to them. If S is vertical, then, by virtue of its saturation with regard to either of the two properties, S is a maximal vertical set. Hence if S is not a maximal vertical set, then S is nonvertical. Then, according to Lemma 5, S is a subset of the graph of some complete Weierstrass set, F. Since $\varphi(F)$ has properties 1) and 2) with regard to which S is saturated, S cannot be a proper subset of $\varphi(F)$. Hence $S = \varphi(F)$.

10. Complete and incomplete analytic multifunctions

The Characterization Theorem suggests the following

Definition I. *A complete analytic multifunction is a nonvertical subset of* C^2 *that is maximal with regard to the property of being a smoothly connected union of denumerably many smooth patches.*

Here, the words *smooth patches* cannot be replaced by *smoothly connected patches* since even the smoothly connected set C^2 is a patch. (Lemma 3 implies only that *smooth* patches are at most 2-dimensional!) On the other hand, according to Lemma 6, the unions of denumerably many smooth patches are identical with the unions of denumerably many regular quasicells.

A complete analytic multifunction S is a *function* if and only if it does not contain any vertical pair of points. If S does contain such a pair, then S contains a pair of regular quasicells that are vertical (i.e. disjoint but having the same Z-projection). In order to characterize complete analytic functions it thus suffices to rule out pairs of vertical quasicells.

Clearly, a complete analytic multifunction S may contain proper subsets that are *maximal subfunctions* of S in the sense of not being proper subsets of any functions contained in S. They are called *branches* of S. Being a subset of a smooth set a branch is of course smooth but it is not necessarily compact. Hence it need not be a continuous function (see Example 3). This is demonstrated e.g., by the branch of the square root consisting of all points (z^2, z) whose amplitude is ≥ 0 and $< \pi$. Two maximal subfunctions clearly may have indenumerable intersections without being identical.

The second half of Lemma 6 can be greatly extended. Not only each smooth patch but (in view of Lemma 10) every smoothly connected union of denumerably many smooth patches, is contained in the graph of a complete Weierstrass set. Conversely, let S be a smoothly connected subset of such a graph. S contains a (not necessarily maximal) regular quasicell, R. One may denote by $\lambda(R)$ the union of R and all (not necessarily maximal) regular quasicells contained in S, and define $\lambda^n(R)$ in analogy to $\kappa^n(R)$ in the proof of Lemma 8. The union V of the $\lambda^n(R)$ for all n is the union of denumerably many regular quasicells contained in S. If the smoothly connected set S contained a point outside of V this point and a point of R would lie in a smooth patch contained in S, which would contradict the construction of V. Hence S is identical with V. One thus arrives at the

Theorem. *In order that a smoothly connected set S in* C^2 *be a subset of the graph of a complete Weierstrass set it is necessary and sufficient that S be the union of denumerably many smooth patches.*

This fact suggests the following concept, which includes both complete and incomplete analytic multifunctions and avoids the element of arbitrariness that has been criticized[8] in the traditional definition of incomplete analyticity following Weierstrass.

Definition II. *An analytic multifunction is a smoothly connected union of denumerably many smooth patches in* C^2.

Each analytic multifunction is a restriction (i.e. a subset) of exactly one complete analytic multifunction with which it is identical if and only if it is a maximal set of its kind.

11. Perspectives

The most immediate problem (to be treated in a subsequent paper[9] is the extension to R^2 of the method herein developed for C^2, where R is the Riemann sphere. One may either include vertical sets as counterparts of constant functions or exclude both vertical and horizontal sets.

Special types of analytic multifunctions might be characterized by studies of the rims of multifunctions.

Finally, the ideas herein developed might be useful in the study of complex multiplace functions (functions "of several complex variables") and in the analysis of quaternion functions.

Addition in the proofs: The ideas here expounded have been further developped by the author in three notes: A characterization of Weierstrass analytic functions. Proc. Nat. Ac. Sci. USA **54**, 1025—1026 (1965); Analytische Funktionen. Weierstrass-Festschrift, Wiss Abh. Arbeitsgemeinschaft f. Forschung Land. Nordrhein-Westfalen **33**, 609—612 (1966); Une caractérisation des fonctions analytiques. C. R. Paris **261**, 4968—4969 (1965).

Professor KARL MENGER
Department of Mathematics
Illinois Institute of Technology
Chicago, Illinois 60616

(Received April 2, 1965)

---

[8] Cf. e.g., L. V. AHLFORS, *Complex Analysis*, New York 1953, p. 210.
[9] Cf. also the notes loc. cit. [4].

Selected Papers on Length

Commentary on Menger's Path Length Papers

Lester Senechal

The determination of the length of a path in a metric space is a variational problem that can be treated by direct methods, without invoking the more extensive machinery of the calculus of variations. In the series of papers [1–10], Menger treated these ideas with increasing generality.

1. Arcs

We first describe the semicontinuity property of the length functional, which is established for arcs in [1] and, in somewhat more detail, in [2]. Let M be a metric space and let the distance between points p, $q \in M$ be denoted by pq. An *arc E* in M is a bicontinuous image in M of a real interval, from which E inherits a linear order. *Arc length* is defined by

$$\lambda(E) = \sup \sum p_{i-1}p_i,$$

where the supremum is taken over all finite sets of points $\{p_0, p_1, \ldots, p_n\}$ of E whose order is consistent with the linear order on E, $n \geq 1$.

Clearly λ is monotone relative to set inclusion, but a simple example of zig-zags approaching a straight line segment shows that λ is not continuous relative to the Hausdorff metric[1] on the subsets of M. However, in [1, 2] Menger shows that λ is lower semicontinuous in the following sense: if the sequence $\{E_n\}$ converges to E in the Hausdorff metric, then

$$\liminf \lambda(E_n) \geq \lambda(E).$$

---

[1] The Hausdorff distance between two sets is the smallest number with the property that, given an element in one of the sets, there is an element of the other whose distance is no greater than that number.

In order to obtain this result, Menger proves a lemma in which he characterizes the length of an arc in a form where the points $\{p_0, p_1, \ldots, p_n\}$ in the definition need not conform to the linear order on E. In fact, let F be a selection of points in any order and define its length by

$$\lambda(F) = \min\left\{ \sum_{k=1}^{n} p_{k-1}^* p_k^* \right\},$$

where the minimum is taken over all permutations $p_0^*, p_1^*, \ldots, p_n^*$ of the points of F. The length of a finite set is, in effect, the length of the *shortest polygon* that can be inscribed in it. He then shows that

$$\lambda(E) = \sup \lambda(F),$$

where the supremum is taken over all finite subsets F of E.

2. Paths

In [2] he goes a step further and extends the definition of λ to paths in the sense of Jordan-Fréchet. Here, for Jordan, a path is a bicontinuous mapping of an interval into a metric space while, for Fréchet, a path is an equivalence class of such mappings, with $f \sim g$ if there is a homeomorphism τ between their domains such that $f = g \circ \tau$. The definition of path invokes the concept of homotopy classes, so that a circle traversed twice represents a path different from a circle traversed just once.[2]

If a path is given by the mapping f on $[a, b]$, then its *basic set* or *trajectory* is the image of $[a, b]$ under f. The length of a path is defined as the supremum of the lengths of finite sets $F = \{t_0, t_1, \ldots, t_n\}$, where $a \le t_0 \le t_1 \le \cdots \le t_n \le b$. Menger shows that the length of a path is at least as great as the length of any arc contained in its basic set. It is obvious that strict inequality can occur: a circle traversed twice has twice the length of its basic set. If the basic set is an arc, then its length agrees with arc length as defined in the preceding section.

In order to extend his theorem on lower semicontinuity, discussed above, Menger defines, in an obvious way, the concept of the length $\lambda(K)$ of an arbitrary continuum K in a metric space, in exactly the same way as was indicated at the end of the preceding section, by first defining λ for finite subsets and then taking the supremum over all finite subsets of K. He then

---

[2] See [6] for a discussion. In [6] and [8] Menger offers an alternative definition of a path that, like Fréchet's definition, does not depend on any particular parameterization.

proves that, if $\{K_n\}$ is any sequence of continua that converges to K in the Hausdorff metric, then

$$\lambda(K) \leq \liminf\{\lambda(K_n)\}.$$

Such considerations are continued in [5], where a path with basic set C is defined as *shortest* through its points if every path whose basic set is a subset of C is at least as long. The theory is further developed for convex and compact metric spaces.[3] The following result is obtained:

Theorem. Let a shortest path with basic set C be given in a convex and compact space and let $\{F_n\}$ be a sequence of finite sets which approach C in the Hausdorff metric. Then the shortest polygons (in the sense indicated at the end of Section 1) that can be inscribed in the F_n approach C and their lengths approach the length of the given shortest path.[4]

3. Path Length Generalized

The publications [7–10] all deal with a generalization of path length in a metric space that is obtained when, instead of using the metric to compute the length of polygons, we employ a function δ that is defined on pairs of points and that satisfies the condition: $\lim p_n = p \Rightarrow \lim \delta(p_n, p) = 0$ and $\lim \delta(p, p_n) = 0$. *Path length*, as before, is defined in terms of inscribed polygons: length exists if $\limsup = \liminf$ for the lengths of all such polygons. *Absolute length* is obtained likewise when we replace the function δ by its absolute value.

The limits superior and inferior in the definition of length allow us to define upper and lower length and upper and lower absolute length for every path. These quantities always exist, but may be infinite.

In [10] it is indicated that, if the metric space is an interval, then path length is an integral in the sense of Burkill and includes, as special cases, the

---

[3] Compactness allows application of a theorem of Hilbert which states that a sequence of paths that are uniformly bounded in length contains a subsequence that is pointwise convergent when each path in the subsequence is parameterized by $\mu \in [0,1]$ where μ gives the location along each of the paths in the subsequence as a proportion of the total length of the path.

[4] In a paper directly following [5] in the Reports of a Mathematical Colloquium, Vol. 2, Menger's student A. N. Milgram took up the case of paths in spaces which are not necessarily convex. His principal theorem states that if M is a continuum which is a subset of the basic set of a path f of finite length, then M is the basic set of a path g whose length is less than twice the length of f.

integrals of Stieltjes and Weierstrass, and that the Weierstrass integral is obtained by setting

$$\delta(x,y) = f\left(x, \frac{y-x}{|y-x|}\right)|y-x|,$$

where the function f is defined for pairs of numbers. In this way, it is also clear that the form of the Weierstrass integral allows for generalization to a metric space, where the function f is then defined for pairs of points in that space.

As we might expect, there can be a severe difference between length and absolute length: we can have length 0 and absolute length ∞. But even when the upper absolute length is finite, path length may fail to exist. Menger develops a necessary and sufficient condition for its existence under the assumption of a finite absolute length. The reader is referred to [10] for details and for the sufficiency of a modification of that condition that ensures semicontinuity for the length function.

References[5]

1. a) Ein Theorem über die Bogenlänge. Anz. Öster. Akad. Wiss. 65 (1928) 264–266; b) Die Halbstetigkeit der Bogenlänge, ibid 65 (1928) 271–281; c) Über eine neue Definition der Bogenlänge. Eine weitere Verallgemeinerung des Längenbegriffes, ibid 66 (1929) 23–25.
2. Untersuchungen über allgemeine Metrik. Vierte Untersuchung. Zur Metrik der Kurven. Mathematischen Annalen 103 (1930) 467–501.
3. Metrische Geometrie und Variationsrechnung. Fundamenta Mathematicae XXV (1935) 441–458.
4. A Theory of Length and its Applications to the Calculus of Variations, Proceedings of the National Academy of Sciences 25 (1939) 474–478.
5. On Shortest Polygonal Approximations to a Curve, Notre Dame University, Reports of a Mathematical Colloquium 2 (1940) 33–39.
6. Définition intrinsique de la notion de chemin, Comptes rendus 221 (1946) 739–741.
7. Stieltjes integrals considered as lengths, Annales de la Société Polonaise de Mathématique XXI (1948) 173–175.
8. What paths have length? Fundamenta Mathematicae XXXVI (1949) 109–118.
9. A topological characterization of the length of paths, Proceedings of the National Academy of Sciences 38 (1952) 66–69.
10. Geometrie générale. Mémorial des Sciences Mathématiques CXXIV (1954) 55–61.

---

[5] The papers [2] and [4] are reprinted in Volume 1 of these Selecta (pp. 333–368 and pp. 399–403, respectively). Both [3] and [4] are discussed by H. Sagan in his Commentary, also in Volume 1.

On Shortest Polygonal Approximations
to a Curve

Karl Menger

Let F be a set consisting of n points of a metric space. If $F^P =$ $[p_1, p_2, \ldots, p_n]$ is a polygon (i.e., a finite ordered set) consisting of the points of F, we set $l(F^P) = \Sigma \, p_i p_{i+1}$, where $p_i p_{i+1}$ denotes the distance from p_i to p_{i+1}. The smallest of the n! numbers $l(F^Q)$ formed for the n! permutations Q of the numbers 1, 2, ..., n will be denoted by $\lambda(F)$. Thus $\lambda(F)$ is *the length of the shortest polygon that can be inscribed into F.*

If A is an arc in a metric space, i.e., a continuous one-to-one image of the closed interval [0, 1], and F any finite subset of A, then we denote by F^A the polygon consisting of the points of F in that order in which we meet them when we traverse A. Obviously $l(F^A) \geqslant \lambda(F)$. Hence the length of A, i.e., the least upper bound of the number $l(F^A)$ for all finite subsets F of A, satisfies the inequality

$$l(A) \geqslant \text{l.u.bd. } \lambda(F)$$

the least upper bound referring to all finite subsets F of A. Clearly, for some finite subsets F of A we may have $l(F^A) > \lambda(F)$. Yet, as the author proved in a previous paper,[1] we have

$$l(A) \leqslant \text{l.u.bd. } \lambda(F), \text{ and hence } l(A) = \text{l.u.bd. } \lambda(F).$$

The length of an arc is the least upper bound of the lengths of the shortest polygons that can be inscribed into finite subsets of the arc.

The theorem was proved for a given arc A by constructing in A for each given $d > 0$ a finite subset F(d) for which

1) $l(F^A(d)) = \lambda(F(d))$, and

2) any two consecutive points of $F^A(d)$ have a distance $\leqslant d$.

On account of the density property 2) we have $l(F^A(d)) > l(A) - \epsilon$ for each prescribed $\epsilon > 0$ provided that d is sufficiently small. Hence by virtue of 1) we have $\lambda(F(d)) > l(A) - \epsilon$ for each $\epsilon > 0$ and each sufficiently small d, from which the theorem follows.

The set $F(d) = [p_1(d), p_2(d), \ldots, p_{n(d)}(d)]$ is constructed in the following way: we establish an order on A, and, for each d call $p_1(d)$ the initial point of A. For $p_2(d)$ we choose the last point which we meet at the distance d from $p_1(d)$ when we traverse A. For $p_{k+1}(d)$ we choose the

---

1. Mathem. Annalen 103 (1930) p. 467.

last point which we meet at the distance d from $p_k(d)$ when we traverse A. On account of the compactness of A this process is easily seen to break off after a finite number of steps, say $n(d) - 1$. For $p_{n(d)}(d)$ we choose the last point of A. For this set $F(d)$ we have $l(F^A(d)) = [n(d) - 2]d + e(d)$ if by $e(d)$ we denote the distance of the last two points of $F(d)$. Since the distance of any two points of $F(d)$, except perhaps that of the last two, is $\geqslant d$, it is clear that each polygon passing through the $n(d)$ points of $F(d)$ has a length $\geqslant [n(d) - 2]d + e(d)$. Hence $l(F^A(d)) = \lambda(F(d))$. Thus the polygon $F(d)$ which obviously has the property 2) has also[2] the property 1).

So far we have spoken about arcs, i.e., continuous one-to-one images of a closed interval. By a continuous curve in a metric space we mean a continuous mapping of a closed interval of real numbers $[a,b]$ on a subset C of the space. If $p = f(t)$ is the point associated with the number t of the interval $[a, b]$, then we denote the continuous curve by C_t. The set C of all points associated with the numbers t of $[a, b]$ will be called the basic set of the curve C_t. By $l(C_t)$, the length of C_t, we mean the least upper bound of the numbers $l(F)$ for all subpolygons of C, i.e., for all ordered finite sets $F = [f(t_0), f(t_1), \ldots, f(t_n)]$ where $t_0 = a$, $t_n = b$, $t_i < t_{i+1}$.

Then as a consequence of the theorem on the length of arcs, we proved (l.c. 1) the following theorem: *If A is an arc and C_f any continuous curve whose basic set contains all the points of A, although perhaps in an entirely different order, then $l(C_t) \geqslant l(A)$.* For, if we had $l(C_t) = l(A) - \varepsilon$ where $\varepsilon > 0$, then for each subpolygon F of C we should have $l(F) \leqslant l(C_t) = l(A) - \varepsilon$. Since each point of A lies in C, we should have, in particular, $l(F) \leqslant l(A) - \varepsilon$ for each subpolygon F of C which is a subset of A, and hence $\lambda(F) \leqslant l(F) \leqslant l(A) - \varepsilon$, in contradiction to our first theorem.

The first remark in the present paper which we wish to add to our former theory supplements the previous theorem by the statement: *If A is an arc and C_t any continuous curve whose basic set contains A as a proper sub-*

---

2. In fact, the polygon $F(d)$ has a stronger property than 1). By a *graph inscribed in a finite set* F, we mean the sum of a set of segments, each joining two points of F. We call the length of a graph the sum of the lengths of the segments which compose the graph. (In a general metric space, we consider pairs of points as segments, and the distance of two points as the length of the segment defined by the two points.) It is clear that each *connected* graph inscribed in the set $F(d)$ contains at least $n(d) - 1$ segments and thus has a length $\geqslant [n(d) - 2]d + e(d)$. Consequently, if we denote by $\kappa(F)$ the length of the shortest connected graph that can be inscribed in a set F we have $\kappa(F(d)) = l(F^A(d))$. It follows that *the length of an arc A is the least upper bound of the lengths of the shortest connected graphs that can be inscribed into finite subsets of the arc.*

By a *graph containing a finite set* F we mean a set which is the sum of a finite number of segments and contains F as a subset. For a finite subset F of a metric space we shall denote by $\iota(F)$ the greatest lower bound of the lengths of all graphs containing F. Mimura (Ergebnisse e. Mathem. Kolloquiums, *4*, Vienna 1932, p. 20) proved that in a metric space the length of an arc A is the least upper bound of the numbers $\iota(F)$ for all finite subsets of A.

set, then $l(C_t) > l(A)$. To prove this remark let p be a point of C—A. Since A is a closed set, the distance d' between p and A is > 0. We call G(d) the set consisting of the points of F(d) and the point p. Any polygon inscribed in the set G(d) contains at least n(d) sides. If d is any number $\leqslant d'$, then each polygon that can be inscribed in G(d) has a length $\geqslant [n(d) - 2]d + e(d) + d' = l(F^A(d)) + d'$. Consequently $l(C) \geqslant l(A) + d' > l(A)$.

In the following, we shall say that a continuous curve C_t is *a shortest path through all its points* if for each continuous curve D_g whose basic set D contains C as a subset, we have $l(D_g) \geqslant l(C_t)$, and for each continuous curve D_g whose basic set has C as a proper subset, we have $l(D_g) > l(C_t)$. We shall say that the continuous curve C_t joining the points a and b is *a shortest path through all its points between a and b* if for each continuous curve D_g joining a and b we have $l(D_g) \geqslant l(C_t)$ provided that D contains C as a subset, the equality sign merely holding if $D = C$. In this terminology, *each arc is a shortest path through all its points, and the shortest path through all its points between the two endpoints of the arc.*

The circle $x = \cos 2\pi t$, $y = \sin 2\pi t$ $(0 \leqslant t \leqslant 1)$ is a shortest path through all its points without being an arc. The circle traversed twice, i.e., the curve $x = \cos 4\pi t$, $y = \sin 4\pi t$ $(0 \leqslant t \leqslant 1)$ is not a shortest path through all its points. Let a be a real number $\geqslant 0$. The continuous curve C_a given by the equations:

$$x = 4t-1,\ y = 0,\ (0 \leqslant t \leqslant \tfrac{1}{4})$$
$$x = 0,\ y = a(4t-1),\ (\tfrac{1}{4} \leqslant t \leqslant \tfrac{1}{2})$$
$$x = 0,\ y = a(3-4t),\ (\tfrac{1}{2} \leqslant t \leqslant \tfrac{3}{4})$$
$$x = 4t-3,\ y = 0,\ (\tfrac{3}{4} \leqslant t \leqslant 1)$$

is a shortest path through all its points between the points $(-1, 0)$ and $(1, 0)$. If $a \leqslant 1$, then C_a is a shortest path through all its points. If $a > 1$, then the polygonal line starting at $(0, a)$, leading to $(0, 0)$, from there to $(-1, 0)$ and then to $(1, 0)$ is a shorter continuous curve whose basic set is identical with that of C_a.

The next question which we shall study is what can be said about the values of λ for arbitrary sequences of finite sets approaching an arc A. So far we know only one family of finite sets which approach the arc A and for which the λ converge toward $l(A)$, viz., the family F(d) consisting of certain subsets of A. In the following we shall prove theorems with much farther reaching implications in this direction.

We shall have to restrict ourselves to the consideration of metric spaces which are convex and complete (as e.g., the vector spaces). The reason for this restriction is that for each finite ordered set $F = [p_1, p_2, \ldots, p_n]$ we shall have to form a continuous curve, called a *polygonal line* through the vertices of F, which is the sum of a segment joining p_1 and p_2, a segment joning p_2 and p_3, ..., a segment joining p_{n-1} and p_n. Now the existence of segments joining any two distinct points is generally guaranteed only by convexity and completeness of the space.[3] In particular, for a finite set F we shall have to form a polygonal line whose length is $= \lambda(F)$.

The concept of approaching which underlies our statements is defined in the following way: We shall say that two sets S and T have a distance $< \delta$ if each point of S has a distance $< \delta$ from some point of T, and each point of T has a distance $< \delta$ from some point of S. The sequence of sets S_1, S_2, ... is said to approach the set T if the distances between S_n and T converge toward 0. This concept of approaching is very weak. It should be noted that if the sequences of finite sets dealt with in the following theorems should converge in some stronger sense, the statements of the theorems would hold a fortiori.

We shall need compactness of the space also because we shall have to make use of the following theorem due to Hilbert: In a compact space each sequence of continuous curves whose lengths are uniformly bounded, contains a convergent subsequence, more precisely, a subsequence converging toward a curve whose length is $\leqslant \lim \inf l_n$, if $l_1, l_2, \ldots$ denote the lengths of the curves C^1, C^2, ... of the subsequence. Here convergence is defined in the following way: for each μ, $0 \leqslant \mu \leqslant 1$, and each n we denote by $p^n(\mu)$ the point of C^n for which the length of the segment of C^n between the initial point of C^n and $p^n(\mu)$ is μl_n. Similarly, we define the point $p(\mu)$ of C_f. In this notation the sequence of curves C^1, C^2, ... is said to converge toward the curve C_f if for each sequence $\mu_1, \mu_2, \ldots$ converging toward $\mu (0 \leqslant \mu \leqslant 1)$ the points $p^n(\mu_n)$ converge toward the point $p(\mu)$.

Remark 1. If C^1, C^2, ... are the basic sets of continuous curves converging toward the curve C_f, and p^1, p^2, ... points of the basic sets C^1, C^2, ..., respectively, converging toward a point p, then p belongs to the basic set C of C_f.

Proof: For each n let μ_n be a number such that $0 \leqslant \mu_n \leqslant 1$ and $p^n = p^n(\mu_n)$. The sequence of numbers $\mu_1, \mu_2, \ldots$ contains a converging subsequence, say, $\mu_{11}, \mu_{12}, \ldots$ Let μ be its limit. Since the curves with the basic sets C^1, C^2, ... converge toward C_f, the points $p^{11}(\mu_{11})$, $p^{12}(\mu_{12})$, ... converge toward the point $p(\mu)$ of C_f. On the other hand they converge toward p. Hence, p lies in the set C.

From Remark 1, we readily obtain:

Remark 2. If C^1, C^2, ... are the basic sets of a sequence of continuous curves converging toward the curve C_f, and D^1, D^2, ... a sequence of subsets of C^1, C^2, ..., respectively, approaching a set D, then D is a subset of C.

---

3. See Mathem. Annalen 100, (1928), p. 87, or Reports of a Math. Colloquium 1 (1939), p. 16.

Remark 3. In a compact and convex metric space, let G^1, G^2, ... be a sequence of finite sets approaching a set S such that $\lim \sup \lambda(G^n) < \infty$. Then there exists a continuous curve D_g such that S is subset of D and $l(D_g) \leqslant \lim \sup \lambda(G^n)$.

Proof: The space being compact and convex, there is, for each n, a polygon Q^n inscribed into G^n for which $l(Q^n) = \lambda(G^n)$. From the assumption $\lim \sup \lambda(G^n) < \infty$ it follows that the lengths of the curves Q^1, Q^2, ... are uniformly bounded. Hence, by virtue of Hilbert's theorem, the sequence Q^1, Q^2, ... contains a subsequence R^1, R^2, ... which converges toward a continuous curve D_g whose length is $\leqslant \lim \inf l(R^n) \leqslant \lim \sup l(Q^n) = \lim \sup \lambda(G^n)$. Since G^1, G^2, ... approach S, from Remark 2 it follows that S is subset of D.

We shall now prove:

Theorem I. In a convex and compact metric space let C_t be a continuous curve which is a shortest path through all its points. Then for any sequence of finite sets F^1, F^2, ... approaching the set C we have $\lim \inf \lambda(F^n) \geqslant l(C_t)$.

We may assume that $\lim \inf \lambda(F^n) < \infty$. The sequence F^1, F^2, ... contains a subsequence, say G^1, G^2, ..., for which the numbers $\lambda(G^1)$, $\lambda(G^2)$, ... converge toward $\lim \inf \lambda(F^n)$. The space being compact and convex, we can, for each n, form a polygon Q^n inscribed into G^n such that $l(Q^n) = \lambda(G^n)$. From Remark 3 it follows that there exists a continuous curve D_g for which C is subset of D and $l(D_g) \leqslant \lim \sup \lambda(G^n) = \lim \inf \lambda(F^n)$. Since by assumption C is a shortest path through all its points, and C is subset of D it follows that $l(D_g) \geqslant l(C_t)$. Hence $l(C_t) \leqslant \lim \inf \lambda(F^n)$.

The next question that we shall study is whether the shortest polygons that we can inscribe into finite sets approaching an arc A do themselves approach the arc. Since these shortest polygons are zigzag lines oscillating between points of the arc it would be conceivable that they do not approach A. However we shall prove:

Theorem II. In a convex and compact metric space let C_t be a continuous curve of finite length which is a shortest path through all its points. If F^1, F^2, ... is a sequence of finite sets approaching C for which $\lim \sup \lambda(F^n) \leqslant l(C_t)$, and if, for each n, P^n is a shortest polygonal line that can be inscribed into F^n, then P^1, P^2, ... approach C.

We assume that P^1, P^2, ... do not approach C, and deduce a contradiction. The assumption implies the existence of a number $\delta > 0$ with the following property: for infinitely many integers n the polygonal line P^n contains a point p^n whose distance from C is $\geqslant \delta$. From the sequence of points so obtained, we can extract a converging sequence of points, say $p^{n(1)}$, $p^{n(2)}$, ... Let p be its limit point. For each k, we denote by $G^{n(k)}$

the finite set consisting of $p^{n(k)}$ and the points $F^{n(k)}$. The sets $G^{n(1)}$, $G^{n(2)}$, ... approach the set $C+p$. Applying Remark 3 to the sets $G^{n(1)}$, $G^{n(2)}$, ... we see that there exists a continuous curve D_g for which $C+p$ is subset of D and $l(D_g) \leqslant \lim \sup \lambda(G^n) \leqslant \lim \sup \lambda(F^n)$. Since by assumption, $\lim \sup \lambda(F^n) \leqslant l(C_t)$ it follows $l(D_g) \leqslant l(C_t)$.

Since D contains C as a proper subset this contradicts the assumption that C is a shortest path through all its points.

Combining the theorems I and II we get a complete survey of the behavior of the shortest polygons inscribed into finite subsets of a curve C which is a shortest path through all its points, in particular of an arc: the lengths of these zigzag lines approach the length of C, and, if C has a finite length, then the zigzag lines themselves approach C. For if F^1, F^2, ... are subsets of the basic set C of a continuous curve C_t, then for each n we have $\lambda(F^n) \leqslant l(C)$, and hence $\lim \sup \lambda(F^n) \leqslant l(C)$. Thus from theorem I it follows that $\lim \lambda(F^n) = l(C)$ provided that C is a shortest path through all its points. Applying theorem II, we obtain:

Theorem III. If in a convex and compact space the continuous curve C_t has a finite length and is a shortest path through all its points, and F^1, F^2, ... is any sequence of finite subsets of C approaching C, then the shortest polygonal lines that we can inscribe into the sets F^1, F^2, ... approach C, and the lengths of these polygonal lines, i.e., the numbers $\lambda(F^1,)$, $\lambda(F^2)$, ..., converge toward $l(C_t)$.

TOPOLOGIE. — *Définition intrinsèque de la notion de chemin.*
Note de M. **Karl Menger**, présentée par M. Élie Cartan.

En cinématique nous considérons des mouvements, c'est-à-dire des transformations continues d'intervalles en sous-ensembles d'espaces distanciés et complets. En topologie, depuis Jordan, nous étudions des trajectoires des mouvements, c'est-à-dire les ensembles des positions d'un mobile. Une idée intermédiaire, résultant d'une abstraction partielle du facteur temporel, est la notion de chemin.

Dans la théorie classique le chemin est une famille maximale de mouvements équivalents. Soit $f[a, b]$ le mouvement ininterrompu associant la position $f(t)$ au moment t de l'intervalle $[a, b]$, la position $f(t)$ n'étant constante au cours d'aucun sous-intervalle de $[a; b]$. Nous disons que $f[a, b]$ est équivalent au mouvement ininterrompu $g[c, d]$ s'il existe une transformation τ de $[a, b]$ en $[c, d]$ telle que pour chaque moment t de $[a, b]$ les points $f(t)$ et $g[\tau(t)]$ soient identiques. En fait, le chemin que nous étudions est toujours présenté comme un mouvement particulier, et chaque théorème doit être complété par la démonstration de son indépendance à l'égard du mouvement particulier choisi parmi les mouvements équivalents.

Bien que Fréchet et Morse aient établi les méthodes pour le choix d'un mouvement représentatif dans chaque chemin, il y a lieu de caractériser le chemin comme notion *sui generis*. On peut établir une solution de ce problème de la manière suivante.

Soient $f[a, b]$ un mouvement ininterrompu donné, $T = \{t_0, t_1, \ldots, t_n\}$ un ensemble de nombres de l'intervalle $[a, b]$ ordonnés selon leur grandeur et tels que $t_0 = a$ et $t_n = b$; à T correspond une chaîne d'ensembles fermés $\{E_1, E_2, \ldots, E_n\}$, où E_k désigne la trajectoire du mobile pendant l'intervalle

---

(⁸) Voir M. Krasner, *Comptes rendus*, 205, 1937. pp. 772-774.

$[t_{k-1}, t_k]$. Par une chaîne nous entendons une suite finie d'ensembles fermés contenant plus d'un point tel que deux élements consécutifs aient toujours au moins un point commun. Appelons $\mathcal{C}(f)$ la famille des chaînes correspondant à tous les ensembles finis T. Évidemment, si f et g sont équivalents, les familles $\mathcal{C}(f)$ et $\mathcal{C}(g)$ sont identiques. On voit que :

1° $\mathcal{C}(f)$ contient des chaînes de normes arbitrairement petites (par norme d'une chaîne on entend le diamètre maximal de ses éléments).

2° Étant données deux chaînes C_1 et C_2 de la famille $\mathcal{C}(f)$, cette dernière contient toujours un amincissement commun de C_1 et C_2. Ici nous disons que la chaîne $C = \{E_1, E_2, \ldots, E_m\}$ est un amincissement de la chaîne $D = \{F_1, F_2, \ldots, F_n\}$ si chaque élément F_i est la somme d'un segment $\{E_{i_k}, E_{i_k+1}, E_{i_{k+1}} - 1\}$ de la chaîne C, où $i_1 = 1$ et $i_{n+1} = m + 1$.

3° La famille $\mathcal{C}(f)$ est saturée par rapport à la propriété 2°, c'est-à-dire qu'il est impossible d'ajouter à $\mathcal{C}(f)$ aucune chaîne sans priver la famille de la propriété 2°.

Nous appelons chemin une famille $\mathcal{C}$ jouissant des propriétés 1°, 2°, 3°. Chaque chemin ainsi entendu, en conjonction avec une famille d'ensembles correspondants finis contenus dans un intervalle $[a, b]$, détermine un mouvement $f[a, b]$ tel que $\mathcal{C} = \mathcal{C}(f)$, deux mouvements $f[a, b]$ et $g[c, d]$ déterminés par le même chemin étant nécessairement équivalents. La détermination de $f[a, b]$ peut être basée sur une suite quelconque de chaînes $C_1, C_2, \ldots$ de la famille $\mathcal{C}$, pourvu que chaque chaîne C_{k+1} soit un amincissement de la chaîne C_k et que les normes de C_k tendent vers o. En conséquence, toute suite de chaînes satisfaisant à ces conditions constitue une seconde définition du chemin, deux suites déterminant le même chemin lorsqu'elles appartiennent à la même famille saturée. Remarquons qu'on peut aussi définir la distance de deux chaînes. En admettant des quasi-chaînes contenant des éléments d'un seul point, on peut inclure des mouvements avec des intervalles de repos.

En faisant correspondre à chaque T le polygone

$$P = \{f(t_0), f(t_1), \ldots, f(t_n)\},$$

nous obtenons une famille $\mathcal{P}$ de polygones. Étant donnés deux polygones P_1 et P_2 de la famille $\mathcal{P}$, cette dernière contient un sous-polygone commun à P_1 et P_2. La famille $\mathcal{P}$ est saturée par rapport à cette propriété, et elle contient des polygones dont les normes par rapport à la famille $\mathcal{P}$ sont arbitrairement petites. Nous appelons norme du polygone $\{p_0, p_1, \ldots, p_n\}$ par rapport à la famille $\mathcal{P}$ la borne supérieure des diamètres des polygones

$$\{q_{k_i} = p_i, q_{k_i+1}, \ldots, q_{k_{i+1}} = p_{i+1}\} \qquad (i = 0, \ldots, n-1)$$

qui forment les segments d'une extension Q de P.

Nous pouvons définir le chemin comme une famille de polygones jouissant des trois propriétés mentionnées.

Cette définition fournit une base entièrement satisfaisante pour une théorie intrisèque de la longueur d'un chemin. Nous appelons longueur de $\mathcal{L}$ la limite (si elle existe) des longueurs des polygones appartenant à $\mathcal{L}$ dont les normes par rapport à $\mathcal{L}$ tendent vers o. Ici la longueur est prise par rapport a une métrique quelconque associant à deux points, comme leur distance, un élément d'un groupe qui est une classe (L), la métrique étant assujettie à la seule condition que la distance de deux points identiques soit toujours l'élément o du groupe. D'autre part le role de la métrique de l'espace distancié est purement topologique.

STIELTJES INTEGRALS CONSIDERED AS LENGTHS

by

KARL MENGER (Chicago).

The Stieltjes integral

$$\int_a^b f(x)\,dg(x) \quad \text{or briefly,} \quad \int_a^b f\,dg$$

is defined as follows: We divide the interval $[a, b]$ into a finite number of intervals

$$a = x_0 < x_1 < \ldots < x_{n-1} < x_n = b.$$

We call max. $(x_{i+1} - x_i)$ the *norm* of the division. Moreover we select a number x_i^* in each interval $[x_i, x_{i+1}]$ and call

$$\sum f(x_i^*)\,[g(x_{i+1}) - g(x_i)]$$

the *Stieltjes sum* associated with the above division and the above selection of points x_i^*. If there exists a number from which the above Stieljes sums differ arbitrarily little provided that the norm of the division is sufficiently small, but regardless of the selection of the x_i^* in $[x_i, x_{i+1}]$, then we call this number the *Stieltjes integral*

$$\int_a^b f\,dg.$$

If in each interval $[x_i, x_{i+1}]$ we select the initial point, i. e., set $x_i^* = x_i$, then we obtain what we may call the *left-side Stieltjes sum*

$$\sum f(x_i)\,[g(x_{i+1}) - g(x_i)]$$

associated with the division of the interval $[a, b]$. If there exists a number from which these left-side Stieltjes sums differ arbitrarily little for every division of sufficiently small norm, then we call this number the *left-side Stieltjes integral*.

In a paper „*What curves have length?*"[1]) we subsumed this left-side Stieltjes integral under a general concept of length. In fact, the above integral is the length of the interval [a, b] derived from the distance

(*) $$\delta(x,y) = f(x)\,[g(y) - g(x)]$$

for every two numbers x, y of [a, b]. The question arises under what conditions a given distance yields a left-side Stieltjes integral as length. The answer is contained in the following remark.

In order that the length of the interval [a,b] derived from the distance $\delta(x,y)$ be a left-side Stieltjes integral with whose integrands the distance is connected by formula (*), it is necessary and sufficient that

1) $$\delta(x,y) \cdot \delta(y,z) + \delta(y,x) \cdot \delta(x,z) = \delta(x,y) \cdot \delta(y,x)$$
for every three numbres x, y, z of [a,b], and

2) *if* $\delta(x,y) = 0$ *and* $\delta(x,z) \neq 0$ *then*
$$\delta(x,z) \cdot \delta(y,w) = \delta(y,z) \cdot \delta(x,w)$$
for every four numbers x, y, z, w of [a,b].

The necessity of the conditions 1) and 2) is an obvious consequence of formula (*). In order to prove that the conditions are sufficient, let $\delta(x,y)$ be a distance satisfying the conditions 1) and 2). If $\delta(x,y) = 0$ for every two elements x, y of [a,b], then the metric is associated with the integral $\int_a^b 0\,dg$ for any function g. If $\delta(x,y) = 0$ does not hold for every x and y, then let x_0 and v_0 be two numbers for wich $\delta(x_0,v_0) \neq 0$. For each u we define

$$g(u) = \delta(x_0, u),$$

$$f(u) = \begin{cases} -\delta(u, x_0)/\delta(x_0, u) & \text{if } \delta(x_0, u) \neq 0, \\ \delta(u, v_0)/\delta(x_0, v_0) & \text{if } \delta(x_0, u) = 0. \end{cases}$$

---

[1]) Fundamenta Mathematicae, t. 35 (in print).

Now let y and z be any two elements. We claim that
$$\delta(y, z) = f(y) \cdot [g(z) - g(y)]$$
where f and g are two functions just defined.

If $\delta(x_0, y) \neq 0$, then
$$f(y) \cdot [g(z) - g(y)] = -\frac{\delta(y, x_0)}{\delta(x_0, y)} \cdot [\delta(x_0, z) - \delta(x_0, y)].$$

The expression on the right side is $\delta(y, z)$ by condition 1).

If $\delta(x_0, y) = 0$, then
$$f(y) \cdot [g(z) - g(y)] = \frac{\delta(y, v_0)}{\delta(x_0, v_0)} \cdot [\delta(x_0, z) - 0].$$

The expression on the right side is $\delta(y, z)$ by condition 2).

What paths have length?

By

Karl Menger (Chicago).

In the classical theory, the length of the curve $y = f(x)$ $(a \leqslant x \leqslant b)$ is determined by computing the integral $\int_a^b \sqrt{1 + f'^2(x)} \, dx$. Geometrically, this means that in determining the length of an arc we really compute the area of a plane domain. The length of the circular arc $y = \sqrt{1 - x^2}$ $(0 \leqslant x \leqslant b)$ is the area of the plane domain $(0 \leqslant x \leqslant b,$ $0 \leqslant y \leqslant 1, \sqrt{1 - x^2})$. If the arc happens to be a quarter of a circle, the domain is not even bounded.

In a series of previous papers [1]), the author has developed a more geometric approach to the problem based on the definition of the length of a path as the limit of the lengths of inscribed polygons which get indefinitely dense in the path. This length was studied in spaces of increasing generality. For instance, when applied to vector spaces our results comprise not only Finsler spaces but spaces with locally Minkowskian metrics in which the indicatrices (or unit spheres) are positive in some directions and negative or zero in others. On each stage we formulated sufficient conditions

---

[1]) [1] Mathematische Annalen **103** (1930), especially pp. 492-501. — [2] Fundamenta Mathematicae **25** (1935), p. 441. — [3] Three notes in the C. R. Paris **201** (1936), p. 705; **202** (1936), p. 1007; **202** (1936), p. 1648. — [4] Ergebnisse eines mathematischen Kolloquiums **8** (1937), p. 1-37. — [5] Proc. Nat. Acad. Sc., **23** (1937), p. 244. — [6] Ibid., **25** (1939), p. 474. — [7] Rice Institute Pamphlets **27** (1940), p. 1-40. — Cf. P a u c, *Les méthodes directes en calcul des variations et en géométrie différentielle*, Hermann, Paris 1941. — In [7], metric methods are also used for the formulation of necessary and sufficient conditions for a line integral to be independent of the path. We add a bibliography of more recent results along these lines: Menger, Proc. Nat. Acad. Sc., **25** (1939), p. 621. — F u b i n i, ibid., **26** (1940), p. 190. — Menger, ibid. **26** (1940), p. 660. — Artin, ibid., **27** (1941), p. 489. — Menger, Reports of a Mathematical Colloquium, 2-nd ser., **2** (1939), p. 45. — Milgram, ibid., **3** (1940), p. 28. — de Pazzi Rochford, ibid., **4** (1940), p. 6.

for the existence as well as the lower semi-continuity of the length. Since lower semi-continuity in a well known way implies the existence of minimizing paths we thus derived existence theorems not only for positively definite but also for semi-definite and indefinite problems of the calculus of variations.

In the course of these studies it became clear that such topological concepts as neighboring paths should not be based on the distance in terms of which the length of polygons and paths is defined. The topological concepts should rather be introduced in terms of a basic topology of the space. Only for the sake of simplicity we described, and continue to describe, this topology in terms of a metric space but we might as well describe it in terms of any topologically equivalent metric [2]).

In the present paper we shall formulate conditions which are not only sufficient but at the same time necessary for the existence of a finite length. On the other hand, we shall not compare the path with neighboring paths. Hence we shall treat the path intrinsically, that is, as a closed interval of real numbers [3]). We shall choose the interval $\mathcal{J} = [0,1]$ and shall describe its topology in terms of the euclidean distance $|x-y|$ of the points x and y.

The lengths of polygons and the length of $\mathcal{J}$ will be derived from a distance $\delta(x,y)$. More precisely, we assume that with every ordered pair of numbers x,y of $\mathcal{J}$ a real number $\delta(x,y)$ is associated, called the *distance from x to y*, which is connected with the underlying topology only by the following one-sided continuity condition [4]):

---

[2]) In [4] and [7] l. c. [1]) we prove that in theorems of the calculus of variations which refer to rectifiable paths or to paths of uniformly bounded lengths, the ordinary length may be replaced by one derived from a more general distance of comparison so that altogether we distinguish *three metrics in the calculus of variation*: the metric which describes the topology and is metrically insignificant; the metric for which we wish to minimize the length; the metric of comparison. In the classical calculus of variations, including Tonelli's theory, one studies only the euclidean metric as the metric describing the topology of the underlying vector space and at the same time as the metric of comparison while the length which we wish to minimize is obtained by a multiplicative distortion of this metric — the integrand being the distorting factor.

[3]) We have formulated an intrinsic definition of the concept of path even if the path is imbedded in a space. Cf. C. R. Paris **221** (1945), p. 739.

[4]) In subsequent papers we shall also admit *complex* distances, in fact, distances belonging to a *normed group*. We shall furthermore weaken the one-sided continuity condition mentioned above.

For each $\delta > 0$ *there exists an* $\eta(\delta) > 0$ *such that* $|x - x'| < \eta(\delta)$ *implies* $|\delta(x, x')| < \delta$.

We shall not assume that, conversely, for points x, x' whose distance $\delta(x, x')$ is small, the number $|x - x'|$ is small. Even if both $\delta(x, x')$ and $\delta(x', x)$ are small or $= 0$, we shall not be able to conclude that $|x - x'|$ is small.

Under this assumption we shall formulate conditions which are both necessary and sufficient in order that $\mathcal{J}$ have a finite length.

With regard to the relation between length and area, our theory reverses the classical point of view. Instead of representing the length as an integral or an area, we may represent integrals and areas as lengths. For instance, we obtain the *integral* $\int_a^b f(x)\, dx$ as the length of the interval $[a, b]$ derived from the distance $\delta(x, y) = f(x) \cdot (y - x)$. In order to obtain the variation of the function $f(x)$ as a length we set $\delta(x, y) = |f(y) - f(x)|$. What we may call the *left-side Stieltjes integral* [5]) $\int_a^b f(x)\, dg(x)$ is derived from the distance $\delta(x, y) = f(x) \cdot (g(y) - g(x))$. The *Weierstrass line integral* [6]) $\int_a^b f(x, \mathring{x})\, dt$ where x is a vector in any vector space, and f is positively homogeneous of degree 1 in $\mathring{x}$, is obtained by setting $\delta(t, t') = f(x(t), x(t') - x(t))$.

We begin by defining the necessary auxiliary concepts. By a *polygon* we mean an ordered set $P = \{x_0, x_1, ..., x_{n-1}, x_n\}$ of elements of $\mathcal{J}$ such that $n \geqslant 1$ and $x_0 < x_1 < ... < x_{n-1} < x_n$. We call $x_1, ..., x_{n-1}$ the *inner* points of P. The polygons P and $Q = \{y_0, ..., y_m\}$ are said to be *exclusive* if the closed intervals $[x_0, x_n]$ and $[y_0, y_m]$ have at most one point in common. The polygon P is called an *end-to-end* polygon if $x_0 = 0$ and $x_n = 1$.

We set

$$vP = \text{Max.} \, (x_{i+1} - x_i), \quad \sigma P = x_n - x_0,$$

---

[5]) cf. the author's paper *The Stieltjes integrals considered as lengths*, Ann. de la Soc. Pol. de Math. t. XXI (1948).

[6]) Concerning Weierstrass integrals cf. [7] l. c. [1]) and, in particular, Pauc's comprehensive presentation in his booklet *Les méthodes directes*.

and call these numbers the *norm* of P and the *span* of P, respectively
By the *length* of P we mean the number

$$\lambda P = \Sigma \delta(x_i, x_{i+1}),$$

by the *absolute length* of P the number

$$\Lambda P = \Sigma |\delta(x_i, x_{i+1})|.$$

We call

$$\chi P = \delta(x_0, x_n)$$

the *chord* of P. Of importance for our purpose is the following number

$$\varkappa P = \frac{\chi P - \lambda P}{|\chi P|} \quad \text{if} \quad \chi P \neq 0.$$

We call $\varkappa P$ the *contraction* of P and complete the definition
by setting

$$\varkappa P = +\infty \quad \text{if} \quad \chi P = 0 > \lambda P,$$
$$\varkappa P = -\infty \quad \text{if} \quad \chi P = 0 < \lambda P.$$

$\varkappa P = 0$ is equivalent with $\lambda P = \chi P$ which, in a euclidean
space, holds if and only if P is an ordered linear polygon. In case
that $\chi P > 0$, the condition $\varkappa P = 1$ is equivalent with $\lambda P = 0$.

We call a sequence of polygons $\{P_n\}$ *distinguished* if each P_n
is an end-to-end polygon and $\lim \nu P_n = 0$. We set

$$\lambda^*\{P_n\} = \lim \sup \lambda P_n \quad \text{and} \quad \lambda_*\{P_n\} = \lim \inf \lambda P_n.$$

We call the least upper bound of the numbers $\lambda^*\{P_n\}$ for all
distinguished sequences the *upper length* of $\mathcal{J}$, the greatest lower
bound of the numbers $\lambda_*\{P_n\}$ the *lower length* of $\mathcal{J}$. We say that $\mathcal{J}$
has a *length* if upper and lower lengths are equal.

We say that $\mathcal{J}$ is of *bounded absolute length* if the least upper
bound, Λ, of the numbers ΛP for all end-to-end polygons is finite.
Clearly, if $\mathcal{J}$ is of bounded absolute length, and has a length $\lambda \mathcal{J}$,
then

$$|\Lambda \mathcal{J}| \leqslant \Lambda.$$

In terms of these concepts we shall prove the following

Theorem. *If $\mathcal{J}$ is of bounded absolute length, then in order
that $\mathcal{J}$ have a finite length it is necessary and sufficient that for each
$\varepsilon > 0$ and $\varkappa > 0$ there exists a $\nu = \nu(\varepsilon, \varkappa) > 0$ such that for each finite
set of exclusive polygons $Q_1, Q_2, ..., Q_k$ whose spans are $< \nu$ and whose
contractions are $> \varkappa$ we have*

(1) $$\Sigma |\chi Q_i| < \varepsilon,$$
(2) $$\Sigma \lambda Q_i > -\varepsilon.$$

Necessity of 1). If Condition 1) is not satisfied, then there exist two numbers ε_0 and $\varkappa_0$, both >0, such that for each n there exists a finite set $Q = \{Q_1^n, Q_2^n, ..., Q_{k_n}^n\}$ of exclusive polygons of spans $<1/n$ and contractions $\geqslant \varkappa_0$ for which

$$\Sigma |\chi Q_i^n| \geqslant \varepsilon_0.$$

For each n, since the span of each Q_i^n is $<1/n$ we can complete the polygons of Q to one polygon Q^n of a norm $<1/n$. The polygon P^n obtained from Q^n by omitting all the inner points of the k_n polygons Q_i^n likewise has a norm $<1/n$. If λ_n is the sum of the lengths of the polygons completing the k_n polygons Q_i^n to Q^n, then

$$\lambda P^n = \lambda_n + \Sigma \chi Q_i^n \quad \text{and} \quad \lambda Q^n = \lambda_n + \Sigma \lambda Q_i^n.$$

Since the contraction of each Q_i^n is $\geqslant \varkappa_0$ we have

$$\lambda Q_i^n < \chi Q_i^n - \varkappa_0 |\chi Q_i^n| \qquad (i = 1, 2, ..., k_n)$$

and thus

$$\lambda Q^n < \lambda_n + \Sigma \chi Q_i^n - \varkappa_0 \Sigma |\chi Q_i^n|.$$

Hence

$$\lambda Q^n < \lambda P^n - \varkappa_0 \Sigma |\chi Q_i^n| < \lambda P^n - \varkappa_0 \cdot \varepsilon_0.$$

We see for each n there exist two polygons P^n and Q^n of norms $<1/n$ whose lengths differ by more than $\varkappa_0 \varepsilon_0$. Thus $\mathcal{J}$ has no finite length.

Necessity of 2). If Condition 2) is not satisfied, then there exist two numbers ε_0 and $\varkappa_0$, both >0, such that for each n there exists a finite set Q of exclusive polygons $\mathcal{C}_1^n, ..., \mathcal{C}_{k_n}^n$ whose spans are $<1/n$, whose contractions are $\geqslant \varkappa_0$ and for which

(*) $$\Sigma \lambda Q_i^n < -\varepsilon_0.$$

Since the necessity of Condition 1) has been established we can assume that

(**) $$\lim_{n \to \infty} \Sigma |\chi Q_i^n| = 0.$$

For each n, we form Q^n and P^n as before and have

$$\lambda P^n = \lambda_n + \Sigma \chi Q_i^n \quad \text{and} \quad \lambda Q^n = \lambda_n + \Sigma \lambda Q_i^n.$$

Thus

$$\lambda Q^n - \lambda P^n = \Sigma \lambda Q_i^n - \Sigma \chi Q_i^n.$$

Fundamenta Mathematicae. T. XXXVI.

8

From (*) and (**) it follows that

$$\lambda Q^n - \lambda P^n < -\varepsilon_0/2 \quad \text{for all large } n,$$

and again $\mathcal{J}$ has no finite length.

Sufficiency. We begin by proving: If Conditions 1) and 2) hold, and thus for every $\varepsilon > 0$ and $\varkappa > 0$ a number $\nu(\varepsilon, \varkappa)$ with the specified properties exists, then for every $\varkappa > 0$ and $\zeta > 0$ the following condition holds:

Condition $C_{\varkappa\zeta}$. For each polygon whose norm is sufficiently small, namely, $< \nu = \nu(\zeta/5, \varkappa)$, there exists a number $\nu' > 0$ with the property that each polygon Q whose norm is $< \nu'$, satisfies the inequality

$$\lambda Q > \lambda P - \varkappa \cdot \Lambda P - \zeta.$$

By assumption, for every finite set of exclusive polygons $Q_1, \dots, Q_k$ whose spans are $< \nu = \nu(\zeta/5, \varkappa)$ and whose contractions are $> \varkappa$, we have

(1')　　　　　　　　$\Sigma |\chi Q_i| < \zeta/5,$

(2')　　　　　　　　$\Sigma \lambda Q_i > -\zeta/5.$

Let P be any polygon whose norm is $< \nu$. Let P be the polygon $\{x_0, \dots, x_n\}$ so that $n+1$ is the number of points of P. For the number $\zeta/5n$ we form the number $\eta(\zeta/5n)$ mentioned in the basic continuity postulate, that is to say, the number for which

($\not\vdash$)　　　　$|x - x'| < \eta$　implies　$|\delta(x, x')| < \zeta/5n.$

If ν' denotes the smaller of the two numbers

$$\eta \quad \text{and} \quad \tfrac{1}{2} \text{Min.} (x_{i+1} - x_i),$$

then we shall prove that ν' has the property claimed in Condition $C_{\varkappa\zeta}$.

In order to prove this contention, let Q be any polygon whose norm is $< \nu'$. Hence

(3)　　　　　　　　$\nu Q < \tfrac{1}{2} \text{Min.} (x_{i+1} - x_i)$

and

(4)　　　　　　　　$\nu Q < \eta = \eta(\zeta/5n).$

For each point x_i of P we denote

by $y_{j(i)}$ the first point of Q which is $> x_i$,
by $y_{k(i)}$ the last point of Q which is $\leqslant x_{i+1}.$

We shall set
$$Q_i = \{y_{j(i)}, y_{j(i)+1}, ..., y_{k(i)}\}$$
and
$$Q_i^* = \{x_i, y_{j(i)}, y_{j(i)+1}, ..., y_{k(i)}, x_{i+1}\}.$$

From (3) it follows that

$$x_i < y_{j(i)} < y_{k(i)} \leqslant x_{i+1} \qquad (i = 0, 1, ..., n-1).$$

Hence each Q_i contains at least two points and thus is a polygon, while each Q_i^* is a polygon containing at least three points. The points $y_{k(i)}$ and $y_{j(i+1)}$ of Q are consecutive and

$$y_{k(i)} \leqslant x_{i+1} < y_{j(i+1)} \qquad (i = 0, 1, ..., n-1).$$

From (4) it thus follows that

(5) $$|\delta(x_i, y_{j(i)})| < \zeta/5n \quad \text{and} \quad |\delta(y_{k(i)}, y_{,(i+1)})| < \zeta/5n.$$

The spans of Q_i and Q_i^* are $\leqslant x_{i+1} - x_i < \nu$. The polygons $Q_0^*, ..., Q_{n-1}^*$ are mutually exclusive. Also the polygons obtained by replacing some (or all) of the Q_i^* by the corresponding Q_i are mutually exclusive. We have

$$\lambda Q_i^* = \delta(x_i, y_{j(i)}) + \lambda Q_i + \delta(y_{k(i)}, x_{i+1}).$$

From (5) it thus follows that

(6) $$\lambda Q_i > \lambda Q_i^* - 2\zeta/5n.$$

If we set $y_{k(0)} = x_0$, then

$$\lambda Q = \Sigma \lambda Q_i + \Sigma \delta(y_{k(i)}, y_{j(i+1)}).$$

Hence by (5)

(7) $$\lambda Q > \Sigma \lambda Q_i - \zeta/5.$$

We shall say that a given polygon Q_i^* is *of the first kind* if $\varkappa Q_i^* > \varkappa$, *of the second kind* if $\varkappa Q_i^* \leqslant \varkappa$. We shall denote summations restricted to polygons of the first kind or the second kind by Σ' and by Σ'', respectively. We have

(8) $$\Sigma \lambda Q_i = \Sigma' \lambda Q_i + \Sigma'' \lambda Q_i.$$

Now polygons of the first kind have spans $< \nu$ and contractions $> \varkappa$. Thus by (1'),

$$\Sigma' |\varkappa Q_i| < \zeta/5,$$

that is to say,

(9) $$\Sigma'|\delta(x_i, x_{i+1}) < \zeta/5$$

and by (2')

(10) $$\Sigma' \lambda Q_i > -\zeta/5.$$

From (6) it follows that

(11) $$\Sigma'' \lambda Q_i > \Sigma'' \lambda Q_i^* - 2\zeta/5.$$

Since for each polygon of the second kind $\varkappa Q_i^* \leqslant \varkappa$ we have

$$\lambda Q_i^* \geqslant \chi Q_i^* - \varkappa \cdot |\chi Q_i^*|.$$

Thus

$$\Sigma'' \lambda Q_i^* \geqslant \Sigma'' \chi Q_i^* - \varkappa \Sigma |\chi Q_i^*| \geqslant \Sigma'' \chi Q_i^* - \varkappa \cdot \Lambda P = \Sigma'' \delta(x_i, x_{i-1}) - \varkappa \cdot \Lambda P =$$
$$= \lambda P - \Sigma' \delta(x_i, x_{i+1}) - \varkappa \cdot \Lambda P.$$

By (9)

$$\Sigma'' \lambda Q_i^* > \lambda P - \zeta/5 - \varkappa \cdot \Lambda P.$$

Combining the last inequality with (7), (8), (10), (11) we conclude

$$\lambda Q > P - \varkappa \cdot \Lambda P - \zeta.$$

Thus conditions $C_{\varkappa\zeta}$ is satisfied.

We conclude the demonstration of the sufficiency of the Conditions (1) and (2) by proving:

If $\mathcal{J}$ is of bounded absolute length and Condition $C_{\varkappa\zeta}$ holds for every $\varkappa$ and ζ, then $\mathcal{J}$ has a finite length.

Let Λ be the finite least upper bound of the numbers ΛP for all polygons. Clearly, the upper length λ^* of $\mathcal{J}$ is finite and there exists a *maximal* sequence of polygons $\{P_n\}$, that is to say, a sequence such that

$$\lim \lambda P_n = \lambda^*.$$

We claim: for every $\varepsilon > 0$ there exists a $\nu' > 0$ such that for each polygon P whose norm is $< \nu'$ we have

$$\lambda P > \lambda^* - \varepsilon.$$

We choose n so large that $\lambda P_n > \lambda^* - \varepsilon/3$. We further choose $\varkappa$ and ζ so that

$$\zeta < \varepsilon/3 \quad \text{and} \quad \varkappa < \varepsilon/3\Lambda.$$

Under these circumstances, Condition $C_{\varkappa\zeta}$ yields our contention.

In subsequent papers we shall formulate conditions which are both necessary and sufficient for $\mathcal{J}$ to have the length ∞, in which case the interval will, of course, be of unbounded absolute length. On the other hand, in connection with the theorem which we have proved above, the question arises as to whether an interval of finite length is not eo ipso of bounded absolute length.

We conclude this paper with an example for the fact *that finite length is compatible with unbounded absolute length*. One is reminded of infinite series which converge without converging absolutely — except that there are more polygons in $\mathcal{J}$ than terms or partial sums in an infinite series.

Our example is based on the following auxiliary function f which is defined for all integers:

$$f(2n) = \tfrac{1}{2} \quad \text{and} \quad (2n+1) = -\tfrac{1}{4} \qquad (n = 0, 1, \ldots).$$

We divide $\mathcal{J}$ into 8 equal segments and define the distances of the end points as follows:

$$\delta\left(\frac{i_1}{8}, \frac{i_1+1}{8}\right) = f(i_1)$$

We further set

$$\delta\left(\frac{i}{8}, \frac{j}{8}\right) = \delta\left(\frac{i}{8}, \frac{i+1}{8}\right) + \delta\left(\frac{i+1}{8}, \frac{i+2}{8}\right) + \ldots + \delta\left(\frac{j-1}{8}, \frac{j}{8}\right) \quad \text{for } i < j.$$

In particular,

$$\delta(0,1) = \tfrac{1}{2} - \tfrac{1}{4} + \tfrac{1}{2} - \tfrac{1}{4} + \tfrac{1}{2} - \tfrac{1}{4} + \tfrac{1}{2} - \tfrac{1}{4} = 1.$$

Next we divide each of the intervals $\left[\dfrac{i_1}{8}, \dfrac{i_1+1}{8}\right]$ into 8 equal segments and define

$$\delta\left(\frac{i_1}{8} + \frac{i_2}{8^2}, \frac{i_1}{8} + \frac{i_2+1}{8^2}\right) = f(i_1) \cdot f(i_2).$$

Again, we set

$$\delta\left(\frac{i_1}{8} + \frac{i_2}{8^2}, \frac{j_1}{8} + \frac{j_2}{8^2}\right) = \delta\left(\frac{i_1}{8} + \frac{i_2}{8^2}, \frac{i_1}{8} + \frac{i_2+1}{8^2}\right) + \ldots + \delta\left(\frac{j_1}{8} + \frac{j_2-1}{8^2}, \frac{j_1}{8} + \frac{j_2}{8^2}\right).$$

Proceeding in this way we set

$$\delta\left(\frac{i_1}{8} + \ldots + \frac{i_{n-1}}{8^{n-1}} + \frac{i_n}{8^n}, \frac{i_1}{8} + \ldots + \frac{i_{n-1}}{8^{n-1}} + \frac{i_n+1}{8^n}\right) = f(i_1) \cdot \ldots \cdot f(i_{n-1}) \cdot f(i_n)$$

and define a distance $\delta(x,y)$ for every two octogonally rational numbers x and y of $\mathcal{I}$ such that $x < y$. We set $\delta(y,x) = \delta(x,y)$ and $\delta(x,x) = 0$. If two octogonally rational numbers differ by less than $1/8^n$, their distances differ by less than $1/2^n$. Hence it is easy to extend the definition of $\delta(x,y)$ to any two numbers x and y of $\mathcal{I}$. The length of each end-to-end polygon is 1. The absolute length of $\mathcal{I}$ is unbounded [7]).

---

[7]) A slight modification of the above construction leads to an arc having *the absolute length* ∞ *and the length* 0. We divide the interval $[0,1]$ into four instead of eight equal parts and define the distances from 0 to $\frac{1}{4}$, and from $\frac{1}{2}$ to $\frac{3}{4}$ to be $\frac{1}{2}$, and the distances from $\frac{1}{4}$ to $\frac{1}{2}$ and from $\frac{3}{4}$ to 1 to be $-\frac{1}{4}$. Iteration of this procedure leads to the indicated result.

Mr. Sheldon L. Levy pointed out that the original example (with divisions into eight parts) can be simplified. It is sufficient to divide the interval $[0,1]$ into three equal parts and to define the distance from 0 to $\frac{1}{3}$ as $\frac{2}{3}$, the distance from $\frac{1}{3}$ to $\frac{2}{3}$ as $-\frac{1}{3}$, and the distance from $\frac{2}{3}$ to 1 as $\frac{2}{3}$. Iteration of this procedure leads to an arc whose absolute length is ∞ and whose length is 1.

Illinois Institute of Technology, Chicago.

A TOPOLOGICAL CHARACTERIZATION OF THE LENGTH OF PATHS.

By Karl Menger

Illinois Institute of Technology

Communicated by Marston Morse, November 8, 1951

Let T be a metrizable topological space. We speak of a universal functional of paths if with every path $\mathcal{P}$ in T a (finite or infinite) number $\lambda\mathcal{P}$ is associated. For any particular metrization of T, the corresponding length of paths is an example of a non-negative universal functional. To different metrizations of T correspond, in general, different lengths. *How are these lengths characterized among the non-negative universal functionals of paths?* In other words, what properties of a functional λ are necessary and sufficient in order that there exist a metrization of T such that, for every path $\mathcal{P}$, the corresponding length is equal to $\lambda\mathcal{P}$? We widen the scope of the problem by admitting metrizations of T for which the distance is non-symmetric.

I. What Is a Path? In the classical theory, a path is obtained from a motion by a process of abstraction. A motion M in T is a continuous association of a point $p_M(t)$ of T with each moment t of a closed interval $[a, b]$. Abstracting from kinematical aspects of the motion, Fréchet called the motion N: $q_N(u)$ ($c \leq u \leq d$) *equivalent* with M if there exists a quasi-similarity between the intervals $[a, b]$ and $[c, d]$ such that $p_M(t) = q_N(u)$ for any two related moments t and u. Here, by a quasi-similarity we mean a set of pairs (t, u) of related moments such that (1) each $t \in [a, b]$ as well as each $u \in [c, d]$ belongs to at least one pair; (2) if (t, u) and (t', u') are two pairs of related moments, then $(t' - t) \cdot (u' - u) \geq 0$.

The *path of* M is the set of all motions which are equivalent to M. *Path* is the path of some motion. One readily defines convergence of paths. The terms initial point and terminal point of a path are self-explanatory. If the terminal point of $\mathcal{P}$ coincides with the initial point of $\mathcal{Q}$, then we call $\mathcal{Q}$ *consecutive* to $\mathcal{P}$, and we denote by $\mathcal{P} + \mathcal{Q}$ the combined path from the initial point of $\mathcal{P}$ to the terminal point of $\mathcal{Q}$.

In an alternate theory, we have more recently[1] defined the path of M by adding geometric specifications to the trajectory of M, i.e., the set of all positions $p_M(t)$ $(a \leq t \leq b)$. This trajectory is the same for any equivalent motion and hence may be called the trajectory of the path. But also nonequivalent motions and different paths may have the same trajectory. While thus the set of points of M is insufficient to characterize the path of M, we may define the latter as *the set of all polygons of* M, that is, the set of finite sequences of points $\{p_M(t_1), p_M(t_2), \ldots, p_M(t_n)\}$ for all ordered sets $\{t_1, t_2, \ldots, t_n\}$ for which $a \leq t_1 < t_2 < \ldots < t_n \leq b$. *Path* is the set of all polygons of some motion.

The sets of polygons which are paths, can be characterized by intrinsic properties without any reference to motions or particular parametrizations. If, for the sake of a simpler formulation, we endow T with a particular metric, a path in T is a set $\mathcal{P}$ of polygons (finite sequences of points of T) with the following properties:

1. *$\mathcal{P}$ contains polygons whose norms relative to $\mathcal{P}$ are arbitrarily small.* By the norm of $P = \{p_1, \ldots, p_n\}$ relative to $\mathcal{P}$ we mean the least upper bound of the diameters of all polygons $\in \mathcal{P}$ from any p_i to the consecutive p_{i+1} $(i = 1, \ldots, n - 1)$, the initial and the terminal point of $\mathcal{P}$ being, if necessary, adjoined to P.

2. *$\mathcal{P}$ contains for every two of its polygons a common refinement.* The polygon P' is called a refinement of P if the sequence P is a subsequence of the sequence P'.

3. *$\mathcal{P}$ is saturated with regard to properties 1 and 2* in the sense that $\mathcal{P}$ is not a proper subset of a set of polygons $\mathcal{P}'$ containing a common refinement for every two of its polygons and such that $\mathcal{P}$ contains polygons of arbitrarily small norms relative to $\mathcal{P}'$ (even though $\mathcal{P}'$ itself may contain polygons of arbitrarily small norms relative to $\mathcal{P}'$).

II. What Is Length? If T is endowed with a particular (not necessarily symmetric) metric, the length $l\mathcal{P}$ of the path $\mathcal{P}$ is defined as the least upper bound of the lengths of all polygons of $\mathcal{P}$, the length of $P = \{p_1, \ldots, p_n\}$ being $\sum d(p_{i-1}, p_i)$. For any metric, the corresponding length of paths is

(A)   *additive:* if $\mathfrak{Q}$ is consecutive to $\mathcal{P}$, then $l(\mathcal{P} + \mathfrak{Q}) = l\mathcal{P} + l\mathfrak{Q}$;

(B)   *lower semi-continuous:* if $\lim \mathcal{P}_n = \mathcal{P}$, then $\lim \inf l\mathcal{P}_n \geq l\mathcal{P}$;

(C)   *regular:* if, for each n, $\mathcal{P}_n$ is a path from p to p_n and $\lim l\mathcal{P}_n = 0$, then $\lim p_n = p$.

The properties A, B, C characterize the lengths among the non-negative universal functionals λ *of paths contained in a compact metrizable space T.* Suppose T and λ are given. First we assume T to be λ-*connected* (i.e., to contain, for any two points p, q, a path $\mathcal{P}$ for which $\lambda \mathcal{P} < \infty$) and *locally* λ-*connected* (i.e., to contain, if $\lim p_n = p$, a path $\mathcal{P}_n$ from p to p_n such that $\lim \lambda \mathcal{P}_n = 0$). If λ has the properties A, B, C, then λ is the length relative to a metric by virtue of which T is *convex* in the sense that, for every two points p and q, T contains at least one path from p to q whose length is $d(p, q)$. We obtain the corresponding distance by setting

$$d(p, q) = g.l.b. \lambda \mathcal{P} \text{ for all paths } \mathcal{P} \text{ from } p \text{ to } q.$$

It can be shown[2] that $d(p, p) = 0$; $d(p, q) > 0$ if $p \neq q$; $d(p, q) + d(q, r) \geq d(p, r)$; and $\lim d(p_n, p) = 0$ if and only $\lim p_n = p$. It remains to be proved that $l\mathcal{P} = \lambda \mathcal{P}$. For any $\epsilon > 0$, let $P = \{p_0, \ldots, p_n\}$ be a polygon of $\mathcal{P}$ such that $\sum d(p_{i-1}, p_i) > l\mathcal{P} - \epsilon$. If $\mathcal{P}_1, \ldots, \mathcal{P}_n$ denote the segments into which P divides $\mathcal{P}$, then $\lambda \mathcal{P} = \sum \lambda \mathcal{P}_i \geq \sum d(p_{i-1}, p_i) > l\mathcal{P} - \epsilon$. Conversely, there exist paths $\mathfrak{Q}_i$ from p_{i-1} to p_i such that $\lambda \mathfrak{Q}_i < d(p_{i-1}, p_i) + \frac{\epsilon}{n}$. Setting $\mathfrak{Q} = \mathfrak{Q}_1 + \ldots + \mathfrak{Q}_n$ we have $\lambda \mathfrak{Q} < \Sigma d(p_{i-1}, p_i) + \epsilon < l\mathcal{P} + \epsilon$. Thus in each neighborhood of $\mathcal{P}$ there exist paths $\mathfrak{Q}$ for which $\lambda \mathfrak{Q} < l\mathcal{P} + \epsilon$. Since λ is lower semicontinuous, $\lambda \mathcal{P} \leq l\mathcal{P}$.

III. Generalizations and Problems. Carathéodory's theorem concerning the compactness of the set of all curves of uniformly bounded length can be generalized as follows.[3] Let λ be any non-negative universal functional of paths contained in a compact metrizable space T. If λ has the properties A, B, C, then, for every $L < \infty$, the set of all paths $\mathcal{P}$ for which $\lambda \mathcal{P} \leq L$, is compact. Using this generalization we prove the following theorem. Let λ be a non-negative universal functional of the paths contained in a compact metrizable space T. *In order that* λ *be the length in a metroid M which* (1) *contains a path of length* $d(p, q)$ *from* p *to* q *whenever* $d(p, q) < \infty$, (2) *can be mapped on T in a one-to-one continuous way and so that* $l\mathcal{P} = \lambda \mathcal{P}$ *for every path, it is necessary and sufficient that* λ *have the properties A, B, C.* Here, by a metroid we mean a set in which every two points have a finite or infinite distance such that $d(p, p) = 0$; $d(p, q) > 0$ if $p \neq q$; $d(p, q) + d(q, r) \geq d(p, r)$. *If T is* λ-*connected, then a metric M can be chosen* (in which all distances are finite); *if T is locally* λ-*connected, then M can be mapped on T topologically.*

Topological characterizations of length whenever T is homeomorphic with a complete metric space would be desirable; similarly, characterizations of length in spaces where the distances satisfy only the condition that $\lim p_n = p$ implies $\lim d(p_n, p) = 0$ (without necessarily being non-negative, triangular, etc.). In such spaces, length is not necessarily a universal functional.[4] Most interesting would be extensions to the area of surfaces in the sense of continuous images of a disk. Such surfaces might be defined as sets of polyhedra.

[1] "Définition intrinsèque de la notion de chemin" *C. R. Acad. Paris*, **221**, 739 (1945).

[2] "Untersuchungen ueber allgemeine Metrik, I," *Mathem. Annalen*, **100**, 96 sq. (1928).

[3] For a slightly weaker generalization cf. "Metric Methods in Calculus of Variations," PROC. NATL. ACAD. SCI., **23**, 247 (1937), and "A Theory of Length and Its Applications to the Calculus of Variations," *Ibid.*, **25**, 476 (1939).

[4] "What Paths Have Length?" *Fund. Math.*, **36**, 109–118 (1949).

Selected Papers on Algebra of Analysis

Commentary on the Algebra of Analysis and Algebra of Functions

A. Sklar

1. Karl Menger's interest in the Algebra of Analysis (which later evolved into the Algebra of Functions)[1] grew out of his experience teaching calculus to large classes of potential naval officers at Notre Dame University during the Second World War. Now throughout his life, Menger was impelled to uncover the basic structure underlying any mathematical area he was involved with, and to express this structure in explicit axiomatic form. So, since calculus is predicated on the behavior of functions under various operations, he resolved to elucidate the fundamental properties of these operations.

He observed that while the operations of addition and multiplication of functions were explicit in the standard treatment of calculus, there was a third, equally if not more important, operation that was never made explicit, but remained hidden under such rubrics as "the chain rule" and "integration by substitution". He called this operation "substitution" and denoted it by simple juxtaposition. (This operation is now more commonly called "composition" and the result of composition is therefore a "composite function".)

Substitution, associative but not generally commutative, is connected to the familiar operations of addition (+) and multiplication (·) by the one-sided[2] distributive laws:

$$(f + g)h = fh + gh \qquad (f \cdot g)h = fh \cdot gh.$$

---

[1] Note that the phrase "Algebra of Functions" already appears in [M1] (in the Table of Contents!).

[2] As Menger notes on p. 1 of [M1], the formulas

$$h(f + g) = hf + hg, \quad h(f \cdot g) = hf \cdot hg,$$

represent not "universal laws", but instead each characterizes "a special class of functions h". Assuming some simple regularity conditions, these classes are, respectively, the (homogeneous) linear functions and the power functions.

All these operations have identity or (as Menger called them) *neutral elements*: the constant function of value zero (denoted by the symbol 0) for addition, the constant function of value one (denoted by the symbol 1) for multiplication. Needing a symbol for the neutral element for substitution, Menger considered using the letter i but discarded that notion because of a conflict with the standard use of i to denote a particular complex number. He chose instead the following letter j, which therefore represents the universal identity function, i.e., the function whose value $j(x)$ is x for every x in the universal domain.

Thus Menger was led to a rather elegant system of postulates for what he called a *tri-operational algebra*. He presented this system in the following form, identical in [M1], [M2], and [M3]:

Commutative Law: $f + g = g + f,$ $\qquad f \cdot g = g \cdot f;$

Associative Law: $(f + g) + h = f + (g + h),$ $(f \cdot g) \cdot h = f \cdot (g \cdot h),$ $(fg)h = f(gh);$

Distributive Law: $(f \cdot g)h = fh \cdot gh,$ $\qquad (f + g)h = fh + gh,$ $(f + g) \cdot h = f \cdot h + g \cdot h;$

Neutral Element: $f + 0 = f,$ $\qquad f \cdot 1 = f,$ $\qquad fj = jf = f,$

$$0 \neq 1 \neq j \neq 0, \qquad 10 = 1;$$

Opposite Element: $f + (-f) = 0.$

In Chapter II of [M1] Menger extends his tri-operational algebra by introducing derivation and antiderivation. On page 31, he postulates a derivative operator D that satisfies the following conditions:

I. $D(f + g) = Df + Dg,$
II. $D(f \cdot g) = f \cdot Dg + g \cdot Df,$
III. $D(fg) = (Df)g \cdot Dg.$

I and II are familiar, while III is usually known in somewhat disguised form as the "chain rule", although mathematicians today will generally recognize what is meant by the phrase "the formula for the derivative of a composite function".[3]

---

[3] Since repeated differentiation is often called for, it is useful to have available a formula for the n-th derivative of a composite function, corresponding to those for the n-th derivative of a sum or product. Several equivalent forms of such a formula exist, of which the oldest and best-known is that of Francesco Faà di Bruno (see [7]). But the most perspicuous form is surely that developed by M. A. McKiernan, who had been a student of Menger's. It appears in [12] in Mengerian notation as

$$D^n(fg) = \sum_{r=1}^{n} (D^r f)g \cdot \sum_{s=0}^{r} \frac{(-1)^{r-1}}{s!(r-s)!} \cdot g^{r-s} \cdot D^n(g^s).$$

Having already defined various functions in Chapter I (including, by way of appropriate functional equations, exponential, logarithmic, and trigonometric functions[4]), Menger proceeds to show how I, II and III enable one to differentiate each of the functions so introduced.

Menger treats antiderivatives in Sections 6–8 of Chapter II in terms of an equivalence relation $\sim$ defined by

$$f \sim g \text{ if and only if } Df = Dg$$

and a symbol S which may be read as "any antiderivative of" and which satisfies the condition

$$D(Sf) = f.$$

There are then the following formulas, corresponding to I, II, III above:

I'. $S(f + g) \sim Sf + Sg$,
II'. $f \cdot g \sim S(f \cdot Dg) + S(g \cdot Df)$,
III'. $fg \sim S((Df)g \cdot Dg)$

Menger shows how these formulas can be used to derive all the standard calculus methods for finding antiderivatives, including those customarily called "integration by substitution" and "integration by parts".

2. As algebraic structures, tri-operational algebras thus constitute a special class of commutative rings. As such, they have many realizations, e.g., rings of polynomials (see the comments of H. Kaiser). Analysis also provides natural examples of tri-operational algebras: e.g., the infinitely differentiable functions defined on the whole real line, or the entire functions in the complex plane.

However, the application of tri-operational algebra to the original area of interest, the functions of elementary calculus, runs into serious complications. These stem from the fact that tri-operational algebra as it stands is not equipped to handle functions with varying domains, or functions whose ranges are not subsets of their domains. Menger was well aware of this (see

---

[4] It should be noted that according to Menger's precise definitions of right, left, and 2-sided inverses on pp. 17–18, *log* is not, as stated on p. 21, a 2-sided inverse of *exp*, but only a left inverse. Similarly, *arctan* is not, as stated on p. 28, a 2-sided inverse of (the equivalent of) the restriction of the tangent to the interval $(-\pi/2, \pi/2)$, but only a right inverse. These slips do not affect the general treatment of the functions involved in the least.

pp. 8–9 in [M1] and p. 48 in [M3]), but apparently at the time was too busy to seek ways out of the difficulties. Later he found such ways,[5] but he was not one to go back and tidy up work long since published. He preferred instead to start afresh and create new work, with perhaps some overlap with the old. Thus, instead of an "Algebra of Analysis, revised edition" or a "Tri-operational Algebra Revisited", we have [M5] and [M6].

It can be said, though, that the innovations in these two papers were foreshadowed earlier: In a footnote on page 98 of [M3], Menger says: "If besides the constant functions with the universal domain we introduced constant functions with restricted domains, we should have to supplement the algebra of functions by an algebra of domains and ranges of functions..." Just such an "algebra of domains and ranges" appears in [M5] and [M6], not as a *supplement to* but as an *intrinsic part* of "the algebra of functions".

Menger's paper [M5] is a revised version of the talk he gave at the Berkeley Symposium on the Axiomatic Method held between December 26, 1957 and January 4, 1958. The paper goes beyond the structure of tri-operational algebra: first, by introducing a partial order relation $\subseteq$ corresponding to the notion of *restriction* of functions (recall that f is a restriction of g if the domain of f is a subset of the domain of g, and $f(x) = g(x)$ for all x in the domain of f) and an empty function ϕ as the minimal element under $\subseteq$ (see Footnote 5); second, by introducing, for each function f the identity function Rf on the domain of f and the identity function Lf on the range of f[6]; and third, by pointing out, almost as an afterthought, that while addition and multiplication of numerical-valued functions can be defined in terms of substitution (by composition of two functions f and g with the respective 2-place functions Sum and Product), substitution cannot be

---

[5] The quickest way would be to add to a tri-operational algebra a universal empty or null element ϕ satisfying

$$\phi + f = \phi \cdot f = \phi f = f\phi = \phi \quad \text{for all } f \text{ in the algebra.}$$

This would give complete closure of the algebra under each of the three operations, so instead of saying, e.g., "the constant function 0 cannot be substituted into the logarithm", we would simply have $\log 0 = \phi$. This would, however, already require some tinkering with the basic identities: thus, instead of $(-f) + f = 0$, one would have, say,

$$(-f) + f = 0f = 0(-f),$$

with the proviso that $0f = \phi$ if and only if $f = \phi$.

[6] In category theory à la Eilenberg-MacLane, the elements corresponding to functions do have identity functions on their domains. But instead of identity functions on their ranges, they have identity functions on their codomains, which are supersets of their ranges; and elements that are equal as functions but have different codomains are considered to be different.

defined in terms of addition and multiplication. This means that in an axiomatic theory of functions, it is appropriate to treat only the single binary operation that Menger called substitution. These points are brought out again in [M6].

3. In reading [M5] and all his later papers in the algebra of functions, one has to keep in mind Menger's conception of what constitutes a genuine function. In common with other mathematicians of his time, Menger felt that a proper function was something that mapped individual numbers or n-tuples of numbers into *single* numbers. The emphasis on numerical domains and ranges could be weakened a bit, though with evident reluctance: note, e.g., the terminological distinction on p. 83 of [M7] (and already on p. 425 of [M6]) between the logical "functors" that map pairs of truth-values into single truth-values, and the isomorphic "functions" one gets on replacing the truth-values True and False by 1 and 0, respectively. But the insistence that what comes out of a "function" must be a *single* entity of the same kind as those that, singly or multiply, go into a function, was never relaxed. Thus Menger could never bring himself to regard, say, the very useful permutation f defined on pairs by $f(x, y) = (y, x)$ as something deserving the name of "function". He would have regarded f as a *pair* of 2-place functions.[7] Similarly for the functions that appear *inter alia*, in recursive function theory (see [18] and [19]) that map single numbers into pairs of numbers: these would be thought of as pairs of 1-place functions. And the identity functions i_n defined on n-tuples by

$$i_n(x_1, \ldots, x_n) = (x_1, \ldots, x_n)$$

would be thought of as n-tuples of n-place functions, and thus ineligible to be elements of algebras of n-place functions for $n \geq 2$. Given this point of view, Menger could not see any way of extending the results he had achieved in [M5] for 1-place functions to n-place functions with $n > 1$. For example, since there is no *identity function*, in his sense, on the domain of an n-place function F for any $n \geq 2$, there can be no element RF in any Menger-style algebra of such functions.

So having reached what evidently seemed to him to be a dead end in the study of 1-place functions, Menger decided, as he says on p. 412 of [M6], "that the real future of the algebra of functions seems to me to lie in the study of the vast field of the so-called functions of several variables – better, *multiplace*

---

[7] He comes closest to making this point of view explicit on pp. 281–282 of [M8].

functions." Accordingly, Menger confined all his later efforts in the algebra of functions to the study of that "vast field."[8] His basic tools in this study were what he ended up calling "parenthesis substitution" and "bracket substitution", defined as follows: For positive integers m, n, suppose that F is an m-place function and $(G_1, \ldots, G_m)$ an m-tuple of n-place functions. Then $F(G_1, \ldots, G_m)$ is the n-place function defined by

$$(F(G_1, \ldots, G_m))(x_1, \ldots, x_n) = F(G_1(x_1, \ldots, x_n), \ldots, G_m(x_1, \ldots, x_n)).$$

If m and n are fixed, then this is evidently an $(m+1)$-ary operation that Menger in Chapter 11 of [M4], called (m, n)-substitution".[9] "Parenthesis substitution" is the combination of these $(m+1)$-ary operations for all $m \geq 1$.

As for "bracket substitution", let F, as before, be an m-place function, but let $(G_1, \ldots, G_m)$ be an m-tuple in which G_1 is an n_1-place function, G_2 an n_2-place function, $\ldots, G_m$ an n_m-place function with $n_1, n_2, \ldots, n_m$ arbitrary positive integers. Set $n = n_1 + \cdots + n_m$. Then $F[G_1, \ldots, G_m]$ is the n-place function defined by

$$F[G_1, \ldots, G_m] = F(G_1(x_1, \ldots, x_{n_1}), \ldots, G_m(x_{n-n_m+1}, \ldots, x_n)).$$

As before, for fixed m, this is an $(m+1)$-ary operation, and "bracket substitution" is the combination of these $(m+1)$-ary operations for all $m \geq 1$. (It should be noted that other people worked with these notions. For example, in [8], Jónsson and Tarski have parenthesis substitution under the name "composition" and bracket substitution under the name "superposition".) Corresponding to the associativity of ordinary composition, the basic property of parenthesis substitution is one that Menger called "superassociativity". It works like this: Let F be an m-place function, $(G_1, \ldots, G_m)$ an m-tuple of n-place functions, and $(H_1, \ldots, H_n)$ an n-tuple of r-place functions. It follows that:

$$(F(G_1, \ldots, G_m))(H_1, \ldots, H_n) = F(G_1(H_1, \ldots, H_n), \ldots, G_m(H_1, \ldots, H_n)).$$

---

[8] He actually confined himself still further: on p. 82 of [M7] he says: "In what follows, only *total* n-place mappings will be considered, i.e., mappings for which the set of the first members of all pairs is the *whole* set X^n". This restriction in fact holds for all papers in the algebra of functions he published after 1961.

[9] People working in multivalued logic and related areas call it "(m, n)-composition": see, e.g., [24].

Bracket substitution has a similar property, also called "superassociativity", which takes the following form:

$$(F[G_1,\ldots,G_m])[H_1,\ldots,H_n] = F[G_1[H_1,\ldots,H_{n_1}],\ldots,G_m[H_{n-n_m+1},\ldots,H_n]],$$

where F is an m-place function, G_i is an n_i-place function for $i = 1,\ldots,m$, and $n = m_1 + \cdots + n_m$. For an arbitrary fixed positive integer n, superassociativity for the $(n+1)$-ary operation of (n, n)-composition (see Footnote 9) is treated in detail in [4] under the name of "the n-substitutive law." In [15], a superassociative operation is called "associative in the first place" ("Associativ in der ersten Stelle"). See the comments of H. Kaiser.

On page 85 of [M7], as part of his discussion of the relation between parenthesis and bracket substitution, Menger introduces certain things he calls "operators". These are unary operations that in various ways transform m-place functions into n-place functions with $n \leq m$. On pp. 86 and 87 he applies certain of these operators to the "functions of 2-valued logic," but makes no further use of them, either in [M7] or in any later paper. To my mind, this is a great pity: he was well enough acquainted with the literature that without much effort he could have seen that the operations admitted by the workers in n-valued propositional calculus are, as stated explicitly in [20], "(repeated) composition which is understood to include the identification and permutation of variables." Now "composition" here means parenthesis and bracket substitution, while "the identification and permutation of variables" can equivalently be done without "variables" by Menger's operators. So Menger could justifiably have said: "Whatever is done in the study of 2- or multi-valued propositional calculus with 'variables' can be done without them." However, what Menger did instead was to go on in [M7] to define "superassociative algebras", and then investigate them with supreme virtuosity, both in [M8], and with H. I. Whitlock in [M9]. But since these algebras contain nothing equivalent to Menger's operators, such staples of "multi-valued logic" as Sheffer functions (see [20]) do not exist in them.[10] Nevertheless, they are interesting algebraic structures in themselves, and have been extensively investigated, both by Menger's students (see [2], [10], [16], [30], [34]) and by mathematicians in the former Soviet Union (see the survey article [23] by Schein and Trohimenko). They have also given rise to interesting applications to projective geometry: see [29] and [5].

---

[10] Note, however, that in appropriate superassociative algebras it is quite possible to speak of other staples of multi-valued logic, in particular "clones" as defined, e.g., on the first page of [24].

4. While writing [M5], Menger got Bert Schweizer and me involved in the work on the algebra of functions.[11] We found the subject interesting and important, and so decided to continue working on it.

Our first results appeared in [25], a series of three papers in the Mathematische Annalen. In the first of these, [25, I], we investigated the consequences of a set of axioms equivalent to that of Menger's in [M5]. In [25, II], we dropped one of the axioms in [25, I] and replaced it with an axiom consisting of two identities which, as we said, "are not implied by our earlier axioms." The resulting set of axioms was thus no longer equivalent to Menger's, though, like his, it still referred to a partial order relation $\subseteq$, an associative binary operation ∘ with an identity element n, and two unary operations R and L. We called any algebraic structure satisfying this revised set of axioms a "function semigroup."

Our investigation of function semigroups culminated in [25, III] with a pair of representation theorems. These showed that under suitable (and rather strong) additional conditions a function semigroup is isomorphic to either a subset of, or the entire set of functions whose domains and ranges are subsets of some fixed set S[12] under the operation of composition.

Nevertheless, even to Bert and me, the three papers in [25] are now mainly of historical interest, since our definitive version of the axioms for an algebraic system modeled on the behavior of functions under composition appeared in [26]. That paper is the basis for all the later work in the algebra of functions by us and some of our students and colleagues: see, e.g., [1, 3, 9, 27, 28, 31, 36].

In [26], we defined a *function system* as a set closed under each of two unary operations R, L and one binary operation ∘ that satisfy the following conditions (in the form given in [31]):

1. The operation ∘ is associative, i.e., for any three elements f, g, h in the set,

$$(f \circ g) \circ h = f \circ (g \circ h).$$

2. For any element f in the set,

   (a) $LRf = Rf$, $RLf = Lf$,
   (b) $Lf \circ f = f = f \circ Rf$.

---

[11] As Menger notes in [M5], Bert and I had some input into the final version of the paper. Among other things, we suggested the notations Rf and Lf which Menger adopted for his right and left neutral functions. In hindsight, D (for domain) would have been better than R (for right), and R (for range) better than L (for left).

[12] Such functions are often called "partial transformations of S", and the semigroup of all such functions under composition is labeled "$PT(S)$": cf. [11], p. 442.

3. For any two elements f, g in the set,

   (a) $L(f \circ g) = L(f \circ Lg)$, $R(f \circ g) = R(Rf \circ g)$,
   (b) $Lf \circ Rg = Rg \circ Lf$,
   (c) $Rf \circ g = g \circ R(f \circ g)$.

It is not hard to check that these identities accurately reflect the behavior of functions, in the following way: The binary operation $\circ$ represents ordinary composition, in which the function $f \circ g$ is defined by: x is in the domain of $f \circ g$ if and only if x is in the domain of g and $g(x)$ is in the domain of f; and $(f \circ g)(x) = f(g(x))$ for all x in the domain of $f \circ g$. And as in [M5], [M6], and [25], Rf represents the identity function on the domain of f and Lf the identity function on the range of f.

Several things may be noted about Axioms 1–3, beginning with the observation that 3(a) is an "algebraization" of the identities.

$$\mathrm{Dom}(fg) = \mathrm{Dom}(g \text{ into } \mathrm{Dom}\, f), \mathrm{Ran}(fg) = \mathrm{Ran}(f \text{ on } \mathrm{Ran}\, g),$$

given by Menger on p. 88 of [M4] and p. 417 of [M6]. In Menger's form, the identities are not amenable to algebraic manipulation, so Menger, though he recognized them as important properties of functions, could not incorporate them into his axioms in [M5].

Axioms 1–3 contain no explicit reference to any partial order. This is because in a function system the partial order $\subseteq$ of restriction can be *defined*, via:

$$f \subseteq g \text{ if and only if } f = g \circ Rf.$$

The axioms also make no mention of any identity element. In fact, a function system need not contain a (universal) identity element. But if it does, then, as we show in [26], "the theory of function systems with an identity element coincides with our previous theory of function semigroups."

We also show in [26] that the class of function systems is broad enough to include inverse semigroups, Brandt semigroups, and (small) categories that are augmented with null elements, as special cases. And of course, any set of functions that is closed under ordinary composition and contains, for every function f in the set, the identity functions on the domain and range of f, is automatically a function system: in this case, we speak of a *concrete* function system.

Now there are function systems that are not concrete function systems nor even (what amounts to the same thing to an algebraist) isomorphic to any concrete function system. While studying related algebraic structures, B. M. Schein proved in [21] (see also p. 51 in [22]) that a necessary and sufficient condition for a function system to be isomorphic to a concrete function system is the following implication:

For any f, g, h in the system, if $f \circ h = g \circ h$, then $f \circ Lh = g \circ Lh$.

Definitive as this result is, Bert and I, being non-algebraists, wanted more: we wanted to characterize concrete function systems explicitly, and not just "up to isomorphism." By adding some rather strong conditions, we were able to do this in [27] for the class of what we called 'intrinsic' function systems. We were able, that is, "to show that each element of an intrinsic function system *is* a function in the ordinary mathematical sense."

Among the added conditions was one that says that for each element f in a function system there is an element g such that $f \circ g = Lf = Rg$. On p. 7 of [26] we had called this condition "the right-subinverse (or RS-) property", and it is stronger than, i.e., implies but is not implied by, Schein's condition.

The uncomfortable strength of the RS-property[13] meant that Bert and I were not completely satisfied with [27]. In the introduction of the paper, we said, "having achieved the characterization it is natural to ask whether the conditions can be weakened. The question remains open..." Having recently returned to this question, I can now report that the conditions *can* be weakened. In fact, in the presence of the other conditions, the RS-property can be replaced by a condition *weaker* than Schein's, namely, the assumption that if f is idempotent (i.e., $f \circ f = f$), then $f \circ Lf = Lf$. This will be the subject of a paper now in preparation.

Leaving characterization questions aside, it is highly advantageous to work with function systems with the RS-property. For example, I showed in [31] that in a function system, the RS-property is equivalent to the statement that every element f in the system admits a *canonical decomposition*, i.e., a representation of the form $f = f_1 \circ f_2$, where $Lf_1 = Lf$, $Rf_1 = Lf_2$, $Rf_2 = Rf$, f_2 is idempotent and f_1 is *invertible*, i.e., there is a unique element f_1^* such that

$$f_1 \circ f_1^* = Lf_1 = Rf_1^*, \; f_1^* \circ f_1 = Lf_1^* = Rf_1.$$

Such canonical decompositions are very useful in dealing with a variety of functional equations of both iterative and non-iterative types.

---

[13] On p. 173 of [25, III] we had already noted that P. Bernays, using different terminology, had shown that in axiomatic set theory the statement "Every function has a right-subinverse" is equivalent to the Axiom of Choice.

To conclude this section, I will note that in [3], D. Cargo determined the form that Green's relations, defined in arbitrary semigroups, take in function systems. And in [1], J. Ballieul applied Cargo's results to determine the number and cardinality of the corresponding equivalence classes in certain finite function systems.

5. While Bert and I were drafting [26], we were simultaneously working to extend Menger's treatment of multiplace functions in [M7] and [M8] to functions whose ranges, as well as domains, could contain n-tuples. Our results in this direction appeared as the 2-part paper [28].

In that paper, we took a concrete rather than an abstract approach and began with a non-empty set S which, to avoid ambiguity, we assumed to be disjoint from any of its Cartesian powers S^n (thus excluding, for example, sets such as $\{a, b, (a,b)\}$). For any two positive integers d, r, we defined a function of *degree d* and *rank r* (and *index d−r*) to be one whose domain is a non-empty subset of S^d and whose range is a subset of S^r. Thus the identity function I_n on S^n is a function of degree n, rank n, index 0. We denoted the set of all functions of degree d and rank r by $\mathcal{F}_{dr}(S)$, and the union of all the sets $\mathcal{F}_{dr}(S)$, together with the empty function ϕ (which has no degree, rank or index), by $\mathcal{F}_\infty(S)$.

In $\mathcal{F}_\infty(S)$, we introduced two unary operations R, L, and three binary operations β, π, σ, as follows: As in a concrete function system, for any F in $\mathcal{F}_\infty(s)$, RF is the identity function on the domain of F, and LF is the identity function on the range of F. Thus $R\phi = L\phi = \phi$, and for any F in $\mathcal{F}_{dr}(S)$, RF is in $\mathcal{F}_{dd}(S)$ and is a restriction of I_d, while LF is in $\mathcal{F}_{rr}(S)$ and is a restriction of I_r. As for the binary operations, we first set

$$\phi\beta F = F\beta\phi = \phi\pi F = F\pi\phi = \phi\sigma F = F\sigma\phi = \phi$$

for any F in $\mathcal{F}_\infty(S)$. Now suppose F is in $\mathcal{F}_{dr}(S)$ and G is in $\mathcal{F}_{es}(S)$. Then $F\beta G$ is the function in $\mathcal{F}_{d+e, r+s}(S)$ specified by: $(x_1, \ldots, x_{d+e})$ is in the domain of $F\beta G$ if and only if $(x_1, \ldots, x_d)$ is in the domain of F and $(x_{d+1}, \ldots, x_{d+e})$ is in the domain of G; and for any such $(d+e)$-tuple $(x_1, \ldots, x_{d+e})$,

$$(F\beta G)(x_1, \ldots, x_{d+e}) = (F(x_1, \ldots, x_d), G(x_{d+1}, \ldots, x_{d+e})).$$

Now set $m = \max(d, e)$. Then an m-tuple $(x_1, \ldots, x_m)$ is in the domain of $F\pi G$ if and only if the d-tuple $(x_1, \ldots, x_d)$ is in the domain of F and the e-tuple $(x_1, \ldots, x_e)$ is in the domain of G. There may be no such m-tuples, in which case $F\pi G = \phi$. If there are such m-tuples, then for any such,

$$(F\pi G)(x_1, \ldots, x_m) = (F(x_1, \ldots, x_d), G(x_1, \ldots, x_e)),$$

and $F\pi G$ is a function of degree $\max(d, e)$ and rank $r + s$.

Having defined β, it becomes possible to define σ 'without variables'. This is not what we did in [28], where we introduced σ as the first and β as the last of our three binary operations. But since a 'variable-free' definition illuminates the relation between σ and ordinary composition, I present it here. If F and G are as above, then

$$F\sigma G = \begin{cases} F \circ (G\beta I_{d-s}), & \text{if } d > s, \\ F \circ G, & \text{if } d = s, \\ F\beta I_{s-d} \circ G, & \text{if } d < s, \end{cases}$$

where $\circ$ represents ordinary composition of functions. It follows that if $F\sigma G \neq \phi$, then the degree of $F\sigma G$ is

$$e + \max(d - s, 0),$$

the rank of $F\sigma G$ is

$$r + \max(s - d, 0) = r - \min(d - s, 0),$$

and the index of $F\sigma G$ is the sum of the indices of F and G. It also follows that $I_1\sigma F = F\sigma I_1 = F$ for all F in $\mathcal{F}_\infty(S)$, so I_1 is the identity element for σ; β and π do not have identity elements.

Thus, σ may be viewed as an extension of ordinary composition, suitably modified to work even when the rank of G is not equal to the degree of F. In [28], we called σ "serial composition", π "parallel composition", and β "bracket composition".

In [28], we showed that each of the three binary operations σ, π, β is associative, and that under σ, R and L, the set $\mathcal{F}_\infty(S)$ is a function semigroup, i.e., a function system with identity. Within this function semigroup, Menger's "parenthesis substitution" and "bracket substitution" appear as very special cases of expressions that involve only binary operations. For example, if F is an m-place function (i.e., a function of degree m, rank 1) and $(G_1, \ldots, G_m)$ are n-place functions (i.e., functions of degree n, rank 1), then Menger's parenthesis substitution $F(G_1, \ldots, G_m)$ can be expressed as $F\sigma(G_1\pi \cdots \pi G_m)$. Similarly, the function $F[G_1, \ldots, G_m]$ resulting from a "bracket substitution" can be expressed as $F\sigma(G_1\beta \cdots \beta G_m)$.

But properly looked at, the results of [28] should serve to augment, rather than diminish, our appreciation of Menger's insights and accomplishments. For example, Menger's feeling that what we call a function of rank $r > 1$ is simply an r-tuple of multiplace functions could be said to be at least partially vindicated by the fact that such a function can be represented in a unique way as a π-composite of r functions of rank 1.

After [28], Bert and I felt that one thing that was needed was a proper axiomatic characterization of the structures that we had called "grammars of functions". A significant step in this direction came in 1974 with the thesis of J. Jurshak [9]. Regretably though, Jurshak's work has not been adequately followed up.

6. These comments would be incomplete without some mention of the influence that Menger had, more through personal contact than through publications, on work in the areas of iteration and functional equations.

In his beautiful paper [6], R. Isaacs, speaking of the functional equation

$$f(f(x)) = g(x), x \in E, \tag{1}$$

where g is a given function from E into E and f is to be determined, says: "our interest in this question arose from the following problem proposed by Menger.[14] Let E be R and $g(x) = a + bx$. There is obviously a linear solution when $b \geq 0$, namely

$$f(x) = a/(1 + b^{\frac{1}{2}}) + b^{\frac{1}{2}}x.$$

Do solutions exist when $b < 0$? The question is answered below."

Isaacs showed that there are indeed solutions to (1) for $b < 0$ and that the solutions are all non-linear and discontinuous. He did this by using the notion of the *orbit* of a function in the form introduced by Kuratowski in a note at the end of [33]. Menger brought Issacs' paper to the attention of Bert and me, and we have ever since considered the notion of orbit to be of fundamental importance in studying the behavior of functions. To cite a single example, the notion of orbit was crucial in [17] where we, together with R. Rice, showed that no complex quadratic polynomial has any iterative roots whatever. But it was not until very recently that a proper "algebraization", in terms of functional equations, of the notion of orbit was achieved: the announcement appears in [32].

---

[14] Presumably in a class or lecture at Notre Dame.

Menger's influence is also apparent in some of the work of M. A. McKiernan, whose paper [12] was mentioned in Footnote 3. His elegant results on series of "iterators" and iterates of functions (see [13], [14], [14']) should also be noted.

D. Zupnik will testify to Menger's influence on his own work, particularly in [35] and [36]. And for my part, I can state that my own work in the area of iteration and functional equations was not only influenced by, but originally arose out of my association with Karl Menger.

References to Menger

[M1] Algebra of Analysis. Notre Dame Mathematical Lectures, Vol. 3, 1944.
[M2] Tri-operational Algebra. Reports of Math. Colloq. Notre Dame, Indiana 5/6 (1944) 3–10.
[M3] General Algebra of Analysis. Reports of Math. Colloq. Notre Dame, Indiana 7 (1946) 46–60.
[M4] Calculus. A Modern Approach, Ginn and Company, Boston, 1955.
[M5] An Axiomatic Theory of Functions and Fluents. The Axiomatic Method With Special Reference to Geometry and Physics. Proceeding of an International Symposium held at the Univ. of California, Berkely, December 26, 1957 – January 4, 1958 (eds.) L. Henkin, P. Suppes, A. Tarski, Studies in Logic and the Foundations of Mathematics, North-Holland Publishing Co., Amsterdam (1959) 454–473.
[M6] The Algebra of Functions: Past, Present, Future. Rendiconti di Matematica 20 (1961) 409–430.
[M7] On Substitutive Algebra and Its Syntax. Zeitschrift für mathematische Logik und Grundlagen der Mathematik 10 (1964) 81–104.
[M8] Superassociative Systems and Logical Functors. Mathematische Annalen 157 (1964) 278–295.
[M9] Two Theorems on the Generation of Systems of Functions. Fundamenta Mathematicae 58 (1966) 229–240 (with H. I. Whitlock).

Other References

1. J. Ballieul: Green's Relations in Finite Function Semigroups. Aequat. Math. 7 (1972) 22–27.
2. P. Calabrese: The Menger Algebras of 2-place Functions in the 2-valued logic. Notre Dame Journal of Formal Logic 7 (1966) 333–340.
3. D. Cargo: Green's Relations in Function Systems and Applications, Ph. D., Dissertation, University of Massachusetts, 1968.
4. R. M. Dicker: The Substitutive Law. Proc. London Math. Soc. (3) 13 (1963) 493–510.
5. D. M. Foley: A Note on Seall's Projective Substitution. J. Geom. 22 (1984) 193–195.
6. R. Isaacs: Iterates of Fractional Order. Canadian J. Math. 2 (1950) 409–416.
7. W. P. Johnson: The Curious History of Faà di Bruno's Formula. Amer. Math. Monthly 109 (2002) 217–234.
8. R. Jónsson, A. Tarski: Boolean Algebras with Operators. Amer. J. Math. 73 (1951) 891–939.

9. J. F. Jurshak: Intrinsic Function Systems and the Axiomatization of Grammars of Functions. Ph. D. Dissertation, Illinois Institute of Technology, 1974.

10. V. Kafka: Axiomatics for Systems of Multiplace Functions. M. S. Thesis, Illinois Institute of Technology, 1965.

11. M. V. Lawson: Semigroups and Ordered Categories. J. Algebra 141 (1991) 442–462.

12. M. A. McKiernan: On the n-th Derivatives of Composite Functions. Amer. Math. Monthly 63 (1956) 331–333.

13. M. A. McKiernan: The Functional Differential Equation $Df = 1/ff$. Proc. Amer. Math. Soc. 8 (1957) 230–233.

14. M. A. McKiernan: Séries d'itérateurs et leurs application aux équations fonctionnelles. C. R. Acad. Sci. Paris 246 (1958) 2331–2334. Le prolongement analytique des séries d'itérateurs; ibid. 246 (1958) 2564–2567.

14′. M. A. McKiernan: On the convergence of series of iterates. Publ. Math. Debrecen 10 (1963) 30–39.

15. W. Nöbauer: Über die Darstellung von universellen Algebren durch Funktionenalgebren. Publ. Math. Debrecen 10 (1963) 151–154.

16. R. Peters: Two Remarks Concerning Menger's and Schultz' Postulates for the Substitutive Algebra of the 2-place Functions in the 2-valued Calculus of Propositions. Notre Dame J. Formal Logic 5 (1964) 125–128.

17. R. E. Rice, B. Schweizer, A. Sklar: When is $f(f(z)) = az^2 + bz + c$? Amer. Math. Monthly 82 (1980) 252–263.

18. J. Robinson: General Recursive Functions. Proc, Amer. Math. Soc. 1 (1950) 703–718.

19. R. M. Robinson: Primitive Recursive Functions. Bull. Amer. Math. Soc. 53 (1947) 925–942.

20. I. G. Rosenberg: On Generating Large Classes of Sheffer Functions. Aequat. Math. 17 (1978) 164–181.

21. B. M. Schein: Restrictive-multiplicative Algebras of Transformations (in Russian). Izvest. Vyss. Ucebn. Zaved. Matem. 4 (1970) 91–102.

22. B. M. Schein: Relation Algebras and Function Semigroups. Semigroup Forum 1 (1970) 1–62.

23. B. M. Schein, V. S. Trohimenko: Algebras of Multiplace Functions. Semigroup Forum, 17 (1979) 1–64.

24. D. Schweigert: On Prepolynomially Complete Algebras. J. London Math. Soc. (2) 20 (1979) 179–185.

25. B. Schweizer, A. Sklar: The Algebra of Functions. (I) Math. Annalen 139 (1960) 366–382; (II) Ibid. 143 (1961) 440–447; (III) Ibid. 161 (1965) 171–196.

26. B. Schweizer, A. Sklar: Function Systems. Math Annalen 172 (1967) 1–16.

27. B. Schweizer, A. Sklar: The Axiomatic Characterization of Functions. Zeitschr. Math. Logik Grundl. Math. 23 (1977) 373–382.

28. B. Schweizer, A. Sklar: A Grammar of Functions. (I) Aequat. Math. 2 (1968) 62–85; (II) Ibid. 3 (1969) 15–43.

29. R. Seall: Connections Between Projective Geometry and Superassociative Algebra. J. Geometry 4 (1974) 11–33.

30. H. L. Skala: The Irreducible Generating Sets of 2-place Functions in the 2-valued Logic. Notre Dame J. Formal Logic 7 (1966) 341–343.

31. A. Sklar: Canonical Decompositions, Stable Functions and Fractional Iterates. Aequat. Math. 3 (1969) 118–129.

32. A. Sklar: A Functional Equation Definition of "Orbit" (abstract). Aequat. Math. 56 (1998) 300.
33. R. Tambs Lyche: Sur l'équation fonctionnelle d'Abel. Fund. Math. 5 (1924) 331–332.
34. H. I. Whitlock: A Composition Algebra for Multiplace Functions. Math. Annalen 151 (1964) 167–178.
35. D. Zupnik: Polyadic Semigroups. Publ. Math. Debrecen 14 (1965) 273–279.
36. D. Zupnik: On Interassociativity and Related Questions. Aequat. Math. 6 (1971) 141–148.

The Influence of Menger's Ideas on the Work of Nöbauer and His School

H. Kaiser

Shortly after the Second World War Wilfried Nöbauer was studying mathematics at the University od Vienna. He asked Edmund Hlawka, the eminent number theorist at the university, to be the supervisor of his doctoral dissertation. Hlawka proposed taking a paper of Eckmann on the uniform distribution of values of the exponential function as a starting point for the research. Nöbauer did, but deviated from the originally intended analytic line of research and pursued instead an algebraic idea in Eckmann's paper. This resulted in a systematic study of permutation polynomials (see below).

Nöbauer wrote an outline of his proposed research and gave it to Hlawka, who generously accepted it. In this way Wilfried Nöbauer became an algebraist. When he was later asked to join the faculty of the Mathematical Institute of the University of Vienna, he discovered that he was the only algebraist there who was active in research. So he studied the papers of his Viennese predecessors in algebra and was influenced by the ideas he found in them. In this way he became acquainted with the work of Karl Menger.

In 1966, Wilfried Nöbauer was appointed Professor of Mathematics at the Vienna University of Technology. As director of what became the "Institute for Algebra and Mathematical Structure Theory" at that University, he gathered around him a group of young mathematicians and together they started to study the "Algebra of Polynomials" in a systematic way. Their study centered around the topics treated in the monograph "Algebra of Polynomials", which was published by H. Lausch and W. Nöbauer in 1973. In this book Menger's ideas in algebra play an important role. The heart of the treatise is a systematic study of the composition of polynomials and polynomial functions. The authors clearly state the influence of Menger's work by writing, on p. 130:

"Menger was the first person who fully realized the significance of the concept of superassociativity, and it was he who introduced selector systems.... Menger also introduced 1-dimensional composition rings and thus axiomatized the composition of functions from a ring into itself."

What follows is a survey of the portions of the work of Wilfried Nöbauer and his colleagues that display the influence of Karl Menger's ideas.

1. The central concept of a composition algebra can be developed in the following way: Let M be a non-empty set, k a positive integer and κ a $(k+1)$-ary operation on M; then κ is called superassociative if κ satisfies the law

$$\kappa\kappa x_0 x_1 \cdots x_k y_1 \cdots y_k = \kappa x_0 \kappa x_1 y_1 \cdots y_k \kappa x_2 y_1 \cdots y_k \cdots \kappa x_k y_1 \cdots y_k$$

$$(x_j, y_t \text{ indeterminates for } j, \; t = 0, 1, \ldots, k).$$

κ is called right-superdistributive with respect to an n-ary operation ω, if κ satisfies the law

$$\kappa\omega y_1 \cdots y_k = \omega, \qquad\qquad\qquad\qquad\qquad \text{for } n = 0;$$
$$\kappa\omega x_1 \cdots x_n y_1 \cdots y_k = \omega\kappa x_1 y_1 \cdots y_k \kappa x_2 y_1 \cdots y_k \cdots \kappa x_n y_1 \cdots y_k, \quad \text{for } n > 0.$$

A family $\{s_1, \ldots, s_k\}$ of elements of M is called a selector system for the $(k+1)$-ary operation κ, if $s_1, \ldots, s_k$ regarded as 0-ary operations on M satisfy the laws

$$\kappa s_i y_1 \cdots y_k = y_i, \quad i = 1, \ldots, k;$$
$$\kappa x_1 s_1 \cdots s_k = x_1.$$

This definition implies that there exists at most one selector system for κ.

For $k = 1$ and $n = 2$ superassociativity reduces to ordinary associativity, right superdistributivity to ordinary right distributivity, and a selector system for κ to an identity for κ.

Let A be an algebra from a class V of algebras of type T with Ω as a family of operations and κ a $(k+1)$-ary operation on A ($k > 0$ an integer). We put $\Omega_1 = \Omega \cup \{\kappa\}$.

The algebra $A = \langle A, \Omega_1 \rangle$ is called a k-dimensional V-composition algebra if

(i) $\langle A, \Omega \rangle$ is an algebra in V,
(ii) κ is superassociative,
(iii) κ is right superdistributive with respect to all $\omega \in \Omega$.

The algebra $A = \langle A, \Omega_1 \rangle$ is called a k-dimensional V-composition algebra with selector system if there is a selector system for κ in A.

Obviously, if V is a variety with respect to Ω, then, for any k, the class of all k-dimensional V-composition algebras is a variety with respect to Ω_1.

Examples of such composition algebras are:

1. Let V be the variety of sets, i.e., $\Omega = \emptyset$, then the k-dimensional V-composition algebras are the so-called k-dimensional superassociative systems. In particular, the 1-dimensional superassociative systems are exactly the semigroups.
2. Let V be the variety of groups. Then the k-dimensional V-composition algebras are called k-dimensional composition groups. In particular, 1-dimensional composition groups are known under the name of "near-rings" (Pilz, 1977).
3. If V is the variety of rings, then the k-dimensional V-composition algebras are called k-dimensional composition rings. 1-dimensional composition rings are also known as composition rings or, in Menger's terminology, tri-operational algebras.
4. If V is the variety of lattices, the k-dimensional V-composition algebras are called k-dimensional composition lattices. For $k = 1$, we simply speak of composition lattices.

Every full function algebra $F_k(A)$ over an algebra A of the variety V is a k-dimensional V-composition algebra if the composition of functions is added to the operations on $F_k(A)$. Since the class of k-dimensional V-composition algebras is a variety, every subalgebra of this composition algebra is also a k-dimensional V-composition algebra. It can be shown that up to isomorphisms, there is no further k-dimensional V-composition algebra. This means that for every V-composition algebra there exists an isomorphic algebra consisting of functions of an algebra in V where the composition is just the composition of functions.

2. In the study of composition homomorphisms the concept of a constant of a k-dimensional V-composition algebra $\langle A, \Omega, \kappa \rangle$ is introduced:

$c \in A$ is called a constant of $\langle A, \Omega, \kappa \rangle$ if

$$\kappa c a_1 \cdots a_k = c \quad \text{for all } (a_1, \ldots, a_k) \in A^k.$$

This definition is motivated by Menger, who defined constants for composition rings.

Superassociative systems have been studied by H. Länger, a member of the Viennese group, in a series of papers (see e.g., Länger, 1976, Länger, 1980).

Near rings are still a subject of intense mathematical research. An exposition of the theory of near rings can be found in the monograph (Pilz, 1977), written by Günter Pilz, another pupil of W. Nöbauer.

Composition rings (Menger's tri-operational algebras) were studied systematically by W. Nöbauer and his pupils. In two special problems in this realm the results obtained turned out to be of interest in modern cryptology. The first problem was that of the permutability of polynomial functions in one variable over the real numbers R with respect to composition. The other problem concerns the study of so called permutation polynomials. A polynomial $f(x)$ over the integers is called a permutation polynomial on the set M of residue classes modulo m, if the mapping induced by $f(x)$ from M into the residue class ring R modulo m is a permutation of M. These polynomials can be used in the RSA – public key cryptosystems (see e.g., Lidl-Müller, 1986, and Lidl-Müller, 1984).

The theory of composition lattices has also been developed by Viennese algebraists (see e.g., Mitsch, 1967).

3. Another field of study of the algebra-group in Vienna, which was heavily influenced by Menger's work, was the study of derivation families with chain rule. The general frame for these investigations is the following:

For a k-dimensional composition algebra $\langle A, \Omega, \kappa \rangle$ where the system Ω of fundamental operations contains at least two binary operations $+$ and $\cdot$ we define:

A system of n unary operations $D_1, \ldots, D_n$ on A is called a family of derivations with chain rule with respect to $+, \cdot$ and κ, if for $x, y, y_1, \ldots, y_k$ in A and $i = 1, \ldots, n$, we have:

(i) $D_i(x + y) = D_i(x) + D_i(y)$,
(ii) $D_i(x \cdot y) = D_i(x) \cdot y + x \cdot D_i(y)$,
(iii) $D_i(\kappa x y_1 \cdots y_k) = \sum_{t=1}^{k} (\kappa D_t(x) y_1 \cdots y_k) \cdot D_i(y_t)$,
       (if all expressions are defined). For $i = 1$ and κ a binary operation, (iii) is the ordinary chain rule.

If $\langle A, \Omega, \kappa \rangle$ has a selector system $\{s_1, \ldots, s_n\}$ the following condition must be fulfilled:

$$\kappa D_i(s_r) y_1 \cdots y_k \text{ is defined for } r = 1, \ldots, n \text{ and } i = 1, \ldots, n.$$

A k-dimensional composition algebra which possesses such a family of derivations with chain rule is called a k-dimensional derivation composition algebra.

The main results in this area concern derivation composition rings (see e.g., Müller, 1968, and Müller, 1975).

4. Another line of research which has grown out of the study of function algebras is the theory of interpolation in universal algebras. It is concerned with the representation of functions (of finite arity) on a universal algebra $\langle A, \Omega \rangle$ with values in A by polynomial functions on any finite subsets of their domains.

Let $\langle A, \Omega \rangle$ be a universal algebra and k a positive integer. A function f: $A^k \to A$ is said to have the interpolation property, if for every finite subset N of A^k there is a polynomial function p over $\langle A, \Omega \rangle$ such that $f(x) = p(x)$ for all $x \in N$. If we denote the set of all k-ary functions over $\langle A, \Omega \rangle$ with the interpolation property by $\mathrm{LP}_k(A)$, the set of all f: $A^k \to A$ by $F_k(A)$, then we say that an algebra $\langle A, \Omega \rangle$ has the interpolation property if and only if $F_k(A) = \mathrm{LP}_k(A)$ for all positive integers k.

This concept has been investigated thoroughly (see e.g., Istinger-Kaiser-Pixley, 1979, Kaiser-Nöbauer, 1982, Fried-Kaiser-Marki, 1982). Classical examples of algebras with the interpolation property are the fields and the Boolean algebra of order 2. A major result of the investigation has been the characterization of algebras with the interpolation property. This has led to the solution of the problem of finding all algebras with the interpolation property in given classes of algebras. For example, in the variety of groups, the elements that have the interpolation property are just the simple non-abelian groups, in the variety of rings, the simple non-zero rings. For a survey of these results and the links with the theory of functional completeness and primality see e.g., Kaiser, 1984.

5. Problems concerning the composition of functions arise in several disciplines, e.g., propositional calculus of many-valued logics, synthesis of automata and universal algebra. A different approach to the study of such problems is the theory of clones. This theory arises naturally if one asks the following question: What functions can be constructed by composition from a given set of (multiplace) functions on a set A with values in A?

We start with a finite set $A = \{0, 1, \ldots, k-1\}$ and we denote the set of all n-ary functions on A with values in A by $O_k^{(n)}$ and let O_k be the join of all $O_k^{(n)}$ from $n = 1$ to ∞.

If f is an n-ary function on A and $g_1, \ldots, g_n$ are t-ary functions on A (with values in A), then we define a t-ary function $f(g_1, \ldots, g_n)$ on A (the so called superposition of $f, g_1, \ldots, g_n$) as follows:

$$f(g_1, \ldots, g_n)(a_1, \ldots, a_t) = f(g_1(a_1, \ldots, a_t), \ldots, g_n(a_1, \ldots, a_t))$$
$$\text{for all } (a_1, \ldots, a_t) \in A^t.$$

For integers $n \geq 1$ and $1 \leq i \leq n$ the i-th n-ary projection on A is the function defined by

$$e_{n,i}(a_1, \ldots, a_n) = a_i \quad \text{for all } (a_1, \ldots, a_n) \in A^n.$$

These projections are just the selectors in the sense of Menger. A set of (multiplace) functions on a fixed set A with values in A is said to be a clone on A if it contains all projections and is closed under superposition.

The set O_k of all functions on A with values in A is a clone, and so is J_k, the set of all projections on A.

The intersection of an arbitrary set of clones on A is again a clone. Therefore it makes sense to speak of the clone generated by a set of multiplace functions on A with values in A.

The connection between clone theory and the universal algebra problems studied by the Viennese algebraists can be established in the following way:

If $A = \langle A, \Omega \rangle$ is a universal algebra we can assign to it two clones in a natural way:

(i) The clone $T(A)$ of all termfunctions on A. This is just the clone generated on A by the system Ω of fundamental operations.
(ii) The clone $P(A)$ of all polynomial functions on A. This is just the clone generated on A by Ω and all unary constant operations.

One of the advantages of clone theory is the fact that one can describe the clone $T(A)$ of the algebra $\langle A, \Omega \rangle$ by "invariants" rather than by the generating set Ω. These invariants are the subuniverses of finite powers of A. If A is finite, then the subuniverses of finite powers of A already determine $T(A)$. The set of all $f \in O_A$ which preserve the subset $\rho \subseteq A^h$ will be denoted by $\mathrm{Pol}_A \rho$. And if S is a set of subsets of powers of A we define $\mathrm{Pol}_A S$ to be the intersection of all $\mathrm{Pol}_A \rho$ for all $\rho \in S$. It is easy to see that, for a given S, $\mathrm{Pol}_A S$ is a clone on A. One very useful result is for example:

For every finite algebra $A = \langle A, \Omega \rangle$ a function on A is a termfunction of A if and only if g preserves all subuniverses of finite powers of A.

For a set Ω of functions on a fixed set A let $\mathrm{Inv}_A \Omega$ denote the family of all subuniverses of finite powers of the algebra $\langle A, \Omega \rangle$. Then, for every set A

the operators

$$\Omega \to \mathrm{Inv}_A\,\Omega \quad \text{and} \quad \mathrm{Pol}\,S \leftarrow S$$

define a Galois correspondence.

On every set A the lattice of subclones of O_A is an algebraic lattice. The atoms of this lattice are called minimal clones on A, and the co-atoms are called maximal clones on A. If A is finite then this lattice is atomic and dually atomic with finitely many coatoms. Hence every proper subclone of O_A is contained in a maximal clone on A.

One of the early deep results of clone theory was the description of all maximal clones on a finite set A by Ivo Rosenberg in 1970 (Rosenberg, 1970).

A detailed exposition of the theory of clones can be found in the monograph (Pöschel-Kalužnin, 1979).

6. In conclusion we mention that, employing ideas and techniques that were developed by Nöbauer et al., G. Dorfer and H. Woracek (1999) have recently proved the famous theorems of Ritt (and more) using purely algebraic arguments, i.e., without any appeal to analytic methods (see also G. Eigenthaler and H. Woracek, 1995).

References

G. Dorfer, H. Woracek: Formal Power Series and Some Theorems of J. F. Ritt in Arbitrary Characteristic. Monatsh. f. Math. 127 (1999) 277–293.

G. Eigenthaler, H. Woracek: Permutable Polynomials and Related Topics. Contributions to General Algebra 9 (1995) 163–182.

M. Istinger, H. Kaiser, A. Pixley: Interpolation in Congruence Permutable Algebras. Colloquium Math. 42 (1979) 219–239.

H. Kaiser, W. Nöbauer: Über interpolierbare Funktionen auf universalen Algebren. Beiträge zur Algebra und Geometrie 12 (1982) 51–55.

E. Fried, H. Kaiser, L. Marki: An Elementary Approach to Polynomial Interpolation in Universal Algebras. Alg. Univ. 15 (1982) 40–57.

H. Kaiser: Interpolation in Universal Algebras. In: P. Burmeister et al. (eds.), Universal Algebra and its Links with Logic, 29–40. Berlin: Heldermann, 1984.

H. Kaiser, W. Nöbauer: Permutation Polynomials in Several Variables Over Residue Class Rings. J. Austral. Math. Soc. 43 (1987) 171–175.

H. Länger: Zur Theorie der superassoziativen Systeme und Menger-Algebren. Ph. D. Diss, Vienna University of Technology, 1976

H. Länger: Commutative Quasi-trivial Superassociative Systems. Fund. Math. 109 (1980) 79–88.

H. Lausch, W. Nöbauer: Algebra of Polynomials. North-Holland, Amsterdam-London, 1973.

R. Lidl, W. B. Müller: On Commutative Semigroups of Polynomials with Respect to Composition. Monatsh. f. Math. 102 (1986) 139–153.

R. Lidl, W. B. Müller: Permutation Polynomials in RSA-Cryptosystems. In: D. Charm (ed.), Advances in Cryptology, 293–301. New York: Plenum, 1984.

H. Mitsch: Trioperationale Algebren über Verbänden. Ph. D. Diss. Univ. Wien, 1967.

W. B. Müller: Derivationen in Kompositionsalgebren Sitzungsber. der Öster. Akad. d. Wiss. Math-nat. Kl. 184 (1975) 239–243.

W. B. Müller: Eindeutige Abbildungen mit Summen-, Produkt- und Kettenregel im Polynomring Monatsh. f. Math. 73 (1969) 354–367.

G. Pilz: Near Rings. North-Holland, Amsterdam-New York-Oxford, 1977.

R. Pöschel, L. A. Kalužnin: Funktionen- und Relationenalgebren. Birkhäuser, Basel-Stuttgart 1979.

I. G. Rosenberg: Über die funktionale Vollständigkeit in den mehrwertigen Logiken, Rozpr. CSAV Rada Mat. Prir. Ved., Praha 80, 4 (1970) 3–93.

NOTRE DAME MATHEMATICAL LECTURES

Number 3

ALGEBRA OF ANALYSIS

by

DR. KARL MENGER

Professor of Mathematics,
University of Notre Dame

NOTRE DAME, INDIANA

1944

TABLE OF CONTENTS

Page

INTRODUCTION   1

I. THE ALGEBRA OF FUNCTIONS OF ONE VARIABLE   4

    1. The Classical Foundation of the
          Theory of Functions   4
    2. Algebra of Functions (Tri-Operational Algebra)   6
    3. The Theory of Constant Functions   10
    4. The Lytic Operations   15
    5. Exponential Functions   18
    6. The Logarithmic Functions   20
    7. The Absolute and the Signum   22
    8. The Power Functions   25
    9. The Trigonometric Functions   28

II. THE ALGEBRA OF CALCULUS   30

    1. The Algebra of Derivatives   30
    2. The Derivation of Exponential Functions . . .   33
    3. The Derivation of Logarithmic Functions . . .   35
    4. Logarithmic and Exponential Derivation   35
    5. The Derivation of the Trigonometric Functions   37
    6. The Foundation of the Algebra of
          Antiderivatives   39
    7. Formulae of the Algebra of Derivation in the
          Notation of Antiderivation   41
    8. The Three Methods of Antiderivation   42

INTRODUCTION

The statements of analysis can be grouped into three classes according to the depth to which the limit concept is used in their formulation and proof.

A first class consists of theorems which are entirely independent of the concept of limit, and deal with approximations. To this group belong graphical and numerical differentiation and integration as well as statements concerning the reciprocity of these two approximative operations.

A second class consists of formulae in whose proofs the concept of limit is used in a mild, so to speak, algebraic, way. This group comprises the bulk of formulae of calculus concerning elementary functions and some formulae concerning all differentiable functions: the rules for the formation of the derivatives of elementary functions, the determination of antiderivatives by substitution and by parts, etc. (Not included in this group is the theorem that each two antiderivatives of the same function differ at most by a constant).

A third group of statements is based on the assumption that in each closed interval each continuous function assumes its maximum. To this group belong the mean value theorem and its applications, of which I mention the Taylor development and its implications concerning maxima and minima, indeterminate forms, and the theorem about the antiderivatives of the same function.

2

In this book, we shall develop the second group of statements from a few assumptions concerning three operations
(addition, multiplication, substitution) and two operators
(derivation and antiderivation). A first part is devoted to
the three operations. A second part deals with the Algebra of
Derivation and Antiderivation. A third part contains a sketch
of the theory of functions of several variables.

In developing the Algebra of Analysis, we shall make use
of the notation of the operator theory. Furthermore, we shall
completely avoid variables in our formulae. These principles
necessitate changes of the current notation most of which result in formal as well as conceptual simplification and
systematization.

In initiating students into calculus, at the present time
one may find it hard to take full advantage of all these
simplifications - not on account of any specific difficulties
inherent to the proposed set-up or the proposed notation, but
because a student of calculus must be enabled to read books on
differential equations, theoretical physics, mathematical
economics, etc., all of which at present are written in the
classical notation. It will take a long time till these applied fields will be presented in a more modern way. In the
meantime, students must acquire not only the knowledge but an
operative grasp of the traditional notation with all its shortcomings from which, in fact, some applied fields, as physical
chemistry, suffer more than mathematics proper. However, a
gradual change of our obsolete notation probably is not only
desirable but unavoidable. The first step in this direction

is undoubtedly an uncompromising exposition of the new ideas
for professional mathematicians and especially teachers of
mathematics. It is one of the aims of this book to
provide the reader with such an exposition.

Several sections of this publication may be helpful in
simplifying the current presentation of calculus even when
our treatment is translated into the classical notation. In
this connection, we mention the Algebra of Antiderivation, the
development of the entire differential calculus from a few
formulae, our treatment of the exponential and tangential
functions on the basis of the functional equations and the
formulae

$$D \exp 0 = 1 \quad \text{and} \quad D \tan 0 = 1,$$

our introduction of the power functions, and the treatment of
antiderivation by substitution.

Apart from these pedagogical aspects, our Algebra of
Analysis seems to open an extended field of research. Many
additional results will be published in more technical papers.
Perhaps one will find the general idea of Algebra of Analysis
related to that of our Algebra of Geometry whose development
the author has outlined in the second lecture in the Rice
Institute Pamphlets, Vol. 27, 1940, p. 41-80.

I. THE ALGEBRA OF FUNCTIONS OF ONE VARIABLE

1. The Classical Foundation of the Theory of Functions.

The classical theory starts with the assumption that a field of numbers is given. That is to say, it starts with a system N of things, called numbers, which we add and multiply according to the well-known laws of a field.

Next, the theory of functions of one variable explicitely defines a function f as the association of a number $f(x)$ with each number x of some subset D_f of N. This set D_f is called the domain of f. The set R_f of all numbers $f(x)$ which f associates with the numbers x of D_f, is called the range of f. A function whose range consists of exactly one number is called constant [*].

The definition of the concept of functions is followed by explicit definitions of the concept of equality of functions and of three binary operations: addition, multiplication, and substitution. We define: $f = g$ if and only if $D_f = D_g$ and $f(x) = g(x)$ for each number x of $D_f = D_g$. (The concept of equality of numbers of the given field is assumed to be known). If $D_f = D_g$, we call sum of f and g [product of f and g] the function which associates the number $f(x) + g(x)$ [the number $f(x) \cdot g(x)$] with each number x of $D_f = D_g$. If R_g is a subset

---

[*] Analyzing the somewhat vague concept of "association" we see that f is a set of ordered pairs of numbers (d,r) such that each element of D_f occurs as the first element of exactly one pair. The second element, r, of the pair (d,r) is that element $f(d)$ of R_f which f "associates" with d.

of D_f, we call f of g the function which associates the number
f(g(x)) with each number x of D_g.

It is an odd fact that in the classical calculus no symbols for the sum of f and g, the product of f and g, and f of
g, are introduced. Using the notation of the calculus of
operators, we shall denote the results of the three operations
by

$$f + g, \qquad f \cdot g, \qquad fg,$$

respectively, so that the numbers associated by these functions
with the number x are

$$(f+g)(x), \qquad (f \cdot g)(x), \qquad (fg)(x),$$

respectively. We shall never omit the dot symbolizing multiplication in order to avoid a confusion of multiplication with
substitution. In this notation the classical definitions of
f + g, f·g, fg read

$$(f+g)(x) = f(x) + g(x), \ (f \cdot g)(x) = f(x) \cdot g(x), \ (fg)(x) = f(g(x)).$$

From these explicit definitions of the classical theory,
one deduces properties of the defined concepts. As two examples we
mention the commutative law for multiplication and the associative law for substitution.

In order to prove f·g = g·f, by virtue of the equality
concept, we have to show that (f·g)(x) = (g·f)(x) for each x.
From the definition of the product of functions we obtain
(f·g)(x) = f(x)·g(x) and (g·f)(x) = g(x)·f(x). From the commutativeness of multiplication in the given field of numbers,
it follows that f(x)·g(x) = g(x)·f(x) which completes the
proof.

In order to prove f(gh) = (fg)h, we have to prove
[f(gh)](x) = [(fg)h](x) for each x. Now from the definition

of substitution it follows that

$$[f(gh)](x) = f[(gh)(x)] = f[g(h(x))]$$
$$[(fg)h](x) = (fg)(h(x)) = f[g(h(x))]$$

which completes the proof.

The commutative law for substitution does not generally hold, as we see, if we set $f(x) = 1 + x$ and $g(x) = x^2$. We have $f(g(x)) = 1 + x^2$ and $g(fx)) = (1 + x)^2$.

2. Algebra of Functions (Tri-Operational Algebra).

In contrast to the classical approach we do not start with a given field of numbers and do not define functions in terms of numbers. In fact, we shall not give any explicit definition of functions, and we shall abolish the dualism between numbers and functions[*].

We start out with a system of things, called functions and denoted by small letters $f, g, \ldots$, and three binary operations: addition denoted by +, multiplication denoted by and substitution denoted by juxtaposition. For these operations we postulate the following laws which in the classical theory (as we have seen in two examples) are deduced from the explicit definitions:

Operation:	Addition	Multiplication	Substitution
Commutative Law:	$f+g = g+f$	$f \cdot g = g \cdot f$	-----------
Associative Law:	$(f+g)+h = f+(g+h)$	$(f \cdot g) \cdot h = f \cdot (g \cdot h)$	$(fg)h = f(gh)$

---

[*] At the same time we rid the foundations of analysis of set theoretical elements which are contained in the explicit definition of functions (see footnote p.4).

Distributive Law:	$(f \cdot g)h = fh \cdot gh$	$(f+g)h = fh+gh$	$(f+g) \cdot h = f \cdot h + g \cdot h$
Neutral Elements:	$f+0 = f$	$f \cdot 1 = f$	$fj = jf = f$
	$0 \neq 1 \neq j \neq 0$		$10 = 1$
Opposite Elements:	$f+(-f) = 0$		

Commutativeness of substitution is not postulated because it does not hold in the classical theory. The three distributive laws listed under

addition   multiplication   substitution

will be called

multiplicative-substitutive, additive-substitutive, additive-multiplicative, distributive laws, respectively, or briefly,

m.s.d. law   a.s.d. law   a.m.d. law.

From the commutative law of multiplication and the a.m.d. law it follows that $h \cdot (f+g) = h \cdot f + h \cdot g$. In absence of a commutative law for substitution the a.s.d. law and the m.s.d. law do not imply

$h(f+g) = hf + hg$ and $h(f \cdot g) = hf \cdot hg$.

In fact, in the classical theory these formulae are not generally true. Each of them represents a functional equality characterizing a special class of functions h.

The neutral elements 0 and 1 in our algebra correspond to the classical functions associating with each x the numbers 0 and 1, respectively. From the commutativity of addition and multiplication in conjunction with the postulates concerning 0 and 1, it follows that $0+f = f$ and $1 \cdot f = f$. For the neutral element of substitution, j, both $jf = f$ and $fj = f$ must be postulated. The element j corresponds to the classical function associating with each number x the number x. Oddly enough, the classical calculus does not introduce a symbol for this fundamental function.

In the same way as one assumes $0 \neq 1$ in defining a field, we assume that the three neutral elements 0, 1, and j are mutually different. Furthermore, we had to postulate that 0 substituted in 1 yields 1 because we shall have to make use of this assumption and, as Rev. F. L. Brown proved by an example, it is independent of the other postulates. (Under certain conditions, the postulate $10 = 1$ could be replaced by the simpler formula $10 \neq 0$).

The three neutral elements can easily be shown to be unique. For instance, if we had also $fj' = f$ for each f, then by applying this formula to $f = j$ we should obtain $jj' = j$. Now $jj' = j'$ by virtue of the neutrality of j. Hence $j = j'$.

With regard to addition we assume the existence of an opposite element to each f. We denote by -f the function which, added to f, yields the sum 0. The formulae which we assumed with regard to addition and multiplication are just those valid in a commutative ring with a unit, i.e., a neutral element of multiplication [*].

In the classical theory the three operations are not universal (i.e., applicable to each pair of functions). E.g., we can form fg only if R_g is part of D_f. Hence we shall not

---

[*] It should be noted that among the formulae concerning addition and substitution, one which would be valid in a non-commutative ring has not been postulated, namely, $h(f + g) = hf + hg$.

postulate in our algebra that the three operations are
universal. We shall interpret our postulates as _formulae_
which are valid if all terms are meaningful. The situation
is the same as in a grupoid in which $f + (g + h) = (f + g) + h$
is true provided that $g + h$, $f + g$, $f + (g + h)$, and $(f + g) + h$
exist.

However, it is worth mentioning that our postulates are
consistent even in presence of the additional assumption
that the three operations are universal. A model satisfying
all these assumptions is the system of all polynomials
$p = c_0 + c_1 \cdot j + c_2 \cdot j^2 + \cdots + c_m \cdot j^m$ with coefficients c_k belonging to a given ring (where j^k is an abbreviation for a
product of k factors j) if sum, product, and substitution are
defined in the ordinary way. That is, if

$q = d_0 + d_1 \cdot j + d_2 \cdot j^2 + \cdots + d_n \cdot j^n$, then

$p + q = (c_0 + d_0) + (c_1 + d_1) \cdot j + (c_2 + d_2) \cdot j^2 + \cdots$

$p \cdot q = c_0 d_0 + (c_0 \cdot d_1 + c_1 \cdot d_0) \cdot j + (c_0 \cdot d_2 + c_1 \cdot d_1 + c_2 \cdot d_0) \cdot j^2 + \cdots$

$pq = c_0 + c_1 \cdot q + c_2 \cdot q^2 + \cdots + c_m \cdot q^m$.

A simple model consisting of four functions $0, 1, j, f$,
is obtained if the three operations are defined by the following tables:

+	0 1 j f		·	0 1 j f			0 1 j f
0	0 1 j f		0	0 0 0 0		0	0 0 0 0
1	1 0 f j		1	0 1 j f		1	1 1 1 1
j	j f 0 1		j	0 j j 0		j	0 1 j f
f	f j 1 0		f	0 f 0 f		f	1 0 f j

In the classical theory we obtain the above system [*) by considering the field consisting of the numbers 0 and 1 (modulo 2) and by calling 0,1,j,f the functions associating with the numbers 0 and 1 the numbers 0,0; 1,1; 0,1; 1,0, respectively. As Rev. F. L. Brown has recently shown, this system is the simplest one satisfying all the postulates since no system consisting of 0,1, and j only satisfies all postulates. However, Father Brown did find a system consisting of the three elements 0,1,j, satisfying all the postulates except the one concerning the existence of a negative element.

3. The Theory of Constant Functions.

We shall now single out a class of functions which we will call constant functions or, briefly, constants. The definition will be in terms of the fundamental operations. From the postulates concerning these operations, it will follow that our constants enjoy the main properties of the constant functions of the classical theory.

We call a function f constant if $f = f0$. If we know of a function that it is constant, then we shall usually denote it by letters c,d,...

From the postulates it follows that $0f = (0+0)f = 0f + 0f$. Adding -(0f) to this equality we obtain the formula

$$0f = 0.$$

In particular $00 = 0$. Hence 0 is a constant. That 1 is a constant, is the content of the postulate $10 = 1$. Since $j0 = 0 \neq j$, we see that j is not a constant. We shall prove

---

[*) It can also be described as the system of all polynomials modulo $j + j^2$ with coefficients belonging to the field 0,1 modulo 2.

the following theorem:

The constants form a ring [*] which is closed with respect to substitution. If c_1 and c_2 are constants, then

$$c_1 + c_2 = c_1 0 + c_2 0 = (c_1 + c_2)0.$$

Similarly, $c_1 \cdot c_2 = (c_1 \cdot c_2)0$. Thus the sum and the product of two constants are constants. From the fact that 0 is a constant, it readily follows that the negative of a constant is a constant. Thus the constants form a ring. Now let c be a constant, and f any function. We have

$$fc = f(c0) = (fc)0.$$

Thus fc is a constant. Using the formula $0f = 0$ we obtain

$$cf = (c0)f = c(0f) = c0 = c.$$

Thus cf is a constant, more specifically, cf is the constant c. This completes the proof of our theorem.

The last formula can also be expressed by saying that if c is a constant, then not only $c0 = c$ but $cf = c$ for each f.

If for a constant c there exists a function c' such that $c \cdot c' = 1$, then this "reciprocal" c' is a constant. For

$$c'0 = 1 \cdot c'0 = (c' \cdot c) \cdot c'0 = c' \cdot (c \cdot c'0) = c' \cdot (c0 \cdot c'0) =$$
$$c' \cdot (c \cdot c')0 = c' \cdot 10 = c' \cdot 1 = c'.$$

We see: if for each constant $c \neq 0$ there exists a reciprocal, then the constants form a field which is closed with respect to substitution. However, the roots of an algebraic

---

[*] Quite accurately, we should say: For the constants all formulae postulated in a commutative ring are valid if they are meaningful. Under additional assumptions and with a sharper definition of constants, we could prove them to form a ring. We should have to call constant a function c such that c0 is defined and $= 0$. We should have to assume that if f0 and g0 are defined, then $(-f)0$, $(f+g)0$, and $(f \cdot g)0$ are defined.

equation with constant coefficients need not be constants. For instance, each of the four functions $0,1,j,1+j$, studied at the end of the last section, satisfies the algebraic equation with constant coefficients $f + f^2 = 0$. The functions j and $1+j$ are not constant.

The definition of equality of two functions in the classical theory is reflected in the following fundamental proposition of our algebra: If $fc = gc$ for each constant c, then $f = g$. If this proposition holds, then we shall speak of a <u>tri-operational algebra with a base of constants</u>.

Clearly, in such an algebra we have $f = g$ if and only if $fc = gc$ for each constant c. Moreover, in order that f be a constant it is necessary and sufficient that $fc = f0$ for each constant c. For from $fc = f0$ it follows that $fc = f0 = f(0c) = (f0)c$. Applying the equality criterion to f and $f0$ we see that f is equal to the constant function $f0$.

A consequence of this last remark is the following first theorem: If for two constants c_0 and c_1 we have $f(c_0 + j) = c_1$, then $f = c_1$. In fact, for each constant c the assumption implies that

$$fc = f(c_0 + (c - c_0)) = f(c_0 + j)(c - c_0) = c_1(c - c_0) = c_1.$$

Another consequence is the following translation theorem: If $f(j + c) = f$ for each constant c, then f is constant. For from the assumption it follows that

$$fc = f(0 + c) = f(j0 + c0) = f[(j + c)0] = [f(j + c)]0 = f0.$$

An Algebra of Functions admits a representation by functions in the classical sense. With each function of our algebra we can associate a function in the classical sense whose domain and whose range are two sets of constants. With the function f of our algebra we can associate the function f^* in the classical sense whose domain consists of those constants which admit substitution into f, and which associates with each such constant c the constant fc. This association of functions in the classical sense with functions of our algebra is readily seen to be a homomorphism. That is to say, we have

$$(f + g)^* = f^* + g^*, \quad (f, g)^* = f^* \cdot g^*, \quad (fg)^* = f^* g^*$$

where addition, multiplication, and substitution of the classical functions on the right sides of these equalities are to be performed in the classical sense. The postulate of a base of constants implies that the above homomorphism is an isomorphism, that is to say, that $f \neq g$ implies $f^* \neq g^*$.

It is to be noted that even in an algebra in which the three fundamental operations are universal and the constants form a field, the postulate of the base of constants need not be satisfied. We obtain an example of the independence of this postulate by considering all polynomials

$$c_0 + c_1 \cdot j + c_2 \cdot j^2 + \cdots + c_m \cdot j^m$$

with coefficients 0 and 1 if addition, multiplication, and substitution are defined in the ordinary sense but modulo 2. There are infinitely many such polynomials but only two constants, viz., 0 and 1. For each polynomial the substitution of 0 and 1 yields either 0 and 0, or 0 and 1, or 1 and 0, or 1 and 1. If we write a polynomial in the form

14

$$p = c_0 + p^{k1} + p^{k2} + \cdots + p^{kn}$$

where c_0 is 0 or 1, then clearly p belongs to one of the following four classes:

Either $c_0 = 0$ and n is even.   Then p0 = 0, p1 = 0,

Or, $c_0 = 0$ and n is odd.   Then p0 = 0, p1 = 1.

Or, $c_0 = 1$ and n is odd.   Then p0 = 1, p1 = 0.

Or, $c_0 = 1$ and n is even.   Then p0 = 1, p1 = 1.

If p_1 and p_2 are two of the infinitely many polynomials belonging to the same class, then for each constant (that is, for c = 0 and c = 1) we have $p_1 c = p_2 c$ and yet $p_1 \neq p_2$. The homomorphic representation of the functions of our algebra by functions in the classical sense which we described above, would lead to the four functions defined in the field modulo 2 mentioned at the end of the preceding section. Each function of the first class would be mapped onto the function representing 0, each function of the second class on the function representing j, each function of the third class on the function representing 1 + j, and each function of the fourth class on the function representing 1.

The following finite example for the same situation may be omitted in a first reading. We consider the polynomials of the preceding example modulo $j + j^4$. That is to say, we set $j + j^4 = 0$. We thus retain a model consisting of 16 polynomials $c_0 + c_1 \cdot j + c_2 \cdot j^2 + c_3 \cdot j^3$ with coefficients $c_k = 0,1$. There are only two constants, 0 and 1, and hence only four possibilities for p0 and p1, as before. Each possibility is realized for a class of four polynomials. E.g., we have p0 = 0 and p1 = 0 for $p_1 = 0$, $p_2 = j + j^2$, $p_3 = j + j^3$, $p_4 = j + j^4$.

This is another example in which each function p, each constant function as well as each of the 14 non-constant functions, satisfies an algebraic equation with constant coefficients, namely, $p + p^4 = 0$, as we see by substituting p into the equality $j + j^4 = 0$.

4. The Lytic Operations.

While we did not postulate universality of the three fundamental operations, we saw that a postulate to this effect would be compatible with our assumptions. Now we are going to introduce a function of a special kind whose very nature, in presence of the other postulates, is incompatible with universality of substitution. We shall call this function rec (an abbreviation for reciprocal) and define it by the equality

$$rec \cdot j = 1.$$

If we substitute 0 in this equality we obtain $(rec \cdot j)0 = 10$ from which, in view of $0 \cdot f = 0$, it follows that

$$0 = rec\ 0 \cdot 0 = rec\ 0 \cdot j0 = (rec \cdot j)0 = 10 = 1.$$

This contradicts the assumption $0 \neq 1$. We see that in presence of the definition of rec we must give up some of our postulates or abandon the universality of substitution by forbidding the substitution of the function 0 into the function rec. We shall follow the latter course.

If the constants form a field, then 0 is the only constant which cannot be substituted into rec. Into rec f we cannot substitute those constants c for which fc = 0. For instance, 1 cannot be substituted into $rec(j - 1)$.

16

It goes without saying that, in the classical notation, rec is the function associating with each number $x \neq 0$ the number $\frac{1}{x}$. From the point of view of domains, the function rec·j is not identical with the function 1. The latter is an extension of the former. For the domain of 1 comprises all numbers; that of rec, and hence of rec·j, all numbers $\neq 0$.

We shall disregard this difference and thus from now on be compelled to interpret our postulates as formulae which are valid whenever their terms are meaningful, and we shall have to interpret each result as a formula admitting those substitutions which are admissible in all terms involved in the derivation of the result from the postulates.

To make the analogy between rec f and -f more conspicuous we shall frequently write neg f instead of -f. This notation is justified since there exists a function neg such that we obtain -f by substituting f into the function neg. This function neg is -j or -1·j. In the classical notation it is the function associating with each number x the number -x.

Instead of postulating the existence of -f for each f it would be sufficient to postulate the existence of a function neg such that j + neg = 0. In view of 0f = 0, jf = f, and the a.s.d. law this postulate implies f + neg f = 0 for each f. We tabulate some analogous facts of the algebra of the functions neg and rec which we shall call the lytic functions with regard to addition and multiplication, respectively.

$$j + neg = 0 \qquad\qquad j \cdot rec = j$$
$$f + neg\ f = 0 \qquad\qquad f \cdot rec\ f = j$$
$$neg\ (f + g) = neg\ f + neg\ g \qquad rec(f \cdot g) = rec\ f \cdot rec\ g$$
$$neg\ neg = j \qquad\qquad rec\ rec = j$$
$$rec\ neg = neg\ rec.$$

In fact, we have

$$rec(f \cdot g) = rec(f \cdot g) \cdot 1 = rec(f \cdot g)(f \cdot rec\ f \cdot g \cdot rec\ g) =$$
$$rec(f \cdot g) \cdot (f \cdot g) \cdot (rec\ f \cdot rec\ g) = 1 \cdot\ rec\ f \cdot rec\ g = rec\ f \cdot rec\ g.$$

Using the formula $rec \cdot rec\ rec = 1$ obtained by substituting rec into $j \cdot rec = 1$ we see

$$rec\ rec = (j \cdot rec) \cdot rec\ rec = j \cdot (rec \cdot rec\ rec) = j \cdot 1 = j.$$

The proof of $neg\ neg = j$ is similar.

From $neg\ f = -1 \cdot f$ it follows that $neg\ f \cdot neg\ g = f \cdot g$. Using this formula we obtain

$$rec\ neg = rec\ neg \cdot (j \cdot rec) = rec\ neg \cdot (neg\ j \cdot neg\ rec) =$$
$$rec\ neg \cdot (neg \cdot neg\ rec) = (rec\ neg \cdot neg) \cdot neg\ rec = 1 \cdot neg\ rec = neg\ rec.$$

We define: f is __even__ if and only if $f\ neg = neg\ f$, and f is __odd__ if and only if $f\ neg = f$. The last of the tabulated formulae can be expressed by saying that rec is odd. Clearly, the product of two even or of two odd functions is even, the product of an odd and even function is odd.

Concerning the lysis of substitution, we mention that if for two functions f and g we have $fg = j$, then we shall call g the right inverse of f, and f the left inverse of g. For instance, the function j is its own right and left inverse since we have $jj = j$. If c_0 and c_1 belong to the ring of constant functions, and c_1 has a reciprocal c_1', then the function $c_1 \cdot j + c_0$ has the function $c_1' \cdot (j - c_0)$ as inverse on either side. In the classical notation, in view of $j(x) = x$, the definition of a pair of inverse functions reads

$$f(g(x)) = x.$$

In other words, f and g are pairs of inverse functions in the classical sense, as log and exp or arctan and tan.

We shall postulate the existence of inverse functions only for special functions f. While there are functions neg and rec satisfying the equations $j + neg = 0$ and $j \cdot rec = 1$ and such that we obtain the negative and the reciprocal of f by substituting f into neg and rec, respectively, there certainly does not exist a function inv satisfying the equation $j \; inv = j$ and such that we could obtain the inverse of f by substituting f into inv. For by virtue of the definition of j we should have $j \; inv = inv$, so that from $j \; inv = j$ it would follow that $inv = j$. But by substituting f into j we obtain f which in general is not the inverse of f. Or we can say: By substituting f into the equality $j \; inv = j$ we obtain $j(inv \; f) = f$ rather than $f(inv \; f) = j$.

A constant function c clearly does not have inverse functions on either side. For, whatever function f may be, cf and fc are constants, thus $\neq j$ since j is not a constant.

If g is a right inverse of f and has itself at least one right inverse, h, then g has only one right inverse, namely, f, and only one left inverse, namely, f. And f has only one right and one left inverse, namely, g. In fact, from $fg = j$ and $gh = j$ it follows that $h = jh = (fg)h = f(gh) = fj = f$, a situation familiar from the axiomatics of group theory.

5. Exponential Functions.

We call the function f an exponential function if and only if
$$f(g+h) = fg \cdot fh \quad \text{and} \quad f \neq 0.$$
We shall denote exponential functions by exp. Thus $\exp(g+h) = \exp g \cdot \exp h$. Substituting a constant c_0 for g, and j for h we see that

$\exp(c_0 + j) = \exp c_0 \cdot \exp j = \exp c_0 \cdot \exp$. If $\exp c_0 = 0$, then it follows that $\exp(c_0 + j) = 0$. If our algebra has a base of constants, then the last formula, by virtue of the first theorem of Section 3, implies that $\exp = 0$ in contradiction to the assumption $\exp \neq 0$. We thus see that $\exp c_0 \neq 0$ for each constant c_0. In further consequence, $\exp f \neq 0$ for each f. For if we had $\exp f = 0$, then we should obtain

$$\exp(fc_0) = (\exp f)c_0 = 0c_0 = 0$$

which, in view of the fact that fc_0 is a constant, contradicts the preceding remark.

If $\exp c_1 = 1$, then $\exp(f + c_1) = \exp c_1 \cdot \exp f = \exp f$ for each f. Conversely, if the constants form a field, from $\exp(c + c_1) = \exp c$ in view of $\exp c \neq 0$ it follows that $\exp c_1 = 1$. Now $\exp(c + 0) = \exp c$. Thus $\exp 0 = 1$. Consequently, $1 = \exp(j + neg) = \exp \cdot \exp neg$ and hence $\exp neg = rec \exp$.

Obviously, in each ring the function 1 is an exponential function. From $\exp 0 = 1$ it follows that 1 is the only constant exponential function. From now on, when talking about exponential functions we shall always mean exponential functions $\neq 1$.

If the constants form a finite field, then no exponential function exists. Let indeed $p \neq 0$ be the characteristic of the field of constants. Since $p - 1$ is the sum of $p - 1$ summands 1, for an exponential function we should have

$$\exp(p - 1) = (\exp 1)^{p-1}.$$

Now in a field of characteristic p we have $c^{p-1} = 1$ for each c. Hence $\exp(p - 1) = 1$. From $(p - 1) + 1 = 0$ it would follow that

$\exp(p-1)\cdot\exp 1 = \exp 0$. Since $\exp 0 = \exp(p-1) = 1$ we should have $\exp 1 = 1$. But then $\exp 2 = \exp 1\cdot\exp 1 = 1$, $\exp 3 = \exp 2\cdot\exp 1 = 1$, etc., hence $\exp c = 1$ for each c.

However, exponential functions do exist in finite rings. For instance, one readily verifies that in the ring of residues modulo 9 the function which under substitution

   of   0,1,2,3,4,5,6,7,8

   yields   1,4,7,1,4,7,1,4,7,   respectively,

is an exponential function. In the infinite ring without divisors of 0 consisting of the numbers $m + n\cdot i$ where m and n are integers and $i^2 = -1$, the function which under substitution of $m + n\cdot i$ yields i^{m+n} is an exponential function. If the constants form the ring of all integers or the field of all rational numbers, then no exponential functions defined for all constants, exist.

6.   The Logarithmic Functions.

We shall now take a step towards the algebra of real functions by assuming, in addition to a base of constants, three postulates about the ring of constants. For the sake of brevity we shall call a constant c a square if there exists a constant $c_1 \neq 0$ such that $c = c_1^2$. Now we postulate for each constant c which is not a divisor of 0:

   1.   If c is a square, then $-c$ is not a square.

   2.   If c is not a square, then $-c$ is a square.

   3.   There exists a constant $1/2$ such that $1/2 + 1/2 = 1$ (and consequently for each c a constant $c/2$ such that $c/2 + c/2 = c$, namely, $c\cdot 1/2$).

Clearly, the product of two squares, as well as the product of two negative squares, is a square. The product of a square and a negative square is not a square. It follows that if a square has a reciprocal, the reciprocal is a square.

Postulate 3 is satisfied in each field of characteristic $\neq 2$. Postulate 1 can be expressed by saying that $c_1^2 + c_2^2 = 0$ implies $c_1 = c_2 = 0$, a weaker form of the postulate for a real field. Postulates 1 and 2 are sufficient to establish in the ring what may be called a multiplicative order: If we call each square "positive", then for each element c of the ring either c is a divisor of 0, or c is positive, or $-c$ is positive, and the product of two positive elements is positive. However, even if the ring is a field, postulates 1 and 2 are not sufficient to order the field (i.e., to guarantee that also the sum of two positive elements is positive) as the field of residues modulo 7 shows if we call 1,2,4 positive. Neither does each ordered field satisfy postulate 2 as the field of all rational numbers shows.

We shall now assume that an exponential function which admits the substitution of each constant, has an inverse on both sides which admits the substitution of each square. We shall call such a function a logarithmic function and denote it by log.

For each constant d, from $\exp d = \exp(d/2 + d/2) = \exp(d/2) \cdot \exp(d/2)$ it follows that each constant $\exp d$ is a square. This fact implies that $\log c$ admits only the substitution of squares. For if $\log c$ is defined, then

$$c = jc = (\exp \log)c = \exp(\log c)$$

and exp d is a square for each d. The same reasoning, in view of exp d $\neq$ 0, shows that log does not admit the substitution of 0. Consequently, the function log $(j \cdot j)$ admits the substitution of each constant $\neq$ 0.

Moreover, we have log 1 = log (exp 0) = (log exp)0 = j0 = 0.

Now let log c_1 and log c_2 be defined. That is, c_1 and c_2 are squares which implies that also $c_1 \cdot c_2$ is a square and log($c_1 \cdot c_2$) is defined. We have

log c_1 + log c_2 = j(log c_1 + log c_2) = log exp(log c_1+ log c_2)

= log(exp log $c_1 \cdot$log exp c_2) = log($c_1 \cdot c_2$).

It follows that

0 = log 1 = log($j \cdot j \cdot$rec$\cdot$rec) = log($j \cdot j$) + log(rec$\cdot$rec)

and hence log(rec$\cdot$rec) = neg log($j \cdot j$), formulae which admit the substitution of each constant $\neq$ 0.

Similarly,

0 = log 1 = log(j$\cdot$rec) = log j + log rec.

However, the last equality admits only the substitution of squares (and of all squares since if c is a square, rec c is a square). Thus the same holds for

log rec = neg log.

7. The Absolute and the Signum.

Under the assumptions of the preceding section one can introduce a function which we shall call the absolute value or, briefly, the absolute, and which we shall denote by abs. We define

abs = exp[$\frac{1}{2} \cdot$log($j \cdot j$)] and abs 0 = 0.

In the classical theory the function corresponding with abs associates with each x the number $|x|$. The function abs

admits the substitution of each constant and is readily seen
to enjoy the following properties:

1. $\text{abs } c_1 \cdot \text{abs } c_2 = \text{abs}(c_1 \cdot c_2)$
2. $\text{abs}^2 = j^2$
3. $\text{abs neg} = \text{abs}$
4. $\text{abs} \neq 0.$

It is easily seen that abs rec = rec abs.

We further define a signum function, denoted by sgn, in
the following way:

$$\text{sgn} = \text{abs} \cdot \text{rec} \quad \text{and} \quad \text{sgn } 0 = 0.$$

In the classical theory the function corresponding with sgn
associates 0 with 0, 1 with each positive, -1 with each negative number.

The function sgn has the following properties:

1. $\text{sgn } c_1 \cdot \text{sgn } c_2 = \text{sgn}(c_1 \cdot c_2)$
2. $\text{sgn}^3 = \text{sgn}$
3. $\text{sgn neg} = \text{neg sgn}$
4. $\text{sgn} \neq 0.$

One readily verifies that sgn yields 1 or -1 according to
whether a square or the negative of a square is substituted.
On this fact one can base another introduction of the assumptions of the preceding section, an introduction which is more
in line with the Algebra of Functions than the postulates 1 and
2 concerning squares.   We can postulate the existence of a
function abs or a function sgn with the four properties mentioned above and define: a constant c which is not divisor of
0, is positive or negative according to whether

abs c = c. or abs c = -c (sgn c = 1 or sgn c = -1).

We remark that the four postulates for sgn are independent. In the field of residues modulo 3 the function s which admits the substitution of all three constants 0,1,-1 and (like the function -j) yields s0 = 0, s1 = -1, s(-1) = 1 satisfies all postulates except the first. In the field of residues modulo 5 the function s which admits the substitution of 0, $\pm 1, \pm 2$ and (like the function j) yields sc = c for each c, satisfies all postulates except the second. In the same field the function s which (like j^2) yields s0 = 0, s1 = s(-1), s2 = s(-2) = -1 satisfies all postulates except the third. In each field the function 0 satisfies all postulates except the fourth.

We have

$$\text{abs exp} = \exp\left[\tfrac{1}{2} \cdot \log(j \cdot j)\right]\exp = \exp\left[\tfrac{1}{2} \cdot \log(\exp \cdot \exp)\right]$$
$$= \exp\left[\tfrac{1}{2} \cdot 2 \cdot \log \exp\right] = \exp j = \exp.$$

For the function log abs, on account of its importance, we shall introduce a special symbol. We shall denote it by logabs. We have

$$\text{logabs} = \log \exp\left[\tfrac{1}{2} \cdot \log(j \cdot j)\right] = \tfrac{1}{2} \cdot \log(j \cdot j).$$

The function logabs admits the substitution of each constant $\neq 0$. It corresponds to the function associating $\log|x|$ with each $x \neq 0$ in the classical theory. We have

$$\text{logabs exp} = (\log \text{abs})\exp = \log(\text{abs exp}) = \log \exp = j,$$
$$\exp \text{logabs} = \exp \log \text{abs} = j \text{ abs} = \text{abs},$$

$\exp(\text{logabs } f + \text{logabs } g) = \exp \text{logabs } f \cdot \exp \text{logabs } g = \text{abs } f \cdot \text{abs } g$. By virtue of $j = \text{sgn} \cdot \text{abs}$ and $\text{sgn}(c_1 \cdot c_2) = \text{sgn } c_1 \cdot \text{sgn } c_2$ it follows that

$$f \cdot g = \text{sgn } f \cdot \text{sgn } g \cdot \exp(\text{logabs } f + \text{logabs } g).$$

The last formula could be used as a definition of multiplication in terms of addition and substitution, in conjunction with the exponential and the signum functions. Algebra of Functions might be developed from postulates about two operations and two particular functions, possibly one particular function.

8. The Power Functions.

We shall now for each constant c introduce a function called the c-th power and denoted by c-po. We define c-po in the same way in which it is defined in the theory of complex functions:

$$c - po = \exp(c \cdot \log).$$

From this definition it follows that c - po admits the substitution of all squares and only of squares. More accurately, we should call the above function the c-th power based on the function exp. However, in some cases we shall see that, for algebraic reasons, power functions are independent of the particular choice of the exponential function used in defining them. For instance

$0 - po = \exp(0 \cdot \log) = \exp 0 = 1$ and $1 - po = \exp(1 \cdot \log) = \exp \log = j$.

If exp' is another exponential function, log' the inverse of exp', and if we define

$$c - po' = \exp'(c \cdot \log'),$$

then, as before, we have

$0-po' = \exp'(0 \cdot \log') = \exp' 0 = 1$ and $1-po' = \exp'(1 \cdot \log') = \exp' \log' = j$.

Moreover, we obtain the following three functional equations for the power functions

$$c_1 - po \cdot c_2 - po = (c_1 + c_2) - po$$

$$c_1 - po \ c_2 - po = (c_1 \cdot c_2) - po$$

$$c - po \ f \cdot c - po \ g = c - po \ (f \cdot g).$$

Proof:

$$c_1\text{-po}\cdot c_2\text{-po} = \exp(c_1\cdot\log)\cdot\exp(c_2\cdot\log) = \exp(c_1\cdot\log + c_2\cdot\log)$$
$$= \exp[(c_1 + c_2)\cdot\log] = (c_1 + c_2)\text{-po},$$

$$c_1\text{-po } c_2\text{-po} = \exp(c_1\cdot\log)\exp(c_2\cdot\log) = \exp[c_1\cdot\log\exp(c_2\cdot\log)]$$
$$= \exp[c_1\cdot(c_2\cdot\log)] = \exp[(c_1\cdot c_2)\log] = (c_1\cdot c_2)\text{-po},$$

$$c\text{-po } f\cdot c\text{-po } g = \exp(c\cdot\log f)\cdot\exp(c\cdot\log g) = \exp(c\cdot\log f + c\cdot\log g)$$
$$= \exp[c\cdot(\log f + \log g)] = \exp[c\cdot\log(f\cdot g)] = c\text{-po}(f\cdot g).$$

From the first of these functional equations it follows that

$$1 = 0\text{-po} = (c + (-c))\text{-po} = c\text{-po}\cdot(-c)\text{-po}, \text{ and hence}$$

$$(-c)\text{-po} = \text{rec } c\text{-po}, \text{ in particular, } (-1)\text{-po} = \text{rec}.$$

Moreover, we have $2\text{-po} = 1\text{-po}\cdot 1\text{-po} = j\cdot j$ and, by induction, we see that for each positive integer n, the n-th power is the product of n factors j. This statement as well as the functional equations for $c\text{-po}$ are independent of the choice of the exponential function exp used in defining the power functions.

From the second functional equation it follows that
$$c\text{-po }\tfrac{1}{c}\text{-po} = \exp(c\cdot\log)\exp(\tfrac{1}{c}\cdot\log) = \exp[c\cdot\log\exp(\tfrac{1}{c}\cdot\log)]$$
$$= \exp[c\cdot j(\tfrac{1}{c}\cdot\log)] = \exp(c\cdot\tfrac{1}{c}\cdot\log) = \exp(1\cdot\log) = \exp\log = j.$$
Thus, $c\text{-po}$ and $\tfrac{1}{c}\text{-po}$ are inverse functions.

In the equation $2\text{-po} = j\cdot j$, the right side admits the substitution of any constant, while the left side admits only the substitution of squares. However, we may consistently extend the definition of $c\text{-po}$ by the following stipulations:

1. $c\text{-po } 0 = 0$

2. If c is a rational number n_1/n_2 with an odd denominator n_2 and a numerator n_1 which is relatively prime to n_2, then the function $c\text{-po}$ is even if the number n_1 is even, and odd if n_1 is odd.

We do not permit the substitution of negative squares into c - po in the remaining cases, that is, if c is a rational number with an even denominator or not rational. We remark that in case that c is a rational number n_1/n_2 whose denominator is even, we have not only

$$n_2- po \ c-po = n_1-po \quad \text{but also} \quad n_2-po \ neg \ c-po = n_1-po.$$

After the above extension our definition includes all the cases covered by the classical theory of power functions.

In case that c is a positive integer we readily see that the extended function c - po is identical with the product of c factors j. For $c = \dfrac{2m}{2n+1}$ or $= \dfrac{2m+1}{2n+1}$ (m and n integers) it is easily seen that the extended c-th powers can be written

$$\frac{2m}{2n+1} - po = exp[\frac{2m}{2n+1} \cdot logabs] = exp[\frac{m}{2n+1} \cdot log(j \cdot j)]$$

$$\frac{2m+1}{2n+1} - po = sgn \cdot exp[\frac{2m+1}{2n+1} \cdot logabs].$$

In operating with integers we have omitted and shall omit the multiplication dot. 2m+1 stands, of course, for $2 \cdot m+1$.

The reader can easily check to which extent our supplementary stipulations concerning the definition of c - po are compatible with the three functional equations for power functions. For instance, the equation $c_1-po \ c_2-po = (c_1 \cdot c_2)-po$ can not be upheld after the extension; in other words, it does not permit the substitution of negative constants. As an example, we mention

$$\tfrac{1}{2} - po \ 2-po = exp[\tfrac{1}{2} \cdot log \ exp(2 \cdot logabs)] = exp \ logabs = abs.$$

Thus the extended function 2 - po and the function $\tfrac{1}{2}$ - po are not inverse. The function $\frac{1}{2}$ - po 2 - po is not = j but = abs which, in fact, is Cauchy's representation of the function abs.

9.   The Trigonometric Functions.

We call f a tangential function if and only if
$$f(g+h) = \frac{fg+fh}{1-fg\cdot fh} \quad \text{and} \quad f \neq 0.$$
We shall denote a tangential function by tan. In the second chapter we shall single out among the tangential functions the ordinary tangent function.

From the definition it follows: If tan g = 0, then tan(f + g) = tan f. Moreover,
$$\tan 0 = \tan(0+0) = \frac{2\cdot\tan 0}{1-\tan 0\cdot\tan 0}.$$
Consequently, if the constant tan 0 is to be real, we have tan 0 = 0. Furthermore, it readily follows from the definition that tan is an odd function.

In this section we shall assume that the constants form the field of real numbers. Moreover, we shall postulate the existence of a smallest constant c > 0 such that tan c = 0. Then tan does not admit the substitution of the constants c/2 and -c/2. For if, say, tan (c/2) were defined, then we should have
$$0 = \tan c = \frac{2\cdot\tan(c/2)}{1-\tan(c/2)\cdot\tan(c/2)}.$$
This equality would imply tan (c/2) = 0 in contradiction to our assumption that c is the smallest number 0 for which tan c = 0. It further follows that tan (c/4) = 1 or -1 since, by virtue of the definition of tan, every other value of tan (c/4) would entail a value for tan (c/2). We shall only admit the

substitution into tan of constants between -c/2 and c/2.

Relative to each tangential function we define a sine and a cosine function in the following way:

$$\sin(2\cdot j) = \frac{2\cdot\tan}{1+\tan^2} \quad \text{and} \quad \cos(2\cdot j) = \frac{1-\tan^2}{1+\tan^2}$$

We obviously have

$$\frac{\sin(2\cdot j)}{\cos(2\cdot j)} = \frac{2\cdot\tan}{1-\tan^2} = \tan(2\cdot j).$$

Substituting $\frac{1}{2}\cdot j$ into the equality, we see that $\tan = \sin/\cos$.

Other useful identities are

$$\sin^2 + \cos^2 = 1 \quad \text{and} \quad 1+\tan^2 = \text{rec } \cos^2.$$

We postulate an inverse of tan on both sides and call it arctan.

II.   THE ALGEBRA OF CALCULUS

1.   The Algebra of Derivatives.

We shall now introduce an operator D associating with a function f a function Df, called the derivative of f. We shall not attempt to formulate criteria as to which functions form the domain of the operator D or as to which constants, if any, may be substituted into Df. In our Algebra of Derivatives, we shall adopt the same point of view as in our Algebra of Functions: We shall derive formulae which are valid in classical calculus provided that the terms involved in the derivation of the formulae are meaningful. In classical calculus, for a given function f and a given constant c, the symbol (Df)c is meaningful if the function $\frac{f - fc}{j - jc}$ has a limit for c. We, too, might define (Df)c in terms of a limit operator, L, and derive the fundamental properties of D from postulates concerning L. But in the present exposition we start out with an undefined operator D subject to a few assumptions connecting D with the Algebra of Functions.

Since D is not a function, the postulates of the Algebra of Functions can not be applied to D. Especially the associative law for substitution does not hold for D. Thus the symbol Dfg is ambiguous. It may mean D(fg) or (Df)g. In order to save parentheses we shall make the convention that the symbol D refers only to the immediately following function or group of functions combined in parentheses. Thus we shall briefly write Dfg for (Df)g and reserve parentheses for the case D(fg).

Three postulates will connect D with the three fundamental operations of the Algebra of Functions:

     I. $D(f + g) = Df + Dg$

    II. $D(f \cdot g) = f \cdot Dg + g \cdot Df$

   III. $D(fg) = Dfg \cdot Dg$

Postulate III replaces the associative law for substitution with respect to D. It states that $D(fg)$ and $(Dg)f$ differ by the factor Dg.

By postulate I we have

$$D0 = D(0 + 0) = D0 + D0.$$

Thus $D0 = 0$. This formula has two important consequences.

By means of it we first see that

$$0 = D0 = D(f + neg\ f) = Df + D\ neg\ f,$$

and thus

$$D\ neg\ f = neg\ Df.$$

Secondly, if c is a constant, that is to say, if $c = c0$, we obtain

$$Dc = Dc0 = D(c0) = Dc0 \cdot D0 = Dc0 \cdot 0 = 0.$$

Postulate II now yields

$$D(c \cdot f) = c \cdot Df + f \cdot Dc = c \cdot Df + 0 = c \cdot Df.$$

We shall call this result the Constant Factor Rule.

In view of $fj = f$ postulate III yields

$$Df = D(fj) = Dfj \cdot Dj = Df \cdot Dj.$$

Hence $Dj = 1$ unless $Df = 0$ for each f which we shall later exclude. Anticipating this development, we shall from now on assume that $Dj = 1$. A frequently used consequence of $Dj = 1$ and $Dc = 0$ is the formula

$$D(j + c) = 1.$$

Applying the formula D neg f = neg Df to f = j we obtain
$$D \ neg = -1.$$
If f is even, that is, if f = f neg, then

Df = D(f neg) = Df neg·D neg = Df neg·-1 = neg Df neg
from which it follows that neg Df = Df neg, or in other words,
that Df is odd. Similarly, one can prove that if f is odd,
then Df is even. Using this fact, we see that

0 = D1 = D(j·rec) = j·D rec + rec·Dj = j·D rec + rec.
It follows that j·D rec = neg rec and D rec = neg (-2)-po.
By virtue of postulate III we conclude further that
$$D(rec \ g) = neg \ (-2)\text{-po} \ g \cdot Dg.$$
By means of postulate II we obtain
D(f·rec g)=f·D(rec g) + rec g·Df = f·neg(-2)-po g·Dg + rec g·Df,
that is, the Quotient Rule
$$D(f \cdot rec \ g) = (g \cdot Df - f \cdot Dg) \cdot (-2)\text{-po} \ g.$$

Let g be a right inverse of f. From fg = j it follows
by virtue of postulate II that
$$Dfg \cdot Dg = Dj = 1, \quad \text{and thus} \quad Dg = rec \ Dfg.$$
If h is a left inverse of f, then hf = j implies that
$$Dhf \cdot Df = Dj = 1, \quad \text{and thus} \quad Dhf = rec \ Df.$$
If h is also a right inverse, then substitution of h into the
last formula yields the preceding formula for the derivation
of a right inverse. For
$$Dh = Dhj = Dhfh = rec \ Dfh.$$

By induction we obtain from the three postulates
$$D(f_1 + f_2 + \cdots + f_n) = Df_1 + Df_2 + \cdots + Df_n$$
$$D(f_1 \cdot f_2 \cdot \ldots \cdot f_n) = p_1 \cdot Df_1 + p_2 \cdot Df_2 + \cdots + p_n \cdot Df_n$$

where p_k denotes the product of the n factors $f_1, f_2, \ldots, f_n$ with the exception of f_k.

$$D(f_1 f_2 \ldots f_n) = Df_1 f_2 \ldots f_n \cdot Df_2 f_3 \ldots f_n \cdot \ldots \cdot Df_{n-1} f_n \cdot Df_n.$$

The second of these rules, for equal factors, yields the formula

$$Df^n = n \cdot f^{n-1} \cdot Df,$$

in particular

$$Dj^n = n \cdot j^{n-1}.$$

This formula in conjunction with postulate I and the Constant Factor Rule enables us to derive each polynomial

$$D(c_0 + c_1 \cdot j + c_2 \cdot j^2 + \cdots + c_m \cdot j^m) = c_1 + 2 \cdot c_2 \cdot j + \cdots + m \cdot c_m \cdot j^{m-1} .$$

We call f an algebraic function, more specifically, an algebraic function belonging to the polynomials $p_0, p_1, \ldots, p_n$, if

$$p_0 + p_1 \cdot f + p_2 \cdot f^2 + \cdots + p_n \cdot f^n = 0.$$

By virtue of the formulae derived in this section we obtain

$$Df = \text{neg}(\sum_{k=0}^{n} f^k \cdot D_{p_k}) \cdot \text{rec}(\sum_{k=1}^{n} k \cdot f^{k-1} \cdot p_k).$$

2. The Derivation of Exponential Functions.

Let exp be an exponential function. We apply the formula $\exp(f + g) = \exp f \cdot \exp g$ to $f = j$ and $g = c$. We obtain

$$D[\exp(j + c)] = D \exp(j + c) \cdot D(j + c) = D \exp(j + c) \cdot 1 = D \exp (j + c).$$

On the other hand

$$D[\exp(j + c)] = D(\exp j \cdot \exp c) = D(\exp \cdot \exp c) = \exp c \cdot D \exp.$$

Thus, $D \exp(j + c) = \exp c \cdot \exp$. Substituting 0 in this equality we obtain

on the left side: $D \exp(j + c)0 = D \exp(0 + c) = D \exp c$

on the right side: $\exp c0 \cdot D \exp 0 = \exp c \cdot D \exp 0.$

Thus $D \exp c = \exp c \cdot D \exp 0$ for each constant c. If we have

34

a base of constants it follows that D exp = exp·D exp 0. We
see that the derivative of an exponential function is a constant multiple of the function.

We shall postulate the existence of an exponential function for which D exp 0 = 1. From now on we shall restrict the
symbol exp to this exponential function defined by the two
postulates

    1. exp(f + g) = exp f·exp g

    2. D exp 0 = 1.

Postulate 2 makes the previous stipulations exp ≠ 0 and ≠ 1
superfluous since D0 = D1 = 0 and thus D00 = D10 = 0 ≠ 1. In
the classical theory, the only differentiable (and even the
only continuous) function satisfying the postulates 1 and 2 is
the function associating e^x with each x.

From the two postulates we have derived that D exp = exp.

In Chapter I we saw in the algebra of the exponential
functions that exp c ≠ 0 for each c. Hence D exp c ≠ 0 for
each c. Thus our postulate 2 concerning the exponential
function implies the existence of a function which, in an algebra with a base of constants, justifies our conclusion
Dj = 1 in the preceding section, in the sense that Djc = 1 for
each constant c. We merely have to apply our previous reasoning to f = exp. From exp j = exp it follows that

        D exp = D(exp j) = D exp j·Dj = D exp·Dj

hence D exp c = D exp c·Djc for each constant c. Since
D exp c ≠ 0, we may multiply both sides of this equality by
rec D exp c and thus obtain Djc = 1 for each constant c. Hence
Dj = 1 if we have a base of constants.

Applying postulate 3 to the formula D exp = exp, we obtain

$$D(\exp f) = D \exp f \cdot Df = \exp f \cdot Df.$$

3. The Derivation of Logarithmic Functions.

By log we shall from now on denote the inverse of the function exp for which D exp 0 = 1 and D exp = exp.

From exp log = j by virtue of postulate III it follows that

$$1 = Dj = D(\exp \log) = D \exp \log \cdot D \log = \exp \log \cdot D \log = j \cdot D \log.$$

Thus, D log = rec.

The function rec on the right side admits the substitution of each constant $\neq$ 0, the function log on the left side the substitution of squares only. Instead of log we shall study the function logabs = log abs which, like rec, admits the substitution of each constant $\neq$ 0.

$$D \text{ logabs} = D[\tfrac{1}{2} \cdot \log(j \cdot j)] = \tfrac{1}{2} \cdot D[\log(j \cdot j)] = \tfrac{1}{2} \cdot D \log(j \cdot j) \cdot D(j \cdot j)$$
$$= \tfrac{1}{2} \cdot \text{rec}(j \cdot j) \cdot 2 \cdot j = \text{rec}(j \cdot j) \cdot j = (\text{rec } j \cdot \text{rec } j) \cdot j = (\text{rec} \cdot \text{rec}) \cdot j$$
$$= \text{rec} \cdot (\text{rec} \cdot j) = \text{rec} \cdot 1 = \text{rec}.$$

Next we compute D abs. We have

$$D \text{ abs} = D(\exp \log \text{ abs}) = D(\exp \text{ logabs}) = D \exp \text{ logabs} \cdot D \text{ logabs}$$
$$= \exp \text{ logabs} \cdot \text{rec} = \text{abs} \cdot \text{rec} = \text{sgn}.$$

We remark that the formulae D log = rec and D abs = sgn by virtue of postulate III entail the formula D logabs = rec.

Applying the last formula and postulate III we obtain

$$D(\text{logabs } f) = D \text{ logabs } f \cdot Df = \text{rec } f \cdot Df.$$

4. Logarithmic and Exponential Derivation.

The formulae at the end of the two preceding sections can also be written as follows:

Df = f·D(logabs f) and Df = rec exp f·D(exp f).
Replacing f in the former formula by a particular function f
is called logarithmic derivation (or differentiation) of f.
Similarly, replacing f in the latter formula by a particular
function f might be called exponential derivation of f.

We apply the former method with benefit whenever logabs f
is simpler than f. As an example of logarithmic derivation,
we treat the power functions. From c - po = exp(c·log) it follows that log c - po = c·log which is indeed simpler than c - po.
We have D(log c - po) = c·D log = c·rec. Hence by the formula
of logarithmic derivation

D c - po = c - po·D(log c - po) = c - po·c·rec = c·(c - 1) - po.

We mention that this formula holds also for the extended
c-th powers in case that c is a rational number with an odd
denominator. For in these cases we obtain

$$D \frac{2m}{2n+1} - po = \frac{2m}{2n+1} - po \cdot D\left\{ \log \exp \left[\frac{2m}{2n+1} \cdot \text{logabs} \right] \right\}$$

$$= \frac{2m}{2n+1} - po \cdot D[\frac{2m}{2n+1} \cdot \text{logabs}] = \frac{2m}{2n+1} \cdot \frac{2m}{2n+1} - po \cdot \text{rec}$$

$$= \frac{2m}{2n+1} \cdot \frac{2(m-n)-1}{2n+1} - po.$$

$$D \frac{2m+1}{2n+1} - po = \frac{2m+1}{2n+1} - po \cdot D\left\{ \text{logabs} \left[\text{sgn} \cdot \exp(\frac{2m+1}{2n+1} \cdot \text{logabs}) \right] \right\}$$

$$= \frac{2m+1}{2n+1} - po \cdot D\left\{ \text{logabs} \exp (\frac{2m+1}{2n+1} \cdot \text{logabs}) \right\}$$

$$= \frac{2m+1}{2n+1} - po \cdot D(\frac{2m+1}{2n+1} \cdot \text{logabs}) = \frac{2m+1}{2n+1} \cdot \frac{2m+1}{2n+1} - po \cdot \text{rec}$$

$$= \frac{2m+1}{2n+1} \cdot \frac{2(m-n)}{2n+1} - po.$$

We see that, in accordance with the general rule, the derivative of the even function $\frac{2m}{2n+1}$ - po is odd, and the derivative
of the odd function $\frac{2m+1}{2n+1}$ - po is even.

As another example we apply logarithmic derivation to the function $f = \exp(j \cdot \text{logabs})$ in the classical theory denoted by x^x. We have logabs $f = j \cdot \text{logabs}$, thus

$$D(\text{logabs } f) = \text{logabs} + j \cdot \text{rec} = \text{logabs} + 1.$$

Hence, $Df = f \cdot D(\text{logabs } f) = \exp(j \cdot \text{logabs}) \cdot (\text{logabs} + 1)$.

In general, for functions starting with the symbol exp the function logabs f is simpler than f, and hence logarithmic derivation is convenient. The same is true for functions f which are products $f_1 \cdot f_2 \cdot \ldots \cdot f_n$ provided that we can find $D(\text{logabs } f_i)$ for $i = 1, 2, \ldots, n$. For $D(\text{logabs } f)$ is the sum of these n functions.

Exponential derivation is convenient whenever exp f is simpler than f. This is the case for functions starting with the symbol log or logabs. As an example, we treat the function $f = \text{logabs}(j + \text{logabs})$. Now, $D(\exp f) = D(j + \text{logabs}) = 1 + \text{rec}$. Hence,

$$Df = \text{rec} \exp f \cdot D(\exp f) = \text{rec}(j + \text{logabs}) \cdot (1 + \text{rec}).$$

5. The Derivation of the Trigonometric Functions.

Let tan be a tangential function, c a constant. From the definition of tan it follows that

$$\tan(j + c) = \frac{\tan j + \tan c}{1 - \tan j \cdot \tan c} = \frac{\tan + \tan c}{1 - \tan \cdot \tan c}$$

By virtue of the quotient rule we obtain

$$D \tan(j + c) = D[\tan(j + c)]$$

$$= \frac{(1 - \tan \cdot \tan c) \cdot D \tan - (\tan + \tan c) \cdot - \tan c \cdot D \tan}{(1 - \tan \cdot \tan c)^2}$$

$$= D \tan \cdot (1 + \tan c \cdot \tan c) \cdot \text{rec}(1 - \tan \cdot \tan c)^2.$$

Substituting 0 we obtain

$$D \tan c = D \tan 0 \cdot (1 + \tan c \cdot \tan c) \cdot \text{rec}(1 - \tan 0 \cdot \tan c)^2.$$

Since tan 0 = 0 we have

D tan c = D tan 0·(1 + tan c·tan c) for each constant c.

If we have a base of constants, then

$$D \tan = D \tan 0 \cdot (1 + \tan^2).$$

We shall now postulate that there is a tangential function tan for which D tan 0 = 1. From now on we shall reserve the symbol tan for this function given by the postulates

1. $\tan(f + g) = \dfrac{\tan f + \tan g}{1 - \tan f \cdot \tan g}$

2. D tan 0 = 1.

For this function we have

$$D \tan = 1 + \tan^2 = \text{rec } \cos^2.$$

In the classical analysis, for each constant a the function tan (a·x) satisfies postulate 1. The function associating tan x with x is the only one which satisfies postulates 1 and 2. In a paper "e and π in Elementary Calculus" (to appear in the near future) we describe how the postulates D tan 0 = 1 and D exp 0 = 1 in conjunction with the functional equations for the tangential and exponential functions yield an intuitive introduction of π and e, as well as a simple development of the "natural" tangential and exponential functions e^x and tan x (x measured in radians).

From tan arctan = j we obtain

1 = D(tan arctan) = D tan arctan·D arctan

= $(1 + \tan^2)$arctan·D arctan = $(1 + j^2) \cdot$ D arctan.

It follows that

$$D \arctan = \text{rec}(1 + j^2).$$

From the definition of the sine function we conclude by virtue of the Quotient Rule

$$2 \cdot D \, \sin(2 \cdot j) = D[\sin(2 \cdot j)]$$

$$= 2 \cdot [(1 + \tan^2) \cdot D \, \tan - 2 \cdot \tan \cdot D \, \tan \cdot \tan] \cdot \text{rec}(1 + \tan^2)^2$$

$$= 2 \cdot (1 - \tan^2) \cdot D \, \tan \cdot \text{rec}(1 + \tan^2)^2 = 2 \cdot (1 - \tan^2) \cdot \text{rec}(1 + \tan^2)$$

$$= 2 \cdot \cos(2 \cdot j).$$

It follows that $D[\sin(2 \cdot j)] = \cos(2 \cdot j)$. Substituting $\frac{1}{2} \cdot j$

we obtain

$$D \, \sin = \cos.$$

Similarly, we arrive at $D \cos = \text{neg sin}$. (It goes without saying that the symbols sin and cos are reserved for the functions defined in terms of the tangential function for which $D \tan 0 = 1$).

6. The Foundation of the Algebra of Antiderivatives.

The Algebra of Antiderivatives is based on an equivalence relation which we shall symbolize by $\sim$, and a right inverse of the operator D which we shall symbolize by S. We shall read the symbol Sf "an antiderivative of f" or "an integral of f" indicating by this expression the multi-valuedness of the operator S in contrast to the uni-valuedness of D. The latter is expressed in the implication

$$\text{If } f = g, \text{ then } Df = Dg$$

which will be of basic importance for the Algebra of Antiderivatives.

Sf is what in the classical analysis is denoted by $\int f(x) dx$ while $f \sim g$ expresses the relation $f'(x) = g'(x)$ for which the classical theory does not introduce a special symbol. Only to some extent $f \sim g$ corresponds to what in classical integral calculus is denoted by $f(x) = g(x) + \text{const}$. As we shall see in this section, $f = g + c$ implies $f \sim g$. However,

our Algebra of Antiderivation neither infers nor postulates
that conversely $f \sim g$ implies $f = g + c$. In classical analysis
the proof of the fact that functions with equal derivatives
differ by a constant, requires deeper logical methods than
the proof of any theorem corresponding to our Algebra of Analysis (see Introduction).

In view of the connection of our antiderivation with
the classical calculus of indefinite integrals, we shall call
f the integrand of Sf.

The two fundamental concepts of the Algebra of Antiderivation are introduced by the postulates:

A. $f \sim g$ if and only if $Df = Dg$

B. $D(Sf) = f$.

No ambiguity will arise if we write postulate B in the form
$DSf = f$ since we shall leave DS undefined. We might, of
course, express postulate B in the form $DS = j$. At the beginning of this section, in calling S a right inverse of D, we
adopted this point of view. But we shall refrain from elabor-
ating on this idea (as in the Algebra of Antiderivates we re-
frained from briefly writing D neg = neg D instead of
D neg f = neg Df) since its consistent extension would necessitate the use of functions of more variables.

From the definition A it follows that the equivalence relation is reflexive, symmetric, and transitive. Since $DO = 0$
and $D1 = 0$, we have $0 \sim 1$. In fact, for each constant c we
have $c \sim 0$. More generally, from the Algebra of Derivatives
it follows that $f + c \sim f$.

Next we consider two fundamental consequences of postulate B. If $Sf \sim g$, then $DSf = Dg$, thus by postulate B, $f = Dg$.

Conversely, if $f = Dg$, then from B it follows that $DSf = Dg$ and hence $Sf \sim g$. We thus see

      C. $Sf \sim g$ if and only if $f = Dg$.

Secondly, we see: If $SDf \sim g$, then $DSDf = Dg$ and from B it follows that $Df = Dg$. Hence $f \sim g$ and $g \sim f$. We thus obtain the result

      D. $SDf \sim f$.

Obviously, this Algebra of Antiderivation solves all the difficulties connected with the multi-valuedness of the operator S. In our formula, Sf stands for any function g for which $Dg = f$. The formulae concerning antiderivatives resulting from our two postulates of the Algebra of Antiderivation express only the equivalence (never the equality) of antiderivatives with functions or other antiderivatives. For instance, from $DO = Dc = 0$ it follows that $SO \sim c$. Clearly, also the classical integral calculus lacks formulae expressing the equality of any antiderivation and any other function.

7. Formulae of the Algebra of Derivation in the Notation of Antiderivation.

The formulae A. - D. of the preceding section enable us to translate each formula of the Algebra of Derivation into a formula about antiderivation. We start translating the postulates I - III of the Algebra of Derivation:

$$Sf + Sg \sim SD(Sf + Sg) \sim S(DSf + DSg) \sim S(f + g).$$
$$f \cdot g \sim SD(f \cdot g) \sim S(f \cdot Dg + g \cdot Df) \sim S(f \cdot Dg) + S(g \cdot Df).$$
$$fg \sim SD(fg) \sim S(Dfg \cdot Dg).$$

We thus obtain

  I' $S(f + g) \sim Sf + Sg$

  II' $f \cdot g \sim S(f \cdot Dg) + S(g \cdot Df)$

  III' $fg \sim S(Dfg \cdot Dg).$

Important is the special case of II' for $f = c$ and $g \sim Sh$. We obtain the Constant Factor Rule

   $c \cdot Sh \sim S(c \cdot h).$

Translating the formulae

  $D \exp = \exp,$ $D \text{ logabs} = \text{rec},$ $D \tan = \text{rec } \cos^2$

we obtain

  $S \exp \sim \exp,$ $S \text{ rec} \sim \text{logabs},$ $S \text{ rec } \cos^2 \sim \tan.$

From $D \ c - po = c \cdot (c - 1) - po$, it follows that

   $S[c \cdot (c - 1) - po] \sim c - po.$

Applying the Constant Factor Rule for $\frac{1}{c}$ (if $c \neq 0$) and replacing $c + 1$ by c, we obtain

   $S \ c - po = \frac{1}{c+1} \cdot (c+1) - po$ if $c \neq 0.$

8. The Three Methods of Antiderivation.

If in formula III' we replace f by Sh we obtain

  III^{*} $Shg \sim S(hg \cdot Dg).$

The formula III^{*} is the source of two methods for the computation of antiderivatives.

The first of these methods consists in applying formula III^{*} read from the right to the left, that is, in the form

  $S(hg \cdot Dg) \sim Shg.$

In words: If the integrand of an antiderivative which we wish to find, can be represented as the product of what results from a function h by substitution of a function g times the derivative of this function g, then we obtain the antiderivative we are looking for, by substituting g into the antiderivative of h. The problem of finding the antiderivative of hg·Dg is thus reduced to the problem of finding the antiderivative of h.

Examples:

$$S(rec \; g \cdot Dg) \sim logabs \; g$$
$$S(tan \; g \cdot Dg) \sim rec \; cos^2 g$$
$$S(exp \; g \cdot Dg) \sim exp \; g, \quad etc.$$

The second method, called antiderivation by substitution, consists in substituting into formula III[*], read from the left to the right, the right inverse of g which we shall denote by g^*. We obtain

$$Sh \; g \; g^* \sim S(hg \cdot Dg)g^*$$

thus

E. $\quad Sh \sim S(hg \cdot Dg)g^*$.

In words: We find the antiderivative of h by substituting into h any function g, multiplying the result by Dg, finding the antiderivative of the product, and substituting into this antiderivative the right inverse of g.

While formula E is correct for each h and g, it is of practical use for given h only if we can find a function g with a right inverse such that $S(hg \cdot Dg)$ is simpler than Sh.

Example:

$$Sh \sim S(h \; tan \cdot D \; tan)arctan.$$

The formula is useful if $S(h \; tan \cdot rec \; cos^2)$ is simpler than Sh. For instance, this is the case if $h = (-\frac{3}{2}) - po(1 + 2 - po)$, in classical notation, $h(x) = (1 + x^2)^{-3/2}$. We have

$$h \tan = (-\tfrac{3}{2}) - \text{po} (1 + \tan^2) = \cos^3,$$

$$h \tan \cdot \text{rec} \cos^2 = \cos,$$

$$S[(-\tfrac{3}{2}) - \text{po} (1 + 2 - \text{po})] \sim S \cos \arctan \sim \sin \arctan.$$

The third method, called antiderivation by parts, consists in an application of formula II', written in the following form

F. $S(f \cdot Dg) \sim f \cdot g - S(g \cdot Df).$

While this formula holds for each f and g, it is of practical use for the computation of an antiderivative Sh only if we succeed in representing h as the product of two functions f and f_1 such that

1) Sf_1 can be found,

2) $S(Df \cdot Sf_1)$ can be found.

If we set $Sf_1 \sim g$, then formula F enables us to compute $S(f \cdot f_1)$:

F'. $S(f \cdot f_1) \sim f \cdot Sf_1 - S(Df \cdot Sf_1).$

While it is immaterial which antiderivative of f_1 we use in the expression on the right side, it is essential that on both places the same antiderivative Sf_1 is selected.

Example:

$$S \text{ logabs} \sim S(\text{logabs} \cdot 1) \sim S(\text{logabs} \cdot Dj)$$
$$\sim j \cdot \text{logabs} - S(j \cdot \text{rec}) \sim j \cdot \text{logabs} - S1 \sim j \cdot \text{logabs} - j.$$

Tri-Operational Algebra

KARL MENGER

1. Classical and Algebraic Development of Geometry and Analysis.

In classical analysis we start with a field of numbers. A function is explicitly defined as a set of pairs of numbers $(x, f(x))$ in which x ranges over what is called the domain of f. Also equality, addition, multiplication, and substitution of functions are explicitly defined. The structure of classical analysis is somewhat comparable to that of classical projective geometry. Veblen and Young start with a set whose elements are called points, and then explicitly define lines and planes as certain sets of points. Also joining and intersecting are introduced in set-theoretical terms.

In contrast to the classical projective geometry, in our Algebra of Geometry[1] we start with one class of undefined elements, called flats, and two binary operations, joining and intersecting, subject to simple postulates. In terms of these operations we introduce the part relation and then define the dimension of flats, that is to say, we single out certain flats called points ("a point is that which has no parts"), and other flats, called lines and planes. Under certain conditions we can associate with each flat a set of points in such a way that these representing sets are the points, lines, and planes of a classical projective geometry and form an isomorphic representation of our algebra of geometry.

In an analogous approach to analysis [2] we start out with one class of entities denoted by $f, g, \ldots$, and three binary operations: addition, multiplication, and substitution, subject to simple postulates. We denote the operations of such a tri-operational algebra by $+$, $\cdot$, and juxtaposition, respectively, and never omit the multiplication dot in order to avoid confusion of multiplication and substitution. We denote the product of f and g by $f \cdot g$, and the element obtained by substituting g and f by fg. We make use of the notation of the theory of operators, and avoid superfluous parentheses.

In terms of the operations we are able to single out a class of elements which we call constant ("a constant is that which cannot be changed by substitution into it"). The constants play a role similar to that of numbers in classi-

---

[1] The development of this algebra of geometry since 1927 is outlined in the author's second lecture in the Rice Institute Pamphlets, 1940, p. 41-80.

[2] An elementary exposition is contained in the pamphlet "Algebra of Analysis," Notre Dame Mathematical Lectures, number 3, 1944. We refer to this booklet by "A. o. A."

3

4

cal analysis. Under certain conditions we can associate with each element of a tri-operational algebra a classical function (i.e., a set of pairs of constants) in such a way that these classical functions form an isomorphic representation of our algebra.

2. The Postulates of the Tri-Operational Algebra.

We assume the following laws:

	Addition	Multiplication	Substitution
Commutative Law	$f+g=g+f$	$f \cdot g = g \cdot f$	....
Associative Law	$f+(g+h)=(f+g)+h$	$f \cdot (g \cdot h)=(f \cdot g) \cdot h$	$f(gh)=(fg)h$
Distributive Laws [3]	$(f \cdot g)h = fh \cdot gh$	$(f+g)h=fh+gh$	$(f+g) \cdot h = f \cdot h + g \cdot h$
Neutral Elements	$f+0=f$	$f \cdot 1 = f$	$fj=jf=f$
	$0 \neq 1 \neq j \neq 0$		$10=1$
Opposite Elements	$f+(-f)=0$		

Here j corresponds, of course, to the function $f(x)=x$ of the classical theory. The formula $10=1$ represents the assumption that 1, the neutral element of multiplication, is not affected by the substitution of 0, the neutral element of addition. That this assumption is independent of the other postulates has been proved by Ferdinand L. Brown.[4]

In applying our postulates to classical functions we have to interpret them as formulae which are valid whenever all terms are meaningful. For instance, the function r defined by $r \cdot j = 1$ and corresponding to the classical function $f(x)=1/x$, does not admit the substitution of 0. The term $r0$ is not meaningful and thus the formula $(r+r)0 = r0 + r0$ not valid. However, our postulates are compatible with the assumption that the three operations are universal in the sense that $f+g$, $f \cdot g$, and fg exist for each f and g. In this paper we shall study tri-operational algebras satisfying this additional assumption.

In such an algebra, with regard to addition and multiplication the elements form a commutative ring with a unit. It should be noted that with regard to addition and substitution the elements do not form a ring since we have not postulated that $f(g+h)=fg+fh$ and this law does not generally hold in a tri-operational algebra. For instance, $1=10=1(0+0)$. If $1(0+0)$ were $=10+10=1+1$, then we should have $1=1+1$ which contradicts the assumption $1 \neq 0$.

In order to obtain examples of tri-operational algebras with universal operations we shall, for each $n>1$, denote by R_n the ring of residues modulo n,

---

(3) We have a distributive law for each pair of fundamental operations. In the column headed by substitution we have listed the ordinary distributive law of algebra of fields or, as we shall say, the additive-multiplicative distributive (a.m.d.) law. Under multiplication we have listed the additive-substitutive (a.s.d.), under addition the multiplicative-substitutive (m.s.d.) law.

(4) Cf. his paper "Remarks concerning tri-operational algebra" in this issue.

and by F_n the set n^n classical functions which are defined on R_n and assume values belonging to R_n. Each element of F_n can also be described as an n-tuple of pairs $(x, f(x))$ of elements of R_n, one pair for each element x of R_n. If for the functions of F_n we define addition, multiplication, and substitution in the classical way, we obtain a tri-operational algebra, denoted by A_n.

For instance, A_2 consists of the four functions f_0, f_1, f_2, f_3 defined for the two elements 0 and 1 of R_2, and assuming values 0 and 1:

$$f_0(0) = 0, \; f_0(1) = 0; \qquad\qquad f_1(0) = 1, \; f_1(1) = 1;$$
$$f_2(0) = 0, \; f_2(1) = 1; \qquad\qquad f_3(0) = 1, \; f_3(1) = 0.$$

Clearly, f_0, f_1, f_2 f_3 play the roles of the elements 0, 1, j, $1+j$, respectively.[5]

Another example of a universal tri-operational algebra is, for each given ring R, the system $P(R)$ of all polynomials over R (i.e., with coefficients belonging to R) for which addition, multiplication, and substitution are defined in the usual way.

3. Constants.

We define: an element c is constant if $c = c0$. While j is not constant, 0 and 1 are. The constants are easily seen[6] to form a ring which is closed with respect to substitution; that is, if c is a constant, then fc and cf are constants for each f. In fact, $cf = (c0)f = c(0f) = c0 = c$. If for each constant $c \neq 0$ there exists an element c' such that $c \cdot c' = 1$, then the constants form a field. For the element c' is a constant since

$$c' = c' \cdot 1 = c' \cdot 10 = c' \cdot (c \cdot c')0 = c' \cdot (c0 \cdot c'0) = c' \cdot (c \cdot c'0) =$$
$$= (c' \cdot c) \cdot c'0 = 1 \cdot c'0 = c'0.$$

If the constants form a field,[7] then the postulate $10 = 1$ can be replaced by the much weaker assumption $10 \neq 0$. First, it is clear that 10 is a constant. For from $0 + 0 = 0$, by the a.s.d. law, it follows that $00 + 00 = 00$. In view of the existence of an element $- 00$ it follows that $00 = 0$ (in words: that 0 is a constant). Now $10 = 1(00) = (10)0$ which proves that 10 is a constant. Assuming that $10 \neq 0$ and that the constants form a field, we are sure of the existence of a constant c' such that $c' \cdot 10 = 1$. Since c' is a constant we have $c' = c'0$ and thus $1 = c' \cdot 10 = c'0 \cdot 10 = (c' \cdot 1)0 = c'0 = c'$. But $c' = 1$ in view of $c' \cdot 10 = 1$ implies $10 = 1$, that is, our former postulate.

The roots of an algebraic equation with constant coefficients need not be

---

[5] Tables for addition, multiplication, and substitution valid in A_2 are contained in "A. o. A", p. 9.

[6] "A. o. A", p. 11.

[7] The totality of (constant and non-constant) elements of a tri-operational algebra cannot be a field. Cf. "A. o. A", p. 15.

6

constant. In A_2 (see section 2) we have $j \cdot j = j$. Thus j satisfies the equation $1 \cdot j + 1 \cdot j^2 = 0$ without being constant.

4. Representations of Tri-Operational Algebras by Classical Functions.

With each element of a tri-operational algebra, A, we may associate a classical function viz., the set of all pairs (c, fc) with c ranging over all constants of A. We thus obtain a homomorphic representation of A since the classical functions representing $f_1 + f_2$, $f_1 \cdot f_2$, and $f_1 f_2$ can be obtained from those representing f_1 and f_2 by classical addition, multiplication and substitution, respectively. The necessary and sufficient condition in order that this representation of A be isomorphic, is that $fc = gc$ for each constant c imply $f = g$. We may express this condition by saying that *the constants form a base of the algebra*. In classical analysis this condition is satisfied by virtue of the explicit definition of the equality of f and g. For a tri-operational algebra the assumption that the constants form a base, is independent of our other postulates. The system $P(R_2)$ of all polynomials over the field of residues modulo 2 contains only two constants, 0 and 1, and yet infinitely many elements.[8] Each of these polynomials is equivalent to one of the functions f_0, f_1, f_2, f_3 of A_2. By this we mean, that each polynomial belonging to $P(R_2)$ assumes the same values on R_2 as one of the four functions f_i of A_2.

Thus tri-operational algebra generalizes classical analysis.

5. The Structure of Tri-Operational Algebras.

For each tri-operational algebra A we denote by CA the ring of constants of A, by $C'A$ the sub-ring of CA consisting of all integers of A, i.e., all elements of the form $1 + 1 + \ldots + 1$. By a *polynomial* of A we mean an element of the form $c_0 + c_1 \cdot j + \ldots + c_n \cdot j^n$ where $c_0, c_1, \ldots, c_n$ belong to CA and j^k stands for a product of k factors j. If $c_0, c_1, \ldots, c_n$ are integers, we shall speak of an *integral polynomial*. Since by adding and multiplying two polynomials (two integral polynomials) as well as by substituting one into the other we obtain a polynomial (an integral polynomial), the polynomials form a tri-operational sub-algebra of A which we shall denote by $\Pi(A)$ and which, in turn, contains the tri-operational sub-algebra of all integral polynomials, denoted by $\Pi'(A)$. The algebra $\Pi(A)$ has the same constants, the algebra $\Pi'(A)$ the same integers as A. If $A = \Pi(A)$ or $= \Pi'(A)$, then we call A a *polynomial algebra* or an *integral polynomial algebra*, respectively. The system $P(R)$ of all polynomials over the ring R is a polynomial algebra. For each algebra A, the coefficients of the polynomials of $\Pi(A)$ and $\Pi'(A)$ belong to the rings CA and $C'A$, respectively. We shall see that $\Pi(A)$ and $\Pi'(A)$ are isomorphic with sub-algebras of $P(CA)$ and $P(C'A)$, respectively.

If a polynomial $p = a_0 + a_1 \cdot j + \ldots + a_n \cdot j^n$ of $\Pi'(A)$ is $= 0$, by substituting $0, 1, \ldots, k$ into p we obtain $k + 1$ linear homogeneous relations between

[8] Cf. "A. o. A", p. 13 sq.

the integers a_0, a_1, ..., a_n. Consequently, if each two integers of $C'A$ are different, then $p=0$ is impossible except when $a_0=a_1=\ldots=a_n=0$. It also follows that if $C'A$ is a field of characteristic k, then $p=0$ is impossible for $n<k$.

If each two integers of A are different, then $C'A$ is isomorphic with the ring R' of all ordinary integers. Each two polynomials of $\Pi'(A)$ with non-identical coefficients are different since otherwise their difference would be a polynomial equal to 0. Hence in this case $\Pi'(A)$ is isomorphic with the ring $P(R')$ of all ordinary integral polynomials with ordinary substituion.

If $C'A$ is a field of characteristic k, then at least the k^k polynomials $a_0+a_1\cdot j+\ldots+a_{k-1}\cdot j^{k-1}$ are different. The example of the integral polynomial algebra $A=P(R_2)$ shows that even if $C'A$ is the field of characteristic 2, each two polynomials of $\Pi'(A)$ may be different.

6. The Fundamental Polynomial.

If $C'A$ is a finite field, and two polynomials of $\Pi'(A)$ with non-identical coefficients are equal, then their difference is $=0$. In the integral polynomial algebra $P(C'A)$ consisting of the polynomials over $C'A$ there exist polynomials of lowest degree which are $=0$. Since $C'A$ is a field, only in one of these polynomials the coefficient of the highest power of j is $=1$. This polynomial will be called the *fundamental* polynomial of $\Pi'(A)$. From the impossibility of $n<k$, proved in the preceding section, it follows that the degree of the fundamental polynomial cannot be lower than the characteristic of the field $C'A$. The polynomial of lowest degree of $P(C'A)$ which can be fundamental in a tri-operational algebra A in which $C'A=R_k$ (or, as we shall briefly say, in an tri-operational algebra *over* R_k') is $-j+j^k$. In the case $k=2$, if $-j+j^2$ is fundamental, then $\Pi'(A)$ is isomorphic with the algebra A_2 considered in section 2.

Let R_k denote the field of characteristic k, and p a polynomial belonging to $P(R_k)$. In order that p be fundamental in some integral polynomial algebra A over R_k it is necessary and sufficient *that for each polynomial* q *over* R_k *the polynomial* pq *obtained by substituting* q *into* p *is divisible by* p. An equivalent condition is that pq be divisible by p for each of the polynomials q *which are residues modulo* p, that is, for the k^n polynomials $q=a_0+a_1\cdot j+\ldots,+a_{n-1}\cdot j^{n-1}$. For if this condition is satisfied, then for each q there exists a polynomial $r(q,p)$ such that $pq=p\cdot r(q,p)$ and $p=0$ implies that $pq=0$. If the condition is not satisfied, then for some polynomial q there exist two polynomials r and s, the latter of a lower degree than p, such that $pq=p\cdot r+s$. The assumption $p=0$ would imply that $pq=0q=0$, thus $s=0$ which contradicts the assumption that p is the polynomial of lowest degree which is $=0$.

From the above characterization of the fundamental polynomials over R_k it follows that *if* p_1 *and* p_2 *are fundamental in two algebras* A_1 *and* A_2 *over* R_k, *respectively, then their product* $p_1\cdot p_2$ *is fundamental in some algebra* A_3 *over* R_k. For

$(p_1 \cdot p_2)q = p_1 q \cdot p_2 q = [p_1 \cdot r(q,p_1)] \cdot [p_2 \cdot r(q,p_2)] = (p_1 \cdot p_2) \cdot r(q,p_1 \cdot p_2).$
We shall see that $p_1 + p_2$ need not be fundamental in any algebra over R_k.

7. Examples of Algebras Over R_2.

We shall study algebras over R_2, the field of residues modulo 2. In order that a binomial $j^m + j^n$ be fundamental it is necessary and sufficient that m and n be powers of 2. The necessity of the condition is seen by substituting $1 + j$. The polynomial $(1+j)^m + (1+j)^n$ is not divisible by $j^m + j^n$ unless m and n are powers of 2. The condition is sufficient for if m and n are powers of 2, then for each polynomial q the polynomial $(j^m + j^n)q = q^m + q^n$ is divisible by $j^m + j^n$. The only polynomials of degrees < 10 which are fundamental in algebras over R_2, besides the six binomials $j + j^2, \ldots, j^4 + j^8$, are

$$(j + j^2)^3, \quad (j + j^2) \cdot (j + j^4), \quad (j + j^2)^2 \cdot (j + j^4).$$

That these polynomials are fundamental follows from the fact that they are products of fundamental binomials. That other polynomials cannot be fundamental is, with three exceptions, seen by substituting $1 + j$ since the resulting polynomials are not divisible by q. The three exceptions are

$$q_1 = j + j^3 + j^5 + j^6,$$
$$q_2 = j + j^2 + j^4 + j^8,$$
$$q_3 = j + j^2 + j^3 + j^4 + j^5 + j^6.$$

For each of these polynomials q_i, the polynomial $q_i(1+j)$ is divisible by q_i. However, $q_1 j^5$ is not divisible by q_1. We prove this as follows: $q_1 = 0$ implies that

$$j^6 = j + j^3 + j^5, \quad j^7 = j^2 + j^4 + j + j^3 + j^5, \quad j^8 = j + j^2 + j^4.$$
Thus $j^{16} = j^2 + j^4 + j^8 = j$. Using this fact we see that

$$q_1 j^5 = j^5 + j^{15} + j^{25} + j^{30} = j^5 + j^{10} = j^5 + j^2 \cdot (j + j^2 + j^4) = j + j^4.$$
Hence $q_1 j^5$ is not divisible by q_1. Similarly, $q_2 j^3$ is not divisible by q_2, and $q_3(1 + j + j^2)$ is not divisible by q_3. It is remarkable that q_2 is not fundamental though it is the sum of two polynomials which are fundamental, namely, $j + j^2$ and $j^4 + j^8$.

8. Algebras Over R_4.

If for a tri-operational algebra A the ring $C'A$ of all integers contains divisors of 0, then A may have a fundamental polynomial and yet contain infinitely many elements. Moreover the degree of the fundamental polynomial may be a divisor of the characteristic of $C'A$. We illustrate these possibilities in the case of tri-operational algebras over R_4.

$P(R_4)$ contains three elements which are equivalent to 0, that is to say, yield 0 after the substitution of each of the four constants 0, 2, 1, -1. These elements are

$$2 \cdot j^2 + 2 \cdot j^3, \quad 2 \cdot j + 2 \cdot j^3, \quad 2 \cdot j + 2 \cdot j^2.$$

The sub-algebra of $P(R_4)$ with the fundamental polynomial $-j+j^4$ thus consists of 64 quadruples of equivalent elements. If we set

$$t_0=0,\ t_1=j^2+j^3,\ t_2=j+j^3,\ t_3=j+j^2,$$

then among the 64 polynomials

$$c_0+c_1\cdot s+t_i\ (i=0,1,2,3)$$

there is exactly one representative of each of these quadruples.

If we set $s=j+j^2+j^3$, then the 16 elements $c_0+c_1\cdot s$ (where c_0 and c_1 are constants) form a tri-operational sub-algebra of $P(R_4)$.

The 256 elements of the sub-algebra of $P(R_4)$ with the fundamental polynomial $-j+j^4$ can be represented in the form

$$c_0+c_1\cdot s+t_i+2\cdot t_j\ (i,j=0,1,2,3).$$

In the tri-operational algebra A_4 consisting of the 4^4 classical functions modulo 4 we find 64 elements forming a tri-operational sub-algebra which is isomorphic with the sub-algebra of $P(R_4)$ having the fundamental polynomial $-j+j^4$. Besides, A_4 contains four elements $u_i(i=0,1,2,3)$ such that $2u_i=t_i(u_0=0)$ and that the 256 elements of A_4 can be represented in the form

$$c_1+c_2\cdot s+t_i+u_j\ (i,j=0,1,2,3).$$

The polynomial $t_3=2\cdot j+2\cdot j^2$ satisfies the following condition: For each polynomial q over R_4 the polynomial $t_3q=2\cdot q+2\cdot q^2$ is divisible by t_3. Hence t_3 is fundamental in an algebra over R_4. This algebra contains infinitely many elements. The degree of t_3 is a divisor of the characteristic 4 of R_4.

9. About Extensions.

If A is a sub-algebra of A', and g' a function of A' not belonging to A, then A' contains all functions $f_1+f_2\cdot g'$ for f_1 and f_2 in A. It is remarkable that *if the constants of A form a field, not all functions of A' can be of the form* $f_1+c\cdot g'$ *where c is a constant.* For if we had

$$j\cdot g'=f_1+c_1\cdot g'\ \text{and}\ g'(1+j)=f_2+c_2\cdot g',$$

then by substituting $j-1$ into the last relation we see that

$$g'=f_2(j-1)+c_2\cdot g'(j-1).$$

Since $f_2(j-1)$ does and g' does not belong to A, we have $c_2\neq0$. Now

$$(j\cdot g')(1+j)=j(1+j)\cdot g'(1+j)=(1+j)\cdot(f_2+c_2g')=f_3+(c_2+c_1\cdot c_2)\cdot g'$$

for some f_3 of A. On the other hand,

$$(j\cdot g')(1+j)=(f_1+c_1g')(1+j)=f_1(1+j)+c_1g'(1+j)=f_4+c_1\cdot c_2g'$$

for some f_4 of A. It would follow that $c_2\cdot g'=f_4-f_3$ which is impossible.

As an example we consider A_2 whose four elements, as far as addition and multiplication are concerned, form the ring of hypercomplex numbers $a_0+a_1\cdot j$

over C_2, where $j^2 = j$. While this ring can easily be extended to a ring of hypercomplex numbers $a_0 + a_1 \cdot j + a_2 \cdot k$ over C_2 where $j \cdot j = j$, $k \cdot k = k$, $j \cdot k = k \cdot j = 0$, according to our theorem it is impossible to extend the tri-operational algebra A_2 to one whose elements would be of the form $a_0 + a_1 \cdot j + a_2 \cdot k$.

We remark that A_2 does admit an extension to a tri-operational algebra whose elements can be written as hypercomplex numbers over A_2, that is, in the form $\sigma_0 + \sigma_1 \cdot l$ where σ_0 and σ_1 belong to A_2. We obtain such an extension by setting $l \cdot l = l$ and defining substitution into σ and l as follows: For any elements σ, τ_0, τ_1 of A_2 we set

$$l(\tau_0 + \tau_1 \cdot l) = (\tau_0 0 + \tau_0 1) \cdot (1 + \tau_1 0) \cdot (1 + \tau_1 1) \cdot l$$
$$\sigma(\tau_0 + \tau_1 \cdot l) = \sigma 0 + (\sigma 0 + \sigma 1) \cdot (\tau_0 + \tau_1 l)$$

where $\sigma 0$ denotes the element of A_2 obtained by substituting 0 into σ. Using these definitions, substitution into $\sigma_0 + \sigma_1 \cdot l$ is defined in accordance with special cases of the a.s.d. and m.s.d laws as follows

$$(\sigma_0 + \sigma_1 \cdot l)(\tau_0 + \tau_1 \cdot l) = \sigma_0 (\tau_0 + \tau_1 \cdot l) + \sigma_1 (\tau_0 + \tau_1 \cdot l) \cdot l(\tau_0 + \tau_1 \cdot l)$$

One can easily verify that this system of 16 hypercomplex numbers with this definition of substitution is a tri-operational algebra. The 16 elements can also be written as hypercomplex numbers

$$a_0 + a_1 \cdot j + a_2 \cdot k + a_2 \cdot l$$

over R_2 if we set

$$k \cdot k = k, \ j \cdot k = k \cdot j = j \cdot l = l \cdot j = k \cdot l = l \cdot k = l \cdot l = l.$$

10. Algebra of Analysis.

We obtain the formulae of differential calculus by assuming an operator D connected with the three operations of the algebra of functions by the laws

1) $D(f+g) = Df + Dg$, 2) $D(f \cdot g) = f \cdot Dg + g \cdot Df$, 3) $D(fg) = (Df)g \cdot Dg$.

The theory of elementary functions can be developed if we postulate an exponential function e and a tangential function t for which

$$e(f+g) = ef \cdot eg, \ (De)0 = 1$$

$$t(f+g) = \frac{tf + tg}{1 - tf \cdot tg}, \ (Dt)0 = 1.$$

For the details of this theory the reader is referred to the booklet *Algebra of Analysis*, Chapter II.

General Algebra of Analysis

KARL MENGER

Introduction

The classical presentation of projective geometry starts with a class of undefined entities, called points. Lines are introduced as undefined sets of points satisfying some basic postulates. Planes are defined as sets of points with certain properties. With the transition to each higher dimension new concepts must be introduced. The fundamental operations of projective geometry, that is, joining and intersecting, are set-theoretically defined. The algebraic properties of these operations are immediate consequences of these definitions on which the classical theory does not elaborate.

In 1927 we developed projective and affine geometry in terms of only one class of undefined entities, which we may call "flats," and of two undefined binary operations, called the joining and intersecting of flats. We postulated simple algebraic properties of these operations, such as commutative, associative, and absorptive laws. On the basis of these assumptions we introduced a part relation for flats. This concept of part enabled us to incorporate into the deductive development of geometry Euclid's famous definition "A point is that which has no parts" and subsequently to define lines, planes and k-dimensional flats. Our Algebra of Geometry removed from the foundations of geometry a great deal of set theory which was explicitly and implicitly contained in the then current developments. It synthesized geometries of different dimensions and various types, as well as non-geometrical theories.[1]

Analysis in some ways presents definite analogies to classical projective geometry. We start out with a class of entities called numbers which, for some purposes of analysis, remain undefined. A function is defined as a set of ordered pairs of numbers (x, y) of a certain type, namely, such that for every two pairs (x_1, y_1) and (x_2, y_2) of the set

$$y_1 \neq y_2 \text{ implies } x_1 \neq x_2.$$

Thus for each x which is the first number of a pair belonging to the set, or, as we briefly say, for each number x of the domain of the function, there is exactly one number y which we shall call $f(x)$, such that the pair $(x, f(x))$ belongs to the set.

The fundamental operations of analysis, that is, addition, multiplication, and substitution, are set-theoretically defined. For instance, by the sum of the

---

1 Jahresber. d. D. Math. Ver., 37, 1928, pp. 309-325. In this country, the algebraic approach has become known under the name of lattice-theoretical foundation of projective geometry.

(46)

two functions (x, y) and (x, y') having the same domain, we mean the function of $(x, y + y')$, that is,

$$(x, f(x)) + (x, g(x)) = (x, f(x) + g(x)).$$

The algebraic properties of these operations are immediate consequences of the definitions on which classical analysis does not elaborate. Addition and multiplication have the same properties as the corresponding arithmetical operations which have been made the objects of extensive studies in modern algebra. In comparison, substitution, the typically analytical operation, has been neglected. The theory of operators contains the first systematic study of this operation.

The transition to functions of more variables is marked by the introduction of new concepts. A function of three variables is a set of ordered triples of numbers (x, y, z) such that for two triples (x_1, y_1, z_1) and (x_2, y_2, z_2) of the set

$$z_1 \neq z_2 \text{ implies either } x_1 \neq x_2 \text{ or } y_1 \neq y_2,$$

in other words, a set of ordered triples $(x, y, f(x, y))$.

The analogy of the situation in analysis with that in the classical projective geometry suggests a development of analysis analogous to our algebra of geometry.[2] We do away with the fundamental distinction between numbers on one hand, and functions defined as certain sets of pairs, triples, . . . of numbers on the other. We start out with one class of undefined entities, called *functions*, and three undefined operations, called *addition, multiplication*, and *substitution*, and denoted by $+$, $.$, and juxtaposition, respectively. We assume that addition and multiplication of functions follow the laws of a ring. About substitution we assume that it is associative and linked to addition and multiplication by the two following distributive laws:

$$(f + g)h = fh + gh, \quad (f \cdot g)h = fh \cdot gh.$$

We shall refer to them as the additive-substitutive and the multiplicative-substitutive (or briefly, a.s.d. and m.s.d.) laws, respectively.

We have $Oh = O$ *and* $(-f)h = -(fh)$ *for every* h. For $Oh = (O + O)h = Oh + Oh$, from which it follows that $Oh = O$. Consequently $fh + -(fh) = O = Oh = (f + -f)h = fh + (-f)h$, from which it follows that $-(fh) = (-f)h$.

Let h be any given element of A. We form the sets $A*h$ and $hA*$ consisting of all elements f of A for which $fh = f$ and $hf = f$, respectively.

Each set $A*h$ *is a ring which, for each of its elements, f, and for each element, g of A, contains gf. Moreover, $fg = f$ if h belongs to $A*g$.* For, if f and f' belong to $A*h$, then so do $f + f'$ and $f \cdot f'$ since $(f + f')h = f + f'$ and $(f \cdot f')h = f \cdot f'$; so do O and $-f$ since $Oh = O$ and $(-f)h = -(fh)$

---

2 Cf. our booklet "Algebra of Analysis," Notre Dame Math. Lectures, 3, 1944, and "Tri-operational Algebra," Reports of a Math. Colloquium, 5-6, 1945. In the sequel we shall refer to these publications as "A.o.A" and "T.O.A."

(47)

$= -f$; and so does gf since $(gf)h = g(fh) = gf$. If h belongs to A*g we have $fg = (fh)g = f(hg) = fh = f$. Similarly we see that

Each set hA*, for each of its elements, f, and for each element, g, of A, contains fg.

The elements of A*O, that is, the functions for which $c = cO$, will be called *constant functions* or *constants* (l.c.[2]). From the above results it follows that A*O is a ring containing gc and cg for every g. In fact, $cg = c$ for every g since O belongs to every A*g. After having done away with numbers as primitive entities in terms of which functions are set-theoretically defined, we thus have singled out algebraically the constant functions. They correspond with the classical constant functions which in turn are in one-to-one correspondence with the class of numbers. For with each number c there corresponds the classical constant function assuming the value c for each x, and each constant function in this way corresponds with exactly one number, viz., the value of the function for every x.

The analogy between the definition of constants in terms of the three operations on one hand, and the definition of points in terms of joining and intersecting on the other, is obvious. In the last section of this paper we shall see how the algebraic point of view synthesizes the theories of functions of different numbers of variables, as it has synthesized geometries of different dimensions.

With regard to the elimination of set theory from the foundations, the analogy between analysis and geometry is only partial. In projective and affine geometry, the algebraic point of view covers all entities normally studied. In the algebra of analysis, we face the following alternative.

The first possible interpretation of our algebra is the *literal* one which implies that *each* two functions can be added, multiplied, and substituted into each other. In this case, we must exclude such functions as the logarithm and the reciprocal into which the constant function O cannot be substituted and which, strictly speaking, cannot even be added to a constant function, since their domains differ from the universal domain of the constant functions.[3]

While from the classical point of view, a class of functions, each two of which can be added, multiplied, and substituted into each other, is very restricted, an algebra of such functions is capable of numerous realizations. Besides the obvious polynomial model there exist both infinite and finite algebras of analysis, the latter ones somewhat reminiscent of the finite projective geometries.

A second way in which the results of our algebra may be interpreted is as *formulae which are valid whenever all terms are meaningful*, a situation com-

---

3 If besides the constant functions with the universal domain we introduced constant functions with restricted domains, we should have to supplement the algebra of functions by an algebra of domains and range of functions and thus face the normal set-theoretical problems of analysis.

(48)

parable to that of a grouppoid in which the associative law holds for those triples of elements for which the operations indicated in the associative formula can be performed.

In both cases we obtain analogues of the formulae of analysis without making use of variables. We have developed ("A . o . A", l. c.²) a fairly complete formulary of Calculus—an uncompromising *Calculus without Variables*. We mentioned l.c.² the affinity of our theory to the theory of operators.

After "A.o.A". and "T.O.A." (l.c.²) had been published it came to our attention that the theory of substitution developed in these publications in some respects resembled also the treatment of substitution contained in the interesting logical work of Schönfinkel[4] and Curry[5]. These authors substitute functions rather than variables into functions and eliminate variables from logic. They introduce an operator I which is neutral with regard to substitution as is the function j in our theory. However, we differ from the logicians in our treatment of constants and of functions of more variables while the tri-operational approach and the systematic development of a formulary of analysis without variables lay outside the domain covered in their papers.[6]

In the present paper, in section 1, we summarize some of our previous results concerning functions of one variable adding remarks concerning the representation of the elements of our algebra by functions in the classical sense. We describe in more detail the concept of the fundamental polynomial of a polynomial algebra. Next we outline theories of functions of rank 2 corresponding to classical functions of two variables (Section 2) and of functions of rank Σ (Section 3). We then develop a general Algebra of Analysis synthesizing all these theories (Section 4).

1. Tri-operational Algebra

We shall speak of a tri-operational algebra if we have a class of entities with three associative operations, such that addition and multiplication are connected with substitution by the a.s.d. and m.s.d. laws, and with each other by the a.m.d. laws

$$(f + g) . h = f . h + g . h, \quad h . (f + g) = h . f + h . g,$$

if addition is commutative and there exist a neutral element O with regard to addition, and a negative —f of each f.[7]

---

4 Math. Annalen 92 (1924) p. 305.

5 Am. Journal of Math. 52 (1930) pp. 508 and 789.

6 Except for the interesting remarks of Curry in Journal of Symbolic Logic, 7 (1942) p. 50.

7 At the suggestion of M. Mannos we drop our original postulates concerning neutral elements with regard to multiplication and substitution. About these "accessory postulates" cf. F. L. Brown's paper, these Reports 7. We will not postulate the commutative law for multiplication in order to cover classical functions of quaternions, nor for substitution since substitution of real functions is not commutative. It should be noted that a tri-operational algebra is not an additive-substitutive ring since we do not postulate $h(f + g) = hf + hg$.

(49)

Each tri-operational algebra, A, is homomorphic with a set of classical functions having the common domain A. With each element f of A we associate the set of ordered pairs (h, fh) of all elements h of A. With

$$f + g, \ O, \ -f, \ f \cdot g, \ fg$$

we associate the sets of pairs

$$(h,(f+g)h), \ (h,O), \ (h,(-f)h), \ (h,(f \cdot g)h), \ (h,(fg)h),$$

respectively. In this way we obtain a homomorphism which we shall call the *natural representation* of A.

A subset A' of A will be called a *basis of isomorphism* of the natural representation if

$$fh' = gh' \text{ for every } h' \text{ of } A', \text{ implies } f = g.$$

If the natural representation is isomorphic, then certainly A itself is a basis of the isomorphism. In the opposite case, there does not exist such a basis. As an example we mention the ring of all polynomials

$$p(x) = a_0 + a_1 x + \ldots + a_n x^n$$

(the coefficients belonging to any ring) with addition and multiplication defined in the customary way while for every polynomial $p(x)$, we consider a_0 as the result of substitution of any polynomial into $p(x)$. This definition satisfies all our assumptions with regard to substitution. The natural representation associates the same classical functions (h, fh) and (h, gh) with every two polynomials f and g having the same constant term.

We take a considerable step toward classical analysis by postulating *that the natural representation is an isomorphism*. However, in classical analysis a much stronger condition is satisfied. The classical definition of equality of functions stipulates that $f = g$ if and only if $f(x) = g(x)$ for every number x belonging to the identical domains of f and g. This definition is reflected by the following postulate:

The set A*O of all constants is a basis of isomorphism.

An example of an algebra not satisfying this postulate while possessing an isomorphic natural representation, is the ring of the polynomials with coefficients mod. 2 and ordinary addition, multiplication, and substitution, provided that two polynomials are considered equal if any only if they have identical coefficients. The only constants are O and 1. Hence each of the infinitely many polynomials belongs to one of four classes according to whether the substitution of O and 1 into the polynomial results in O, O or O, 1 or 1, O or 1, 1. Yet, for any two different polynomials, f and g, there exists a polynomial h such that $fh \neq gh$.

Let R by any given ring of constants, i.e., a subring of A*O. We shall call a function p a *generalized polynomial* in R if, for every function f of A,

(50)

the function pf belongs to the ring generated by adjoining f to R, that is to say, if

$$pf = a_0 + a_1 . f + \ldots + a_n . f^n$$

for some elements $a_0, a_1, \ldots, a_n$ of R, where f^k is an abbreviation for a product with k factors f. We shall call p a *polynomial* in R if the number n and the $n + 1$ elements a_i of R are the same for every f of A.

We take a further step towards the classical situation by assuming the existence of *neutral elements*, 1 and j, with regard to multiplication and substitution, respectively,

$$1 . f = f . 1 = f \text{ and } jf = fj = f \text{ for every } f.$$

In presence of a function j, a function p is a polynomial in R if and only if $p = a_0 + a_1 . j + \ldots + a_n . j^n$ for some integer n and $a_0, \ldots, a_n$ belonging to R, and each generalized polynomial is a polynomial.

If $j = O$, then we have $f = jf = Of = O$ for every f. Hence $j \neq O$ unless A consists of O only. If $j \neq O$, then $jO = O \neq j$. While we thus see that j is not constant, our postulates do not imply that 1 is constant.[8] If in the ring consisting of O, 1, j, $1 + j$ for which $f + f = O$ and $f . f = f$, we define substitution in such a way that $(1 + j) (1 + j) = O$, then we obtain an algebra in which O is the only constant, j the only polynomial. In particular, we obtain $1O = O$, $11 = 1j = 1$, $1(1 + j) = 1 + j$.

By an *integer* of a tri-operational algebra A we shall mean an element which is the sum of a finite number of ones, $1 + 1 \ldots + 1$, or the negative of such a sum. In presence of the postulate $1O = 1$, the integers form a subring, I, of the ring of all constants of A. The ring I either contains infinitely many different elements in which case I is isomorphic with the ring $I\infty$ of all ordinary integers, or a finite number, n, of different elements in which case I is isomorphic with the ring I_n of the residues $O, 1, \ldots, n-1$ mod. n. The ring I_n is a field if and only if n is a prime number.

The polynomials of A with integer coefficients form a subring, Q, of the ring of all polynomials of A. If each two polynomials with non-identical coefficients are different, then Q is isomorphic with the ring of all ordinary polynomials over I, two polynomials being considered as equal if and only if they have identical coefficients. If two polynomials with non-identical coefficients are equal, then Q contains a polynomial $a_0 + a_1 . j + \ldots + a_k . j^k$ which is $= O$ though $a_k \neq O$. Then I must be a finite ring.

Let us now study a given algebra, A, in which I is the field I_P for some prime number p. Then let n denote the minimum degree of a polynomial which is $= O$ although not all of its coefficients are $= O$. Exactly one polynomial of degree n and highest coefficient 1 is $= O$, say,

$$f = a_0 + a_1 . j + \ldots + a_{n-1} . j^{n-1} + j^n.$$

---

[8] Cf. F. L. Brown, Reports of a Math. Coll. 5-6, (1944) p. 13.

(51)

We call f the *fundamental polynomial* of A. We have $n \geqslant p$ since by substituting the constants $O, 1, \ldots, p-1$ into f we obtain p relations between the $n+1$ integers $a_0, a_1, \ldots, a_{n-1}, 1$.

For each polynomial g, the polynomial fg is divisible by f. For if for some g we have $fg = p \cdot f + q$ where q is of lower degree than f, then from $f = O$ it follows that $fg = O$ and $p \cdot f = O$, and thus $q = O$. Since by assumption, f is a polynomial of lowest degree which is $= O$, it follows that all coefficients of q are O, thus $fg = p \cdot f$.

On the other hand, if the polynomial f with coefficients belonging to I_P (the highest coefficient being $= 1$) has the property that, for every polynomial g with coefficients belonging to I_P, the polynomial fg is divisible by f, then we can exhibit a tri-operational algebra A, in which the integers form the field I_P and f is the fundamental polynomial. Let A be the ring of the p^n polynomials.

$$p = c_0 + c_1 \cdot j + \ldots + c_{n-1} \cdot j^{n-1}$$

with coefficients $c_0, c_1, \ldots, c_{n-1}$ belonging to I_P which are the residues modulo f. If we define substitution in the normal way, namely,

$$pq = c_0 + c_1 \cdot q + \ldots + c_{n-1} \cdot q^{n-1}$$

for every q, then A is a tri-operational algebra. For A is a ring in which substitution is associative and the a.s.d. and m.s.d. laws hold. It thus remains for us to show:

If $q_1 \equiv q_2$ (mod. f), then $q_1 q \equiv q_2 q$ and $q q_1 \equiv q q_2$ (mod. f) for every q.

From $q_2 = q_1 + p \cdot f$, by the a.s.d. and m.s.d. laws, it follows that $q_2 q = q_1 q + pq \cdot fq$. Since fq is divisible by f, we have

$$q_2 q \equiv q_1 q \quad (\text{mod. } f)$$

We further see that for every k

$$q_2{}^k = (q_1 + p \cdot f)^k \text{ and } q_2{}^k \equiv q_1{}^k \text{ (mod. f).}$$

If $q = b_0 + b_1 \cdot j + \ldots + b_{n-1} \cdot j^{n-1}$, then

$qq_1 = b_0 + b_1 \cdot q_1 + \ldots + b_{n-1} \cdot q_1{}^{n-1},$

$qq_2 = b_0 + b_1 \cdot q_2 + \ldots + b_{n-1} \cdot q_2{}^{n-1},$

and hence $qq_1 \equiv qq_2$ (mod. f).

Clearly f is the fundamental polynomial in this tri-operational algebra A.

We thus see: For each field I_P, *in order that the polynomial*

$$f = a_0 + a_1 \cdot j + \ldots + a_{n-1} \cdot j^{n-1} + j^n$$

with coefficients belonging to I_P be the fundamental polynomial of some tri-operational algebra in which the integers form the field I_P, it is necessary and sufficient that

$$fg \equiv O \quad (\text{mod. } f)$$
$$(52)$$

for every polynomial g with coefficients belonging to I_P or, what is equivalent, for each of the p^n residues modulo f.[8a]

A polynomial f has this property if and only if[9] f is the least common multiple of polynomials of the form $(jp^n - j)^m$ for various positive integral values of n and m.

We obtain an Algebra of Calculus[10] by assuming the existence of an operator D which is connected with the three operations by the following laws

$$D(f + g) = Df + Dg$$
$$D(f \cdot g) = f \cdot Dg + g \cdot Df$$
$$D(fg) = (Df)g \cdot Dg$$

and such that we do not have $Df = O$ for every f.

In a polynomial tri-operational algebra with integers forming the field I_P such an operator exists if and only if[11] the fundamental polynomial of the algebra is a power of a polynomial which is fundamental in another algebra with integers forming the field I_P.

2. Functions of Rank 2

In what follows we shall develop various algebraic aspects of functions of more variables. We begin by outlining a theory of what we call functions of rank 2. These entities correspond to functions of two variables.

Let A_2 be a class of elements f, g, . . . , such that with every f and every ordered pair g_1, g_2 an element $f(g_1, g_2)$ of A_2 is associated. We shall call the elements of A_2 *functions of rank* 2, and refer to $f(g_1, g_2)$ as the function obtained by *substituting* g_1, g_2 into f. We postulate that this substitution is *associative*, that is to say, that

$$[f(g_1, g_2)] \ (h_1, h_2) = f[g_1(h_1 h_2), \ g_2(h_1, h_2)].$$

Let f', f'' be a given ordered pair of functions. We call f an *invariant* of the pair f', f'' if

$$f(f', f'') = f.$$

We shall denote the set of all invariants of f',f'' by $A_2{}^*(f',f'')$. The set of all invariants is closed in the following sense. If f_1 and f_2 belong to $A_2{}^*(f', f'')$, then so does $h(f_1, f_2)$ for every function h of A_2. For we have

$$[h(f_1, f_2)] \ (f', f'') = h[f_1(f', f''), f_2(f', f'')] = h(f_1, f_2).$$

Next we postulate the existence of a *universal invariant* which we shall denote by O, that is to say, a function of O such that

$$O(g_1, g_2) = O \text{ for every pair } g_1, g_2.$$

[8a] For an ideal-theoretical approach of these and similar problems cf. M. Mamos' "Ideals in Tri-Operational Algebras I" Thesis Note 7.

[9] Cf. A. N. Milgram, *"Saturated Polynomials,"* Reports of a Math. Coll. 7.

[10] "A. of A." pp. 30-44.

[11] Cf. J. C. Burke, Reports of a Math. Coll. 7.

We call a function of rank 2 *constant* if and only if it belongs to $A_2^*(O,O)$. From the above remark it follows that if c_1 and c_2 are constant, then so is $h(c_1,c_2)$ for every h. Moreover, we see: If c is constant, we have

$$c(g_1,g_2) = [c(O,O)] \, (g_1,g_2) = c[O(g_1,g_2), O(g_1,g_2)] = c(O,O) = c$$

for every pair g_1,g_2. In other words, each constant is a universal invariant.

We denote by A' and A'' the sets of all functions f such that for every g_1 and g_2

$$f(g_1,g_2) = f(O,g_2) \text{ and } f(g_1,g_2) = f(g_1,O),$$

respectively. The functions belonging to A' (to A'') correspond to the functions of two variables which do not depend upon the first (the second) variable. Both classes A' and A'' are closed with regard to substitution. If f_1 and f_2 belong to A', then so does $h(f_1, f_2)$ for every h.

Obviously, each constant belongs to A' and A''. Conversely, if f belongs to A' and A'', then $f(g_1,g_2) = f(O,O)$ for every g_1,g_2. For from $f(g_1,g_2) = f(O,g_2)$ and $f(g_1,g_2) = f(g_1,O)$ it follows that $f(O,g_2) = f(g_1,O)$ for every g_1 and g_2, in particular, for $g_1 = O$. Hence $f(O,g_2) = f(O,O)$ and thus $f(g_1,g_2) = f(O,O)$. But the preceding postulates do not imply that a function belonging to A' and A'' is necessarily constant. E.g., if A_2 consists of only two elements O and 1, and we have $f(g_1,g_2) = O$ for every f, g_1, g_2, then 1 belongs to A' and A'', and yet $1 \neq 1(O,O)$.

We approach the classical situation by assuming the existence of a *neutral pair* of functions j_1, j_2 such that

$$f(j_1, j_2) = f \text{ for every } f.$$

In presence of such a neutral pair, *a function f is constant if* and only if it belongs to both A' and A''.* For if in the preceding reasoning we let g_1, g_2 be the neutral pair, we obtain $f = f(j_1, j_2) = f(O,O)$.

A further step is the assumption of the existence of two *selectors* j_1' and j_2' such that

$$j_1'(g_1, g_2) = g_1 \text{ and } j_2'(g_1, g_2) = g_2$$

for each g_1, g_2. Under this assumption A'' is identical with $A_2^*(j_1', O)$, the set of all functions f for which $f = f(j_1',O)$, and also with the set $A_2^*(j_1',g_1)$ for every function of g_1 of A''. We further see that if f belongs to A', then $f(j_2', j_1')$ belongs to A'', and vice versa.

In classical analysis, the neutral pair consists, of course, of selectors. We have $j_1 = j_1'$ and $j_2 = j_2'$.

---

* This theorem is used in finding particular solutions of homogeneous partial differential equations, e.g., the Laplace equation $z_{xx} + z_{yy} = 0$. If a solution z is the product of two functions X and Y which are independent of the second and the first variable, respectively, then from $X''/X = Y''/Y$ we conclude that these two quotients are constants.

(54)

With regard to addition and multiplication we may follow one of two ways. Either we may proceed as we did in tri-operational algebra, and postulate *the existence of two binary operations*, denoted by $+$ and $.$, *by virtue of which A_2 is a ring with a unity*. In this case we have to assume two distributive laws connecting the two binary operations with the substitution:

the a.s.d. law $(f + g)(h_1, h_2) = f(h_1, h_2) + g(h_1, h_2)$.

the m.s.d. law $(f . g)(h_1, h_2) = f(h_1, h_2) . g(h_1, h_2)$.

Or we may postulate *the existence of two functions s and p with properties corresponding to the assumptions about a ring with a untiy, viz.*,

$$s(f, s(g, h)) = s(s(f, g), h) \text{ and } p(f, p(g, h)) = p(p(f, g), h)$$
$$s(f, O) = s(O, f) = f \text{ and } p(f, 1) = p(1, f) = f,$$
$$s(f, -f) = s(-f, f) = O,$$
$$p(f, s(g, h)) = s(p(f, g), p(f, h)),$$
$$p(s(g, h), f) = s(p(g, f), p(h, f)).$$

In this way, addition and multiplication are reduced to the level of individual functions such as j_1, j_2, O, and we obtain a *purely substitutive algebra*. The two distributive laws connecting substitution with addition and multiplication are consequences of the associative law for substitution. The element O of the ring which is neutral with regard to addition, can easily be shown to be a universal invariant, i.e., to satisfy the condition $O(g_1, g_2) = O$ for every pair g_1, g_2 .Hence if A_2 is assumed to be a ring we need not postulate the existence of a universal invariant.

If A_2 has been made a ring by either set of assumptions, then we can prove that A'' *is a tri-operational algebra* with regard to the following binary operations

$$f + g = s(f, g), \quad f . g = p(f, g), \quad fg = f(g, O), \quad j = j_1.$$

A' is an isomorphic tri-operational algebra.

We see: *if A_2 is a class with*

1) an associative substitution of pairs $f(g_1, g_2)$ admitting a neutral pair of selectors,

2) ring operations which are defined either by virtue of postulated addition and multiplication satisfying the a.s.d. and m.s.d. laws or by virtue of the existence of two proper elements s and p,

then A_2 determines a tri-operational algebra with which the subsets A' and A'' are isomorphic.

An example of a finite system satisfying the postulates for a set A_2 of functions of rank 2 is the set of 16 elements

$$O, 1, j_1, j_2, j_1 + 1, j_2 + 1, s = j_1 + j_2, s + 1,$$
$$(55)$$

$p = j_1 . j_2, \ p+1, \ p+j_1, \ p+j_2, \ p+j_1+1, \ p+j_2+1, \ p+s, \ p+s+1.$
This set is isomorphic with the 16 functions of two variables ranging over the residues mod. 2. For every element f we have
$$f + f = O, \quad f . f = f.$$

In classical analysis two differential operators, ∂_1 and ∂_2, are given which are connected with the three operations in the following way

1) $\partial_1 s = \partial_2 s = 1$
2) $\partial_1 p = j_2, \ \partial_2 p = j_1$
3) $\partial_1 [f(g_1, g_2)] = s[p(\partial_1 f(g_1, g_2), \partial_1 g_1), \ p(\partial_2 f(g_1, g_2), \partial_2 g_2)]$
4) $\partial_1 j_2$ belongs to A', $\partial_2 j_1$ to A''.

Under these assumptions one can prove that for each function f of A', the function $\partial_1 f$ belongs to A' and $\partial_2 f = O$. Consequently, in the tri-operational algebra A' the operator $\partial_1 f$ satisfies the postulates which we assumed for the differential operator in a tri-operational algebra.

3. Functions of Rank Σ

Obviously, the concepts developed in the preceding section can be generalized so as to comprise entities corresponding to functions of a finite number or denumerably many variables. We shall study more general entities, called functions of rank Σ, for every (finite or infinite, ordered or unordered) set Σ. In the present section we shall outline a "semi-algebraic" theory of such entities.

Let Σ be a given set. We start with a class, A, of elements denoted by f, g, ... Let M be the set of all mappings of Σ into A. Each element F, G, ... of M is an association of an element f^a of A with each element a of Σ. More precisely, each element of M is a set of pairs (a, f^a) with a ranging over Σ, and each f^a belonging to A. By [F] we shall mean the range of the mapping F, that is, the set of all elements of A associated by F with some element of Σ.

We assume that with each element f of A, and each element G of M, an element fG of A is associated. Under this assumption we call the elements of A *functions of rank* Σ, and fG the function obtained by *substituting* G into f. If Σ consists of the numbers 1, 2, ..., ρ, then we briefly speak of functions of rank ρ. In the preceding section we dealt with functions of rank 2.

Since each element of M is a function in the classical sense, i.e., a set of pairs (a, f^a), the present theory is a partial relapse into set-theory. However, we adopt the set-theoretical point of view only with regard to the places at which we substitute into f, and not with regard to domains of variables which are being substituted. In a classical introduction of functions of rank Σ, first of all we should have to have a set Σ of domains, that is, an association of a set R^a with each element a of Σ. Then a function of rank Σ would be defined as a set of sets.

$$(x^1, \ldots, x^a, \ldots; f(x^1, \ldots, x^a, \ldots))$$

(56)

where for each a the elements x^a range over R^a. In contrast, the theory outlined in the present section is an *analysis without variables*. These remarks will explain why we called our theory semi-algebraic.

Since in the set-theoretical part of the theory, that is, in the part dealing with the elements of M, all the mappings have the same domain, namely Σ, no set-theoretical difficulties will arise in connection with addition, multiplication, and substitution of mappings. If F and G are elements of M, then we define $F + G$ and $F \cdot G$ as those mappings of Σ into A which with each element a of Σ associate the functions $f^a + g^a$ and $f^a \cdot g^a$, respectively. We define FG as that mapping of Σ into A which with each element a associates the function $f^a G$.

We assume that substitution is associative, that is,

$$(fg)H = f(GH).$$

By virtue of this assumption, the substitution in M is associative, that is to say,

$$(FG)H = F(GH).$$

Moreover, we have

$$(F + G)\,H = FH + GH \quad \text{and} \quad (F \cdot G)\,H = FH \cdot GH.$$

Hence M is a tri-operational algebra. We obtain interesting finite examples of such tri-operational algebras in the following way. Let $A_n{}^m$ denote the classical functions of m variables ranging over the residues mod.n, that is, the functions associating a number $f(x_1, x_2, \ldots, x_m)$ of the range $0, 1, \ldots, n-1$ with each of the n^m ordered m-tuples $x_1, x_2, \ldots, x_m$ of numbers $0, 1, \ldots, n-1$. There are (n to the power n to the power m) such functions. $n-1$ of these functions are constant. We may denote them by $0, 1, \ldots, n-1$. We further denote the function $f(x_1 x_2, \ldots, x_m) = x_l$ by j_l. Then we have $f(j_1, j_2, \ldots, j_m) = f$ for every f. Let $M_n{}^m$ be the tri-operational algebra of the ordered m-tuples of functions of $A_n{}^m$. The m-tuples

$$(0, 0, \ldots, 0), \quad (1, 1, \ldots, 1), \quad (j_1, j_2, \ldots, j_m)$$

are the neutral elements of $M_n{}^m$. The n^m elements $(c_1, c_2, \ldots, c_m)$ where the c_i are constants of $A_n{}^m$, are the constants of $M_n{}^m$.

If $n = m = 2$, then $A_2{}^2$ is the system of the 16 functions discussed in the preceding section, which can be obtained as sums of $0, 1, j_1, j_2, p$. The tri-operational algebra $M_2{}^2$ consists of the 256 pairs of such functions. The constants of $M_2{}^2$ are the four elements

$$(0, 0), \quad (0, 1), \quad (1, 0), \quad (1, 1).$$

They form a ring. Since we have $f + f = f$ and $f \cdot f = f$ for each element f of $A_2{}^2$, we have

$$(f, g) + (f, g) = (f, g) \quad \text{and} \quad (f, g) \cdot (f, g) = (f, g)$$

for every element (f, g) of $M_2{}^2$. In particular, for the neutral element of $M_2{}^2$ with regard to substitution we have $(j_1, j_2) \cdot (j_1, j_2) = (j_1, j_2)$. Hence $M_2{}^2$ contains only the following 16 polynomials

(57)

$$(c_1, c_2) + (d_1, d_2) \cdot (j_1, j_2) = (c_1 + d_1 \cdot j_1, c_2 + d_2 \cdot j_2).$$

where c_1, c_2, d_1, d_2 are O or 1. The other 240 elements of $M_2{}^2$ are not polynomials. For instance, (s, O), (s, s), (p, j_1) are not polynomials.

We say that the function f is an *invariant* of the element F' of M if $f = fF'$. We denote the set of all invariants of F' by $A*F'$.

Each function of invariants of F' is an invariant of F', that is to say, if F is an element of M such that each element f of [F] belongs to $A*F'$, then so does hF for every function h of A. For we have

$$hF = h(FF') = (hF)F'.$$

For each subset A_1 of A, and each given element F' of M, we denote by A_1F' the set of the functions f_1F' for all elements f_1 of A_1. If $F'F' = F'$, then AF' is a subset of $A*F'$. For from $F'F' = F'$ it follows that

$$f_1F' = f_1(F'F') = (f_1F')F'.$$

We postulate the existence of a universal invariant O such that $OF = O$ for each F of M. Denoting the mapping of Σ on [O] by O, we call f *constant* if it belongs to $A*O$, that is, if $fO = f$. If C is an element of M such that [C] is a set of constants, then fC is a constant for every f. Moreover, every constant c is a universal invariant, that is to say, we have $cF = c$ for each element F of M.

If we assume that the elements of A form a ring and that the distributive laws $(f + g)H = fH + gH$ and $(f \cdot g)H = fH \cdot gH$ hold, then the O of the ring is automatically a universal invariant.

Let T be a given subset of Σ. If G is any element of M mappings a of Σ on g^a, then we shall denote by G^T the element of M mapping a on

g^a if a belongs to T,
O if a belongs to $\Sigma - T$.

We call A^T the set of all elements f of A such that $f = fG$ for every element G of M. This set A^T is isomorphic with an algebra of functions of rank T. It follows that the algebra A^T is isomorphic with the algebra A^U for every subset U of Σ which can be brought into a one-to-one correspondence with T.

In presence of a neutral element J of M such that $fJ = f$ for every f of A, a function is constant if and only if it belongs to A^T for every subset T of Σ. Moreover, in this case A is isomorphic with a set of classical functions of Σ variables each ranging over A. The "natural" representation in case that Σ consists of the numbers 1, 2, ..., ρ associates with each f the set of ordered $(\rho + 1)$ triples

$$f_1, f_2, \ldots, f_\rho, f(f_1, f_2, \ldots, f_\rho).$$

If no neutral element $J = (j_1, j_2, \ldots, j_\rho)$ exists, this representation is a homomorphism but not necessarily as isomorphism.

A step toward classical analysis is the assumption of such a neutral element J, and of selectors, that is, an element j^a for each a of Σ such that $j^aF = f^a$ for every F of M. In classical analysis, J is a mapping of Σ on selectors.

(58)

4. Algebra of Analysis

We shall now outline a purely algebraic theory which synthesizes the semi-algebraic theories of functions of rank Σ for the various sets Σ by summarizing the common features of these theories.

First of all, for every set Σ, the set M of all mappings of Σ into a ring R is a ring. More specifically, M is the direct product of a set Σ of rings each of which is isomorphic with R, or, as we shall briefly say, M is the Σ-th power of the ring R. Since in the algebra of functions of rank Σ, the set A of all functions is a ring, we see that for every Σ the set M is a ring. More specifically, in the case of rank Σ the ring M is the Σ-th power of the ring A.

Next, for every set Σ we defined a substitution FG for each element G of M into each element F of M. We proved this substitution to be associative. Moreover, it obviously satisfies the a.s.d. and m.s.d. laws

$$(F + G) H = FH + GH, \quad (F \cdot G) H = FH \cdot GH.$$

Hence, for every Σ the set M is a tri-operational algebra.

In the third place, while the ring A is not a tri-operational algebra since no binary substitution is defined in A, the ring A and the tri-operational algebra M are connected by a substitution. For each element f of A and each element G of M there is an element fG in A. This substitution (which enabled us to define the substitution in M) is associative in the sense that $f(GH) = (fG)H$, and connected with the operations of the ring A by the distributive laws

$$(f + g) H = fH + gH \text{ and } (f \cdot g) H = fH \cdot gH.$$

Finally we see that the operations of the tri-operational algebra M are related to the ring operations in A and to the substitution of elements of M into elements of A. For every g of A we denote by F^g that element of M for which $\{F^g\}$ consists of g only. In other words, F^g is the mapping of Σ on the set consisting of the single element g. This mapping of A into M is an isomorphism with regard to the ring operations

$$F^{g+h} = F^g + F^h, \quad F^o = O, \quad F^{g \cdot h} = F^g \cdot F^h.$$

Moreover, for each g of A, and H of M we have

$$F^g H = F^{gH}.$$

In the last equality, $F^g H$ denotes the element of M obtained by substituting H into F^g while F^{gH} is the element of M corresponding with the element gH of A.

We also notice the following accidental features.

If A contains an element 1 which is neutral with regard to multiplication, then so does M. The mapping of Σ on the set containing only 1 is this neutral element of M. Without any danger of confusion we may denote this element of M by 1 in the same way as we denote by O the element of M for which the set $[O]$ contains only the element O of A.

(59)

If M contains a neutral element for the substitution into the elements of A, that is, an element J such that fJ = f for every f, then the tri-operational algebra M contains an element which is neutral with regard to right-side substitution, namely, the same J, that is to say, we have FJ = F for every F of M. Conversely, if M contains a neutral element with regard to right-side substitution, such that FJ = F, then fJ = f for every f of A. for

$$F^f = F^f J = F^{fJ}$$

and since the association of F^g with g is one-to-one it follows that fJ = f.

We shall speak of a *Generalized Algebra of Analysis* RT if we have .

1) a ring R

2) a tri-operational algebra T,

3) a substitution of each element G of T into each element f of R satisfying the associative and distributive laws

f (GH) = (fG) H,

(f + g) H = fH + gH, (f . g) H = fH . gH.

4) a one-to-one mapping of R on a subset T^R of T associating with each g of R an element F^g of T such that

$F^{g+h} = F^g + F^h$, $F^0 = O$, $F^{g.h} = F^g . F^h$, $F^g H = F^{gH}$.

We shall speak of an *Algebra of Analysis* if

5) the ring R contains an element which is neutral with regard to multiplication, that is, a unit 1 such that 1 . f = f . 1 = f for every f of R.

6) the tri-operational algebra T contains a neutral element with regard to substitution, that is, an element J such that JF = FJ = F for every F of T.

This concept incorporates the common features of the algebras of functions of rank Σ for the various sets Σ. If $T = T^R$ and $F^g = g$ for every g, then the algebra of analysis TR is isomorphic with the tri-operational algebra T.

(60)

AN AXIOMATIC THEORY OF FUNCTIONS AND FLUENTS

KARL MENGER

Illinois Institute of Technology, Chicago, Illinois, U.S.A.

The topic of this paper is a theory of some basic applications of mathematics to science. Part I deals with concepts of pure mathematics such as the logarithm, the second power, and the product, and with substitutions in the realm of those functions. Part II is devoted to scientific material such as time, gas pressure, coordinates — objects that Newton called *fluents*. Part III formulates articulate rules for the interrelation of fluents by functions. Properly relativized, the latter play that connective role for which Leibniz originated the term *function*.

I. FUNCTIONS

Explicitly, a real function with a real domain — briefly, a *function* — may be defined as a class of consistent ordered pairs of real numbers. Here and in the sequel, two ordered pairs of any kind are called *consistent* unless their first members are equal while their second members are unequal. If each pair $\in f_1$ (that is, belonging to the function f_1) is also $\in f_2$ — in symbols, if $f_1 \subseteq f_2$ — then f_1 is called a *restriction* of f_2; and f_2 an *extension* of f_1. The *empty* function (including no pair) will be denoted by $\emptyset$. The class of the first (the second) members of all pairs $\in f$ is called the *domain* of f or dom f (the *range* of f or ran f). If ran f includes exactly one number, then f is said to be a *constant* function.

The following typographical convention [1] will be strictly adhered to:

roman type for numbers; *italic type for functions*.

For instance, the logarithmic function — briefly, *log* — is the class of all pairs (a, *log* a) for any a $>$ 0. The constant function consisting of all pairs (x, 0) for any x will be denoted [2] by O. The following are examples of a formula and a general statement, respectively: *log* e $= 1$, and $\emptyset \subseteq f$ for any f. Here, 0, 1, e as

---

[1] Cf. Menger [10] referred to in the sequel as *Calculus*.

[2] Symbols for constant functions that are more elaborate than italicized numerals, such as c_1 and c_0, must be used in order to express certain laws; e.g., that $c_{0+1} = c_0 + c_1$.

well as *log*, *O*, and *ø* are designations of specific entities, while a, x, and *f* are *variables* (i.e., symbols replaceable with the designations of specific entities according to the respective legends) — *number variables* or *function variables* as indicated typographically.

The intersection of any two functions is a function; e.g., that of *cos* and *sin* is the class of all pairs $((4n + 1)\pi/4, (-1)^n/\sqrt{2})$ for any integer n. The union of *cos* and *sin*, however is not a function. From the set-theoretical point of view, *functions do not constitute a Boolean algebra* [3]. But any two functions have a sum, a difference, a product, and a quotient provided $\dfrac{f_1}{f_2}$ is defined as the class of all pairs (x, q) such that $(x, p_1) \in f_1$, $(x, p_2) \in f_2$ and $\dfrac{p_1}{p_2} = q$ for some p_1 and p_2 — a definition that dispenses with any reference to zeros in the denominators. For instance,

$$cot = \frac{1}{tan}, \; \frac{f}{0} = \text{ø, and } \frac{f_1}{f_2} \cdot f_2 \subseteq f_1 \text{ for any } f, f_1, \text{ and } f_2.$$

Moreover, any function f_2 may be substituted into any function f_1, the result f_1f_2 (denoted by mere juxtaposition, whereas multiplication will always be denoted by a dot!) being the class of all pairs (x, z) such that $(x, y) \in f_2$ and $(y, z) \in f_1$ for some y. The *identity function*, i.e., the class of all pairs (x, x) for any x — an object of paramount importance — will be denoted [4] by *j*. Its main property is bilateral neutrality under substitution:

(1) $$jf = f = fj \text{ for any } f.$$

For each *f*, there is a *bilaterally inverse* function [5], Inv *f*, which is the largest class of pairs (x, y) such that $(y, x) \in f$ and that, under substitution, $f \text{ Inv } f \subseteq j$ and $(\text{Inv } f)f \subseteq j$. For instance, $\text{Inv } j^3 = j^{\frac{1}{3}}$ and $\text{Inv } exp = log$. If j_+^2 is the class of all pairs (x, x^2) for any $x \geq 0$, then

$\text{Inv } j_+^2 = j^{\frac{1}{2}}$, $\text{Inv } j^{\frac{1}{2}} = j_+^2$; similarly, $\text{Inv } j_-^2 = -j^{\frac{1}{2}}$, $\text{Inv } -j^{\frac{1}{2}} = j_-^2$. But $\text{Inv } j^2$ consists of the single pair (0, 0); and $\text{Inv } cos = $ ø, while Inv *f*

---

[3] For this reason, the postulational theories of binary relations, which are based on Boolean algebra (cf. especially, McKinsey [8] p. 85 and Tarski [22] p. 73), are inapplicable to functions.

[4] Cf. *Calculus.* p. 74 and pp. 99–105. Cf. also Menger [11] and [12].

[5] Cf. *Calculus.* pp. 91–95, where Inv *f* is denoted by *j///f*. The fertility of this concept of inverse functions has been brought out by M. A. McKierman's interesting and promising studies on operators. Cf. McKiernan [6], [7].

is a branch of *arccos* if $f \subseteq cos$ and dom f is the interval $[n\pi, (n+1)\pi]$, for some integer n.

In the traditional literature, the identity function has remained anonymous — one of the symptoms for the neglect of substitution in analysis. The usual reference—" the function x" — and the symbol x are complete failures in basic assertions. Even in order to assert substitutive neutrality, concisely expressed in (1), analysts are forced to introduce a better symbol than x — an ad hoc name of the identity function, say h — and then must resort to an awkward implication: If $h(x) = x$ for each x, then $h(f(x)) = f(x) = f(h(x))$ for any f and any $x \in$ dom f.

The overemphasis on additive-multiplicative processes, which is characteristic of mathematics in the second quarter of this century, becomes particularly striking in passing from theories of functions based on explicit definitions to postulational theories — theories of rings of functions, of linear function spaces, etc., which stress those properties that functions or entities of any kind share with numbers. One of the few exceptions doing justice to substitution is the *trioperational algebra of analysis* [6]. In it, functions are (undefined) elements subject to three (undefined) operations. With regard to the first two, denoted by $+$ and $.$, the elements constitute a ring including neutral elements, 0 and 1. The third, called substitution and denoted by juxtaposition, is associative and right-distributive with regard to the ring operations [7]:

(2) $(f + g)h = fh + gh$ and $(f.g)h = fh.gh$ for any f, g, h.

For many purposes, it is important to postulate a neutral element j satisfying (1).

Trioperational algebra has interesting applications to rings of polynomials as well as non-polynomials [8] but does not apply to the realm of *all* functions, even though the three operations can be defined for any two functions. The only ring postulate that is not generally satisfied is that, for each g, there exist an f such that $f + g = 0$. For instance, $- log + log$ is not 0, but rather the restriction of 0 consisting of all $(x, 0)$ for $x > 0$. Only $f + log \subseteq 0$ has solutions (namely, any $f \subseteq - log$). What narrowed

[6] Cf. Menger [13], [14].

[7] In keeping with the traditional attitude toward substitution, the laws (2) are hardly ever mentioned even though they are as important in analysis as is the multiplicative-additive distributive law.

[8] Cf. especially Milgram [18] p. 65, Heller [4] and Nöbauer [19].

the scope of trioperational algebra in its original form was the fact that it did not take the relation $\subseteq$ into account.

A more satisfactory postulational approach to functions may be based on the following idea of a *hypergroup*: a set $\mathscr{G}$ satisfying six postulates:

I. $\mathscr{G}$ is partially ordered by a relation $\subseteq$. For some purposes it is convenient to assume that $\mathscr{G}$ includes a (necessarily unique) *minimal* element $\varnothing$, such that $\varnothing \subseteq \gamma$ for each γ; or, even further, that $\mathscr{G}$ is *atomized* in the sense that (1) for each $\gamma \neq \varnothing$, at least one $\alpha \subseteq \gamma$ is an *atom* (i.e., such that $\alpha' \subset \alpha$ if and only if $\alpha' = \varnothing$); (2) $\gamma_1 \subseteq \gamma_2$ if and only if each atom $\subseteq \gamma_1$ is also $\subseteq \gamma_2$. For other purposes, $\mathscr{G}$ may be assumed to be *intersectional*, i.e., to include, for any two elements γ_1 and γ_2, a maximal element $\subseteq \gamma_1$ and $\subseteq \gamma_2$ — an *intersection*, $\gamma_1 \cap \gamma_2$.

II. In $\mathscr{G}$, there is an *associative operation*, $\circ$.

III. $\mathscr{G}$ includes a *bilaterally and absolutely neutral element*, ν, such that $\gamma \circ \nu = \gamma = \nu \circ \gamma$ for any γ.

Clearly, ν is unique. The connection between $\subseteq$, $\circ$, and ν is established in the following postulate that simplifies the author's original development and is due to Prof. A. Sklar.

IV. $\gamma \subseteq \delta$ if there exists an element $\nu' \subseteq \nu$ such that $\nu' \circ \delta = \gamma$ and if and only if there exists an element $\nu'' \subseteq \nu$ such that $\gamma = \delta \circ \nu''$.

It readily follows that $\circ$ is *bilaterally monotonic*; that is to say, $\gamma_1 \subseteq \gamma_2$ implies $\gamma_1 \circ \gamma \subseteq \gamma_2 \circ \gamma$ and $\gamma \circ \gamma_1 \subseteq \gamma \circ \gamma_2$ for any $\gamma_1, \gamma_2, \gamma$. If there is a minimal element, then $\varnothing$ may be a *bilateral strict annihilator*:

$$\varnothing \circ \gamma = \varnothing = \gamma \circ \varnothing \text{ for each } \gamma.$$

Moreover, if $\nu_1 \subseteq \nu$ and $\nu_2 \subseteq \nu$, then $\nu_1 \circ \nu_2 \subseteq \nu_1$ and $\subseteq \nu_2$; thus, if $\mathscr{G}$ is intersectional, $\nu_1 \circ \nu_2 \subseteq \nu_1 \cap \nu_2$.

V. For each γ, there exist two *unilaterally and relatively neutral elements*, Lγ and Rγ (the *left-neutral* and the *right-neutral* of γ) such that:
1) L$\gamma \circ \gamma = \gamma = \gamma \circ$ Rγ;
2) L$(\gamma_1 \circ \gamma_2) \subseteq$ Lγ_1 and R$(\gamma_1 \circ \gamma_2) \subseteq$ Rγ_2 for each γ_1 and γ_2;
3) if $\mu \subseteq \nu$, then L$\mu \subseteq \mu$ and R$\mu \subseteq \mu$.

Clearly, L$\gamma \subseteq \nu$ and R$\gamma \subseteq \nu$ for each γ. If L$\gamma = \gamma$ and/or R$\gamma = \gamma$, then $\gamma \subseteq \nu$. If $\gamma \subseteq \nu$, then L$\gamma = \gamma =$ Rγ. Hence LL$\gamma =$ RL$\gamma =$ Lγ for every γ. Moreover, L$\gamma = \varnothing$, R$\gamma = \varnothing$, and $\gamma = \varnothing$ are equivalent. If $\gamma \subseteq \delta$, then L$\gamma \subseteq$ Lδ and R$\gamma \subseteq$ Rδ. If χ is an *annihilator* is the sense that $\gamma \circ \chi =$ Lχ and $\chi \circ \gamma =$ Rχ for each γ, and if $\chi \subseteq \nu$, then $\chi = \varnothing$. Moreover, Lγ and Rγ are charaterized among the elements $\subseteq \nu$ by the following minimum property:

If $\mu \subseteq \nu$, then $\mu \circ \gamma = \gamma$ implies $L\gamma \subseteq \mu$, and $\gamma \circ \mu = \gamma$ implies $R\gamma \subseteq \mu$. It follows that $L\gamma$ and $R\gamma$ are unique for each γ. If $\circ$ is commutative, then $L\gamma = R\gamma$ for each γ.

It will suffice, here, bypassing *unilaterally opposite* elements, to postulate finally

VI. For each γ, there is a *bilaterally opposite element* Op γ such that Op $\gamma \circ \gamma \subseteq R\gamma$ and $\gamma \circ$ Op $\gamma \subseteq L\gamma$ for each γ,

and which, if one sets Op $\gamma \circ \gamma = R'\gamma$ and $\gamma \circ$ Op $\gamma = L'\gamma$, has the following minimax property:

1) if $\delta \circ \gamma \subseteq R\gamma$ and $\gamma \circ \delta \subseteq L\gamma$, then $\delta \circ \gamma \subseteq R'\gamma$ and $\gamma \circ \delta \subseteq L'\gamma$;

2) if $\delta \circ \gamma = R'\gamma$ and $\gamma \circ \delta = L'\gamma$, then Op $\gamma \subseteq \delta$.

Op γ is unique for each γ, and $L'\gamma \circ \gamma = \gamma \circ R'\gamma$, which might be called $C\gamma$, the *core* of γ. If $\mu \subseteq \nu$, then Op $\mu = R'\mu = L'\mu = C\mu = \mu$. For each atom, Op α is an atom, and $C\alpha = \alpha$. Additional assumptions would guarantee that

Op $\delta \circ$ Op $\gamma \subseteq$ Op$(\gamma \circ \delta)$; Op Op $\gamma \subseteq \gamma$; Op Op Op $\gamma =$ Op γ for each γ, δ. However, $\gamma \subseteq \delta$ does not imply Op $\gamma \subseteq$ Op δ.

An element γ of a hypergroup will be called *right-elementary* (or *left-elementary*) if $\delta \subset \gamma$ implies $R\delta \subset R\gamma$ (or $L\delta \subset L\gamma$). Each atom is bilaterally elementary — briefly, *elementary*. If $\mathcal{G}$ is commutative and γ is unilaterally elementary, γ is elementary. If $\mathcal{G}$ is atomized, then ϱ is right-elementary if and only if $\varrho \circ \mu = \varrho$ implies $\mu \subseteq \nu$. If, in contrast, $\varkappa \circ \gamma \subseteq \varkappa$ for each γ, then $\varkappa$ may be called a *left-annihilator*; and each $\varkappa'$ that is $\subseteq \varkappa$, a *leftquasiannihilator*. Clearly, $\varkappa' \circ \gamma$ and $\gamma \circ \varkappa'$ are left-quasi-annihilators for any γ. If each element of $\mathcal{G}$ is right-elementary (or elementary), then $\mathcal{G}$ will be said to be *right-elementary* [9] (or *elementary*).

With regard to addition as well as multiplication, the set of all functions is a commutative elementary hypergroup. The universal neutrals are 0 and 1; the relative neutrals of f are, as it were, vertical projections of f on 0 and 1, respectively; the opposites of f are $-f$ and $\dfrac{1}{f}$.

With regard to substitution, the set of all functions is a (non-commutative) right-elementary hypergroup. The universal neutral is j. The relative neutrals, Rf and Lf, correspond to dom f and ran f, respective-

---

[9] Prof. B. Schweizer proposes to call δ a *right-neutralizer* of γ if $\gamma \circ \delta \subseteq \nu$ and $L\delta \subseteq R\gamma$, and to say that δ is (1) *maximal* if $\gamma \circ \delta' \subseteq \gamma \circ \delta$ for each right-neutralizer δ' (2) *saturating* if $\gamma \circ \delta = L\gamma$. One might then postulate that each element γ, on either side, has at least one maximal neutralizer or at least one saturating neutralizer.

ly.[10] Thus the contrast between functions (classes of pairs of numbers) and their domains and ranges (classes of numbers) disappears. Op f is Inv f. The left annihilators $\neq \emptyset$ are what may be called *universal constant functions*; the left-quasiannihilators $\neq \emptyset$ are the *constant functions*[11].

Another example of a hypergroup is the set of all binary relations in some universal set with regard to what logicians call the *relative product*[12]. The universal neutral is the identity relation, while the relative neutrals again correspond to domains and ranges. Op γ is a restriction — in general, a *proper* restriction — of the converse of the relation γ.

Geometrically, the situation may be interpreted in a set (a *"plane"* consisting of *"points"*) that is decomposed into mutually disjoint subsets (*"vertical lines"*). *"Simple"* sets, i.e., sets having at most one point in common with each vertical line, are the counterpart of functions. This vertical simplicity corresponds to right-side elementariness of functions. Substitution can be illustrated if, secondly, the plane is decomposed into disjoint subsets (*"horizontal lines"*) each of which has exactly one point in common with each vertical line; and if, thirdly, there is given a *"diagonal"* set having exactly one point in common with each vertical line as well as with each horizontal line. The diagonal corresponds to j; each horizontal line, to a universal constant function; the vertical (the horizontal) projection of a simple set f on the diagonal, to Rf (to Lf); the points, to atoms. Fig.1, p. 460, based on the assumption of ordinary vertical and horizontal lines and a straight diagonal, j, shows a simple plane construction[13] of the result of substituting g into f. For any point a in the set g, move horizontally to j, then vertically to f, and finally horizontally back to the vertical line through a. The set of all points thus obtained is

[10] In contrast to groups and hypergroups, a Brandt *groupoid* (Mathematische Annalen, vol. 96) only permits the composition of *some* elements. In „categories" (i.e., essentially, groupoids) of mappings of groups on groups, MacLane calls the one-side identities of a mapping its domain and range.

[11] In a self-explanatory way, one can say that functions constitute a commutative elementary *hyperfield* with regard to addition and multiplication (with the multiplicative annihilator 0) and (non-commutative) right-elementary hyperfields with regard to addition and substitution as well as multiplication and substitution. The functions may also be said to constitute a *trioperational hyperalgebra*.

[12] Cf., e.g., McKinsey [8] and Tarski [22].

[13] Cf. *Calculus* pp. 89 ff. The traditional postulational theory of binary relations is inapplicable to functions (cf. [3]). On the other hand, the plane construction of functional substitution here described may, as Prof. M. A. McKiernan observed, be utilized for binary relations instead of the 3-dimensional construction proposed by Tarski [22] pp. 78, 79.

fg. In the figure, *f* has the shape of an exponential curve; *g*, that of $-j^2$; hence *fg*, that of the probability curve.

Notwithstanding the analogy (brought out in the concept of a hypergroup) of addition and multiplication with substitution, the latter has a definite primacy. In an atomized non-commutative hypergroup $\mathscr{G}$, any binary operation $\times$ (such as $+$ and .), defined in the class of all atoms $\subseteq \nu$, may be extended to any two elements γ' and γ'' of $\mathscr{G}$ by defining $\gamma' \times \gamma''$ as the minimum element including all atoms α such that there

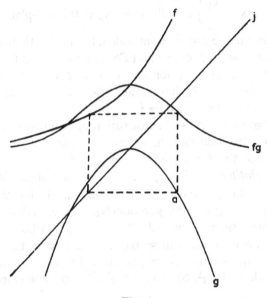

Fig. 1

exist two atoms $\alpha' \subseteq \gamma'$ and $\alpha \subseteq \gamma''$ satisfying the following conditions:

$$R\alpha = R\alpha' = R\alpha'' \text{ and } L\alpha = L\alpha' \times L\alpha''.$$

(This is essentially how the arithmetical operations are extended from numbers to functions.) Moreover, Neg *f* and Rec *f*, the negative and the reciprocal of *f*, are obtainable by substituting *f* into $-j$ and j^{-1}, respectively; and, as will be shown presently, even $f + g$, $f \cdot g$, and $\dfrac{f_1}{f_2}$ can be obtained from 2-place functions *S*, *P*, and *Q* by substitution. In contrast,

[14] In fact, no 2-place function yields *fg* even by substitution of *f* and *g*. Cf. *Calculus*, p. 304.

there are no functions of any kind which, for each f and g, would yield fg or even Inv f by additions and multiplications. [14] Beyond any question, in the realm of functions substitution is the operation par excellence.

For any integer $m \geq 1$, a class of consistent pairs whose second members are real numbers while their first members are sequences of m real numbers is called, briefly, an m-*place function* [15] and will be designated by a capital italic, except where lower case italics emphasize the 1-place character of functions such as those treated in what precedes. The class Q of the pairs $\left((x_1, x_2), \dfrac{x_1}{x_2} \right)$ for all x_1 and $x_2 \neq 0$ is a 2-place function. So are sum and product, S and P, from which, because of their associativity, an m-place sum and product, S^m and P^m, can be derived for each $m > 2$. Of particular importance is, for any two integers $1 \leq i \leq k$, the i-th k-place *selector function* $I_i{}^k$, that is, the class of all pairs $((x_1, \ldots, x_k), x_i)$ for any $x_1, \ldots, x_k$. Clearly, $I_1{}^1 = j$.

There are two main types of substitution of sequences of m functions into an m-place function (which, for $m = 1$, coincide with one another and with the substitution as defined on p. 455):

a) *product substitution*: $F^m[G_1, \ldots, G_m]$, whose domain is a subset of the Cartesian product, dom $G_1 \times \ldots \times$ dom G_m, which is the class of all sequences $(\gamma_1, \ldots, \gamma_m)$ for any $\gamma_1 \in$ dom $G_1, \ldots, \gamma_m \in$ dom G_m.

b) *intersection substitution*: $F^m(G_1, \ldots, G_m)$, whose domain is $\subseteq$ dom $G_1 \cap \ldots \cap$ dom G_m. Unless $G_1, \ldots, G_m$ have the same place-number, that intersection is empty and $F^m(G_1, \ldots, G_m) = \emptyset$; e.g., $P(j, S) = \emptyset$, while $P(I_1{}^2, S)$ and $P(I_2{}^2, S)$ are non-empty. Clearly, $Q(f_1, f_2) = \dfrac{f_1}{f_2}$, as defined on p. 455; and $I_i{}^m(G_1, \ldots, G_m) = G_i$ for any m functions of the same placenumber. A simple generalization of the plane construction described on p. 459 to the 3-dimensional space [16] yields $F^2(G_1{}^2, G_2{}^2)$.

Traditionally, $P(j^2, log)$, $P[j^2, log]$, $P(P, S)$, $P[P, S]$, $P(I_2{}^2, S)$, $P[j^2, S]$ are referred to as the functions $x^2 \log x$, $x^2 \log y$, $xy(x + y)$, $xy(u + v)$, $y(x + y)$, $x^2(y + z)$, respectively.

Either type of substitution can be extended to a realm of sequences

---

[15] One might introduce numbers as 0-plane functions.

[16] Cf. Menger [15] p. 224. Recently, S. Penner in his Master's thesis at Illinois Institute of Technology has extended the geometric axiomatics of substitution, outlined on p. 459 of the present paper, from 1-place to m-place functions in the $m + 1$-dimensional space.

of functions. With each sequence, besides the number of functions in it, called the sequence-number, a placenumber will be associated. Either substitution of a second sequence into a first presupposes that *the sequence-number of the second be equal to the place-number of the first.*

a) An s-place *array* of r functions is a sequence such that the sum of the r places-numbers is s; for instance $\Phi_r{}^s = [F_1{}^{m_1}, \ldots, F_r{}^{m_r}]$, where $s = m_1 + \ldots + m_r$. Product substitution, defined by

$$\Phi_r{}^s\Psi_s{}^t = [F_1{}^{m_1}[G_1, \ldots, G_{m_1}], \ldots, F_r{}^{m_r}[G_{s-m_{r-1}+1}, \ldots G_s]],$$

clearly is associative and admits unilateral neutrals:

$$\Phi_r{}^s[\Psi_s{}^t X_t{}^u] = [\Phi_r{}^s\Psi_s{}^t]X_t{}^u \text{ and } |_r{}^r\Phi_r{}^s = \Phi_r{}^s = \Phi_r{}^s|_s{}^s,$$

where, for any $k \geq 1$, $|_k{}^k = [j]^k$, an array of k functions j.

b) An s-place *throw* of r functions is a sequence such that all r place numbers are s; for instance, $F_r{}^s = (F_1{}^s, \ldots, F_r{}^s)$. Intersection substitution, defined by

$$F_r{}^s G_s{}^t = (F_1{}^s(G_1{}^t, \ldots, G_s{}^t), \ldots, F_r{}^s(G_1{}^t, \ldots, G_s{}^t)),$$

is associative and admits unilateral neutrals. Let $I_k{}^k$ be the k-place throw of all k-place selector functions in the natural order, and let $(I_k{}^k)^h$ denote the k-place throw of hk functions forming a chain of h throws $I_k{}^k$. Then

$$F_r{}^s(G_s{}^t H_t{}^u) = (F_r{}^s G_s{}^t)H_t{}^u \text{ and } I_r{}^r F_r{}^s = F_r{}^s = F_r{}^s I_s{}^s.$$

By mean sof intersection substitution, the array $\Phi_r{}^{rs}$ and the throw $F_r{}^s$ with the same components $F_1{}^s, \ldots, F_r{}^s$ can be connected: $F_r{}^s = \Phi_r{}^{rs}(I_s{}^s)^r$.

Commutativity and associativity of addition and the distributive law can be expressed in the formulae:

$$S = S(I_2{}^2, I_1{}^2); \ S[j, S] = S[S, j]; \ P[S, j] = S[P, P](I_1{}^3, I_3{}^3, I_2{}^3, I_3{}^3).$$

The existence of right-neutrals has the following simple

Corollary. *Every non-empty function of any number of places lends itself to substitutions (of both types) with non-empty results.*

For any $k \geq 1$, the k-place throws of k functions form a hypergroup by intersection substitution. More generally, throws as well as arrays of functions constitute what might be called *hypergroupoids* — a concept that will be studied elsewhere.

Both types of substitution can be extended to n-ary relations. For instance, if P is a class of sequences of $n + 1$ elements; and if $\Pi_1, \ldots, \Pi_n$ are classes of (not necessarily consistent) ordered pairs, then $P[\Pi_1, \ldots, \Pi_n]$ will denote the class of all sequences $(\alpha_1, \ldots, \alpha_n, \gamma)$ such that for some $\beta_1, \ldots, \beta_n$:

$(\alpha_1, \beta_1) \in \Pi_1, \ldots, (\alpha_n, \beta_n) \in \Pi_n$ and $(\beta_1, \ldots, \beta_n, \gamma) \in P$.

In what precedes, only *real* functions have been considered, but all statements (including the following remarks) remain valid if one selects a ring (or, where division is involved, a field) and writes *element of the ring (the field)* instead of *real number*.

The definitions of arithmetical operations for functions (addition, etc.) merely presuppose classes of consistent pairs *whose second members are real numbers*. The nature of the first members plays no role. Operating on functions with disjoint domains, however, yields $\emptyset$; for instance, $j\_^2 + log = \emptyset$ and $j.S = \emptyset$. Hence, for some results in a class of functions to be non-empty, it is necessary that *some domains be non-disjoint* [17]. With this proviso, the arithmetical operations may be extended to what I will call *functors*—classes of consistent quantities, if *quantity* is any ordered pair whose second member is a number [18]. Of course only functors whose domains consist of mathematical entities are objects of pure mathematics. Mathematical functors that are not functions have been called *functionals*; e.g., the class $\int_0^1$ of all pairs $(f, \int_0^1 f)$ for any integrable function f.

Substitution presents an altogether different situation. If the result $f_1 f_2$ is non-empty it is so because the first member of the pair $(y, z) \in f_1$ is the second member in a pair $(x, y) \in f_2$; in other words, because functions are classes of pairs *whose first and second members are of like nature* [19]. A similar reason accounts for substitutions of sequences of functions into functions of several places. In view of the corollary on p. 462, *the only functors that lend themselves to substitution with some non-empty results are the functions*. Calling every class of consistent quantities a "function" (which has been proposed) thus epitomizes overemphasis on addition and multiplication as well as supreme disregard for the paramount operation in the realm of functions — substitution.

II. FLUENTS

The objects of science and geometry to which Newton referred as fluents and which he and his successors have treated with supreme virtuosity

---

[17] Functions of the same place-number, and even throws, satisfy this condition, and actually lend themselves to meaningful addition and multiplication.

[18] Cf. *Calculus*, Chapter VII.

[19] What that common nature of the elements *is* plays no role in the definition of substitution. For any set S, one may consider classes of consistent pairs of elements of S (self-mappings of S) and define substitution. Examples include n-place throws of n functions.

have not, in the classical literature, ever been defined either explicitly or, by postulates, implicitly. There are of course scientific procedures determining, for instance, $p\gamma_0$, the gas pressure in atm. of a specific instantaneous gas sample γ_0, corresponding to arithmetical definitions of *log* 2. But the function *log* (even though its definition on p. 454 presupposes the understanding of *log* x for any x) must be distinguished from the numbers *log* x as well as from the class ran *log*. Similarly, p — in the sequel, fluents as well as 1-place functions will be designated by lower case italics — must be distinguished from the numbers $p\gamma$ as well as from ran p (the class of all those numbers). The fluent p is the class of all pairs $(\gamma, p\gamma)$ for any instantaneous gas sample γ.

Besides this (as it were, objective) pressure p, there is, for any observer A, a fluent p_A, *the gas pressure in atm. observed by* A, which is the class of all pairs $(\alpha, p_A\alpha)$ for any act α of A's reading a pressure gauge calibrated in atm., where $p_A\alpha$ denotes the number — *the pure number*, say, 1.5 — read by A as the result of α.

> Thus extramathematical features (such as "denomination" and "dimension") that are often attributed to the values of p and p_A are, as it were, absorbed in the definitions of these fluents. Their values being pure numbers, also ran p and ran p_A are *objects of pure mathematics*. In contrast, dom p and dom p_A and, therefore, p and p_A themselves are *extramathematical objects*. The definition of an entire fluent adds to the knowledge of its values the idea of a class — a class that is highly significant in some physical laws and, in fact, indispensable if intuitive understanding (however efficient) of those laws is to crystallize in articulate formulations.

Differentiation between p, on the one hand, and the numbers $p\gamma$ or the class ran p, on the other, however slight the difference may appear, is at variance with the entire traditional literature on fluents inasmuch as the latter is at all articulate. McKinsey, Sugar, and Suppes [20] introduce time as a class of numbers (clock readings) and Artin [21] takes a similar position (whereas, from the point of view here expounded, t_A, for an observer A, is the class of all pairs $(\tau, t_A\tau)$ for any act τ of clock reading performed by A). Courant says explicitly [22] that Boyle's law deals only with the *values* of p and v and not with those quantities themselves. All that physics supplies, he emphasizes, are the classes of values of p and v.

In fact, Courant mentions p as an example of a *variable* (a symbol that

[20] Cf. McKinsey, Sugar and Suppes [9].
[21] Cf. Artin [1], p. 70.
[22] Cf. Courant [3], p. 16.

may be replaced with the designation of any element of a class of numbers), thereby illustrating another error pervading the traditional literature: the identification of fluents with what herein is called number variables, and the indiscriminate use of the term *variable* as well as the same (italic) type for both.

Yet — and this is a mere hint of the actual gulf separating the two — number variables *may* be interchanged, whereas fluents (e.g., abscissa and ordinate along a curve in a Cartesian plane, x being the class of all pairs $(\pi, x\pi)$ for any point π on the curve) *must not*. For instance, the class of all (x, y) such that $y = x^2$ is the same as the class of all (y, x) such that $x = y^2$, whereas the parabola $y = x^2$ and the parabola $x = y^2$ are different curves.

The confusion is enhanced by the use of the term variable, thirdly, for symbols that are replaceable with the designations of any element of some well-defined class of fluents or of classes of consistent quantities — in other words, for *fluent variables* or *c.c.q. variables*; e.g., for u in the statement $\dfrac{d \, sin \, u}{du} = cos \, u$ for any c.c.q. u that is continuous on (the limit class) dom u. Here, u may be replaced with the designation of the time [23] or the abscissa or even a continuous functional (as is $\int_0^1$ in the realm of continuous functions whose limit is defined by uniform convergence), but nor with the designation of a number. One has

$$\frac{d \, sin \, t}{dt} = cos \, t, \quad \frac{d \, sin \, x}{dx} = cos \, x \text{ and even } \frac{d \, sin \int_0^1}{d\int_0^1} = cos \int_0^1,$$

whereas $\dfrac{d \, sin \, 1}{d1} = cos \, 1$ is nonsense.

The literature also contains allusions to fluents that avoid confusing them with either classes of numbers or number variables. But those allusions (usually to "variable numbers") are inarticulate beyond recognition. For instance Russell [24], Tarski [25], and other logicians in discussing number variables have repeatedly criticized the misconception of numbers that are capable of various values; and indeed, there are no numbers that are both 0 and 1, nor, as some one put it, numbers that have different values on weekdays and on Sundays. What logicians seem to overlook, however, is the fact that many obscure allusions to "variable numbers" do not refer to number variables in the logico-mathematical

[23] Strictly speaking, the domain of a fluent is not a limit class. In a model, however, according to the concluding remarks of the present paper, t and s may be assumed to be *continuous classes of consistent quantities on domains that are limit classes*. Cf. *Calculus*, pp. 220–225.

[24] Cf. Russell [20], p. 90.

[25] Cf. Tarski [21], pp. 3, 4.

sense, but rather represent utterly confused references to Newton's fluents. A fluent (without of course being a variable number) may indeed assume both the value 1 and the value 0. In fact, it may (as does, e.g., the admission fee in $ to certain art galleries) assume the value 1 on weekdays and the value 0 on Sundays.

In the broadest sense, a *fluent* may be defined as *a class of consistent quantities with an extramathematical domain* — the c.'s c.q. with mathematical domains being functions and functionals. Fluents such as the class *h* of all pairs (F, *h*F) for any Frenchman F, where *h*F is F's height in cm. (studied in biology and sociology), are sometimes called *variates*; their domains, *populations*.

Clearly, not every quantity, as defined on p. 463, is *interesting*; nor is every fluent *significant*, even if its elements are interesting quantities — think of the union of the height in the population of France and the weight in the population of Italy. Nor, for that matter, is every function and every functional important. While the general theory, of course, provides the scheme for handling *all* fluents, it is up to the individual investigator to apply it to some of the countless cases that are theoretically or practically significant.

Some critics of the theory here expounded have suggested that its basic idea, the concept of fluent, has always been known, viz., under the name of "real function" and, moreover, follows the pattern of Kolmogoroff's well-established concept of random variables — r.v.'s. Besides overextending the use of the term function (see p. 463), those critics seem to overlook: (1) that what is essential in the theory is the formulation of definitions for the (heretofore only intuitively used) concepts that Newton called fluents — definitions that are at variance with their traditional treatment, which ignores classes of pairs altogether (see p. 464); (2) that scientific fluents and r.v.'s lack one another's very characteristics and are, if anything, complementary rather than parallel concepts. [26] If Δ is a physical die, then the (*extramathematical*) class t_Δ of all pairs $(\delta, t\delta)$ for any act δ of rolling Δ is *an experimental fluent but not a r.v.* — not even if an additive functional ("probability") is defined for the 2^6 subsets of ran $t_\Delta = \{1, \ldots, 6\}$ (i.e., the class S of all possible outcomes of rolling Δ). On the other hand, in presence of such a probability functional on the subsets of S, any (*purely mathematical*) function having S as *domain* is *a r.v. but not a scientific fluent*; e.g., the function f for which $f(1) = \sqrt{7}$, $f(2) = \pi + e$, ..., $f(6) = \cos 2 + \log 5$. By their definitions, r.v.'s lack connections with experiment and observation. Again, scientific fluents such as t_Δ, gas pressure, and time lack the characteristic of r.v.'s, since the definition of a reasonable probability on subsets of their *domains* is completely out of the question. (What should be the probability of an act of rolling Δ, or of a gas sample or of an act of clock reading? Only in the *range* of a scientific fluent can one define frequency, relative frequency and, perhaps, probability.) (3) That even some-

[26] Cf. Menger [17], pp. 222–223.

one referring to all functors as "functions" cannot escape the use of a special term (say, "functions in the strict sense") referring to those functors whose domains consist of numbers or sequences of numbers. For (because of their substitutive properties, *not shared by any other functors*) these functors play a special role, and therefore are omnipresent in science as well as in mathematics. While in the light of the conceptual clarifications, terminological questions are quite insignificant, it does seem most appropriate to call *fluents* what Newton called fluents, and *functions*, what Leibniz called functions.

The union of non-identical fluents with the same domain is not a fluent. From a set-theoretical point of view, *fluents do not constitute a Boolean algebra*. One of the few positive formal properties of fluents is the possibility of substituting them into 1-place functions: $\log p$ is the class of all pairs $(\gamma, \log p\gamma)$ for any sample γ — a definition analogous to that of $\log \cos$. But while also the cosine of the logarithm is a c.c.q. $\neq \emptyset$, the pressure of the logarithm is empty. Every function permits *some* non-empty substitutions, whereas a fluent (like a functional) permits *none*.

Attempts have been made to dodge the problem of articulately connecting various fluents by *defining* some of them as functions of others [27]. Yet, even if in a gallery a sign declares that admission costs $ 1 on weekdays and is free on Sundays, the concept of admission fee cannot very well be said to be defined as a function of the time. Someone unfamiliar with that concept will not grasp it by reading the sign while, on the other hand, the concept is comprehensible to persons ignorant of the days of the week. Actually, admission fee might (for operative purposes) be defined as the class a of all pairs (A, aA) for any act of admitting a visitor, where aA is the amount in $ charged during A. The sign, comprehensible only to those who know a and t, stipulates how the two are connected.

By substitutions into 2-place functions S, P, etc., significant addition, multiplication etc. of fluents can be defined, provided that their domains are non-disjoint — the only condition for arithmetical operations on c.'s c.q. to be non-empty (see p. 463). For instance, $P(p, v)$, the result of intersection substitution of p and v (whose common domain is the class of all γ) is $p.v$. But a slight change in the point of view raises difficulties. What, in view of the fact that dom p_A and dom v_A consists of acts of different (manometric and volumetric) observations, is the meaning of $p_A.v_A$? Since Boyle, it has become traditional to associate with that symbol (if only intuitively, i.e., without explicit definitions) the class $((\pi, \beta), p_A\pi.v_A\beta)$ for any two simultaneous acts π and β that A directs to

[27] Cf. the references in footnotes 20 and 21.

the same object; thus $p_A \cdot v_A$ denotes the restriction of $P[p_A, v_A]$ to the class Γ of all pairs of *simultaneous and co-objective* acts $\in$ dom $p_A \times$ dom v_A. It thus appears that in operating on fluents, besides referring to the elements of their several domains, one may well have to relativize the operations to certain pairings of those domains. Such relativizations are imperative in formulating — articulately formulating — relations between fluents.

III. RELATIVE CONNECTIONS OF FLUENTS BY FUNCTIONS

Consider Boyle's law for gas undergoing an isothermal process — in proper units, $v = \dfrac{1}{p}$. If all that physics supplied were the values of v and p or the classes of those values, then Boyle might have discovered his law upon being presented with a bag containing cards each indicating a value of p, and another bag informing him of the values of v. But why, in that situation, should Boyle have paired each number in the first bag just with its reciprocal in the second rather than, say, with its square root? As a matter of fact, Boyle did not primarily pair numbers at all. Pairing numbers is what mathematicians do in defining functions. What Boyle actually paired were observations pertaining to the same object; and he discovered that

(3) $v\gamma = \dfrac{1}{p\gamma}$ for any inst. gas sample γ at the fixed temperature.

This statement is comparable to

(4) $cot\ \mathrm{x} = \dfrac{1}{tan\ \mathrm{x}}$ for any number x that is not a multiple of $\pi/2$,

with v and p corresponding to *cot* and *tan*; and the sample variable γ, to the number variable x.

Unfortunately, the classical literature has done all that was possible to conceal the existing analogies. Besides, it has simulated a parallelism between v, p and x by indiscriminately referring to them as "variables" and using the same (italic) type for all of them (whereas the functions are usually denoted by cot and tan). In an attempt to mask the confusion between fluents and number variables, a contradiction in terms comparable to "enslaved freeman" was coined: "dependent variable". Finally, the true analogues of $v = 1/p$, formulae such as $cot = 1/tan$ (connecting two functions just as Boyle's law connects two fluents) are anathema, and only the corresponding statements about numbers, such as (4) are admitted. [28]

For an observer A, Boyle's law takes the form

(5) $$v_A\beta = \frac{1}{p_A\pi} \text{ for any two acts } (\pi, \beta) \in \Gamma.$$

Relativizing connections of two fluents to a class Γ of pairs of simultaneous co-objective acts is very natural though not logically cogent. At any rate, since Galileo and Boyle, such (tacitly understood) relativizations have become second nature to physicists, who have transplanted them, as matters of course, even to quantum mechanics — a field where they are rather problematic. In $v = 1/p$, the pairing is altogether hidden.

On the level of general statements about fluents, however, the need for *explicit* relativizations is evident. The question "Is $w = 1/u$?" for any two fluents is incomplete. Certainly it does not necessarily refer to the entire class dom $u \times$ dom w; that is to say, it does not necessarily mean "Is *each* value of w the reciprocal of *each* value of u?" In this sense, for an affirmative answer it would be necessary that both u and w were constant fluents. The question thus must refer to some *subset* of dom $u \times$ dom w. But to *which* subset? No particular subset of the Cartesian product of any two (especially disjoint) sets is or can be "natural". The intended subset must be *specified*. Such a relativization is necessary in order to make the question complete.

In the broadest sense, the connection of a class of consistent quantities w with another c.c.q. v relative to a set $\Pi \subseteq$ dom $u \times$ dom w by the function f is described in the following basic definition:

$$w = fu(\text{rel. } \Pi) \text{ if and only if } (\alpha, \beta) \in \Pi \text{ implies } w\beta = fu\alpha.$$

Here, the consequent might be replaced with: $(u\alpha, w\beta) \in f$. For instance, (3) results if Π is the class I of all pairs (γ, γ). The connection of *functions* by functions in traditional analysis is relative to restrictions of j. If j' is the restriction of j to numbers that are not multiples of $\pi/2$, then (4) subsumes under the general scheme:

$cot = j^{-1} \, tan$ (rel. j') since cot y $= j^{-1} \, tan$ x for any (x, y) $\in j'$.

---

[28] It is not unusual to write, e.g.: if $f = 1/g$, then $g = 1/f$ for any two functions f and g (thus dispensing with number variables). But, in violation of automatic substitutive procedures, the function variables f and g are replaced, e.g., by *tan* x and *cot* x, and not in the traditional literature by *tan* and *cot*.

Clearly, $w = fu$ (rel. Π) implies $u = \text{Inv } fw$ (rel. conv. Π); and

$w = fv$(rel. Π) and $v = gu$ (rel. P) imply $w = fgu$ (rel. ΠP).

It is now clear why functions have been defined as on p. 454, and "multi-valued" functions have been strictly excluded. If the latter were admitted, then, relative to *every* pairing, *every* fluent would be a function of *every* other fluent. The question "Is w a function of u rel. Π?", which is so important in science (e.g., in thermodynamics), would be deprived of any meaning. However, for any 2-place function F, one may define:

(6) $F(u, w) = 0$ (rel. Π) if and only if $(\alpha, \beta)\in\Pi$ implies $F(u\alpha, w\beta) = 0$.

Of course only if $F(u\alpha, w\beta) \neq 0$ for *some* $(\alpha, \beta)\in$ dom $u \times$ dom w (especially, if $F \neq 0$) does (6) establish a connection between u and w.

The most general connection of a functor w with n functors $v_1, \ldots, v_n$ relative to $\text{P} \subseteq \text{dom } v_1 \times \ldots \times \text{dom } v_n \times \text{dom } w$ by the n-place function G is given by:

$w = G[v_1, \ldots, v_n]$ (rel. P) if and only if

$(\beta_1, \ldots, \beta_n, \gamma) \in \text{P}$ implies $w\gamma = G(v_1\beta_1, \ldots, v_n\beta_n)$.

The chain rule reads as follows:

$w = G[v_1, \ldots, v_n]$ (rel. P) and $v_i = F_i[u_{i,1}, \ldots, u_{i,m}]$ (rel. Π_i) imply

$w = G[F_1, \ldots, F_n][u_{1,1}, \ldots, u_{n,mn}]$ (rel. P$[\Pi_1, \ldots, \Pi_n]$).

The rate of change of w with, say, v_n rel. P (keeping $v_1, \ldots, v_{n-1}$ unchanged) is a fluent with the domain P, which must not be confused with the n-th place partial derivative D_nG, which is an n-place function with a domain $\subseteq$ dom G. While the two symbols are frequently misrepresented as synonyms, the concepts are connected [29] by the formula:

$$\left(\frac{\partial w}{\partial v_n}\right)_{v_1, \ldots, v_{n-1}} = D_nG(v_1, \ldots, v_n) \text{ (rel. P).}$$

But it is important to note that the rate of change of w with v_n rel. P may well exist without w being a function of $v_1, \ldots, v_n$ rel. P. An analogous distinction is necessary between the cumulation of w with v_n and the n-th place partial integral of G.

From the preceding exposition of the material, based on explicit definitions, there emerge the outlines of its axiomatic treatment. A group

[29] Cf. *Calculus*, Chapter XI, especially pp. 306–315 and 332–341 and Menger [16].

I of postulates has to be devoted to *partial order* in a realm of undefined entities (called n-ary relations) in which there are two operations, *intersection* and *Cartesian multiplication*, subject to postulates of group II. In terms of these operations, associative substitutions are introduced (group III). Union of relations plays a small role, if any, and certainly none in that important subclass of relations whose elements are called *classes of consistent pairs* (group IV), because in the realm of c.'s c.p. union cannot in general be defined. Of particular significance among c.'s c.p. are *selector* and *identity relations* (group V) which, as has been illustrated in the realm of the 1-place functions, play the roles of domains of c.'s c.p. At this point, the class of all real numbers (or, if one pleases, a field or ring) enters the picture. By means of it, consistent classes of quantities or *functors* can be singled out (group VI) and, among them, *functions* constituting a hypergroupoid. Selector relations that are functions are the all-important selector functions, including the identity function *j*. What precedes is a basis for treating the connection of one functor with n other functors by means of an n-place function relative to an $n + 1$-ary relation between their domains, as well as a functional interrelation of m functors relative to an m-ary relation.

Clearly, such an axiomatic theory represents the most general treatment of *models* in the sense in which this term is used in science, especially, in social sciences. An analogy appears with postulational geometry, which deals with undefined elements, called points and lines for the sake of a suggestive terminology, while all that is assumed about them is that they satisfy certain assumptions. Subsequently, they are compared with observable objects, e.g., in the astronomical space, with cross hairs and light rays. Models are formulated in terms of functor variables — undefined classes of consistent quantities. called, say, time and position or pressure and volume and denoted by t and s or p and v, for the sake of a suggestive terminology, while all that is assumed about them is that, relative to undefined pairings of their domains, those functors are interrelated by certain functions. Subsequently, an observer A compares them with observed fluents (t_A and s_A or p_A and v_A) relative to specified pairings of the domains of the latter. He trusts that, within certain limits of accuracy, the statements concerning the undefined functors in the model will be verified by known connections between the observed fluents — some of them, he hopes, by previously unknown connections [30].

[30] The ideas here outlined seem to supplement the existing theory on concept formation in empirical science; cf. Carnap [2] and Hempel [5].

As far as the general theory of fluents is concerned, the prediction may be ventured that indiscriminate uses of the term "variable" and of nondescript letters x will give way to more careful distinctions; and that references to domains of fluents as well as to pairings of those domains, once introduced, will be permanently incorporated in the articulate formulations of scientific laws.

Acknowledgements

The author is grateful to Professors M. A. McKiernan, B. Schweizer, and A. Sklar for valuable suggestions in connection with this paper, and to the Carnegie Corporation of New York for making it possible to devote time to the development of the material.

Bibliography

[1] ARTIN, E., *Calculus and Analytic Geometry*. Charlottesville 1957, 126 pp.

[2] CARNAP, R., *The methodogical character of theoretical concepts*. In Minnesota Studies in Philosophy of Science, vol. 1, Minneapolis 1956.

[3] COURANT, R., *Differential and Integral Calculus*, vol. 1,

[4] HELLER, I., *On generalized polynomials*. Reports of a Mathematical Colloquium 2nd. ser., issue 8 (1947), pp. 58–60.

[5] HEMPEL, C. G., *Fundamentals of concept formation in the empirical sciences*. International Encyclopedia of Unified Science, vol. 2 no. 7 Chicago 1952.

[6] McKIERNAN, M. A., *Les séries d'itérateurs et leurs applications aux équations fonctionelles*. Comptes Rendus Paris, vol. 246 (1958), pp. 2331–2334.

[7] ——, *Le prolongement analytique des séries d'itérateurs*. Comptes Rendus Paris, vol. 246 (1958), pp. 2564–2567.

[8] McKINSEY, J. C. C., *Postulates for the calculus of binary relations*. Journal of Symbolic Logic, vol. 5 (1940), pp. 85–97.

[9] ——, SUGAR, A. C. and P. SUPPES, *Axiomatic Foundations of classical particle mechanics*. Journal of Rational Mechanics and Analysis, vol. 2 (1953), pp. 253–272.

[10] MENGER, K., *Calculus. A Modern Approach*. Boston 1955, XVIII + 354 pp.

[11] ——, *The ideas of variable and function*. Proceedings of the National Academy, U.S.A., vol. 39 (1953) pp. 956–961.

[12] ——, *New approach to teaching intermediate mathematics*. Science, vol. 127 (1958) pp. 1320–1323.

[13] ——, *Algebra of Analysis*. Notre Dame Mathematical Lectures, vol. 3, 1944. 50 pp.

[14] ——, *Tri-operational algebra*. Reports of a Mathematical Colloquium, 2nd series, issue 5–6 (1945) pp. 3–10 and issue 7 (1946) pp. 46–60.

[15] ——, *Calculus. A Modern Approach*. Mimeographed Edition, Chicago 1952, XXV + 255 pp.

[16] ——, *Rates of change and derivatives*. Fundamenta Mathematicae, vol. 46 (1958), pp. 89–102.

[17] ——, *Random variables and the general theory of variables*. Proceedings of the 3rd Berkeley Symposium on Mathematical Statistics and Probability, vol. 2, Berkeley 1956, pp. 215–229.

[18] MILGRAM, A. N., *Saturated polynomials*. Reports of a Mathematical Colloquium, 2nd series, issue 7 (1946), pp. 65–67.

[19] NÖBAUER, W., *Über die Operation des Einsetzens in Polynomringen*. Mathematische Annalen vol. 134 (1958) pp. 248–259.

[20] RUSSELL, B., *The Principles of Mathematics*. Vol. 1. Cambridge 1903, XXIX + 534 pp.

[21] TARSKI, A., *Introduction to Logic*. New York 1941, XVIII + 239 pp.

[22] ——, *On the calculus of relations*. Journal of Symbolic Logic, vol. 6 (1941) pp. 73–89.

The algebra of functions:
past, present, future [*]

by K. MENGER (Chicago, Illinois)

RIASSUNTO. — *Viviamo, si dice, nell'età dell'algebra e delle teorie astratte assiomatiche. Ma uno dei concetti matematici più importanti — il concetto di funzione — non è stato nè « algebrizzato » nè assiomatizzato. Così come nelle definizioni astratte di gruppo e di corpo gli algebristi pongono assiomi corrispondenti a qualche proprietà essenziale dei numeri, in questa Conferenza mi propongo di precisare talune proprietà essenziali di funzioni reali, e di scegliere qualcuna fra esse che possa servire come fondamento di una teoria astratta — un'algebra di funzioni.*

Mi occuperò qui di funzioni reali dette tradizionalmente funzioni « di una variabile » (ogni tale oggetto dell'analisi pura è di fatto un insieme di coppie ordinate di numeri reali di un certo tipo, cioè un insieme tale che, se (x, y) e (x', y') vi appartengono, allora da x = x' segue y = y'). Ma lo sviluppo della teoria aprirà anche prospettive di estensioni alle funzioni dette « di più variabili ».

I. The beginnings. The past few decades have witnessed the development of extensive theories of vector spaces and rings, applicable to the family $\mathcal{F}_S$ of all real functions having a set S as their common domain. These theories deal with, and to some extent generalize, the *addition* of such functions and their *multiplication* by scalars and by each other. Addition and multiplication in $\mathcal{F}_S$ are associative and commutative operations analogous to the addition and multiplication of numbers. Just as there are neutral numbers, 0 and 1, there are neutral functions: the constant functions of the values 0 and 1. They are defined as the results of associating, with every element of S, the numbers 0 and 1, respectively, or, in other words, as the sets of all pairs (x, 0) and (x, 1) for each element x of S. I will denote these functions by *0* and *1*. (More generally, all functions will be designated in *italics*, while all references to

(*) Conferenza tenuta il 29 aprile 1961 presso l'Istituto Nazionale di Alta Matematica.

numbers will be printed in roman type (^{1})). Thus

(1) $0 + f = f = f + 0$ and $1 \cdot f = f = f \cdot 1$ for any function f.

Moreover, addition and multiplication are connected by a distributive law

(2) $(f + g) \cdot h = f \cdot h + g \cdot h$ for any functions f, g, h.

Some difficulties arise with regard to reciprocals. Whereas in the realm of numbers only 0 lacks a reciprocal, in $\mathcal{F}_S$ there are, besides the constant function 0, many functions without reciprocals. If f is any function belonging to $\mathcal{F}_S$ and including at least one pair $(x_0, 0)$ for some element x_0 of S (that is to say, if f assumes the value 0 at all), then a reciprocal of f with a domain including x_0 does not exist; in other words, then there is no reciprocal of f in $\mathcal{F}_S$. This is why the family $\mathcal{F}_S$ is not a field. Apart from this difference, however, the addition and multiplication of functions resemble the corresponding operations for numbers. In particular, each function in $\mathcal{F}_S$ as well as each number has exactly one negative; that is to say, for each f in $\mathcal{F}_S$ there exists exactly one function $-f$ such that

(3) $-f + f = 0,$

whence $\mathcal{F}_S$ is an additive-multiplicative ring.

In contrast to numbers, however, real functions lend themselves to an operation that is much more closely related to the very idea of function than are addition and multiplication — the *substitution* of functions into functions. This operation must not be confused with the substitution of, say, y *for* x in the formula $x + 1 = 1 + x$, which is nothing but a simple replacement resulting in $y + 1 = 1 + y$. Substituting the 3rd root *into* (not *for*!) the 5th root results in an altogether different function: the 5th root of the 3rd root, that is, the 15th root; in symbols: $\sqrt[3]{} \; \sqrt[5]{} = \sqrt[15]{}$.

Numbers do not lend themselves to substitution; nor can one, if S consists of arcs, surfaces or the like, substitute real functions

---

(^{1}) Only in quoting the classical literature shall I adhere to the rather bizarre traditional convention: to print, *regardless of meaning*, one-letter-symbols in italics and multi-letter symbols in roman type, e.g., e, x, $f(x)$, $\log x$, $\arctan x$, etc.

even into functions belonging to $\mathcal{F}_S$. Only if the domain of f consists of *real numbers* or of (finite or infinite) *sequences of real numbers* does f lend itself to the substitution of real functions. Thus, in traditional real analysis, substitution is confined to functions in the comparatively restricted sense in which this word is used by most scientists and by the numerous mathematicians who, following Volterra and Hadamard, reserve it for the results of associating numbers *with numbers or sequences of numbers*, and refer to arc length, area, and similar associations of numbers with arcs, surfaces and the like by a different (specially created!) term, namely, as *functionals*. Functionals do not lend themselves to substitution any more than numbers. There is no area of the arc length [2].

Functions whose domains are sets of single numbers, however (e.g., roots, powers, and trigonometric functions) can be substituted into one another. I will denote the result of substituting g into f by fg. (In the symbol for the product $f \cdot g$ the dot will never be omitted!) This associative, if in general non-commutative, operation possesses a neutral element, namely, the set of all pairs (x, x) for any x, which I will call the *identity function* and denote by j. Thus, in analogy to (1),

(1') $fj = f = fj$ for any function f.

In the realm of functions, substitution truly is the operation par excellence [3]. Yet, compared to addition and multiplication,

[2] Fréchet's classical synthesis of functions such as roots and trigonometric functions, on the one hand, and functionals such as arc length, on the other, has been immensely fertile as far the former functions (in the strict sense) and functionals are analogous, that is, with regard to limits, continuity properties, addition and multiplication. For example, if the domain S is a compact space, then every lower semicontinuous association of numbers with the elements of S attains its minimum regardless of whether S consists of numbers or pairs of numbers or curves or functions. But substitution, which promises to play a role of paramount importance in the realm of real functions in the restricted sense of this word, escapes that analogy, since functionals do not lend themselves to the substitution of either real functions or functionals into them.

[3] For a simple graphical construction of the result of substituting g into f see my Note « Frammenti piani auto-duali e relative sostituzioni. *Rend. Acc. Lincei* ser. 8, vol. 30, 1961, p. 713. That paper generalizes the graphical substitution of linear functions (i.e., of non-vertical lines) to projective planes and outlines a foundation of projective geometry on that substitution.

substitution has in classical analysis unfortunately been treated as a step-child. No symbol has been introduced for the neutral function, although the identity function is of the same paramount importance for analysis as 0 and 1 are for arithmetic. Traditionally it is referred to by its value for x as « the function x ». Two distributive laws involving substitution, which are at least as important for analysis as is the additive-multiplicative law (2), usually are not even mentioned: an *additive-substitutive* and a *multiplicative-substitutive-distributive law*:

$$(2')\ (f+g)h = fh + gh \ \text{ and } \ (f \cdot g)h = fh \cdot gh \ \text{ for any functions } f, g, h.$$

In the early 1940's, I developed an abstract theory of functions[4] by postulating a tri-operational algebra, more specifically, by assuming a set $\mathcal{F}$ of elements (whose nature remains undefined) and three operations, called addition, multiplication, and substitution. The elements will be designated in *italics*. The results of applying the said operations to the elements f and g (called their sum, product, and transform) may be denoted by

$$f+g \quad \text{or} \quad \mathrm{S}(f, g), \quad f \cdot g \quad \text{or} \quad \mathrm{P}(f, g), \quad fg \quad \text{or} \quad \mathrm{T}(f, g),$$

respectively. The operations were assumed to be associative, to admit mutually unequal neutral elements, denoted by $0, 1$, and j satisfying (1) and (1'), and to be connected by three distributive laws (2) and (2'). It was further assumed that addition and multiplication are commutative, and that each element has a negative satisfying (3). In contrast to classical analysis, where the neutral functions are explicitly defined as classes of pairs of numbers, in the algebra of functions, $0, 1$ and j are elements of unspecified nature, given and uniquely determined by the assumption that they satisfy (2) and (2'): $0 + f = f = f + 0$ or $\mathrm{S}(0, f) = f = \mathrm{S}(f, 0)$ for each element f, etc.

The simplest finite tri-operational algebra is the quadruple of elements $0, 1, j$, and n for which the operations are defined by

---

[4] Cf. *Algebra of Analysis*. Notre Dame Mathem. Lectures, vol. 3, 1944 and « Tri-Operational Algebra », *Reports of a Mathem. Colloquium*, 2nd ser., issue 5-6, 1945, p. 3 and issue 7, 1946, p. 46.

the following tables:

S	0 1 j n
0	0 1 j n
1	1 0 n j
j	j n 0 1
n	n j 1 0

P	0 1 j n
0	0 0 0 0
1	0 1 j n
j	0 j j 0
n	0 n 0 n

T	0 1 j n
0	0 0 0 0
1	1 1 1 1
j	0 1 j n
n	1 0 n j .

In an abstract tri-operational algebra $\mathcal{F}$, au element c is called a *constant* if and only if $cf = c$ for each element f of $\mathcal{F}$. For example, 0 and 1 are constants, while j is not. If c and d are constants, then so are $c + d$, $c \cdot d$ as well as fc and cf for each f in $\mathcal{F}$.

Examples of tri-operational algebras include *polynomials*, i.e., functions $c_0 + c_1 \cdot j + ... + c_n \cdot j^n$ for any non-negative integer n, where the c_i belong to a field of constant functions.

With each element g of $\mathcal{F}$, one can associate the set Φ_g of all pairs (f, gf) for each f in $\mathcal{F}$. No matter what may be the nature of the element g of the abstract algebra, the associated set Φ_g is a mapping of $\mathcal{F}$ into $\mathcal{F}$. If $g_1 \neq g_2$, then $\Phi_{g_1} \neq \Phi_{g_2}$, since Φ_{g_1} includes the pair $(j, g_1 j)$, which is (j, g_1), while Φ_{g_2} includes the pair $(j, g_2 j)$, which is (j, g_2). I used the functions Φ_g as representations of the elements g of the abstract algebra $\mathcal{F}$.

Let Γ_g denote the set of all pairs $(c, g c)$ for any constant c in $\mathcal{F}$. Each Γ_g is a function having the set of all constants as its domain. The sets Γ_g, which are of course subsets of the corresponding Φ_g, are of particular interest in case that $\Gamma_{g_1} \neq \Gamma_{g_2}$ for any two elements $g_1 \neq g_2$. It this case, the constants are said to constitute a *substitutive base* of $\mathcal{F}$, and the Γ_g may be used as a representation of $\mathcal{F}$. If, in particular, $\mathcal{F}$ is the additive-multiplicative-substitutive algebra of the real functions whose domains consist of real numbers, then each constant function corresponds to a number (viz., the value of the function), and Γ_g is essentially equal to g.

Tri-operational algebra gave rise to some interesting algebraic studies [5]. Analysts, however, continue to speak about « the function $\log x$ », « the function $f(x)$ », and « the function x », thus symboli-

---

(5) See various papers in *Reports of a Mathem. Colloquium*, issue 7, 1946, especially, A. N. MILGRAM « Saturated Polynomials », p. 65. Cf. also W. NÖBAUER, *Mathem. Annalen*, 134, 1958, p. 248 and his study of rational functions forthcoming in Sitzungsber. Wiener Akad. Wissensch.

cally identifying functions and values of functions while rejecting a symbol for the identity function. This idiosyncrasy is hard to understand. For, in the first place, symbols for the identity function are coming more and more into use in topology, operator theory, recursive logic, and in the theory of combinators (where they were first introduced by Schönfinkel). In the second place, the identification of functions and function values (first criticized by Frege, Peano, E. H. Moore, and others) is becoming more and more inconvenient in analysis itself. How keenly some mathematicians feel those difficulties and how hard they try to overcome them by creating symbols such as $x \to f(x)$ for f, and $x \to x$ for the identity function is evident from a recent book (^{6}), which lists all those attempted remedies — with the single exception of the truly simple solution proposed in the algebra of functions : Denote the identity function and the constant function of value 1 (the functions *themselves* and not numerical values of theirs !) by symbols such as j and *1*, and designate the other functions by the x-free parts, such as *log* and f, of their traditional symbols (^{7}).

In order to arrive at a proper appreciation of this solution one merely has to try to express the fundamental law (1') or even its special case $jj = j$ in terms of those other proposed remedial symbols, e.g., in the above-mentioned arrow notation or in terms of Church's well-known lamdas :

$$\lambda_x f(x) \quad \text{for } f \quad \text{and} \quad \lambda_x x \quad \text{for } j.$$

Only the formula $jj = j$ seems to be comparable in structure as well as in simplicity with the analogous classical formulae $0 + 0 = 0$ and $1 \cdot 1 = 1$ for numbers or the formulae $0 + 0 = 0$ and $1 \cdot 1 = 1$ for functions.

One of the obstacles that undoubtedly stood in the way of tri-operational algebra was the wide-spread belief that only functions *with the same domain* have a sum and a product and that a function g can be substituted into a function f only if the range (Italian :

(^{6}) J. L. KELLEY, « *General Topology* », 1958, p. 12.

(^{7}) I recently have tried to illustrate the defectiveness of the traditional symbolism of analysis (which lacks symbols for the identity, the square, and the cube function) by describing a world whose arithmetical vocabulary lacks symbols for the numbers 1, 2, 3. Cf. « Gulliver in the Land without One, Two, Three », Mathem. Gazette vol. 43, 1959, p. 241.

codominio) of g is equal to, or at least a subset of, the domain of f. It seemed impossible in the realm of real functions to substitute the constant function -1 into the square root or the logarithmic function. But then, how can $\mathcal{F}$ include -1 as well as $\sqrt{}$, and fg for every f and g? Apparently, a tri-operational algebra cannot include the square root; in other words, the theory seemed to be practically confined to polynomials and rational functions.

While the said belief will be seen to be unfounded, for other reasons it became apparent that the development of a satisfactory algebra of functions actually required some additional ideas.

II. **The present algebra of functions.** The difficulties had to be attacked from two sides : a) by improving the treatment of the explicitly defined real functions, and b) by introducing into the abstract theory a relation not considered in the original tri-operational algebra.

a) With regard to explicitly defined functions, onecan overcome the misconception that addition, multiplication, and substitution could be performed only for *certain* pairs of functions, by stressing ([8]) the following explicit definitions of those operations for *every* pair f, g of real functions whose domains consist of real numbers. The family of all these functions will be denoted by $\mathcal{R}$.

$f + g$ (or $f \cdot g$) is the set of all pairs (x, y) of real numbers such that, for some numbers p and q, the pair (x, p) belongs to f, and (x, q) belongs to g, while $y = p + q$ (or $y = p \cdot q$). For example, $1 + log$ is the set of all (x, y) such that, for some p and q, the pair (x, p) belongs to 1 (thus p $= 1$), and (x, q) belongs to log (thus $x > 0$ and q $= log$ x, while $y = p + q$; in other words, $1 + log$ is the set of all pairs (x, $1 + log$ x) for any x > 0. Similarly, $0 \cdot log$ is the set of all pairs (x, 0) for any x > 0. The function $0 \cdot log$ thus is not equal to 0; it is a subset of the function 0, namely, the restriction of 0 to the domain of log, that is, to the set of all positive numbers.

If f and g are functions whose domains are disjoint, then, according to these definitions, $f + g$ and $f \cdot g$ are the empty set of pairs of numbers, which I will call the *empty function* and denote by $\varnothing$. For example, if 1_N is the restriction of the function 1 to the set of all negative numbers, that is, the set of all pairs (x, 1) for any x < 0, then $1_N + log = \varnothing$. In every study of sets, sooner

([8]) « An axiomatic Theory of Functions and Fluents » in *The Axiomatic Method*, ed. Henkin et al., Amsterdam, 1959, p. 454.

10

or later, the need for the consideration of the empty set makes its-elf felt. That in the traditional literature the empty function has not played a great role, if any, is due to the fact that, in the classical analysis, functions have not explicitly been treated as sets.

NEG f, the negative of f, may be defined as the set of all (x, y) such that (x, p) belongs to f and $y + p = 0$. The preceding is tantamount to the usual definition of $-f$. But the traditional difficulties concerning reciprocals and quotients disappear. For any function belonging to $\mathcal{F}$, the reciprocal REC f may be defined as the set of all (x, y) such that, for some number p, the pair (x, p) belongs to f and $y \cdot p = 1$. If $f = 0$, then no such pair (x, p) exists. Hence REC $0 = \varnothing$. Thus even 0 has a reciprocal. The quotient f/g may be defined as the product $f \cdot$ REC g without any references to the zeros of the denominator.

The result of substituting g into f is defined as the set fg of all pairs (x, z) such that there exists a real number y such that (x, y) belongs to g, and (y, z) belongs to f. For example, if $f = log$ and $g = -j$, then $log(-j)$ is the set of all (x, z) such that, for some real number y, the pair (x, y) belongs to $-j$ (thus $y = -x$), and (y, z) belongs to log (thus $z = log\ y = log(-x)$); in other words, $log(-j)$ is the set of all (x, z) such that $-x > 0$ and $z = log(-x)$ — briefly, it is the set of all pairs (x, $log(-x)$) for any $x < 0$. Similarly, $log(-1)$ is seen to be the empty function. Thus -1 may well be substituted into log (and into the square root) and the result is not different from that of certain additions, such as $1_N + log$.

In the traditional way, let Dom f (and Ran f) be the domain (and the range) of the function f, that is, the set of all numbers x (of all numbers y) such that a pair (x, y) belongs to f. Clearly, Dom $(f + g)$ as well as Dom $(f \cdot g)$ is the (set-theoretical) intersection of Dom f and Dom g, while there are no simple connections between Ran $(f + g)$ or Ran $(f \cdot g)$ with Ran f and Ran g. It is possible, however, to describe both Dom (fg) and Ran (fg) in terms of the following concepts [9]. If f is a function belonging to $\mathcal{R}$, and A and B are two sets of numbers, then denote the set of all pairs (x, y) belonging to f and such that x belongs to A (that y belongs to B)

by f on A (by f into B).

_____

[9] Cf. the author's book *Calculus. A Modern Approach*, Boston 1955 p. 180.

One then can formulate the following laws ([10]) :

(4) $\text{Dom}(fg) = \text{Dom}(g \text{ into } \text{Dom}f)$, $\quad \text{Ran}(fg) = \text{Ran}(f \text{ on } \text{Ran}g)$.

Summarizing one can say: Addition, multiplication, and substitution can be performed for *every* pair of functions belonging to $\mathcal{R}$. The three operations are associative and have neutral elements *0, 1, j* satisfying (1) and (1'). Addition and multiplication are also commutative; and the three distributive laws (2) and (2') hold. Why then is the family $\mathcal{R}$ not a tri-operational algebra?

The answer is that the explicitly defined function NEG f does not necessarily satisfy (3), and that in fact for some functions f belonging to $\mathcal{R}$ no $-f$ satisfying (3) exists. Exactly *one* such function $-f$ exists if and only if f is *inextensible*; that is to say, if Dom f is the set of all numbers. For example, *0, 1*, and *j* have negatives : $-0 \ (= 0)$, -1, and $-j$. There does not, however, exist a function $-log$ such that $-log + log = 0$, since $\text{Dom}(f + log)$ is a subset of Dom log, and hence a proper subset of Dom *0*, for any f in $\mathcal{R}$. Similarly, only if g is an inextensible function and Ran f does not include the number 0 does there exist a function REC g (and exactly one such function) for which REC $g \cdot g = 1$. If g is not inextensible, then all that can be said is that there is more than one function f_1 and more than one function f_2 such that

$$f_1 + g = 0 \text{ on Dom } g \quad \text{and} \quad f_2 \cdot g = 1 \text{ on (Dom } g - 0_g),$$

where 0_g is the set of all x such that (x, 0) belongs to g.

Since there are difficulties about inverting multiplication and even addition, the well-known problems connected with the inverse of (non-commutative!) substitution are not surprising. I will call a function belonging to $\mathcal{R}$ *universal* if its range is the set of all numbers; and I will call g *essential* if for any two pairs (x, y) and (x', y') belonging to g, from y = y' it follows that x = x'. In order that for a function g there exist a function INV g such that ([11])

(3') $\qquad\qquad\qquad$ INV $g \ g = j = g$ INV g

([10]) loc. cit. ([9]), p. 88.

([11]) The following convention for saving parentheses is very useful: *What is within the reach of a symbol in bold-face letters is the immediately following function symbol.* For example, INV $g \ g = ($INV $g) \ g$ and D $fg = ($D$f) \ g$, for the results of substituting g into INV g and into Df. This convention will be consistently adhered to in all that follows. The inverse of $g \ g$ and the derivative of $f \ g$ will be denoted by INV $(g \ g)$ and D $(f \ g)$.

it is necessary and sufficient that g be inextensible, universal, and essential, in which case INV g is the set of all pairs (x, y) such that (y, x) belongs to g, while

$$\text{Dom INV } g = \text{Ran } g \qquad \text{and} \qquad \text{Ran INV } g = \text{Dom } g.$$

The three conditions are satisfied, e.g., for the function j^3, for which INV $j^3 = j^{1/3}$; and for $\sqrt[3]{j + \sqrt{j^2 + 1}} + \sqrt[3]{j - \sqrt{j^2 + 1}}$, whose inverse is $\dfrac{1}{2}(j^3 + 3j)$.

Suppose that g is essential but not inextensible (say, $g = log$ or tan on $[-\pi/2, \pi/2]$) or not universal (say, $g = exp$ or $arctan$) or neither inextensible nor universal (say, $g = arcsin$, whose domain and range are $[-1, 1]$ and $[-\pi/2, \pi/2]$, respectively). Then no function satisfying $(3')$ exists, since for each function f at least one of the functions fg or gf is $\neq j$. Even the substitution of log into the exponential function results only in j_P, the restriction of j to the set of all positive numbers.

Suppose that g is not essential (say, $g = j^2$ or cos or the inextensible universal function $j^3 - j$). Then there are at least two unequal functions f_1 and f_2 such that

$$f g_1 = g f_2 = j \text{ on Ran } g,$$

while $fg = j$ for no function f. E.g., if $g = j^2$, then $j^2 j^{1/2} = j^2(-j^{1/2}) = j$ on Ran j^2, while even the substitution of j^2 into $j^{1/2}$ yields a result $\neq j$, namely, the function assuming for each x the value $|x|$ or abs x.

Each function g possesses, however, a (perhaps empty) restriction that I call [12] the *core* of g, for which there exists a unique inverse in a sense to be defined later. Co g is the set of all pairs (x, y) belonging to g such that g does not include a pair (x', y) for $x' \neq x$. Each essential function is equal to its core. The core of the function $j \cdot exp$ (assuming the value $x \cdot e^x$ for any x) is the restriction of that function to the class of all numbers ≥ 0. The core of j^2 consists of the single pair $(0, 0)$; and Co $(j^2 - 1)^2 = \varnothing$.

Summarizing one can say that it is the impossibility of strictly inverting substitution, and even addition (and not limitations con-

---

[12] Cf. the author's *Calculus. A Modern Approach*, 1955, p. 91 and loc. cit (8), p. 458.

cerning the feasibility of the operations themselves) that keeps $\mathcal{R}$ from being a substitutive group, a multiplicative group, and an additive group (as well as from constituting a tri-operational algebra).

b) The new step that had to be taken in the abstract theory ([13]) was a widening of its scope by including in the foundation a relation of partial order that reflects the restriction-extension relation of functions belonging to $\mathcal{R}$. That relation in $\mathcal{F}$ will be denoted by $\subseteq$. It will remain undefined in the abstract theory. In applications to $\mathcal{R}$, the formula $f \subseteq g$ will mean that f is a *restriction* of g, and g is an *extension* of f or, in other words, that the set of pairs f is a subset of the set of pairs g. In $\mathcal{R}$, one has $\varnothing \subseteq f$ for each f, while in the abstract theory it is not necessary to postulate the existence of a minimum element $\varnothing$.

In another respect, however, it has seemed convenient to narrow the scope of the theory. Instead of addition, multiplication, and substitution, only one associative operation with a neutral element will be considered in $\mathcal{F}$. The result of the operation applied to f and g will be designated by

$$f \circ g \quad \text{or} \quad O(f, g),$$

the neutral element by n. Thus

$$(1_0) \quad n \circ f = f = f \circ n \quad \text{or} \quad O(n, f) = f = O(f, n) \quad \text{for each } f.$$

The theory thus will represent a joint treatment of the three operations considered in $\mathcal{R}$; and one may replace $\circ$ and n with $+$, 0 or with $\cdot$, 1 or with juxtaposition and j. In fact, the theory has still other applications.

The connection between the operation and the partial order is *bilateral monotony* of $\circ$ with regard to $\subseteq$:

$f \subseteq g$ implies $f \circ h \subseteq g \circ h$ and $h \circ f \subseteq h \circ g$ for any f, g, h.

So far, we have assumed that $\mathcal{F}$ is a (not necessarily commutative) partially ordered semi-group with a neutral element. While n is *universally* (i.e., for every f) bilaterally neutral, it will further be postulated ([14]) that for each f there exist two *individual* unilateral

---

([13]) Cf. loc. cit. ([8]), p. 457.
([14]) Loc. cit. ([8]), p. 457.

neutrals, a right-side neutral $\mathrm{R}f$ and a left-side neutral $\mathrm{L}f$, satisfying the following conditions (the third of which might be weakened):

(α) $\mathrm{L}f \circ f = f = f \circ \mathrm{R}f$ for any f;

(β) $\mathrm{L}(f \circ g) \subseteq \mathrm{L}f$ and $\mathrm{R}(f \circ g) \subseteq \mathrm{R}g$ for any f and g;

(γ)   If $f \subseteq n$, then $\mathrm{L}f = \mathrm{R}f = f$.

When applied to substitution in $\mathcal{R}$, the function $\mathrm{R}f$ (the function $\mathrm{L}f$) corresponds to the sets of all pairs (x, x) such that f includes a pair (x, y) (a pair (y, x)). These functions thus in some ways play the role of Dom f (of Ran f), which, however, are sets of *numbers* in contrast to f, $\mathrm{R}f$ and $\mathrm{L}f$, which are sets of *pairs of numbers*. The algebra of functions overcomes that dualism and operates exclusively with undefined elements of one kind.

How can the negative of a function, for example NEG *log*, be characterized in terms of addition and the partial order relation of restriction? While there is no function f such that $f + log = 0$, there are functions such that $f + log \subseteq 0$. But this condition also holds for every *restriction* of $- log$, even for $\varnothing$. In order to rule out proper restrictions of $- log$, one must consider functions f such that $f + log$ is the *most extended* possible restriction of 0, that is, the set O_P of all (x, 0) for each x > 0. But this maximum condition also holds for each *extension* of $- log$, e.g., for the function assuming the value log x or 1 according as x > 0 or x < 0 (i.e., the settheoretical union of the functions log and 1_N). What actually characterizes the negative of log is the following *minimax* property [15]: — *log is the most restricted function which, when added to log, yields the most extended restriction of 0.* Analogous minimax conditions characterize NEG g, REC g, and INV g for any g. The inverse of g can be defined by a minimax property characterizing the set of all pairs (x, y) such that (y, x) belongs to the core of g.

For simple postulates guaranteeing that each element f of an abstract algebra $\mathcal{F}$ possess opposite elements satisfying minimax conditions, the reader is referred to a very interesting paper by Schweizer and Sklar [16]. In it, the authors develop a definition of OP f which, when applied to the three operations in $\mathcal{R}$, yields

---

[15] Loc. cit. (8), p. 458.
[16] *Mathem. Annalen* 139, 1960, p. 366.

NEG f, REG f, and INV f. Besides, they elaborate the algebra of functions in other directions, e.g., by studying idempotent and constant elements.

More recently, Schweizer and Sklar[17] found a very interesting representation of an abstract algebra $\mathcal{F}$. They represent each element g by a mapping closely related to Φ_g introduced in tri-operational algebra. Substitution and restriction in the realm of these mappings are isomorphic with the operation ∘ and the relation $\supseteq$ in $\mathcal{F}$. This isomorphism demonstrates that the postulates of an abstract algebra, as formulated by Schweizer and Sklar, actually reflect all essential properties of functions. Another interesting, if to my mind less natural, representation of $\mathcal{F}$, likewise based on relations (4), was given by Johnson[18], who represents each g by a mapping of sets of subsets of $\mathcal{F}$ on subsets of $\mathcal{F}$, thereby obtaining in general only a homomorphism.

III. **Open problems.** Some problems concerning the algebra of the so-called functions of one variable (cf. Riassunto) — better, 1-*place functions* — are still unsolved. Yet the real future of the algebra of functions seems to me to lie in the study of the vast field of the socalled functions of several variables — better, *multi-place functions*.

(A) ON 1-PLACE FUNCTIONS. In the algebra of 1-place functions, I believe that so far not enough use has been made of the relation of *proper restriction* : $f \subset g$ in the sense of $f \subseteq g$ but $f \neq g$. For example, one might postulate that $f \subset g$ implies $R f \subset R g$. In the theory of opposite functions, it might be useful to define g as *essential* if $f \subset g$ implies $L f \subset L g$ for any f.

Just as the mono operational algebra of functions studied on the preceding pages generalizes and extends the concept of *groups*, one might consider partially ordered systems with two or three operations each of which satisfies the postulates of an algebra of functions and which are connected by distributive laws — generalizations and extensions of the concepts of *rings*, *fields*, and *tri-operational algebras* in the original sense (i.e., without the consideration of a relation of partial order).

Before multi-valued complex functions can be attacked from the point of view of algebra, the development of an *algebra of single-*

(17) *Mathem. Annalen* 143, 1961, p. 440 and Notices Amer. Math. Soc. 7, 1960, p. 956.

(18) *Mathem. Annalen* 142, 1961, p. 317.

valued complex functions must be completed. It is in progress from the standpoint of the explicit theory of those functions as well as from the abstract point of view. In both theories, functions are denoted by symbols of their own instead of by their values for z or for x + iy. For example, there are functions j and *con*, the identity and the conjugate. Explicitly, they are defined as the sets of all pairs (z, z) or (x + iy, x + iy) and (z, $\bar{z}$) or (x + iy, x — iy), respectively, for any complex number z or any two real numbers x, y. Postulationally, they are introduced by assumptions such as (1') and

$$con\ con = j \neq con, \quad con\,(f+g) = con\,f + con\,g,$$

$$con\,(f \cdot g) = con\,f \cdot con\,g \quad \text{for any } f \text{ and } g.$$

The functions

$$\frac{1}{2}\,(j + con) = re \quad \text{and} \quad \frac{1}{2}\,(j - con) = im,$$

traditionally called « the functions x and y », are real-valued. This fact can be expressed in the formulas

$$re\ re = re,\ re\ im = im, \quad im\ re = im\ im = 0.$$

Also $im\,(j \cdot con) = 0$, where $j \cdot con = re^2 + im^2 = abs^2$.

Of great importance is the introduction of the limit concept in function algebras. This problem will be discussed later.

(B) ON MULTI-PLACE FUNCTIONS. Explicitly, a 2-place function is defined as a set of ordered pairs whose first members are ordered pairs of numbers while the second members are numbers, and such that if ((x, y), z) and ((x', y'), z') belong to the set, then z $\neq$ z' implies (x, y) $\neq$ (x', y'), that is, x $\neq$ x' and-or y $\neq$ y'. If ((x, y), z) belongs to the function F, then one sets z = F(x, y). By Ran F (by Dom F) is meant the set of all numbers z (of all ordered pairs (x, y)) for which some ((x, y), z) belongs to F.

Important examples inlude :

α) *Constant* 2-place functions, whose ranges consist of a single number, e.g., the function $1^{(2)}$ such that $1^{(2)}$ (x, y) = 1 for any x and y ; in other words, the set of all pairs ((x, y), 1) for any pair (x, y) ;

β) *Selector functions* I and J, whose values for any x and y are $I(x, y) = x$ and $J(x, y) = y$ (traditionally, they are referred to as « the functions x and y »);

γ) *Sum* and *Product* functions S and P such that $S(x, y) = x + y$ and $P(x, y) = x \cdot y$ for any x and y.

It goes without saying that these and the following ideas can be generalized to p-place functions for any positive integer p.

Each 2-place function G may be substituted into any 1-place function f, the result being the 2-place function fG consisting of all pairs $((x, y), z)$ for which there exists a number u such that $((x, y), u)$ belongs to G, and (u, z) belongs to f. Clearly, $jG = G$ for any function G.

Moreover, G lends itself to substitutions of ordered pairs of functions:

a) of pairs of p-place functions F_1, F_2, for any p, whereby the subtitution results in a p-place function $G(F_1, F_2)$ consisting of those pairs $((x_1, \dots, x_p), z)$ for which there exist numbers u_1 and v_1 such that $((x_1, \dots, x_p), u_1)$ and $((x_1, \dots, x_p), u_2)$ belong to F_1 and to F_2, respectively, and $((u_1, u_2), z)$ belongs to G. The domain of $G(F_1, F_2)$ is a subset of the set-theoretical intersection of Dom F_1 and Dom F_2. Clearly,

(5) $G(I, J) = G$, $I(F_1, F_2) = F_1$, $J(F_1, F_2) = F_2$ for any G, F_1, F_2.

b) of a p-place function M and a q-place function N for any p and q (which may or may not be equal numbers!), whereby the substitution results in a $(p + q)$-place function $G[M, N]$ consisting of the pairs $((x_1, \dots, x_p, x_{p+1}, \dots, x_{p+q}), z)$ for which there exist numbers u and v such that $((x_1, \dots, x_p), u)$ and $((x_{p+1}, \dots, x_{p+q}), v)$ belong to M and N, respectively, and $((u, v), z)$ belongs to G. The domain of $G[M, N]$ is a subset of the Cartesian product of Dom M and Dom N.

Accordingly, there are two substitution algebras of multi-place functions which I call *intersection* and *product substitution* and symbolize by the use of parentheses and brackets, respectively [19].

Some examples of formulae of the substitutive algebra of functions follow. $I_k^{(p)}$ will denote the kth n-place selector for which

(19) Loc. cit. (8) p. 461. For details concerning intersection substitution, cf. the book *Calculus*, loc. cit. (9).

$I_k^{(p)}(x_1, ..., x_p) = x_k$ for each $x_1, .., x_p$ (thus $I_1^{(2)} = I$, $I_2^{(2)} = J$, and $I_1^{(1)} = j$).

$$S(P, I) = P(I, S(J, 1^{(2)})), \qquad P(J, I) = P.$$

The last formula expresses the *commutativity* of multiplication. An expression of *associativeness* in intersection substitution requires the introduction of 3-place selectors:

$$P(I_1^{(3)}, P(I_2^{(3)}, I_3^{(3)})) = P(P(I_1^{(3)}, I_2^{(3)}), I_3^{(3)}).$$

The more remarkable is the following elegant expression of associativeness in product substitution:

$$P[P, j] = P[j, P].$$

The *additive-multiplicative distributive* law ((2) requires 3-place selectors even if product substitution is used:

$$P(S(I_1^{(3)}, I_2^{(3)}), I_3^{(3)}) = S(P(I_1^{(3)}, I_3^{(3)}), P(I_2^{(3)}, I_3^{(3)})) \quad \text{and}$$

$$P[S, j] = S[P, P](I_1^{(3)}, I_3^{(3)}, I_2^{(3)}, I_3^{(3)}).$$

The analogues of $L f$ and $R f$ in the algebra of 1-place functions are

1) a 1-place function $L G \subseteq j$ such that [20]

$$L G G = G, \text{ and } f \subset L G \text{ implies } f G \subset G;$$

2) two 2-place functions $R_1 G \subseteq 1$ and $R_2 G \subseteq J$ such that

$$G(R_1 G, R_2 G) = G \text{ while from } F_1 \subset R_1 G \text{ and/or}$$

$$F_2 \subset R_2 G \text{ it follows that } G(F_1, F_2) \subset G.$$

The introduction of these functions overcomes the traditional trialism of 2-place functions, their domains, and their ranges — sets of triples of numbers, of pairs of numbers, and of single numbers, respectively. Explicitly, $R_1 G$ can be defined as the set of all $((x, y), x)$ such that $((x, y), z)$ belongs to G for some z.

---

[20] $L G G = (L G) G$ according to the convention about the reach of bold-face symbols. Cf. [11].

Two right-inverses of G can be defined as follows : $\text{INV}_1\, G$ is the set of all $((x, y), z)$ such that $((z, y), x)$ belongs to G provided that this set is a 2-place function; that is to say, provided that, if $((u, v), w)$ and $((u', v'), w')$ belong to G and $u \neq u'$, then $v \neq v'$ and/or $w \neq w'$. The said condition and an analogue concerning $\text{Inv}_2\, G$ are satisfied if $G = (2I + 1^{(2)}) \cdot J^3$. Then

$$\text{INV}_1\, G = \frac{1}{2}\,((J/I^3) - 1^{(2)}) \quad \text{and} \quad \text{INV}_2\, G = \sqrt[3]{J/(2I + 1^{(2)})}.$$

In abstract 2-place algebra, inverses can be characterized by minimax properties. They satisfy modifications of the postulates

$$G\,(\text{INV}_1\, G, J) = \text{R}_1\, G \quad \text{and} \quad G\,(I, \text{INV}_2\, G) = \text{R}_2\, G.$$

(C) APPLICATIONS TO THE CALCULUS OF PROPOSITIONS. The simplest non-trivial substitutive 2-place algebra consists of the sixteen functions each of which is a quadruple F of pairs $((x, y), F(x, y))$, where $x, y, F(x, y) = 0, 1$ (and of the restrictions of these functions). I shall denote those functions, which are isomorphic with the sixteen functors of the calculus of propositions, by

$$1, A, B, C, D, E, I, J \text{ and } 1' (= 0), A', B', C', D', E', I', J'.$$

$1, 0, I, J$ are the 2 place constant functions and the selectors. The values of $A - E$ are given by

	(0,0)	(0,1)	(1,0)	(1,1)	corresponding to the logical functor
A	1	1	1	0	incompatability
B	1	0	0	0	joint rejection
C	1	1	0	1	implication
D	1	0	1	1	
E	1	0	0	1	equivalence

and $F'(x, y) = 1 - F(x, y)$ for each functor F and $x, y = 0, 1$.

Applying to this example the method of substitutional algebra one obtains not only the traditional formulae of the calculus of propositions in a new form - in fact, a *formulary of the calculus of propositions without propositional variables* (to be developed in a forthcoming paper) — but also new insights into the structure of the algebra of propositions, and new points of view on the classification and even the very nature of classical results. There is, e. g., the famous theorem that all sixteen functors can be expressed in terms of the functor A alone (as well as in terms of B alone). For example, the classical expression of the implication C in terms of A reads

$$C(x, y) = A(A(x, y), x) \qquad \text{for any} \quad x, y = 0, 1.$$

This expression for C, however, makes essential use of variables. From the standpoint of substitutional algebra, the isolated x should be replaced by the value $I(x, y)$ of the selector I, whereupon the formula can be written in the following form that is free of variables :

$$C = A(A, I).$$

But in this way C is expressed in terms of A and I, and not in terms of A alone. Nor is it possible to express C and eleven other functors substitutionally in terms of A alone. In fact, for any functor F there are at most four functors that can be expressed in terms of F alone.

(D) THE PRIMACY OF SUBSTITUTION IN THE ALGEBRA OF FUNCTIONS. The remark that, in the realm of functions, substitution is the operation par exellence and more profoundly connected with the idea of fuctions than are addition and multiplication can now be substantiated.

Just as certain associations of numbers with numbers and with ordered pairs of numbers lead to explicit definitions of 1-place and 2-place functions, certain associations of functions with functions and with ordered pairs of functions lead to explicit definitions of 1-place and 2-place *operators*. Operators studied on the preceding pages (and herein consistenly denoted by bold face letters) include a) the 2-place operators S, P, T associating with any ordered pair of function f, g the functions $S(f, g) = f + g$, etc. ; b) the 1-place operators NEG, REC, and INV associating with any f the function NEG f, etc. E.g., one may define REC as the class of all pairs $(f, $ REC $f)$ for any function f; and P as the class of all pairs $((f, g), $ P $(f, g))$ for any f and g. But there is a profound difference between S, P, NEG, and REC, on the one hand, and T and INV, on the other. There are 2-place functions, viz., S and P, and 1-place functions, viz., $-j$ and j^{-1}, with the following property : Substitution of any

f, g into those functions, yields the results of applying the corresponding operators to f and g; for example, $P(f, g) = P(f, g)$ and REC
$f = j^{-1}f$ for any f and g. But there is no 2-place function T such
that ([21])

(6) $T(f, g) = T(f, g) = fg$ for each f and g.

For suppose that there were such a function T satisfying (6). For
any number d, let c_d be the constant 1-place function of value d.
In particular, $c_d 0 = d$ for any d. Substitution of c_b into c_a yields
c_a for any a, b. By (6),

$$T(c_a, c_b) = T(c_a, c_b) = c_a c_b = c_a \qquad \text{for any a and b.}$$

In particular, the 1-place function $T(c_a, c_b)$ and c_a would assume
equal values for 0. But $c_a 0 = a$ and $T(c_a, c_b) 0 = T(c_a 0, c_b 0) =$
$= T(a, b)$. Hence $T(a, b) = a = I(a, b)$ for each a, b and, consequently, $T = I$. But $T(j, j^2) = j j^2 = j^2$, whence from (6) it would
follow that $T(j, j^2) = j^2$, whereas, by (5), $I(j, j^2) = j$.

Neither can the operator INV be reduced to substitution into a
1-place function even in the realm of linear functions. For suppose
there were a function v such that

(7) $vf = \text{INV}\, f$ for each linear function f.

Since $\text{INV}\,(cj) = \dfrac{1}{c} j$ for each $c \neq 0$, from (7) it would follow that

(8) $v(cj) = \dfrac{1}{c} j$ for each $c \neq 0$.

In particular, $v(cj)$ and $\dfrac{1}{c} j$ would assume equal values for 1. Con-
sequently, $v(c) = \dfrac{1}{c}$ for each $c \neq 0$, whence $v = j^{-1}$. But $j^{-1}(cj) = \dfrac{1}{c} j^{-1}$,
in contradiction to (8).

These remarks demonstrate that the operators T and INV are
of a much deeper nature than are P and REC.

([21]) Cf. the book *Calculus*, loc. cit ([9]), p. 304.

Examples of 1-place operators that can be reduced to substitution include, besides NEG and REC, also the additive and multiplicative operators S_g and P_g (which play a great role in the theory of linear differential equations) associating with any f the functions

$$S_g f = g + f \qquad \text{and} \qquad P_g f = g \cdot f,$$

respectively, and, for each g, the important operators T_g and $_gT$ for which

$$T_g f = fg \qquad \text{and} \qquad _gTf = gf \qquad \text{for any } f.$$

Examples of 1-place operators that cannot be reduced to substitution include, besides INV, the *derivative* D, the *Laplace transform* LAP, and the *iterator* I. The latter plays a great role in McKiernan's remarkable investigations on operators in the spirit of the algebra of functions ([22]).

Based on a substitutive algebra of functions, the theory of operators assumes a truly algebraic character, as demonstrated in a reformed differential calculus ([23]) whose fundamental formulae (connecting D with the operations of the tri-operational algebra in $\mathcal{R}$) read:

$$D(f+g) = Df + Dg, \quad D(f \cdot g) = Df \cdot g + f \cdot Dg, \quad D(fg) = Dfg \cdot Dg;$$

$$D\,0 = 0, \qquad D\,1 = 0, \qquad D\,j = 1.$$

The algebra can be extended to partial derivatives of multi-place functions. There are, moreover, axiomatizable algebras of linear operators such as LAP, but also of non-linear operators such as the Legendre transform.

(E) REMARKS ON LIMITS. A sequence of numbers $x_1, x_2, \ldots$ may be looked upon as a function whose domain is the set of integers in their natural order (or an infinite subset of this set) and for which $f(1) = x_1, f(2) = x_2, \ldots$ From this point of view, the limit, being the result of associating numbers with functions, is a *functional* and consequently does not lend itself to the substitution of real functions (not even of integer-valued functions !). But the

---

([22]) *C. R. Acad. Sci. Paris* vol. 246, 1958, **p.** 2331 and p. 2564.
([23]) *Cf. Calculus, A Modern Approach*, loc. cit. ([9]). Chapters VIII and IX.

limit may also be considered as a (symmetric) function *Lim* with infinitely many places that are somehow identifiable.

According to Fréchet's original postulates for abstract limit classes, the domain of that function *Lim* includes $(x_2, x_3, \dots)$ if it includes $(x_1, x_2, x_3, \dots)$ as well as $(d, d, \dots)$ for any d, and *Lim* $(d, d, \dots) = d$.

Being a function, *Lim* lends itself to substitution. Intersection substitution of p place functions $F_1, F_2, \dots$ into *Lim* results in a (perhaps empty) p-place function *Lim* $(F_1, F_2, \dots)$, called the *pointwise limit* and consisting, if p = 1, of all pairs $((x_1, x_2, \dots), z)$ for which there exist numbers $y_1, y_2, \dots$ such that (x_i, y_i) belongs to F_i and $((y_1, y_2, \dots), z)$ belongs to *Lim*.

In contrast, the *uniform limit* is an ∞-place operator which associates a (perhaps empty) function LIM UN $\cdot (F_1, F_2, \dots)$ with F_1, $F_2, \dots$ Even if p = 1, this operator cannot be reduced to an ∞-place function. For suppose that G were an ∞-place function such that

$$(9) \qquad G(f_1, f_2, \dots) = \text{LIM}(f_1, f_2, \dots) \qquad \text{for each } f_1, f_2, \dots$$

For each constant function c_d, from LIM UN $\cdot (c_d, c_d, \dots) = c_d$ it would follow that $G(c_d, c_d, \dots) = c_d$ and hence $G(d, d, \dots) = d$ for any d. If for each positive integer p the set of all pairs (x, x^p) for $0 \le x \le 1$ is denoted by k_p, it thus would follow further that

$$G(k_1, k_2, \dots)0 = 0 \qquad \text{and} \qquad G(k_1, k_2, \dots)1 = 1,$$

wherefore the domain of the 1-place function $G(k_1, k_2, \dots)$ would include 0 and 1, while LIM UN $\cdot (k_1, k_2, \dots) = \varnothing$.

In $\mathcal{R}$, the function *Lim* is bilaterally distributive with regard to addition and multiplication, e.g.,

$$Lim(f_1 \cdot g_1, f_2 \cdot g_2, \dots) = Lim(f_1, f_2, \dots) \cdot Lim(g_1, g_2, \dots).$$

Of great importance is the left-distributivity of *Lim* with regard to substitution :

$$Lim(f_1 g, f_2 g, \dots) = Lim(f_1, f_2, \dots)g \qquad \text{for any } g, f_1, f_2, \dots$$

On the other land, the relation

$$Lim(fg_1, fg_2, \dots) = f\, Lim(g_1, g_2, \dots) \qquad \text{for each } f, g_1, g_2, \dots$$

characterizes continuous functions f.

Accordingly, in an abstract algebra of functions wherein a ∞-place function *Lim* has been somehow axiomatically introduced, one may postulate

$$Lim\ (f_1 \circ g,\ f_2 \circ g,\ ...\ ) = Lim\ (f_1,\ f_2,\ ...\ ) \circ g \quad \text{for each} \quad g, f_1, f_2, ... ,$$

and one may define f as *continuous* in a purely algebraic way by the condition

$$Lim\ (f \circ g_1,\ f \circ g_2,\ ...\ ) = f \circ Lim\ (g_1,\ g_2,\ ...\ ) \quad \text{for each} \quad g_1, g_2, ...$$

[*Entrata in Redazione il 14 settembre 1961*]

Zeitschr. f. math. Logik und Grundlagen d. Math.
Bd. 10, S. 81—104 (1964)

ON SUBSTITUTIVE ALGEBRA AND ITS SYNTAX

by KARL MENGER in Chicago, Illinois (U.S.A.)

Introduction

In the calculus of propositions, meaningful expressions in ŁUKASIEWICZ' parentheses-free frontal notation were characterized in 1932 as strings of symbols satisfying certain formal conditions.[1] A string made up of references[2] to binary and unitary logical functors (e.g., C for implication, and n for negation) and to propositions or truth-values is well-formed if and only if the number of references to propositions or truth-values exceeds the number of the binary symbols in the entire string without exceeding it in any of its initial segments, and the string terminates in a proposition symbol. This theorem (which clearly also applies to arithmetical expressions written in ŁUKASIEWICZ' notation) later was generalized[3], has been rediscovered[4], and has found its way into compendia and textbooks[5].

These characteristic conditions concern relations between symbols of unlike types in the string: references to functors and references to propositions or symbols for operations and symbols for numbers. Hence they are inapplicable to the algebra of functions which I have tried to develop during the past two decades[6]. For the principal formulae of this algebra are free of references to the elements of the domains of the functions; and the expressions in those formulae include, besides parentheses and commas, only symbols of one type, namely, references to the functions themselves.

---

[1] MENGER, „Eine elementare Bemerkung über die Struktur logischer Formeln". Ergebnisse e. math. Kolloquiums 3 (1932) 22—24. According to ŁUKASIEWICZ, C. R. Soc. Sci. Let. Varsovie III 24 (1932) these conditions were also discovered by JAŚKOWSKI.

[2] By *references* to elements of a certain kind we mean symbols designating specific elements or variables, i.e., symbols that may be replaced by such designations.

[3] SCHRÖTER, „Axiomatisierung der Fregeschen Aussagenkalküle". Leipzig, 1943, and GERNETH Bull. Amer. Math. Soc. 54 (1948), 803.

[4] P. HALL, "Some word problems". J. Lond. Math. Soc. 33 (1958), 482—496.

[5] CHURCH, "Introduction to Mathematical Logic, I". Princeton 1956, p. 85, and ROSENBLOOM, "The Elements of Mathematical Logic", Chapter IV. New York 1950.

[6] MENGER, "Trioperational Algebra". Reports of a Math. Coll. 5—6 (1944), 3—10; "General Algebra of Analysis". ibid. 7 (1946), 46—60; "Axiomatic Theory of Functions and Fluents" in "The Axiomatic Method" ed. HENKIN et al., Amsterdam 1959, p. 454—473; "The Algebra of Functions: Past, Present, Future". Rend. Mat. Roma 20 (1961), 400—430; "A Group in the Substitutive Algebra of the Calculus of Propositions". Arch. Math. 13 (1962), 471—478.

6 Ztschr. f. math. Logik

Chapter I of the present paper begins with a brief description of the algebra of functions and in fact, more generally, of mappings and their compositions, illustrated by applications to the 2-valued calculus of propositions. It further contains definitions of abstract systems that include the families of mappings as special cases.[7]) In Chapter II, three kinds of parentheses-free notations for the algebra of mappings and for the general abstract systems are developed and characterizations of the well-formed expressions are formulated. These syntactic studies are perhaps warranted by the role that the algebra of compositions or substitutions of mappings seems to be destined to play in applications (e.g., to automata, networks, and linguistic studies) as well as in pure mathematics. For years, in numerous lectures I have ventured the prediction that the future of large parts of algebra as well as of analysis belongs to the theory of the substitutions of functions and mappings.

Chapter 1. Substitutive function algebras and their generalizations

1. Two kinds of substitutions

Underlying the algebra of explicitly defined substitutions is a set X, which will be called the *base set*. Its elements will be referred to as *basic elements*. The set of all ordered m-tuples of basic elements will be denoted by X^m ($m = 1, 2, \ldots$). Calling two ordered pairs of elements of any kind *consistent* unless their first members are equal and their second members are unequal we define an m-*place mapping* as a set of consistent pairs whose first members belong to X^m while their second members are elements of X. In what follows, only *total* m-place mappings will be considered, i.e., mappings for which the set of the first members of all pairs is the *whole* set X^m; but, for the sake of brevity, they will be referred to as m-place mappings[8]).

If F is an m-place mapping and $x_1, \ldots, x_m$ is an element of X^m, then the second member of the pair belonging to F whose first member is $x_1, \ldots, x_m$ will, in the usual way, be called the *value* of F for $x_1, \ldots, x_m$ and will be denoted by $F(x_1, \ldots, x_m)$. In classical analysis, especially in the case of functions, this symbol for a value of the mapping also serves as an (often inconsistently used) designation of the mapping F itself. In the sequel, mappings and their values will be consistently distinguished and the distinction will be stressed by the adherence to the following typographical convention[9]): references to mappings are in *italics*; references to basic elements are in roman type.

---

[7]) Important results about these abstract systems are contained in a paper by H. I. WHITLOCK forthcoming in the Mathematische Annalen.

[8]) The recent development of the algebra of 1-place functions (beyond the original trioperational algebra) began with the introduction into the theory of a partial order concept which has made it possible to extend the theory to nontotal functions. The most important contributions to this development are the papers by SCHWEIZER and SKLAR "The Algebra of Functions, I and II". Math. Ann. **139** (1959/60), 366—382, and **143** (1961), 440—447.

[9]) This typographical convention was introduced in MENGER, "Calculus. A Modern Approach". Boston 1955.

Mappings with m places for which X consists of real numbers are called m-place real *functions* while, if X consists of propositions or of truth-values, some logicians refer to them as *logical functors*. The same term is sometimes used if integers serve as truth-values, e.g., 1 for True, and 0 for False, even though in this case the mappings may also be considered as real functions.

For every positive integer n, the following n-place mappings are of paramount importance: the *selectors* I_k^n, where k is any positive integer $\geq n$; and the *constant* mapping C_x^n of value x, where x is any element of X. They are defined by

$$I_k^n(x_1, \ldots, x_n) = x_k \quad \text{and} \quad C_x^n(x_1, \ldots, x_n) = x \quad \text{for every} \quad x_1, \ldots, x_n \quad \text{in} \quad X^n.$$

Instead of I_1^1, we shall also write j.

The fundamental operations of the algebra are *compositions* of mappings. For any positive integer m, a composition of m + 1 mappings may be looked upon either as an *application* of an m-place mapping F to an ordered m-tuple $G_1, \ldots, G_m$ of mappings or as a *substitution* of the m-tuple into F. (As the following definitions demonstrate, substituting an m-tuple *into* F does of course not mean that the m-tuple will be substituted *for* F in the sense of replacing it.) The basic nature and wide applicability of the theory are apparent from the fact that every n-ary operation may be considered as a substitution into a specific n-place mapping. E.g., addition and multiplication of two functions, which classical and modern analysis and algebra have greatly overemphasized at the expense of substitution, are nothing but substitutions into two 2-place functions, sum and product, respectively; and in the same way the operations of Boolean algebra can be reduced to substitution — the operation that actually is at the core of those branches of mathematics.

Two kinds of substitution, which in the case m = 1 are identical, will be distinguished[10]). *Parenthesis substitution* into an m-place mapping is defined only for m-tuples of mappings *with one and the same place-number*. The result of parenthesis substitution of the m-tuple $G_1, \ldots, G_m$ of n-place mappings into the m-place mapping F is defined as the n-place mapping $F(G_1, \ldots, G_m)$ for which

$$(1) \quad \langle F(G_1, \ldots, G_m)\rangle (x_1, \ldots, x_n) = F(G_1(x_1, \ldots, x_n), \ldots, G_m(x_1, \ldots, x_n))$$

for every $x_1, \ldots, x_n$ in X^n. Here and in the sequel, $\langle \rangle$ serves as a grouping symbol. *Bracket substitution* of any ordered m-tuple $G_1, \ldots, G_m$ of mappings (say, with the place-numbers $n_1, \ldots, n_m$, respectively) into the m-place mapping F results in the $(n_1 + \cdots + n_m)$-place mapping $F[G_1, \ldots, G_n]$ for which

$$(2) \quad \langle F[G_1, \ldots, G_m]\rangle (x_1, \ldots, x_{n_1 + \cdots + n_m})$$
$$= F(G_1(x_1, \ldots, x_{n_1}), \ldots, G_m(x_{n_1 + \cdots + n_{m-1} + 1}, \ldots, x_{n_1 + \cdots + n_m}))$$

for every $x_1, \ldots, x_{n_1 + \cdots + n_m}$ in $X^{n_1 + \cdots + n_m}$. In classical analysis, the fundamental difference between the two types of substitution has been expressed, not too clearly, by writing, e.g., $F(G(x, y), H(x, y))$ for the function $F(G, H)$ — traditionally all

---

[10]) Cf. loc. cit. [9]) and the third and fourth paper loc. cit. [6]).

6*

single-letter symbols including x and y are italicized — and $F(G(t, u), H(v, w))$ or something equally arbitrary for the function $F[G, H]$. Formulae such as (2) can be further improved by setting

(3) $$F[x_1, \ldots, x_m] = F(x_1, \ldots, x_m)$$

for every m-place mapping F and every element of S^m thereby eliminating parentheses from the formulary of bracket substitution.

The principal property of these substitutions, called loc. cit.[6]) *superassociativity*, can be described in the following two equalities:

(4) $$\langle F(G_1, \ldots, G_m)\rangle (H_1, \ldots, H_n) = F(G_1(H_1, \ldots, H_n), \ldots, G_m(H_1, \ldots, H_n))$$

for any m-place mapping F, any n-place mappings $G_1, \ldots, G_m$, and any mappings $H_1, \ldots, H_n$ having one and the same place-number;

(5) $$\langle F[G_1, \ldots, G_m]\rangle[H_1, \ldots, H_{n_1 + \cdots + n_m}]$$
$$= F[G_1[H_1, \ldots, H_{n_1}], \ldots, G_m[H_{n_1 + \cdots + n_{m-1} + 1}, \ldots, H_{n_1 + \cdots + n_m}]$$

for any m-place mapping F, any mappings $G_1, \ldots, G_m$ and, if the place-number of G_i is n_i, any $n_1 + \cdots + n_m$ mappings $H_1, \ldots, H_{n_1 + \cdots + n_m}$.

Moreover,

(6) $F(I_1^m, \ldots, I_n^m) = F$ and $F[I_1^1, \ldots, I_1^1] = F$ for every m-place mapping F, $I_k^m(G_1, \ldots, G_m) = G_k$ for every m-tuple of mappings of the same place-number,

(7) $C_x^m(G_1^n, \ldots, G_m^n) = C_x^n$ and $C_x^m(G_1^{n_1}, \ldots, G_m^{n_m}) = C_x^{n_1 + \cdots + n_m}$.

Mapping equalities, e.g., (4)—(7), of course express the equality of mappings as sets of pairs. Two (total!) mappings M and N are equal if and only if (a) their place-numbers are equal, and, if that common place-number is p, then (b) $M(x_1, \ldots, x_p) = N(x_1, \ldots, x_p)$ for every $x_1, \ldots, x_p$ in X^p.

In the sequel, the place-number of a mapping M will be indicated in a superscript to M. *But the superscript may be omitted where the place-number is appearent from the number of mappings substituted into M.*

Contrary to what one might presume at first glance, parenthesis substitution is more general than bracket substitution.

Remark 1. *Parenthesis substitution cannot be expressed in terms of bracket substitution.* E.g., if F^2 is any 2-place mapping, and G^1 and H^1 are 1-place mappings, then $F(G^1, H^1)$, for example $F(j, j)$, is a 1-place mapping, whereas every bracket substitution somehow involving F^2 yields a mapping with a place-number ≥ 2. (In Section 3, this fact will be reconsidered.)

Remark 2. *Bracket substitution can be expressed in terms of parenthesis substitution by the use of selectors. One has*

$$F[G_1^{n_1}, G_2^{n_2}, \ldots, G_m^{n_m}] = F(G_1(I_1^n, \ldots, I_{n_1}^n), G_2(I_{n_1+1}^n, \ldots, I_{n_1+n_2}^n), \ldots,$$
$$G_m(I_{n_1 + \cdots + n_{m-1} + 1}^n, \ldots, I_n^n)),$$

where $n = n_1 + n_2 + \cdots + n_m$.

Using this last law one readily derives the following *mixed superassociative laws* connecting the two kinds of substitution:

$$\langle F[G_1^{n_1}, \ldots, G_m^{n_m}] \rangle \, (H_1^p, \ldots, H_{n_1 + \cdots + n_m}^p)$$

(8)
$$= F(G_1(H_1^p, \ldots, H_{n_1}^p), \ldots, G_m(H_{n_1 + \cdots + n_{m-1} + 1}^p, \ldots, H_{n_1 + \cdots + n_m}^p)),$$

$$\langle F(G_1^n, \ldots, G_n^n) \rangle \, [H_1^{p_1}, \ldots, H_n^{p_n}] = F(G_1[H_1^{p_1}, \ldots, H_n^{p_n}], \ldots, G_n[H_1^{p_1}, \ldots, H_n^{p_n}]).$$

It should be noted that there are also important *operators* whose results can be expressed in terms of parenthesis substitution and selectors. One of them is the *diagonalization* operator T_δ associating with every m-place mapping F^m the 1-place mapping

$$T_\delta \, F^m = F^m(I_1^1, \ldots, I_1^1).$$

From this, by (8) one infers

(9)
$$T_\delta \langle F(G_1, \ldots, G_m) \rangle = F(T_\delta \, G_1, \ldots, T_\delta \, G_m) = T_\delta \langle F[G_1, \ldots, G_m] \rangle.$$

Another such operator is the *reductor* of modul n, by which is meant the operator T_ϱ^n associating, for every positive integer m, with every m n-place mapping F^{mn} the n-place mapping

$$T_\varrho^n \, F^{mn} = F^{mn}(I_1^n, \ldots, I_n^n, I_1^n, \ldots, I_n^n, \ldots, I_1^n, \ldots, I_n^n).$$

Remark 3. *Parenthesis substitution can be expressed in terms of bracket substitution and reductors. One has*

(10)
$$F(G_1^n, \ldots, G_m^n) = T_\varrho^n \langle F[G_1^n, \ldots, G_m^n] \rangle.$$

More generally, for every function φ mapping the integers $1, \ldots, m$ on the integers $\varphi 1, \ldots, \varphi m$ where $1 \leq \varphi i \leq m$, and for every integer $n \geq \mathrm{Max}(\varphi 1, \ldots, \varphi m)$, one may define an operator T_φ^n associating with every m-place mapping F^m the n-place mapping

$$T_\varphi^n \, F^m = F(I_{\varphi 1}^n, \ldots, I_{\varphi m}^n).$$

These operators include in particular *permuters* obtained if φ is a 1-1-function. Transformers admit of an associative composition but are not in general associatively connected with substitutions.

2. Applications to the functors of 2-valued Logic

In the case of 2-valued logic, X consists of two elements, say, 1 and 0 or T and F. We shall denote the four 1-place functors and eight of the 2-place functors as follows:

	j	n	i	o
1	1	0	1	0
0	0	1	1	0

	1	A	B	C	D	E	I	J
1,1	1	0	1	1	1	1	1	1
1,0	1	1	1	0	1	0	1	0
0,1	1	1	1	1	0	0	0	1
0,0	1	1	0	1	1	1	0	0

F' will be the 2-place functor $n F$ for which $F'(x_1, x_2) \neq F(x_1, x_2)$ for every x_1, x_2 in X^2, where F stands for any of the functors $1, A, \ldots, J$. Clearly, A is incompatibility; B is nonexclusive disjunction; C is implication; E is equivalence; A' is conjunction. The functors j, I, and J are the selectors I_1^1, I_1^2, and I_2^2, respectively, which traditionally remain anonymous and have often been confused with "variables". The functors 1 and $1'$ are the constant 2-place functors of the values 1 and 0, while i and o are the constant 1-place functors. Instead of $1'$ it is often convenient to write 0. (Note the typographical difference between the functor symbols $0, 1$ and the symbols 0, 1 for the elements of X.)

The substitutive functor algebra seems to introduce new points of view in the calculus of propositions[11]. Classical results and methods are resolved into theorems and procedures of various degrees. As an example, consider the famous result that all unitary and binary functors can be expressed in terms of incompatibility (usually denoted by a stroke). Algebraic statements concerning A include the following theorems, where $\mathfrak{F}(F_1, \ldots)$ and $\mathfrak{F}[F_1, \ldots]$ denote the sets of all functors obtainable from the functors $F_1, \ldots$ by parenthesis and bracket substitution, respectively.

$$\mathfrak{F}(A) = \{A, A', 1, 0\};$$

that is to say, *the only 2-place functors that can be expressed in terms of A alone by parenthesis substitution are:*

$$A' = A(A,A); \quad 1 = A(A,A(A,A)) = A(A',A); \quad 0 = A(A(A,A(A,A)),$$

$$A(A,A(A,A)) = A(1,1);$$

and $\mathfrak{F}[A] = \{A, A[A,A], A(A,A[A,A]], \ldots\}.$

All functors in $\mathfrak{F}[A]$ have even place-numbers. For example $A[A,A]$ is the 4-place functor assuming the value 1 for x_1, x_2, x_3, x_4 if and only if $x_1 = x_2 = 1$ and/or $x_3 = x_4 = 1$.

By diagonalization and bracket substitution one can obtain from A fourteen of the sixteen 2-place functors, viz., all functors except E and E'.

First, one can obtain the four 1-place functors by the use of (9). Indeed,

$$T_\delta A = A(j,j) = n;$$

$$T_\delta \langle A[A,A] \rangle = A(T_\delta A, T_\delta A) = A(n,n) = j;$$

$$T_\delta \langle A[A,A[A,A]] \rangle = A(n,j) = i;$$

$$T_\delta \langle A[A[A,A[A,A]],A[A,A[A,A]]] \rangle = A(i,i) = o.$$

[11]) Cf. the last paper loc. cit. [6]) and MENGER, "Function Algebra and Propositional Calculus" in "Self-Organizing Systems 1962". Spartan Books, Washington, as well as MENGER and SCHULTZ, "Postulates for the substitutive Algebra of the 2-place functors in the 2-valued calculus of propositions" forthcoming in Notre Dame J. f. Symb. Log. 1963.

Bracket substitution of the 1-place functors into A yields nine other 2-place functors, as evident from the following bracket substitution table of A.

$A[\ ]$	j	n	i	o
j	A	C	I'	1
n	D	B	I	1
i	J'	J	0	1
o	1	1	1	1

which summarizes sixteen results: $A[j,j] = A$, $A[j,n] = C$, and so on. One obtains A', B', C', and D' by substituting A, B, C, and D into n. But E and E' are not among the results. Nor can they be obtained by bracket substitution of 1-place functors into any of the fourteen other functors. Indeed, the bracket substitution tables of these functors can be derived from the table for A by virtue of the superassociative law (4). For example, since $C = A[j,n]$ and $I' = A[j,i]$, one finds

$$C[n,i] = A[j,n][n,i] = A[n,o] = 1 \quad \text{and} \quad I'[n,j] = A[j,i][n,j] = A[n,i] = I.$$

Clearly, E and E' cannot appear as results. Nor can E and E' be obtained by dia gonalization of the functors $B, C, \ldots$ or of functors obtainable from them by bracket substitution. Since the diagonalization of functors obtainable from A alone yields all four 1-place functors, nothing can be added by the diagonalization of the functors $B, C, \ldots$ or of functors resulting from substituting into each other some of the fourteen functors already obtained. This concludes the proof of the theorem that only the functors $\neq E$ and E' can be obtained.

The bracket substitution table of E, which plays an exceptional role, follows. It demonstrates that *only E', I, I', J, J', 1, and 0 are obtainable from E and 1-place functors by bracket substitution.*

$E[\ ]$	j	n	i	o
j	E	E'	I	I'
n	E'	1	I'	I
i	J'	J	1	0
o	J	J'	0	1

The only 2-place functor obtainable from E by diagonalization and bracket substitution is $E[i,i] = 1$.

By bracket substitution, diagonalization, and reduction, all 1-place and all 2-place functors can be obtained from A. In view of the preceding theorem it suffices to show that E and E' can be obtained. Since $E = A(A,B)$, by the use of (8) one obtains

$$E = \mathbf{T}_\varrho^2 \langle A[A,B]\rangle, \quad \text{where} \quad B = A[n,n] \quad \text{and} \quad n = \mathbf{T}_\delta A.$$

3. Superassociative Systems

In this section, $\mathfrak{A}$ will mean a set with each of whose elements, F, there is associated a positive integer, denoted by $\varrho\,F$ and called the *rank* of F. $\mathfrak{A}$ will be called a *superassociative system of the first kind* if, with every element F of $\mathfrak{A}$ of rank m and every ordered m-tuple of elements $G_1, \ldots, G_m$ of one and the same rank, an element $F(G_1, \ldots, G_m)$ of rank $\varrho\,G_m$ is associated in such a way that the superassociative law (4) is satisfied. $\mathfrak{A}$ will be called a *superassociative system of the second kind* if with F and every ordered m-tuple $G_1, \ldots, G_m$ an element of rank $\varrho\,G_1 + \cdots + \varrho\,G_m$ is associated in such a way that the law (5) is satisfied. If $\mathfrak{A}$ is a superassociative system of both kinds and the laws (8) are satisfied then $\mathfrak{A}$ will be called a *mixed superassociative system*. For many purposes it is important to assume the existence of elements I_k^n such that the laws (6) hold.

An element C of rank 1 may be called constant under substitution or, briefly, *constant* if $C\,F = C$ for every F of rank 1, and $C\,G_1 = C\,G_2$ for every pair of elements G_1 and G_2 of equal rank. Setting $C\,G = C^m$, where $m = \varrho\,G$ (and $= \varrho\,C^m$, so that $C = C^1$) one has (7), whence C^m might be called *constant of rank* m.

The following example, due to H. I. WHITLOCK shows that, in a system of explicitly defined functions, an element that is constant under substitution need not be a constant function in the sense of a function assuming only one value. If X is the triple $\{0, 1, 2\}$, define for every positive integer n three functions I^n, F^n, and G^n as follows:

$$I^n(x_1, \ldots, x_n) = x_1,$$

$$F^n(0, x_2, \ldots, x_n) = 0, \quad F^n(1, x_2, \ldots, x_n) = F^n(2, x_2, \ldots, x_n) = 1;$$

$$G^n(0, x_2, \ldots, x_x) = 0, \quad G^n(1, x_2, \ldots, x_n) = G^n(2, x_2, \ldots, x_n) = 2.$$

One readily verifies that this system of functions is closed under substitution and that, for every n, both F^n and G^n are constant of rank n, even though each of them assumes two values. With regard to substitution, this system is isomorphic with the system of the functions I^n, 1^n, 2^n, for all positive integers n, where 1^n and 2^n are the constant functions of value 1 and 2, respectively. This isomorphism demonstrates the impossibility of defining constant functions in terms of substitution.

In this abstract theory one may also admit elements of rank 0 into which nothing can be substituted. Their counterpart in the theory of mappings would be the basic elements. Hence, if $\varrho\,G = m$ and $\varrho\,K_1 = \ldots = \varrho\,K_m = 0$, then $G(K_1, \ldots, K_m)$ is an element of rank 0 corresponding to a value of G. If one postulates that for every element K of rank 0 there exists a constant function C_K of rank 1 such that $C_K\,K' = K$ for every element K' of rank 0, then, setting $C_K\,G = C_K^m$, one has

$$C_K^m(K_1, \ldots, K_m) = K$$

for every m elements $K_1, \ldots, K_m$ of rank 0. Extending Remark 2 (p. 84) one can express every result of a bracket substitution as a result of parenthesis substitution even in presence of elements of rank 0.

$$F[G_1^{s_1}, \ldots, G_m^{s_m}] = F(H_1, \ldots, H_m)$$

if one sets $s_0 = 0$, $s_1 + \cdots + s_m = s$, and for $i = 1, \ldots, m$

$$H_i = \begin{cases} G_i(I_{s_1 + \cdots + s_{i-1} + 1}^s, \ldots, I_{s_1 + \cdots + s_i}^s) & \text{if } s_i > 0 \\ C_K^s & \text{if } s_i = 0 \text{ and } G_i^0 = K. \end{cases}$$

Parenthetically it may be mentioned that, in 2-valued logic, $F(j, j)$ or $F(I_1^1, I_1^1)$, which is not equal to the result of any bracket substitution involving F and functions of positive place-numbers, can in presence of 0-place functions be expressed as the result of a bracket substitution by the use of the 2-place disjunction B and conjunction A' and the 3-place selectors or by the 3-place function

$$B(A'(I_1^3, n\, I_3^3),\ A'(I_2^3, I_3^3)).$$

If this function be ad hoc denoted by G, then

$$F(j, j) = G[F(0, 0), F(1, 1), j],$$

where $F(0, 0)$ and $F(1, 1)$ are 0-place functions, viz., values of F. More generally, $F(G^1, H^1)$ can be obtained by the bracket substitution of

$$[F(0, 0), F(0, 1), F(1, 0), F(1, 1), G0, G1, H0, H1, j]$$

into a 9-place function, defined in analogy to BOOLEAN canonical forms in terms of multi-place conjunctions and disjunctions. (The preceding remark is due to KARL MENGER jr. and H. I. WHITLOCK.)

In the abstract theory, one might postulate $G[K_1, \ldots, K_m] = G(K_1, \ldots, K_m)$ if $\varrho K_1 = \ldots = \varrho K_m = 0$, in analogy to (3).

4. Systems with one generator

One of the main problems of the abstract theory of superassociative systems is the study of the sets $\mathfrak{A}(F_1, \ldots)$ and $\mathring{\mathfrak{A}}[F_1, \ldots]$ substitutively generated by certain elements of $\mathfrak{A}$. Some examples and the abstract characterization of the 1- and 2-place functors of the 2-valued calculus of propositions follow.

The associative system $\mathfrak{A}(n, i)$, *where n and i are elements of rank 1 for which*

$$n\, n\, n = n, \quad n\, n\, i = i, \quad i\, n = i, \quad i\, i = i \quad \text{and} \quad n\, i \neq i$$

is isomorphic with the system of the four 1-place functors of the 2-valued logic if one defines $n\, n = j$ *and* $n\, i = o$.

An element F of rank m will be called *idempotent* if $F(F, \ldots, F) = F$. Clearly, if F is idempotent, then $\mathfrak{A}(F) = \{F\}$. Of the sixteen 2-place functors, the following six are idempotent: *1*, *0*, *I*, *J*, *A'*, *B*.

In the realm of elements of rank 2, if F is not idempotent, set

$$F(F, F) = G, \quad F(F, G) = H_1, \quad F(G, F) = H_2, \quad F(G, G) = K.$$

Clearly, $G(F, F) = K$. One further readily proves

If $K = F$, then $H_1(F, F) = H_2(G, G) = H_2$ and $H_2(F, F) = H_1(G, G) = H_1$; then, furthermore, $H_1 = G$ if and only if $H_2 = F$, and $H_1 = F$ if and only if $H_2 = G$.

Indeed, $H_1(F, F) = F(F, G) (F, F) = F(G, K) = F(G, F) = H_2$. And $H_1 = G$ then implies that $H_2 = H_1(F, F) = G(F, F) = K = F$.

It follows that *a system* $\mathfrak{A}(F)$ *consisting of two elements, namely, F and G, is of one of the following six types:*

a) $K = F = H_1$ and $H_2 = G$;

b) $K = F = H_2$ and $H_1 = G$;

c) $K = G = H_1$ and $H_2 = F$;

d) $K = G = H_2$ and $H_1 = F$;

e) $K = G$ and $H_1 = H_2 = F$;

f) $G = K = H_1 = H_2$.

Eight of the sixteen functors are of the types a)—e), namely,

the functors	J'	I'	C and D'	D and C'	E and E'
are of type	a)	b)	c)	d)	e)

Type f) is not represented in the 2-valued logic. In 3-valued logic, where the domain of each 2-place functor consists of the nine pairs (x, y), $(x, y = 0, 1, 2)$, type f) is represented by the functor F for which $F(x, y) = 0$ unless $x = y = 2$ and $F(2, 2) = 1$, while $G(x, y) = 0$ for every x, y.

If $\mathfrak{A}(F)$ consists of *three* elements, then from $K = F$ and $H_1 = H_2$, say $= H$, it follows that $F(H, H) = H$. Indeed,

$$F(H, H) (F, F) = F(F(F, G), F(F, G)) (F, F) = F(F(G, K), F(G, K)) =$$
$$= F(F(G, F), F(G, F)) = F(H, H).$$

But since $F(F, F) \neq F$ and $G(F, F) \neq G$, from $F(H, H) (F, F) = F(H, H)$ it follows that $F(H, H) \neq F, G$ and thus $= H$.

No 2-place functor of 2-valued logic generates a system consisting of three elements. But the tables

$F(\,)$	F	G	H
F	G	H	H
G	H	F	H
H	H	H	H

and

$F'(\,)$	F'	G'	H_2'
F'	G'	G'	G'
G'	H_2'	G'	F
H_2'	G'	G'	H_2'

are materialized by the following functors of the 3-valued logic:

	F	G	H	F'	G'	H_2'
0,0	0	0	0	0	0	0
0,1	0	0	0	2	2	2
0,2	0	0	0	1	0	2
1,0	0	0	0	0	0	0
1,1	2	1	0	0	0	0
1,2	0	0	0	0	0	0
1,3	0	0	0	0	0	0
2,1	0	0	0	0	0	0
2,2	1	2	0	2	2	2

If $\mathfrak{A}(F)$ consists of *four* elements, then from $K = F$ and $H_1 = H_2 = H$, if one sets $F(H, H) = L$, it follows that $F(L, L) = H$ and that the equalities in each of the following six pairs are equivalent:

$$F(F, H) = F \quad \text{and} \quad F(G, H) = G;$$
$$F(F, H) = G \quad \text{and} \quad F(G, H) = F;$$
$$F(F, H) = H \quad \text{and} \quad F(G, H) = H;$$
$$F(F, L) = F \quad \text{and} \quad F(G, L) = G;$$
$$F(F, L) = G \quad \text{and} \quad F(G, L) = F;$$
$$F(F, L) = H \quad \text{and} \quad F(G, L) = H.$$

One readily sees furthermore that *all systems* $\mathfrak{A}(F)$ *consisting of four elements such that* $K = F$, $H_1 = H_2 \, (= H)$, *and satisfying the relations*

g) $F(F, H) = G = F(H, F)$, $F(F, L) = H = F(L, F)$, $F(H, L) = H = F(L, H)$

are isomorphic. In the 2-valued logic, A and B' generate such systems.

This completes the proof of the following **theorem**.

The sixteen systems each of which is generated by a 2-place functor of the 2-valued logic can be abstractly characterized as follows: Six of them, generated by the idempotent functors 1, 0, I, J, A', *and* B, *consist of a single element.* $\mathfrak{F}(I')$, $\mathfrak{F}(J')$, $\mathfrak{F}(C)$ *as well as* $\mathfrak{F}(D')$, $\mathfrak{F}(D)$ *as well as* $\mathfrak{F}(C')$, *and* $\mathfrak{F}(E)$ *as well as* $\mathfrak{F}(E')$ *are systems of two elements characterized by the relations* a), b), c), d), *and* e), *respectively.* $\mathfrak{F}(A)$ *and* $\mathfrak{F}(B')$ *are systems of four elements characterized by the relations* g).

Chapter II. The syntax of the algebra of mappings and of its abstract generalizations

5. Ambiguous strings of functor symbols

Since, by definition, $F^1(G^n)$ and $F^1[G^n]$ are equal, one may write $F^1 G^n$ — briefly $F G^n$ — for either of those expressions. The symbols $F \langle G H^p \rangle$ and $F \langle G H^p \rangle$ denote results of differently defined operations even though, by the superassociative law (1), which for $m = n = 1$ expresses associativity, the results themselves are equal so that, from the point of view of the designation of functions, one may unambiguously write $F\,G\,H^p$.

On the other hand, superassociativity notwithstanding, as simple a string as $M^2\,N^1\,P^1\,Q^1$ is ambiguous even with regard to the designated functions unless it be interspersed with symbols for substitution (such as brackets or parentheses) and grouping symbols (such as $\langle\,\rangle$ and commas). In bracket substitution, it might mean $M^2[N^1\,P^1, Q^1]$ or $M^2[N^1, P^1\,Q^1]$; and in parenthesis substitution it admits of three interpretations, the third being $\langle M^2(N^1, P^1)\rangle\,Q^1$. As another example, consider the string $A\,A\,A\,A\,C$ which, interlaced with parentheses, occurs in the following three equalities:

$$A(A, A(A, C)) = C, \quad A(A(A, A), C) = A, \quad \langle A(A, A)\rangle\,(A, C) = I'.$$

Are there unambiguous descriptions of the results of substitution that are free of grouping symbols as well as of pincer-like double symbols (such as parentheses and brackets) indicating substitution? What are the rules for well-formed unambiguous expressions?

6. A partial synthesis of substitutive and logico-arithmetical strings

Let there be given an alphabet of script letters $a, b, c, \ldots, x, y, z$, each letter with the numeral of an integer ≥ -1, called the *weight* of the letter, as a superscript. The weight of a string $\mathfrak{S}$ of such letters will be defined as the sum of the weights of all letters in $\mathfrak{S}$. A string $\mathfrak{S}$ will be called *balanced* if its weight is negative while the weight of each initial segment of $\mathfrak{S}$ is non-negative. Clearly, the weight of a balanced string is -1.

If $\mathfrak{S}$ is balanced, then the number of letters with negative weights exceeds the sum of the weights of the other letters in the entire string (namely, by 1) while this inequality does not hold for any initial segment of $\mathfrak{S}$. The following remarks are readily proved.

1. An *elementary* string, i.e., a string consisting of a single letter is balanced if and only if that letter is of negative weight. In every nonelementary balanced string, the weight of every terminal segment (in particular, of the terminal letter) is negative while that of the initial letter is nonnegative.

2. A balanced string $\mathfrak{S}$ remains balanced if symbols of weight 0 that may occur in $\mathfrak{S}$ are deleted or if further symbols of weight 0 are introduced anywhere in $\mathfrak{S}$, except at the end (since the terminal letter in a balanced string must be of weight -1).

3. A balanced string $\mathfrak{S}$ remains balanced if any balanced segment of $\mathfrak{S}$ is replaced by any balanced string. In other words, if $\mathfrak{S}'$, $\mathfrak{S}''$, and $\mathfrak{S}_i\,\mathfrak{S}'\,\mathfrak{S}_t$ are balanced, then so is $\mathfrak{S}_i\,\mathfrak{S}''\,\mathfrak{S}_t$.

4. Two nonidentical balanced segments of a string are either disjoint or one of them is a noninitial segment of the other.

5. In a string $\mathfrak{S}$ of negative weight, every letter is the initial element of exactly one balanced segment. This segment will be called the *interval* (in $\mathfrak{S}$) of that letter.

6. If the balanced string $\mathfrak{S}$ begins with a letter of nonnegative weight m, say l^m, then there are m + 1 uniquely determined balanced segments $\mathfrak{S}_0, \mathfrak{S}_1, \ldots, \mathfrak{S}_m$ such that

$$\mathfrak{S} = l^m \, \mathfrak{S}_0 \, \mathfrak{S}_1 \ldots \mathfrak{S}_m.$$

The m + 1 segments $\mathfrak{S}_i$ will be called the *components* of $\mathfrak{S}$ as well as of l^m.

7. Every balanced (proper) segment $\mathfrak{S}'$ of a balanced string $\mathfrak{S}$ is a component of exactly one balanced segment $\mathfrak{S}''$ of $\mathfrak{S}$ (namely, of the shortest of the balanced segments of $\mathfrak{S}$ that include $\mathfrak{S}'$ as a proper segment; $\mathfrak{S}''$ may be $\mathfrak{S}$ itself).

It is often convenient to distinguish the letters of negative weight from the others typographically or at least by omitting their superscript as in the following two examples of balanced strings (followed by their components):

(11) $\qquad\qquad l^2 m^1 u v n^0 w p^2 x y z; \quad m^1 u v, n^0 w, p^2 x y z.$

(12) $\qquad\qquad l^3 a m^2 b u v n^1 e w p^3 q x y z; \quad a, m^2 b u v, n^1 e w, p^3 q x y z.$

Balanced strings admit of two important applications.

I. In what may be called *the logico-arithmetical or functor interpretation*, one replaces every script letter of weight m by the symbol for an (m + 1)-place mapping (especially, a logical functor or a function) while each letter of weight -1 is replaced by a reference to a basic element, which might be called a 0-place mapping. E.g., if in (11) one replaces l^2, m^1, n^0, p^2 by italics F^3, G^2, H^1, K^3 denoting mappings with the superscribed place-numbers, and replaces the letters without superscripts by references in roman type to basic elements, one obtains the string

(11') $\qquad\qquad F^3 \, G^2 \, \mathrm{u} \, \mathrm{v} \, H^1 \, \mathrm{w} \, K^3 \, \mathrm{x} \, \mathrm{y} \, \mathrm{z},$

which designates what is traditionally denoted by

(11'') $\qquad\qquad F(G(\mathrm{u}, \mathrm{v}), H(\mathrm{w}), K(\mathrm{x}, \mathrm{y}, \mathrm{z})).$

For a string $\mathfrak{S}$ consisting exclusively of references to mappings of place-numbers 2, 1, and 0, the weight is equal to the number of binary symbols in the string. A logical string $\mathfrak{S}$ thus is balanced if and only if the number of proposition symbols exceeds the number of binary symbols in $\mathfrak{S}$ while this inequality does not hold for any initial segment of $\mathfrak{S}$, and 2) the terminal element of $\mathfrak{S}$ is not a unitary symbol (and hence is the reference to a proposition). But this is the condition characteristic of well-formed strings in ŁUKASIEWICZ' notation mentioned at the beginning of the introduction. More generally, in its logico-arithmetical interpretation a string $\mathfrak{S}$ is well-formed if and only if $\mathfrak{S}$ is balanced; that is to say, if in $\mathfrak{S}$, but not in any initial segment of $\mathfrak{S}$, the number of references to basic elements exceeds the sum of the place-numbers of all functions (of nonnegative weight) diminished by the number of those functions; or, in still other words, *if the number of letters exceeds the sum of their place-numbers in $\mathfrak{S}$ but not in any initial segment of $\mathfrak{S}$.*

II. In *the substitutive or operator interpretation*, one replaces every script letter of weight -1 by a reference (in italic type) to a mapping while every script letter of positive weight is replaced by an operator symbol (say, a bold-face letter with the superscript m) for the substitution of an m-tuple of mappings into the mapping whose symbol immediately follows the operator symbol. In the sequel, this interpretation will be carried out by writing

$$\mathbf{P}^m \, F \, G_1 \ldots G_m \quad \text{for} \quad F(G_1, \ldots, G_m), \quad \text{and} \quad \mathbf{B}^m \, F \, G_1 \ldots G_m \quad \text{for} \quad F[G_1, \ldots, G_m].$$

(The letters $\mathbf{P}$ and $\mathbf{B}$ should be reminiscent of parenthesis and bracket substitution.)

If a, b, e, and g in (12) are replaced by the symbols F, G, H, and K for mappings with the place-numbers 3, 1, 2, and 3, respectively, and $u, \ldots, z$ are replaced by the mapping symbols $U, \ldots, Z$ while each letter with positive weight is replaced by an equally weighted $\mathbf{B}$, then (12) reads

12') $$\mathbf{B}^3 \, F \, \mathbf{B}^2 \, G \, U \, V \, \mathbf{B}^1 \, H \, W \, \mathbf{B}^3 \, K \, X \, Y \, Z$$

designating the function

(12'') $$F[G[U, V], H \, W, K[X, Y, Z]].$$

If the mappings $U, \ldots, Z$ have the place-number 0 and are replaced by $u, \ldots, z$, then, by (3), one may replace $\mathbf{B}^3$, $\mathbf{B}^2$, $\mathbf{B}^1$, $\mathbf{B}^3$ in (12') by $\mathbf{P}^3$, $\mathbf{P}^2$, $\mathbf{P}^1$, $\mathbf{P}^3$, and the brackets in (12'') by parentheses. One obtains (11''). In these examples, the operator interpretation of (12) and the functor interpretation of (11) coincide, which illustrates the following general remark.

8. *For every well-formed logico-arithmetical string, there is an equivalent balanced string in operator interpretation.* Indeed, the latter can be obtained from the former if each symbol (of nonnegative weight m) for an (m + 1)-place functor is replaced by an operator symbol of weight m + 1 followed by the functor symbol, and if all symbols in the logico-arithmetical string are assigned the weight -1.

The operators $\mathbf{B}^0$ and $\mathbf{P}^0$ would indicate that, into a 0-place mapping, a 0-tuple (that is, nothing) is being substituted. One thus might interprete $\mathbf{B}^0$ and $\mathbf{P}^0$ as operators acting on 0-place elements somewhat like 1-place functions act on numbers. Or one might altogether dispense with 0-place operators and assume that every operator has a *positive* weight while assigning the weight 0 (rather than -1) to every mapping. In this case, a string $\mathfrak{S}$ of operator and mapping symbols would have to be called balanced *if, in $\mathfrak{S}$ but not in any initial segment of $\mathfrak{S}$, the number of mapping symbols exceeds the weight* (which would be equal to the sum of the weights of the operator symbols).

Still another possibility would be to assign the weight 0 to all mappings and to write $\mathbf{P}^{m+1} \, F \, G_1 \ldots G_m$ for $F(G_1, \ldots, G_m)$, and $\mathbf{B}^{m+1} \, F \, G_1 \ldots G_m$ for $F[G_1, \ldots, G_m]$; that is to say, to give the weight n to a substitutive operator representing an (n − 1)-ary operation. In this case, a string $\mathfrak{S}$ would be balanced *if the number of letters exceeds the sum of their weights in $\mathfrak{S}$, but not in any initial segment of $\mathfrak{S}$.*

7. Beyond logico-arithmetical strings

Two facts make it necessary to go beyond the logico-algebraic mapping strings.

1) In such strings, all elements of lowest weight (propositions, numbers, and elements of X) are considered as being of one type, whereas *the elements of lowest weight in the algebra of mappings and their abstract generalizations are differentiated by unlike place-numbers or ranks*. The property of being balanced, which characterizes well-formed logico-arithmetical strings, is, therefore, insufficient to characterize the well-formed expressions of the substitutive algebra. E.g., if F is 1-place or 3-place and if K, L, M have unequal place-numbers, then $\mathbf{B}^2 F G H$ and $\mathbf{P}^3 F K L M$ are balanced without being what will be called well-formed strings in operator interpretation.

2) The converse of Remark 8 is invalid. *There are well-formed substitutive strings for which no equivalent logico-arithmetical strings exist.* A simple example is

(13) $\mathbf{B}^6 \mathbf{B}^3 F G H K \mathrm{u v w x y z}$ with the seven components $\mathbf{B}^3 F G H K$, u, $\ldots$, z,

where F, G, H, K are mappings with the place-numbers 3, 2, 1, 3, respectively. (13) describes a value of the 6-place mapping $F[G, H, K]$, namely

(13″) $\langle F[G, H, K] \rangle (\mathrm{u}, \ldots, \mathrm{z})$.

Applying (2) to (13″) one obtains (11″), for which an equivalent logico-arithmetical string does exist, namely, (11′). But (13″) cannot be written as such a string.

In (4), (5), and (8), only the expressions on the right sides can be written as logico-arithmetical strings, not those on the left sides. Hence, *in terms of logico-arithmetical strings, superassociativity of composition cannot be described in a single formula. It can only be expressed in an implication such as*

$$\text{if } E \, \mathrm{u v} = F \, G_1 \, \mathrm{u v} \, G_2 \, \mathrm{u v}, \quad \text{then}$$

$$E \, H_1 \, \mathrm{x y} \, H_2 \, \mathrm{x y} = F \, G_1 \, H_1 \, \mathrm{x y} \, H_2 \, \mathrm{x y} \, G_2 \, H_1 \, \mathrm{x y} \, H_2 \, \mathrm{x y}.$$

Here, the antecedent amounts to the introduction ad hoc of the symbol E for the composite mappings. This step is necessary because of the lacuna in the traditional symbolism, which lacks a notation for that composite mapping.

8. Well-formed strings in the substitutive algebra

The following concepts, which are insignificant for logico-arithmetical strings, are important in the substitutive interpretation.

Of the $m + 1$ components (defined in Remark 6) of a balanced string beginning with l^m, we shall call $\mathfrak{S}_0$ the *scope* of l^m, and $\mathfrak{S}_1, \ldots, \mathfrak{S}_m$ the *adjunct* components of l^m and of $\mathfrak{S}$.

A balanced segment $\mathfrak{S}'$ of a balanced string $\mathfrak{S}$ will be called *bound* or *free* (in $\mathfrak{S}$) according as $\mathfrak{S}'$ does or does not belong to the scope of a script letter. E.g., the last letter of $\mathfrak{S}$ as well as $\mathfrak{S}$ itself is free in $\mathfrak{S}$ whereas the first italic in a nonelementary string is bound. The following proper segments are free in (12′):

$$U, V, \mathbf{B^2}\, G\, U\, V, W, \mathbf{B^1}\, H\, W, X, Y, Z, \mathbf{B^3}\, K\, X\, Y\, Z$$

while F, G, H, and K are bound. In (13), the letters u, v, . . ., z are free while F and $\mathbf{B^3}\, F\, G\, H\, K$ are bound.

The remainder of this section is devoted to balanced strings consisting of 1) boldface operator symbols $\mathbf{B}$ and $\mathbf{P}$ with superscribed positive (!) weights; and 2) capital letters in italic type (which will be understood to be of weight -1) to each of which a nonnegative integer, called its *rank*, is associated whose numeral will often appear as a superscript.

If $\mathfrak{S}$ is a balanced string, the *rank* $\varrho\,\mathfrak{S}$ will be defined recursively as follows:

a) if $\mathfrak{T}$ is elementary and consists, say, of F^m, then $\varrho\,\mathfrak{T} = \varrho\,F^m = m$;

b) if $\mathfrak{T} = \mathbf{B}^m\,\mathfrak{T}_0\,\mathfrak{T}_1 \ldots \mathfrak{T}_m$, then $\varrho\,\mathfrak{T} = \varrho\,\mathfrak{T}_1 + \cdots + \varrho\,\mathfrak{T}_m$;

c) if $\mathfrak{T} = \mathbf{P}^m\,\mathfrak{T}_0\,\mathfrak{T}_1 \ldots \mathfrak{T}_m$, then $\varrho\,\mathfrak{T} = \varrho\,\mathfrak{T}_m$.

A string $\mathfrak{S}$ will be said to be *well-formed* if

1) $\mathfrak{S}$ *is balanced;*

2) *the weight of every operator symbol is equal to the rank of its scope;*

3) *for every* $\mathbf{P}^m$, *the* m *adjunct components have one and the same rank.*

The following remarks are easily proved.

9. *Every balanced segment of a well-formed string is well-formed.*

10. *If in well-formed string $\mathfrak{S}$ a free balanced segment $\mathfrak{T}$ is replaced by a well-formed string of rank* m, *then the resulting string is well-formed and of rank $\varrho\,\mathfrak{S}$ if and only if* m $= \varrho\,\mathfrak{T}$. The sufficiency follows from the definitions of rank and well-formedness. The necessity can be proved by repeated application of the remark that if, in the well-formed string $\mathfrak{P} = \mathbf{Q}^m\,\mathfrak{P}_0\,\mathfrak{P}_1 \ldots \mathfrak{P}_m$ (where $\mathbf{Q} = \mathbf{B}$ or $\mathbf{P}$) an adjunct component is replaced by a well-formed string of unlike rank, then the resulting string either is not well-formed (namely, if $\mathbf{Q}^m = \mathbf{P}^m$ and m > 1) or has a rank $\neq \varrho\,\mathfrak{P}$.

By a *substitution* within the well-formed string $\mathfrak{S}$ (more specifically, by the $\mathbf{Q}$-substitution of $\mathfrak{U}_1, \ldots, \mathfrak{U}_n$ into $\mathfrak{T}^n$ within $\mathfrak{S}$) we mean the replacement of a free balanced segment $\mathfrak{T}^n$ (of rank n) by the string $\mathbf{Q}^n\,\mathfrak{T}^n\,\mathfrak{U}_1 \ldots \mathfrak{U}_n$, where $\mathfrak{U}_1, \ldots, \mathfrak{U}_n$ are balanced strings, and $\mathbf{Q} = \mathbf{B}$ or $\mathbf{P}$. This substitution will be said to be *total* if $\mathfrak{T}^n = \mathfrak{S}$; *left-elementary*, if $\mathfrak{T}^n$ is an elementary string; *right-elementary*, if $\mathfrak{U}_1, \ldots, \mathfrak{U}_n$ are n elementary strings; *elementary*, if it is both left- and right-elementary.

In order that the result of the $\mathbf{Q}$-substitution of $\mathfrak{U}_1, \ldots, \mathfrak{U}_n$ *into* $\mathfrak{T}^n$ *within* $\mathfrak{S}$ *be a well-formed string it is necessary and sufficient that* $\mathfrak{U}_1, \ldots, \mathfrak{U}_n$ *be well-formed and that, if* $\mathfrak{T}^n$ *is contained in an adjunct component of some* $\mathbf{P}$, *then*

$$\varrho\,\mathfrak{U}_1 = \ldots = \varrho\,\mathfrak{U}_n = \text{n} \quad if \quad \mathbf{Q} = \mathbf{P},$$

$$\varrho\,\mathfrak{U}_1 + \cdots + \varrho\,\mathfrak{U}_n = \text{n} \quad if \quad \mathbf{Q} = \mathbf{B}$$

(*thus* $\varrho \, \mathfrak{U}_1 = \ldots = \varrho \, \mathfrak{U}_n = 1$ *if* $\mathbf{Q} = \mathbf{B}$ *and all* $\varrho \, \mathfrak{U}_i > 0$). *If the result* $\mathfrak{S}^*$ *is well-formed, then* $\varrho \, \mathfrak{S}^* = \varrho \, \mathfrak{S}$ *if* $\mathfrak{T}^n$ *is contained in an adjunct component of some* $\mathbf{P}$, *whereas otherwise*

$$\varrho \, \mathfrak{S}^* = \begin{cases} \varrho \, \mathfrak{S} - n + \varrho \, \mathfrak{U}_1 + \cdots + \varrho \, \mathfrak{U}_n & \textit{if} \quad \mathbf{Q} = \mathbf{B} \\ \varrho \, \mathfrak{S} - n + r & \textit{if} \quad \varrho \, \mathfrak{U}_1 = \ldots = \varrho \, \mathfrak{U}_n = r \quad \textit{and} \quad \mathbf{Q} = \mathbf{P}. \end{cases}$$

The necessity of the condition follows from Remark 10. Its sufficiency and the formulae for $\varrho \, \mathfrak{S}^*$ follow from the definitions of well-formed strings and their ranks.

In order that a string $\mathfrak{S}$ *consisting of letters* $\mathbf{B}$ *and* $\mathbf{P}$ *and italics designate the result of iterated substitutions within* $\mathfrak{S}$ *it is necessary and sufficient that* $\mathfrak{S}$ *be well-formed.* $\mathfrak{S} = \mathbf{Q}^n \, \mathfrak{P}_0 \, \mathfrak{P}_1 \ldots \mathfrak{P}_n$ is the result of substituting $\mathfrak{P}_1, \ldots, \mathfrak{P}_n$ into $\mathfrak{P}_0$ within $\mathfrak{S}$. Since $\mathbf{Q}^m \, F^m \, G_1 \ldots G_m$ is, by definition, the result of a substitution, the assertion follows by induction according to the number of bold-face letters in $\mathfrak{S}$. The substitutions by which $\mathfrak{S}$ can be obtained in this way are, of course, in general neither total nor semi-elementary.

Pure $\mathbf{B}$-*strings* (i.e., strings including no $\mathbf{P}$) or *homogeneous* $\mathbf{P}$-*strings* (i.e., strings all of whose bold-face letters are $\mathbf{P}$ of one and the same weight) are among those that can be built up by right-elementary substitutions. In such a string $\mathfrak{S}$, if it is nonelementary, let $\mathbf{Q}^n$ be the last bold-face letter that is immediately preceded by an italic. (If there is no such letter, let $\mathbf{Q}^n$ be the first letter in $\mathfrak{S}$.) The interval of $\mathbf{Q}^n$ clearly is of the form $\mathbf{Q}^n \, \mathfrak{P}_0 \, H_1 \ldots H_n$ for n elementary strings $H_1, \ldots, H_n$; and their $\mathbf{Q}$-substitution into $\mathfrak{P}_0$ is right-elementary. But the inhomogeneous $\mathbf{P}$-string $\mathbf{P}^2 \, F^2 \, \mathbf{P}^3 \, G_1^3 \, K_1^2 \, K_2^2 \, K_3^2 \, G_2^2$ and the mixed string $\mathbf{P}^2 \, F^2 \, \mathbf{B}^2 \, L_1^2 \, M_1^2 \, M_2^3 \, L_2^5$ cannot be obtained by right-elementary substitutions since $\mathbf{P}^2 \, F^2 \, G_1^3 \, G_2^2$ and $\mathbf{P}^2 \, F^2 \, L_1^2 \, L_2^5$ are not well-formed.

It is easy to characterize the substitutive operator strings that have logico-arithmetical mapping counterparts.

In order that a well-formed string $\mathfrak{S}$ *consisting of references to operators of positive weight and mappings — basic elements being considered as* 0-*place mappings — possess an equivalent logico-arithmetical string it is necessary and sufficient that* 1) *the scope of every operator symbol, if any, be elementary and* 2) *all free mapping symbols have the place-number* 0. The equivalent string can be obtained by omitting the operator symbols and transferring the superscribed weight of every operator symbol to the mapping symbol in its scope as indication of the place-number of the mapping. (This is what one has to do, e.g., in the transition from (12′), (11′).) Two simple well-formed strings violating conditions 1) and 2), respectively, are

$$\mathbf{P}^2 \, \mathbf{P}^2 \, F^2 \, G_1^2 \, G_2^2 \, H_1^1 \, H_2^1 \quad \text{and} \quad \mathbf{P}^2 \, F^2 \, G_1^2 \, G_2^1.$$

9. An irredundant method of coding results of substitution

A well-formed string such as $\mathbf{B}^2 \, F^2 \, G_1^2 \, G_2^1$ includes redundant information because the weight of $\mathbf{B}^2$ and the place-number of F^2 in its scope must be equal. Hence $\mathbf{B}^2 \, F \, G_1^2 \, G_2^1$ would be sufficient. Clearly, the ranks of free italics in a string $\mathfrak{S}$ (such as G_1^2 and G_2^1 in the example) cannot be inferred from the weights of the bold-face letters in $\mathfrak{S}$. But neither can even the ranks of some bound italics. E.g., the string $\mathbf{B^3 \, B^2} \, F \, G_1 \, G_2 \, H_1 \, H_2 \, H_3$ designates $\langle F[G_1, G_2] \rangle [H_1, H_2, H_3]$ and implies that $\varrho \, G_1 + \varrho \, G_2 = 3$. but it fails to determine the place-numbers of G_1 and G_2 — which of them is 2 and which is 1. Yet, according to the superassociative law, the designated function is equal to $F[G_1[H_1, H_2]. \, G_2 \, H_3]$ in one case, and to $F[G_1 \, H_1, \, G_2[H_2, H_3]]$, in the other. Hence omission of the superscribed ranks may create ambiguities.

Irredundancy can, however, be achieved by omitting the superscribed weights of the bold-face symbols.

First consider a nonelementary string of the form

$$\mathbf{Q_u \, Q_{u-1}} \cdots \mathbf{Q_2 \, Q_1} \, F_1 \, F_2 \ldots F_{v-1} \, F_v \, ,$$

where $u, v \geq 1$ and each $\mathbf{Q_i}$ is either $\mathbf{B}$ or $\mathbf{P} (1 \leq i \leq u)$. For typographical reasons, the ranks of the italics have not been superscribed. In syntactic discussions, each italic F_i may be replaced by the numeral, say $\varrho(i)$ for its rank or place-number. The string thus may be written in the form

$$\mathfrak{B} = \mathbf{Q_u} \cdots \mathbf{Q_1} \, \varrho(1) \ldots \varrho(v),$$

which will be called a *block*. With $\mathfrak{B}$ we associate two integer-valued functions, φ and σ. having an initial segment of $\{0, 1, \ldots, u\}$ or this entire set as their domain. They are defined as follows.

$$\varphi(0) = 0, \quad \sigma(0) = 1;$$

and, for $1 \leq k \leq u$,

$$\varphi(k) = \varphi(k - 1) + \sigma(k - 1)$$

provided $\sigma(k - 1)$ is defined and > 0; otherwise $\varphi(k)$ remains undefined;

$$\sigma(k) = \varrho(\varphi(k)) \quad \text{or} \quad \varrho(\varphi(k - 1) + 1) + \cdots + \varrho(\varphi(k))$$

according as $\mathbf{Q_k} = \mathbf{P}$ or $\mathbf{B}$, provided that $\varphi(k)$ is defined and $\leq v$; otherwise $\sigma(k)$ remains undefined.

Clearly, $\varphi(1) = 1$ and $\sigma(1) = \varrho(1)$:

$$\varphi(2) = 1 + \varrho(1) \quad \text{and} \quad \sigma(2) = \varrho(1 + \varrho(1)) \quad \text{or} \quad \varrho(2) + \cdots + \varrho(1 + \varrho(1))$$

according as $\mathbf{Q_1} = \mathbf{P}$ or $\mathbf{B}$, provided $1 + \varrho(1) \leq v$, while $\varrho(2)$ remains undefined if $1 + \varrho(1) > v$; and so on.

The nonelementary block $\mathfrak{B}$ will be called *well-formed* if and only if

   I. $\varphi(u)$ is defined;

  II. $\varphi(u) = v$;

 III. $\varrho(i) = \varrho(\varphi(k))$ for each i and k such that

$$1 \leq k \leq u, \quad \varphi(k - 1) < i \leq \varphi(k), \quad \text{and} \quad \mathbf{Q_k} = \mathbf{P}.$$

Condition I implies that v is not too small. Condition II, which might be replaced by $\varphi(u) \geq v$, implies that v is not too large.

It should be noted that $\mathfrak{B}$ may include elements of rank 0; that is to say, that $\varrho(i)$ may be 0 for some i. If, in a well-formed block, $\varrho(i) = 0$ for $\varphi(k-1) < i \leq \varphi(k)$, then $\sigma(k) = 0$ so that $\varphi(k+1)$ is undefined. It follows that $k = u$. Such a block thus terminates in $\sigma(u-1)$ zeros while $\varrho(i) > 0$ for $1 \leq i \leq \varphi(u-1)$.

If $\mathfrak{B}$ is well-formed, we define $\varrho(\mathfrak{B}) = \sigma(u)$. By induction according to u, one proves without difficulty that *the block* $\mathfrak{B}$ *designates the result of successive complete substitutions* (u *in number*) *into* F_1 *if and only if* $\mathfrak{B}$ *be well-formed. The place-number of the designated function is* $\sigma(u)$, *that is,* $\varrho(\varphi(u-1)+1) \div \cdots \div \varrho(v)$ *according as* Q_u *is* **P** *or* **B**.

Now let there be given any nonelementary string $\mathfrak{S}$ consisting of bold-face symbols **B** and **P** and of numerals (indicating the ranks of italics). There are positive integers $t, u_1, \ldots, u_t, v_1, \ldots, v_t$ such that, if

$$s_j = u_1 + \cdots + u_j \qquad (1 \leq j \leq t),$$

then

$$\mathfrak{S} = Q_{s_t} \cdots Q_{s_{t-1}+1} \varrho_t(1) \cdots \varrho_t(v_t) Q_{s_{t-1}} \cdots Q_{s_{t-2}+1} \varrho_{t-1}(1) \cdots \varrho_{t-1}(v_{t-1}) \cdots$$
$$\cdots Q_{s_2} \cdots Q_{s_1+1} \varrho_2(1) \cdots \varrho_2(v_2) Q_{s_1} \cdots Q_1 \varrho_1(1) \cdots \varrho_1(v_1).$$

First, define the functions φ_1 and σ_1 associated with the (not necessarily well-formed) block

$$\mathfrak{B}_1 = Q_{u_1} \cdots Q_1 \varrho_1(1) \cdots \varrho_1(v_1).$$

If $\sigma_1(u_1)$ is defined, that is, if $\varphi_1(u_1)$ is defined and $\leq v_1$, then the function ϱ_2 can be extended by setting

$$\varrho_2(v_2 + 1) = \sigma_1(u_1)$$

and, if $v_1 > \varphi_1(u_1) + 1$,

$$\varrho_2(v_2 + w) = \varrho_1(\varphi_1(u_1) + w - 1) \quad \text{for} \quad 2 \leq w \leq v_1 + 1 - \varphi_1(u_1).$$

Next one, may define the functions φ_2 and σ_2 associated with the block

$$\mathfrak{B}_2 = Q_{u_1+u_2} \cdots Q_{u_1+1} \varrho_2(1) \cdots \varrho_2(v_1 + v_2 + 1 - \varphi_1(u_1)).$$

If $\sigma_2(u_2)$ is defined, that is, if $\varphi_2(u_2)$ is defined and $\leq v_2 + v_1 + 1 - \varphi_1(u_1)$, then one can extend the function ϱ_3 and define φ_3 and σ_3.

If

$$\mathfrak{B}_j = Q_{s_j} \cdots Q_{s_{j-1}+1} \varrho_j(1) \cdots \varrho_j(v_1 + \cdots + v_j + j - 1 - \varphi_1(u_1) - \cdots - \varphi_{j-1}(u_{j-1}))$$

has been obtained, then one can extend the function ϱ_{j-1} by setting

$$\varrho_{j+1}(v_{j+1} + 1) = \sigma_j(u_j)$$

and, if $v_1 + \cdots + v_j \geq \varphi_1(u_1) + \cdots + \varphi_j(u_j) - j + 1$,

$$\varrho_{j+1}(v_{j+1} + w) = \varrho_j(\varphi_j(u_j) + w - 1) \quad \text{for} \quad 2 \leq w \leq v_1 + \cdots + v_j + j - \varphi_1(u_1) -$$
$$- \cdots - \varphi_j(u_j);$$

and one can define φ_{j+1} and σ_{j+1}.

Proceeding in this way one arrives at

$$\mathfrak{B}_t = \mathbf{Q}_{s_t} \ldots \mathbf{Q}_{s_{t-1}-1} \varrho_t(1) \ldots \varrho_t(v_t) \varrho_t(v_t + 1) \ldots \varrho_t(v_1 + \cdots + v_t + t - 1 - $$
$$- \varphi_1(u_1) - \cdots - \varphi_{t-1}(u_{t-1}))$$

and can define φ_t and σ_t.

$\mathfrak{B}$ will be called well-formed if and only if

I. $\varphi_t(u_t)$ is defined;

II. $v_1 + \cdots + v_t = \varphi_1(u_1) + \cdots + \varphi_t(u_t) + t - 1$:

III. $\varrho_j(i) = \varrho_j(\varphi_j(k))$ for each i. j. k such that

$$1 \leqq j \leqq t. \quad 1 \leqq k \leqq v_1 + \cdots + v_j + j - 1 - \varphi_1(u_1) - \cdots - \varphi_j(u_j),$$
$$\varphi_j(k - 1) < i < \varphi(k), \quad \text{and} \quad \mathbf{Q}_{s_j - k} = \mathbf{P}.$$

Condition I implies that the blocks $\mathfrak{B}_1, \ldots, \mathfrak{B}_t$ are not too short; Condition II implies that $\mathfrak{B}(= \mathfrak{B}_t)$ is not too long.

10. Functor trees

This section is devoted to another coding of balanced strings that are free of consecutive letters of nonnegative weight. A substitutive interpretation of such a string is free of consecutive bold-face letters and thus possesses an equivalent logico-arithmetical string. The symbolism to be developed altogether dispenses with operator symbols and intersperses the mapping symbols with finite sequences of positive integers — hereinafter briefly called *sequences*.

If $\sigma = i_1, \ldots, i_m$ is such a sequence and if $1 \leqq k < m$, we set $\sigma_k = i_1 \ldots i_k$ and write $\sigma = \sigma_k i_{k-1} \ldots i_m$. We call every σ_k an *ancestor* of σ, and refer to σ as a *descendent* of σ_k, and as the *immediate* descendent of σ_{m-1}.

An order relation. $<$ and $>$ (read: *precedes* and *follows*) will be defined by the stipulations that

a) every sequence precedes all its descendents;

b) if $i_{k-1} \neq i'_{k-1}$, then

$$\sigma = i_1 \ldots i_k i_{k-1} \ldots i_{k-m} < \quad \text{or} \quad > \sigma' = i_1 \ldots i_k i'_{k-1} \ldots i'_{k+m'}$$

according as $i_{k-1} < i'_{k-1}$ or $i_{k-1} > i'_{k-1}$.

A set T of sequences will be called an *index tree* provided that. if $i_1 \ldots i_m$ $(m > 2)$ belongs to T, then so does the sequence $i_1 \ldots i_{m-1}$ and that all sequences in T have the same first element. If this element is i_1, then T will be called an i_1-tree. If T includes only one sequence (which then necessarily consists of i_1 alone) T will be called *elementary*.

A sequence σ will be called *interior* or *terminal* (in T) according as T does or does not include at least one descendent of σ. The number of immediate descendents (in T) of an interior sequence σ will be denoted by $N(\sigma)$ and called the *offspring number* of σ (in T).

The sequence $\chi = i_1 \ldots i_m$ $(m > 1)$ will be called *characteristic* (in T) if every descendent of χ_{m-1} is terminal and $i_m = 1$. Moreover, if T is elementary, the sequence i_1 will be called *characteristic* (in T). Clearly, for a sequence σ in T to be interior it is necessary and sufficient that σ be an initial segment of a characteristic sequence.

Let $\chi^r = i_1^r \ldots i_{m_r}^r$ $(1 \leqq r \leqq t)$ be all the characteristic sequences of T labelled (if $t > 1$) in such a way that $\chi^{r-1} < \chi^r$ for $2 \leqq r \leqq t$. There is an integer p_r such that

$$\chi_{p_r-1}^{r-1} = \chi_{p_r-1}^r \quad \text{and} \quad i_{p_r}^{r-1} < i_{p_r}^r$$

Clearly, $1 < p_r < m_r$ and

(14) $\qquad N(\chi_s^r) \geqq i_{s+1}^r \quad \text{for} \quad 1 \leqq r \leqq t \quad \text{and} \quad p_r \leqq s_r \leqq m_r - 1.$

We now enumerate all sequences belonging to T in their order of precedence. The first (uncharacteristic) interior sequences are $\chi_1^1, \chi_2^1, \ldots, \chi_{m-1}^1$ and, if we set $p_1 = 1$, the following $i_{s+1}^1 - 1$ uncharacteristic terminal sequences between χ_s^1 and χ_{s+1}^1 are

(15¹) $\qquad \chi_s^1 1, \chi_s^1 2, \ldots, \chi_s^1 i_{s+1}^1 - 1 \qquad (p_1 \leqq s \leqq m_1 - 2).$

Because of $i_{m_1}^1 = 1$, the sequence $\chi_{m_1-1}^1$ is followed by χ^1.

If $t > 0$, then χ^{r-1} for $1 < r \leqq t$ is followed by the uncharacteristic terminal sequences

(16ʳ) $\qquad \chi_s^{r-1} i_{s+1}^{r-1} + 1, \chi_s^{r-1} i_{s+1}^{r-1} + 2, \ldots, \chi_s^{r-1} N(\chi_s^{r-1})$

for $s = m_{r-1} - 1, m_{r-1} - 2, \ldots, p_r$ (in this order!), which are altogether

$$N(\chi_{m_{r-1}}^{r-1} - 1) - i_{m_{r-1}}^{r-1} + \cdots + N(\chi_{p_r}^{r-1}) - i_{p_r}^{r-1}$$

uncharacteristic terminals sequences. They, in turn, are followed by the $i_{p_1}^r - i_{p_r}^{r-1} - 1$ uncharacteristic terminal sequences

(17ʳ) $\qquad \chi_{p_r-1}^{r-1} i_{p_r}^{r-1} + 1, \chi_{p_r-1}^{r-1} i_{p_r}^{r-1} + 2, \ldots, \chi_{p_r-1}^{r-1} i_{p_r}^r - 1.$

Of course, if in (14) for some s the equality sign holds and/or $i_{p_r}^r = i_{p_r}^{r-1} + 1$, then the corresponding sets (16ʳ) and (17ʳ) are empty. (17ʳ) is followed in T by the interior sequence $\chi_{p_r}^r$. Between χ_s^r and χ_{s+1}^r for $p_r \leqq s \leqq m_r - 2$, there are the $i_{s+1}^r - 1$ uncharacteristic terminal sequences

(15ʳ) $\qquad \chi_s^r 1, \ldots, \chi_s^r i_{s-1}^r - 1 \qquad (p_r \leqq s \leqq m - 2, 2 \leqq r \leqq t).$

Finally, if we set $p_{r+1} = 1$, then T includes the following

$$N(\chi_{m_t-1}^t) - i_{m_t}^t + \cdots + N(\chi_{p_t,1}^t) - i_{p_{t-1}}^t + 1$$

uncharacteristic terminal sequences

(16ᵗ) $\chi_s^t i_{s+1}^t + 1, \chi_s^t i_{s+1}^t + 2, \ldots, \chi_s^t N(\chi_s^t)$ for $s = m_t - 1, m_t - 2, \ldots, p_{t+1}$ (in this order!).

With each nonelementary index tree T we can associate a string X(T) consisting of the t characteristic sequences χ^r ($1 \leqq r \leqq t$) of T (including altogether $m_1 + \cdots + m_r$ integers) and the $\sum (m_r - p_r)$ offspring numbers of its interior sequences, arranged as follows:

$$N(\chi_{p_1}^1) N(\chi_{p_1+1}^1) \ldots N(\chi_{m_1-1}^1) \chi^1 \ldots N(\chi_{p_t}^t) N(\chi_{p_t+1}^t) \ldots N(\chi_{m_t}^t) \chi^t.$$

In this way, one obtains an element of the set $\mathfrak{X}$ of all strings X consisting of sequences $\sigma^1, \ldots, \sigma^t$ of positive integers and single positive integers, $c_1^r, \ldots, c_{m_r-p_r}^r$ preceding σ^r ($1 \leqq r \leqq t$) and satisfying the following conditions.

A. *X includes at least one sequence having at least two members.*

B. *All sequences in X begin with one and the same number and terminate in 1. No sequence in X is an initial segment of another sequence in X.*

C. *If* $t > 1$, *then* $\sigma^1 < \sigma^2 < \cdots < \sigma^t$.

In view of B, it follows from C that. if $t > 1$, then, for every $r > 1$ and $\leqq t$, there is a number $p_r \geqq 2$ such that

$$c_{p_r-1}^{r-1} = c_{p_r-1}^r \quad \text{and} \quad c_{p_r}^{r-1} < c_{p_r}^r.$$

We furthermore set $p_1 = 1$.

D. *If the sequence* σ^r *has exactly* m_r *members, then X includes exactly* $m_r - p_r$ *single numbers between* σ^{r-1} *and* σ^r.

E. *If* $1 < r \leqq t$ *and* $1 \leqq s \leqq m_r - p_r$, *then the* s-*th number between* σ^{r-1} *and* σ^r *is not less than the* $(p_r + s)$-*th member of* σ^r.

(E reflects inequality (14).)

Conversely, for every string X belonging to the set $\mathfrak{X}$, there is exactly one index tree T for which

$$\chi^r = \sigma^r \quad \text{and} \quad N(\chi_{p_r+s}^r) = c_s^r \quad (1 \leqq r \leqq t, \ 1 \leqq s \leqq m_r - p_r).$$

Indeed, the numbers in σ^1 are sufficient for the determination of the (uncharacteristic) sequences preceding σ^1. After the sequence σ^{r-1} has been determined, the numbers $c_{p_{r-1}}^{r-1}, \ldots, c_{m_{r-1}}^{r-1}$ satisfying condition E make it possible to fill in all the (uncharacteristics) sequences, if any, immediately following σ^{r-1}, while the numbers $i_{p_r}^r, \ldots, i_{m_r}^r$ are sufficient for the determination of the remaining (uncharacteristic) sequences preceding σ^r. This procedure may be continued until σ^t is reached. Then the numbers $c_{m_t-1}^t, \ldots, c_1^t$ make it possible to determine all uncharacteristic sequences, if any, following σ^t.

It is of course understood that there is no doubt as to where every member of the string begins and terminates. Otherwise, separators would have to be inserted in order to clarify that 1) each of the multi-digit numerals, if any, describes one number; 2) certain segments of numerals each describe one sequence. Knowing where each sequence terminates one can, by mere counting, determine where each sequence begins. E.g., if in a string there is only one termination mark, then it must be at the end of the string, and the number of numerals in the string is odd; if it is $2m - 1$, then σ^1 includes the last m numerals.

For example, from the string

$$4 \quad 2 \quad 111 \quad 1 \quad 121 \quad 3 \quad 2 \quad 1431$$

one easily reconstructs the tree including the following sequences

$$1, 11, 111, 112, 12, 121, 13, 14, 141, 142, 143, 1431, 1432.$$

An index tree will be called *homogeneous* if all its interior sequences have the same offspring number. If that number is n, the tree will be said to be of *order* n. A non-elementary tree of order n thus is a set, T, of sequences such that

α) the first member in all sequences in T is one and the same positive integer i_1 while each of the following members is one of the numbers $1, \ldots, n$;

β) if $k > 1$ and $i_1 \ldots i_{k-1} i_k$ belongs to T, then so do $i_1 \ldots i_{k-1}$ and the $n - 1$ sequences $i_1 \ldots i_{k-1} m$ for $m = 1, \ldots, i_k - 1, i_k + 1, \ldots, n$.

Clearly, a homogeneous index tree can be reconstructed from its order number, n, and its characteristic sequences. The number of noncharacteristic sequences between χ^{r-1} and χ^r is

$$(m_{r-1} - p_r) n - (i^{r-1}_{p_r} + \cdots + i^{r-1}_{m_{r-1}}) + (i^r_{p_r} + \cdots + i^r_{m_r}).$$

Knowing, e.g., that 1121 and 1221 are the only characteristic sequences of a tree T of order 2 one finds that T is the set of the following sequences:

$$1, 11, 111, 112, 1121, 1122, 12, 121, 122, 1221, 1222.$$

The mere knowledge of an order number and a string of numerals without fragmentation into sequences (say, by commas or spaces) is not sufficient for the reconstruction of the tree. E.g., the string

$$1 \quad 1 \quad 1 \quad 1 \quad 2 \quad 1 \quad 1 \quad 2 \quad 1$$

may be broken up into

$$1 \quad 1 \quad 1; \quad 1 \quad 2 \quad 1 \quad 1 \quad 2 \quad 1 \quad\quad \text{and} \quad\quad 1 \quad 1 \quad 1 \quad 1 \quad 2 \quad 1; \quad 1 \quad 2 \quad 1$$

and furthermore describes the single characteristic sequence of a third tree.

If with every sequence of an index tree T a mapping or an element of an abstract substitutive system is associated — with every interior sequence a mapping whose place-number or rank is equal to the offspring-number of the sequence — then result will be called a *mapping tree* (on T) of the *first* or *second* kind according as the substitutions are understood to be parenthesis or bracket substitution. In order to describe such a mapping tree of wither kind, one may insert in the string describing T (1) every mapping symbol corresponding to an interior or characteristic sequence, σ or χ, immediately preceding $N(\sigma)$ and χ, respectively, (2) the other mapping symbols in the order of the noncharacteristic sequences.

It is easy to characterize well-formed strings of this kind. The logico-arithmetical string (11'), written in the form (12') with weighted operators, may be described as the following mapping tree of the second kind:

$$F\,3\,G\,2\,\mathrm{u}\,111\,\mathrm{v}\,H\,1\,\mathrm{w}\,121\,K\,3\,\mathrm{x}\,131\,\mathrm{y}\,\mathrm{z}\,.$$

If the place-numbers of the mappings are superscribed, one may of course dispense with the offspring-numbers of the index tree and write

$$F^3\,G^2\,\mathrm{u}\,111\,\mathrm{v}\,H^1\,\mathrm{w}\,121\,K^3\,\mathrm{x}\,131\,\mathrm{y}\,\mathrm{z}\,.$$

Similarly, the expression for 0 in terms of A obtained on p. 86, which can be written with unweighted operators in the form

$$P\,A\,P\,A\,A\,P\,A\,A\,A\,P\,A\,A\,P\,A\,A\,A$$

may be described as the homogeneous mapping tree of order 2

$$A\,A\,A\,A\,A\,1121\,A\,A\,A\,A\,A\,1221\,A\,.$$

(Eingegangen am 4. Juli 1963)

MENGER, K.
Math. Annalen 157, 278—295 (1964)

Superassociative Systems and Logical Functors

By

KARL MENGER in Chicago*

1. Introduction

Mathematicans often refer to our time as the age of algebra and of axiomatic theories. Paradoxically, however, the branch of mathematics which for three centuries has been at the very core of mathematics — analysis — has been neither 'algebraicized' nor axiomatized.

Algebraists continue the great tradition of the theories of groups and fields in studying monooperational and biooperational systems mostly shaped after addition and multiplication — sometimes even far-fetched generalizations in preference to the most fertile ideas of analysis.

Analysts are developing the ideas that originated in the 17th century beyond anything that the founders of the theory could dream, but have not yet filled the gaps that those great men left by omitting names and symbols for the most important objects of their studies — the identity, the selectors, and the constant functions. To this day, those paramount functions are often symbolically identified with the values that the functions assume for x or for $x_1, \ldots, x_n$, e. g., in speaking about "the function x which, for every number x, assumes the value x" and, correspondingly, even about "the functions $\log x$ and $f(x)$", instead of achieving uniformity by referring to the latter as the functions $\log$ and f, and introducing a symbol for the identity function[1]). This procedure blocks the algebraization of analysis and the automation, in the spirit of our time, of the operations with functions, especially, of the substitution of functions into functions, which is the operation par excellence and yet, in comparison with addition and multiplication, has been rather neglected[2]).

For just as one needs symbols, 0 and 1, for the constant functions of the values 0 and 1 in order to write

$$0 + f = f + 0 = f \text{ and } 1 \cdot f = f \cdot 1 = f,$$

---

* I wish to thank Dr. E. F. STUEBEN and Mr. H. I. WHITLOCK for valuable suggestions and Miss A. LIPSON for work on a I.B.M. 7090 proving the consistency of some superassociative systems. All three were supported by the Office of Naval Research (grant Nonr (G)-0003-64) during the summer 1963.

[1]) An analogue of analysis lacking symbols for the most important functions would be an arithmetic without symbols for the numbers 1, 2, 3. Cf. [4], [5].

[2]) For example, analysts often speak of *the* distributive law — referring to the additive-multiplicative distribution, while they rarely even mention the additive-substitutive and multiplicative-substitutive distributive laws,

$$(f + g)h = fh + gh \text{ and } (f \cdot g)h = fh \cdot gh,$$

emphasized in [6]—[9]. Here, juxtaposition of function symbols indicates substitution; see footnote[2].

one needs a symbol, say j, for the identity function in order to write the analogous formula
$$jf = fj = f,$$
where substitution is denoted by juxtaposition of function symbols[3]); and one needs a symbol, say I_k^n, for the selector of the k-th of n places in order to write
$$I_k^n(F_1, \ldots, F_n) = F_k \text{ and } F(I_1^n, \ldots, I_n^n) = F.$$
Without the use of symbols such as j and I_k^n, these important facts cannot possibly be expressed each in a single formula. The lack of such a symbol is also at the root of the unnatural notation $\dfrac{d}{dx} F(x, G(x, y))$ for the so-called total x-derivative of a function of which the classical symbolism can only express the value assumed for x, y. If D_i denotes the i-th place partial derivative, then abbreviating I_1^2 to I one can express the intended idea, concisely and unambiguously[4]), by writing $D_1(F(I, G))$. This function is equal to
$$D_1 F(I, G) \cdot D_1 I + D_2 F(I, G) \cdot D_1 G, \text{ that is, } D_1 F(I, G) + D_2 F(I, G) \cdot D_1 G,$$
since $D_1 I$ is a constant function of value 1. More generally, $D_1 I_k^n = I^n$ or 0^n according as i = k or i $\neq$ k. These are minor examples of the systematic algebraization of which analysis is capable if functions and their values are clearly distinguished; cf. [9]. Another consequence of denoting a function F into which n-tuples of functions may be substituted or, as we shall say, an n-*place function* F by $F(x_1, \ldots, x_n)$ is the traditional misnomer[5]) "function of n variables."

[3]) Juxtaposition is used in writing $F(G, H)$ for the result of substituting the pair of functions (G, H) into F. The result of substituting single functions is sometimes indicated by a circle sandwiched between the function symbols, as in $j \circ f = f \circ j = f$. Such a symbol is superfluous if multiplication is consistently denoted by a dot, as in $1 \cdot f = f \cdot 1$.

[4]) The first parentheses in $(D_1 F)(I, G)$ distinguishing it from $D_1(F(I, G))$ can be dispensed with if one consistently adheres to the convention that *what is within the reach of an operator symbol is only the immediately following symbol for a function for which the operator is meaningful*. Derivatives such as $D_1(F(I_2^2, I_1^2))$ can hardly be expressed in any single classical symbol.

[5]) For an ideal gas, the temperature t is a function *of* the pressure p and the volume v; in proper units, $t = p \cdot v$ or $t = P(p, v)$. But the product P *itself*, which connects t with the other two fluents, is the set of all pairs $((x, y), x \cdot y)$, where x and y stand for any two numbers. It is not a function *of fluents* (such as p and v, often called variables), nor *of variables* in the toto caelo different sense in which this word is used in logic (such as the letters x and y in the definition of P), nor in fact *of anything*, just as a number in number theory is not a number of apples or of centimeters or of anything. The connection of t with p and v by the function P means, if γ stands for any sample of an ideal gas or for any state of such a gas, that
$$t\gamma = P(p\,\gamma, v\,\gamma) \text{ or that } ((p\,\gamma, v\,\gamma), t\,\gamma) \text{ belongs to } P \text{ for any } \gamma.$$
Not only the formulae of physics but also those of geometry have to be reinterpreted; e.g., in the geometry of spheres,
$$v = \frac{4}{3} \pi r^3 \text{ means } v\sigma = \frac{4}{3} \cdot \pi \cdot r^3 \sigma \text{ for any sphere } \sigma.$$
Here σ is a sphere variable while the radius r and the volume v are fluents. (Cf. 3, 4, 9, 10.) Physical fluents can be added and multiplied but do not lend themselves to the principal operation in the realm of functions: One can not substitute functions or other fluents into a fluent. A discrimination between fluents and functions from the point of view of type theory has been developed in the interesting paper [2] by J. J. MEHLBERG.

20*

Even logicians treat the selectors and constant functors in the calculus of propositions as step children, referring to them by their values (without introducing symbols for those functors themselves) and thereby neglecting important distinctions. For example, the classical result that every 2-place functor can be expressed in terms of incompatibility is capable of a refinement, if not in need of a certain correction. Let, in LUKASIEWICZ consistently frontal (and therefore parenthesis-free) notation, Cpq denote the implication p → q, and let Apq denote the incompatibility of p and q, traditionally designated by Sheffer's stroke p|q. It can be readily proved that Cpq if and only if ApApq or, in the notation of analysis, $C(p, q) = A(p, A(p, q))$. This formula determines the value of C for any p and q in terms of p, q, and A. It may, therefore, be regarded as an extensional definition of C. Yet, just as classical analysis lacks a symbol for the function $F(I, G)$, it is impossible to write down, in terms of A, the functor to which C is supposed to be equal[6]). This can be done only if one introduces into the formula a symbol, say I, for the first selector, for which $I(p, q) = p$ for every p and q, by writing

$$C(p, q) = A(I(p, q), A(p, q)) .$$

Then it is clear that the functor to which C is being equated is $A(I, A)$. But this is an expression in terms of A and I, and not in terms of A alone. Nor is it possible to express, for the functor C, anything but its value Cpq in terms of p, q, and A alone. More can be done, however, for the negation of A, which is the conjunction A', and the constant functors 1 and 0. But these functors are the only ones that can themselves be expressed in terms of A alone, namely as follows:

$$A' = A(A, A), 1 = A(A, A(A, A)), 0 = A(A(A, A), A(A, A)) .$$

Thus, however satisfactory the extensional definition of C in terms of A may be, it must be admitted that A', 1 , and 0 are more closely related to A than are the other twelve functors – a distinction that has been neglected in the classical literature.

A trioperational algebra of analysis (an axiomatic description of systems of functions with addition, multiplication, and substitution), developed in a series of papers [1], [6], [7], [8], [13] in the early 1940's, attracted only limited attention[7]). The theory was essentially confined to functions having one and the same domain. Progress was made by adjoining, to the postulates concerning operations, assumptions about a partial order reflecting the fact that a function may be a restriction of another function, which then is an

---

[6]) Even the connection of scientific fluents by functions has an analogue in the connection of what SEALL [18] calls *truth-valued fluents* by logical functors. A truth-valued fluent s is defined by associating a truthvalue $s\alpha$ (in the 2-valued logic, 0 or 1) with each object or situation α of a certain kind. The functor A actually is the quadruple $((x, y), A(x, y))$ for all x, y in $\{0, 1\}$; and $r = A(p, q)$ means that $r\alpha = A(p\alpha, q\alpha)$ or that $((p\alpha, q\alpha), r\alpha)$ belongs to A.

[7]) More recently, NÖBAUER has elaborated trioperational algebra in a series of interesting papers [14].

extension of the first. In this way it became possible to consider dom f and ran f (the domain and the range of f) themselves as functions, namely, as restrictions of the identity function (cf. [10]). In a series of interesting papers [16], SCHWEIZER and SKLAR have developed this theory for 1-place functions. The main problems of the axiomatic theory of functions—and one may venture to predict: the main topic of algebra during the next decades—lie, however, in the substitutive algebra of multiplace functions. There are two kinds of such substitutions (cf. [7]—[11]): *intersection or parenthesis substitution* associating with each m-place function F and each m-tuple of n-place functions $G_1, \ldots, G_m$ an n-place function $F(G_1, \ldots, G_m)$ whose domain is the intersection, dom $G_1 \cap \cdots \cap$ dom G_m; and *product or bracket substitution* associating with F and each m-tuple of functions $G_1, \ldots, G_m$ of various place numbers $n_1, \ldots, n_m$ an $(n_1 + \cdots + n_m)$-place function $F[G_1, \ldots, G_m]$ whose domain is the cross product, dom $G_1 \times \cdots \times$ dom G_m. Traditionally, the distinction is indicated, somewhat obscurely, by writing $F(G_1(x_1, \ldots, x_n), \ldots, G_m(x_1, \ldots, x_n))$ and

$$F(G_1(x_1, \ldots, x_{n_1}), \ldots, G_m(x_{n_1 + \cdots + n_{m-1} + 1}, \ldots, x_{n_1 + \cdots + n_m}))$$

for the results of the two types of substitution. The basic property of both operations is *superassociativity* (cf. [8], [10]), which in the case of parenthesis substitution reads

$$(F(G_1, \ldots, G_m))(H_1, \ldots, H_n) = F(G_1(H_1, \ldots, H_n), \ldots, G_m(H_1, \ldots, H_n))$$

for all n-tuples of functions $H_1, \ldots, H_n$ of one and the same place number. An important paper by SCHWEIZER and SKLAR [17] outlines even more general schemes[8]), while the abstract theory of parenthesis substitution as a generalization of the theory of semigroups has been developed by H. I. WHITLOCK [19].

The present paper deals with abstract concepts reflecting parenthesis substitution in the realm of 2-place functions although some ideas can readily be extended to multiplace algebras. The simplest systems with one and two generators will be enumerated, and the last section is devoted to an important system with three generators.

The ordered n-tuples of n-place functions lend themselves to an associative substitution. E. g., for n = 2,

$$((F_1, F_2)(G_1, G_2))(H_1, H_2) = (F_1, F_2)((G_1, G_2)(H_1, H_2)) .$$

In the case of the 2-valued calculus of propositions, there are 256 functor pairs; and they constitute a 1-place substitutional algebra, which can easily be seen to be isomorphic to the semigroup of the 256 functions mapping the set $\{0, 1, 2, 3\}$ into itself. Why then should one investigate superassociative

---

[8]) POST [15] has developed an algebra of the functors of 2-valued logic, which amounts to the use of bracket substitution in conjunction with an operation that may be called *place identification* (cf. [8] and [12]). Unfortunately, POST presents this important work in terms of an obscure notion of variables. A purely algebraic way of writing the formula $C = A(I, A)$ in bracket substitution with place identification is $C = A[j, A]_{(1,2;3)}$, where the subscript indicates that in the 3-place function $A[j, A]$ the first and the second places are to be identified.

systems rather than simply apply the theory of semigroups to the ordered n-tuples of n-place functions ? The answer is that such an approach would not lead to characterizations of special functors and of the systems they generate. The following study of superassociative systems, however, results, e. g., in postulational characterizations of the incompatibility A as well as of the system generated by A. In particular, the system with three generators in the last section represents an axiomatization of the substitutive algebra of all 2-place functors of the 2-valued calculus of propositions — a system that requires three generators, namely, A (i. e., the negation of conjunction) or the negation of disjunction and two other functors, e. g., the selectors.

2. General concepts

A system with superassociative 2-place substitution, briefly, a *superassociative 2-place system* is a set $\mathfrak{S}$ such that every ordered triple X, Y, Z of elements of $\mathfrak{S}$ determines an element $X(Y, Z)$ of $\mathfrak{S}$, and that

(I) $(X(Y, Z))(U, V) = X(Y(U, V), Z(U, V))$ for every X, Y, Z, U, V in $\mathfrak{S}$. $X(Y, Z)$ will be called the result of *substituting* the ordered pair (Y, Z) into X or of *applying* X to the pair (Y, Z).

For any two elements X and Y of $\mathfrak{S}$, let X/Y be the element $X(Y, Y)$, which will be said to result from X by *diagonalization* with Y. From the Superassociativity Law (I) it follows that diagonalization is associative; that is to say,

(II) $(X/Y)/Z = X/(Y/Z)$ for every X, Y, Z in $\mathfrak{S}$ (so that one may write $X/Y/Z$) and that diagonalization is connected with substitution by the laws

(III) $(X(U, V))/Z = X(U/Z, V/Z)$ and $X/(Y(U, V)) = (X/Y)(U, V)$.

For any subset $\mathfrak{T}$ of $\mathfrak{S}$, the set $\mathfrak{S}(\mathfrak{T})$ *generated* by $\mathfrak{T}$ is the smallest set $\mathfrak{F}$ including (1) all elements of $\mathfrak{T}$, and (2) $X(Y, Z)$ for any three elements X, Y, Z belonging to $\mathfrak{F}$. If $\mathfrak{T}$ consists of the elements $F_1, \ldots, F_n$, then we also write $\mathfrak{S}(F_1, \ldots, F_n)$ instead of $\mathfrak{S}(\mathfrak{T})$. If a system $\mathfrak{S}$ includes elements $F_1, \ldots, F_n$ such that $\mathfrak{S} = \mathfrak{S}(F_1, \ldots, F_n)$, then $F_1, \ldots, F_n$ will be called a *system of generators* of $\mathfrak{S}$. A system $\mathfrak{S}$ with n generators $F_1, \ldots, F_n$ is completely determined if the elements $F_1(X, Y), \ldots, F_n(X, Y)$ are known for every X and Y in $\mathfrak{S}$. For each F_i, the elements $F_i(X, Y)$ can be arranged in a matrix, in which X is the row index and Y the column index. This matrix will be called the *substitution table* of F_i. Hence one can say that *a system is determined by the substitution tables of its generators*. $G(X, Y)$, for any elements of $\mathfrak{S}$, can be found by expressing G in terms of the generators and substituting (X, Y) into them.

An element X of a system $\mathfrak{S}$ will be called *idempotent* if $X/X = X$. Clearly, $\mathfrak{S}(F) = \{F\}$ if and only if F is idempotent.

Idempotent elements will also be said to be of *order* 1. Setting $X^{[1]} = X/X$ and $X^{[n+1]} = (X^{[n]})/X$ we call X of *order* n if $X = X^{[n]} \neq X^{[m]}$ for $m < n$.

If $\mathfrak{S}(F)$ contains more than one element, then $F/F \neq F$. In the sequel, in this case, we set

(1) $$F/F = G.$$

An element X of $\mathfrak{S}$ will be called *constant* in $\mathfrak{S}$ if $X(Y, Z) = X$ for every Y and Z in $\mathfrak{S}$; and X will be called *symmetric* in $\mathfrak{S}$ if $X(Y, Z) = X(Z, Y)$ for every Y and Z in $\mathfrak{S}$. If every element of $\mathfrak{S}$ is symmetric, then $\mathfrak{S}$ will be called symmetric. Symmetry of F is sufficient for the symmetry of $\mathfrak{S}(F)$.

If $F(F, G) = F(G, F)$, then $\mathfrak{S}(F)$ will be called *basically symmetric*. If F is of order 2, then from what has been said it follows:

(2) $$F/G = F \text{ and } G/F = F .$$

Moreover, because of (III),

(3) $$X/G = X, (X(F, G))/F = X(G, F) \text{ and } (X(G, F))/F = X(F, G)$$

for every X in $\mathfrak{S}(F)$. Consequently, $X(F, G) = F$ or G if and only if $X(G, F) = G$ or F, respectively.

With H. I. WHITLOCK [19] we call a mapping π of a system $\mathfrak{S}$ on a system $\mathfrak{S}^*$ a *permutomorphism* if for any three elements X, Y, Z of $\mathfrak{S}$

(4) $$\pi(X(Y, Z)) = \pi X(\pi Z, \pi Y) .$$

If such a mapping exists, $\mathfrak{S}$ and $\mathfrak{S}^*$ will be called *permutomorphic*.

3. The logical functors

As an example of a superassociative system, consider the 2-place functors of the 2-valued calculus of propositions; in other words, since True and False may be represented by 1 and 0, the set of the 16 functions whose domains consist of the pairs (0,0), (0,1), (1,0), (1,1) and which assume the values 0 and/or 1. We denote these functions by

	1	A	B	C	D	E	I	J	1'	A'	B'	C'	D'	E'	I'	J'
(0,0)	1	1	0	1	1	1	0	0	0	0	1	0	0	0	1	1
(0,1)	1	1	1	1	0	0	0	1	0	0	0	0	1	1	1	0
(1,0)	1	1	1	0	1	0	1	0	0	0	0	1	0	1	0	1
(1,1)	1	0	1	1	1	1	1	1	0	1	0	0	0	0	0	0

Following LUKASIEWICZ, we denote implication by C. Clearly, B is disjunction; E is equivalence; A is incompatibility (traditionally expressed by Sheffer's stroke); A' is conjunction; I and J are the selectors; 1 and $1'$, instead of which one may write 0, are the constant functors. In contrast to the numerals 1 and 0, the functor symbols 1 and 0 are italicized.

By $F(G, H)$ we mean the function with the value $F(G(\mathrm{x}, \mathrm{y}), H(\mathrm{x}, \mathrm{y}))$ for any x, y. The following six functors are readily seen to be idempotent: $1, 0, I, J, B$, and A'. More generally, in order that a 2-place function F be idempotent it is necessary and sufficient that $F(\mathrm{k}, \mathrm{k}) = \mathrm{k}$ for every element k that is a value of F.

$\mathfrak{S}(A) = \{A, A', 1, 0\}$ (see Introduction) and $\mathfrak{S}(B') = \{B', B, 0, 1\}$. As examples of substitution tables, we mention those of A in $\mathfrak{S}(A)$ and of B' in $\mathfrak{S}(B')$.

A	A	A'	1	0
A	A'	1	A'	1
A'	1	A	A	1
1	A'	A	0	1
0	1	1	1	1

B'	B'	B	0	1
B'	B	0	B	0
B	0	B'	B'	0
0	B	B'	1	0
1	0	0	0	0

That is to say, $A(A, A) = A'$, $A(A, A') = 1, \ldots, B'(1, 1) = 0$.

From the tables it is clear that A and B' are elements of order 2; and that the systems $\mathfrak{S}(A)$ and $\mathfrak{S}(B')$ are isomorphic. Since all entries in the table for A belong to the set $\{A, A', 1, 0\}$ it follows that this quadruple is all that A generates. The substitution tables of A' and B can easily be derived from those of A and B', respectively. 1 and 0 are readily shown to be constant in $\mathfrak{S}(A)$ and $\mathfrak{S}(B')$.

At the same time it is clear that none of the 16 functors generates a system containing more than 4 elements. In fact, each of the remaining functors, $I', J', C, C', D, D', E, E'$, generates only a functor pair. E. g., $\mathfrak{S}(I')$ consists of $I' = I'(I, I')$ and $I = I'/I = I'(I', I)$.

The substitution table of A in the realm of all 16 functors includes 256 entries which may be summarized in the following formulae, where $A(X; Y_1, \ldots, Y_n) = Z$ and $A\{U_1, U_2, U_3\} = Z$ stand for the n formulae $A(X, Y_i) = Z$ ($i = 1, \ldots, n$) and the three formulae $A(U_1, U_2) = A(U_2, U_3) = A(U_3, U_1) = Z$, respectively.

$A(X, Y) = A(Y, X)$ for every X and Y;

$A(X, X) = A(X, 1) = X'$, $A(X, X') = A(0, X) = 1$ for every X, where $(X')' = X$;

$A(A'; B, C, D, E, I, J) = A\{E, J, I\} = A(B, E) = A(C, I) = A(D, J) = A$;

$A(B'; A, C, D, E, I', J') = A\{E, I', J'\} = A(A, E) = A(C, J') = A(D, I') = B$;

$A(C'; A, B, D, E', I, J') = A\{E', I, J'\} = A(D, E') = A(A, I) = A(B, J') = C$;

$A(D'; A, B, C, E', I', J) = A\{E', I', J\} = A(C, E') = A(B, I') = A(A, J) = D$;

$A\{A, C, I'\} = I$, $A\{A, D, J'\} = J$, $A\{A, B, E'\} = E$;

$A\{B, D, I\} = I'$, $A\{B, C, J\} = J'$, $A\{C, D, E\} = E'$;

$A\{A', B', E'\} = A\{A', C', I'\} = A\{A', D', J'\} = A\{B', C', J\}$
$$= A\{B', D', I\} = A\{C', D', E\} = 1.$$

4. Pairs with one generator

$\mathfrak{S}(F) = \{F, G\}$ if and only if all three elements $F(F, G)$, $F(G, F)$, $F(G, G)$ belong to the set $\{F, G\}$. In this case, F will be said to *generate a pair*.

Clearly, if F is not of order 2, then $F/G = G/F = G$. Moreover, from $F(F, G) = F$ or G it follows that

$F(G, F) = F(F/F, G/F) = (F(F, G))/F = F/F$ or G/F, respectively .

If F is of order 2, then $G/F = F$ and one can say: $F(F, G) = F$ or G if and only if $F(G, F) = G$ or F, respectively. This rules out two of the eight a priori possible substitution tables for F.

Theorem 1. *There are six mutually nonisomorphic superassociative pairs with one generator. Exactly two of them have a generator of order 2. These two pairs, which are basically unsymmetric, are isomorphic to $\mathfrak{S}(I')$ and $\mathfrak{S}(J')$. Of the other pairs, two are basically unsymmetric, and two are symmetric. Of the unsymmetric pairs, one is isomorphic to $\mathfrak{S}(C)$ and $\mathfrak{S}(D')$, the other, to $\mathfrak{S}(C')$ and $\mathfrak{S}(D)$. One of the symmetric pairs is isomorphic to $\mathfrak{S}(E)$ and $\mathfrak{S}(E')$. In these five pairs, F is nonconstant, while G is constant or nonconstant according as F is not of order 2 or is of order 2. The sixth pair satisfies $F(X, Y) = G(X, Y) = G$ for every X, Y in F, G. (Thus G is constant while F may be called pseudoconstant.) The last pair is not isomorphic to a system generated by a functor of 2-valued logic. $\mathfrak{S}(I')$ and $\mathfrak{S}(J')$ are permutomorphic and so are $\mathfrak{S}(C)$ and $\mathfrak{S}(D)$.*

It follows that the 8 of the 16 functors which generate pairs can be characterized by either one of the following sets of Postulates:

F	$\mathfrak{S}(F)$ contains at most 2 elements satisfying the relations:	$\mathfrak{S}(F)$ contains at least 2 elements satisfying the aforementioned equalities and the following (instead of the inequality):	Consequences
I'	$(F/F)/F = F \neq F(F, F/F)$	$F(F, F/F) = F/F$	$F(F/F, F) = F$
J'	$(F/F)/F = F \neq F(F/F, F)$	$F(F/F, F) = F/F$	$F(F, F/F) := F$
C and D'	$(F/F)/F = F/F = F(F, F/F) \neq F(F/F, F)$	$F(F/F, F) = F$	—
D and C'	$(F/F)/F = F/F = F(F/F, F) \neq F(F, F/F)$	$F(F, F/F) = F$	—
E and E'	$(F/F)/F = F/F \neq F(F/F, F) = F(F, F/F)$	$F(F/F, F) = F(F, F/F) = F$	—
—	$F/F = F/F/F = F(F, F/F) = F(F/F, F) \neq F$		—

The last pair is isomorphic to the system generated by the functor of the 3-valued logic which assumes the value 1 for (2,2) and the value 0 for every other pair (x, y), where x, y = 0, 1, 2.

This still leaves $\mathfrak{S}(A)$ and $\mathfrak{S}(B')$ to be characterized.

5. Triples with one generator

If F is of order 2 and $\mathfrak{S}(F)$ contains at least 3 elements, then both $F(F, G)$ and $F(G, F)$ are $\neq F$ and G. In this case, we will set

(5) $$F(F, G) = H .$$

Lemma 1. In a basically symmetric system $\mathfrak{S}(F)$ containing at least three elements, $F \neq F/H \neq G$.

$$F/H = F/(F(F, G)) = F/(F(G, F)) = F/(F(F/F, G/F)) = (F/H)/F .$$

whereas, according to (1) and (2), $F \neq F/F$ and $G \neq G/F$.

Lemma 2. If F is of order 2 and $F/H = H$, then $H/F \neq F, G$.

$F/H = H$ implies $(F/H)/F = H/F$ and hence $F/(H/F) = H/F$. This rules out $H/F = F$ or G, as in the proof of Lemma 1.

If F generates a triple $\{F, G, H\}$, and $\mathfrak{S}(F)$ is basically symmetric, then by Lemma 1, $F/H = H$. A basically symmetric system $\mathfrak{S}(F)$ containing at least three elements, F, G, and H will be said to be *of the first* or *second kind* according as $F/H = H$ or $\neq H$. By Lemma 2, *every basically symmetric triple is of the first kind.* $\mathfrak{S}(A)$ and $\mathfrak{S}(B')$ are symmetric quadruples of the second kind.

If F is of order 2 and generates $\mathfrak{S}(F) = \{F, G, H\}$, then, by (3) and (5), $F(F, G) = F(G, F) = H$. Thus (in contrast to pairs) *every triple generated by an element of order 2 is basically symmetric (and hence of the first kind).*

If F is of order 2 and $\mathfrak{S}(F)$ is a triple of the first kind, then, by Lemma 2, $H/F = H$. Moreover, $X(F, H) = F$ or G or H by (3) implies $X(G, H) = G$ or F or H; and $X(H, F) = F$ or G or H implies $X(H, G) = G$ or F or H, respectively, for every element X in $\{F, G, H\}$.

Hence

Theorem 2. *There are 9 different triples generated by an element of order 2. For each of them,*

$$F/F = G, F/G = F, F/H = H, F(F, G) = F(G, F) = H, F(G, H) =$$
$$= (F(F, H))/F, F(H,G) = (F(H, F))/F.$$

Each of the 9 triples is determined by a choice of M and N from $\{F, G, H\}$ and by setting $F(F, H) = M$ and $F(H, F) = N$. The triple corresponding to the choice M, N and the triple corresponding to the choice N, M are permutomorphic. The 3 triples corresponding to the choices M, M are symmetric; that is to say, $X(Y, Z) = X(Z, Y)$ for X, Y, Z in $\{F, G, H\}$.

Miss A. LIPSON showed on an I. B. M. 7090 that the 9 mentioned triples are indeed consistent, and that there are altogether 485 triples with one generator: Besides the 9 triples described in Theorem 2, there are

6 triples for which $F/G = H$ and $F/H = F$, each triple being determined by the choices of $F(F, G)$ and $F(G, F)$;

　　4 triples for which $F/G = H$ and $F/H = G$;

　　25 triples for which $F/G = F/H = H$;

　　36 triples for which $F/G = F/H = G$;

405 triples for which $F/G = G$ and $F/H = H$. Each one of these last triples is determined by a choice of the six elements $F(X, Y)$ for $X \neq Y$ from $\{F, G, H\}$ subject only to the condition that at least one of the elements $F(F, G)$ and $F(G, F)$ must be H.

Addition in the proofs: Mr. WHITLOCK has proved that there are 253 non-permutomorphic triples with one generator.

6. On quadruples

It will now be assumed that $\mathfrak{S}(F)$ contains at least 4 elements.

Lemma 3. If F is of order 2, and if $\mathfrak{S}(F)$ contains, besides F, G, and $H = F(F, G)$, an element K such that $F/H = K$, then

a) $F/K = H$. Proof: $F/K = F/(F/H) = (F/F)/H = G/F(F, G) = (G/F)(F, G) = F(F, G)$.

b) $G/H = H$ and $G/K = K$. Proof: $G/H = (F/F)/H = F/K$ and G/K
$= (F/F)/K = F/H$.

c) $H/F = F(G, F)$. Proof: $H/F = (F(F, G))/F = F(F/F, G/F)$.

d) $H/H = F(K, H)$. Proof: $H/H = (F(F, G))/H = F(F/H, G/H)$.

e) $H/K = F(H, K) = H(F, G)$. Proof: $H/K = (H/F)/H = (F(G, F))/H$
$= F(G/H, F/H) = F(H, K)$ by b); $H(F, G) = (F(F, G)) (F, G) = F(H, K)$.

In the same way, one proves that if $H = F(G, F)$ and $F/H = K$, then

c') $H/F = F(F, G)$.

d') $H/H = F(H, K)$.

e') $H/K = F(K, H) = H(G, F)$.

Combining these results one can say

Lemma 4. If F is of order 2, and $\mathfrak{S}(F)$ is basically symmetric, and $F/H = K$,
then

$$F/K = H, F(H, K) = F(K, H) = H/H = H/K = H(F, G) = H(G, F),$$
$$H/F = H, K/F = K, K/H = K/K.$$

Lemma 5. Under the assumptions of Lemma 4, $F(H, K) \neq F, G, K$.

$F = F(H, K)$ (which is $= H/H$) would imply that $F = H/H = H/(H/F)$
$= (H/H)/F = F/F$; and $F(H, K) = G$ would lead to a similar contradiction.
$H/H = K$ would imply

$$H = F/K = F/H/H = K/H = (F(H, K))/H = F(H/H, K/H) = F(K, H) = K.$$

Lemma 6. Under the assumptions of Lemma 4, if either $F(F, H)$ or
$F(H, F)$ is F or H or K, then $F(H, K) \neq F \neq F(K, H)$.

$F(F, H) = F$ or H would imply $F(F/K, H/K) = F/K$ or H/K, respectively.
In presence of $H = F(H, K)$ (which is $= H/K$ by Lemma 4), this would entail
$F(H, H) = H$, which is a contradiction. $F(F, H) = K$ would imply $F(F/H,$
$H/H) = K/H$. In presence of $H = F(K, H)$ (which is $= H/H$), this would
entail $F(K, H) = (F/H)/H = F/(H/H) = F/H = K$, which is a contradiction.

Lemma 7. Under the assumptions of Lemma 4, if either $F(F, K)$ or
$F(K, F)$ is F or G or K, then $F(H, K) \neq H \neq F(K, H)$.

Contradictions result from $F(F, K) = F$ or K by diagonalization with H,
and from $F(F, K) = G$ by diagonalization with K.

Lemma 8. Under the assumptions of Lemma 4, if $\mathfrak{S}(F)$ contains more than
4 elements, then there is at least one element Z such that Z/F does not belong
to the set

$$\mathfrak{F} = \{F, G, H, K\}.$$

If $\mathfrak{S}(F)$ contains at least 5 elements, then there is at least one element not
belonging to $\mathfrak{F}$ among

$$F(F, H), F(F, K), F(H, F), F(K, F), F(H, K).$$

If for such an element, Z, we had $Z/F = F$ or G or H or K, then $(Z/F)/F$ would
be G or F or H or K, respectively. But $Z/(F/F) = Z$ by (3), which is a contradiction.

Now under the assumptions of Lemma 4 let $\mathfrak{S}(F)$ contain at most 4 elements.
Then $\mathfrak{S}(F) = \mathfrak{F}$. From Lemma 5 it follows that $F(H, K) = H$. But then from

Lemmas 6 and 7 it follows that $F(F, H) \neq F, H, K$ and $F(F, K) \neq F, G, K$; that is to say,

$$F(F, H) = F(H, F) = G \text{ and } F(F, K) = F(K, F) = H .$$

Consequently, $F(F/F, H/F) = F(H/F, F/F) = G/F$, that is, $F(G, H) = F(H, G) = F$ and similarly, $F(G, K) = F(K, G) = H$. Hence

Theorem 3. *There is exactly one basically symmetric system of the second kind generated by an element of order 2 and containing at most 4 elements. It is a (symmetric) quadruple isomorphic with $\mathfrak{S}(A)$ and $\mathfrak{S}(B')$. The elements G, H, K correspond to $A', 1, 0$ and to $B, 0, 1$, respectively.*

In order that $\mathfrak{S}(F)$ be isomorphic with $\mathfrak{S}(A)$ it is necessary and sufficient that

I. $\mathfrak{S}(F)$ *contains at most four elements.*

II. $F/F/F = F.$

III. $F(F, F/F) = F(F/F, F).$

IV. $F/(F(F, F/F)) \neq F(F, F/F).$

Being of the second kind the system $\mathfrak{S}$ contains at least 2 elements. Being basically symmetric and generated by an element of order 2 the system contains at least 3 elements according to Theorem 1. Being of the second kind and generated by an element of order 2 the system $\mathfrak{S}$ is not a triple according to Theorem 2. Hence $\mathfrak{S}$ contains at least 4 elements. Comparison of the remarks preceding Theorem 3 with the substitution table for A establishes the isomorphism of $\mathfrak{S}(F)$ with $\mathfrak{S}(A)$.

In view of the preceding remarks, Postulate I may be replaced by the assumption

I_{min}. $\mathfrak{S}(F)$ *is the system with the least possible number of elements satisfying the other postulates.*

7. Independence questions and alternative postulates

Postulates I—IV *of the preceding section are independent.* $\mathfrak{S}(F) = \{F, G, H, K\}$, where F has the substitution tables

F	F	G	H	K	F	F	G	H	K	F	F	G	H	K
F	G	H	F	G	F	G	H	F	K	F	G	H	K	K
G	H	H	K	G	G	K	F	H	G	G	H	F	K	K
H	H	K	K	F	H	F	H	K	G	H	K	K	H	K
K	G	K	F	F	K	K	G	F	H	K	K	K	K	K

satisfies the postulates except II, III, IV, respectively. The first system, in which F is of order 4, is due to H. I. Whitlock. The last two are examples of the quadruples generated by an element of order 2 whose consistency was established by Miss A. Lipson. Among these systems, 64 satisfy Postulate III without being basically symmetric, while 448 are basically symmetric and of the first kind.

The independence of Postulate I can be demonstrated by the quintuple $\mathfrak{S}(F) = \{F, G, H, K, Z\}$ for which $F(X, Y) = Z$ for all pairs X, Y except

$$F/F = G, F/G = F, F(F, G) = F(G, F) = F/K = H, F/H = K .$$

E. STUEBEN has constructed, for any integer $n > 5$, a system of n elements satisfying Postulates II, III, IV.

Postulate I may be replaced by the following I* which we break up into five assumptions:

I*. *At least one member of each of the following five pairs belongs to the set* $\mathfrak{F} = \{F, G, H, K\}$:

I a. $F(F, H), F(G, H)$; I a'. $F(H, F), F(H, G)$;

I b. $F(F, K), F(G, K)$; I b'. $F(K, F), F(K, G)$;

I c. $F(H, K), F(K, H)$.

In view of $H/F = H$ and $K/F = K$, each of these five assumptions implies that both members of the pair under discussion belong to $\mathfrak{F}$. Hence I*, II, and IV imply Postulate I.

The eight postulates I a, I a', I b, I b', I c, II, III, IV *are independent.*

The independence of II, III, and IV follows from the same examples that demonstrated their independence from Postulate I. The independence of I c is demonstrated by the quintuple $\mathfrak{S}(F) = \{F, G, H, K, V\}$ with the substitution table

F	F	G	H	K	V
F	G	H	M_1	M_2	V
G	H	F	M_1	M_2	V
H	M_3	M_3	K	V	V
K	M_4	M_4	V	H	V
V	V	V	V	V	V

for any of the 16 choices of M_1, M_2, M_3, M_4 from the set $\{H, K\}$.

The following sextuples $\mathfrak{S}(F) = \{F, G, H, K, V, W\}$ satisfy all postulates except I a and I b, respectively.

F	G	H	K	V	W
F	G	H	V	H	H	V
G	H	F	W	H	W	H
H	G	F	K	H	W	V
K	H	H	H	H	H	H
V	H	F	F	H	W	H
W	G	H	G	H	H	V

F	G	H	K	V	W
F	G	H	G	V	G	H
G	H	F	F	V	F	H
H	G	F	K	H	K	H
K	H	H	H	H	H	H
V	G	F	K	H	W	V
W	H	H	H	H	V	V

Systems that are permutomorphic with the preceding ones demonstrate the
independence of Ia' and Ib'.

8. Idempotent pairs

Turning to systems with several generators we call, for any integer n, the
n elements $X_1, \ldots, X_n$ *independent* if none of them belongs to the set generated
by the $n-1$ other elements. The set $\{X_1, \ldots, X_n\}$ will be called *idempotent*
if the $X_1, \ldots, X_n$ are independent and $\mathfrak{S}(X_1, \ldots, X_n) = \{X_1, \ldots, X_n\}$.
If $n = 1$, then $\{X_1\}$ is idempotent if and only if X_1 is idempotent. Each element
of an idempotent set is idempotent, since $X_i(X_i, X_i) = X_k(k \neq i)$ would
contradict the independence of X_k. On the other hand, if F and G are idempotent
functions on $\{0, 1, 2\}$ such that $F(0, 1) = 0$, $G(0, 1) = 2$, $F(0, 2) = 1$, then
$F(F, G) \neq F, G$, and the set $\{F, G\}$ is not idempotent.

In order to obtain a survey of all idempotent pairs $\{X, Y\}$ (one of which
will play a great role in the sequel) we describe either component Z of such a
pair by the ordered triple

$$[Z]: Z(Z, Z'), Z(Z', Z), Z(Z', Z'),$$

where Z stands for either one of X and Y, and Z' stands for the other one.
$Z(Z, Z) = Z$ in both cases since X and Y are idempotent.

Of the eight a priori possible sequences $[Z]$, three turn out to be impossible.
E.g., from $X(X, Y) = Y$, $X(Y, X) = X$, $X(Y, Y) = X$ it would follow that

$$Y = X(X, Y) = (X(X, X))(X, Y) = X(X(X, Y), X(X, Y)) = X(Y, Y) = X.$$

This rules out $[Z] = Z', Z, Z$. Similar contradictions arise from the assumptions

$$[Z] = Z, Z', Z \text{ and } [Z] = Z', Z', Z.$$

This leaves five types of components of idempotent pairs:

If $[Z] =$	then Z will be said to be of
Z, Z, Z	of Type 0 or *constant*
Z, Z', Z'	of Type 1 or of the type of I
Z', Z, Z'	of Type 2 or of the type of J
Z, Z, Z'	of Type 3
Z', Z', Z'	of Type 4

If Z is of type 4, then Z' is constant. E. g., $Z'(Z, Z') = Z'$ because

$$Z'(Z, Z') = (Z(Z, Z'))(Z, Z') = Z(Z(Z, Z'), Z'(Z, Z')) = Z(Z', Z'(Z, Z')),$$

whence $Z'(Z, Z') \neq Z$.

Theorem 4. *There are twelve types of idempotent pairs. Nine of them can be
illustrated by the fifteen pairs of unlike idempotent 2-place functors* $0, 1, I, J, B, A'$
of the 2-valued logic.

Types	0,0	0,1	0,2	0,3	0,4	1,2	1,3	2,3	3,3
Functor Pairs	0,1	$0,I$ $1,I$	$0,J$ $1,J$	$0,B$ $1,A'$	$0,A'$ $1,B$	I,J	I,A' I,B	J,A' J,B	A',B

The second and third pairs are permutomorphic, and so are the seventh and eighth. It will further be noted that I and J are always of the types 1 and 2, respectively, but that B and A' are of type 3 with one exception, namely, in conjunction with 1 and 0, respectively, in which cases they are the only illustrations of elements of type 4.

There are no pairs of unlike idempotent elements of the type 1,1 and 2,2 in 2-valued logic. There are, however, such pairs of functors in the 3-valued logic. For example, set

$I(x, y) = x$ for every x, y = 0, 1, 2.

$I^*(x, y) = x$ for every x, y except 0, 2; and $I^*(0, 2) = 1$.

One readily proves that $I(F, G) = F$ and $I^*(F, G) = F$ for $F, G = I, I^*$. In a similar way, one constructs an idempotent pair of the types 2, 2.

If F in F, G is of type 4, then, as one readily shows, $G(F, G) = G(G, F) = G$. Thus G is of type 0 or 3. The latter occurs in the 3-valued logic, e.g. if $F(0, 1) = F(1, 0) = F(1, 1) = 1$ while $F(x, y) = 0$ otherwise, and $G(x, y) = 1$ except for $G(0, 0) = 0$.

9. An octuple with two generators

We now assume that F is an element of order 2 generating a basically symmetric system of the second kind (i. e., satisfies Postulates II, III, IV). $\mathfrak{S}(F)$ thus includes the set $\mathfrak{F}$. We adjoin to $\mathfrak{S}(F)$ an element L that is idempotent and connected with F by one more postulate VI; and we make an assumption I^* replacing Postulate I. Thus we have, in addition to II, III, IV, the following postulates:

V. $L/L = L$.

VI. $L(F, L) = F$.

I*. $\mathfrak{S}(F, L)$ contains at most eight elements or $\mathbf{I}_{min}$.

Postulates I^*, $II - VI$ for F and L are *consistent* since they are satisfied by the functors A and I as well as by the functors B' and I.

Definition. Let $\mathfrak{T}$ be the sequence of elements

$$F, G, H, K, L, L_2 = F/L, L_3 = F(F, L), L_4 = F/L_3 .$$

Lemma 9. $G/F = L/F = F, F/F = L_2/F = G, H/F = L_3/F = H, K/F = L_4/F = K$

$L/F = L/(L(F, L)) = (L/L)(F, L) = L(F, L) = F$;

$L_2/F = (F/L)/F = F/(L/F) = F/F = G$;

$L_3/F = (F(F, L))/F = F(F/F, L/F) = F(G, F) = H$;

$L_4/F = (F/L_3)/F = F/(L_3/F) = F/H = K$.

Lemma 10. $L(F, G) = F$, $L_2(F, G) = G$, $L_3(F, G) = G$, $L_4(F, G) = F$.

Two diagonalizations of Postulate VI with F yield $L(G, F) = G$ and $L(F, G) = F$. It follows that $L_2(F, G) = (F/L) (F, G) = F/(L(F, G)) = F/F = G$. Furthermore, $L_3(F, G) = (F(F, L)) (F, G) = F (F(F, G), L(F, G)) = F(H, F) = G$ and $L_4(F, G) = (F/L_3) (F, G) = F/(L_3(F, G)) = F/G = F$.

Lemmas 9 and 10 have two important immediate consequences.

Corollary 1. The elements X/F and $X (F, G)$ belong to $\mathfrak{F}$ for any X and Y in $\mathfrak{T}$.

Corollary 2. Any two elements of $\mathfrak{T}$ yield unlike results either when diagonalized with F or when applied to (F, G) or in both cases.

From Corollary 2 it follows that the eight elements of $\mathfrak{T}$ are distinct. Hence, Postulate I* implies that $\mathfrak{S}(F, L) = \mathfrak{T}$. If the subsystem of $\mathfrak{S}(F, L)$ generated by F contained an element not in $\mathfrak{F}$, then, by Lemma 8, it would contain an element V such that V/F would not belong to $\mathfrak{F}$. By Lemma 9, such a V would be different from the elements of $\mathfrak{T}$ and thus would be a ninth element of $\mathfrak{S}(F, L)$, contrary to Postulate I*. The subsystem of $\mathfrak{S}(F)$ of $\mathfrak{S}(F, L)$ thus satisfies Postulates I–IV, and one can obtain $F(X, Y)$ from Theorem 3 for any two elements X and Y of $\mathfrak{F}$.

Therefore, in view of Corollary 1, Theorem 3 can be used in determining, for every X and Y in $\mathfrak{T}$,

$(F(X, Y))/F$, which is $F(X/F, Y/F)$, and $(F(X, Y)) (F, G)$, which is $F(X(F, G), Y(F, G))$. By Corollary 2, $F(X, Y)$ can be determined for every X and Y in $\mathfrak{T}$. E. g., in order to determine $F(G, L_4)$, observe that

$(F(G, L_4))/F = F(G/F, L_4/F) = F(F, K) = H$, whence $F(G, L_4) = H$ or L_3;

$(F(G, L_4)) (F, G) = F(G(F, G), L_4(F, G)) = F(K, F) = H$, whence $F(G, L_4) = F$ or H. Since the Postulates are consistent, it follows that $F(G, L_4) = H$. In this way, one readily establishes the entire substitution table of F.

In a similar way, one can determine the table for L after $L(X, Y)$ has been determined for every X and Y in $\mathfrak{F}$. The latter determination can be based on

Lemma 11. $L(X, Y) = X$ if both X and Y belong to $\{H, K\}$.

$L/H = L/(F/K) = (L/F)/K = F/K = H$ and similarly, $L/K = K$. From Theorem 3 and Lemma 4 it follows that $H/H = H/K = H$ and $K/H = K/K = K$. By diagonalizing Postulate VI with K and H one obtains $L(H, K) = H$ and $L(K, H) = K$.

If any element of $\mathfrak{F}$ is diagonalized with H or K or if (F, G) is substituted into it, the result is either H or K. Hence, from Lemma 11, one obtains:

Lemma 12. $(L(X, Y))/H = L(X/H, Y/H) = X/H$, $(L(X, Y))/K = X/K$, $(L(X, Y)) (F, G) = L(X(F, G), Y(F, G))$, for every X and Y in $\mathfrak{F}$ and thus, by Corollary 1, for every X and Y in $\mathfrak{T}$.

Lemma 13. Two elements of $\mathfrak{T}$ yield the same results when diagonalized with F if and only if they yield the same results when diagonalized with H and when diagonalized with K.

Indeed, for F and L_2 the results of diagonalization with H and K are K and H, respectively; for G and L, they are H and K; for H and L_3, they are H and H; and for K and L_4, they are K and K.

From Lemma 13 and Corollary 2 one obtains

Corollary 3. Any two elements of $\mathfrak{T}$ yield unlike results in at least one of the following cases: when diagonalized with H or when diagonalized with K or when applied to (F, G).

From Lemma 12 it thus follows that $L(X, Y) = X$ for every X and Y in $\mathfrak{T}$. Summarizing, one can formulate

Theorem 5. *There is exactly one system* $\mathfrak{S}(F, L)$ *containing at most eight elements satisfying Postulates* II—VI. *The substitution table for* F *is*

F	F	G	H	K	L	L_2	L_3	L_4
F	G	H	G	H	L_3	L	L	L_3
G	H	F	F	H	F	H	F	H
H	G	F	K	H	L_2	L	L_4	L_3
K	H	H	H	H	H	H	H	H
L	L_3	F	L_2	H	L_2	H	F	L_3
L_2	L	H	L	H	H	L	L	H
L_3	L	F	L_4	H	F	L	L_4	H
L_4	L_3	H	L_3	H	L_3	H	H	L_3

and $L(X, Y) = X$ *for every* X *in* $\mathfrak{S}(F, L)$. *The system is isomorphic to* $\mathfrak{S}(A, I)$ *and* $\mathfrak{S}(B', I)$, *the element* $L_2 = F/L$ *corresponding to* I' *while* $L_3 = F(F, L)$ *and* $L_4 = F/L_3$ *correspond to* C, C' *and* D', D, *respectively*.

We now prove the following extension of Lemma 8.

Lemma 14. If F and L satisfy Postulates II—VIII and $\mathfrak{S}(F, L)$ contains more than 8 elements, then $\mathfrak{S}(F, L)$ contains an element Z such that Z/F does not belong to $\mathfrak{T}$.

Certainly $\mathfrak{S}(F, L)$ contains a ninth element Z^* such that

$$Z^* = F(U, V) \text{ and/or } Z^* = L(X, Y)$$

where U, V, X, Y belong to $\mathfrak{T}$. In either case, because of $X/G = X$ for every element X of $\mathfrak{T}$, we have $Z^*/G = Z^*$ and hence $(Z^*/F)/F = Z^*$. It follows that Z^*/F is an element $Z (= Z^*$ or $\neq Z^*)$ for which Z/F does not belong to $\mathfrak{T}$.

The independence of the postulates for L from I—IV is seen from the following examples:

V. For X, Y in $\{F, G, H, K\}$, define $F(X, Y)$ as in Theorem 3; and set

$$F(L, X) = F(X, L) = F(X, G),$$

$$L(X, Y) = G(X, Y), L(X, L) = L(L, X) = G(X, G), L(L, L) = L.$$

Then V is not satisfied since $L(F, L) = K$. The element L is something of an alter ego of G without, however, belonging to $\mathfrak{S}(F)$.

VI. Set $F = A$ and $L = D$. Then VI is not satisfied since $D(D, D) = 1$. But $D(A, D) = A$. It is interesting to note that $\mathfrak{S}(A, D) = \mathfrak{S}(A, J)$, which is isomorphic to the system $\mathfrak{S}(F, M)$ to be discussed, in Theorem 6.

Postulates II—VI yield only a small part of the substitution table of L. The proof that L_2, L_3, L_4 are different from each other and from F, G, H, K, L

remains valid. But F/L_2 may well be a nineth element, as can be seen in 3-valued logic by defining $F(2, 2) = 0$ and $F(x, y) = 2$ otherwise, and $L(x, y) = x$ for every x, y in $\{0, 1, 2\}$.

In analogy to Theorem 5, one proves

Theorem 6. *There is exactly one system* $\mathfrak{S}(F, M)$ *containing at most 8 elements satisfying Postulates* II, III, IV *and*

VII. $M/M = M$.

VIII. $M(M, F) = F$.

$M(X, Y) = Y$ *for every* X *and* Y *in* $\mathfrak{S}(F, M)$. *The substitution table of* F *is analogous to that in Theorem 5. The system* $\mathfrak{S}(F, M)$ *is isomorphic to* $\mathfrak{S}(A, J)$, $\mathfrak{S}(B', J)$, *the element* $M_2 = F/M$ *corresponding to* J' *while* $M_3 = F(F, M)$ *and* $M_4 = F/M_3$ *correspond to* D, D' *and* C', C, *respectively.*

10. Postulates for the substitutive algebra of the 2-place functors in 2-valued logic

We now consider a system with three generators, $\mathfrak{S}(F, L, M)$ satisfying Postulates II–VIII. The postulates are consistent since they are satisfied by the functors A, I, and J.

Definition. Let $\mathfrak{V}$ be the sequence

$$F, G, H, K, L, L_2, L_3, L_4, M, M_2, M_3, M_4,$$
$$N_1 = F(L_2, M_2), N_2 = F/N_1, N_3 = F(L_3, M_3), N_4 = F/N_3.$$

The following is an extension of Corollary 2 in the preceding section.

Lemma 15. From Postulates II–VIII it follows that

If X is in $\{G, L, M, N_1\}$ $\{F, L_2, M_2, N_2\}$ $\{H, L_3, M_3, N_4\}$ $\{K, L_4, M_4, N_3\}$
then $X/F =$ F G H K ;
if X is in $\{L, L_4, M_2, M_3\}$ $\{L_2, L_3, M, M_4\}$ $\{F, H, N_1, N_3\}$ $\{G, K, N_2, N_4\}$
then $X(F, G) =$ F G H K

Since no two elements of $\mathfrak{V}$ have like results under both the diagonalization and the substitution, the 16 elements of $\mathfrak{V}$ are distinct. The assumption that $\mathfrak{S}(F, L, M)$ contains at most 16 elements implies

$$\mathfrak{S}(F, L, M) = \mathfrak{V}.$$

If either of the subsystems of $\mathfrak{S}(F, L, M)$ generated by F and L or F and M contained a ninth element, then by Lemma 14 the subsystem would contain an element V such that V/F would not belong to $\mathfrak{F}$. Since for each element X of $\mathfrak{V}$ the element V/F does belong to $\mathfrak{F}$, it follows that V would be a seventeenth element of $\mathfrak{S}(F, L, M)$ against the assumption.

It furthermore follows that in the subsystems $\mathfrak{S}(F, L)$ and $\mathfrak{S}(F, M)$ the substitution tables of Theorems 5 and 6 are valid. Since X/F and $X(F, G)$ belong to $\mathfrak{F}$ for every X and Y in $\mathfrak{V}$, one has

$$(L(X, Y))/F = L(X/F, Y/F) = X/F$$

and

$$(L(X, Y)) (F, G) = L(X(F, G), Y(F, G)) = X(F, G),$$

whence $L(X, Y) = X$ for every X and Y in $\mathfrak{V}$. Similarly, $M(X, Y) = Y$.

$(F(X, Y))/F$ and $(F(X, Y))$ (F, G) can be determined from Theorems 5 and 6, and they determine $F(X, Y)$ for every X and Y in $\mathfrak{V}$.
Summarizing, one can formulate

Theorem 7. *There is exactly one system* $\mathfrak{S}(F, L, M)$ *containing at most 16 (i. e., the minimum number of) elements and satisfying Postulates* II–VIII. *In it,*

$$L(X, Y) = X \text{ and } M(X, Y) = Y \text{ for every } X, Y \text{ in } \mathfrak{S}(F, L, M).$$

The substitution table of F *is isomorphic to that of* A, *the elements*

$$F, G, H, K, L, L_2, L_3, L_4, M, M_2, M_3, M_4, N_1, N_2, N_3, N_4$$

corresponding to

$$A, A', 1, 0, I, I', C, C', J, J', D, D', B, B', E', E,$$

respectively.

References

[1] MANNOS, M.: Ideals in tri-operational algebra. Reports Math. Coll., 2nd series, Notre Dame 7, 73—79 (1946).

[2] MEHLBERG, J. J.: A classification of mathematical concepts. Synthese 14, 78—86 (1962).

[3] MENGER, K.: A counterpart of Occam's Razor in pure and applied mathematics. Synthese 12, 415—428 (1960) and 13, 331—349 (1961).

[4] — Gulliver in the land without One, Two, Three. Math. Gaz. 43, 241—250 (1959).

[5] — Gulliver's return to the land without One, Two, Three. Am. Math. Monthly 67, 641—648 (1960).

[6] — Tri-operational algebra. Reports Math. Coll., 2nd ser., Notre Dame 5—6, 3—10 (1945).

[7] — General algebra of analysis. Reports Math. Coll., 2nd ser., Notre Dame 7, 46—60 (1946).

[8] — Algebra of analysis. Notre Dame Math. Lect. 3 (1944).

[9] — Calculus. A Modern Approach. Boston: Ginn 1955.

[10] — An axiomatic theory of functions and fluents. In The Axiomatic Method, 454 to 473, ed. Henkin et al. Amsterdam: North-Holland Publ. Co. 1959.

[11] — Algebra of functions: past, present, future. Rend. Mat. Roma 20, 409—430 (1961).

[12] — Function algebra and propositional calculus. In Self-Organizing Systems 1962, 525—532. Ed. Yovits et al. Washington: Spartan Books.

[13] MILGRAM, A. N.: Saturated polynomials. Reports Math. Coll., 2nd ser., Notre Dame 7, 65—68 (1946).

[14] NÖBAUER, W.: Über die Operation des Einsetzens in Polynomringen. Math. Ann. 134, 248—259 (1958) and Funktionen auf kommutativen Ringen. Math. Ann. 147, 166—175 (1962).

[15] POST, E.: The two-valued iterative systems of mathematical logic. Princeton Univ. Press (1941).

[16] SCHWEIZER, B., and A. SKLAR: The algebra of functions. Math. Ann. 139, 366—382 (1960); 143, 440—447 (1961) and a forthcoming third paper.

[17] — A mapping algebra with infinitely many operations. Coll. Math. 9, 33—38 (1962).

[18] SEALL, R. E.: Truth-valued fluents and qualitative laws. Philos. Sci. 30, 36—41 (1963).

[19] WHITLOCK, H. I.: A composition algebra for multiplace functions. Math. Ann. 157, 167—178 (1964).

(Received December 16, 1963)

21*

FUNDAMENTA
MATHEMATICAE
LVIII (1966)

Two theorems on the generation of systems of functions

by

Karl Menger and H. Ian Whitlock (Chicago) *

This paper deals with two basic questions about multiplace functions ("functions of several variables") defined on a finite set $N_m = \{1, ..., m\}$. How many functions can k functions generate by composition, and how many functions are needed to generate by composition all p-place functions?

The essential feature of the paper is its algebraic approach to the subject matter in contrast to the traditional treatment of functions in logic [1]. Consider e.g. the functions over N_2. By composition, the two basic logical functions, negation and disjunction, do not generate more than eight functions, namely, the four 1-place functions, four of the sixteen 2-place functions and none of the higherplace functions (see Example 2). All that Sheffer's stroke (herein denoted by a frontal A) generates are four of the 2-place functions. The traditional statement that $A(x, y)$ also generates e.g. the 1-place negation $n(x)$ is based on the fact that $n(x) = A(x, x)$. But in so saying one substitutes x for y; and similarly one substitutes $A(y, z)$ for y in saying that $A(x, y)$ generates $A(x, A(y, z))$. Substitution of an expression for a variable, however, is not the composition of functions. Nor is it possible to obtain any 1-place of 3-place function from A by compositions.

From our strictly algebraic point of view, we prove that the maximum number of functions that k functions can generate depends upon k but (except for trivial limitations) is independent of the placenumbers of the functions (Corollary 2 of Theorem I). At least p functions are necessary (Corollary 3 of Theorem I), and p properly chosen functions are sufficient (Theorem II), to generate all p-place functions for p > 1 with one important exception: the 2-place functions over N_2. Thus while three functions are needed to generate all the 2-place func-

---

* Theorem I and its Corollaries are due to the first author, Theorem II is the work of the second.

[1] Another algebraic approach to the study of multiplace functions is the Marczewski abstract algebra which, however, stresses the domains of the functions rather than their composition.

tions of the 2-valued logic, only two functions are necessary for all other finite-valued logics; and three functions are sufficient to generate all 3-place functions of the 2-valued as well as of all other finite-valued logics.

Let S be a set, and p a natural number. By a p-*place function over* S we mean ([2]) a mapping of S^p (the set of all ordered p-tuples of elements of S) into S. In other words, a p-place function over S is a set of pairs (T, s) containing, for each ordered p-tuple T of elements of S exactly one pair where s is an element of S. If the p-place function is denoted by F, and (T, s) belongs to F, then we write, as is customary, $s = F(T)$. The set of all elements of S that are second members of pairs (T, s) belonging to F is called the *range* of F—briefly, ran$F$.

If F is a p-place function over S, and $F_1, ..., F_p$ are q-place functions over S for some natural number q, then $F(F_1, ..., F_p)$ will denote the q-place function consisting of the pairs $(U, F(F_1(U), ..., F_p(U)))$ for all elements U of S^q. This q-place function is said to be the result of the *composition* of F with $F_1, ..., F_p$ or of the *application* of F to $F_1, ..., F_p$ or of the *substitution* of $F_1, ..., F_p$ into F. (Even in the case p = 1, substitution of a function F_1 *into* F has of course nothing whatever to do with the substitution of F_1 *for* F in the sense of replacing F by F_1 or y by x). Clearly,

$$\operatorname{ran} F(F_1, ..., F_p) \subseteq \operatorname{ran} F.$$

The main property of the operation just defined is what we have called *superassociativity* ([3]):

$$(F(F_1, ..., F_p))(G_1, ..., G_q) = F(F_1(G_1, ..., G_q), ..., F_p(G_1, ..., G_q))$$

for any p-place function F, any q-place functions $F_1, ..., F_p$, and any functions $G_1, ..., G_q$ of one and the same place-number—all of them over S.

If G is a set of functions over S (not necessarily of the same place-number), then the smallest set of functions containing G (as a subset) that is closed under substitution will be called the set *generated* by G and denoted by $\mathfrak{S}G$. Thus $\mathfrak{S}G$ is the smallest set including 1) G as a subset, and 2) the function $F(F_1, ..., F_p)$ if the p-place function F

([2]) Cf. H. I. Whitlock [7]. It will be noted that we herein adhere to the typographical convention introduced by the senior author of this paper: All references to functions are in *italic* type (e.g., *F, A, I, O, n, deg, Max*); all references to numbers and to elements of the domains and ranges of functions are in lower case roman type (e.g., p, s, x, y); sets of numbers or subsets of domains and ranges are denoted by capital letters in roman type (e.g., S, T); sets of functions in **bold face** (e.g., **G, F**).

([3]) Cf. K. Menger [3] and the bibliography in that paper.

and the functions $F_1, ..., F_p$ having one and the same place-number belong to the set. Clearly, each function in $\mathfrak{S}G$ must have the same place-number as one of the functions in G. A set of functions will be called *homogeneous* if all its elements have the same place-number. If G is homogeneous, then so is $\mathfrak{S}G$.

EXAMPLE 1. If $S = \{0, 1\}$, consider the homogeneous set $G = \{A, I\}$ where I is the first 2-place *selector* assuming the value $I(x, y) = x$ for every pair (x, y) in S^2, and A is *incompatibility*, for which $A(0, 0) = A(0, 1) = A(1, 0) = 1$ and $A(1, 1) = 0$. The set $\mathfrak{S}\{I, A\}$ consists of the eight functions $I, A, A' = A(A, A), I' = A(I, I), C = A(I, A), C' = A(C, C), 1 = A(A, A')$ and $1' = 0 = A(1, 1)$. Here, C is the *implication*, for which $C(0, 0) = C(0, 1) = C(1, 1) = 1$ and $C(1, 0) = 0$; 1 is the *constant* 2-place function of value 1, for which $1(x, y) = 1$ for every (x, y) in S^2; and $F'(x, y) = 1 - F(x, y)$ for each (x, y) in S^2, where $F = A, I, C, 1$. The function A', for which $A'(x, y) = Min(x, y)$ is the *conjunction*; and 0 is the other constant function.

EXAMPLE 2. If $S = \{0, 1\}$, consider the nonhomogeneous set $\{n, B\}$, where n is the 1-place function for which $n(x) = 1 - x$ for $x = 0, 1$, and B is the *disjunction*, for which $B(x, y) = Max(x, y)$. Setting $nB = B'$ and $nn = j$ one readily verifies that

$$\mathfrak{S}\{n\} = \{n, j\}, \quad \mathfrak{S}\{B\} = \{B\}, \quad \mathfrak{S}B' = \{B', B, 1, 0\},$$
$$\text{and} \quad \mathfrak{S}\{n, B\} = \{n, j, o, i, B, B', 1, 0\}.$$

Here, $o = B'(n, j) = 0(n, n)$ and $i = B'(o, o) = no$ are the *constant* 1-place functions over S, and j is the identity function over $\{0, 1\}$, for which $j(x) = x$ for $x = 0, 1$.

REMARK 1. *For every p-place function F in $\mathfrak{S}G - G$, there is, for some natural number q, a q-place function, G, in G, and p-place functions $F_1, ..., F_q$ in $\mathfrak{S}G$ such that $F = G(F_1, ..., F_q)$.*

If G is given we first associate, with some elements F of $\mathfrak{S}G$, a natural number, called the *degree* of F relative to G. We get $deg(F, G) = 1$ if and only if F belongs to G. If the elements of degree $\leqslant n$ are defined, let F be a p-place function such that $deg(F, G)$ is not $\leqslant n$. We set $deg(F, G) = n+1$ if there exist 1) a function in G, say a q-place function G, and 2) p-place functions $F_1, ..., F_q$ of degree $\leqslant n$ such that $F = G(F_1, ..., F_q)$.

Parenthetically we remark that $deg(F, G)$ expresses a relation between F and the set G (and not a property of F, nor even a relation between F and $\mathfrak{S}G$). Relative to $G = \{I, A\}$ in Example 1, I and A have the degree 1; $I', A',$ and C, the degree 2; C' and 1, the degree 3; and 0 has the degree 4. Relative to $\{I, A, 1\}$, the degree of 1 is 1, and that of 0 is 2. Relative to $\{C, A\}$ the degree of 1 is 2, and that of 0 is 3, since $C(C, C) = 1$ and $A(1, 1) = 0$. Set $\mathfrak{S}\{I, A\} = \mathfrak{S}\{I, A, 1\} = \mathfrak{S}\{A, C\}$.

The subsequent proof of Remark 1 pertains to one and the same set G of functions over the same S, whence $deg(F, G)$ will be abbreviated to $degF$. The set F, of all functions in $\mathfrak{S}G$ that have a finite degree relative to G is a subset of $\mathfrak{S}G$ which 1) contains the subset G, and 2) is closed under substitution. We prove that, more precisely,

$$degF(F_1, ..., F_p) \leqslant degF + Max(degF_1, ..., degF_p)$$

for any p-place function F and any p-tuple of functions $F_1, ..., F_p$ having one and the same place-number. Indeed, this inequality clearly holds if $degF = 1$. Assume its validity for all F of a degree $\leqslant$n, and suppose that $degK = $ n+1. By the definition of degree, there exist a function in G, say a q-place function G, and functions $H_1, ..., H_q$ in $\mathfrak{S}G$ whose degrees are $\leqslant$n such that $K = G(H_1, ..., H_q)$. By the superassociative law,

$$K(F_1, ..., F_p) = \big(G(H_1, ..., H_q)\big)(F_1, ..., F_p)$$
$$= G\big(H_1(F_1, ..., F_p), ..., H_q(F_1, ..., F_p)\big).$$

Since, by the inductive assumption, the inequality holds for each of the functions $H_i(F_1, ..., F_p)$, it holds for $K(F_1, ..., F_p)$.

By definition, $\mathfrak{S}G$ is the smallest set with properties 1) and 2). Hence $F = \mathfrak{S}G$. In other words, each function $\mathfrak{S}G$ has a finite degree relative to G. This clearly entails Remark 1.

COROLLARY. *If F belongs to $\mathfrak{S}G$, then* ran$F \subseteq$ ran$G$ *for some function G in G.* (This function G need not have the same place-number as F.)

REMARK 2. *If T and T' are two elements of S^p such that $G(T) = G(T')$ for each p-place function G belonging to the set G, then $F(T) = F(T')$ for each p-place function F belonging to $\mathfrak{S}G$.*

In view of Remark 1, the proof by induction is straight forward.

If G is a homogeneous set of, say p-place, functions, then by a substitutive base—briefly, a *base*—of G we mean a subset B of S^p with the following properties: 1) for each T in S^p, there exists an element T' in B such that $G(T) = G(T')$ for all G in G; 2) if T' and T'' are two elements of B, then $G(T') \neq G(T'')$ for at least one function G in G. In other words, a base of G is a minimal subset of S^p with property 1). By Remark 2, a base of a homogeneous set G is also a base of $\mathfrak{S}G$.

Thus *the set of all* p-*place functions with one and the same base* B *is closed with respect to substitution into* (i.e. left-side composition with) *functions as well as with respect to the application to* (i.e. right-side composition with) *functions having the base* B.

If F is a *constant* p-place function over S, then any single element of S^p constitutes a base of F. Now let S be $\{0, 1\}$. The base of each non-constant p-place function F consists of exactly two elements of S^p: one for which F assumes the value 0, and for which F assumes the value 1. Any base of the function A in Example 1 necessarily contains $(1, 1)$ and any one of the three other pairs. A base of $\{A, F\}$ consists of $(1, 1)$ and one or two of the other pairs, for any F. Any one other pair in conjunction with $(1, 1)$ constitutes a base of $\{A, I\}$. The bases of $\{A, C\}$ are $\{(1, 1), (1, 0), (0, 0)\}$ and $\{(1, 1), (1, 0), (0, 1)\}$.

One furthermore readily proves

REMARK 3. *Let E be the set $\{I, I', J, J', E, E'\}$ where J is the second 2-place selector, for which $J(x, y) = y$ and $E(x, y) = 1$ or 0 according as $x = y$ or $x \neq y$. A pair of 2-place functions $\{F, G\}$ over $\{0, 1\}$ has a base including all four pairs if and only if F and G belong to E without constituting one of the pairs $\{I, I'\}$, $\{J, J'\}$, $\{E, E'\}$.*

If $(G_1, \dots, G_k)$ is an ordered k-tuple of p-place functions, then by the *range* of the k-tuple—briefly, ran $(G_1, \dots, G_k)$—we mean the subset of S^k consisting of all k-tuples $(s_1, \dots, s_k)$ for which there exists an element T of S^p such that $s_i = G_i(T)$ $(i = 1, \dots, k)$. Clearly, $\mathrm{ran}(G_1, \dots, G_k)$ is a subset of the Cartesian product, $\mathrm{ran}\,G_1 \dots \times \mathrm{ran}\,G_k$. For example,

$$\mathrm{ran}(A, C) = \{(0, 1), (1, 0), (1, 1)\},$$

$$\mathrm{ran}(A, C, I) = \{(0, 1, 1), (1, 0, 1), (1, 1, 1)\},$$

$$\mathrm{ran}(A, C, I) = \{(1, 1, 0), (1, 0, 1), (0, 1, 1)\}.$$

REMARK 4. *Any base of a homogeneous ordered set G, is in a one-to-one correspondence with ran G. The power of any such set $G = \{G_1, \dots, G_k\}$ does not exceed the product of the powers of $\mathrm{ran}\,G_1, \dots, \mathrm{ran}\,G_k$.*

What is the maximum number of functions that a set G of k functions over one and the same set S can generate? What is the minimum number of p-place functions generating all p-place functions over S?

First consider the case where $G = \{F\}$. Let B be a base of F and let r be the number of elements in ran F and the number of elements in B. The values of any function H generated by F determine H, whence $\mathfrak{S}F$ includes at most r^r functions, regardless of the place-number of F.

Next consider the case where $G = \{F_1, F_2\}$ and F_1 and F_2 have the same place-number, say p. Let r_1 and r_2 be the numbers of elements in their ranges. According to the Corollary of Remark 1, the range of each function in $\mathfrak{S}\{F_1, F_2\}$ is a subset of either ranF_1 or ranF_2. According to Remark 2, a base of $\{F_1, F_2\}$ is also a base of $\mathfrak{S}\{F_1, F_2\}$. If b is the number of elements in such a base, then at most r_1^b and r_2^b functions are generated whose ranges are $\subseteq \mathrm{ran}F_1$ and $\subseteq \mathrm{ran}F_2$, respectively. If r_{12} is the number of elements in the intersection of ranF_1 and ranF_2, then the maximum number of functions in $\mathfrak{S}\{F_1, F_2\}$ is $r_1^b + r_2^b - r_{12}^b$, regardless of the place number p. By induction, one readily sees that, if $r_{i_1,...,i_h}$ is the number of elements in the intersection of $\operatorname{ran}F_{i_1}, ...,$ $\operatorname{ran}F_{i_h}$, then the maximum number of functions in $\mathfrak{S}\{F_1, ..., F_k\}$ is

$$\sum_{h=1}^{k} \sum_{i_1...i_h} (-1)^{h+1} r^b_{i_1,...,i_h}.$$

Suppose now that $F_1, ..., F_k$ are all the p-place functions in G and that G also contains functions $G_{k+1}, ..., G_t$ having the range numbers $r_{k+1}, ..., r_t$ but place-numbers $\neq p$. The range of a function in $\mathfrak{S}G$ is a subset of one of the t functions in G. According to Remark 2, all p-place functions in $\mathfrak{S}G$ have the same base as $\{F_1, ..., F_k\}$. The total number of functions with those b base elements whose ranges are subsets of one of the t ranges of the functions in G is a sum like the one above, the only difference being that the summation of h ranges from 1 to t instead of from 1 to k. We can express this result in

THEOREM I. *If G is a set of t functions, $F_1, ..., F_t$, having s different place-numbers, $p_1, ..., p_k$, let b_k be the number of base elements of the set of all p_k-place functions in G; and, for any h such that $1 \leqslant h \leqslant t$ and any set of h functions $F_{i_1}, ..., F_{i_h}$, let $r_{i_1,...,i_h}$ denote the number of elements in the intersection of $\operatorname{ran}F_{i_1}, ..., \operatorname{ran}F_{i_h}$. Then the number of p_k-place functions in $\mathfrak{S}G$ does not exceed*

$$a_k = \sum_{h=1}^{t} \sum_{i_1,...,i_h} (-1)^{h+1} r^{b_k}_{i_1,...,i_h}.$$

where one summation extends over all sets $\{i_1, ..., i_h\}$ of he of the numbers $1, ..., t$, and the other summation extends over the numbers h from 1 to t.

The number of functions in $\mathfrak{S}G$ does not exceed $\sum_{k=1}^{s} a_k$.

It should be noted that *the place-numbers $p_1, ..., p_s$ themselves do not enter into the upper bounds given above for the numbers of functions generated.* An obvious limitation for the number of p-place functions that can be generated is of course *the number of all p-place functions.* This number is less than the given upper bound when there are too many generators.

In Example 2, we have $t = s = 2$; $p_1 = 1$, $b_1 = 2$; $p_2 = 2$, $b_2 = 2$; $r_1 = r_2 = r_{12} = 2$. Hence, according to Theorem I, the number of functions in $\mathfrak{S}\{n, B\}$ is at most $(2^2 + 2^2 - 2^2) + (2^2 + 2^2 - 2^2)$. It actually is 8.

COROLLARY 1. *If G is a homogeneous set of k functions with b base elements, and the ranges of the functions in G, which contain $r_1, ..., r_k$ elements, are disjoint, then $\mathfrak{S}G$ includes at most $r^b_1 + ... + r^b_k$ functions.*

We now come to the most important special cases of Theorem I. They concern the sets $S = N_m = \{1, ..., m\}$ for some natural number m.

The range of each function contains at most m elements. The base of any homogeneous set of k functions contains at most m^k elements. Hence

COROLLARY 2. *A homogeneous set of k functions over N_m generates at most m^{m^k} functions, regardless of their place-number. If G is a set of functions over N_m having s different place-numbers, $p_1, ..., p_s$, and if k_i is the number of p_i-place functions in G, then $\mathfrak{S}G$ includes at most $\sum_{i=1}^{s} m^{m^{k_i}}$ functions.*

COROLLARY 3. *A set G generating all p-place functions over N_m includes at least p functions.*

We now turn to the questions whether there actually exist sets of k functions over N_m that generate m^{m^k} functions, and whether p functions are sufficient to generate all p-place functions over N_m.

Obviously, the lower bound, p, stipulated in Corollary 3 is unsharp in three simple cases:

a) $m = 1$ and $p > 1$. There is only one single p-place function over N_m for each p.

b) $m > 1$ and $p = 1$. No single function generates the full semi-group of 1-place functions over N_m. Two or three functions are needed (cf. Piccard [4]) to generate those m^m functions according as $m = 2$ or $m > 2$.

c) $p = m = 2$. Three functions are needed (cf. Menger [2]) to generate all 2-place functions over N_2. If one considers $S = \{0, 1\}$ instead of N_2, then from Remark 3 one readily concludes: Unless the pair of 2-place functions $\{F, G\}$ is a subset of the set E, it has a base of at most three elements and, therefore, cannot generate more than eight functions. If $\{F, G\}$ is a subset of E, then $\mathfrak{S}\{F, G\} \subseteq \mathfrak{S}E$, and $\mathfrak{S}E$ is easily seen to consist of eight elements: the six functions in E and the constant functions 1 and 0. All sixteen 2-place functions over S are indeed generated by some triples of functions, e.g., by $\{A, I, J\}$.

Except for these cases, however, the lower bound, p, stipulated in Corollary 3 will now be proved to be sharp.

In the proof, we shall make extensive use of the *selectors*. For any two natural numbers, m and p, there are p such p-place functions over N_m. Where m is kept fixed, the k-th p-place selector over N_m will be denoted by $I_k^{(p)}$ and is defined for $1 \leqslant k \leqslant p$ by

$$I_k^{(p)}(x_1, ..., x_p) = x_k \quad \text{for any} \quad x_1, ..., x_p \text{ in } N_m.$$

In some cases, we shall continue to write I and J for $I_1^{(2)}$ and $I_2^{(2)}$ over N_m, respectively. The single 1-place selector $I_1^{(1)}$ is the identity function j over N_m. It has the fundamental property that any function F (of any number of places) remains unchanged upon substitution into j; that is to say, $jF = F$ for any F.

If $(F_1, ..., F_k)$ is an ordered k-tuple of k-place funcions over N_m, then, according to Remark 4, $\mathrm{ran}(F_1, ..., F_k)$ and any base of the set $F = \{F_1, ..., F_k\}$ consist of equally many elements of N_m^k. If the (unique) base of F includes all k^m elements of N_m^k, then the set F will be called *perfect*. If F is perfect, then $\mathrm{ran}\,(F_1, ..., F_k)$ (as well as the range of F in any order) is a permutation of the base of F. For any k-place function H, we set $H(F_1, ..., F_k) = H(F_1, ..., F_k)^1$ and define

$$(F_1, ..., F_k)^{r+1} = \left(F_1(F_1, ..., F_k)^r, ..., F_k(F_1, ..., F_k)^r\right).$$

Clearly, k^m iterations of the permutation of the k^m elements yield the identical permutation; that is to say,

$$(F_1, ..., F_k)^{k^m} = (I_1^{(k)}, ..., I_k^{(k)}).$$

Since each component of each $(F_1, ..., F_k)^r$ belongs to $\mathfrak{S}F$ we thus have

LEMMA 1. *If F is a perfect set of k-place functions over N_m, then $\mathfrak{S}F$ includes all k-place selectors, $I_i^{(k)}$ $(1 \leqslant i \leqslant k)$.*

EXAMPLE 3. Consider the triple (F, I_1, I_2) of 3-place functions over N_2, where

$$F(1,1,1) = F(1,2,2) = F(2,1,1) = F(2,2,1) = 2,$$

$$F(2,2,2) = F(1,2,1) = F(2,1,2) = F(1,1,2) = 1;$$

$$I_1(x, y, z) = x, \quad I_2(x, y, z) = y \quad \text{for each} \quad (x, y, z) \text{ in } N_2^3.$$

The set $\{F, I_1, I_2\}$ is easily seen to be perfect. (F, I_1, I_2) produces a cyclical permutation of the triples

$$(1,1,1), (2,1,1), (2,2,1), (2,2,2), (1,2,2), (2,1,2), (1,2,1), (1,1,2).$$

Hence $(F, I_1, I_2)^8 = (I_1, I_2, I_3)$, where $I_3(x, y, z) = z$. It follows that I_3 belongs to $\mathfrak{S}F$. Indeed, $I_3 = I_2(F, I_1, I_2)^7 = I_1(F, I_1, I_2)^6 = F(F, I_1, I_2)^5$.

A classical theorem in Boolean algebra asserts that each "function of p variables $x_1, ..., x_p$" over $\{0, 1\}$ can be represented in two (so-called *normal*) forms: as a sum of products and as a product of sums. The first half states, more precisely, that each $F(x_1, ..., x_p)$, except the function assuming only the value 0, is, for some number k, where $1 \leqslant k \leqslant 2^p$, the sum of k products of the form $y_1 \cdot ... \cdot y_p$, where, for each $i = 1, ..., p$, one has $y_i = x_i$ or $y_i = 1 - x_i$. This theorem can be expressed in terms of the functions n, A', and B, mentioned in Examples 1 and 2, and the p-place selectors $I_i^{(p)}$, which we shall denote, briefly, by $I_1, ..., I_p$. It is convenient to set $n^1 = n$, $n^2 = nn = j$. Hence $n^k F = nF$ or $= F$ according as $k = 1$ or 2. If $(i_1, \ldots, i_p)$ is an ordered p-tuple of numbers belonging to $\{1, 2\}$, then we define a p-place function

$$P_{i_1, \ldots, i_p} = A'\Big(\ldots\big(A'\big(A'(n^{i_1}I_1, n^{i_2}I_2), n^{i_3}I_3\big), \ldots, n^{i_p}I_p\big)\Big).$$

Each such function corresponds to one of the products $y_1 \cdot \ldots \cdot y_p$ and obviously belongs to $\mathfrak{S}\{n, A', I_1, \ldots, I_p\}$.

The classical theorem asserts that, for any p-place function F over $\{0, 1\}$ there exist k ordered p-tuples $(i_{h1}, \ldots, i_{hp})$ where $1 \leqslant h \leqslant k$, for some k $(1 \leqslant k \leqslant 2^p)$ such that

$$F = B\Big(\ldots\big(B\big(B(P_{i_{11}, \ldots, i_{1p}}, P_{i_{21}, \ldots, i_{2p}}), P_{i_{31}, \ldots, i_{3p}}\big), \ldots, P_{i_{k1}, \ldots, i_{kp}}\big)\Big).$$

Hence F belongs to $\mathfrak{S}\{n, A', B, I_1, \ldots, I_p\}$.

Post [5] generalized the theorem just mentioned to the set $\mathfrak{S}_m^p$ of all p-place functions over N_m. We assume N_m to be ordered according to $1 < 2 < \ldots < m$, and define

$$A'(x, y) = Min(x, y), \quad B(x, y) = Max(x, y), \quad n(x) = x+1$$
$$\text{for} \quad 1 \leqslant x \leqslant m-1 \text{ and } n(m) = 1.$$

We set $n^{k+1} = nn^k$ for $1 \leqslant k \leqslant m-1$. Clearly, $n^m = j$, where $jF = F$ for each F over N_m. The functions $P_{i_1, \ldots, i_p}$ are defined as in the classical theorem, but for all ordered p-tuples $(i_1, \ldots, i_p)$ of numbers $1, \ldots, m$. Any p-place function F over N_m, except the constant p-place function of value 1, can be expressed, just as in the classical case, in terms of B and k functions $P_{i_{h1}, \ldots, i_{hp}}$ $(1 \leqslant i \leqslant k)$ for some k such that $1 \leqslant k \leqslant m^p$.

We now prove

LEMMA 2. *Let* p, r, *and* m *be natural numbers* >1. *Then the set* $\mathfrak{S}_m^p$ *of all* p-*place functions over* N_m *is a subset of*

$$\mathfrak{S}^* = \mathfrak{S}\{N_r^*, A_r^*, B_r^*, I_1^{(p)}, \ldots, I_p^{(p)}\}$$

where N_r^*, A_r^*, *and* B_r^* *are* r-*place functions defined as follows:*

$$N_r^* = nI_1^{(r)}, \quad A_r^* = A'(I_1^{(r)}, I_2^{(r)}), \quad B^* = B(I_1^{(r)}, I_2^{(r)}).$$

Let F be a function belonging to $\mathfrak{S}_m^p$. According to Post's Theorem, F belongs to $\mathfrak{S} = \mathfrak{S}\{n, A', B, I_1, \ldots, I_p\}$, where I_1 is an abbreviation for $I_1^{(p)}$. We prove that F belongs to $\mathfrak{S}^*$ by induction on the degree of F relative to the set $\mathfrak{S}$. If $degF = 1$ then, being a p-place function, F is one of the selectors I_1 and therefore belongs to $\mathfrak{S}^*$. Only if $p = 2$, also $degA' = degB = 1$; but in this case $A' = A_r^*(I_1, I_2, I_2, \ldots, I_2)$ and $B = B_r^*(I_1, I_2, I_2, \ldots, I_2)$. For $p > 2$, assume that all p-place functions of a degree $\leqslant n$ relative to $\mathfrak{S}$ belong to $\mathfrak{S}^*$, and let F be a p-place function of degree $n+1$. Clearly, F is either nK or $A'(K, L)$ or $B(K, L)$ for two functions K and L of a degree $\leqslant n$. But

$$nK = N_r^*(K, K, ..., K), \quad A'(K, L) = A_r^*(K, L, ..., L),$$
$$B(K, L) = B_r^*(K, L, ..., L).$$

In any case, F thus belongs to $\mathfrak{S}^*$.

An immediate consequence is

LEMMA 3. *For any two natural numbers,* p *and* r, *if* $\mathfrak{S}\{F_1, ..., F_k\}$ $= \mathfrak{S}_m^r$, *then* $\mathfrak{S}_m^p \subseteq \mathfrak{S}\{F_1, ..., F_k, I_1^{(p)}, ..., I_p^{(p)}\}$.

The set $\mathfrak{S}\{F, I_1, I_2\}$ in Example 3 contains, as has been shown, all three 3-place selectors. As one readily verifies, $N_3^* = F(I_1, I_1, I_1)$ and, if one sets $K = F(I_1, I_2, I_2)$ and $L = N_3^*(I_1, I_2, I_2)$, $M = N_3^*(I_2, I_1, I_1)$, then $A_3 = N_3^*(K, K, K)$ and $B_3 = F(L, M, M)$. Thus also N_3^*, A_3^*, and B_3^* belong to $\mathfrak{S}\{F, I_1, I_2\}$. From Lemma 2 it follows that this set is $\mathfrak{S}_2^3$. We thus have established the case $p = 3$ of

LEMMA 4. *For each* $p > 2$, *there exist* p *functions generating* $\mathfrak{S}_2^p$.

Assume $p > 3$, and consider

$$G = \{F(I_1, I_2, I_3), I_1, I_2, I_4, ..., I_p\},$$

where I_1 is an abbreviation of $I_1^{(p)}$. As one easily verifies, G is perfect. Hence, by Lemma 1, $\mathfrak{S}G$ includes all p-place selectors (also I_3). We further show that $\mathfrak{S}G$ includes N_p^*, A_p^*, and B_p^*. Setting $F^* = F(I_1, I_2, I_3)$ one can verify that $N_p = F^*(I_1, I_1, ..., I_1)$. Setting $K = F^*(I_1, I_2, ..., I_2)$ $L = N^*(I_1, I_2, ..., I_2)$ and $M = N_p^*(I_2, I_1, ..., I_1)$, one furthermore verifies that

$$A_p^* = N_p^*(K, K, K, ..., K) \quad \text{and} \quad B_p^* = F^*(L, M, ..., M).$$

By Lemma 2, $\mathfrak{S}_2^p \subset \mathfrak{S}G$, which completes the proof of Lemma 4.

EXAMPLE 4. Consider the 2-place function F over N_3 defined by

$$F(1, 1) = F(1, 2) = F(1, 3) = 2 ;$$
$$F(2, 2) = F(3, 1) = F(3, 2) = 3 ;$$
$$F(3, 3) = F(2, 1) = F(2, 3) = 1 .$$

Martin [1] has proved that this function F in conjunction with I and J generates all 3^{3^2} 2-place functions over N_3. If we define G by $G(x, y)$ $= F(y, x)$ for each (x, y) in N_3^2, then the set $\{F, G\}$ is easily seen to be perfect and, therefore, by Lemma 1, includes I and J. It follows that $\mathfrak{S}\{F, G\} = \mathfrak{S}_3^2$.

EXAMPLE 5. For any $m > 3$, consider the 2-place function F defined by

$$F(i, i) = i+1 \quad \text{for} \quad 1 \leqslant i \leqslant m-1, \quad F(m, m) = 1,$$

$$F(1, 2) = F(1, 4) = 2, \quad F(2, 3) = F(2, 4) = 1,$$

$$F(x, y) = x \quad \text{for all other pairs (x, y) in } N_m^2.$$

Set $F = F_1$ and $F(F_k, F_k) = F_{k+1}$, and define 1-place functions g_k over N_m for $1 \leqslant k \leqslant m$ by setting $g_k(i) = F_k(i, i)$ for $1 \leqslant i \leqslant m$. Since $g_1(i) \equiv i+1$ (mod. m) it is clear that these m functions g_k are the m cyclical permutations of $(1, 2, ..., m)$. From the definition of F, one further sees that $t = F(g_m, g_1)$ is the transposition interchanging 1 and 2, and that $h = F(g_m, g_2)$ has the values $h(1) = h(2) = 1$ and $h(k) = k$ for $3 \leqslant k \leqslant m$. It is well known that g_1, t, and h generate all m^m functions in $\mathfrak{S}_m^1$. Since the functions $T = F(F_m, F_1)$ and $H = F(F_m, F_2)$ belong to $\mathfrak{S}\{F\}$ and

$$F(i, i) = g_1(i), \quad T(i, i) = t(i), \quad H(i, i) = h(i) \quad \text{for} \quad 1 \leqslant i \leqslant m,$$

it is clear that, for each function u in $\mathfrak{S}_m^1$, the set $\mathfrak{S}\{F\}$ contains a function U such that $U(j, j) = u$; that is to say, $U(i, i) = u(i)$, for $1 \leqslant i \leqslant m$. Hence $\mathfrak{S}\{F\}$ includes m^m mutually different functions.

We now define G by setting $G(x, x) = F(x, x)$,

$$G(1, 2) = 3, \quad G(2, 3) = 2, \quad \text{and}$$
$$G(x, y) = y \quad \text{for all other (x, y) in } N_m^2.$$

It is easy to verify that the set $\{F, G\}$ is perfect. Hence $\mathfrak{S}\{F, G\}$ includes the 2-place selectors, I and J.

Słupecki [6] proved, for every natural number $m > 2$, an important theorem which may be formulated as follows. If H is any 2-place function over N_m which, for no f in $\mathfrak{S}_m^1$ is equal to either fI or fJ, then

$$\mathfrak{S}_m^2 \subset \mathfrak{S}\{g_1, t, h, H, I, J\}.$$

In Example 5, $\mathfrak{S}\{F, G\}$ includes I and J and U for every u in $\mathfrak{S}_m^1$. By induction on the degree of 2-place functions relative to $\{g_1, t, h, F, I, J\}$ one sees that $\mathfrak{S}\{F, G\} = \mathfrak{S}_m^2$.

We abbreviate $I_1^{(p)}$ to I_1 and define, for each $p > 2$,

$$F_p = F(I_1, I_2), \quad G_p = G(I_1, I_2),$$

where F and G are the functions studied in Example 5 if $m > 3$, and the functions in Example 4 if $m = 3$. In any case, the set

$$\mathcal{G} = \{F_p, G_p, I_3, ..., I_p\}$$

is readily seen to be perfect, whence $\mathfrak{S}\{F_p, G_p, I_2, ..., I_p\}$ includes all p-place functions. Now set

$$\mathcal{G}^* = \{F, G, I_1, ..., I_p\}.$$

By an inductive proof similar to that of Lemma 2, we see that the p-place functions in $\mathfrak{S}G$ and in $\mathfrak{S}G^*$ are the same. Hence $\mathfrak{S}G = \mathfrak{S}_m^p$. We thus have the first half of

THEOREM II. *If* $m > 1$ *then, except for the case* $m = p = 2$, *the bounds given in Corollaries 3 and 2 of Theorem I are sharp; that is to say, there are* p *functions generating all* p-*place functions over* N_m; *and there exists a homogeneous set of* k *functions generating* m^{m^k} *functions.* More specifically, *there exists a homogeneous set of* p *functions including* $p-2$ *selectors and generating* $\mathfrak{S}_m^p$; *and there exists a homogeneous set of* k *functions including* $k-2$ *selectors and generating* m^{m^k} *functions.*

In order to obtain a homogeneous set of k functions generating the maximum number of functions, for any place-number $p > k$, consider a homogeneous set F of k functions, $F_1, ..., F_k$, generating $\mathfrak{S}_m^k$. (F may be so chosen as to include $k-2$ selectors.) For $p > k$, set $F_i^{(p)} = F_i(I_1^{(p)}, I_2^{(p)})$ and

$$G = \{F_1^{(p)}, ..., F_k^{(p)}\}.$$

The number of functions in $\mathfrak{S}F$ and in $\mathfrak{S}G$ is the same. Thus the k functions in G generate m^{m^k} functions.

Addition in the proofs. In Remark 1, a function F in G is not necessarily obtainable by substituting functions belonging to $\mathfrak{S}G$ into a function belonging to G. Cf. [2] p. 291 for an example of two functions F and G in the 3-valued logic such that $F(X, Y) = G(X, Y) = G \neq F$ for each X, Y in $\{F, G\}$.

References

[1] N. M. Martin, *The Sheffer functions of the 3-valued logic*, J. Symb. Logic 19 (1954), pp. 45-51.

[2] K. Menger, *Algebra of functions: past, present, future*, Rend. Math. Roma 20 (1961), pp. 409-430.

[3] — *Superassociative systems and logical functors*, Math. Annalen 157 (1964), pp. 278-295.

[4] S. Piccard, *Sur les fonctions définies dans les ensembles finis quelconques*, Fund. Math. 24 (1935), pp. 298-301.

[5] E. L. Post, *Introduction to a general theory of elementary propositions*, Amer. J. Math. 43 (1921), pp. 163-185.

[6] J. Słupecki, C. R. Soc. Sci. Let. Varsovie, Cl. III, 32 (1939).

[7] H. I. Whitlock, *A composition algebra for multiplace functions*, Math. Annalen 157 (1964), pp. 167-178.

Reçu par la Rédaction le 20. 4. 1965

Selected Papers on Didactics, Variables, and Fluents

Commentary on Didactics, Variables, Fluents

B. Schweizer and A. Sklar

In the introduction to Part IV of his "Selected Papers in Logic and Foundations, Didactics, Economics" [M11] Menger says: "The realization of the need for a clarification of the basic concepts and a reform of the didactics [of mathematics] first came to me in the early 1930's in Vienna – but not in connection with the Circle. It was rather when teaching a course in calculus at the University that I was vexed by finding that the traditional treatment forced me to dodge some difficult questions. A few years later in America, ... some of these questions were raised by beginners whose initiation into calculus was my responsibility..."

Menger responded to these questions in several ways. On the one hand, he began to develop what became the "Algebra of Analysis" and "Algebra of Functions". On the other, he began to try to improve both the way calculus and related areas of mathematics were presented to students and the way in which calculus was used as a scientific tool. These efforts inevitably led him to investigate and sort out the various notions lumped together under the omnipresent term "variable". In particular, he noted that what were often called "physical variables" or "variable quantities" had nothing but the name in common with variables *per se*. In view of their importance, he decided that these miscalled "variable quantities" deserved a separate name: he finally settled, for better or worse, on the name "fluent".

Menger's writings on the subject span more than a quarter century. During this time his scope expanded, his insights deepened, and he sometimes changed his mind. The reprinted items that follow this commentary–([M1], the excerpt from [M2], [M3], [M5] and [M6]) and [M10], which is reprinted in Volume 1, pp. 575–593 – constitute a small but important representative sample of his work in the intertwined areas of calculus, "variables", and "fluents".

1. Variables

At the beginning of "To the...general reader" in [M2], Menger notes that Weierstrass, in his 1880 lectures, had defined a 'variable' as a "symbol that stands for any number or any element of a certain class of numbers." Since he is shortly going to speak about other types of variables ('fluent variables' and 'function variables'), Menger refers to variables as defined by Weierstrass as 'numerical variables.' He defines 'function variables' as "symbols that stand for any function or any element of a certain class of functions." As for 'fluent variables', he says that they are symbols "that may be replaced by...elements of a certain class" of fluents.

It is clear from all this that a general definition of 'variable' is lurking in the background, but no such general definition appears in [M2]. Indeed, it seems that the only place in all of Menger's writings where there is such a definition is in Section 2 of [M3], where he says:

> "More generally, if a formula contains a letter and is accompanied by a legend concerning the replacement of the letter by the designations of elements of a certain class, then logicians and mathematicians refer to the letter as a *variable*, and call the said class the *scope* of the variable."

(Actually, this is giving "logicians and mathematicians" as a group too much credit. In real logico-mathematical life, things that do not fit the definition are often called 'variables', while things that do fit the definition are either overlooked entirely, or are called anything *but* 'variables'. We shall give some examples below.)

In the definition, the word "formula" is to be taken in the widest possible sense. Also, the word "letter" should be replaced, in each of its occurrences, by the word "symbol". It should be noted, moreover, that there is no requirement in the definition that the scope of a variable be given explicitly: Although it is true in almost all cases that the scope of a variable is known to be a subset of a particular given set, it may not be known *which* subset. For example, an existential statement involving a variable is simply the assertion that the scope of the variable is non-empty. And "solving an equation" generally means determining at least one, or possibly all, the elements in the scope of a variable called an "unknown" of the equation.

Probably the most common use of variables is in definitions, particularly of functions. Consider, for example, the following definition of a sequence of polynomial functions on the reals:

$$C_0(x) = 2 \text{ and } C_1(x) = x \text{ for all real } x, \tag{1}$$

$$C_{n+1} = C_1 \cdot C_n - C_{n-1} \text{ for all positive integers } n. \tag{2}$$

In (1), x is a variable whose scope is R, the set of all real numbers. In (2), there are 3 related variables: n, whose scope is the set of positive integers, $n-1$, whose scope is the set of all non-negative integers, and $n+1$, whose scope is the set of integers ≥ 2. In (1), we could replace 'x' by 'y' or 't' or 'ξ', or indeed, most anything *except* 'C'.

In (2), if we replace '$n-1$' by 'm', then (2) becomes the completely equivalent statement

$$C_{m+2} = C_1 \cdot C_{m+1} - C_m \text{ for all non-negative integers } m. \tag{2'}$$

The symbol 'C' in (1) and (2) is of course not a variable: it is the designation of a particular sequence (i.e., a function whose domain is the set of non-negative integers) whose values are particular (polynomial) functions.

Here are other examples of definitions involving variables:

A binary operation on a (non-empty) set S is a function whose domain is $S \times S$ and whose range is a subset of S.

A binary operation $*$ on a set S is *associative* if

$$(x * y) * z = x * (y * z) \text{ for all elements } x, y, z \text{ of } S.$$

In the first definition 'S' is a variable whose scope is the very large set (or "class" if you like) of non-empty sets. There are five variables in the second definition: S, $*$, x, y, z. Notice that we cannot properly fix the scopes of $*$, x, y, z until we replace S by the designation of some particular non-empty set.

Returning to [M2], we note that, immediately after introducing the notion of "numerical variables", Menger says: "In calculus, numerical variables may be used or they may, as will be shown herein, be dispensed with." He does *not* say, here or anywhere else, that *all* variables may be dispensed with in calculus, nor, in any of his writings, does he say that his version of calculus is "variable-free."[1] For instance, his elegant and totally explicit formulation of the chain rule, namely

$$D(f \circ g) = (Df \circ g) \cdot Dg,$$

contains the function variables "f" and "g", but no numerical variables.

---

[1] Others are looser in their use of the expressions "variable-free" or "without variables." If the fundamental objects in some theory are certain functions, and the theory is treated without use of variables whose scopes are domains or ranges of those functions, then such treatment, even though replete with variables over the functions themselves, is called "variable-free": see, e.g. [13]. In this much looser sense, Menger's treatment of calculus, and his later treatment of propositional calculus could be called "variable-free."

One example, outside calculus, showing that numerical variables can "be dispensed with", concerns the so-called "general polynomial equation of degree n". This is generally written in the form

$$x^n + a_1 x^{n-1} + \cdots + a_n = 0$$

(or something similar), in which the variable x is called the "unknown", and the variables $a_1, \ldots, a_n$ are called "coefficients" or "parameters".[2] Menger spoke on this subject in 1967 at the Fifth Annual Meeting on Functional Equations in Waterloo, Ontario, and the following abstract of his talk appeared in Aequationes Mathematicae, Vol.1, 1968, p. 281:

> "The so called general algebraic equation of degree n with the parameters $a_1, \ldots, a_n$ is the problem to find, for any values of the parameters, all numbers x such that $x^n + a_1 x^{n-1} + \cdots + a_n = 0$. Let $I_k^{(n)}$ be the k-th n-place selector function assuming for any n-tuple $(x_1, \ldots, x_n)$ the value x_k. The general algebraic equation may be considered as the functional equation asking for all n-place functions F such that
>
> $$F^n + I_1^{(n)} \cdot F^{n-1} + \cdots + I_n^{(n)} \equiv O^{(n)},$$
>
> where $O^{(n)}$ is the constant n-place function of value 0. One may study the implications of this point of view for functional as well as algebraic equations."

2. Non-Variables

It is now commonplace for people using probability theory to note that "a 'random variable' is neither random nor variable." But this attitude was far from common in the early 1950's, particularly among working statisticians. Consequently Menger felt that he had to devote the first part of his talk at the Third Berkeley Symposium on Mathematical Statistics and Probability [M3] to an explanation of what a mathematical variable was (so leading to the definition quoted in the preceding section). He could then proceed to discuss "random variables" as things that were not variables, and in particular, to show that in practice there were two different kinds of such things: "probabilistic random variables" as defined by Kolmogorov in [7] and "statistical random variables" as rather loosely defined by Wald [14] and [15] (see the note in Section 2 of the Commentary on Probabilistic

---

[2] Menger discusses "unknowns", "parameters", and other things in his paper [M9], which is reprinted as Chapter 10 of [M11].

Geometry). Menger tightened up Wald's definition and observed that the basic difference between the two kinds of random variables is that "a probabilistic r.v. has a f.p.s [fundamental probability space] as its *domain*, whereas a statistical r.v. has a f.p.s. as its *range*."

This difference means that, unlike two or more probabilistic r.v.'s with the same domain, two or more statistical r.v.'s, even with the same domain, have no built-in joint distribution – any joint distribution has to be either assumed or inferred from prior results.[3] This is not a drawback in actual statistical practice, where assumptions, in particular of independence, are routine, but for theoretical purposes it is simpler to work with probabilistic r.v.'s alone. Nevertheless, the distinction between probabilistic and statistical r.v.'s is latent even in the most highly polished theoretical works: it may be glimpsed, e.g., in the following words of M. Loève from his excellent text ([8], p. 174):

> "It is important to realize fully that measurements of a stochastic variable are relative to the induced pr[obability] space; the original pr[obability] space is but a mathematical fiction. Yet it is basic, for it permits the use of a 'common frame of reference' for the families of stochastic variables we investigate – the families of sub σ-fields of events they induce on the original pr[obability] space."

And there are indications that forthcoming developments in probability theory may see the re-emergence into prominence of the distinctions so clearly elucidated by Menger in [M3]: in this connection, see [12].

Before turning to [M5], we note that, while random variables, of whatever stripe, are not variables, variables whose scopes are sets of random variables abound in works of probability theory. Such "r.v. variables" (as Menger calls them in [M3], presumably to avoid the odd-looking phrase "random variable variables") are commonly denoted by capital letters, especially X and Y.

Now any person encountering mathematics will come across the letters 'x' and 'y', lower case and capital, used over and over again, in a great variety of ways. This variety (usually!) offers no difficulty to the experienced mathematician, but can be, and more often than not is, thoroughly confusing to the beginner. So Menger, in [M5], sorted out some (by no means all!) of the uses of 'x' and 'y', and proposed ways to alleviate the confusion. Most of Menger's examples (e.g., numerical variables, operators, function variables) are straightforward and need not be discussed in detail. Example II. "The Identity

---

[3] This is an important distinction which, strangely, Menger fails to make.

Function" will be treated in the next section, and Examples VIII and X "Specific Fluents" and "Fluent Variables" in Section 5. Here we will only discuss Example V "Indeterminates."

In Example V of [M5], although Menger carefully distinguishes "a polynomial form containing one letter" from the look-alike "polynomial function", he does not do the same with what are often called "polynomials in several variables" as against "polynomial functions of several variables." We indicate how to fill this gap by considering the case where "several" means two. Then what is called "a polynomial in two variables" is a function whose domain is the set of all ordered pairs of non-negative integers and whose range is a subset of a field (or slightly more generally, of a commutative ring with 1), and which has the value 0 (the zero of the field or ring) for all but a finite subset of the domain. This statement is then supplemented by a definition of how two such polynomials are to be added or multiplied. Then one shows that all such polynomials can be expressed in terms of (a) constant polynomials, i.e., those whose value is 0 at all pairs except the pair (0, 0), (b) the specific polynomial which has the value 0 at all pairs except the pair (1, 0), where it has the value 1 (the unit of the ring or field), and (c), the specific polynomial which has the value 0 at all pairs except the pair (0, 1), where it has the value 1. The first specific polynomial mentioned can be, and often is, labeled 'x', and the second 'y'. Thus Menger's example x^2-9y^2 is a polynomial whose value is (the real integer) 1 at the pair (2, 0), whose value is -9 at the pair (0, 2), and whose value is 0 at all other pairs. Similarly, $3xy$ is the polynomial whose value is 3 at the pair (1, 1) and 0 at all other pairs. Clearly, 'x' and 'y' are not "variables" of any sort (and since they are quite specific objects, calling them "indeterminates" is a historical misnomer). And certainly, the polynomial x^2-9y^2 is quite different from the function P, which can be defined, using 'x' and 'y' in this case as numerical variables, by $P(x,y) = x^2-9y^2$.

3. Menger's "Calculus: A Modern Approach"

We have included Chapter 1 of the textbook [M2] in these Selecta because that chapter introduces the two major innovations that underlie Menger's treatment of the calculus: his "mini" or "pre-limit calculus" and his provision of explicit symbols for the most basic (and previously nameless) real functions.

Menger's mini-calculus can be concisely described in standard mathematical terms as a theory of integration of step-functions and differentiation of continuous piecewise-linear functions.[4] Note that each of these two sets of functions is closed under linear combinations. More significantly, note that the integral of a step-function is a continuous piecewise-linear function, and that the derivative of a continuous piecewise-linear function is a step-function. Thus in this setting, not only are all the most basic results of the integral and differential calculus separately valid, but also the fundamental relationships between integration and differentiation (the so-called "Fundamental Theorem of the Calculus") — and all without recourse to any limit processes whatever.

Once this mini-calculus is developed, it can be used, first, to provide approximations to integrals and derivatives of more general functions (Menger does this in Chapters 2 and 3 of [M2], and then, after limits are introduced (via Chapter 5) as the basis for the development of "full" calculus (Chapters 6, 8, and 9).

Of course, since Menger is writing an elementary textbook, he does not introduce his mini-calculus as we have done above. Instead, he devotes Chapter 1 to an exposition, in geometric form, of what amounts to a particularly simple special case of his mini-calculus (a "mini-mini-calculus" so to speak).[5] This deals with the integrals of constant functions (or rather, with the graphs, referred to as "horizontal lines" of such functions) and with derivatives (of the graphs, referred to as "non-vertical lines") of linear functions. Since Menger always thinks of a function as a set of pairs of numbers, and since a curve in a coordinatized plane can also be considered a set of pairs of numbers, the two versions are interchangeable, and Menger consistently uses the same notation for a function and its graph.

In Section 1 of Chapter 1, Menger introduces the symbol j for "the line for which $jx = x$ for any x", so that j, as already noted in "To the Instructor and General Reader", denotes the identity function. This allows him, a few lines later, to write $cj + c'$, where c and c' are numerical variables, for a general "non-vertical straight line", therefore for a general linear function. When this is supplemented by the introduction, in Section 3 of Chapter 2, of the notation j^c for the cth power function (so that j^2 denotes the square and $j^{1/2}$, with a different domain, the square root), then this is the extent of Menger's 'new' notation for functions. But this

---

[4] In Section 1 of Chapter 2 of [M2] Menger takes care of the fact that there will generally be points where a continuous piecewise-linear function has no derivative by, in effect, noting the existence of (and difference between) left- and right-derivatives at such points.

[5] The complete mini-calculus is developed at the beginning of Chapter 2.

seemingly small innovation is vital, for at one stroke it removes most, if not all of the ambiguities that plagued, and still plague, the standard presentation of calculus, and converts the subject into a pure and unambiguous calculus of functions.

Besides these major innovations, there are many other felicities in [M2]. One that deserves mention is Menger's simple and perspicuous technique of graphical composition of two real functions (pp. 89–90). (Interestingly, this technique is well-known in the special case of *iteration*, i.e., composition of a function with itself, but seems to be still quite unknown for the general case.)

In spite of favorable reviews of an earlier, mimeographed version of [M2] (see [1] and [4]), Menger had great difficulty in finding a publisher for the finished work. When it finally appeared, in a rather meager format, it was met by almost total incomprehension and some open hostility. Many teachers of calculus received, at second- or third-hand, an impression which Menger felt had been deliberately spread and which can be paraphrased as follows: "Here's a man who claims that the way I teach is totally wrong and who says he has the right way. Yet all he does is replace the letter x by the letter j, and he makes a big noise about it. He may have done good work in the past, but now..." The effect of this reception on Menger is discussed in the Introduction to these Selecta.

Nevertheless, from the beginning there have been a few who have appreciated Menger's contribution and have tried to extend it. Thus, at the end of his review referred to above, I. T. Adamson wrote:

> "Every teacher of the Calculus must read [Menger's book]. And he will be thick-skinned indeed who, having read it, does not seriously rethink his lectures; the thinner-skinned reader will rewrite his – and the reviewer confidently believes he will rewrite them à la Menger."

Adamson did a bit of rewriting himself: his paper [2] begins by saying: "The aims of this note are first, to present the theory of 'change of variable' in multiple integrals in the spirit of Menger's book on the calculus..."

Those people who have actually taught calculus à la Menger have generally been favorably impressed by the results and have tried to acquaint other teachers with them. In particular, J. Harkin, beginning with [6] and continuing to the present, has been engaged in spreading knowledge of Menger's approach to his fellow-teachers. And such attempts are not confined to mathematicians: see [5]. But we give the last word in this section to the well-known mathematician A. Rapoport. At the end of a review [10] directed to non-mathematicians, he wrote:

"Whether Menger's reforms will revolutionize the teaching of the calculus and the writing of calculus textbooks depends on the outcome of the battle between semantic awareness and loyalty to established routines. In the long run, semantic awareness usually wins, but many short-term victories fall to the entrenched faces of conformity. Pending the final victory, Menger's *Calculus* is strongly recommended to all brave teachers willing to blaze new paths, to confront the raised eyebrows of department chairmen, and to defend their convictions of what constitutes good teaching of mathematics."

4. Differentials, Rates of Change and Other Fluents

The development of calculus without numerical variables, together with the consistent use of derivative and antiderivative operators, D and D^{-1}, and the elimination of the appendage "$(x)dx$" in $\int_a^b f(x)dx$, revolutionizes the subject. First of all, it has profound effects on pedagogy. And the natural extension of this development provided by the introduction of partial derivative operators (as in Chapter 11 of [M2]) permits far-reaching simplifications and clarifications of other areas of mathematics, most notably the calculus of variations: see the following commentary by L. Senechal and B. Schweizer on "A Mengerian Tour Along Caratheodory's Royal Road."

On the other hand, this development seemingly divorces the subject from its application to science. This apparent separation naturally and ineluctably led Menger to the question of reuniting theory and practice. Thus Chapter 7 and a good part of Chapter 11 of [M2] are devoted to a discussion of the applications of the calculus to (physical) science. The underlying philosophical viewpoint of this discussion can be traced back to Menger's Vienna days, and in particular, to his participation in the Vienna Circle (see [M12]), but the actual content, at least in large part, only as far back as the work embodied in [M1].

In that paper, Menger points out the ambiguities that arise in the standard presentation of thermodynamics[6] and proceeds to show how these ambiguities could be eliminated. Since a major, though not the only, source of the ambiguities arises from the use of all-too-easily misinterpreted differential expressions, Menger's major recommendation, in Section 8 of the paper, is that "physicists in presenting the elements of the theory should refrain from referring to the poorly understood concept of a general differential altogether".

---

[6] Ambiguities which, we are told on good authority, persist in textbooks to the present day.

To anyone who has ever suffered through a course in classical thermodynamics, Menger's analysis is a revelation. But of course, it is not only in thermodynamics that difficulties with differentials arise. These difficulties were so prevalent that in the early 50's, C. B. Allendoerfer, then editor of the American Mathematical Monthly, in effect commissioned three different mathematicians, namely: M. K. Fort Jr., C. G. Phipps and H. J. Hamilton, to write about them. The resulting articles appeared in Volume 59 of the journal and are reprinted in [3]. In an accompanying editorial note (also reprinted in [3]), Allendoerfer writes:

"The best description of a differential that I have been able to formulate is: The presence of a differential in a mathematical expression is a sign to the reader that this expression is obtained by a limiting process from a second expression in which the differentials dx, dy, etc. are replaced by finite increments Δx, Δy, etc. The nature of this limiting process depends on the particular expression involved and must be inferred from the context" [shades of G. Berkeley's "ghosts of departed quantities"].

This was in 1952. The situation was not improved 15 years later, when one could come across an article titled "Manipulations with differentials made respectable" [9]; and it remains unimproved today. Thus it would appear that in advocating the elimination of differentials in elementary mathematics, Menger was prescient. It should also be said that while he did not object to the use of differentials in more advanced mathematics, where they could be precisely defined (and sometimes actually are), he preferred the use of other techniques when available.

Returning to [M2], we note that the applications to science in the book are treated in terms of notions called "quantities", "consistent classes of quantities (c.c.q's)" or "fluents", and certain types of c.c.q's called "variable quantities (v.q.'s)". These are defined as follows:

A *quantity* is an ordered pair whose first member (the *object* of the quantity) can be anything whatever, and whose second member (the *value* of the quantity) is a (real) number. Two quantities are *equal* if they are equal as ordered pairs, i.e., have the same object and the same value; they are *consistent* unless they have the same object but different values, in which case they are *inconsistent*. A *c.c.q.* (consistent class of quantities) or *fluent* is a set of pairwise consistent quantities. The *domain* of a fluent is the set of its objects; its *range* is the set of its values. A fluent whose domain "consists of objects of scientific study" is a *variable quantity* or *v.q.*

According to Menger, ordinary real functions are special types of fluents, but definitely not v.q.'s. Given two fluents w and u, a *pairing* of Dom w with

Dom u (in that order) is a one-to-one correspondence Π between Dom w and a subset of Dom u. A fluent w is a function f of a fluent u, written $w = fu$, relative to a pairing Π if for every a in Dom w, wa, the value of w at a, is given by $wa = f(u(\Pi a))$.

This machinery is carried over into [M6], except that the terms "consistent class of quantities" and "variable quantity" are suppressed, so only the terms "quantity" and "fluent" appear. The later paper also makes clear something that is fudged in [M2], i.e., that for certain purposes we want the domains of some fluents to be *limit classes*, i.e., spaces in which there is an appropriate notion of limits. Now let w be such a fluent and suppose there is a pairing Π of Dom w with Dom u. For $b \neq a$ in Dom w (and assuming $u(\Pi b) \neq u(\Pi a)$) form the quotient

$$\frac{wa - wb}{u(\Pi a) - u(\Pi b)}$$

If this expression has a limit as $b \to a$ in Dom w, then this limit is the *rate of change at a of w with respect to u* (*relative to* Π) and it is denoted by $(dw/du)_{\Pi}a$; it is also the value at a of the *fluent* $(dw/du)_{\Pi}$. If $w = fu$ relative to Π and f is differentiable, then it is not difficult to show that $(dw/du)_{\Pi}$ is (essentially) $(Df)u$, where Df is the usual derivative of f.

5. Later Work on Variables and Fluents

For inclusion in his "Selected Papers" [M11], Menger chose the works that represent his last words on variables and fluents, namely: (1) The paper "Variables, constants, fluents" [M9] which, as he emphatically pointed out, "supersedes all preceding papers on variables"; (2) a combined version of the two "Counterparts of Occam's razor" papers [M7, M8]; (3) the second half of the paper [M10] (reprinted in Volume 1, pp. 575–593) which is, essentially, a recapitulation from the perspective of logical positivism of the main points of various earlier papers. In addition, he included the classic "Why Johnny hates math" [M4] which should be must-reading for every teacher of mathematics, and somewhat abridged versions of his "Gulliver" papers, whose pointed satire is – sad to say – almost as valid today as it was a half-century ago.

Although the paper [M9] succinctly summarizes the contents of previous papers on the subject, it also shows that Menger's analysis is incomplete and open to some criticism. Thus, the discussion of parameters is rather weak; and when illustrating the notions of equality, consistency and inconsistency of quantities, Menger says "If Smith weighs 160 pounds and has a bank account of \$160, then . . . the following three quantities are not only consistent, but *equal* because they have the same object and the same value: (Smith, 160), (Smith, his weight in pounds), (Smith, his account in dollars)". But no one would want to say that a number, a weight, and a bank account were the same thing because in certain units they happened to have the same value. Similarly, if a given volume of fluid were to exhibit, in appropriate units, equal pressures and temperatures through a particular stretch of time, no one would want to say that "pressure" and "temperature" were the same thing. Such objections were already raised by E. W. Adams in his "comments on Menger's 'Variables, constants, fluents'", which, together with Menger's "Rejoinder to Adams" are part of the paper [M9] but are not reproduced in [M11]. In his rejoinder (which is briefly summarized in [M11]), Menger steadfastly maintains that "the value of fluents, as I have defined them, are numbers – pure numbers and not so-called *denominate numbers* such as 3 feet . . .". The reason for this is of course the fact that Menger wants to use functions to combine the values of quantities; but it does seem to us that this comes close to laying applied mathematics upon a Bed of Procrustes.

In the papers [M7] and [M8], Menger considers his study of variables from a more philosophical point of view. 'Occam's Razor' or 'Law of Parsimony' is the maxim which states that "It is vain to do with more what can be done with fewer" or, equivalently, "Entities are not to be multiplied beyond necessity".[7] Menger's counterpart to the Law of Parsimony is what he calls a 'Law against Miserliness' which states that "It is vain to try to do with fewer what requires more" or, equivalently, "Entities must not be reduced to the point of inadequacy"; and as the counterpart to Occam's razor he introduces his prism or comb. In [M7] and [M8] he shows how these tools can be used to separate and/or untangle the multitudinous uses of the word "variable".

Now, a razor cuts a thing down to its barest essentials; but a prism or a comb works in the opposite way: the higher the resolution of the prism or the finer the teeth of the comb, the more constituent elements one finds. And thus, the more Menger looked, the more he found; the more he found, the more he wrote; and the more he wrote, the harder it became for readers to keep up with him. Add to this the fact that the times were (and still are) such that there is little reward (particularly for young scholars) for investigating matters such as those described in this chapter, and it is small wonder that Menger's elaborate machinery has found no takers, neither among mathematicians, scientists nor philosophers of science. Nevertheless, Menger deserves much credit for

---

[7] Cf. the discussion of William of Occam in Chapter 14 of [11].

venturing to undertake such a task and the scholarly community owes him a great debt. Anyone who takes the time to read some of his papers on variables and related matters will come away with deep and valuable insights into the semantics of mathematical symbolism and terminology, as well as with a more profound understanding of the relationship between pure mathematics and the way it is applied in real-world situations.

6. Conclusion

We can find no better ending to this commentary than the following excerpt from a letter to Karl Menger, written by the well-known logician-philosopher Ernest Nagel and dated November 9, 1960. (The book referred to is an expanded version of the booklet "Basic Concepts of Mathematics" that was never completed.)

"I hope you will not think me presumptuous in saying, despite my being just a philosopher rather than a professional mathematician, that I regard the book as a contribution partly to 'popular education', but more importantly to a needed intellectual housecleaning of central mathematical notions and/or mathematical language – a housecleaning from which advanced students as well as beginners could greatly profit. I have naturally been familiar with some of the ambiguities and unclarities to which you call attention in standard presentations of algebra, geometry and function theory; but until I read your discussion I did not quite realize the full enormity of the confusions and logical sins that continue to be committed in current textbooks in those subjects. The distinctions you introduce in connection with the miscellany of divers uses to which the term 'variable' is frequently put seem to me most valuable; and if you will permit me to say so, you show rare courage in exhibiting the persistent 'double talk' of which many mathematicians (along with many physicists) are guilty."

This letter is over 40 years old: but the comments are as timely and valid today as they were then!

References to Menger

M1. The Mathematics of Elementary Thermodynamics. American Journal of Physics 18 (1950) 89–103.

M2. Calculus. A Modern Approach, Ginn and Company, Boston, 1955.

M3. Random Variables from the Point of View of a General Theory of Variables, Proceedings of the Third Berkeley Symposium on Mathematical Statistics and Probability (1954–55), L. M. LeCam, J. Neyman (eds.), University of California Press, Berkeley 2 (1956) 215–229.

M4. Why Johnny Hates Math. The Mathematics Teacher 49 (1956) 578–584.

M5. What Are x and y? The Mathematical Gazette 40 (1956) 246–255.
M6. Rates of Change and Derivatives. Fundamenta Mathematicae 46 (1958) 89–102.
M7. A Counterpart of Occam's Razor in Pure and Applied Mathematics. Ontological Uses, Synthese 12 (1960) 415–428.
M8. A Counterpart of Occam's Razor in Pure and Applied Mathematics. Semantic Uses, Synthese 13 (1961) 331–349.
M9. Variables, Constants, Fluents. Current Issues in the Philosophy of Science, H. Feigl, G. Maxwell (eds.), New York, 1961, 304–318.
M10. Mathematical Implications of Mach's Ideas: Positivistic Geometry, The Clarification of Functional Connections. In: Ernst Mach, Physicist and Philosopher, Boston Studies in the Philosophy of Science VI, R. S. Cohen, R. J. Seeger (eds.), Reidel Dordrecht, 1970, 107–125.
M11. Selected Papers in Logic and Foundations, Didactics, Economics, Vienna Circle Collection, Vol. 10, Reidel, Dordrecht, 1979.
M12. Reminiscences of the Vienna Circle and the Mathematical Colloquium, Vienna Circle Collection, Vol. 20, L. Golland, B. McGuinness, A. Sklar (eds.), Kluwer, Dordrecht, 1994.

Other References

1. I. T. Adamson: Review of K. Menger: Calculus, a Modern Approach. Mathematical Gazette 39 (1954) 245–250.
2. I. T. Adamson: Transformation of integrals. American Mathematical Monthly 65 (1958) 590–596.
3. T. Apostol et al. (eds.): Selected Papers on Calculus. Mathematical Association of America, 1968.
4. W. L. Duren Jr.: Review of K. Menger: Calculus, a Modern Approach. Bull. Amer. Math. Soc. 61 (1955) 185–187.
5. T. Erber: Modern Calculus Notation Applied to Physics. Eur. J. Phys. 15 (1994) 111–118.
6. J. B. Harkin: The Pre-limit Calculus of K. Menger. Math. Teachers J. 23 (1973) 71–75.
7. A. N. Kolmogorov: Grundbegriffe der Wahrscheinlichkeitsrechnung, Springer-Verlag. Translated as: Foundations of the Theory of Probability. 2nd Ed., Chelsea 1956, 1933.
8. M. Loève: Probability Theory 1, 4th Ed., Springer-Verlag, 1977.
9. M. E. Munroe: Manipulations with Differentials Made Respectable. In: K. O. May (ed.), Lectures in Calculus. Holden-Day, 1967, 127–144.
10. A. Rapoport: Review of K. Menger: Calculus, a Modern Approach. ETC. 12 (1955) 137–144.
11. B. Russell: A History of Western Philosophy, Simon and Schuster, New York, 1945.
12. A. Sklar: Random Variables, Distribution Functions and Copulas. In: L. Rüschendorf, B. Schweizer, M. D. Taylor (eds.), Distributions with Fixed Marginals and Related Topics. CMS Lecture Notes 28 (1996) 1–14.
13. A. Tarski, S. Givant: A Formalization of Set Theory Without Variables. AMS Colloquium Publ. 41 (1987).
14. A. Wald: On the Principles of Statistical Inference, Notre Dame Mathematical Lectures. No. 1 (1942).
15. A. Wald: Sequential Analysis, Wiley, 1947.

A Mengerian Tour[1] Along Caratheodory's Royal Road

L. Senechal and B. Schweizer

Introduction

It is ironic that Caratheodory's work on the foundations of thermodynamics is better known today [1, 2] than his equally elegant formulation of the calculus of variations, once called "Königsweg" by specialists. This formulation is set forth in Chapters 12 and 13 of Caratheodory's book [3] and also discussed in detail in [4]. The seminal idea for this development is Hilbert's invariant integral (which in [5] is likened to the gauge transformation of electrodynamics), from which perspective the calculus of variations is so formulated that the Hamilton-Jacobi theory and Hamilton's canonical equations are developed in parallel with the Beltrami and Euler equations. Another striking feature is that sufficient conditions for extremization arise naturally and immediately, in contrast with the historical development, following Jacobi and Weierstrass, in which sufficiency is achieved much less directly [6].

Although he published nothing more on the calculus of variations after 1940, Menger's interest in the subject stayed with him throughout the remainder of his life. He lectured on it regularly, usually dividing a year-long course into two parts, the first devoted to the classical theory, the second to direct metric methods. As he lectured on the classical theory, he became more and more enamoured of Caratheodory's approach. He made additional simplifications, streamlined the notation, and put the matter in the context of his ideas on the algebra of functions and operators. In the end-product, briefly outlined below, superfluous variables are stripped away and the essential ideas take on a most elegant and simple form.

---

[1] This article is based on notes taken by the first-named author in a series of lectures given by Menger at the University of Arizona during the Fall Semester of 1964.

Functional Notation

The symbol $I^{(n)}$ denotes the identity mapping on $\mathbb{R}^n$. If the dimension is clear from the context, the superscript can be be omitted. The n coordinate (or *selector*) functions of $I^{(n)}$ are $I_1^{(n)}, \ldots, I_n^{(n)}$ or $I_1, \ldots, I_n$ if the dimension n is understood. In two dimensions, I_1, I_2 are denoted by I, J and in one dimension I_1 is denoted by j. Thus, for example, the function whose value for (x, y) is $\tan(x/\sqrt{x^2 + y^2})$ can be denoted by $\tan(I/\sqrt{I^2 + J^2})$. The advantages of Menger's functional notation will quickly become apparent as we pursue his axiomatized presentation of Caratheodory's formulation.

Preliminaries

The variational problem we consider is that of minimizing the integral

$$\int_a^b G(j, f, Df) \qquad (*)$$

for piecewise continuously differentiable functions f which satisfy the boundary conditions $f(a) = c$, $f(b) = d$ and are thus coterminal. We will furthermore assume that G is defined and continuously twice differentiable on $[a, b] \times \mathbb{R}^2$, and that G is *positively regular*[2] in the sense that the second order partial derivative $D_{33}G$ is strictly positive on that same set. These are severe restrictions, but encompass many of the classical problems such as the brachistochrone and minimal surface area of revolution problems. It will be obvious, furthermore, that the requirement on the domain of G can be relaxed and is made here only for simplicity's sake.

The Caratheodory Relations

Lemma. Suppose there exists a two-place function P such that

$$G(I, J, P) = 0,$$
$$D_3 G(I, J, P) = 0.$$

Then $(*)$ is minimized by every solution of the differential equation $Df = P(j, f)$, on $[a, b]$.

---

[2] The designation *regular* was made by Hilbert in 1900, but the importance of the concept had already been recognized by Legendre as early as 1786.

Proof. From the first of the equations above it is clear that the integral (∗) has value 0 for any solution of the differential equation, so suppose that g is defined on $[a, b]$ but does not satisfy the differential equation. Then there exists a number $c \in [a, b]$ such that $Dg(c) \neq P(c, g(c))$. The convexity of G with respect to its third place implies that the one-place function $D_3G(c, g(c), j)$ is increasing, so that the one-place function $G(c, g(c), j)$ achieves a minimum at $P(c, g(c))$, where its derivative is 0. Therefore

$$G(c, g(c), Dg(c)) > G(c, g(c), P(c, g(c))) = 0,$$

so that the integral (∗) is positive for g. □

Clearly, the conditions of the lemma are much too restrictive to be of interest in themselves. This situation changes radically if we subtract from the integrand of (∗) the exact form

$$D_1 T(j, f) + D_2 T(j, f) \cdot Df,$$

where T, like P, is a continuously differentiable two-place function. The integral of this quantity corresponds to Hilbert's invariant integral and is constant on families of coterminal functions f. For the new integrand, the hypothesis of the lemma yields

$$G(I, J, P) = D_1 T + D_2 T \cdot P, \qquad \text{(CR1)}$$

$$D_3 G(I, J, P) = D_2 T. \qquad \text{(CR2)}$$

We call these the *Caratheodory relations.* They provide far-reaching general sufficient conditions for minimization.

Theorem. If continuously differentiable two-place functions P, T exist that satisfy the Caratheodory relations, then each solution of the differential equation $Df = P(j, f)$ minimizes the integral (∗) relative to coterminal functions.

Proof. Define the three-place function

$$G^* = G - D_1 T(I_1, I_2) - D_2 T(I_1, I_2) \cdot I_3.$$

Then

$$D_3 G^* = D_3 G - D_2 T(I_1, I_2)$$

and

$$D_{33}G^* = D_{33}G,$$

so that G^* is positively regular and the Caratheodory relations imply that the conditions of the lemma are satisfied by G^*. □

The Beltrami and Euler Equations

We solve for D_1T, D_2T in the CR's. From the equality of the mixed second-order partial derivatives of T we obtain the *Beltrami equation*:

$$D_2G(I,J,P) = D_{13}G(I,J,P) + D_{23}G(I,J,P) \cdot P$$
$$+ D_{33}G(I,J,P) \cdot (D_1P + D_2P \cdot P). \qquad \text{(B)}$$

Since it is a first-order partial differential equation for the two-place function P, it can be reduced to a system of two first order ordinary differential equations or, equivalently, to a second order ordinary equation. In fact, if $Df = P(j, f)$, then it follows from (B) that

$$DD_3G(j,f,Df) = D_2G(j,f,Df), \qquad \text{(E)}$$

which is the *Euler equation*.

Thus the Beltrami equation for P gets replaced by an equivalent second-order ordinary equation: one-parameter families of solutions of the Euler equation which fill a simply-connected region without overlap determine particular choices for solutions P of (B).

Once we have a P that satisfies (B) we can find T by integrating the second CR relation. Integrability is assured by the very nature of the Beltrami equation as an exactness condition. In this way we see that, for the hypotheses under which we are working, one-parameter families of solutions of the Euler equation which fill a simply-connected region without overlap consist entirely of minimizing functions.

The Hamilton-Jacobi Equation

The positive regularity condition on G implies that D_3G is invertible with respect to its third place, so that we can solve (CR2) for P as a function of I, J, D_2T. If we let $S = Inv_3D_3G$, then

$$P = S(I, J, D_2T).$$

Substituting into (CR1) we obtain

$$G(I, J, S(I, J, D_2T)) = D_1T + D_2T \cdot S(I, J, D_2T).$$

If we define

$$H = -G(I_1, I_2, S) + I_3 \cdot S,$$

then we have

$$D_1T + H(I, J, D_2T) = 0, \qquad \text{(H-J)}$$

which is the *Hamilton-Jacobi equation*. The *Hamiltonian function H* is the *Legendre transformation* of G with respect to the third place of G. The Legendre transformation is involutory, so that G is the Legendre transformation of H with respect to the third place of H.

Once we have a solution T of (H-J) we can compute the corresponding P as follows. We have

$$\begin{aligned} D_3H &= -D_3G(I_1, I_2, S) \cdot D_3S + S + I_3 \cdot D_3S \\ &= S, \end{aligned}$$

since $D_3G(I_1, I_2, S) = I_3$ by the definition of S. From the second CR relation we have therefore

$$D_3H(I, J, D_2T) = S(I, J, D_2T) = P.$$

The equation $P = D_3H(I, J, D_2T)$ is called the *T-subsidiary equation*.

Example. Let g be a continuous and positive one-place function and let $G = g(I_2) \cdot \sqrt{1 + I_3^2}$. Clearly $D_{33}G > 0$, so G is positively regular. We obtain

$$S = \frac{I_3}{\sqrt{g^2(I_2) - I_3^2}},$$

and

$$H = -\sqrt{g^2(I_2) - I_3^2}.$$

The Hamilton-Jacobi equation is

$$D_1 T - \sqrt{g^2(J) - (D_2 T)^2} = 0,$$

and the T-subsidiary equation is

$$P = \frac{D_2 T}{\sqrt{g^2(J) - (D_2 T)^2}}.$$

Particular solutions of the former are obtained by setting $D_1 T$ equal to a constant[3] two-place function C and by computing

$$D_2 T = \pm \sqrt{g^2(J) - C^2}.$$

Solutions to the latter are then computed as

$$P = \pm C^{-1} \sqrt{g^2(J) - C^2}.$$

Hamilton's Canonical Equations

The partial differential H-J equation can, like the B equation, be reduced to equivalent first-order ordinary equations. First we note that

$$D_2 H = -D_2 G(I_1, I_2, S) - D_3 G(I_1, I_2, S) \cdot D_2 S + I_3 \cdot D_2 S$$
$$= -D_2 G(I_1, I_2, S).$$

Applying D_2 to the H-J equation, we obtain

$$D_2 H(I, J, D_2 T) + D_3 H(I, J, D_2 T) \cdot D_{22} T + D_{21} T =$$
$$= D_2 H(I, J, D_2 T) + P \cdot D_{22} T + D_{12} T = 0.$$

Let f satisfy $Df = P(j, f)$ and set $g = D_2 T(j, f)$. Then, by substituting (j, f) into the two places of the above functions, we obtain further

$$D_2 H(j, f, D_2 T(j, f)) + P(j, f) \cdot D_{22} T(j, f) + D_{21} T(j, f) =$$
$$= D_2 H(j, f, g) + Dg = 0.$$

---

[3] The first-named author has shown that by making a different choice for $D_1 T$ it is possible to obtain the conjugate point theory of Jacobi for the class of examples considered here.

Combined with the T-subsidiary equation, we thus have *Hamilton's canonical equations*

$$Df = D_3H(j,f,g),$$
$$Dg = -D_2H(j,f,g).$$

By its definition, the first component f of a solution of these equations is also a solution of the Euler equation.

Extensions and Applications

All of the above considerations can be extended in a straightforward way to vector-valued functions. We need only replace the positiveness for the partial derivative $D_{33}G$ in the definition of positive regularity by positive definiteness of the Hessian of the vector-valued second partial derivative.

Menger, in his lectures, initiated a treatment of analytical mechanics based on the Caratheodory framework, but this must be regarded as an unfinished task.

References

1. S. Chandrasekhar: An Introduction to the Study of Stellar Structure, Dover Publications, New York, 1957.
2. W. G. V. Rosser: An Introduction to Statistical Physics, Halsted Press, New York, 1982.
3. C. Caratheodory: Variationsrechnung und partielle Differentialgleichungen erster Ordnung, Teubner, Leipzig and Berlin, 1935.
4. M. Giaquinta, S. Hildebrandt: Calculus of Variations, vol. 1, Springer-Verlag, Berlin, 1996.
5. H. Rund: The Hamilton-Jacobi Theory in the Calculus of Variations, D. Van Nostrand, London, 1966.
6. H. Sagan: Introduction to the Calculus of Variations, McGraw-Hill, New York, 1969.

TO THE INSTRUCTOR AND GENERAL READER*

What Is A Variable?

The conceptual and semantic clarification of calculus is centered on the analysis of the hitherto obscure general term "variable," which is resolved into an extensive spectrum of well-defined meanings.

The only clear (if one-sided) definition heretofore formulated goes back to Weierstrass who, in his celebrated lectures in the 1880's, defined it as a symbol that stands for any number or any element belonging to a certain class of numbers. Bertrand Russell, who at the turn of the century investigated the various aspects of variables probably more thoroughly than anyone before him, said: "Variable is perhaps the most distinctly mathematical of all notions; it is certainly also one of the most difficult to understand ... and in the present work [*The Principles of Mathematics*, 1903] a satisfactory theory as to its nature, in spite of much discussion, will hardly be found." In fifty years this situation has not been improved.

In this book a solution of the problem is attempted by distinguishing and making precise various equally important uses of the term "variable" in pure and applied calculus. Some of these variables differ from each other as profoundly as do trigonometric and geometric tangents. But whereas no one has, on account of a flimsy equivocation, confused tangents of angles and tangents of curves, this book seems to be the first to maintain clear distinctions between the following three concepts:

I. Variables according to Weierstrass, herein called *numerical variables*, as x and y in

(1) $x^2 - 9y^2 = (x + 3y) \cdot (x - 3y)$ for any two numbers x and y.

Here, as throughout this book, the numerical variables are printed in roman type. Without any change of the meaning, x and y may be replaced by any two non-identical letters, e.g., by a and b or by y and x; that is to say, two numerical variables may be interchanged:

$y^2 - 9x^2 = (y + 3x) \cdot (y - 3x)$ for any two numbers y and x

is tantamount to (1). In calculus, numerical variables may be used or they may, as will be shown herein, be dispensed with.

II. Variables or variable quantities in the sense in which scientists use these terms; for instance, *t*, the time; *s*, the distance traveled (in chosen units); *x* and *y*, the abscissa and ordinate in a physical or postulational plane (relative to a chosen frame of reference); etc. These "variables" are defined and thoroughly discussed in Chapter VII under

---

*The beginner should start on page 1 or the one facing it.

the names of *consistent classes of quantities* and – reviving Newton's terminology – of *fluents*. They are herein consistently denoted by letters in italic type. Fluents cannot be dispensed with in formulas expressing scientific laws, such as Galileo's

(2) $s = 16t^2$.

Nor can they be interchanged:

$$2x + 3y = 5 \quad \text{and} \quad 2y + 3x = 5$$

are different lines. (If, on the other hand, in pure analytic geometry, the first of these two lines is defined as

the class of all pairs (x, y) of numbers such that 2x + 3y = 5,

where x and y are numerical variables, then

the class of all pairs (y, x) of numbers such that 2y + 3x = 5

is an equivalent definition.)

III. Variables in the sense of u and w in statements such as

(3)   If $w = 16u^2$, then $\dfrac{dw}{du} = 32u$ for any two fluents, u and w.

These "variables" belong to a third type, first explicitly introduced by the author in 1952 (see Bibliography). They are herein referred to as *fluent variables*, since they partake in characteristics of numerical variables as well as of fluents. In (3), u and w may be replaced by any two elements of a certain class – but not by two numbers. If u were replaced by 3, and w by 144, then the antecedent $144 = 16 \cdot 3^2$ would be valid, and yet the consequent $\dfrac{d\,144}{d\,3} = 32 \cdot 3$, utterly nonsensical. What u and w in (3) may be replaced by are fluents, such as t and s regarding a motion, or x and y along a plane curve:

(3′)   If $s = 16t^2$, then $\dfrac{ds}{dt} = 32t$;

(3″)   If $y = 16x^2$, then $\dfrac{dy}{dx} = 32x$.

About the Notation

Obviously, the preceding clarifications have nothing to do with notation. But it may be pointed out that, in the literature, the equivocations of the term "variable" are accentuated by the use of the same letters (mainly, x and y in italic type): (1) as numerical variables; (2) to denote specific consistent classes of quantities or fluents, namely abscissa and ordinate; (3) as fluent variables; indeed, (3″) often stands for (3).

In formulas such as

$$(4) \qquad \mathbf{D} \sin x = \cos x \quad \text{and} \quad \int_0^{\pi/2} \cos x \, dx = [\sin x]_0^{\pi/2},$$

x can be simply shed; and the formulas may be replaced by

$$(4') \qquad \mathbf{D} \sin = \cos \quad \text{and} \quad \int_0^{\pi/2} \cos = [\sin]_0^{\pi/2}.$$

In fact, there is a trend toward actually shedding x after the letters f and g — symbols that stand for any function or any element of a certain class of functions and which, therefore, have been called *function variables*. In several more advanced books on analysis one can read

$$(5) \qquad \text{If } \mathbf{D}f = g, \quad \text{then} \quad \int_a^b g = [f]_a^b.$$

Only the second formula in (5) contains numerical variables (namely, a and b) and g may be replaced by any function that is continuous between a and b.

Why then is x in (4) usually retained, and, even in more advanced books, reintroduced in applications of (5) to specific functions? The reason is one of the great curiosities in the history of modern analysis — comparable only to the lack of a symbol for the number zero in ancient arithmetic. The function that is perhaps more important than any other — the *identity function* assuming the value x for any x — lacks a traditional symbol. Therefore, in contrast to (4), in

$$(6) \qquad \mathbf{D} \log x = 1/x \text{ for } x > 0, \quad \mathbf{D}x^3 = 3x^2, \quad \mathbf{D}(\tfrac{1}{2}x^2) = x,$$

x cannot be shed. One obviously would not write

$$\mathbf{D} \log = 1/\,, \qquad \mathbf{D}\,^3 = 3\,^2, \quad \mathbf{D}(\tfrac{1}{2}\,^2) = \,,$$

and this is why, for the sake of uniformity, x is also retained in (4). Traditionally, the identity function is referred to by its value for x as "the function x" — incidentally, a fourth current meaning of the italic letter x.

There is, however, another way to preserve uniformity, namely, by rescuing the identity function from anonymity. If one denotes it by j (so that $j x = x$ and $j^n x = x^n$ for any x), then (6) reads

$$(6^*) \qquad \mathbf{D} \log x = j^{-1} x \text{ for } x > 0, \quad \mathbf{D}j^3 x = 3j^2 x, \quad \mathbf{D}(\tfrac{1}{2}j^2)x = j x;$$

and in (6*), x can be shed as in (4):

$$(6') \quad \mathbf{D} \log = j^{-1} \text{ (on the class P of all positive numbers)},$$

$$\mathbf{D}j^3 = 3j^2, \quad \mathbf{D}(\tfrac{1}{2}j^2) = j.$$

The formulas thus obtained (or, if one pleases, the results of calculus "in this notation") connect the functions themselves rather than their values and are of algebraic beauty – as it were, *streamlined*. Yet it is a matter of taste which one prefers: (4') and (6') or (4) and (6). But it is a fact (proved at the end of Chapter IV) and not a mere matter of taste that only if the identity function is granted the same status that *log* and *cos* have enjoyed for centuries can calculus be (to use another term that is in vogue) completely *automatized*. Automation is impossible "in the classical notation." In the automatized calculus one obtains specific results from general theorems by replacing function variables by the designation of specific functions, and numerical variables by the designations of specific numbers. For instance, from (5) one obtains:

$$\text{If } \mathbf{D} \ log = j^{-1}, \text{ then } \int_a^b j^{-1} = [log]_a^b.$$

Since, according to (6'), actually $\mathbf{D} \ log = j^{-1}$ on the class P, one concludes that $\int_a^b j^{-1} = [log]_a^b$, where the numerical variables a and b may be replaced by any two numbers in the class P. For instance,

$$\int_1^2 j^{-1} = [log]_1^2 \ (= log \ 2 - log \ 1 = log \ 2).$$

A symbol for the identity function was introduced, as early as 1924, by the Russian logician Schönfinkel and has been adopted by H. B. Curry in his work on combinatoric logic. But whereas in topology extensive use has been made of a symbol for the identity mapping, that for the identity function had not, before the publication of the booklet "Algebra of Analysis" (see Bibliography), found its way into mathematical analysis – in particular not even into Curry's own mathematical papers.

To the specialist, perhaps more interesting is the automation of the theory of what, traditionally, is called the theory of functions of two variables. Most important are the functions assuming for every pair (x, y) the values x and y. Traditionally, they are referred to by their values as "the function x" – a fifth meaning of x – and "the function y." In an automatized calculus, they cannot remain anonymous. In Chapter XI they are called the *selector functions* and are denoted by I and J so that

I (x, y) = x and J (x, y) = y for any x and y.

Chapter XI contains the theories of partial derivatives and partial integrals. Calculus thus developed as an algebra of functions synthesizes the spirit of the oldest logic and that of the most modern machine age.

Extensions of the ideas here expounded to *complex functions*, to the theory of *operators*, to *random variables*, etc. will be given in other publications.

Applications to Science

Perhaps the most important automation is that of the applications of calculus to science. An example is the application of derivatives of functions to rates of change of one fluent with respect to another. The scheme of these applications is the following statement: If w is connected with u by the function f, then the rate of change $\dfrac{d\,w}{d\,u}$ is connected with u by the derivative $\mathbf{D}\,f$; or

(7) $$\text{If } w = f(u), \text{ then } \frac{d\,w}{d\,u} = \mathbf{D}\,f(u).$$

Traditionally, "derivative" and "rate of change" are considered as synonyms. But the former operator associates a function with every function of a certain kind; the latter, a fluent with certain pairs of fluents. If, in (7), the function variable f is replaced by sin, the result is

$$\text{If } w = sin\ u, \text{ then } \frac{d\,w}{d\,u} = \mathbf{D}\ sin\ u.$$

Hence, by virtue of $\mathbf{D}\ sin = cos$,

(8) $$\text{If } w = sin\ u, \text{ then } \frac{d\,w}{d\,u} = cos\ u.$$

(Incidentally, (3) results from (7) by virtue of $\mathbf{D}(16\,j^2) = 32\,j$.) Replacing the fluent variables u and w by the time and the position of a linear oscillator and calling v its velocity one obtains

(8') $$\text{If } s = sin\ t, \text{ then } \frac{d\,s}{d\,t} = cos\ t \text{ or } v = cos\ t.$$

If in physics one ascertains the validity of $s = sin\ t$ for an oscillator, then he can conclude that for this oscillator $v = cos\ t$.

The contributions of mathematics ($\mathbf{D}\,sin = cos$) and of physics ($v = \dfrac{d\,s}{d\,t}$ and $s = sin\ t$) are completely separated, whereby the roles of these two sources of insight into nature are greatly clarified. (7) — it may be repeated — is the general scheme for applying the calculus of derivatives to science. There is an analogous and, in fact, more important scheme for the application of integral calculus to consistent classes of quantities. In Chapter XI, these ideas are extended to partial rates of change and partial derivatives, and are applied to thermodynamics thereby preparing the basic mathematical tools needed in physical chemistry.

On the basis of pure observations alone, all that physics can claim about a falling object is that s equals $16t^2$ *within certain limits* and that v equals $32t$ *within certain limits*. What is the logical connection between these two statements? In Chapter VII it will be shown that, if $\sim$ denotes "is equal within certain limits," then

$v \sim 32t$ (in presence of proper initial conditions) implies $s \sim 16t^2$;

but that $s \sim 16t^2$ does not, conversely, imply $v \sim 32t$; in fact, $s \sim 16t^2$ *does not permit any inference whatever concerning v.* In the realm of strictly observational statements, inferences "by integration" are highly significant, whereas inferences "by differentiation," while often extremely important heuristically, are not of logical character.

Calculus is a theory of pure functions just as arithmetic is a theory of pure numbers. There are analogies between the two in the realms of applications as well as of pure theory, both on the specific and on the general levels.

	CALCULUS	LEVEL	ARITHMETIC
A P P L I E D	If $s = \sin t$, then $\dfrac{ds}{dt} = \cos t$	Specific	If 1 ft. = 12 in., then 3 ft. = 36 in.
	$\dfrac{d \sin u}{d u} = \cos u$ for any fluent u	General	$3 \cdot 12\,a = 36a$ for any object a
P U R E	**D** $\sin = \cos$	Specific	$3 \cdot 12 = 36$
	D $f^2 = 2f \cdot$ **D** f for any function f	General	$(a + 1)^2 = a^2 + 2a + 1$ for any number a.

But there is a difference in attitude. No arithmetician stoops to references to objects, whereas some analysts refuse to raise their eyes above the level of fluent variables. Yet the transition to the pure level does not in any way jeopardize the applicability of arithmetic to feet and inches or of calculus to time and position, nor does the shedding of numerical variables in calculus. The formulas

$$\mathbf{D}\ \sin x = \cos x \text{ for any x and } \frac{d \sin u}{d u} = \cos u \text{ for any } u,$$

which are often considered as having the same meaning, really belong to altogether different realms (one using a numerical variable, the other a fluent variable), but both follow from **D** $\sin = \cos$, – the first by evaluation, the second by virtue of (7). Nothing is lost by writing simply **D** $\sin = \cos$ or $\sec = 1/\cos$. In fact, as will be shown in Chapter VII, it is the latter formulas (and not those obtained by inserting numerical variables) that are the perfect analogue of the traditional formulas

$$\frac{ds}{dt} = \cos t \text{ and } v = 1/p \text{ in physics.}$$

Students' Questions

Within the traditional conceptual frame of calculus it is exceedingly difficult to answer questions pertaining to some classical definitions, statements, and formulas such as the following:

The function x is the function assuming for any x the value x; or the class of all pairs (x, x).

The line $2x + 3y = 5$ is the class of all pairs (x, y) such that $2x + 3y = 5$, which is the same as the class of all pairs (y, x) such that $2y + 3x = 5$, but quite different from the line $2y + 3x = 5$.

In algebra, $\dfrac{x^2 - 1}{x - 1} = x + 1$, whereas in calculus, the functions $\dfrac{x^2 - 1}{x - 1}$ and $x + 1$ are not identical.

For falling objects, $s = s(t) = s(v)$.

The expressions

$$\int_a^b f(x)\,dx \quad \text{and} \quad \int_a^b f(t)\,dt$$

are not only numerically equal but have precisely the same meaning. On the other hand, if p denotes force, then

$$\int_a^b p\,ds \quad \text{and} \quad \int_a^b p\,dt,$$

and, if one writes $p = p(s) = p(t)$, then even

$$\int_a^b p(s)\,ds \quad \text{and} \quad \int_a^b p(t)\,dt$$

have not only different meanings (work and impulse) but are, in general, numerically different.

Questions concerning such statements naturally arise in the minds of the very best students. Some of them may be not only disappointed but discouraged in the pursuit of mathematical studies if they do not receive clear answers. Following the approach described in this book, teachers can readily answer these and similar questions.

CHAPTER I

THE TWO BASIC PROBLEMS OF CALCULUS
AND THEIR SOLUTIONS FOR STRAIGHT LINES

In Fig. 1 there appears a horizontal line O and a simple curve f. A curve is said to be *simple* if it intersects no vertical line (i.e.,no line perpendicular to O) in more than one point. The letter S is not a simple curve — a fact illustrated by any $-sign; neither is a circle. The upper half and the lower half of a circle, taken separately, are simple curves.

On the line O, a point 0 and a point 1 have been chosen. The segment between them is called a linear unit. The point which is 2 (or any number, a) units to the right of 0 is called the point 2 (the point a). Vertical distances are measured in the linear unit between 0 and 1 turned through a right angle; areas, in square units such as that marked above the segment from 0 to 1.

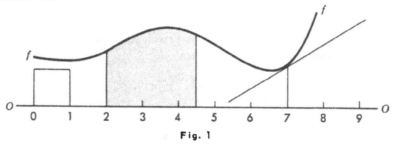

Fig. 1

A domain has been shaded. It is bounded by a portion of the curve f, the segment from 2 to 4.5 on O, and two lateral vertical segments. WHAT IS THE AREA OF THIS DOMAIN? In other words, HOW MANY SQUARE UNITS ARE CONTAINED IN THIS DOMAIN? This question is a typical example of the first basic problem to which this book is devoted. The number in question will be denoted by the symbol $\int_{2}^{4.5} f$ (read: integral of f from 2 to 4.5). If a and b are any two points on O belonging to the projection of f, then

$$\int_{a}^{b} f \quad \text{(read: the } integral \text{ of } f \text{ from a to b)}$$

denotes the area of the domain above O and below f from a to b — briefly, *the area under f from a to b*.

In Fig. 1, at the point of f above 7, the tangent to that curve has been drawn. HOW STEEP IS THAT TANGENT? In other words, HOW MANY

1

UNITS DOES THE TANGENT RISE VERTICALLY FOR EACH HORI-
ZONTAL UNIT? This question is an example of the second basic problem to be treated in this book. The number in question is called the
derivative of f at 7 and will be denoted by $Df7$. If x is any point on the
projection of f on O, then

$$Dfx \quad \text{(read: the derivative of } f \text{ at } x)$$

denotes the steepness or slope of the tangent to f above the point x –
briefly, *the slope of f at x.*

Approximate solutions of area problems are as old as geometry. For a
few special curves, Archimedes (about 250 B.C.) found precise areas and
slopes, but in a cumbersome way. The systematic study of the two problems began about 1700, when Newton in England and Leibniz in Germany
discovered an exact method applicable to many curves. Their ingenious
theory has become known as CALCULUS.

In the attempt to understand and control Nature man is faced with innumerable scientific questions and engineering problems that can be reduced to the determination of areas under simple curves and slopes of
tangents. Since these geometric problems can be solved by calculus,
calculus has become one of the most powerful of scientific tools and a
major factor in the development of modern civilization.

Before the basic problems for curves like f in Fig. 1 are attacked, they
will be solved for the most primitive curves imaginable, namely, for
straight lines; more specifically, the area under a horizontal line and the
slope of a nonvertical line will be determined. In these simple situations,
first numerical and then graphical solutions of the problems are presented;
finally the solution is given that is based on the idea of Newton and
Leibniz.

1. NUMERICAL SOLUTIONS

a. **Area.** In Fig. 2(a), the horizontal line is drawn that is .5 units above
the line O; it is denoted by .5; in Fig. 2(b), the horizontal line $-.5$ is .5
units below O. More generally, if c is any number, the horizontal line that
is c units above the line O will be called *the line c* – briefly, c. Clearly,
O is the line at the altitude 0. (.5, $-.5$, c for lines are italic type.)

The domain below the line .5 and above O from 2 to 5 is shaded. The

domain is a rectangle of height .5 and of base 5 – 2. Thus $\int_2^5 .5$ (which

denotes the area of this domain) is .5 (5 – 2) = 1.5 square units. More
generally,

If c is the horizontal line of the altitude c, then $\displaystyle\int_a^b c = c \cdot (b-a)$, for any a, b, and c. For instance,

$$\int_1^3 2.5 = 2.5 \cdot (3-1) = 5; \quad \int_1^3 0 = 0; \quad \int_2^5 (-.5) = (-.5) \cdot (5-2) = -1.5.$$

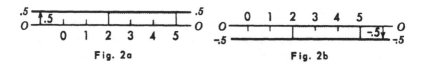

Fig. 2a   Fig. 2b

Fig. 2(b) illustrates the last equality. The shaded rectangle of 1.5 square units below O and above the line $-.5$ is said to be -1.5 square units above O and below the line $-.5$.

That the area of this rectangle is negative cannot be proved (nor, of course, disproved) just as it cannot be proved (or disproved) that its altitude is negative. Both statements express conventions concerning the use of the words area and altitude. The conventions are universally adopted because they are useful in distinguishing between the altitudes of the shaded rectangles in Fig. 2(a) and (b), and between the areas of these two rectangles. Without these conventions both rectangles would have the altitude .5 and the area 1.5.

b. Slope. In Fig. 3, l denotes the straight line that has the altitude 1 above the point 2, and the altitude 2.5 above the point 5 on O. These facts are conveniently expressed in the formulas

$$l\,2 = 1 \quad \text{and} \quad l\,5 = 2.5.$$

More generally, the altitude of l above any point a on O — briefly, the *altitude of l at a* — will be denoted by $l\,a$.

The difference $l\,5 - l\,2$ is called the *rise* of l from 2 to 5, and is denoted by $[l]_2^5$. More generally, if a and b are any two numbers, then $l\,a$ and $l\,b$ are numbers and $[l]_a^b = l\,b - l\,a$.

By the slope of l at 2 (which, according to what was said on p. 2, will be denoted by $\mathbf{D}\,l\,2$) is meant the ratio of the rise of l from 2 to any other point, say 5, to the distance on O from 2 to that point; in a formula,

$$\mathbf{D}\,l\,2 = \frac{[l]_2^5}{5-2}.$$

Clearly, $\dfrac{[l]_2^5}{5-2} = \dfrac{1.5}{3} = .5$. Hence $\mathbf{D}\,l\,2 = .5$.

From similar triangles in Fig. 3 it is clear that the same ratio is obtained if 5 is replaced by 4 or 6; $\dfrac{[l]_2^4}{4-2} = \dfrac{[l]_2^6}{6-2} = .5$.

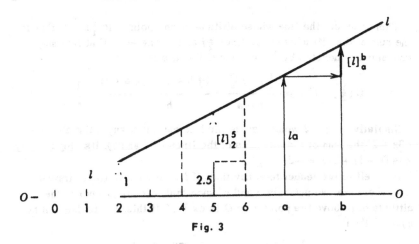

Fig. 3

Moreover, it is clear that, for any two different points a and b,

$$\frac{[l]_a^b}{b-a} = \frac{[l]_2^b}{b-2} = \frac{[l]_2^5}{5-2} = .5.$$ Hence, for any a on O and for any b ≠ a,

$$D l a = \frac{[l]_a^b}{b-a} = .5.$$ Similarly one sees:

If *m* is any nonvertical straight line and a is any point on O, then

$$D m a = \frac{[m]_a^b}{b-a} = \frac{mb-ma}{b-a}, \quad \text{for any } b \neq a, \text{ and}$$

$$D m x = D m a, \quad \text{for any point x on } O.$$

For instance, let *j* denote (*as it will throughout this book*) the line having the altitude 0 at 0, the altitude 2 at 2, the altitude −3 at −3 — more generally, the line which, for any number x has the altitude x above the point x on O; in a formula, *j* is the line for which

$$j x = x \quad \text{for any x.}$$

Then

$$D j a = \frac{[j]_a^b}{b-a} = \frac{jb-ja}{b-a} = \frac{b-a}{b-a} = 1 \quad \text{for any a and any b} \neq a.$$

Next consider the line having, for any number x, the altitude $\frac{1}{2}x$ above the point x. This line (which, incidentally, is the line l in Fig. 3) may be

denoted by $\frac{1}{2}j$. Its slope at any a is $D(\frac{1}{2}j)a = \dfrac{[\frac{1}{2}j]_a^b}{b-a} = \dfrac{\frac{1}{2}b-\frac{1}{2}a}{b-a} = \dfrac{1}{2}.$

Clearly, $D(\frac{1}{2}j)x = \frac{1}{2}$ at any x.

Then consider the line whose altitude at any point x is $\frac{1}{2}$x + 1, that is, the sum of the altitudes of the lines $\frac{1}{2}j$ and 1, wherefore that line may be conveniently denoted by $\frac{1}{2}j + 1$. Its slope at a is

$$\mathsf{D}\,(\tfrac{1}{2}j+1)\,a \;=\; \frac{[\tfrac{1}{2}j+1]_a^b}{b-a} \;=\; \frac{(\tfrac{1}{2}b+1)-(\tfrac{1}{2}a+1)}{b-a} \;=\; \frac{1}{2}\,.$$

Similarly, $-3j + 2$ would denote the line having at any x the altitude $-3x + 2$ (the sum of the altitudes of the lines $-3j$ and 2). Its slope at any x is $\mathsf{D}\,(-3j+2)\,x = -3$.

It is left to the reader to verify that, if l is any nonvertical straight line, there are two numbers c and c′ such that, for any number x, the altitude of l above the point x on O is cx + c′. If this line is denoted by $cj + c'$, then

$$\mathsf{D}\,(cj+c')\,x \;=\; c \quad \text{for any x on } O.$$

EXERCISES

INTRODUCTORY NOTE. On the preceding pages, numerals and letters in roman type (0, 1, 7, a, c, x, etc.) designate numbers or points on the line O; whereas numerals and letters in *italic* type designate simple curves (the horizontal lines O, $.5$, c, etc., the nonvertical lines l, j, etc., the curve f). *This typographical distinction will be maintained throughout this book.* If in longhand (on paper or on a blackboard) the distinction between roman and italic type should be difficult, one may denote

numbers and points on O by $O, 1, 7, a, c, x$;

simple curves by $\underline{O}, \underline{.5}, \underline{c}, \underline{l}, \underline{j}, \underline{f}$ (underlined).

If one avoids j and f as designations of numbers, then he may omit the underlining in the designation of the line j and the curve f. But one must not omit it in the symbols for the horizontal lines O, $.5$, and c, since 0, .5, and c designate numbers.

The letters x and y will be avoided as designations of altitudes of horizontal lines. The reason is that lines with the altitudes x and y would have to be denoted by x and y, whereas we wish to reserve the letters x and y (in italics) for concepts of analytic geometry discussed in Chapter VII.

Finally, it should be noted that, in this book, the symbols

$$\mathsf{D}\,3,\quad \mathsf{D}\,c,\quad \int_2^4 3,\quad \int_a^b c$$

remain undefined, and therefore meaningless. A number has neither a slope, nor an area, nor a weight, nor a color, ...

1. On cross-section paper, draw figures illustrating

(a) $\displaystyle\int_2^5 .5 = \int_5^8 .5 = \int_{-1}^2 .5 = \int_{-4}^{-1} .5 = 1.5$;

(b) $\displaystyle\int_0^3 (-4) = \int_{-2}^1 (-4) = \int_{-4}^{-1} (-4) = -12$;

(c) $\displaystyle\int_1^3 0 = 0$ and $\displaystyle\int_1^1 3 = 0$.

2. Show that $\displaystyle\int_a^b c = \int_{a+k}^{b+k} c$ and $\displaystyle\int_a^b c + \int_b^{b'} c = \int_a^{b'} c$.

*3. Apply the formula $\displaystyle\int_a^b c = c(b-a)$ to cases where b is less than a

(that is, the point b on 0 is to the left of the point a). For instance, what

are $\displaystyle\int_5^2 .5, \int_3^1 2.5, \int_3^1 0, \int_5^2 (-.5)$? When is $\displaystyle\int_a^b c$ positive and

when negative?

4. Determine the numbers c_1 and c_0 in such a way that $c_1 j + c_0$ is the
line having

(a) the altitude 1 at -1, and the altitude 5 at 1;

(b) the altitudes a′ at a and b′ at b, where a, b, a′, b′ are any four given
    numbers (a $\neq$ b);

(c) the slope 3 and the altitude 2 at 1;

(d) the altitude 4 at both 1 and 3;

(e) the altitude 0 at a, and the slope c;

(f) the altitude c at a, and the rise r from a to b.

5. On cross-section paper, draw figures illustrating $D(2j+3)a = 2$ for
(a) a = 1.5 using b = 3; (b) a = -3 using b = -1; (c) a = -1 using
b = 2. Then illustrate $D(2j+3)(-1) = 2$ using any point b.

6. Show that $D(-2j+3)2 = D(-2j+4)2 = D(-2j-1)2$ and illustrate
this equality on cross-section paper.

2. GRAPHICAL SOLUTIONS

a. Slope and Slope Line. The slope of a nonvertical line can be determined graphically. No knowledge of arithmetical division (used in the

numerical solution of the slope problem) is presupposed. However, parallel lines have to be drawn (an operation not needed in the numerical solution).

In Fig. 4, the slope of the line l at a is determined as follows: Through the point a on O, the segment m parallel to l is drawn, whose projection on O has unit length, that is, extends from a to a + 1. The triangles in Fig. 4 are similar. In the large triangle, the height is $[l]_a^b$, and the base is b − a. In the small triangle, the height is $m(a + 1)$ (that is, the altitude of m at a + 1), and the base is (a + 1) − a = 1. Hence the proportion

$$\frac{[l]_a^b}{b-a} = \frac{m(a+1)}{1}.$$

The quotient on the right side is equal to $m(a + 1)$. The quotient on the left side is equal to the slope $D\,l\,a$. Consequently

$$D\,l\,a = m(a + 1).$$

Hence the following

Graphical Solution of the Slope Problem for Nonvertical Lines. *To find the slope of l at a, construct m as indicated, and measure the altitude of m at a + 1.*

In Fig. 4 there appears also the horizontal line of altitude $D\,l\,a$, which passes through the terminal point of m. This line will be called

the *slope line* of l or the *derivative* of l − briefly, $D\,l$.

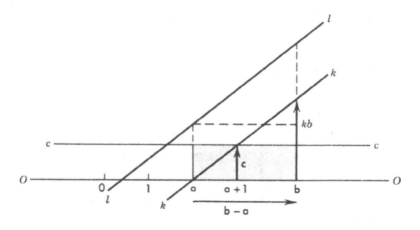

Fig. 4

The horizontal line $D\,l$ has the same altitude at any point of O, just as the line l has the same slope at any point. The slope line of the horizontal line 3 is the line O; in a formula, $D\,3 = O$. More generally, $D\,c = O$ for any horizontal line c.

Since, if l is a line, $l3$ and la (for any a) are numbers, $D(l3)$ and $D(la)$ are undefined symbols. Therefore $Dl3$ cannot mean anything but the slope of l at 3 or the altitude, $(Dl)3$, at 3 of the slope line Dl. This is why, from the outset, symbols such as $Dl2, D(\frac{1}{2}j)x$, etc. have been used and could be used without any ambiguity.

b. Area and Area Lines. $\displaystyle\int_a^b c$, the area of the shaded domain under the horizontal line c from a to b in Fig. 5, can also be determined graphically. Draw the line k joining the point a on O to the point on the line c above the point a + 1 on O. The right triangles in Fig. 5 whose hypotenuses lie on the line k are similar. In the small triangle, the height is c linear units,

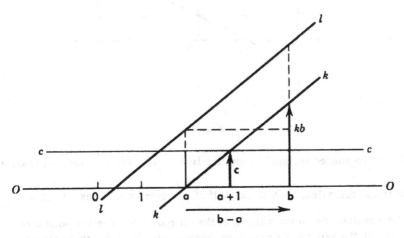

Fig. 5

and the base is $(a + 1) - a = 1$ unit. In the large triangle, the height is kb linear units (the altitude of the line k at b), the base is $b - a$ units. Because of the similarity of the triangles, the four numbers kb, $b - a$, c, and 1 are in the proportion $\dfrac{kb}{b-a} = \dfrac{c}{1}$. The quotient on the right is equal to c. If both sides of the equality $\dfrac{kb}{b-a} = c$ are multiplied by the number $b - a$, the result is $kb = c \cdot (b - a)$. Since $c \cdot (b - a) = \displaystyle\int_a^b c$, it follows that

$$\int_a^b c = kb;$$

that is to say, the number of square units in the shaded rectangle is equal to the number of linear units in the altitude of k above the point b. Thus,

to find $\int_a^b c$, construct the line k as indicated and measure its altitude at b. The situation is particularly clear on cross-section paper. E.g., in

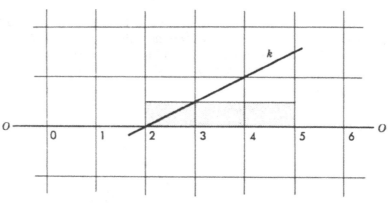

Fig. 6

Fig. 6 the shaded rectangle under the line .5 from 2 to 5 obviously consists of three rectangles, each of one half of a square unit. Thus $\int_2^5 .5 = 1.5$. The construction yields a line k of the altitude 1.5 above the point 5 on O. Thus, in the vertical segment from that point to the line k there are as many linear units as there are square units in the shaded rectangle. In other words, the number of linear units in that segment is equal to the product of the numbers of linear units in the two perpendicular sides of the shaded rectangle.

The age in years of a father, say 35, may be equal to the product of the ages in years, 7 and 5, of his children, and there is nothing paradoxical about this fact. Of course, it must not be expressed by saying that the father is the product of the two children. Nor should one say that the vertical segment from 5 to k is the product of two perpendicular sides of the rectangle.

The line k in Fig. 6 has not only the altitude $\int_2^5 .5$ at 5, but also the altitudes $\int_2^4 .5$ and $\int_2^3 .5$ at 4 and 3 respectively. In fact, for any number b between 0 and 5, the altitude of k at b equals the area under .5 from

2 to b. (Even at 2, its altitude is 0, as is $\int_2^2$ .5, the area of a rectangle whose base is 0.) For this reason, the line k in Fig. 6 will be called *the 2-area line* of the horizontal line *.5* or *the 2-integral* of *.5*, denoted by

$\int_2$ .5; in a formula,

in Fig. 6, $k = \int_2 .5.$

Similarly, in Fig. 5, the altitude of the line k at any point b' that one might mark on O would be equal to the area under c from a to b', wherefore k is called the a-area line (or the a-integral) of c; in a formula,

$k = \int_a c.$ Clearly, k is the *only* line whose altitudes are thus related

to the areas under c.

Since k a = 0, the altitude kb' of k at any point b' is equal to the rise of k from a to b'. In a formula,

$$[k]_a^{b'} = \int_a^{b'} c \text{ and, more generally, } [k]_x^{x'} = \int_x^{x'} c \text{ for any x and x'.}$$

But the last equality holds also for some lines other than k, for instance for the line l (parallel to k) in Fig. 5, and, in fact, for any line that is parallel to k. Indeed any such line has the same rise from x to x' as has k. Any such line is therefore called *an area line* or *an integral* of c. The a-area line of c is the one that has the altitude 0 at a (or crosses the line

O at a). The line l in Fig. 5 is the .5-area line of c or $\int_{.5} c.$ For any

number h, there is an area line of c having the altitude h at a.

The preceding results yield the following rule:

Graphical Solution of the Area Problem for Horizontal Lines. *To find the area under the horizontal line c from a to b, construct an area line as indicated in Fig. 5 (either k or any line parallel to k) and measure its rise from a to b.*

The reader may wonder why, besides k, area lines parallel to k are even considered. Those lines will play an important role in Chapter II.

EXERCISES

7. Which of the following are numbers and which are lines?

$$\int_1 3, \quad \mathbf{D}\,c\,2, \quad \int_1^2 5, \quad \mathbf{D}\,j\,3, \quad \mathbf{D}\,3, \quad \sqrt[3]{5}, \quad \mathbf{D}\,(2j), \quad 4, \quad [2j]_3^5 \,.$$

8. On cross-section paper construct the slope and the slope line of l, if:

(a) $l\,1 = 1.5$ and $l\,1.5 = 2.75$ (b) $l\,4 = 1$ and $l\,6.5 = -2.75$

(c) $l\,1 = -2$ and $l\,1.5 = -.75$ (d) $l\,1 = -2$ and $l\,3 = .5$

(e) $l\,2 = 1$ and $l\,3.5 = 1$ (f) $l = 2j + 1$

(g) $l = -3j + 2$ (h) $l = .5$.

Check the results by the numerical method. Then draw nonvertical lines and determine their slopes graphically without measuring the altitudes of the lines anywhere. Only one mensuration is indispensable in each case: that of the altitude of the slope line.

9. What can be said about the slope line of a line which (a) has the rise 0 from a to b? (b) has the rise r from a to a + 1? (c) has the rise 1 from a to b? Show that a straight line l is a simple curve if and only if l is not vertical.

10. For what lines l is (a) $\mathbf{D}\,l = l$; (b) $\mathbf{D}\,l$ parallel to l? (c) Prove that if $l = cj + c'$ (see p. 5), then $c = \mathbf{D}\,l\,0$ and $c' = l\,0$.

11. On cross-section paper construct

(a) $\displaystyle\int_2^{3.5} 2$ (Count the number of unit squares or halves of unit squares

under the line 2 from 2 to 3.5, and measure the rise of the area line from 2 to 3.5.)

(b) $\displaystyle\int_2^{2+\sqrt{2}} \sqrt{2}, \quad \int_{1-\sqrt{2}}^1 \sqrt{2}, \quad \int_{-2-\sqrt{2}}^{-2} \sqrt{2}.$ Here, $\sqrt{2}$ denotes the

positive square root of 2. (In these cases, the rectangles cannot be broken up into fractions of unit squares as the rectangles in Fig. 6. But the results of the constructions can be checked by the numerical method.)

(c) The area lines $\displaystyle\int_{\sqrt{2}} \sqrt{2}$ and $\displaystyle\int_{1-\sqrt{2}} \sqrt{2}.$

(d) The areas under portions of horizontal lines without measuring the altitudes of these lines or the lengths of the portions. Only one mensuration is indispensable in each case: that of the rise of an area line.

(e) $\displaystyle\int_1^3 O$ and several area lines of O. How are the lines $\displaystyle\int_1 O$ and

$\displaystyle\int_2 O$ related? Is there an area line of O having the altitude 3 at 2?

12. (a) Prove $\displaystyle\int_a c = c(j - a)$.

(b) Show that, if l is any area line of the horizontal line c, there is a number b such that $l = cj + b$.

(c) If $c \neq 0$, where has the area line of c whose altitude at a is h the altitude 0 ?

(d) Find the particular area line of .5 having the altitude 3 at 2; and those of -2 having the altitudes $-.5$ at 1, and 3 at 1.

(e) Show that $\displaystyle\int_0^a 3c = 3\int_0^a c$, for every horizontal line c and every point a.

⋆13. Construct an area line l of the horizontal line 1.5 and determine its rise from 2 to 4.5.

(a) Without changing the point 0 on O, replace the original unit of length by a new unit half as long; that is to say, choose the former point .5 as 1. The horizontal line 1.5 now has the altitude 3; the former points 2 and 4.5 are now 4 and 9. Using the new unit, construct an area line l' of the horizontal line 3. How is the rise of l' from 4 to 9 in the new linear unit related to the rise of l from 2 to 4.5 in the old? How is the area in the new square unit related to the area in the old?

(b) Without changing the point 0, replace the original linear unit by one twice as long and repeat the constructions.

(c) How does, in general, a change of the linear unit affect (1) the rise of an area line of a given horizontal line between two points? (2) the area of the domain under the horizontal line between the two points?

⋆14. Let l be the line for which $l\,2 = 1$ and $l\,2.5 = 4$. Construct the slope line $\mathbf{D}\,l$. Then change the unit as in the preceding exercise and construct the slope line for the same line l. What is, in general, the effect of a change of the unit on the slope and the slope line of a nonvertical line?

⋆15. To determine the area under the horizontal line 400 from 1 to 3 graphically, one will want to measure vertical segments in a shorter linear unit than horizontal segments. Areas, then, are measured in unit rectangles, whose horizontal and vertical sides have the length of one respective unit. Choose a horizontal unit of $1''$ and a vertical unit of $.01''$, construct an area line of the line *400*, and show that

(a) the area $\displaystyle\int_1^3 400$ is $400(3-1)$ unit rectangles;

(b) the rise of the area line from 1 to 3 is 800 vertical units.

Then construct the line $200j$ whose rise from 3 to 5 is 400. Construct its slope line and show that the altitude of $\mathbf{D}(200j)$ is 200 vertical units.

*16. Construct graphically $\displaystyle\int_5^2 .5, \quad \int_3^1 0, \quad \int_5^2 (-.5)$, and check the results by the numerical method.

*17. What is the meaning of those of the following symbols that have been defined? (The others will remain undefined throughout this book.) $\mathbf{D}(2j)$, $\mathbf{D}(\tfrac{1}{2}j)3$, $\mathbf{D}\,2$, $\mathbf{D}\,2$, $\mathbf{D}\,j2$, $\mathbf{D}(j2)$, $\mathbf{D}(\mathbf{D}j)$, $\mathbf{D}(\mathbf{D}j)3$,

$\mathbf{D}(\mathbf{D}j3)$, $\displaystyle\int_0 3$, $\displaystyle\int_1 5$, $\mathbf{D}\int_3^4 j$, $\mathbf{D}\int_3 j$.

3. THE IDEA OF NEWTON AND LEIBNIZ

a. The Relation between Slope and Area. Slope and area seem to be totally unrelated. Newton's teacher, Barrow, probably was the first to see clearly that, in fact, these concepts are very closely related — somewhat like square and square root — and this discovery is the very core of Calculus.

If n is a number different from 0, there is exactly one positive number that is called the square of n. For instance, the square of 3 is 9. If m is a positive number there are two numbers that are called the square roots of m, either root the negative of the other. For instance, the square roots of 9 are 3 and -3.

(i) *The square of a square root of a positive number m is m.*

(ii) *A square root of the square of a number n ($\neq 0$) is either n or $-n$.*

If l is a nonvertical line, there is one horizontal line that is called the slope line or derivative of l. For instance, $\mathbf{D}(3j+3) = 3$. If c is any horizontal line, there are infinitely many lines that are called area lines or integrals of c, each of them parallel to any other. For instance, the horizontal line *3* has the area lines $3j+2$, $3j$, $3j-6$; in fact, for any number d, the line $3j+d$ is an area line of *3*.

Theorem I. *The slope line of any area line of a horizontal line* c *is* c; *or, the derivative of any integral of* c *is* c; in a formula,

(I) If c is any horizontal line, then $\mathbf{D} \int_a c = c$, for any a.

Theorem II. *Any area line of the slope line of a nonvertical line* l *is parallel to or identical with* l; *or, any integral of the derivative of* l *is parallel to* l *or identical with* l.

Indeed, if l is the line $cj + d$, then $\mathbf{D}\, l$ is the horizontal c; and if l' is any area line of c, then there is a number d' such that $l' = cj + d'$. The line l' is parallel to l, if $d \neq d'$, and identical with l, if $d' = d$.

For instance, $\frac{1}{2}j$ has the altitude 2.5 at 5. Its slope line, $\mathbf{D}\,(\frac{1}{2}j)$, is the horizontal line $\frac{1}{2}$. The 5-area line of the latter is that line $\frac{1}{2}j + d'$ which has the altitude 0 at 5, that is, the line $\frac{1}{2}j - 2.5$. It is 2.5 units below $\frac{1}{2}j$; in a formula, $\int_5 \mathbf{D}\,(\frac{1}{2}j) = \frac{1}{2}j - 2.5.$

More generally, $\int_a \mathbf{D}\, l$, the a-area line of $\mathbf{D}\, l$, has the altitude 0 at a.

The line l has the altitude $l\,a$ at a. Hence $\int_a \mathbf{D}\, l$ is $l\,a$ units below l; in a formula,

(II) If l is any nonvertical line, then $\int_a \mathbf{D}\, l = l - l\,a$ for any a.

The relation between square and square root can also be expressed as follows:

(iii) *If* m *is a root of* n, *then* n *is the square of* m.

(iv) *If* n *is the square of* m, *then* m *is a root of* n.

Correspondingly, the following theorems can easily be verified:

Theorem III. *If* l *is an area line (or an integral) of* c, *then* c *is the slope line (or derivative) of* l.

Theorem IV. *If* c *is the slope line (or derivative) of* l, *then* l *is an area line (or integral) of* c.

An inspection of Fig. 7 will not reveal whether c was given and the

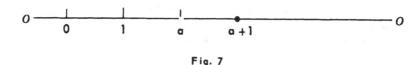

Fig. 7

a-area line of c was constructed or l was given and the slope line of l was constructed. The figure illustrates both

$$\int_a c = l \qquad \text{and} \qquad \mathbf{D}\, l = c;$$

and this duplicity of the meaning of Fig. 7 is a striking expression of the reciprocity of slope and area.

EXERCISES

18. Verify that similar relations of reciprocity exist between

(a) the tangents (in the sense of trigonometry) of angles that are not odd multiples of right angles, on the one hand, and the arctangents of numbers, on the other;

(b) the women who are mothers of women, on the one hand, and daughters, on the other.

b. The Calculus Solution of the Area Problem. There are two numbers having the square 9. They are called the square roots of 9. One of them is > 0 and is denoted by $\sqrt{9}$, the other one is < 0 and is denoted by $-\sqrt{9}$. Clearly $\sqrt{9} = 3$ and $-\sqrt{9} = -3$.

Some statements are valid for either root; e.g., the fourth power of either root equals 81; the absolute value of either root lies between 2 and 4; etc. For some purposes it is therefore convenient to have a symbol such as $\pm\sqrt{9}$ that stands for either root. In terms of the latter symbol

$$(\pm\sqrt{9})^4 = 81 \qquad \text{and} \qquad 2 < |\pm\sqrt{9}| < 4.$$

On the other hand, not being a specific number, $\pm\sqrt{9}$ is neither > 2 nor $\not> 2$. Obviously, only $\sqrt{9} > 2$, whereas $-\sqrt{9} \not> 2$.

There are infinitely many straight lines having as their common slope line or derivative the horizontal line *3*. They are called the *antiderivatives* of *3*. Each of them has the altitude 0 somewhere: one, and only one, at 0; one, and only one, at 2; and, for any number a, one and only one at a. These lines will be denoted by

$$D_{(0,0)}^{-1}\,3, \quad D_{(2,0)}^{-1}\,3, \quad D_{(a,0)}^{-1}\,3,$$

respectively. One readily verifies that

$$D_{(0,0)}^{-1}\,3 = 3j, \quad D_{(2,0)}^{-1}\,3 = 3j - 6, \quad D_{(a,0)}^{-1}\,3 = 3j - 3a.$$

More generally, for any three numbers a, b, and c, the antiderivative of the horizontal line c (of altitude c) which has the altitude b at a, will be denoted by $D_{(a,b)}^{-1}$ c. (Of course, many of these lines coincide; see Exercise 20b.)

Since, by definition of antiderivatives, *3* is the slope line of the line $D_{(a,0)}^{-1}$ *3*, it follows from Theorem IV that $D_{(a,0)}^{-1}$ *3* is an area line of *3*. Having the altitude 0 at a, $D_{(a,0)}^{-1}$ *3* must be the a-area line of *3*; in a formula,

$$D_{(a,0)}^{-1}\,3 = \int_a 3 \text{ and, more generally, } D_{(a,0)}^{-1}\,c = \int_a c.$$

Theorems III and IV can be summarized in the following

Theorem V. *Any area line of c is an antiderivative of c, and any antiderivative of c is an area line of c.*

In view of Theorem V, the graphical solution of the area problem, which calls for the construction of area lines, can be restated; and this restatement is the

Calculus Solution of the Area Problem for Horizontal Lines. *To find the area under the horizontal line c from a to b, find any antiderivative of c and determine its rise from a to b.*

Some statements are valid for any antiderivative of *3;* for instance, the derivative of any antiderivative of *3* is *3*. For some purposes it is therefore convenient to have a symbol that stands for any antiderivative of *3*. A convenient symbol is $D^{-1}\,3$. For instance,

$$D\,(D^{-1}\,3) = 3 \text{ and, for any c, } D\,(D^{-1}\,c) = c.$$

On the other hand, not being a specific line, $D^{-1}\,3$ has at 2 neither the altitude 0 nor an altitude $\neq 0$. Only $D_{(2,0)}^{-1}3$ has the altitude 0 at 2, whereas $D_{(1,0)}^{-1}3$ has an altitude $\neq 0$ at 2; in formulas,

$$D_{(2,0)}^{-1}3\,2 = 0 \quad \text{and} \quad D_{(1,0)}^{-1}3\,2 \neq 0.$$

$D^{-1}\,3\,2$, and more generally, D^{-1} ca, for any c and any a, will remain undefined and therefore meaningless symbols.

The Calculus Solution of the area problem for horizontal lines, proved in this section, can be condensed in the formula

$$\int_a^b c = [D^{-1} c]_a^b \text{ for any horizontal line c and any two numbers a and b.}$$

This simple formula contains the essence of Newton's and Leibniz' idea, the application of which to more complicated curves has marked an epoch in the history of mathematics and science.

EXERCISES

19. Verify the Calculus Solution of the area problem for
(a) a horizontal line below O;

*(b) the lines .5 and $-.5$ from 5 to 2 (see Exercise 16). Note that the result of the construction for the area always agrees with the sign rule: the product of the altitude and the base is positive if the factors are both positive or both negative, and negative if one factor is positive and one negative.

(c) Verify Theorem II for a line l with negative slope.

20. Show that
(a) $D^{-1}_{(0,b)} c \neq D^{-1}_{(0,b')} c$ for any two numbers b and $b' (\neq b)$;

(b) $D^{-1}_{(a,b)} c = D^{-1}_{(a',b')} c$, if and only if $ac - b = a'c - b'$;

(c) $D^{-1}_{(a,b)} c = b + \int_a^{} c$ for any two numbers a and b and for any hori-

zontal line c;

(d) For any two numbers a and $a' (\neq a)$, is $D^{-1}_{(a,0)} c = D^{-1}_{(a',0)} c$?
(Hint. Reconsider Exercise 11(e).)

(e) Analyze the meaning of the formulas

$$DD^{-1}_{(a,b)} c = c \quad \text{and} \quad DD^{-1} c = c, \text{ on the one hand,}$$

$$D \int_a^{} c = c, \text{ on the other.}$$

One of the main prerequisites for the understanding of calculus is the realization that the first formulas are immediate consequences of basic definitions, whereas the last expresses a profound discovery.

21. If $\int^b c$ denotes the line having the altitude $\int_a^b c$ at any point a

on O, show that $\mathbf{D} \int^b c = -c.$

*22. Show that the Calculus Solution of the area problem has the follow-
ing counterpart: *The slope of a nonvertical line l can be found by determining the altitude of a horizontal line having l as an area line.* In contrast to the calculus solution of the area problem, however, the above
solution of the slope problem is of little importance.

The Mathematics of Elementary Thermodynamics

KARL MENGER

Illinois Institute of Technology, Chicago, Illinois

THE purpose of this paper[1] is to bridge the gap between calculus as taught in elementary mathematics courses, and calculus as applied in the first courses on thermodynamics. Many beginners are baffled by two discrepancies.

(A). In his calculus course, the student has, at best, acquired the idea of the "complete" differential df of a function f of several variables. In thermodynamics he learns that the quantity of heat dq, transferred to a gas, is $pdv+du$, where p denotes the pressure, v the volume, and u the internal energy. He is told that dq is an "incomplete" differential. What is an incomplete differential? The answer given in most textbooks of thermodynamics (that it is a differential which is not complete) is incomprehensible to one not familiar with the concept of a general differential, and many presentations of this latter concept are obscure. When clarified, it proves to belong to those ideas which, by virtue of their very simplicity, are beyond a real understanding of beginners (cf. Sec. 8). Some physicists suggest that, in view of the incompleteness of the differential, the quantity of heat should be denoted by δq, dq, or Q. This suggestion is good but, obviously, not a substitute for the much needed definitions of the terms used in the basic laws of thermodynamics. In the absence of such definitions, most beginners who give any thought to the equation $dq = pdv+du$ interpret it as follows: If the volume and the energy of a gas undergo small changes Δv and Δu, then the quantity of heat Δq, which has to be transferred to the gas, is approximately equal to $p\Delta v+\Delta u$. As simple examples show, this interpretation is incorrect no matter what we mean by "approximately equal" (cf. Sec. 2).

(B). In a *mathematical* statement, two functions are denoted by the same symbol if, and

only if, they are equal: that is, if and only if: (1) they have the same domain, and (2) with each element of this domain they associate the same number or, as we say, the same "function value." For instance, the function associating with each number t, the number $16t^2$, and the function associating with each number s, the number $16+16(s+1)(s-1)$, may be denoted by the same symbol, say F, and we have

$$F(t) = 16t^2$$

and $F(s) = 16+16(s+1)(s-1)$. In *physics*, two functions are denoted by the same symbol if the meaning of the function values is the same. For instance, the formulas $v=v(t)$, $v=v(s)$ express the fact that the velocity of a falling body is determined by the time t elapsed since the release of the body, as well as by the distance s the body has traversed. Since $v=gt$ and $v=\sqrt{(2gs)}$, the mathematician would use different symbols for the two functions and would write $v=f_1(t)$ and $v=f_2(s)$. In thermodynamics, the student who takes seriously what he has learned in mathematics and in physics runs into difficulties with regard to almost every function. He should see at least one function treated in both notations with a discussion of their relative merits (cf. Secs. 4 and 5).

A corollary of the difference mentioned in (B) is the need for different symbols for partial derivatives in physics and mathematics. The difficulties are enhanced by the use of the same symbol for functions and numbers (variables).

We shall present the first law of thermodynamics without any reference to the concept of a differential (1) as a relation between time rates of change (Sec. 1) and (2) using a Stieltjes integral (Sec. 6). Also the second law will be analyzed without any essential reference to differentials (Sec. 10). An Appendix contains a definition of the thermodynamic equality of quasi-static processes.

1. The Differential Form of the First Law

We begin with a presentation of the first law that presupposes nothing but the idea of the

---

[1] In preparing the final draft of this paper the author has profited by suggestions from Burton D. Fried and Leo A. Schmidt of the Illinois Institute of Technology, and from Dr. Eric Lype of the Armour Research Foundation. Part of the content which was presented in lectures at the University of Notre Dame, is incorporated in the Master's thesis of Vincent J. Cushing (1946). An abstract of a paper "The use of differentials in thermodynamics" was published in *Am. Math. Monthly* **56**, 210 (1949).

89

derivative (df/dt) or $f'(t)$ of a function of one variable $f(t)$. We shall study a thermodynamical process II of a homogeneous substance for which we can determine the following five functions of the time:

$v(t)$, the volume at the moment t;
$p(t)$, the pressure at the moment t;
$\theta(t)$, the temperature at the moment t;
$u(t)$, the internal energy at the moment t;
and
$q(t)$, the quantity of heat transferred to the substance between an initial moment t_0 and the moment t.

We shall call such a process *quasi-static*. An example is the gradual change of volume, temperature, and energy of an ideal gas through the transfer of heat under constant pressure. On the other hand, if by opening a shutter in the container we let the gas suddenly expand into an originally empty neighboring container of equal size, the gas undergoes a process which is not quasi-static. No uniform pressure can be determined while the gas is expanding. The volume is doubled during the process but, at its intermediate stages, the gas is not homogeneous since in the original container the density drops from its initial value δ to $\delta/2$ while in the neighboring container the density rises from 0 to $\delta/2$. Once and for all, we point out that *all processes studied in this paper, except those considered in Statement C of Sec. 10, are quasi-static.*

We call the process II *differentiable* if the functions $v(t)$, $u(t)$, and $q(t)$ possess continuous derivatives. For a differentiable process, the first law may be introduced as the following relation between $p(t)$ and the time rates of change of the functions $v(t)$, $u(t)$, and $q(t)$

$$(dq/dt) = p(dv/dt) + (du/dt)$$
or
$$q'(t) = p(t)v'(t) + u'(t) \qquad (1)$$

at every moment t.

In calculus, the student has learned that if the function $f(t)$ has a continuous derivative, then, for every number t and every number Δt, we have

$$f(t+\Delta t) - f(t) \sim_{\Delta t} f'(t)\Delta t,$$

where the symbol $\sim_{\Delta t}$ indicates that the expressions on the left and on the right side are approx-imately equal in the following sense. If $|\Delta t|$ i sufficiently small, then the difference between $f(t+\Delta t)-f(t)$ and $f'(t)\Delta t$ is not only small—thi is obvious, since both $f(t+\Delta t)-f(t)$ and $f'(t)\Delta$ themselves are small—but small in compariso with $|\Delta t|$; that is, by choosing $|\Delta t|$ sufficientl small (but $\neq 0$) we can make

$$\left| \frac{f(t+\Delta t) - f(t) - f'(t)\Delta t}{\Delta t} \right|$$

as small as we please.

Applying this theorem to the three derivative $q'(t)$, $v'(t)$, and $u'(t)$ we see that Eq. (1) i equivalent to

$$q(t+\Delta t) - q(t) \sim_{\Delta t} p(t)[v(t+\Delta t) - v(t)] + [u(t+\Delta t) - u(t)]. \qquad (2$$

Hence the expositor may either start with Eq (1) and derive Eq. (2) or start with Eq. (2) an derive Eq. (1).

2. A Misinterpretation of the Differential Form of the First Law

In formula (1) which connects $p(t)$ and th time rates of change of the functions $v(t)$, $u(t)$ and $q(t)$, every term is clearly and unambiguousl defined. Traditionally, however, the first law i written in the form of an equality of differential

$$dq = pdv + du. \qquad (3$$

This last equality is clear and unambiguous onl when considered as an abbreviation for Eq. (1) in other words, when supplemented by the re mark that dq, dv, and du are differentials wit regard to the time of the functions of on variable $q(t)$, $v(t)$, and $u(t)$. This point is rarel emphasized. In lieu of the above remark, mos books present those discussions of the incom pleteness of the differential dq which we men tioned in the introduction to this paper. Experi ence shows that, in spite of those discussions many beginners misinterpret formula (3) b translating it into the following statement:

If the volume and the energy of a gas underg sufficiently small changes Δv and Δu, then th quantity of heat that has to be transferred to th gas is approximately equal to $p\Delta v + \Delta u$. Mor precisely, Δq and $p\Delta v + \Delta u$ differ by a quantit which is small in comparison with Δv and Δu, or

$$\Delta q \sim _{\Delta v,\,\Delta u} p\Delta v + \Delta u. \qquad (4)$$

While no textbook makes this erroneous claim, few books warn against it. Since, under these circumstances, the misunderstanding frequently occurs, it seems advisable to show explicitly that the above statement (4) is neither a logical consequence of Eqs. (1) and (2) nor generally valid for thermodynamical processes. This can be shown by means of simple examples:

(a) *A mathematical example.*—Let $F(x)$, $A(x)$, $B(x)$, and $C(x)$ be four functions satisfying the relation

$$F'(x) = A'(x) + C(x) \cdot B'(x) \qquad (5)$$

for every x. In view of what we saw in Sec. 1, from Eq. (5), it follows that if $|\Delta x|$ is sufficiently small, then

$$\Delta F \sim _{\Delta x} \Delta A + C(x)\Delta B.$$

But the relation (5) does by no means imply that for small changes ΔA and ΔB of $A(x)$ and $B(x)$, the change of $F(x)$ will differ from $\Delta A + C\Delta B$ little, let alone by a quantity which is small in comparison with ΔA and ΔB. If, for instance,

$$F(x) = -(2x-1)^3/6, \quad A(x) = B(x) = x^2 - x,$$
$$C(x) = -2x,$$

then relation (5) holds and yet $F(x)$ may undergo large changes although $A(x)$ and $B(x)$ change little or not at all. For instance,

for $x = 0$ we have

$$F(0) = 1/6, \quad A(0) = 0, \quad \text{and} \quad B(0) = 0;$$

for $x = 1$ we have

$$F(1) = -1/6, \quad A(1) = 0, \quad \text{and} \quad B(1) = 0.$$

Hence, as x changes from 0 to 1, the net change

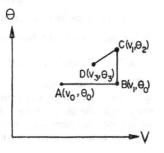

FIG. 1. Thermodynamic process applied to an ideal gas, in which, after appreciable change of q the initial volume and temperature are almost restored.

of F is $-\frac{1}{3}$, while the net change of both A and B is 0. If x changes from 0 to a number close to 1, then the change of F is almost $-\frac{1}{3}$, while the net changes of A and B are almost 0. It is true that the change of x from 0 to 1 is relatively large and that for smaller changes of x, such as from 0 to $\frac{1}{2}$, the net changes of A and B are appreciable. It is furthermore true that if x changes from 0 to a number which is sufficiently close to 0, then the difference between ΔF and $\Delta A + C\Delta B$ is small in comparison with $[(\Delta A)^2 + (\Delta B)^2]^{\frac{1}{2}}$. But in the erroneous statement (4) one compares the net change Δq merely with the net changes ΔA and ΔB and completely ignores the quantity Δt. This is where the error lies, for the quantity Δt plays an essential role in the correct formula (2).

(b) *A physical example.*—We study one mole of an ideal gas for which, at every moment, $pv = R\theta$. The internal energy u remains constant, if the volume is changed while the temperature is kept constant. Morever, u depends linearly upon θ, if either volume or pressure is kept constant. We have

$$u - u_0 = c_v(\theta - \theta_0) \quad \text{and} \quad u - u_0 = c_p(\theta - \theta_0),$$

respectively, where u_0 is the energy at the temperature θ_0 and where the constants c_v and c_p satisfy the equality $c_p - c_v = R$.

Every process in which, after an appreciable change of q, the initial volume and temperature of the gas are restored or almost restored, disproves statement (4). We carry out in detail a simple instructive example due to Mr. Burton D. Fried. One mole of an ideal gas, with the initial volume, temperature, and pressure v_0, θ_0, and p_0, is contained in a cylinder of the type employed in a Carnot cycle, and undergoes the following process Π_1:

(1) First, the gas expands isothermally to the volume $v_1 > v_0$.

(2) Then, the gas is heated at constant volume to the temperature $\theta_2 = v_1\theta_0/v_0$. When it reaches this temperature, the pressure assumes its original value p_0.

(3) Finally, the gas is cooled and its volume decreased under constant pressure. Let v_3 and θ_3 denote the final volume and temperature,

In a (v, θ)-plane, the process is represented by a triangular path from $A = (v_0, \theta_0)$ via $B = (v_1, \theta_0)$ and $C = (v_1, \theta_2)$ to $D = (v_3, \theta_3)$ (see Fig. 1).

Computing Δq in the traditional way we see that $d\theta = 0$ on AB; $dv = 0$ on BC; and $dp = 0$ on CD. Hence,

$$\Delta q = \int_{ABCD} dq = \int_A^B p\,dv + \int_B^C c_v\,d\theta$$
$$+ \int_C^D (p\,dv + c_p\,d\theta)$$
$$= \int_{v_0}^{v_1} \frac{R\theta_0}{v}\,dv + \int_{\theta_0}^{\theta_2} c_v\,d\theta + \int_{v_1}^{v_3} p_0\,dv + \int_{\theta_2}^{\theta_3} c_p\,d\theta$$
$$= R\theta_0 \ln(v_1/v_0) + c_v(\theta_2 - \theta_0)$$
$$+ p_0(v_3 - v_1) + c_p(\theta_3 - \theta_2).$$

Now,

$$c_v(\theta_2 - \theta_0) + c_p(\theta_3 - \theta_2) = (c_p - R)(\theta_2 - \theta_0)$$
$$+ c_p(\theta_3 - \theta_2) = c_p(\theta_3 - \theta_0) - R(\theta_2 - \theta_0)$$

and

$$R(\theta_2 - \theta_0) = R\theta_0(v_1 - v_0)/v_0 = p_0(v_1 - v_0).$$

Hence,

$$\Delta q = p_0 \Delta v + c_p \Delta \theta + R\theta_0 \ln(v_1/v_0) - 2p_0(v_1 - v_0),$$

where we have set $\theta_3 - \theta_0 = \Delta \theta$ and $v_3 - v_0 = \Delta v$.

If we choose v_3 and θ_3 closer and closer to v_0 and θ_0, then Δv and $\Delta \theta$ approach 0, while the corrective term

$$\Delta q - p_0 \Delta v - c_p \Delta \theta = R\theta_0 \ln(v_1/v_0) - 2p_0(v_1 - v_0)$$

not only fails to become small in comparison with $\Delta \theta$ and Δv but does not become small at all. For the corrective term is independent of v_3 and θ_3 and, as is easily seen, negative provided that $v_1 > v_0$.

It may be of interest to describe the process II_1, by five functions of the time $v(t)$, $\theta(t)$, $p(t)$, $u(t)$, and $q(t)$. Corresponding to the three stages of the process, each of these functions is defined in three time intervals:

$$v(t) = \begin{cases} v_1(t) & \text{for} \quad t_0 \leq t \leq t_1 \\ v_1 & \text{for} \quad t_1 \leq t \leq t_2 \\ v_3(t) & \text{for} \quad t_2 \leq t \leq t_3, \end{cases}$$

where $v_1(t)$ and $v_3(t)$ are functions satisfying the conditions

$$v_1(t_0) = v_0, \quad v_1(t_1) = v_1,$$
$$v_3(t_2) = v_1, \quad v_3(t_3) = v_3;$$
$$\theta(t) = \begin{cases} \theta_0 & \text{for} \quad t_0 \leq t \leq t_1 \\ \theta_1(t) & \text{for} \quad t_1 \leq t \leq t_2 \\ p_0 v_3(t)/R & \text{for} \quad t_2 \leq t \leq t_3, \end{cases}$$

where $\theta_1(t)$ is a function satisfying the condition

$$\theta_1(t_1) = \theta_0, \quad \theta_1(t_2) = \theta_2 = v_1 \theta_0 / v_0;$$
$$p(t) = R\theta(t)/v(t) = \begin{cases} R\theta_0/v_1(t) & \text{for} \quad t_0 \leq t \leq t_1 \\ R\theta_1(t)/v_1 & \text{for} \quad t_1 \leq t \leq t_2 \\ p_0 & \text{for} \quad t_2 \leq t \leq t_3; \end{cases}$$

and

$$u(t) = \begin{cases} u_0 & \text{for} \quad t_0 \leq t \leq t_1 \\ u_0 + c_v \theta_1(t) & \text{for} \quad t_1 \leq t \leq t_2 \\ u_0 + c_v \theta_2 + c_p p_0 v_3(t)/R & \text{for} \quad t_2 \leq t \leq t_3. \end{cases}$$

It follows that

$$p\frac{dv}{dt} + \frac{du}{dt} = \begin{cases} (R\theta_0/v_1(t))v_1'(t) + 0 & \text{for} \quad t_0 \leq t \leq t_1 \\ 0 + c_v \theta_1'(t) & \text{for} \quad t_1 \leq t \leq t_2 \\ p_0 v_3'(t) + c_p p_0 v_3'(t)/R & \text{for} \quad t_2 \leq t \leq t_3 \end{cases}$$

Hence, by Eq. (1),

$$\frac{dq}{dt} = \begin{cases} R\theta_0 v_1'(t)/v_1(t) & \text{for} \quad t_0 \leq t \leq t_1 \\ c_v \theta_1'(t) & \text{for} \quad t_1 \leq t \leq t_2 \\ p_0(1 + c_p/R)v_3'(t) & \text{for} \quad t_2 \leq t \leq t_3 \end{cases}$$

and by integration

$$q(t) = \begin{cases} q(t_0) + R\theta_0 \ln[v_1(t)/v_0] & \text{for} \quad t_0 \leq t \leq t_1 \\ q(t_0) + R\theta_0 \ln(v_1/v_0) \\ \quad + c_v[\theta_1(t) - \theta_0] & \text{for} \quad t_1 \leq t \leq t_2 \\ q(t_0) + R\theta_0 \ln(v_1/v_0) + c_v(\theta_2 - \theta_0) \\ \quad + p_0(1 + c_p/R)[v_3(t) - v_1] & \text{for} \quad t_2 \leq t \leq t_3 \end{cases}$$

In particular

$$q(t_3) = q(t_0) + R\theta_0 \ln(v_1/v_0) + c_v(\theta_2 - \theta_0)$$
$$+ p_0(1 + c_p/R)(v_3 - v_1)$$

Transforming the last two terms we find

$$\Delta q = p_0 \Delta v + c_p \Delta \theta + R\theta_0 \ln(v_1/v_0) - 2p_0(v_1 - v_0).$$

3. What Is and What Is Not a Function?

The last example also shows that if we merely know the net changes which the volume and the energy (or the temperature) of a gas undergo, we cannot even approximately guess how much heat has been transferred to the gas during the process. For in the process Π_1, even if the net changes Δv, $\Delta \theta$, and Δu are as small as we please, Δq is appreciable. But, if Π_2 denotes the expansion of the volume of the gas under constant pressure from v_0 to v_3 (in the diagram, Π_2 would be represented by the straight segment AD), then we obtain

$$\Delta q = \int_{AD} (p_0 dv + c_p d\theta)$$

$$= p_0 \Delta v + c_p \Delta \theta = p_0(1 + c_p/R)\Delta v.$$

Thus, for Π_2, if Δv is sufficiently small, Δq is as small as we please.

We thus see that Δq is not a function of Δv and $\Delta \theta$. Here, by "function" we mean the association of exactly one number with each element of some set. (This set is called the *domain* of the function.) If the elements of the set are numbers, pairs of numbers, triples of numbers, . . . we speak of a function of one, two, three, . . . variables. (Some authors call the functions in this sense *one-valued* or *single-valued* functions.) Moreover, we see that q is not a function of v and θ. For, if in Π_1 we set $v_3 = v_0$, then Δv, $\Delta \theta$, Δp are precisely zero, and yet

$$\Delta q = R\theta_0 \ln(v_1/v_0) - 2p_0(v_1 - v_0) < 0.$$

The beginning student of physics would easily grasp this situation, if in studying the concept of a function he had been shown examples of associations of numbers with numbers which do not fall under the definition. The temperature of a room is a function of the time, and the temperature of a man in the room is a function of the time; but the temperature of the man is not a function of the temperature of the room, since (at different moments) the man's temperature may be different although the room temperature is the same. Neither is the room temperature a function of the man's temperature. Even if, during a particular time interval, the man's temperature should be the same whenever the room temperature is the same, we still should not call the former a function of the latter since a little reflection would reveal the coincidental character of the particular situation.[2] Unfortu-

nately, many mathematics books fail to present such examples.[3]

On the contrary, numerous books on calculus define, at the outset, a function as the association of one *or more* numbers with each element of a set (domain). According to this definition (of what sometimes is called a *many-valued* function), every measurable quantity varying in time is a function of every other such quantity; e.g., a man's temperature would indeed be a function of the room temperature. For this very reason, the general concept of a many-valued function is physically useless; moreover, it is also unimportant in the elements of mathematics. In calculus, we differentiate and integrate exclusively one-valued functions and even in the theory of complex functions we study only a very limited class of many-valued functions.

While thus the introduction of the general concept of a many-valued function does not serve any useful purpose, it creates in beginners the disposition to those misunderstandings which are particularly harmful in thermodynamics. Mathematicians ought to limit elementary discussions strictly to one-valued functions, but supplement the definition with examples not falling under the concept.

On the other hand, physicists ought to refrain from using such symbols[4] as $(\partial q/\partial v)_\theta$ which are meaningless except under the assumption that q is a function of v and θ—and this, as we have seen and as these physicists themselves point out, is not the case.

4. An Equation of State and the Energy

We shall now study a class of processes. Usually the processes of this class are called *quasi-static processes of a substance satisfying an equation of state*. Mathematically, these processes are characterized by a relation between volume, pressure, and temperature (which are said to

---

[2] Cf. the author's lecture notes "The concept of a function," edited by B. D. Fried (1948).

[3] Cooley, Gans, Kline, and Wahlert in their *Introduction to mathematics* (Houghton Mifflin, 1937), a book with many merits, claim (p. 262) that the marriages in New York are a function of the imports of Siam from the United States. The authors substantiate this claim by representing the marriages and imports as two functions of the time. By coincidence, whenever the imports differ so do the marriages. But even so, we should consider the situation as an example *not* falling under the concept of a function.

[4] H. Margenau and G. M. Murphy, *The mathematics of physics and chemistry* (Van Nostrand, 1943), p. 11.

characterize the *state* of the substance). The relation (equation of state) is assumed to hold for all quasi-static processes of the substance and to be of such a nature that every two of the quantities v, p, and θ determine the third. More precisely, we assume the existence of three functions

$$V(p, \theta), \quad P(v, \theta), \quad \Theta(v, p)$$

possessing continuous first partial derivatives which are so related that the following equalities hold

$$
\begin{aligned}
V(P(v, \theta), \theta) &= v \quad \text{for every } v \text{ and } \theta, \\
V(p, \Theta(v, p)) &= v \quad \text{for every } v \text{ and } p, \\
P(V(p, \theta), \theta) &= p \quad \text{for every } p \text{ and } \theta, \\
P(v, \Theta(v, p)) &= p \quad \text{for every } p \text{ and } v, \\
\Theta(V(p, \theta), p) &= \theta \quad \text{for every } \theta \text{ and } p, \\
\Theta(v, P(v, \theta)) &= \theta \quad \text{for every } \theta \text{ and } v.
\end{aligned}
$$

For every process of the class, at every moment t we have

$$v(t) = V(p(t), \theta(t)), \quad p(t) = P(v(t), \theta(t)),$$
$$\Theta(v(t), p(t)) = \theta(t)$$

and hence

$$v'(t) = \frac{\partial V}{\partial p} p'(t) + \frac{\partial V}{\partial \theta} \theta'(t) \qquad (6)$$

as well as similar formulas for $p'(t)$ and $\theta'(t)$. Here $(\partial V/\partial p)$ and $(\partial V/\partial \theta)$ denote the functions of t obtained by substituting $p(t)$ and $\theta(t)$ into the partial derivatives of $V(p, \theta)$.

We shall furthermore assume that the energy is determined by volume, pressure, and temperature or, as we say, that u is a function of the state. In Sec. 2 we saw that, even for one and the same process, the quantity of heat need not be determined by volume, pressure, and temperature. In other words, q is not a function of the state.

In view of the equation of state, every function of the state, in particular u, is determined by any two of the three quantities v, p, and θ. This example illustrates the difference, mentioned in Paragraph (B) of the introduction, between the notations of the physicist and the mathematician.

(1) Since, in general, the way u depends upon v and p is totally different from the ways u depends upon v and θ, or upon p and θ, the mathematician, being exclusively interested in the way the function values depend upon the independent variables, would denote the three functions by three different symbols, say by

$$U_1(v, p), \quad U_2(v, \theta), \quad U_3(p, \theta)$$

with the understanding that we have

$$U_3(p, \theta) = U_2(V(p, \theta), \theta)$$
$$\text{for every } p \text{ and } \theta \quad (7_{\text{math}})$$
$$U_3(p, \theta) = U_1(V(p, \theta), p)$$
$$\text{for every } p \text{ and } \theta \quad (8_{\text{math}})$$

and similar formulas connecting any U_i to any U_j ($j \neq i$). The mathematician would denote the six partial derivatives of the three functions by

$$\frac{\partial U_1}{\partial v}, \quad \frac{\partial U_1}{\partial p}, \quad \frac{\partial U_2}{\partial v}, \quad \frac{\partial U_2}{\partial \theta}, \quad \frac{\partial U_3}{\partial p}, \quad \frac{\partial U_3}{\partial \theta}.$$

By differentiating the relations (7_{math}), (8_{math}) etc., he would obtain the relations

$$\frac{\partial U_3}{\partial \theta} = \frac{\partial U_2}{\partial v} \frac{\partial V}{\partial \theta} + \frac{\partial U_2}{\partial \theta} \qquad (9$$

$$\frac{\partial U_3}{\partial p} = \frac{\partial U_2}{\partial v} \frac{\partial V}{\partial p}, \quad \text{etc.} \qquad (10$$

(2) The physicist, being primarily interested in the physical meaning of the function values, uses the same letter to denote the three functions and writes

$$u = u(v, p) = u(v, \theta) = u(p, \theta). \qquad (7_{\text{phys}}$$

In order to avoid ambiguities which would be disturbing from his as well as the mathematician's point of view, the physicist denotes the six partial derivatives of u by

$$\left(\frac{\partial u}{\partial v}\right)_p, \left(\frac{\partial u}{\partial p}\right)_v, \left(\frac{\partial u}{\partial v}\right)_\theta, \left(\frac{\partial u}{\partial \theta}\right)_v, \left(\frac{\partial u}{\partial p}\right)_\theta, \left(\frac{\partial u}{\partial \theta}\right)_p$$

respectively. Accordingly he expresses the relations (9) and (10) in the form

$$\left(\frac{\partial u}{\partial \theta}\right)_p = \left(\frac{\partial u}{\partial v}\right)_\theta \left(\frac{\partial V}{\partial \theta}\right)_p + \left(\frac{\partial u}{\partial \theta}\right)_v \qquad (9_{\text{phys}}$$

and

$$\left(\frac{\partial u}{\partial p}\right)_\theta = \left(\frac{\partial u}{\partial v}\right)_\theta \left(\frac{\partial V}{\partial p}\right)_\theta, \qquad (10_{\text{phys}}$$

c., where $(\partial V/\partial\theta)_p$ and $(\partial V/\partial p)_\theta$ could, without ambiguity, be replaced by $(\partial V/\partial\theta)$ and $(\partial V/\partial p)$, respectively.

The function of the time $u(t)$ is related to the three functions u of two variables as follows: at every moment t, we have

$$(t) = u(v(t),\, p(t)) = u(v(t),\, \theta(t))$$
$$= u(p(t),\, \theta(t)) \quad (11_{phys})$$
$$(t) = U_1(v(t),\, p(t)) = U_2(v(t),\, \theta(t))$$
$$= U_3(p(t),\, \theta(t)). \quad (11_{math})$$

Accordingly, we have, for instance,

$$\frac{du}{dt} = \left(\frac{\partial u}{\partial v}\right)_\theta \frac{dv}{dt} + \left(\frac{\partial u}{\partial \theta}\right)_v \frac{d\theta}{dt}, \quad (12_{phys})$$

$$u'(t) = \frac{\partial U_2}{\partial v} v'(t) + \frac{\partial U_2}{\partial \theta} \theta'(t), \quad (12_{math})$$

and two similar equalities for U_1 and U_3.

Once the coordination of the terminologies used in physics and mathematics is established, it is only fair for the physicist to insist on the merits of his own notation in the field of thermodynamics. For in developing this theme the physicist has to represent the energy as a function of many pairs of quantities besides (v, p), $v, \theta)$, and (p, θ). The mathematician would have to introduce further symbols, such as $u_4, u_5, \cdots$ which would be unintuitive unless a definite order of all thermodynamical quantities were generally adopted.

It may be pointed out, furthermore, that in applying the concept of a function to geometry the mathematician himself often uses the notation of the physicist. He describes the equation of the motion of a point in the plane by $x = X(t)$, $y = Y(t)$ and denotes the curvature of the path at the moment t by $\kappa(t)$. Let $S(t)$ be the length of the path traversed between an initial moment 0 and t. For given s, let $T(s)$ be the moment at which $S(T(s)) = s$. Now most geometers write $\kappa(s)$ instead of $\kappa(T(s))$, although the way in which the curvature depends upon s is, in general, totally different from the way in which the curvature depends upon t. However, it seems doubtful that one could develop a consistent, purely logical syntax of this physical-geometrical functional notation, except with regard to partial derivatives.

5. The Specific Heat

Traditionally, the first application of the first law is made in computing the specific heat. For a given differentiable process Π we call $(q'(t)/\theta'(t))$ the specific heat at the moment t. We have

$$\frac{q'(t)}{\theta'(t)} \sim_{\Delta t} \frac{q(t+\Delta t) - q(t)}{\theta(t+\Delta t) - \theta(t)}.$$

From the first law, Eq. (1), we obtain

$$\frac{q'(t)}{\theta'(t)} = \frac{p(t)v'(t) + u'(t)}{\theta'(t)}. \quad (13)$$

(a) *Constant volume.*—First we consider a process, Π_{v_0} during which the volume is constant, $v(t) = v_0$, and thus $v'(t) = 0$. For every such process Π_{v_0} belonging to the class studied in Sec. 4, from Eqs. (13) and (12) in view of $v'(t) = 0$ we obtain

$$c_v(t) = \left(\frac{q'(t)}{\theta'(t)}\right)_v = \begin{cases} (\partial u/\partial\theta)_v \\ \qquad \text{in the physical notation,} \\ (\partial U_2/\partial\theta)(v_0,\, \theta(t)) \\ \qquad \text{in the mathematical notation.} \end{cases}$$

Here the last term indicates that in $(\partial U_2/\partial\theta)$ we substitute v_0 for v, and $\theta(t)$ for θ.

We see that, if for a process Π_{v_0} at two different moments the temperature has the same value (the volume has automatically the same value v_0), then at these moments the specific heat will be the same. But much more can be said. Even if the substance with the energy function $u(v, \theta)$ undergoes different processes, Π_{v_0} and $\Pi_{v_0}*$, with constant volume v_0, then whenever the temperatures are equal, so are the specific heats—in other words, the specific heat for constant volume is a function of the state.

We can also express this fact by saying that there exists a function $c(v, \theta)$ such that, if a substance with the energy function $u(v, \theta)$ undergoes any process Π_{v_0}, then at every moment t the specific heat is

$$c_{v_0}(t) = c(v_0,\, \theta(t)).$$

(b) *Constant pressure.*—Let Π_{p_0} denote a process of our substance for which $p(t) = p_0$. Then

$p'(t) = 0$ and from Eqs. (13), (6), and the analog of (12) for $U_2 = u(p, \theta)$ it follows that

$$c_{p0}(t) = \frac{q'(t)}{\theta'(t)} = p_0 \frac{\partial V}{\partial \theta}(p_0, \theta(t)) + \frac{\partial U_2}{\partial \theta}(p_0, \theta(t))$$

$$= p_0\left(\frac{\partial V}{\partial \theta}\right)_p + \left(\frac{\partial u}{\partial \theta}\right)_p.$$

Using Eq. (9_{phys}) we obtain

$$c_{p0}(t) = p_0\left(\frac{\partial V}{\partial \theta}\right)_p + \left(\frac{\partial u}{\partial v}\right)_\theta\left(\frac{\partial V}{\partial \theta}\right)_p + \left(\frac{\partial u}{\partial \theta}\right)_v.$$

We see also that c_{p0} is a function of the state. Again, conditions can be formulated under which c_p does not change during the process, and assumes the same value for all processes Π_p within a certain range. Clearly,

$$c_p - c_v = [p + (\partial u/\partial v)_\theta](\partial V/\partial \theta)_p.$$

6. An Integral Form of the First Law

The form Eq. (1) of the first law is confined to differentiable processes for which the volume, the energy, and the quantity of heat are differentiable functions of the time. We shall now present an integral form of the first law which does not presuppose differentiability of any of the functions, and thus is applicable to a much wider class of processes.

We shall base this formulation on the concept of the Stieltjes integral of a function $f(x)$ with respect to a function $g(x)$ in an interval, $a \leq x \leq b$, denoted by

$$\int_a^b f(x)dg(x).$$

Due to its fundamental role in pure as well as applied mathematics, this concept has lately been included in several elementary textbooks.[5] Its main applications are to moments in mechanics and statistics. As we shall see, Stieltjes integrals are also useful in thermodynamics.

The idea of a Stieltjes integral is so simple that it can easily be presented to a beginning student of theoretical physics. In order to define

---

[5] D. V. Widder, *Advanced calculus* (Prentice-Hall, 1947), Chap. V.

$\int_a^b f(x)dg(x)$ we divide the interval $a \leq x \leq b$ int smaller intervals

$$a = x_0 < x_1 < \cdots < x_k < x_{k+1} < \cdots$$
$$< x_{n-1} < x_n = b. \quad (14$$

Let λ denote the length of the longest of thes intervals, that is, the largest of the number $x_{k+1} - x_k$. In each of the small intervals we choos a number $x_k \leq x_k^* \leq x_{k+1}$. Then we form th so-called Stieltjes sum

$$\sum_{k=0}^{n-1} f(x_k^*)[g(x_{k+1}) - g(x_k)]$$

which is an approximation to $\int_a^b f(x)dg(x)$. I order to obtain a better approximation we form Stieltjes sums for subdivisions of the interva $a \leq x \leq b$ which are finer, that is, for which λ i smaller. The integral $\int_a^b f(x)dg(x)$ is defined a the limit, whenever it exists, of the Stieltje sums as λ approaches zero. In the special cas that $g(x) = x$ the integral is the ordinary integra $\int_a^b f(x)dx$. In the special case that $f(x) = 1$ fo $a \leq x \leq b$ we readily see that $\int_a^b 1dg(x) = g(b) - g(a)$ for every function $g(x)$. If $g(x)$ is differ entiable, one can easily prove that

$$\int_a^b f(x)dg(x) = \int_a^b f(x)g'(x)dx.$$

But, the Stieltjes integral also exists for non differentiable functions $g(x)$ and even for som discontinuous functions $g(x)$; for instance, it exists for functions having only a finite numbe of maxima and minima in the interval $a \leq x \leq b$ provided that $f(x)$ is not too discontinuous.

The function $g(x)$ is said to be of *bounded variation* in the interval $a \leq x \leq b$, if there exists a finite number V such that for every subdivision (Eq. (14)) of the interval we have

$$|g(x_1) - g(x_0)| + \cdots + |g(x_{k+1}) - g(x_k)|$$
$$+ \cdots + |g(x_n) - g(x_{n-1})| \leq V.$$

It can be shown that every continuous function having only a finite number of maxima and minima in an interval is of bounded variation in this interval. The converse of this theorem is not true. The continuous function

$$g(x) = \begin{cases} x^2 \sin(1/x) & \text{for} \quad 0 < x \leq 1 \\ 0 & \text{for} \quad x = 0 \end{cases}$$

n be proved to be of bounded variation in the interval $0 \leq x \leq 1$, although it has infinitely many maxima and minima in the interval.

It can be shown that $\int_a^b f(x) dg(x)$ exists whenever $f(x)$ is continuous and $g(x)$ of bounded variation in the interval $a \leq x \leq b$. The integral form of the first law reads

$$q(t_1) - q(t_0) = u(t_1) - u(t_0) + \int_{t_0}^{t_1} p(t) dv(t). \quad (15)$$

If $v(t)$ is differentiable, then the last integral is $= \int_{t_0}^{t_1} p(t) v'(t) dt$. We can divide equality (15) by $-t_0$. If also $q(t)$ and $u(t)$ are differentiable, and we let t_1 approach t_0, then we obtain the differential form, Eq. (1), of the first law for $t = t_0$. But from what was said about Stieltjes integrals it follows that the integral form of the first law is also applicable to processes which are not differentiable, even to processes in which the volume undergoes discontinuous changes, provided that at the moments of these explosions the pressure is continuous.

7. Expansion and Contraction

The remark contained in this section is due to Mr. Leo A. Schmidt. We call a process Π_m an *expansion*, if $a \leq t_1 \leq t_2 \leq b$ implies $v(t_1) \leq v(t_2)$ and $v(a) < v(b)$; a *contraction*, if $a \leq t_1 < t_2 \leq b$ implies $v(t_1) \geq v(t_2)$ and $v(a) > v(b)$. The subscript in Π_m indicates that, in both cases, $v(t)$ is what is called a *monotonic* function. If $g(x)$ is monotonic for $a \leq x \leq b$, then the Stieltjes integral $\int_a^b f(x) dg(x)$ satisfies the following mean value theorem: There exists a number x^* between a and b such that

$$\int_a^b f(x) dg(x) = f(x^*)[g(b) - g(a)].$$

From the integral form (15) of the first law it thus follows for Π_m that

$$q(t_1) - q(t_0) = p(t^*)[v(t_1) - v(t_0)] + u(t_1) - u(t_0)$$

for some t^* between t_0 and t_1, or briefly, that

$$\Delta q = p^* \Delta v + \Delta u.$$

If we set $p(t_0) = p$, it follows that

$$\Delta q - p \Delta v - \Delta u = (p^* - p) \Delta v.$$

Now, if we denote by Δp_{max} the maximum change

that the pressure undergoes during the process Π_m then we have

and

$$|p^* - p| \leq \Delta p_{max}$$

$$\left| \frac{\Delta q - p \Delta v - \Delta u}{\Delta v} \right| \leq \Delta p_{max}.$$

In words: *In an expansion or contraction, if the change in pressure is sufficiently small, the difference between Δq and $p \Delta v + \Delta u$ is as small in comparison with Δv as we please.*

8. What is a Differential?

We shall discuss the case of three variables from which the transition to two or one, as well as to four or more variables, is easy.

Let f be a function of three variables which in a domain D admits three continuous first partial derivatives. For every triple (x, y, z) belonging to the domain D, and every triple of numbers $(\Delta x, \Delta y, \Delta z)$, we form the number

$$\Lambda_f(x, y, z; \Delta x, \Delta y, \Delta z) = \frac{\partial f}{\partial x}(x, y, z) \Delta x$$

$$+ \frac{\partial f}{\partial y}(x, y, z) \Delta y + \frac{\partial f}{\partial z}(x, y, z) \Delta z.$$

If we set $\Delta r = [(\Delta x)^2 + (\Delta y)^2 + (\Delta z)^2]^{\frac{1}{2}}$, then, as is taught in the elements of calculus,

$$\Lambda_f(x, y, z; \Delta x, \Delta y, \Delta z) \sim_{\Delta r} f(x + \Delta x, y + \Delta y, z + \Delta z) - f(x, y, z).$$

That is, for small Δr the difference between Λ_f and Δf is not only small—this is obvious because for small Δr both Λ_f and Δf themselves are small—but small in comparison with Δr. In other words, by choosing Δr sufficiently small we can make

$$\left| \frac{\Delta f - \Lambda_f}{\Delta r} \right| \quad (16)$$

as small as we please. [The statement mentioned in Sec. 1 about a function $f(t)$ is the analog for functions of one variable of our last statement about $f(x, y, z)$.]

For every triple (x, y, z) belonging to D the following *difference function* of f at (x, y, z), $\Delta f = f(x + \Delta x, y + \Delta y, z + \Delta z) - f(x, y, z)$, is a function (and, in general, a complicated function) of Δx, Δy, Δz, while $\Lambda_f(x, y, z; \Delta x, \Delta y, \Delta z)$, which we call the *differential* of f at (x, y, z), is a linear function of Δx, Δy, Δz.

The significance of the above result lies in the fact that, for sufficiently small Δr, the inaccuracy we incur by replacing the complicated difference function by the linear differential is as small as we please in comparison with Δr.

Traditionally, the last fact is symbolized as follows

$$df = \frac{\partial f}{\partial x} dx + \frac{\partial f}{\partial y} dy + \frac{\partial f}{\partial z} dz. \quad (17)$$

In presenting Eq. (17) to the beginning mathematician and physicist one cannot emphasize too strongly that this formula is neither more nor less than a symbol for the possibility of approximating Δf by Λ_f in the way explained in the preceding paragraph. It is the conviction of the author that it would be best not to present to beginners the symbolic expression Eq. (17) at all.

Now let us assume that in a domain D of the (x, y, z)-space three continuous functions, $P(x, y, z)$, $Q(x, y, z)$, and $R(x, y, z)$, are given. For every triple (x, y, z) belonging to D and every triple of numbers $(\Delta x, \Delta y, \Delta z)$ we may form the number

$$\Lambda_{P, Q, R}(x, y, z; \Delta x, \Delta y, \Delta z) = P(x, y, z)\Delta x$$
$$+ Q(x, y, z)\Delta y + R(x, y, z)\Delta z.$$

In this way for given P, Q, and R, we have defined a function of six variables. We can also say: given P, Q, and R, for every triple (x, y, z) belonging to D, we have defined a linear function of $(\Delta x, \Delta y, \Delta z)$. If, in particular, P, Q, and R are the partial derivatives of a function $f(x, y, z)$

$$P = \partial f/\partial x, \quad Q = \partial f/\partial y, \quad R = \partial f/\partial z,$$

then $\Lambda_{P, Q, R}(x, y, z; \Delta x, \Delta y, \Delta z)$ is what we have denoted by $\Lambda_f(x, y, z; \Delta x, \Delta y, \Delta z)$ and called the differential of f. For this reason, if P, Q, and R are any three given continuous functions, $\Lambda_{P, Q, R}$ has received the rather unfortunate name of a (general) *differential* while the differential Λ_f of a function is called a *complete* or *exact* differential.

The traditional symbols

$$P(x, y, z)dx + Q(x, y, z)dy + R(x, y, z)dz$$

for the (general) differential, shaped after the symbols on the right side of Eq. (17), are, as we shall see, even more unfortunate. At this place we just emphasize that in our definition of $\Lambda_{P, Q, R}$ the numbers Δx, Δy, Δz are in no way restricted and, in particular, need not be small.

Summarizing we can say: Λ_{f_x, f_y, f_z} is the differential of f. The differential $\Lambda_{P, Q, R}$, for three given functions P, Q, and R with the domain D, is the association of the linear function $\Lambda_{P, Q, R}(x, y, z; \Delta x, \Delta y, \Delta z)$ of $(\Delta x, \Delta y, \Delta z)$ with every (x, y, z) belonging to D.

For instance, if we have

$$P(x, y, z) = 3x^2 + yz + 2, \quad Q(x, y, z) = 0,$$
$$R(x, y, z) = 4x$$

for the domain of all triples (x, y, z), then with $(0, 0, 0)$, $(1, 1, 1)$, and $(3, 2, 1)$ we associate the linear functions

$$\Lambda_{P, Q, R}(0, 0, 0; \Delta x, \Delta y, \Delta z) = 2\Delta x,$$
$$\Lambda_{P, Q, R}(1, 1, 1; \Delta x, \Delta y, \Delta z) = 6\Delta x + 4\Delta z,$$
$$\Lambda_{P, Q, R}(3, 2, 1; \Delta x, \Delta y, \Delta z) = 31\Delta x + 12\Delta z.$$

The concept of a differential is thus exceedingly simple. But it is of that kind of simplicity and abstractness which baffles beginners. The usual symptom of their lack of understanding is their response to the above definition of $\Lambda_{P, Q, R}$ with the question "But what *is* a differential?" They will not ask this question with regard to the differential of a function f. For in the mind of the beginner, Λ_f gets its real meaning from the fact that it is approximately equal to Δf, the difference between values of f. This point of view is, of course, incorrect. *By definition*, Λ_f is a linear function of $(\Delta x, \Delta y, \Delta z)$ for every (x, y, z), precisely as is $\Lambda_{P, Q, R}$; and it is a *theorem* that, if Λ_f is so defined, then Λ_f and Δf differ by a quantity which is as small as we please in comparison with $\Delta r = [(\Delta x)^2 + (\Delta y)^2 + (\Delta z)^2]^{\frac{1}{2}}$ provided that Δr is sufficiently small. But the tie between Λ_f and Δf is what really interests the beginner. No similar tie exists between $\Lambda_{P, Q, R}$ and more familiar concepts. The result is that $\Lambda_{P, Q, R}$ is not understood by beginners with the exception of a small minority.

The traditional expression, $Pdx + Qdy + Rdz$, for the differential adds to the confusion. For in the symbolic equality Eq. (17) we write dx, dy, and dz in order to indicate that the inaccuracy incurred by replacing Δf by Λ_f gets smaller and naller in comparison with $|\Delta x|$, $|\Delta y|$, and $|\Delta z|$ s these quantities get sufficiently small. Hence ne expression $Pdx+Qdy+Rdz$ suggests that ome inaccuracy gets smaller or, at any rate, nat something happens to $P\Delta x+Q\Delta y+R\Delta z$ rhen $|\Delta x|$, $|\Delta y|$, and $|\Delta z|$ get sufficiently small. As a matter of fact, however, in general nothing articular happens under these circumstances.

The conclusion which the author draws from hese facts is *that physicists in presenting the lements of the theory should refrain from referring o the poorly understood concept of a general 'ifferential altogether.*

How such references can easily be avoided in presenting the first law has been shown in Sec. 1 where we introduced the law as the relation (1) between time derivatives, and in Sec. 6 where we presented the integral form Eq. (15).

After the beginner has really understood the first law in these forms which he is prepared to understand, the physicist may, for historical reasons, wish to present the differential form Eq. (3). In doing so he ought to emphasize the remark made in Sec. 2 that the differentials are meant with respect to the time, and he should present examples, such as those of Sec. 2, which forestall misinterpretations of the differentials in other directions. For

$$p\Delta v+\Delta u \quad \text{and} \quad [p+(\partial u/\partial v)]\Delta v+(\partial u/\partial \theta)\Delta \theta$$

are general differentials in two variables v, u and v, θ, respectively. For instance, the latter differential can be written in the form

$$A(v, \theta)\Delta v+B(v, \theta)\Delta \theta,$$

where

$$A(v, \theta)=P(v, \theta)+(\partial u/\partial v)(v, \theta)$$

and $B(v, \theta)=(\partial u/\partial \theta)(v, \theta)$, $P(v, \theta)$ representing the pressure as a function of volume and temperature. In general, the differential $A(v, \theta)\Delta v+B(v, \theta)\Delta \theta$ is not the differential of a function $f(v, \theta)$. In their futile attempts to connect this differential with more familiar concepts beginners are misled by the unfortunate form Eq. (3) of the first law into the misinterpretations discussed in Sec. 3 which cannot be prevented by simply replacing the symbol dq by dq, δq, or Q.

9. A General Scheme for the Elimination of Differentials

More easily comprehensible to the beginner than the differential $\Lambda_{P, Q, R}(x, y, z; \Delta x, \Delta y, \Delta z)$ is the simple idea of the triple of functions $(P(x, y, z), Q(x, y, z), R(x, y, z))$ because it may be interpreted as the association of a vector with each point. With the point (x, y, z) we associate the vector with the components $P(x, y, z)$, $Q(x, y, z)$, and $R(x, y, z)$.

The possibility of replacing the idea of the differential $\Lambda_{P, Q, R}$ by that of the vector field (P, Q, R) is most obvious in mechanics. A force field in the three-dimensional space is given by the association of a vector $(P(x, y, z)$, $Q(x, y, z)$, $R(x, y, z))$ with each point (x, y, z) of a domain.

The differential with the classical symbol

$$P(x, y, z)dx+Q(x, y, z)dy+R(x, y, z)dz \quad (18)$$

means that for every point (x, y, z) we define a linear function of $(\Delta x, \Delta y, \Delta z)$; that is, that, for every point (x, y, z), with every vector from (x, y, z) to $(x+\Delta x, y+\Delta y, z+\Delta z)$ we associate a number, namely, the scalar product of this vector with the vector $(P(x, y, z), Q(x, y, z), R(x, y, z))$. The physical meaning of this scalar product is the work we should do by moving a unit mass from (x, y, z) to $(x+\Delta x, y+\Delta y, z+\Delta z)$ along the straight line, if the force vector everywhere along this path were identical with the force vector at the initial point (x, y, z) which, in general, is, of course, not the case.

Now, in spite of the fact that in some respects the concept of work is more basic than that of force, one will probably hesitate to introduce the beginner to the theory of force fields by taking the differential Eq. (18) as the starting point. One will much rather begin with the force field and then define the work W done by moving a unit mass along the path $x=x(t)$, $y=y(t)$, $z=z(t)$, $(t_0 \le t \le t)$ as the Stieltjes integral

$$\int_{t_0}^{t_1} [P(x(t), y(t), z(t))dx(t)$$

$$+Q(x(t), y(t), z(t))dy(t)$$

$$+R(x(t), y(t), z(t))dz(t)]$$

corresponding to the integral form Eq. (15) of the first law. For differentiable motions, we may relate the time derivatives of W and the components of the velocity,

$$W'(t) = P(x(t), y(t), z(t))x'(t)$$
$$+ Q(x(t), y(t), z(t))y'(t)$$
$$+ R(x(t), y(t), z(t))z'(t)$$

corresponding to the form Eq. (1) of the first law.

The general procedure for eliminating differentials $\Lambda_{P, Q, R}$ is to concentrate on the triple of functions $(P(x, y, z), Q(x, y, z), R(x, y, z))$. For instance, we shall not say that the differential $Pdx + Qdy + Rdz$ is complete or exact, or, in other words, that there exists a function f such that $df = Pdx + Qdy + Rdz$. Instead we shall say that the triple of functions (P, Q, R) is *exact* if, and only if, there exists a function f such that

$$P = \partial f/\partial x, \quad Q = \partial f/\partial y, \quad R = \partial f/\partial z.$$

Similarly, we shall say, the triple (P, Q, R) is *integrable* if, and only if, there exists a function $s(x, y, z)$ such that the triple (sP, sQ, sR) is exact. The beginner can be expected to know that every pair of functions of two variables $(P(x, y), Q(x, y))$, but not every triple of functions of three variables, is integrable.

10. The Second Law

The second law, in its traditional form, is the conjunction of various, more or less independent, statements. We shall mention the three main mathematical consequences of the second law.

(a) *The existence of the entropy.*—There exists a function of the state, $S(v, \theta)$, for which

$$\frac{\partial S}{\partial v} : \frac{\partial S}{\partial \theta} = \left[P(v, \theta) + \left(\frac{\partial u}{\partial v} \right)_\theta \right] : \left(\frac{\partial u}{\partial \theta} \right)_v.$$

In the terminology of the preceding section, statement (a) contends that the pair of functions, $P(v, \theta) + (\partial u/\partial v)_\theta$, $(\partial u/\partial \theta)_v$ is integrable or, in the classical terminology, that the differential

$$[P(v, \theta) + (\partial u/\partial v)_\theta]dv + (\partial u/\partial \theta)_v d\theta$$

is integrable. As we mentioned at the end of the preceding section, every pair of two functions of two variables and every differential in two variables is integrable. Hence, in the case of processes which can be characterized by two functions $v(t)$ and $\theta(t)$, statement (a) is provable and thus does not impose any limitations on the course of nature.

The situation changes, if we study more complicated processes, for instance, processes in which each state is characterized by four numbers v_1, θ_1, v_2, and θ_2 describing volume and temperature of two substances. In this case, the corresponding quartet of functions or, in the classical terminology, the differential

$$\left[P_1(v_1, \theta_1) + \left(\frac{\partial u_1}{\partial v} \right)_{\theta_1} \right] dv_1 + \left(\frac{\partial u_1}{\partial \theta_1} \right)_{v_1} d\theta_1$$
$$+ \left[P_2(v_2, \theta_2) + \left(\frac{\partial u_2}{\partial v_2} \right)_{\theta_2} \right] dv_2 + \left(\frac{\partial u_2}{\partial \theta_2} \right)_{v_2} d\theta_2 \quad (19)$$

need not be and, in general, is not integrable. For instance, one readily proves that the differential

$$\frac{R\theta_1}{v_1} dv_1 + c_1 d\theta_1 + \frac{R\theta_2}{v_2} dv_2 + c_2 d\theta_2 \quad (20)$$

corresponding to two ideal gases, is not integrable. For the assumption that there exists a function $S(v_1, \theta_1, v_2, \theta_2)$ such that

$$\frac{\partial S}{\partial v_1} : \frac{\partial S}{\partial \theta_1} : \frac{\partial S}{\partial v_2} : \frac{\partial S}{\partial \theta_2} = \frac{R\theta_1}{v_1} : c_1 : \frac{R\theta_2}{v_2} : c_2$$

leads to a contradiction. However, in the case of two substances, the analog of statement (a) postulates the integrability only if the two substances have the same temperature, that is, if $\theta_1 = \theta_2 = \theta$. The differential Eq. (20) reads in this case

$$\frac{R\theta}{v_1} dv_1 + \frac{R\theta}{v_2} dv_2 + (c_1 + c_2)d\theta \quad (21)$$

and Eq. (21) is indeed integrable. For, let S be the function

$$S(v_1, v_2, \theta) = R \log v_1 + R \log v_2 + (c_1 + c_2) \log \theta. \quad (22)$$

Then we have

$$\frac{\partial S}{\partial v_1} = \frac{R}{v_1}, \quad \frac{\partial S}{\partial v_2} = \frac{R}{v_2}, \quad \frac{\partial S}{\partial \theta} = \frac{c_1 + c_2}{\theta}$$

and, thus, obviously

$$\frac{\partial S}{\partial v_1} : \frac{\partial S}{\partial v_2} : \frac{\partial S}{\partial \theta} = \frac{R\theta}{v_1} : \frac{R\theta}{v_2} : c_1 + c_2.$$

ence the differential Eq. (21) is compatible ith proposition (a).

One easily sees that, for instance, the differ- tial $\theta dv_1 + dv_2 + d\theta$ is incompatible with propo- tion (a). Hence proposition (a) precludes the ossibility of processes corresponding to the bove differential.

We finally mention the obvious fact that for very pair of functions of two variables there xist *many* functions whose partial derivatives re proportional to the functions of the pair. or instance, if

$$\frac{\partial S}{\partial v} : \frac{\partial S}{\partial \theta} = P(v_1\theta) + \left(\frac{\partial u}{\partial v}\right)_\theta : \left(\frac{\partial u}{\partial \theta}\right)_v,$$

hen, for every function f of one variable if $(v, \theta) = f(S(v, \theta))$, we have

$$\frac{\partial T}{\partial v} : \frac{\partial T}{\partial \theta} = \frac{\partial S}{\partial v} : \frac{\partial S}{\partial \theta}.$$

imilarly, if a triple of functions or a differential n three variables, such as Eq. (21), is integrable, here exist many functions S satisfying the equired conditions. One of them is selected ccording to physical principles and called the ntropy.

(b) *The nature of the integrating denominator.—* he ratio

$$(\partial S/\partial v) : [P(v, \theta) + (\partial u/\partial v)_\theta]$$

which, by statement (a), is equal to the ratio $\partial S/\partial \theta) : (\partial u/\partial \theta)_v$, is independent of v. If we lenote this ratio by $1/N(\theta)$, then $N(\theta)$ is called n *integrating denominator* of the pair of functions $^2(v, \theta) + (\partial u/\partial v)_\theta$, $(\partial u/\partial \theta)_v$ or of the correspond- ng differential.

While the exactness of this pair of functions of wo variables, and hence the existence of *some* ntegrating denominator, is provable (as we mphasized in discussing statement (a), the xistence of a *particular* denominator, namely, of denominator which is independent of v (asserted n statement (b)), is a law of nature. This law recludes, for instance, the possibility that

$$P(v, \theta) + (\partial u/\partial v)_\theta = v \text{ and } (\partial u/\partial \theta)_v = e^v.$$

Statement (b) is even more restrictive with egard to processes in which two or more sub- tances (at the same temperature θ) are involved.

But we see that the differential Eq. (21) satisfies the condition of statement (b) since $N(\theta) = \theta$ is an integrating denominator.

From statements (a) and (b) in conjunction with the first law we obtain

$$\frac{\partial S}{\partial v}(v(t), \theta(t))v'(t) + \frac{\partial S}{\partial \theta}(v(t), \theta(t))\theta'(t)$$

$$= \frac{P(v(t), \theta(t)) + (\partial u/\partial v)_\theta}{N(\theta(t))}v'(t)$$

$$+ \frac{(\partial u/\partial \theta)_v\theta'(t)}{N(\theta(t))} = \frac{q'(t)}{N(\theta(t))}.$$

If for every moment t we set

$$s(t) = S(v(t), \theta(t)),$$

we have

$$s'(t) = q'(t)/N(\theta(t))$$

and

$$s(t_1) - s(t_0) = \int_{t_0}^{t_1} \frac{dq(t)}{N(\theta(t))}.$$

The last Stieltjes integral might be used as a definition of $s(t)$ even in cases when $q(t)$ is a not differentiable function.

As a corollary, we deduce: If Π_{Q0} is an adia- batic quasistatic process for which $q(t) = q_0$ during the period $t_0 \leq t \leq t_1$, then $s(t_1) - s(t_0) = 0$. In words: *During an adiabatic quasi-static process, the entropy remains constant.* This statement is also valid for processes in which two or more substances are involved.

(c) *The monotony of the entropy.* During an adiabatic generalized process Γ the entropy never decreases.

Since we have just seen that during an adia- batic quasi-static process the entropy remains constant we first illustrate statement (c) by a nonquasi-static process during which the entropy increases; secondly, we mention a process during which the entropy decreases and which, conse- quently, by statement (c) is impossible in nature.

We let an ideal gas undergo the nonquasi-static process mentioned in Sec. 1 in such a way that both containers are adiabatically isolated from the rest of the world. Then we have

$$v(t_0) = v_0, \quad v(t_1) = 2v_0,$$
$$q(t) = q_0 \text{ for } t_0 \leq t \leq t_1.$$

Experience shows that the temperature also remains constant during this process

$$\theta(t) = \theta_0 \quad \text{for} \quad t_0 \leq t \leq t_1.$$

In analogy to formula (22), for the entropy of an ideal gas characterized by its volume v and its temperature θ, we obtain expression $S(v, \theta) = R \log v + c \log \theta$. If $s(t)$ denotes the entropy of the gas at the moment t, we thus have

$$s(t_0) = S(v_0, \theta_0), \quad s(t_1) = S(2v_0, \theta_0)$$
$$s(t_0) = S(v_0, \theta_0) = R \log v_0 + c \log \theta_0$$
$$s(t_1) = S(2v_0, \theta_0) = R \log(2v_0) + c \log \theta_0,$$

thus

$$s(t_0) < s(t_1).$$

A nonquasi-static process precluded by statement (c) is the reverse of the above expansion, that is, a spontaneous adiabatic contraction of a gas to half of its initial volume. Since in this case

$$s(t_0) = S(2v_0, \theta), \quad s(t_1) = S(v_0, \theta_0)$$

we should have

$$s(t_0) > s(t_1)$$

which is impossible by statement (c).

Appendix. Which Processes are Thermodynamically Equal?

Let M be a continuous *motion* of a particle in the (x, y)-plane. We describe M by two continuous functions $X(t)$ and $Y(t)$, both defined for the interval $t_0 \leq t \leq t_1$ where, for every t of this interval, $X(t)$ denotes the abscissa, $Y(t)$ the ordinate of the particle at the moment t. By the *trace* of the motion M we shall mean the set (or locus) of all points (x, y) traversed by the moving point; that is, (x, y) belongs to the trace of M if, and only if, there exists at least one number t such that $t_0 \leq t \leq t_1$ and $x = X(t)$ and $y = Y(t)$.

Now let M^* be a continuous motion of a particle in the same plane, described by the two functions $X^*(u)$ and $Y^*(u)$ both defined for $u_0 \leq u \leq u_1$. The motions M and M^* are *identical* if, and only if, they occur during the same time interval and, at every moment, the positions of the two particles are identical; that is, if, and only if, $t_0 = u_0$, $t_1 = u_1$ and $X(s) = X^*(s)$, $Y(s) = Y^*(s)$ for every s of the interval.

Besides this strict identity, we are also interested in weaker relations between M and M^*. For instance, we might call M and M^* *kinematically equal* if they differ only inasmuch as they occur at different time intervals but become identical if one of the clocks used in describing the motions, say the second, is advanced by c time units; that is, if and only if $t_0 = u_0 + c$, $t_1 = u_1 + c$, $X^*(u) = X(u + c)$, and $Y^*(u) = Y(u + c)$ for every u between u_0 and u_1.

We might call M and M^* *kinematically similar* if they become identical when the second clock, at the moment it is accelerated in the ratio $b:1$ and then advanced by time units; that is to say, if $t_0 = bu_0 + c$, $t_1 = bu_1 + c$, $X^*(u) = X(bu + c)$, and $Y^*(u) = Y(bu + c)$ for every u between u_0 and u_1.

When shall we say that M and M^* are *geometrically equal* or, as it is also expressed, that M and M^* determine the same *path*? A *necessary* condition is, of course, that M and M^* have the same trace. But this condition is not sufficient. For the mere trace of a motion does not, for instance, determine the length of the path traversed, and certainly this length is geometrically relevant. For example if we merely know that the trace of a motion is the set of all points on the circle $x^2 + y^2 = r^2$, then we know that the length of the path is at least $2\pi r$. But it is $2n\pi r$, if the moving particle traverses the circle n times according to the equations $x = r \cos nt$, $y = r \sin nt$ for $0 \leq t \leq 2\pi$. In fact any number $\geq 2\pi r$ is the length of the path of some motion having the above circle as its trace.

A *sufficient* condition for geometric identity is kinematic equality or similarity. But even similarity is by no means necessary. If, for instance, M_1 is described by

$$x = r \cos\sqrt{t}, \quad y = r \sin\sqrt{t} \quad \text{for} \quad 0 \leq t \leq 4\pi^2$$

and M_1^* is described by

$$x = r \cos u, \quad y = r \sin u \quad \text{for} \quad 0 \leq u \leq 2\pi,$$

then M_1 and M_1^* are not kinematically similar although they are equal in every geometrical respect.

These remarks show that the geometric equality of motions or the equality of their paths is less than their kinematic similarity and more than the mere identity of their traces.[6]

In the last example as in the case of kinematically similar motions we see that the time intervals $0 \leq t \leq 4\pi^2$ and $0 \leq u \leq 2\pi$ during which the motions take place, are monotonically related in such a way that in related moments the positions of the particles are identical. The relation is established by the function $t = u^2$ which amounts to a (nonlinear) distortion of the time scale satisfying the condition of monotony. For if the moment u' precedes the moment u'', then the moment $t' = u'^2$ precedes the moment $t'' = u''^2$.

Next we consider the following two motions along the X axis: M_2 described by the functions

$$X(t) = \begin{cases} 2t & \text{for } 0 \leq t \leq 1/2 \\ 1 & \text{for } 1/2 \leq t \leq 1, \end{cases} \quad Y(t) = 0 \text{ for } 0 \leq t \leq 1,$$

---

[6] A path can thus be defined either as *a class of motions* (namely, as the class of all motions which are geometrically equal to any motion of the class) or as *a trace with some additional information* (namely, information from which we can infer the way the trace has been traversed). This intrinsic definition of a path has been developed in the author's note "Définition intrinsèque de la notion de chemin," *Comptes Rendus* 221, 739 (1945). In what follows, we apply the first definition to thermodynamical processes. But also our intrinsic definition can be applied to thermodynamics.

d M_2^* described by the functions

$$X(u) = \begin{cases} \frac{3}{2}u & \text{for } 0 \le u \le 1/3 \\ 1/2 & \text{for } \frac{1}{3} \le u \le 2/3, \\ \frac{3}{2}u - \frac{1}{2} & \text{for } \frac{2}{3} \le u \le 1 \end{cases} \quad Y(u) = 0 \text{ for } 0 \le u \le 1.$$

he first particle moves from $(0, 0)$ to $(1, 0)$ and rests ere for one-half time unit; the second moves from $(0, 0)$ o $(\frac{1}{3}, 0)$, rests there for one-third time unit, and then oves to $(1, 0)$. We call $\frac{1}{2} \le t \le 1$ and $\frac{1}{3} \le u \le \frac{2}{3}$ *rest intervals* f the two motions. In the case of M_2 and M_2^* it would e impossible to establish a one-to-one relation or a strictly onotonic relation between the time intervals in such a ay that at related moments the positions of the moving articles are identical. Yet M_2 and M_2^* are equal in every eometric respect. For the differences regarding the speed r the rest intervals are kinematic rather than geometric. he motions M_2 and M_2^* become identical if the clocks y means of which we describe the motions are changed in he following way. We let the first clock stop between $= \frac{1}{2}$ and $t = 1$ so that the entire motion takes place between $= 0$ and $t = \frac{1}{2}$. The second clock is decelerated in the ratio :4 between $u = 0$ and $u = \frac{1}{3}$ so that the first part of M_2^* akes place between 0 and $\frac{1}{4}$; between $u = \frac{1}{3}$ and $u = \frac{2}{3}$ we et it stop; between $u = \frac{2}{3}$ and $u = 1$ we let it resume its ecelerated rate so that the third part of M_2^* takes place etween $\frac{1}{4}$ and $\frac{1}{2}$. If the clocks are adjusted in this way, t every moment the positions of the particles are identical. ince we have achieved this identity by merely changing he time scales without ever reversing either clock we hall consider M_2 and M_2^* as geometrically equal.

In general, we shall say that two motions

M given by $X(t)$, $Y(t)$ for $t_0 \le t \le t_1$,
M^* given by $X^*(u)$, $Y^*(u)$ for $u_0 \le u \le u_1$,

re geometrically equal if, and only if, we can relate the moments between t_0 and t_1, and the moments between u_0 nd u_1 in a nonrecurrent way and so that for every pair (t, u) of related moments we have

$$X(t) = X^*(u) \quad \text{and} \quad Y(t) = Y^*(u).$$

Here we say that the moments are related in a *nonrecurrent* way if

(1) every moment between t_0 and t_1 is related to at least one moment between u_0 and u_1; and every moment between u_0 and u_1 is related to at least one moment between t_1 and t_2;

(2) if t, u as well as t', u' are pairs of related moments, then the numbers $t - t'$ and $u - u'$ are not of opposite signs; that is, it never occurs that one of them is positive and the other one negative.

One readily derives from this definition that if t, u and t', u' are two pairs of related moments such that $u < u'$, then $t \le t'$ and that from $t < t'$ it follows that $u \le u'$. One further sees that t_0, u_0 as well as t_1, u_1 is a pair of related moments. If one moment of a rest-interval of either motion is related to a moment of the other motion, then every moment of the rest-interval is related to the same moment of the other motion. One can easily prove that for geometrically equal motions the paths traversed have the same length.

Now let Π and Π^* be two thermodynamical processes. For the sake of simplicity we shall assume that Π is determined by two functions $v(t)$ and $\theta(t)$ for $t_0 \le t \le t_1$ and Π^* by two functions $v^*(t^*)$ and $\theta^*(t^*)$ for $t_0^* \le t^* \le t_1^*$. The extension of the subsequent remarks to processes whose characterization requires more than two functions presents no more difficulties than the extension of what has been said about motions in the plane to motions in a space of three or more dimensions.

We say that Π and Π^* are thermodynamically equal, if the motions in the (v, θ)-plane described by

$$v(t), \theta(t) \quad \text{for } t_0 \le t \le t_1$$

and

$$v^*(t^*), \theta^*(t^*) \quad \text{for } t_0^* \le t^* \le t_1^*$$

are geometrically equal.

One readily proves that if the processes Π and Π^* are thermodynamically equal, then

$$\int_{t_0}^{t_1} P(v(t), \theta(t)) dv(t) = \int_{t_0^*}^{t_1^*} P(v^*(t^*), \theta^*(t^*)) dv^*(t^*).$$

Some scientists regard an interest in the history of their subject as mere antiquarianism, and it may be that the very remote past consists largely of mistakes to be avoided. But it deserves to be remembered that the history of any scientific discipline intimately determines the current modes of investigation. The frames of reference which appear eligible at any given epoch, the instruments accepted as respectable, and the types of "fact" taken to have evidential value are historically conditioned. To pretend otherwise is to claim for human reason, as manifested in scientific progress, a universality and fixity it has never manifested.—MAX BLACK, "The Definition of Scientific Method," *Science and Civilization*, Edited by Robert C. Stauffer (The University of Wisconsin Press, Madison, 1949).

RANDOM VARIABLES FROM THE POINT OF VIEW OF A GENERAL THEORY OF VARIABLES

KARL MENGER

ILLINOIS INSTITUTE OF TECHNOLOGY

1. Introduction

In his great book *Sequential Analysis*, Wald defines (see p. 5 in [1]) a random variable as a variable x such that "for any given number c a definite probability can be ascribed to the event that x will take a value less than c." As a first example of a random variable, Wald mentions the outcome x of the experiment of weighing an object selected at random from a lot of n known objects. He calls x a random variable "since a probability can be ascribed to the event that x will take a value less than c, for any given c." If n_c is the number of objects in the lot whose weight is less than c, that probability is n_c/n. On page 11, Wald says that "statistical problems arise when the distribution function of a random variable is not known and we want to draw some inference concerning the unknown distribution function on the basis of a limited number of observations." He then mentions, as an example, the random variable x assuming the value 0 if a unit selected from a completely unknown lot of products is nondefective, and the value 1 if the unit is defective.

In 1947, I submitted to Wald the following two observations: (1) the concept "variable" on which the notion of random variable is based (see p. 5 in [1]) does not appear to be that of a numerical variable, the only one then clearly defined; (2) the statement and example on page 11 seem to be at variance with the definition of random variables on page 5.

I believe that I carry out Wald's intentions by saying that he fully agreed with both remarks and expressed the hope to clarify the statistical concept of random variables at a later occasion. His untimely death in 1950, after the completion of his fundamental book on statistical decision functions (in which he essentially retained the treatment of random variables of *Sequential Analysis*) prevented him from carrying out this plan.

For the past few years I have tried to analyze the ideas behind the general term "variable"—a term that, in spite of its frequent and heretofore indiscriminate use, has never been introduced by a comprehensive definition (either explicitly, in terms of other concepts, or implicitly, by postulates). As a result of these studies [2], [3], [4], and especially [5], it appears that there is not one comprehensive concept of variable. The underlying material has been resolved into an extensive spectrum of concepts. That array begins in mathematical logic; it traverses algebra, analysis, the various types of geometry, and physical science; it touches social science, and it ends in statistics. Some of those concepts have only one common bond—the name variable. In content, they differ about as much as do the tangent of an angle in trigonometry and the tangent to a curve in geometry. But whereas no one has ever confused the latter two ideas because of a flimsy equivoca-

215

tion, the equivocal use of the term "variable" has indeed resulted in confusion. Some of those ambiguities account for the obscurity in the foundations of pure analysis, others for the lack of articulate rules concerning certain applications of analysis to science. "Variables" will be discussed in sections 2–5.

The term "random," which is widely used in statistics and the theory of probability, is in a condition that very much resembles that of "variable." It will be analyzed in section 6.

In sections 7–11, these results are applied to random variables. Various types must be distinguished even within the realm of these "variables." In the last section, the relation between random variables in statistics and in the theory of probability will be shown in a new light.

2. Logico-mathematical variables

In pure mathematics, the only concepts of variable that possess clear traditional definitions are of the type of the so-called *numerical variables*. The latter concept may be illustrated as follows. The formulas

(1)
$$\frac{2}{3^2-1} = \frac{1}{3-1} - \frac{1}{3+1}, \qquad \frac{2}{e^2-1} = \frac{1}{e-1} - \frac{1}{e+1}, \cdots,$$

and countless other formulas can be synthesized in one formula containing a letter and accompanied by, as it were, a legend with directions concerning the use of the letter. Such a general statement is

(2)
$$\frac{2}{x^2-1} = \frac{1}{x-1} - \frac{1}{x+1},$$

where x may be replaced by the designation of any number $\neq 1$ and -1. Usually, the legend is abbreviated to "for any number x $\neq$ 1 and -1." The class of all numbers $\neq 1$, -1 is called the *scope* of the numerical variable x in (2). If x is replaced by 3, e, and other numbers, specific statements result, namely, (1) and other formulas.

The statement

(3) $(x + 3y)^2 = x^2 + 6xy + 9y^2$ for any x and any y

involves two numerical variables. The meaning of (2) and (3) is not changed if x is replaced by any other letter (except e), or x and y by any two nonidentical letters, say, by a and b or by a and y, and even if they are interchanged as in

(4) $(y + 3x)^2 = y^2 + 6yx + 9x^2$ for any x and any y .

If the scope of a numerical variable consists of a single number, then the letter designates a specific number as, for instance, does e.

In this paper, all numbers and numerical variables are printed in roman type, while italics are reserved for functions, function variables to be defined presently, and concepts introduced in subsequent sections.

Analysis abounds in laws of the following type:

(5) $D(f \cdot g) = f \cdot Dg + g \cdot Df,$

where f and g may be replaced by the designations of any two differentiable functions, such as *log*, *cos*, etc. The letters f and g in (5) are called *function variables* each having the class of all differentiable functions as its scope. The following statement,[1] which is equivalent to (5), involves numerical as well as functional variables:

(6) $$D(f \cdot g) x = f x \cdot D g x + g x \cdot D f x$$

for any f, g, and any x for which f, g, and $f \cdot g$ have derivatives.

More generally, if a formula contains a letter and is accompanied by a legend concerning the replacement of the letter by the designations of elements of a certain class, then logicians and mathematicians refer to the letter as a *variable*, and call the said class the *scope* of the variable.

In examples (2) to (6), each replacement results in a statement. In such a case, I shall say that the variable is used *indicatively*. One of the numerous other uses of variables is illustrated in the following definition.

(7)   Let *log* be the class of all pairs $(x, log\ x)$ for any $x > 0$.

If, in the formal part of (7), x is replaced, say, by 3, the result is an element of the class defined by (7), namely, the pair $(3, log\ 3)$, and not a statement. I shall refer to this use of x as *conjunctive*.

3. Consistent classes of quantities

The transition from logico-mathematical variables to the objects of statistics (the height, the weight, etc.) and to what scientists call variables (the mass of radium, the time, etc.) is a step into a different world. This contrast is what I had in mind when comparing "variables" in various fields with "tangents" in the fields of trigonometry and geometry.

A simple scheme that is useful in the mathematical treatment of scientific and statistical material is supplied by the following concept of *quantity:* an ordered pair in which the second member (the *value* of the quantity) is a number, whereas the first member (the *object* of the quantity) may be anything. I shall call two quantities *consistent* unless their objects are equal, and their values are unequal. For instance, if A is a certain resident of Chicago, then (A, 69) is a quantity. Here 69 may be A's height in inches. If B is another resident of Chicago, (B, 71) and (B, 169), where 71 may be B's height, 169 his weight in pounds, are two inconsistent quantities, though either is consistent with the quantity (A, 69).

Consider the class of all pairs $(C, h C)$ for any resident C of Chicago, where $h C$ denotes C's height in inches. I shall designate this class by h. It is what statisticians study under the name of the height in inches within the population of Chicago. Any two quantities belonging to h are consistent. I shall call a class with this property a *consistent class of quantities*—briefly, c.c.q.

In the definition of h, the letter C is a logico-mathematical variable (used conjunctively) whose scope is the population of Chicago. Replacing C by A or B, one obtains the quantities $(A, h A) = (A, 69)$ and $(B, h B) = (B, 71)$ belonging to h. But the class h, as a whole, is totally different from a numerical variable. It is not a symbol that may be replaced by designations of specific numbers. It is a class of pairs whose

---

[1] The typographical convention mentioned above (that is, the use of roman type for numbers and numerical variables, and of italics for functions and function variables) makes it unnecessary to say in the legend to (6): for any "functions" f, g and any "number" x . Moreover, without any danger of confusion, $f(x)$ can be abbreviated to fx.

second elements are numbers; and it is a specific class, just as 3 is a specific number and *log* is a specific function. That *h* and, in fact, *any object of statistical studies is totally different from numerical variables* is the first point that I wish to stress in this paper.

Another example of a consistent class of quantities is *w*, the weight in pounds of the population of San Francisco—the class of all pairs (D, *w* D) for any resident D of San Francisco, where *w* D denotes the weight in pounds of D. Also the union or set-theoretical sum of the classes *h* and *w* is a consistent class of quantities, though it is not likely ever to be practically significant.

Among the highly significant notions covered by the concept "consistent classes of quantities" are all those to which scientists refer as "variables" and "variable quantities." (In this connection, I shall avoid the former term to forestall confusion with logico-mathematical variables.) That *this vast material actually comes under the heading of consistent classes of quantities* is the **second point** that I wish to emphasize. Since those concepts are not usually defined as classes of ordered pairs, the preceding remark will be illustrated by two examples.

Consider decaying radioactive substances and let *m* be the mass in grams. This mass is a consistent class of quantities whose objects are instantaneous specimens of the substance or substances under consideration, that is, pieces of the material at a definite instant. If *m* μ denotes the mass in grams of the specimen μ, then *m* may be defined as the class of all pairs (μ, *m* μ) for any instantaneous specimen μ.

If *P* is a specific swinging pendulum, denote by *a* the angle in radians between the pendulum and a vertical line. The variable quantity *a* is the class of all pairs (σ, *a* σ) for any state σ of P, where *a* σ is the said angle in the state σ.

The class of all objects—all specimens, all states, etc.—is called the *domain* of the variable quantity; the class of all values of a variable quantity is referred to as its *range*.

A c.c.q. whose domain consists of numbers (that is, a class of ordered pairs of numbers) is called a *function*—by mathematicians as well as scientists. An example is the logarithmic function or the function *log* as defined in (7). More generally, for any positive integer p, a consistent class of quantities whose objects are ordered p-tuples of numbers is universally called a p-*place function*. Examples include the maximum, the sum, the average, and weighted means of ordered p-tuples of numbers.

While some mathematicians propose to call any c.c.q. a function—even the mass *m*, a function whose domain is the class of specimens—all physicists and many mathematicians refer to *m* (by itself) as a variable quantity and not as a function—a term that they reserve for *log*, the exponential function, the sum, etc. The difference is purely terminological and, therefore, utterly unimportant.

However, even mathematicians who refer to the height *h* and the mass *m* as functions need a specific term for what every scientist calls functions. They may refer to *log*, to the exponential function, etc., as "functions in the narrow sense of the word" or in some other cumbersome way. But they must refer to them *somehow*, because—and this has nothing to do with terminology, but is a fact and, I believe, a rather significant fact, and the **third point** that I wish to stress in this paper—*there is an important difference between functions (such as log and exp) and other consistent classes of quantities*. Not only are functions the only c.c.q.'s for which derivatives and integrals can be defined—even the most general theories of integration are inapplicable to the height *h* and the mass *m*— but *functions are the only c.c.q.'s into which other c.c.q.'s can be substituted*. One can define the logarithm of the mass and the logarithm of the cosine, but there is no mass of the height nor a mass of the logarithm.

In discussions, algebraists usually deny the significance of the last point. The point was questioned also in the discussion following the presentation of this paper. Opponents claim that even the mass lends itself to substitutions, namely, of "functions" whose values are instantaneous specimens. And indeed, if with each block β of pitchblende one associates the specimen $\mu(\beta)$ of radium that can be extracted from β, then he can substitute this "function" into m and thereby define the class of all pairs $[\beta, m \mu(\beta)]$ as the mass of radium in blocks of pitchblende. But he has not substituted a consistent class of quantities, since the values of a c.c.q. are numbers and not specimens of radium. Moreover, since Galileo, most quantitative laws of science are in terms of substitutions such as that of m into log. These laws express the connections between scientific c.c.q.'s by means of functions, for example, the connection of the time with the mass by the function log.

The applicability of numbers and functions to the most diverse quantities and c.c.q.'s is the very reason for the omnipresence of the former in science, whereas specimens of radium and m occur only in certain branches of physics, and inhabitants of Chicago and h are studied only in sociology. What some modern algebraists, in a spirit of hyperformalism, seem to overlook is the specific role of mathematics as a universal tool—a role which is epitomized in the characterization of functions among consistent classes of quantities.

4. Constant variable quantities. Observations. Consistent classes of pairs

In contrast to numbers, variable quantities can be divided into those that are constant and those that are not. A partial realization of this distinction is probably what has prompted some analysts to contrast "constants" and "variables," even though in the absence of an explicit definition of variable quantities they usually express that distinction in an obscure way.

I shall call a c.c.q. *constant* if its range consists of one single element, and nonconstant otherwise. For instance, the mileage as well as the speed of a specific car is constant while the car is parked. At least the former is nonconstant while the car is moving. In a parked car, the range of the speed consists of the single number 0; that of the mileage consists of the value attained when the car was parked.

In particular, of course, there are constant and nonconstant functions. The constant function *3* (designated by an italic) of the value 3 (in roman type) is the class of all pairs (x, 3) for any number x.

Useful in a mathematical treatment of statistical and probabilistic material is also the following concept of *observation:* a pair in which the first member is an *act* of observation, and the second member is the *result* of that act. Essentially, this concept is due to von Mises who refers to an act as "Beobachtung" and to the result as "Merkmal" (characteristic).

If the result is a number, for example, a scale mark, then the observation is a quantity. But there are acts of observation the results of which are p-tuples of numbers, for example, the p numbers of points observed when p dice have been rolled. There are also acts with altogether nonnumerical results, such as head and tail observed when a coin has been tossed.

The domain of m is also the domain of other variable quantities: an instantaneous specimen μ has also a volume, a temperature, etc. In contrast, an act of observation has, in general, only one result. (If the mass of μ is being observed on scales calibrated in different units, then the results are different, but so also are the acts.) Hence any two observations are consistent.

With most scientific variable quantities, such as m and a, there correspond classes of observations. For instance, there is a class m^* of *mass observations* in grams consisting of all pairs $(\beta, m^*\beta)$ for any act β directed to a scale (calibrated in grams) on which a specimen of a decaying substance is being weighed, where $m^*\beta$ denotes the result of the act β.

Various acts of observation $\beta_1, \beta_2, \cdots$ may be directed to the same instantaneous specimen μ; for example, acts of various observers or, if the decay is a quasi-stationary process, successive acts of the same observer. The results $m^*\beta_1, m^*\beta_2, \cdots$ may well be unequal, in which case the class of all quantities (μ, r) for any specimen and any result r of an act directed to μ is not consistent. In fact, each specimen μ gives rise to a variable quantity m^*_μ of all mass observations directed to μ, and the value $m\mu$ of m is somehow derived from the variable quantity m^*_μ.

As Dr. M. A. Woodbury pointed out in the discussion of the present paper, the preceding remarks apply to a certain extent also to functions. For instance, with the function *log* there corresponds a variable quantity *logarithmic computation:* the class of all pairs each consisting of an act of computing the logarithm of a number and the result of that act.

Clearly, the *actual domain* of a class of observations consists of a finite number of acts, the *potential domain* (including the acts that may still be carried out) may be either finite or what I shall call *indefinite*. Similarly, one has to distinguish the *actual range* (that is, the class of all actual results) and the *potential range* (including the possible results of acts belonging to the domain). In a class of at most five observations of a die (and in some larger classes) the range consists of at most five numbers, the potential range of six. The potential range is what von Mises called "Merkmalmenge" (mistranslated into "sample space"—neither are the elements samples nor is, in general, the range a space). The actual range of the class of all observations directed to the gravitational acceleration at a certain place consists essentially of one number. The acceleration at the particular place is constant.

Classes of observations are special *consistent classes of pairs*. A c.c.p. is a class of pairs (of any kind) not containing two pairs whose first members are equal and whose second members are unequal.

5. C.C.P. variables

Just as (3) is valid for any two numbers, and (5) for any two functions, the following statement is valid for any two consistent classes of quantities of a certain kind:

(8) $$\text{If } w = log\ u, \text{ then } \frac{\mathrm{d}w}{\mathrm{d}u} = \frac{1}{u}$$

for any two c.c.q.'s w and u, if the range of u consists of positive numbers.

The letters u and w in (8) are symbols accompanied by a legend according to which u and w may be replaced by the designations of specific c.c.q.'s. For instance, one may replace u by m, the mass present in a chemical reaction, and w by the time; or u by r, the distance from a certain charge, and w by p, the potential due to the charge. One thus obtains the statements: If $t = log\ m$, then $dt/dm = 1/m$; and if $p = log\ r$, then $dp/dr = 1/r$. In other words, u and w are logico-mathematical variables—but variables whose scopes consist of c.c.q.'s. I therefore shall call u and w in (8) *c.c.q. variables*. (Such variables must not be replaced by numbers. If, in (8), u were replaced by 1, and w by 0, the implication would be nonsensical even though the antecedent would be valid.)

The confusion in the literature is epitomized in the traditional misstatement that the following formulas (often presented without legends) have the same meaning:

(9) $$\frac{d\,log\,u}{du} = \frac{1}{u}$$ for any c.c.q. u whose values are positive ;

(10) $$D\ log\ x = \frac{1}{x}$$ for any number $x > 0$.

Actually, (9) deals with the rate of change of the c.c.q. variable $log\ u$ with u, and (10) with the value for x of the derivative of the function log. (If w is connected with u by the function f, then the rate of change of w with u is connected with u by Df, the derivative of f.)

It is clear how c.c.p. variables are to be defined.

The development of the idea of c.c.q. variables is the **fourth point** that I wish to emphasize in concluding the discussion of "variables" in analysis and science. Traditionally, the term "variable" has been used indiscriminately in the sense of

(1) numerical variable;

(2) consistent class of quantities (in particular, for scientific variable quantity);

(3) c.c.q. variable.

6. Various meanings of the term "random"

The following descriptions include five of the most important uses of the term "random" in statistics.

(A) Randomness is often attributed to *samples* from a population, that is, to subclasses of a class. Analysis reveals that the subclasses cannot be divided into those that are, and those that are not, random samples or even potential random samples, since in many cases every subclass is a possible random sample. Randomness in connection with samples is a property of selections of subclasses from the population rather than a property of subclasses. Its rigorous treatment (of which no example is known to me) would present exceedingly difficult problems, since a sound theory would have to be formulated in terms of (a) acts of selection, (b) aims of selections, (c) information available—as shown by the following simple examples. One cannot regard the selection of those who are over six feet tall from the people passing a busy corner as a random selection in a study of height or even weight, though one might regard it as such for other purposes, for example, possibly in a study of income. Neither would the former purposes be served by the selection of every one-hundredth passer-by if one should know that arrangements have been made as a result of which the 100th, 200th, 300th, $\cdots$ passer-by would be over six feet tall. Without such knowledge, however, one might regard the choice of every one-hundredth as a random selection.

(B) The word "random" is used with regard to certain *sequences*, namely, to irregular sequences of the type of von Mises' collectives (a concept made precise and consistent by Wald's relativization to a definite set of principles of selection) and to *sample numbers*.

(C) Numerous references are made to random *events* and, in a related sense, to random *experiments*. Random and nonrandom events in the sense in which these words are used by a man on the street will be contrasted in a simple example. Suppose that a man were offered a game based on the following understanding. Unobserved by the man, two coins are tossed. A reliable friend of the man inspects the outcome. If no tail turns up, the friend will say "zero"; if at least one tail appears, he will, at his discretion, say either "one" or "two." The man on the street will attribute random character to the two events "zero" and *either* "one" *or* "two." But he will not refer to any of the four events "one," "two," *either* "zero" *or* "one," and *either* "zero" *or* "two" as random events. The formulation of general principles guiding the man on the street in attributing random character to some events and not to others is an important (and, to my knowledge, unsolved) problem—important, because it might suggest postulates for the implicit definition of random events.

(D) Randomness is attributed to *processes*, such as Brownian motion and diffusion. The partly obscure relations of the meanings (A), (B), (C) with each other and with (D) will not here be investigated. But agreement seems to exist in the literature as to the relation between (D) and the following type (E). Random processes are generally considered as classes or families of random variables.

(E) Thus the consideration of the various uses of the terms "variable" and "random" has finally led to the concept discussed in the introduction to this paper. In the literature, *random variables* (or, briefly, r.v.'s), in contrast to the types (A) and (C) of randomness, are introduced by definitions. Besides the one quoted in the introduction, there is the well-known measure-theoretical definition of a r.v. But are these two definitions equivalent? And do they cover all that statisticians actually study under the name of r.v.'s? For reasons discussed in the following sections, I do not believe that these questions can be answered affirmatively.

A r.v. in either sense will be seen to be a consistent class of pairs, and consequently general statements that are valid for *any* r.v. may conveniently be expressed in terms of c.c.p. variables. With numerical variables, r.v.'s have, notwithstanding frequent remarks to the contrary [6], [7], nothing whatever to do.[2]

7. Probabilistic and statistical random variables

The fifth point here to be stressed is the fact that, *in the literature, two different concepts have been studied under the name of "random variable."* A r.v. as defined by Kolmogorov and others in the *theory of probability* (that is, a measurable function on what Neyman calls a fundamental probability set, briefly, f.p.s.) is not identical with a r.v. as envisaged by Wald and others in *statistics* (that is, a consistent class of pairs for which a distribution function is known). A r.v. of either type is a consistent class of pairs that is related to a fundamental probability set, but with the basic difference that a probabilis-

---

[2] This point was clearly recognized by Halmos who, in his excellent treatment of probability, says, "A random variable is a function, a function whose numerical values are determined by chance...in other words, a function attached to an experiment" (see p. 188 in [8]). However, I fail to see the basis of Halmos' factual statement, "Ever since rigor has come to be demanded in mathematical definitions, it has been recognized that the word 'variable,' particularly a variable whose values are 'determined' somehow or other, means in precise language a function" (see p. 187 in [8]). On the contrary, the classical treatises by de la Vallée Poussin, and G. H. Hardy, as well as recent books by R. Courant and A. A. Albert, take positions that are the direct opposite of what Halmos calls the recognized point of view.

tic r.v. has a f.p.s. as its *domain*, whereas a statistical r.v. has a f.p.s. as its *range*. More precisely, the two concepts can be defined as follows:

A *probabilistic r.v.* is a triple $\mathfrak{R}, p, v$ consisting of

(1) a σ-field, $\mathfrak{R}$, of sets (called *random events*) including a maximal set R_{max}, that is, a set that contains each element of $\mathfrak{R}$ as a subset;

(2) a σ-additive function p (called *probability*) whose domain is $\mathfrak{R}$ and which assumes the value 1 for R_{max};

(3) a consistent class of quantities v whose domain is R_{max} and which has the following property: If $R_{v, x}$ denotes the class of all elements a of R_{max} such that $va < x$, then $R_{v, x}$ belongs to $\mathfrak{R}$, for any number x. (This property is also expressed by saying that the event that v assumes a value less than x is a random event.)

The function p_v assuming the value $p_v(x) = p(R_{v, x})$ for any number x is called the *probabilistic distribution function* of the random variable $\mathfrak{R}, p, v$.

A *statistical r.v.* is a triple $w, \mathfrak{S}, q$ consisting of

(1) a consistent class of pairs w;

(2) a σ-field $\mathfrak{S}$ of subsets of Ran w (the potential range of w) that includes Ran w itself (the elements of $\mathfrak{S}$ are called *statistical random events*);

(3) a σ-additive function q (called *statistical probability*) whose domain is $\mathfrak{S}$ and which assumes the value 1 for Ran w.

An example of a probabilistic r.v. is the triple $\mathfrak{T}, r, f$,where $\mathfrak{T}$ is the field consisting of the four sets: the empty set, $\{0\}$, $\{1, 2\}$, and $\{0, 1, 2\}$ (the maximal set); r is the additive function whose values for the afore-mentioned random events are 0, 1/4, 3/4, and 1, respectively; and f is the consistent class of quantities (0, 0), (1, 1), and (2, 1), that is, the function assuming the value $x \cdot (3 - x)$ for any x in $\{0, 1, 2\}$.

If r is replaced by the function r' assuming the values 0, 1/3, 2/3, and 1, one obtains a different probabilistic r.v. A third is $\mathfrak{T}_1, r_1, f$, where $\mathfrak{T}_1$ consists of all eight subsets of $\{0, 1, 2\}$ and r_1 is the additive function assuming the values 1/4, 1/2, and 1/4 for the sets $\{0\}, \{1\}, \{2\}$, respectively.

If j is the c.c.q. consisting of the three pairs (0, 0), (1, 1) and (2, 2), that is, the identity function on $\{0, 1, 2\}$, then neither $\mathfrak{T}, r, j$ nor $\mathfrak{T}, r', j$ is a probabilistic r.v. Indeed, the event that j assumes a value $< 3/2$ is the class $\{0, 1\}$, which does not belong to $\mathfrak{T}$. However, the triple $\mathfrak{T}_1, r_1, j$ is a probabilistic r.v.

An example of a statistical r.v. is the triple $a, \mathfrak{T}, r$, where a is the outcome of the game described in section 6 (C), that is, the class of all pairs $(a, a\,a)$ for any act a of the man listening to the announcements "zero," "one," and "two" of his friend, where $a\,a$ is the result of a, and the announcements are labeled 0, 1, and 2, respectively. The triples a, $\mathfrak{T}, r'$ and $a, \mathfrak{T}_1, r_1$ are different statistical r.v.'s.

Let h be the function assuming the values 0, 1, and 2 for all numbers between $-\pi$ and $-\pi/3$, between $-\pi/3$ and $\pi/3$, and between $\pi/3$ and π, respectively. If $\mathfrak{U}$ is the class of all Borel subsets of the sum of the three open intervals, and t is the Borel-Lebesgue measure, then $h, \mathfrak{U}, t$ is a statistical r.v., while $\mathfrak{T}, r, h$ is a probabilistic r.v. (just as $a, \mathfrak{T}, r$ is a statistical r.v., while $\mathfrak{T}, r, f$ is a probabilistic r.v.).

On a roulette that is divided into three sectors each subtending $2\pi/3$ radians, one can play a game such that for its outcome, g, the triple $g, \mathfrak{U}, r'$ (where r' has its previous meaning) is a statistical r.v. Any game of dice or cards in conjunction with a class of random outcomes and their probabilities gives rise to a statistical r.v.

8. Qualitative random variables

Neyman and other outstanding statisticians, after defining what in the preceding section has been called probabilistic random variables, emphasize that the most interesting problems concerning r.v.'s arise in so-called hypothetical cases, where the probability and distribution functions are unknown. For the same reasons that prompted observation (2) mentioned in the introduction, I cannot quite accept this point of view. How can one, after having defined r.v. relative to a σ-field of sets as well as to a σ-additive function p (even if p enters only into the definition of the distribution function of the random variable) discuss random variables for which no function p is known?

One must not answer that even where p is unknown, this function exists. This answer would be mathematically significant only if the assumption of the mere existence of a σ-additive function on a σ-field $\Re$ imposed restrictions on $\Re$. This, however, the assumption does not do. There exist σ-additive functions (assuming the value 1 for R_{max}) on every σ-field with a nonvacuous maximal set. An example is, if s denotes any element of R_{max}, the function assuming for a set R in $\Re$ the value 1 or 0 according to whether R does or does not contain s. Hence no restriction whatever is imposed on $\Re$ by assuming that a probability function exists.[3]

It rather appears—and this is the **sixth point** that I wish to emphasize—that *statisticians frequently study* (especially in problems to which they attach particular interest) *a more rudimentary concept than random variables*, namely, mere pairs $\Re,u$ or $w,\mathfrak{S}$ rather than triples including also a σ-additive function on the σ-field of sets. I shall refer to such pairs as *qualitative random variables*. Particularly important are statistical qualitative r.v.'s.

Let g be the class of all pairs $(a, g\,a)$, for any act a of reading the points that turn up in a game of dice, where $g\,a$ is the result of the act a. Let $\mathfrak{S}$ be the class of all subsets of Ran g, for instance, if the game is with one die, the class of the 2^6 subsets of $\{1, 2, 3, 4, 5, 6\}$. If I am told that the die is loaded but not given any other information (in particular, no records of past outcomes), then I may well decide to consider g, $\mathfrak{S}$ a qualitative statistical r.v.

If a, $\mathfrak{T}$ and j have the same meaning as in section 7, and $\mathfrak{T}'$ denotes the class of all subsets of $\{0, 1, 2\}$, then $a,\mathfrak{T}'$ and $\mathfrak{T},j$ are not even qualitative r.v.'s.

9. R.V. variables

After what has been said in section 5, two examples will sufficiently illustrate the concept of r.v. variables.

If $\Re$ is the σ-field of all subsets of Dom u, then $\Re$, u is a qualitative probabilistic r.v. for any c.c.q. u. If Dom u is finite, and p is the additive function assuming equal values for any two subsets consisting of a single element of Dom u ("elementary events"), then $\Re$, p, u is a probabilistic r.v. Similarly, if $\mathfrak{S}$ is the σ-field of all subsets of Ran w, then w, $\mathfrak{S}$ is a qualitative statistical r.v. In the preceding statements, u and w are c.c.p. variables.

Also the following somewhat curious general statement deals with concepts that belong to the category of r.v. variables. *The probability of denumerably many incidents all*

[3] This point was brought out in a discussion with Dr. R. Seall and Mr. M. McKiernan. The assumption that a probability function *of a certain kind* exists may impose restrictions of $\Re$. For instance, the existence of a function p assuming infinitely many values obviously presupposes that $\Re$ contains infinitely many elements.

of which give rise to random events is itself a probabilistic random variable. More precisely, assume that $\mathfrak{R}$ includes all subsets of R_{max}, and that R_{max} is denumerable. Then, for any σ-additive function p on $\mathfrak{R}$, one can define a function p^* on R_{max} assuming for any "incident" (that is, for any element r of R_{max}) the same value that p assumes for the corresponding "elementary event" (that is, the set $\{r\}$). The triple $\mathfrak{R}$, p, p^* is a probabilistic r.v.

10. Four types of statistical studies concerning consistent classes of pairs

The definition and the treatment of probabilistic r.v.'s in the literature have been considerably more precise and lucid than those of statistical r.v.'s. The reason is that, in the case of the former, nothing conceptually different has to be added to the ideas of σ-field and σ-additive function on which mathematicians have concentrated for a long time. In contrast, statistical r.v.'s have domains whose elements (such as acts of observation or physical objects) are, as it were, one step closer to reality, and concerning elements of this kind pure and even applied mathematics have been rather inarticulate. Yet an analysis reveals that most statistical studies actually deal with those domains and their elements.

(1) In the case of some variable quantities, all a safety engineer wishes to ascertain is the maximum and/or minimum value. (The difference between these two numbers is what most statisticians call the "range" of the variable quantity.) Such studies do not require references to the domains of the variable quantities.

(2) The typical statistical investigations of individual c.c.p.'s, which are concerned with frequencies, cannot be confined to the ranges and do require references to the domains of the c.c.p.'s. For instance, all that can be said about the range R of a class of observations in relation to an element r of its potential range is either that R does, or that R does not, contain r. (In any class or set, any element has, as it were, the frequency 1 or 0.) It cannot be correctly said (although it sometimes *is* said) that R contains r, say, 3 times. The frequency 3 is associated with r if and only if the domain of the class of observations contains exactly three acts with the result r. Thus all significant processes of counting (and, in more complicated cases, measuring) in statistical investigations are performed within the *domains* of c.c.p.'s. Subsequently, the numerical results of the processes are paired with the elements of the *ranges*, whereby the latter are transformed into what might be called *weighted classes*. Oddly enough, no traditional term (such as *weighted* and *relatively weighted* class) exists for the concept of a class with a frequency or relative frequency defined for the elements, in spite of the paramount importance of this concept in statistics and, incidentally, in algebra.

(3) Typical order statistics (the theory of runs, etc.) are sequences of observations or quantities. It was in his study of sequences of a certain kind ("collectives") that von Mises introduced the distinction between acts and results. Just for sequences of pairs, however, frequency studies do not really require references to the first members of the pairs. Consider, for instance, the sequences of observations

(11) $(a_1, r_1), (a_2, r_2), \cdots, (a_n, r_n).$

All frequency studies concerning the sequence (11) can be based on the sequence $r_1, r_2, \cdots, r_n$ of the results. One associates the frequency 3 with the result r if there are exactly three indices, say, 2, 3, and 7, such that $r_2 = r_3 = r_7 = r$. Thus the indices (which constitute, as it were, the domain of the *sequence* and are attached to the results as well as to the acts) may take over the role of the acts (which constitute the domain of the *class of pairs* belonging to the sequence).

(4) In investigations into the correlation of two c.c.p.'s and the regression curve of a variable quantity w on a variable quantity v, references to the domains of the c.c.p.'s and to Dom v and Dom w are absolutely essential. What one primarily pairs are elements of those domains. Subsequently one studies pairs of elements of the ranges and pairs of values of v and w, induced by that primary pairing.

Summarizing one can say that few statistical investigations—essentially only those in order statistics—can dispense with references to the domains of c.c.p.'s. *In frequency studies concerning nonsequential c.c.p.'s, for instance, concerning one variable quantity as well as connections between two variable quantities, references to the domains are indispensable.* This is the **seventh** point that I wish to emphasize.

The remarks about correlation and regression curves bring out an important **eighth point.** *The questions as to what is the correlation coefficient or the regression line of w on v are unanswerable since they are incomplete.* In problems of this kind, references not only to the elements of Dom v and Dom w but to specific pairs of elements (one element belonging to Dom v, and one to Dom w) are indispensable. *A definite pairing of the domains (or of subclasses of the domains) must be given, and only relative to that pairing can the afore-mentioned questions be investigated.*

Pairings are mentioned by statisticians. To physicists, since Galileo and Boyle, pairing of *simultaneous* acts of observation has become second nature and is tacitly understood. If m^* is weight observation, say, with regard to a piece of radium that in 1900 weighed 1 gram, and t^* is time observation, then Dom m^* consists of acts of scale readings, and Dom t^* of acts of calendar readings. In the formula

(12) $$t^* = 1900 - 2.3 \cdot 10^3 \, log \, m^*,$$

it is perfectly, if implicitly, understood that the values of t^* and m^* for simultaneous scale and calendar readings are being connected. Formula (12) is an abbreviation for

(13) $$t^* \tau = 1900 - 2.3 \cdot 10^3 \, log \, m^* \beta \,,$$

for any pair (β, τ) of simultaneous acts of scale and calendar readings.

Formula (12) does not connect *any* value of t^* and *any* value of m^*. Hence it is completely misleading when some mathematicians claim [9] that laws such as (12) deal only with the values of quantities, and not with quantities themselves; that t^* and m^* are numerical variables, and the like. If m^* and t^* were numerical variables, they could be interchanged as can x and y in (3).

But if P and P_1 are two pendulums, and a and a_1 are their angles with a vertical line (compare section 4), then the relation between a and a_1 relative to the class Σ of all pairs (σ, σ_1) of simultaneous states may not be of particular significance. One may be more interested in the class Σ' of pairs (σ, σ_1'), where σ_1' is the state of P_1, one quarter of a period after σ_1. The results of pairing a and a_1 relative to Σ' and Σ are quite different. It may well be that $a_1 = -a$ relative to Σ', while a_1 is not at all a function of a relative to Σ and a different function of a relative to another pairing Σ'' of the domains.

The omission of references to the pairing of the domains and even to the domains altogether accounts for the lack of articulate rules concerning the application of analysis to science.

The need for a pairing of the domains *as one of the data* becomes perfectly obvious on the level of c.c.p. variables. For, what pairing of Dom v and Dom w—of one abstract set with another abstract set—is "natural"? To be applicable to radioactivity as well as to sociology, general statistical statements must contain not only c.c.p. variables, such as v and w, that may be replaced, say, by m^* and t^* in one case, and both by the height of men in Chicago, in the other. They must contain also a pairing variable Π that may be replaced by the pairing, say, according to the simultaneity in one case, and according to the father-son relation, in the other. That this procedure has been neglected in the literature is due to the lack of an explicit definition and of a clear treatment of c.c.q. variables.

11. What are statistical random events?

Modern statisticians have derived ingenious methods for the translation of a frequency record concerning the domain of a qualitative statistical random variable w, $\mathfrak{S}$ into a definition of a probability function q on $\mathfrak{S}$. Theories guide the statistician in choosing q.

Less attention seems to have been paid to the problem of defining $\mathfrak{S}$, even though guiding principles in this respect would be equally desirable.

Suppose the man on the street plays a long series of games with his friend, as described in section 7 (C). If, at the outset, he is told that the coins are unbiased, then, on the basis of past experience with unbiased coins, he will assume the probabilities $1/4$ and $3/4$ for 0 and 1 or 2, respectively. If he is told that the coins are somehow biased without being given further information, he will set up a frequency record on the basis of which a statistician will advise him. But concerning the outcomes 1 and 2 which, according to his understanding with his friend, are entirely left to the latter's discretion, no frequency record will be significant. Even if the friend should seem to speak the truth (which would give the probabilities $1/2$ and $1/4$ to 1 and 2, respectively) or if he should use random sample numbers to simulate probabilities $3/8$ for both 1 and 2, the man on the street will not make predictions even on averages of future outcomes since he knows that his friend may change his policy at any time.

If the friend plays the same game also with a second person, who was told (rightly or wrongly) that the friend would always speak the truth, then the same frequency record that was insignificant to the first man will appear to be significant to the second.

12. The connection between probability and statistics

Many authors have likened the relation between probability theory and statistics to that between postulational and physical geometry. Postulational geometry and probability are based on unproven assumptions concerning undefined concepts, whereas physical geometry and statistics deal with phases of reality that approximately satisfy the postulates. If chalk dots and chalk streaks on a blackboard are called "points" and "lines," then Euclidean geometry, dealing with undefined points and lines is approximately valid. If the outcome of rolling a die is called "incident" and the relative frequency of an outcome in a long series of trials "probability" of the incident, then the theory of a certain probabilistic random variable (whose domain consists of six undefined elements called "incidents") is approximately valid.

Besides postulational and physical geometry, there exists a third theory dealing with points and lines, namely, pure analytic geometry. There (in contrast to physical geometry), points, lines, etc. are *defined*, but (in contrast to physical geometry) they are defined without any reference to reality, namely, as ordered pairs of numbers, as classes of such pairs satisfying linear equations, etc. As a concluding ninth point, I wish to emphasize that, besides the postulational probability theory and statistics, *there exists a third theory dealing with incidents, events, probability, etc. In it* (in contrast to the postulational theory) *events and probabilities are defined;* but (in contrast to statistics) *they are defined without any reference to reality.*

I shall illustrate this idea in the case of the weak law of large numbers: The probability is close to 1 that, in a sufficiently large sequence of independent trials of the same kind, the relative frequency of success is very close to the a priori probability of success in each trial. The purely mathematical background of the (more colorful than lucid) traditional formulation of this important law is the following:

FIRST COMBINATORIC LAW OF LARGE NUMBERS. Let B be a finite set, A a subset of B, and let a and b denote the numbers of elements in A and B. For any sequence Γ_k of k elements belonging to B, let $a(\Gamma_k)$ denote the relative frequency in Γ_k of elements belonging to A, that is to say, the number of those elements in Γ_k divided by k. For any two numbers, c and c', call $F_k(c, c')$ the number of all sequences Γ_k for which

$$(14) \qquad\qquad c \leqq a\,(\Gamma_k) \leqq c',$$

so that, in particular,

$$(15) \qquad\qquad F_k\,(0,\,1)\ = b^k \text{ for any } k = 1,\,2,\,\cdots.$$

Then, for any pair of positive numbers x and y (no matter how small), there exists a number $N(x, y)$ such that

$$(16) \qquad k > N\,(x,\,y) \text{ implies } F_k\left(\frac{a}{b} - x,\ \frac{a}{b} + x\right) > (1 - y) \cdot b^k.$$

An example of such a number $N(x, y)$ is Cantelli's number

$$(17) \qquad\qquad \frac{2}{x^2}\,log\,\frac{4}{x^2 y} + 2\,.$$

In words: *If* k *is sufficiently large, then among the* b^k *sequences of* k *elements of B those in which the relative frequency of the elements belonging to A differs from* a/b *by less than* x *have a relative frequency* $> 1 - y$.

To the chalk dots in physical geometry and the undefined elements called points in postulational geometry, there corresponds in analytic geometry the purely mathematical definition of a point as an ordered pair of numbers. Similarly, to the repeated trials and observed frequencies in statistics and the undefined elements called events and probability in the postulational probability theory, there correspond in the combinatoric probability theory the following purely mathematical definitions.

A *trial* is an element of a finite set B. A *success* is an element of a subset A of B. The *a priori probability* of success in a trial is the ratio of the numbers of elements in A to that of the elements in B; a sequence of independent trials is a sequence of elements of B. An *event* is a class of such sequences having the same length. The *probability* of an event is the ratio of the number of sequences in the class to the total number of sequences of the same length. Clearly, any two sequences of the same length are considered as equally probable.

That the first combinatoric theorem differs from the traditional law is clear not only from the absence in the former of references to reality but from the fact that the combinatoric theorem covers only cases with *rational a priori probability*, whereas in the traditional law the a priori probability might well be $1/\sqrt{2}$ or $1/e$. However, one can obtain a Second Combinatoric Theorem covering any (rational or irrational) a priori probability p. For this purpose a Poisson setup is needed; that is to say, a sequence of finite sets $B_1, B_2, \cdots$ and a sequence of respective subsets $A_1, A_2, \cdots$ must be given with a_n/b_n converging to p.

Clearly, the combinatoric method can be extended to the multinomial case, in which each set B_n has k^m mutually disjoint and jointly exhaustive subsets $A_n^1, \cdots, A_n^m$—the same m for any n. The method can further be extended to sampling theory, confidence limits and, to a certain extent, to continuous distributions.

REFERENCES

[1] A. WALD, *Sequential Analysis*, New York, John Wiley and Sons, 1947.

[2] KARL MENGER, "The ideas of variable and function," *Proc. Nat. Acad. Sci.*, Vol. 39 (1953), pp. 956–961.

[3] ———, "On variables in mathematics and in natural sciences," *Brit. Jour. Phil. Sci.*, Vol. 5 (1954), pp. 134–142.

[4] ———, "Variables de diverses natures," *Bull. des Sci. Math.*, Vol. 78 (1954), pp. 229–234.

[5] ———, *Calculus. A Modern Approach*, Chicago, Illinois Institute of Technology Bookstore, 1953; final edition, Boston, Ginn and Co., 1955.

[6] J. F. KENNEY, *Mathematics of Statistics*, Part I, 2d ed., New York, D. Van Nostrand Company, 1947.

[7] M. G. KENDALL, *The Advanced Theory of Statistics*, Vol. I, 4th ed., London, Griffin, 1948.

[8] P. R. HALMOS, *Measure Theory*, New York, D. Van Nostrand Company, 1950.

[9] R. COURANT, *Differential and Integral Calculus*, Vol. I, New York, Interscience, 1951.

WHAT ARE x AND y?

By Karl Menger
Illinois Institute of Technology

Introduction [1]

1. *Consistent Classes of Quantities.* An ordered pair whose second member (or *value*) is a number, while its first member (or *object*) may be anything, will be referred to as a *quantity*. By a *consistent class of quantities*—briefly, c.c.q.— we mean a class of quantities that does not contain two quantities with equal objects and unequal values. Reviving Newton's term, we will refer to c.c.q.'s also as *fluents*. The class of all objects (of all values) of the quantities belonging to a fluent is called the *domain* (the *range*) of the fluent.

Given a plane and a unit of length, any pair $(\kappa, r(\kappa))$ consisting of a circle κ and its radius $r(\kappa)$ is a quantity. The class of all these pairs for any circle κ is a c.c.q.—the radius r. The following three classes, for various reasons, are not c.c.q.'s : that of all pairs $((r(\kappa), \kappa)$ for any circle κ ; that of all pairs $(\kappa, \zeta(\kappa))$ for any κ, where $\zeta(\kappa)$ is the center of the circle κ ; that of all pairs of numbers.

If, for a positive integer n, the objects of all quantities belonging to a fluent are n-tuples of numbers [2] then the fluent is called an *n-place function*. An example of a 2-place function is the class of all pairs $((a, b), a)$ for any two numbers a and b ; another is the class of all pairs $((a, b), b)$. For obvious reasons, these functions might be called the 2-place *selectors* ; the first will be denoted by I, the second by J ; their values for (a, b) by $I(a, b)$ (which equals a) and $J(a, b)$ (which equals b). Examples of 1-place functions are the class of all pairs (c, c) for any number c, and the class of all pairs $(x, \log x)$ for any $x > 0$. They are called the *identity* and the *logarithmic function*. The former will herein be denoted by j, its value for c, by $j(c)$, which equals c.

A class of pairs of functions that does not contain two pairs with equal first and unequal second members is called an *operator*. An example is the class of all pairs (f, Df) for any everywhere differentiable function f, where Df denotes the derivative of f.

2. *Variables.* By a *variable* we mean a symbol that, in some context and according to definite stipulations, may be replaced by the designation of any element of a certain class (called the *scope* of the variable). In the preceding definitions, the letters c and x are *numerical variables*, i.e., symbols that in the pairs mentioned may be replaced by numerals such as 3, and then yield pairs such as $(3, 3)$ and $(3, \log 3)$ belonging to the identity and the logarithmic functions, respectively. In those definitions, the scope of c is the class of all numbers, that of x is the class of all positive numbers. The letter f in the definition of the operator is a *function variable*, i.e., a variable whose scope is a class of functions.

It is common to all variables that they may be replaced by any other symbol of the same kind without any change of the meaning. For instance, the logarithmic function may be defined as the class of all pairs $(c, \log c)$ for any $c > 0$; and the operator mentioned, as the class of all pairs (g, Dg) for any everywhere differentiable function g.

3. *Remarks.* Some mathematicians (if hardly any scientists) call all fluents " functions ". But only fluents whose domains consist of numbers or systems of numbers have the power to connect c.c.q.'s. The logarithm connects the radius r with log r, which is the class of all pairs $(\kappa, \log r(\kappa))$ for any circle κ ; similarly, the area a with log a ; the exponential function with the identity function ; the function cos with log cos ; etc. The radius r lacks the

power to connect two fluents. There is no radius of the area nor a radius of the cosine. Nor can fluents other than those of the type of log be differentiated or integrated : $dr/d\kappa$ and $\int r\, d\kappa$ are meaningless. Hence a special name for those fluents whose domains consist of numbers or systems of numbers is practically indispensable, and no name for them is more appropriate than " functions ".

Many scientists and some mathematicians call fluents " *variable quantities* " —often simply " *variables* ". The latter usage is quite unfortunate. It certainly has not been conducive to the maintenance of a clear and sharp distinction between what above have been called fluents (specific classes !) and variables (replaceable symbols !). Yet this distinction is of the utmost importance in pure and applied mathematics if terms and symbols are to be used according to articulate rules.

A Question with a Dozen Answers.

Two symbols are ubiquitous in mathematical writings—the letters [3] x and y. What is their meaning? The answer altogether depends upon the context.

I. *Numerical Variables.* In the following statements, definitions, and problems, x and y are numerical variables.

(1) The class of all pairs $(x, x+1)$ for any x contains an element not belonging to the

the class of all pairs $\left(x, \dfrac{x^2-1}{x-1} \right)$ for any $x \neq 1$,

namely, the pair $(1, 2)$; and one element, namely, $(2, 3)$, not belonging to the class of all pairs $\left(x, \dfrac{x^2-x-2}{x-2} \right)$ for any $x \neq 2$.

(2) In the realm of integers mod. 2, the class of all pairs (x, x) for any x, and the class of all pairs (x, x^2) for any x, are equal.

(3) Let j^{-1} denote the class of all pairs (x, x^{-1}) for any $x \neq 0$.

(4) Let L denote the class of all pairs (x, y) such that $2x+3y=5$. Obviously equivalent is the definition of L as the class of all pairs (y, x) such that $2y+3x=5$. The class L' of all pairs (x, y) such that $2y+3x=5$ is different from L.

(5) $x^2 - 9y^2 = (x+3y) \cdot (x-3y)$ for any x and y ;
$y^2 - 9x^2 = (y+3x) \cdot (y-3x)$ for any x and y (or any y and x).
(The two (or three) preceding statements are identical in meaning.)

(6) $(x+yi)^2 = x^2 - y^2 + 2xyi$ for any x and y.

(7) $\sin 2x = 2 \sin x \cos x$ for any x.

(8) $\mathbf{D} \sin x = \cos x$ for any x.

(9) $\mathbf{D} \log x = j^{-1}(x)$ for any $x > 0$.

(10) If $F(x, y) = x^2 y^5$ for any x and y,
then $\mathbf{D}_1 F(x, y) = 3x^2 y^5$ and $\mathbf{D}_2 F(x, y) = 5x^3 y^4$ for any x and y.

(11) Find all numbers x such that $x^2 - 1 = 0$.

(12) Find all pairs of numbers x, y such that $2x+3y=5$ and $x+y=3$.

As these examples show, numerical variables are put to a variety of uses. Theorems I(5)–(10) stipulate that a replacement of the variables by numerals yields a formula expressing a valid connection of specific numbers ; e.g., I(5) yields $5^2 - 9 \cdot 7^2 = (5+3 \cdot 7) \cdot (5-3 \cdot 7)$. In each definition involved in I(1)–(3), such a replacement results in an element of the class defined ; e.g., $(3, 3^{-1})$ belongs to the function j^{-1}. In I(4), replacement of x and y in (x, y) by numerals yields an element of L if and only if it transforms the formula $2x+3y=5$ into a valid connection of specific numbers ; e.g., $(4, -1)$ is, and $(-1, 4)$ is not, an element of L. In a fourth way x and y are used in problems such as (11) and (12). These uses might be roughly described as *indicative, conjunctive, conditional,* and *imperative.*

II. *The Identity Function.* In the following statements, x designates the identity function, j, defined in the Introduction.

(1) The function $x + 1$ is an extension of the (non-identical) functions

$$\frac{x^2 - 1}{x - 1} \quad \text{and} \quad \frac{x^2 - x - 2}{x - 2}.$$

(2) In the realm of integers mod. 2, the functions x and x^2 are identical. Either consists of the two pairs $(0, 0)$ and $(1, 1)$.

(3) The function $x^2 - 1$ is a polynomial, and $\mathbf{D}(x^2 - 1) = 2x$.

(4) $\dfrac{d \sin x}{dx} = \cos x$.

(5) The class of all pairs $(f, x.f)$ for any function f is an operator.

II(1), (2) are equivalent with I(1), (2), since the function $x + 1$ is the class of all pairs $(x, x + 1)$, and so on. Yet in none of the examples II is x a numerical variable. That the function $3 + 1$ is an extension of the function $\dfrac{3^2 - 1}{3 - 1}$ is utter nonsense, as is the expression $\dfrac{d \sin 3}{d3}$. That also in II(4) the letter x may be interpreted as the designation of the identity function is seen by defining for any two functions f and g the *rate of change* of f with regard to g as the function $\dfrac{df(x)}{dg(x)} \left(\text{briefly, } \dfrac{df}{dg} \right)$ assuming the value

$$\lim_{g(x) \to g(a)} \frac{f(x) - f(a)}{g(x) - g(a)}$$

for any number a for which this limit exists. Then $\dfrac{d \sin x}{dx}$ is the rate of change of the sine with the identity, $\dfrac{d \sin x}{dj(x)} \left(\text{briefly, } \dfrac{d \sin}{dj} \right)$.

III. *The Selector Functions.* In the following examples, x and y designate the 2-place selector functions, I and J, defined in the Introduction.

(1) If a function of x and a function of y are equal, then the two functions are constant; in symbols, if $f(x) = g(y)$, then there exists a constant function c such that $f = g = c$.

(2) $\dfrac{\partial x^3 y^5}{\partial x} = 3x^2 y^5$ and $\dfrac{\partial x^3 y^5}{\partial y} = 5x^3 y^4$.

(3) In the pure analytic geometry of the Cartesian plane, where points are defined as ordered pairs of numbers, and curves as certain classes of such pairs, $2x + 3y = 5$ and $2y + 3x = 5$ are different straight lines.

Clearly, (1) has the following meaning : If $f(I(a, b)) = g(J(a, b))$ for any a and b, then there exists a number c such that $f(t) = g(t) = c$ for any t. In (2), in contrast to I(10), the letters x and y cannot be replaced by numerals. That they may be interpreted as the selector functions is seen as follows. For any three 2-place functions F, G, and H, let $\left(\dfrac{\partial F}{\partial G} \right)_H$ denote the 2-place function assuming the value

$$\lim_{\substack{(x, y) \to (a, b) \\ H(x, y) = H(a, b)}} \frac{F(x, y) - F(a, b)}{G(x, y) - G(a, b)}$$

for any pair (a, b) for which this limit exists. If this function $\left(\dfrac{\partial F}{\partial G} \right)_H$ is called the *partial rate of change* of F with regard to G keeping H constant, then the traditional expression $\dfrac{\partial F}{\partial x}$ designates the partial rate of change of F with regard to I keeping J constant; similarly, $\dfrac{\partial F}{\partial y}$ designates $\left(\dfrac{\partial F}{\partial J}\right)_I$.

(3) expresses the same fact as I(4), namely, that L and L' are different. But whereas, in the definition of L, x and y may be interchanged without affecting the meaning, in the traditional symbol for this line, as III(3) shows, x and y must not be interchanged. In the latter context they designate specific 2-place functions, in the former they are numerical variables.

IV. *Real-Valued Complex Functions.* In the following statements, x is the class of all pairs $(a+bi, a)$ for any a and b; and y is the class of all pairs $(a+bi, b)$.

(1) $z = x + yi$.

(2) $\dfrac{1}{3}\dfrac{dz^3}{dz} = (x + yi)^2$.

Here, the letter z designates the identity function in the realm of complex numbers. (In the formula $D \sin z = \cos z$, it may be interpreted as this identity function or, in presence of the stipulation " for any z," as a numerical variable whose scope is the class of all complex numbers.)

V. *Indeterminates.* The following formulae (without any qualifying or explanatory legends) belong to the theory of polynomial and rational forms.

(1) $\dfrac{x^2 - 1}{x - 1} = \dfrac{x^2 - x - 2}{x - 2} = \dfrac{x + 1}{1}$.

(2) In the realm of integers mod. 2, the polynomial forms $(x + 1) \cdot (x - 1)$ and $x^2 + 1$ are equal; the forms x and x^2 are considered as unequal.

(3) $x^2 - 9y^2 = (x + 3y) \cdot (x - 3y)$.

It will be noted that the statements V(1), (2) about rational forms are (for opposite reasons) quite unparallel to the statements II(1), (2) about the corresponding rational functions. In contrast to a function, a form is not a class of pairs of numbers and thus is not meant to be evaluated for any specific argument. In contrast to a numerical variable, the letter x in a form is not supposed to be replaced by specific numerals. Just as a complex number may be considered as an ordered pair of real numbers, a polynomial form containing one letter is completely characterized by the sequence of its coefficients ; that is to say, the form may be regarded as a sequence of numbers belonging to a given field. A rational form is an ordered pair of such sequences. Forms thus are hypercomplex numbers of a certain kind that are equated, added, and multiplied according to well-known laws. For instance, if, in a self-explanatory way, the rational forms in (1) are denoted by

$$\dfrac{(1, 0, -1)}{(1, -1)} , \quad \dfrac{(1, -1, -2)}{(1, -2)} , \quad \dfrac{(1, -1)}{(1)} ,$$

then the first two are equal because $(1, 0, -1) \cdot (1, -2) = (1, -1) \cdot (1, -1, -2)$. Operating with such sequences becomes more perspicuous if to each element of the sequence an indicator of its positions is appended. It is customary to suffix such an indicator as a quasi-exponent and to use as its quasi-base the letter x. This choice is motivated by the parallelism between rational forms and rational functions (the form x corresponding to the identity function x), even though this parallelism is very incomplete, as shown by the contrast between V(1), (2) and II(1), (2). In a form, each letter x thus is, as it were, a holder of a place card (the quasi-exponent) describing the position of a co-efficient. Such a letter in a form is called an indeterminate.

VI. *Parts of Operational Symbols.* The symbols $\dfrac{d}{dx}$, $\dfrac{\partial}{\partial x}$, and $\dfrac{\partial}{\partial y}$ are often interpreted as synonyms of D, D_1, and D_2. In this case, the first x is a part of the symbol for the derivative just as D is a part of the synonymous symbol D—not more and not less. In the last two symbols, the letters x and y indicate which of two operations is intended : that pertaining to the first place in a 2-place function, or that pertaining to the second place. Of course, adopting this interpretation one attributes to the two letters x in the expression $\dfrac{d}{dx} \sin x$ totally discrepant meanings since the second x is either a numerical variable or the identity function, whereas in II(4) both x were interpreted as the identity function.

VII. *Operators.* The class mentioned in II(5) is often referred to as the operator x, for instance, in

(1) $(D^2 + xD + 1)J_0 = 0$.

Here J_0 is the 0-th Bessel function, and 1 designates the identity operator, that is, the class of all pairs (f, f) for any function f.

VIII. *Specific Fluents (Abscissa and Ordinate).* In contrast to pure analytic geometry, mentioned in III(3), the points and lines in a physical or postu-lational plane are not arithmetically defined. They are physical objects in the former (e.g., chalk dots and streaks on a blackboard), and undefined elements (called points and lines) satisfying certain assumptions, in the latter. In any physical or postulational plane, one may choose a Cartesian frame of reference consisting, essentially, of three non-collinear points, o, ξ, and η, called origin, first, and second unit point, respectively. Relative to such a frame, one can, by a well-known procedure, associate two numbers, which we denote by $x\pi$ and $y\pi$, with each point π in the plane. They are called the coordinates of π relative to the chosen frame. In particular,

$$x_0 = y_0 = y\xi = x\eta = 0 \quad \text{and} \quad x\xi = y\eta = 1.$$

In this way, two consistent classes of quantities are defined whose common domain is the class of all points in the plane under consideration ;
  the *abscissa*, x, is the class of all pairs $(\pi, x\pi)$ for any point π ;
  the *ordinate*, y, is the class of all pairs $(\pi, y\pi)$ for any point π.
In defining these fluents, use has been made of a point variable : the letter π, which may be replaced by the designation of any specific point whereby one obtains a specific pair belonging to the fluent considered ; e.g., replacing π by ξ and o in the definition of x one obtains the pairs $(\xi, x\xi) = (\xi, 1)$ and $(o, x_0) = (o, 0)$ belonging to x. The latter pair belongs also to y.
  In this sense x and y are used in the following statements in physical or postulational geometry.

(1) The lines $2x + 3y = 5$ and $2y + 3x = 5$ are different.

(2) If $y = \sin x$, then $\dfrac{dy}{dx} = \cos x$.

Here, $2x + 3y = 5$ and $y = \sin x$ designate the classes of all points π such that $2x\pi + 3y\pi = 5$ and $y\pi = \sin x\pi$, respectively. The rate of change of the fluent y with regard to the fluent x is defined as the fluent assuming the value

$$\lim_{x\rho \to x\pi} \frac{y\rho - y\pi}{x\rho - x\pi}$$

for any point π on the sine curve, ρ being confined to the same curve.

IX. Function Variables. In the following examples, y is a function variable put to various uses. Its scope in (1) and (2) is the class of all twice differentiable functions.

(1) If $D^2y + y = 0$, then y is a periodic function.

(2) Find y such that $D^2y + y = x$ and $y(0) = 1$.

The letter x in (2) as well as in $\dfrac{d^2y}{dx^2} + y = x$ designates the identity function; in $y''(x) + y(x) = x$ for any x, it is a numerical variable.

X. Fluent Variables. In the following examples, x and y are variables whose scopes are classes of fluents; that is to say, x and y are symbols that may be replaced by the designations of specific fluents.

(1) $\dfrac{d\sin x}{dx} = \cos x$ for any fluent x.

(2) If $y = \sin x$, then $\dfrac{dy}{dx} = \cos x$.

(3) If $y = \cos x$, then $\displaystyle\int_{\alpha_0}^{\alpha} y\, dx = \sin x\alpha - \sin x\alpha_0$.

In (2) and (3), x and y must not be replaced by numbers. For $\pi/2$ and 1 the antecedent in (2) is valid and yet the consequent is utterly nonsensical. However, one may replace x by the time t, and y by the position s of a linear oscillator, in which case (2) yields a statement about the rate of change of s with regard to t, namely,

if $s = \sin t$, then $\dfrac{ds}{dt} = \cos t$.

Here, t is defined as the class of all pairs $(\tau, t\tau)$ for any act τ of reading a clock, and s as the class of all pairs $(\sigma, s\sigma)$ for any act σ of reading the position of the oscillator, where $t\tau$ and $s\sigma$ denote the numbers read as the results of the respective acts. If for any act τ of clock reading, $\Gamma\tau$ designates the *simultaneous* act of reading the position, then the rate of change of s with regard to t is defined as the fluent assuming for any act τ_0 the value

$$\lim_{t\tau \to t\tau_0} \frac{s(\Gamma\tau) - s(\Gamma\tau_0)}{t\tau - t\tau_0}.$$

In physics, this rate of change is identified with, or defined as, the *velocity* of the oscillator.

More generally, let x and y be any two fluents, defined as classes of pairs $(\alpha, x\alpha)$ and $(\beta, y\beta)$, and let Γ be a pairing of an element $\Gamma\alpha$ belonging to the domain of y with any element α of the domain of x. One can introduce the rate of change of y with regard to x relative to the pairing Γ, namely, as the fluent $\dfrac{dy}{dx}$ assuming the value

$$\lim_{x\alpha \to x\alpha_0} \frac{y(\Gamma\alpha) - y(\Gamma\alpha_0)}{x\alpha - x\alpha_0}$$

for any α_0 in the domain of x for which this limit exists. In X(2) such a pairing was (as it is customary) tacitly taken for granted. In the example of the oscillator, the pairing was by *simultaneity*; in VIII(2), it was tacitly understood that with each point in the domain of x the *same* point in the domain of y should be associated. Also $\dfrac{df}{dg}$, as defined in II, subsumes under $\dfrac{dy}{dx}$ provided that with each number x in the domain of g the *same* number $\varGamma x = x$ in the domain of f is paired (in other words, that $\varGamma = j$). With this understanding, X(1) yields II(4) if the fluent variable x is replaced by j.

Relative to such a pairing $\varGamma$, one may also define the *cumulation* of y with regard to x from α_0 to α, denoted by $\displaystyle\int_{\alpha_0}^{\alpha} y \, dx$, namely, as the number to which for various n and various sequences $\alpha_0, \alpha_1, \ldots, \alpha_{n-1}, \alpha_n = \alpha$ the product sums

$$y(\varGamma\alpha_0) \cdot (x\alpha_1 - x\alpha_0) + \ldots + y(\varGamma\alpha_{n-1}) \cdot (x\alpha_n - x\alpha_{n-1})$$

are as close as he pleases provided that the largest of the numbers

$$|\, x\alpha_k - x\alpha_{k-1}\,| \quad (k = 1, \ldots, n)$$

is sufficiently close to 0.

XI. *Sundries.* The meaning of x and y in formulae and expressions such as $dy = \sin x \, dx$ and $M(x, y) \, dx + N(x, y) \, dy$ will not here be discussed since the essential problem in these cases lies in the interpretation of the symbol d. Neither will be the numerous uses of x and y in vector algebra, vector analysis, etc. be elaborated on, since they are usually accompanied by stipulations that rule out the danger of confusion with the meanings discussed in this paper. The various meanings of x and y as so-called *random variables* are parallel to some of those herein listed and are summarized elsewhere [4].

XII. *Dummies.* In all previous examples, x and y have clearly defined, if totally discrepant, meanings. In the following formulae they have no meaning whatsoever.

(1) $\displaystyle\int_0^1 \cos x \, dx = \sin 1 - \sin 0.$

(2) $\displaystyle\int_a^{a+d} f(x) \, dx = \int_a^{a+d} f(y) \, dy$ for any two numbers a and d, and any function f that is continuous in the closed interval between a and $a + d$.

The simplest definition of the integral in (1) is

$$\int_0^1 \cos x \, dx = \lim_{n \to \infty} \frac{1}{n} \cdot \left[\cos \frac{1}{2n} + \cos \frac{3}{2n} + \ldots + \cos \frac{2n-1}{2n} \right].$$

More generally,

$$\int_a^{a+d} f(x) \, dx = \lim_{n \to \infty} \frac{d}{n} \cdot \left[f\left(a + \frac{1}{2n} \, d \right) + f\left(a + \frac{3}{2n} \, d \right) + \ldots + f\left(a + \frac{2n-1}{2n} \, d \right) \right].$$

Since the expression on the right side is free of x it is not surprising that there is a trend toward shedding the meaningless reference to x in the integral and toward writing simply $\displaystyle\int_a^{a+d} f.$

The English Language and the Mathematical Symbolism.

" [Old English] lacked an adequate system of pronouns and ambiguities were multiplied in Middle English when *hē* ' he ', *hēo* ' she ', and Anglian *hēo* ' they ' . . . became identical in pronunciation. . . . The listener or reader [had to] gather the meaning from the context," says S. Potter in *Our Language*, and he continues : " That is why Middle English adopted and adapted these structural words [' they ', ' them ', and ' their '] from Scandinavian to supply its needs. Then, as now, intelligibility was a strong determining factor." " But although the *th*-forms must . . . be reckoned a great advantage to the language," says O. Jesperson in *Growth and Structure of the English Language*, " it took a long time before the old forms were finally displaced." But eventually the new forms were generally adopted, for, as Potter puts it, "when men find that their words are imperfectly apprehended they naturally modify their speech and they deliberately prefer the unambiguous form."

In cases I and X of the traditional mathematical symbolism, the letter x, serving as a numerical and as a fluent variable, plays the roles of, as it were, mathematical " pronouns " that differ from one another as much as do " he " and " they ". Moreover, in Cases II, III, IV, VII, and VIII, x has been found playing the roles of five (more or less unrelated) mathematical " nouns ", in Cases V and VI the roles of (totally unrelated) suffixes, and in Case X no role whatsoever. Is it surprising that even excellent teachers of mathematics find " that their words are imperfectly apprehended " by some beginners? This often deplored situation is usually attributed to the profundity of mathematics, the rareness of mathematical talent in students, and the like, even though it might well be accounted for, and certainly is seriously aggravated by, the fact that ubiquitous mathematical symbols are ambiguous, nay, if the neologism be permitted, duodeciguous.

The physicist, operating with a number of fluents that exceeds the number of letters in the alphabet, is forced to use, in different contexts, the same letter in different meanings, e.g., t for the temperature as well as for the time ; but no physicist has ever confused temperature and time. The equivocations in mathematics are more subtle and insidious. Indeed, which traditional treatises on analysis do clearly distinguish between the numerical variable in I(8), the identity function in II(4), and the fluent variable in X(1)?

The author has " naturally modified " mathematical expressions by introducing structural symbols that reflect the conceptual variety behind the letters x and y. Intelligibility has been a " strong determining factor " in the new approach to pure and applied analysis l.c. [1], where he "deliberately prefers unambiguous forms." The ten principal points of this clarification can be summarized as follows.

A. The designations of numbers (0, 1, 3, e, i, ...) and numerical variables (x, y, a, b, c, ...) are printed in roman type ; the designations of specific functions (*log*, *cos*, the constant functions *0*, *1*, *3*, ...) and function variables in italics ; operators in bold face. For instance, I(8) reads I(8') **D** sin x $=cos$ x for any x.

B. The following statements involving the specific functions j, I, and J are equivalent to the laws I(7)–(10) concerning any number belonging to a certain scope :

I(7') $sin\ (2j) = 2\ sin\ .\ cos.$
(The dot indicating multiplication must not be omitted. Mere juxtaposition means substitution, as in $sin\ (2j)$.)
I(8'), (9') **D** $sin = cos$ and **D** $log = j^{-1}$ (on the class of all numbers >0).
I(10') If $F = I^3 . J^5$, then $D_1 F = 3I^2 . J^5$ and $D_2 F = 5I^3 . J^4$.
Either of the following formulae, which are equivalent with I(5),
I(5') $I^2 - 9J^2 = (I + 3J) . (I - 3J)$ and $J^2 - 9I^2 = (J + 3I) . (J - 3I)$
can be obtained from the other by substituting J in the first, and I in the second place. But the formulae I(5') are not identical in meaning.

C. The class of all x such that $x^2 - 1 = 0$ consists of all numbers for which the functions $j^2 - 1$ and 0 assume equal values. Since $j^2 - 1 = 0$, by itself, is a false statement, the class will be denoted by $\{j^2 - 1 = 0\}$. Similarly, L and L' in I(3) are $\{2I + 3J = 5\}$ and $\{2J + 3I = 5\}$, respectively. (Here, 5 is the constant 2-place function of value 5 ; it is denoted by $5^{(2)}$ where there is any danger of confusing it with the constant 1-place function 5 of value 5.) In pure Cartesian plane geometry, L and L' are called lines, and

I(4') $\{2I + 3J = 5\} \neq \{2J + 3I = 5\}$.

D. The real-valued complex functions in IV might be denoted by re and im. In the realm of complex numbers

$$\text{IV}(1'),\ (2')\quad j = re + i\,im \quad \text{and}\quad \tfrac{1}{2}Dj^2 = (re + i\,im)^2.$$

This is one of the many cases where intelligibility of the symbolism is achieved by following the way mathematicians *talk*. Who would orally refer to the x of a complex number rather than to its real part? Only in *writing* do mathematicians resort in this case as in so many others to the letters x and y that are so heavily fraught with connotations.

E. The italic letters x and y are reserved for specific fluents : Cartesian coordinates in a physical or postulational plane. The line consisting of all points π such that $2x\pi + 3y\pi = 5$ can, without the use of point variables, be written $\{2x + 3y = 5\}$. Clearly,

$$\text{VIII}(5')\quad \{2x + 3y = 5\} \neq \{2y + 3x = 5\}.$$

F. In view of the specific meaning of x and y, it is wise to avoid the use of x and y as function variables ; and it is necessary to refrain from using x and y as fluent variables whose scopes include coordinates (just as one must avoid the use of e as a numerical variable whose scope includes the specific number e). The letters u, v, w, ... and f, g, h, ... are convenient fluent and function variables—symbols that may be replaced, e.g., by the time t and the ordinate y ; and the functions cos and j^2, respectively. The letters a, b, c, ... are convenient as variables whose scopes consist of constant functions—symbols that may be replaced e.g., by the constant functions 0 and 3.

G. If g is a function, let c_g denote the constant operator of value g (that is, the class of all pairs (f, g) for any function f) just as, if b is any number, c_b (instead of b) might denote the constant function of value b (that is, the class of all pairs (x, b) for any number x).

Operators are objects of three operations (addition, multiplication, and substitution) just as are functions. The operators that are neutral with regard to these operations are c_0, c_I, and j, that is, the classes of all pairs $(f, 0)$, (f, I), and (f, f) for any function f, respectively. [5]

The operator defined in (II 5), which plays a great role in quantum mechanics, clearly is the product of c_j (the constant operator of value j) and the identity operator j, since

$$(c_j.\ \text{j})f = c_j f.\ \text{j}f = j\ .f. \quad \text{Similarly,}\ (c_j.\ \text{D})f = c_j f.\ \text{D}f = j\ .Df.$$

Hence, VII (1) is rendered by

$$\text{VII}(1')\quad (D^{\text{II}} + c_j.\ \text{D} + \text{j})J_0 = 0 \quad \text{or}\quad (D^{\text{II}} + c_j.\ \text{D} + \text{j})J_0(\text{x}) = 0.$$

Roman type has been used for the number 0, italic type for the constant function 0 of value 0. It will further be noted that roman numerals in the exponents distinguish substitutive iteration from multiplicative iteration which is traditionally expressed by arabic numerals in exponents. By the same token, one might distinguish the functions $sin^{-\text{I}} = arcsin$ and $sin^{-1} = csc$.

H. As a bearer of the ordinal number (as " basis of the exponent ") indicating the position of the coefficient in a polynomial or rational form, one may use an asterisk, $*$; in an n-ary form: $*_1, \ldots, *_n$; in a binary form also : $*$ and $\dagger$. One has

in arithmetic : $\dfrac{x^2 - 1}{x - 1} = x + 1$ for any number $x \neq 1$;

in the algebra of forms : $\dfrac{*_2 - 1}{* - 1} = * + 1$;

in analysis : $\dfrac{j^2 - 1}{j - 1} = j + 1$ (on the class of all numbers $\neq 1$).

I. Since x and y mean abscissa and ordinate, $\dfrac{d}{dx}$, $\dfrac{\partial}{\partial x}$ and $\dfrac{\partial}{\partial y}$ mean the rates of change with regard to these specific fluents. More generally, the operators $\dfrac{d}{d}$ and $\dfrac{\partial}{\partial}$ associate a fluent with two fluents ; the operators D, D_1, and D_2 associate a function with one function.

J. Again following verbatim the mathematician's oral expression, one may denote the area from 0 to 1 under the cosine curve by $S_0{}^1$ cos, just as one may write $S_a{}^{a+d}f$. Similarly, in presence of a symbol for the identity function, one may write $S_a{}^b$ cos $(2j)$ and $S_1{}^2 j^{-1}$. [6]. If a and b are two numbers, then $S_a{}^b f$ associates a number with any function f of a certain kind, whereas $\displaystyle\int_{\alpha_0}^{\alpha} w \, du$ associates a number with two fluents u and w, if the domain of the former includes α_0 and α.

Clearly, instead of the symbols here presented, other notations might be used to differentiate visibly between meanings of x and y that are worlds apart. But it is believed that only the maintenance in some form of strict distinctions between those various meanings makes it possible to present algebra, analytic geometry, and pure as well as applied analysis as a system of formulae connected by articulate rules. [7]   K.M.

REFERENCES

1. This paper elaborates on ideas expressed in the Appendix to the author's book *Calculus. A Modern Approach*, Ginn and Co., Boston 1955.

2. Where there is no indication to the contrary, " number " in this paper means *real* number.

3. Traditionally, x and y are printed in italic type as are all single letters used as mathematical symbols.

4. Cf. the author's paper *Random Variables and the General Theory of Variables* to be published in the Proceedings of the Third Berkeley Symposium on Mathematical Statistics and Probability, University of California, 1955.

5. Since substitution of operators (universally denoted by mere juxtaposition) is often called multiplication, the identity operator j is usually denoted by 1.

6. After submitting the present paper, I found that Dr. I. T. A. C. Adamson (*Mathematical Gazette*, September 1955) has suggested a similar symbol for definite integrals in his interesting review of my 1953 notes on calculus, edited in a greatly enlarged and improved form in the book l.c.[1].

1868. Dürer gained much by his visit to Venice. He had begun a translation of Euclid, and had been to Bologna to continue his study of mensuration. He had developed from grandiose structural compositions to a more thoughtful kind of painting.—Pierre Descargues, *Dürer*. [Per Mr. E. H. Lockwood].

1869. A STORM AT SEA.

Mr. Jolter was far from being unconcerned at the uncommon motion of the vessel, the singing of the wind, and the uproar he heard above him ; . . . the poor governor's heart died within him, and he shivered with despair. His recollection forsaking him, he fell upon his knees in the bed, and fixing his eyes upon the book which was in his hand, began to pronounce aloud with great fervour, " The time of a complete oscillation in the cycloid is to the time in which a body would fall through the axis of the cycloid DV, as the circumference of a circle is to its diameter."—Smollett : *Peregrine Pickle*, Chap. 35—Saintsbury Ed., II, p. 2. (What was Mr. Jolter's book?) [Per Prof. H. G. Forder.]

FUNDAMENTA
MATHEMATICAE
XLVI (1958)

Rates of change and derivatives

by

Karl Menger (Chicago)

In Memory of Alexander Wundheiler

1. Introduction. Ever since Lagrange initiated a new epoch in pure analysis by defining the derivative of a function, the logical clarity of applied mathematics has suffered from a confusion of those derivatives with the rate of change of one variable quantity with respect to another. Yet a mere count of the ideas involved in the two concepts clearly demonstrates that the situations studied in pure and in applied mathematics are basically unlike. The derivative associates a function with one function; for instance, the cosine function with the sine function. The rate of change associates a variable quantity with two variable quantities; for instance, the velocity with the distance travelled and the time.

This paper is devoted to the clarification of that difference and to the formulation of articulate rules coordinating the two situations. For the past 200 years, their synthesis has been immensely successful. In fact, the application of the derivative to rates of change in the physical universe has been a contribution of paramount importance to the development of science. But that application has been based on intuitive manipulations rather than on a conceptual foundation.

2. Fluents. One of the principal sources of shortcomings in the traditional literature is the lack of an adequate treatment of variable quantities. Many who use those words fail to explain them by any (either explicit or implicit) definition. Others give a definition — quantities capable of assuming various values — without, however, (either explicitly or implicitly) defining quantity. Still others, finally, confuse variable quantities with number variables such as the letters x and c in the general statement:

$$x - c^2 = (\sqrt{x} + c)(\sqrt{x} - c) \text{ for any number } x \geqslant 0 \text{ and any number } c.$$

("Number" here and in the sequel means: real number.) In order to forestall the latter (particularly obnoxious) confusion, in the present paper I will altogether avoid the words *variable quantity* and replace

them by the brief term that Newton used in reference to distance travelled, time, velocity, and the like, namely, *fluents*.

For any class A, a *fluent* with the domain A is a class of ordered
pairs in each of which the first member is an element of A, and the second
member is a number — a class that is free of what I will call inconsistent
pairs, that is, pairs whose first members are equal while their second
members are unequal. (Cf. [2]-[5], and [6], especially Chapter VII.) (The
reason for calling these classes fluents with the domain A rather than
real functions on A will become apparent in Section 3.)

If A is the population of Warsaw and, for any inhabitant a of that
city, ha denotes a's height in cm, then the class h of all pairs (a, ha) is
a fluent with the domain A.

The observed position on a straight line S of a particle moving
along S (more precisely, its directed distance in cm from a certain point
on S, called origin) is a fluent if s is defined as the class of all pairs $(\sigma, s\sigma)$
for any act σ of reading the scale mark opposite the particle, where $s\sigma$
denotes the number read on the scale as the result of the act σ.

For a particle at rest relative to the origin, the range (i. e., the class
of all values) of s consists of only one number. Such a fluent is said to
be *constant* (^{1}).

3. Functions. The cosine function is a fluent if it is defined in
the customary way as the class of all pairs $(x, cos\,x)$ for any number x.
Any fluent whose domain is a class of real numbers will be called a *function*.
The restriction of the word function to fluents of a special type
(namely, to the fluents studied in pure mathematics, whose domains
consist of numbers (^{2})), agrees with the use of the term by Leibniz, who
introduced it in order to describe certain connections between Newton's
fluents. One may ask what function in Leibniz' sense connects s with
the time t; whether s is the sine of t; whether s is the logarithm of some
other given fluent; and so on. But one cannot intelligently ask what
fluent other than a function in Leibniz' sense connects s with t; whether
s is the pressure of t; and whether the position is the temperature of
some other given fluent.

(^{1}) The existence of constant fluents shows that the definition of fluents as quantities capable of assuming various values in absence of a definition of quantity is not
only incomplete but in some ways too restrictive. For, a constant fluent must by no
means be confused with the number that is its only value.

(^{2}) More generally, functions include fluents whose domains consist, for some
positive integer n, of ordered n-tuples of numbers; of infinite sequences of numbers;
or of mutually similar well-ordered sets of numbers. Fluents whose domains are functions have been called *functionals*. Fluents, in turn, are special cases of *mappings* —
classes of mutually consistent pairs of any kind.

4. The substitution of fluents into functions. Let u be any fluent, that is, for some class A (called the *domain* of u and denoted by $\text{Dom}\,u$),

the class of all pairs (a, ua) for any $a \in A$;

and let f be any function whose domain is the class of all numbers. Under these assumptions,

the class of all pairs $\big(a, f(ua)\big)$ for any $a \in A$

is a fluent with the domain A that may be called the *result of substituting* u *into* f, and that may be denoted by $f(u)$. If each product of two fluents is designated by placing a dot between the symbols for the factor fluents (as in $f \cdot g$), then one may dispense with the parentheses in the symbols $f(u)$ and $f(ua)$ for results of substitutions and evaluations; and one may write ([3])

fu is the class of all pairs (a, fua) for any $a \in A$.

It will be noted that in the preceding definition of the fluent fu, the letter f serves as a function variable while the letter u is what may be called a *fluent variable* ([4]); that is to say, u may be replaced with designations of specific fluents (e. g., the time t) just as f may be replaced with designations of specific functions (e. g., the cosine function), each such replacement yielding the definition of the result of substituting a specific fluent into a specific function (for instance, $\cos t$).

If f is a function whose domain does not include all numbers, then fu may be defined as

the class of all pairs (a, fua) such that $a \in \text{Dom}\,u$ and $ua \in \text{Dom}\,f$.

([3]) Mere juxtaposition of the symbols for a function and a fluent is also the *traditional* designation of the result of substituting the fluent into the function where the latter has a *multi*-letter symbol, as in $\cos t$ and $\log p$ for the cosine of the time and the logarithm of the pressure. No one writes $\cos(t)$ or $\log(p)$. Only when a fluent is substituted into a function with a *single*-letter symbol, such as a Bessel function or a Legendre polynomial, parentheses are in traditional use; and so they are in $f(u)$, where a single letter serves as a function variable. In view of the utter mathematical irrelevance of the number of letters in a symbol for a function, any rational standardization should unify the two cases.

([4]) It will be noted that u *is not a number variable.* No meaningful definition results if, in the definition of fu, the letter f is replaced with a function symbol, and u with a numeral. This is a first indication of the dangers in confusing fluents or fluent variables with number variables.

It is easy to describe the domain and the range of the fluent fu if, for any fluent v, any class D, and any class C of numbers, one considers the following two classes:

v (on D), the class of all pairs in v whose 1st member belongs to D;

v (into C), the class of all pairs in v whose 2nd member belongs to C.

In terms of these concepts (which are also useful in other connections) one readily proves:

(1)
$$\mathrm{Dom}(fu) = \mathrm{Dom}\, u \ (\text{into } \mathrm{Dom}f),$$
$$\mathrm{Ran}(fu) = \mathrm{Ran}f \ (\text{on } \mathrm{Ran}\, u)$$

for any function f and any fluent u.

The following two remarks will clarify the roles of the variables in formulae (1) and in the definition of fu:

1. f is a function variable but not a general fluent variable. Indeed, f may be replaced with the designation of the cosine function but not (if non-vacuous results are to be expected) with the designations of human height or observed position (as defined in Section 2). While the cosine of any fluent u is defined and satisfies formulae (1), the position of a fluent u on a line S remains undefined, and $\mathrm{Dom}\, u$ (into $\mathrm{Dom}\, s$) is a vacuous symbol for each fluent u.

2. u is a fluent variable but not a number variable. True, upon replacement of u with, say, 3, the symbol fu happens to remain meaningful (namely, to yield the value that f assumes for 3). But formulae (1) and the very definition of fu become nonsensical if u is replaced with the symbol 3, since the number 3 has no domain ([5]) nor does it assume a value for a.

5. Relative equality of fluents. Fluents being classes, it is clear when two fluents are equal, when u is an extension of v, and when u is a restriction of v. For instance, the fluents v (on D) and v (into C) are proper restrictions of v if D (or, more generally, the intersection of D and $\mathrm{Dom}\, u$) is a proper subclass of $\mathrm{Dom}\, v$, and if C (or the intersection of C and $\mathrm{Ran}\, v$) is a proper subclass of $\mathrm{Ran}\, v$.

([5]) Of course, the number 3 must not be confused with a constant fluent of value 3 (cf. ([1])). The constant function 3 of value 3 is the class of all pairs of numbers $(x, 3)$ for any number x. The reader will note that this paper adheres to the typographical convention introduced in the authors book *Calculus. A Modern Approach* [6]. Symbols referring to numbers, to *functions*, and to **operators** are printed in roman, in *italic*, and in **bold face** type, respectively. This standardization greatly simplifies the reading of formulae, and saves some parentheses and a great deal of verbiage.

In pure mathematics, these concepts are satisfactory with regard to the fluents studied, that is, with regard to functions. Two functions, f and g, are equal if and only if any pair belonging to either function also belongs to the other, which is the case if and only if $\operatorname{Dom} f = \operatorname{Dom} g$ and $fx = gx$ for any element x of that common domain.

In science, however, it is often necessary to relate fluents with disjoint domains — fluents which, mathematically speaking, are unequal. Yet a workable and useful definition of scientific equality of v and u is given, namely, relative to a certain subclass Π of the Cartesian product $\operatorname{Dom} u \times \operatorname{Dom} v$. I will write

$$v = u \ (\text{rel. } \Pi) \qquad \text{or} \qquad v \underset{\Pi}{=} u$$

if and only if $(\alpha, \beta) \in \Pi$ implies $v\beta = u\alpha$.

For instance, let s' be the observed position of a particle moving along a straight line S' (more precisely, the particle's directed distance from an origin on S'), defined as the class of all pairs $(\sigma', s'\sigma')$ for any act σ' of reading the scale mark opposite the particle. Physicists frequently compare the fluent s' with s (as defined in Section 2) relative to the class Γ_1 of all pairs of simultaneous acts (σ, σ'). The fluents s and s' are equal relative to this class Γ_1 if and only if

$$(\sigma, \sigma') \in \Gamma_1 \qquad \text{implies} \qquad s'\sigma' = s\sigma \,.$$

Again, physicists compare s and s' relative to the class Γ_2 of pairs (σ, σ') of acts that are respectively simultaneous with equal readings on two timers (set in motion at different instants and possibly calibrated in different units).

Since Galileo and Boyle, most comparisons of fluents in classical physics have been based, in some way or other, on *simultaneity of acts of observation or of physical states*. The comparison of functions, expressed in the mathematical concept of equality (as defined at the beginning of Section 5) is implicitly based on *equality of numbers*; more precisely, $f = g$ means equality relative to the class of pairs of equal numbers belonging to the intersection of $\operatorname{Dom} f$ and $\operatorname{Dom} g$. In contrast, statisticians use a great variety ([6]) of subclasses of the Cartesian products of the domains (or, as many statisticians say, various *pairings* of the *populations* of the *variates*).

6. Properties of relative equality. For any class Π of pairs, let Π^* denote the class of all pairs (β, α) such that (α, β) belongs to Π. For any two classes Π and P, let $P\Pi$ denote the class of all pairs (α, γ)

([6]) In comparing the height h and the weight w in the population of Warsaw, one may study the class Π of all pairs (α, α) for any inhabitant α; or the class Π_1 of all pairs (α, β), where β is the father of α; or the class Π_2 of all pairs of twins, and so on.

for which there exists an element β such that $(a, \beta) \in \Pi$ and $(\beta, \gamma) \in P$. Finally, let $I(C)$ for any class C denote the class of all pairs (γ, γ) such that $\gamma \in C$. It then is clear that relative equality has the following properties:

Symmetry. If $v = u$ (rel. Π), then $u = v$ (rel. Π^*).

Transitivity. If $v = u$ (rel. Π) and $w = v$ (rel. P), then $w = u$ (rel. $P\Pi$). Here, $\Pi \subseteq \mathrm{Dom}\,u \times \mathrm{Dom}\,v$ and $P \subseteq \mathrm{Dom}\,v \times \mathrm{Dom}\,w$, wherefore $P\Pi \subseteq$ $\subseteq \mathrm{Dom}\,u \times \mathrm{Dom}\,w$.

Reflexivity. $u = u$ $\big(\mathrm{rel.}\,I(\mathrm{Dom}\,u)\big)$.

Relative equality of a fluent with another fluent that is the result of a substitution is of paramount importance in applications of analysis to science:

$$w = fu \ (\mathrm{rel.}\,\Pi) \quad \text{or} \quad w \underset{\Pi}{=} fu$$

if and only if $(a, \beta) \in \Pi$ implies $w\beta = fua$.

From the transitivity of relative equality it follows that

if $v = fu$ (rel. Π) and $w = gv$ (rel. P), then $w = gfu$ (rel. $P\Pi$).

7. Rate of change. From this point on, it will be assumed that the domain of any fluent studied is a limit class. There is no difficulty about defining when a fluent is *continuous*. The idea of rate of change is more complex. The velocity of a moving particle, the slope of a curve, and similar examples suggest that the rate of change of a fluent w with respect to a fluent u is itself a fluent. Consequently, its domain must be clearly defined. Moreover, it will appear that an articulate definition of that rate of change must be relative to a subclass Π of $\mathrm{Dom}\,u \times \mathrm{Dom}\,w$; that is to say, that one must define a fluent

$$\frac{dw}{du} \ (\mathrm{rel.}\,\Pi) \quad \text{or} \quad \underset{\Pi}{dw/du}$$

a relativization comparable to that presupposed by a scientifically workable definition of equality.

The salient point of the theory to be expounded is the following. The domain of $\underset{\Pi}{dw/du}$ is a subclass of that class Π; in a formula,

$$\mathrm{Dom}\,(\underset{\Pi}{dw/du}) \subseteq \Pi .$$

Clearly, since $\mathrm{Dom}\,u$ and $\mathrm{Dom}\,w$ are limit classes, so are (in a natural way) $\mathrm{Dom}\,u \times \mathrm{Dom}\,w$ and $\mathrm{Dom}\,(\underset{\Pi}{dw/du})$.

To facilitate matters, it will be assumed that $\mathrm{Dom}\,u$ and $\mathrm{Dom}\,w$ are endowed with a continuous semi-metric, by which I mean that a number $d(a', a)$ be associated with any two elements a' and a of $\mathrm{Dom}\,u$ in such a way that $\lim d(a_n, a) = 0$ if and only if $\lim a_n = a$; and that a similar definition applies to w. These auxiliary metrics as such will be of no significance and may be replaced with any two other metrics that preserve the limits.

For any number $d > 0$, the pair (ξ, η) of $\mathrm{Dom}\, u \times \mathrm{Dom}\, w$ is said to be a d-*neighbor* of the pair (a, β) if $d(\xi, a) < d$ and $d(\eta, \beta) < d$. Two pairs (a, β) and (ξ, η) are said to be u-*discriminating* if $ua \neq u\xi$. A pair $(a, \beta) \,\epsilon\, \Pi$ will be called Π-*normal* if, for each number $d > 0$, the class Π contains a u-discriminating d-neighbor of (a, β).

If (a, β) is Π-normal and c is a (finite) number, then I will call c the value of dw/du for (a, β), and I will write

$$dw/du(a, \beta) = c$$
$$\underset{\Pi}{}$$

if and only if the following is satisfied.

Condition c. For each positive integer n, there exists a number $d_n > 0$ such that

$$\left| \frac{w\eta - w\beta}{u\xi - ua} - c \right| < \frac{1}{n}$$

for each u-discriminating d_n-neighbor $(\xi, \eta) \,\epsilon\, \Gamma$ of (a, β).

Since (a, β) is Π-normal, clearly no two unequal numbers c and c′ can be values of dw/du for (a, β). Accordingly, the said rate of change may be defined as the class of all pairs

$$\left((a, \beta), \lim_{\substack{\xi \to a \\ \eta \to \beta \\ (\xi, \eta) \epsilon \Pi}} \frac{w\eta - w\beta}{u\xi - ua} \right)$$

for any Π-normal pair (a, β) for which the limit exists; that is to say, such that Condition c is satisfied for some number c.

Clearly, the value of the rate of change for (a, β) depends only upon the values of u and w for the members of neighbor pairs $\epsilon \Pi$. The limit may exist for each pair $(a, \beta) \,\epsilon\, \Pi$, or for some but not all of these pairs, or for none. Accordingly, $\mathrm{Dom}(dw/du)$ may be Π, or a proper subclass of Π, or the vacuous class.

The following remark bears out what has been said in Section 1: The rate of change is a binary operator, and the words "the rate of change of a fluent" remain undefined. The reader should beware of mistaking the quotients

$$\frac{u\xi - ua}{d(\xi, a)}$$

for a possible basis of a unitary rate of change (or fluxion) of the fluent u. For, as far as u is concerned, the auxiliary distance might well be replaced by its double, thereby reducing each of the said quotients to its half.

8. General remarks (⁷). For any class C and any number p, the constant fluent of value p with the domain C will be denoted by p (on C). For any fluent u, the element a will be called u-*changing* if $a \in \mathrm{Dom}\,u$ and for each positive integer n there exists an element $\xi_n \in \mathrm{Dom}\,u$ such that $d(\xi_n, a) < 1/n$ and $u\xi_n \neq ua$. The class of all u-changing elements will be denoted by $\mathrm{Ch\,Dom}\,u$ (Ch for characteristic). In this notation,

$$\frac{du}{du}\big(\text{rel. } I(\mathrm{Ch\,Dom}\,u)\big) = 1\,(\text{on } \mathrm{Ch\,Dom}\,u), \text{ for any fluent } u\,.$$

Here, $I(\mathrm{Ch\,Dom}\,u)$ (see Section 5) is the class of all pairs (a, a) for any $a \in \mathrm{Ch\,Dom}\,u$; and the equality is relative to the class of all pairs $\big(a, (a, a)\big)$ for any $a \in \mathrm{Ch\,Dom}\,u$. With the same definition of equality, also the following more general formula is self-explanatory.

$$\frac{d(pu+q)}{du}\big(\text{rel. } I(\mathrm{Ch\,Dom}\,u)\big) = p \text{ (on } \mathrm{Ch\,Dom}\,u),$$

for any fluent u and any two numbers p and q .

It will be noted that the preceding results do not presuppose that u have a limit anywhere, let alone that u be continuous. If $\mathrm{Dom}\,\lim u$ denotes the class of all $a \in \mathrm{Dom}\,u$ such that $\lim_a u$ exists, then, as one readily proves,

$$\frac{du^2}{du}\big(\text{rel. } I(\mathrm{Ch\,Dom}\,u)\big) = u + \lim u, \text{ for any fluent } u\,.$$

The domain of the fluent on the right side is the class of all elements belonging to both $\mathrm{Ch\,Dom}\,u$ and $\mathrm{Dom}\,\lim u$. The fluent assumes the value $ua + \lim_a u$ for each a belonging to that domain. The equality is relative to the class of pairs $\big(a, (a, a)\big)$ for any a in that domain. Any polynomial of u can be treated similarly.

If u is continuous at a, and (a, β) belongs to $\mathrm{Dom}(dw/du)_\varPi$, then w is continuous at β in the following restricted sense: For any positive integer n there exists a number $d'_n > 0$ such that $|w\eta - w\beta| < 1/n$ for each u-discriminating d'_n-neighbor (ξ, η) of (a, β) in $\varPi$.

(⁷) Most of the following remarks were suggested to the author by his colleague Prof. Abe Sklar.

9. The derivative of a function. Df is defined as the class of all pairs

$$\left(a, \ \lim_{\substack{x \to a \\ x \,\epsilon\, \mathrm{Dom}f}} \frac{fx - fa}{x - a}\right) \text{ for any } a \,\epsilon\, \mathrm{Dom}f \text{ such that the limit exists.}$$

This definition can be rephrased in such a way as to subsume under the definition of a rate of change. For this purpose it is necessary to have at one's disposal a symbol for the class of all pairs (x, x) for any number x. I will use ([8]) the letter j, and refer to this traditionally symbolless class as the *identity function*.

The class j plays a twofold role in rephrasing the definition of Df. Firstly, j is the fluent with respect to which the rate of change of the fluent f will be considered. Secondly, j (on $\mathrm{Dom}f$) (that is, the class ([9]) of all pairs (x, x) for any $x \,\epsilon\, \mathrm{Dom}f$) is the class relative to which I will consider the rate of change of f with respect to j.

In view of the fact that a pair of numbers (x, y) belongs to j (on $\mathrm{Dom}f$) if and only if $x \,\epsilon\, \mathrm{Dom}f$ and $y = x$, it is clear that

$$\frac{df}{dj}(\text{rel. } j \text{ (on } \mathrm{Dom}f)) \quad \text{or} \quad \underset{j \,(\text{on } \mathrm{Dom}f)}{df/dj}$$

is the class of all pairs

$$\left((a, a), \ \lim_{\substack{x \to a \\ x \,\epsilon\, \mathrm{Dom}f}} \frac{fx - fa}{jx - ja}\right) \text{ for any } \quad a \,\epsilon\, \mathrm{Dom}\,Df,$$

that is, for any $a \,\epsilon\, \mathrm{Dom}f$ such that the limit exists.

The domain of this rate of change and $\mathrm{Dom}\,Df$ are disjoint classes, since the latter consists of numbers and the former of pairs of equal numbers (a, a) such that $a \,\epsilon\, \mathrm{Dom}\,Df$. But except for this fact, the rate of change of f with respect to j is equal to Df. The difference between them is comparable to that between the rational number $\frac{3}{1}$ and the integer 3. Hence

THEOREM I. *The derivative of a function f is essentially the rate of change of f with respect to j and relative to j (on $\mathrm{Dom}f$). In a formula,*

$$Df \underset{\Pi}{=} \frac{df}{dj}(\text{rel. } j \text{ (on } \mathrm{Dom}f))$$

where the equality is relative to the class Π of all pairs $((x, x), x)$ for any $x \,\epsilon\, \mathrm{Dom}\,Df$.

([8]) Cf. [1]. Occasionally, I have used the letter I for the identity function. But it seems preferable to reserve italic capitals for functions of several places and to designate one-place functions by non-capital letters. Also the symbol id has been suggested for the identity function.

([9]) Of course, j (on $\mathrm{Dom}f) = I \,(\mathrm{Dom}f)$ if $I(C)$ is defined as in Section 6.

10. On functionally connected fluents. Suppose that $w \underset{\Pi}{=} fu$ and that (a, β) is Π-normal. Assume further that

$$dw/du(a, \beta) = c \quad \text{and} \quad Dfua = c'.$$

Since (a, β) is Π-normal, there exist arbitrarily close u-discriminating neighbor pairs (ξ, η) of (a, β). For each sufficiently close pair of this kind,

$$\frac{w\eta - w\beta}{u\xi - ua}$$

is arbitrarily close to c. In view of $w \underset{\Pi}{=} fu$, this means that

$$\frac{fu\xi - fua}{u\xi - ua}$$

is arbitrarily close to c. If u is continuous at a, then the said difference quotient is also close to c'; wherefore $c = c'$. The preceding reasoning shows that, if u is continuous at a, the existence of $Dfua$ implies the existence of $dw/du(a, \beta)$.

In order to infer, conversely, from the existence of the latter number that of $Dfua$, one must assume that, for any number $x \in \text{Dom} f$ that is sufficiently close to ua, there is an element $\xi' \neq a$ and an element η' such that (ξ', η') is a close neighbor pair of (a, β) belonging to Π and that $u\xi' = x$. I will say that u is *Darboux-continuous* at a if u satisfies the following condition: For each positive integer m, each ξ such that $d(\xi, a) < 1/m$ and $u\xi \neq ua$, and each number x between $u\xi$ and ua, there exists an element ξ' of $\text{Dom} u$ such that $d(\xi', a) < 1/m$ and $u\xi' = x$. I will call Π *quasi-continuous* at (a, β) if for each positive integer m there exists a positive integer n such that $d(\xi, a) < 1/n$ implies the existence of an element η such that $(\xi, \eta) \in \Pi$ and $d(\eta, \beta) < 1/m$. In this terminology, the preceding results can be summarized as follows:

THEOREM II. *If $w \underset{\Pi}{=} fu$ and u is continuous at a and $Dfua$ exists, then so does $dw/du(a, \beta)$. If u is Darboux-continuous at a, if Π is quasi-continuous at (a, β), and if $dw/du(a, \beta)$ exists, then so does $Dfua$. In either case,*

$$dw/du(a, \beta) = Dfua.$$

In the preceding theorem, f is only a function variable and, therefore, must not be replaced with designations of fluents other than functions. In contrast, u and w are general fluent variables that may, in particular, be replaced with the designations of functions — for instance, continuous nowhere differentiable functions.

If f is replaced with a designation of the sine function, then, in view of $\boldsymbol{D}\sin = \cos$, Theorem II yields the following

COROLLARY. *If* $w \underset{\Pi}{=} \sin u$, *then* $dw/du \underset{\Pi}{=} \cos u$.

Here, the fluent variables u and w may be replaced with designations of specific fluents, e. g., the time t and the position s of a certain oscillator, and Π with the designation of a specific class of pairs, e. q., Galileo's class Γ of pairs of simultaneous acts of clock readings and mark readings. Or u and w may be replaced with the abscissa x and the ordinate y along a sine curve in the Cartesian plane, and Π with the class I of pairs of equal points on that curve. In this way one obtains:

If $s \underset{\Gamma}{=} \sin t$, then $ds/dt \underset{\Gamma}{=} \cos t$.

If $y \underset{I}{=} \sin x$, then $dy/dx \underset{I}{=} \cos x$.

But while u and w are fluent variables, it is perfectly clear that they are not number variables. Replacing in the corollary u and w with designations of numbers one obtains false implications. If u and w are replaced with π and 0, respectively, in the resulting implication

$$\text{if } 0 = \sin\pi, \text{ then } d0/d\pi = \cos\pi$$

the antecedent is valid while the consequent is nonsensical.

11. Reciprocal rates of change. The element (a, β) of $\Pi \subseteq$ $\subseteq \mathrm{Dom}\,u \times \mathrm{Dom}\,w$ will be called Π-*binormal* if for each n the class Π contains an $1/n$-neighbor (ξ, η) of (a, β) such that $u\xi \neq ua$ and $w\eta \neq w\beta$. Clearly, if (a, β) is Π-binormal, then (β, a) is Π^*-binormal.

THEOREM III. *If* (a, β) *is* Π-*binormal and the numbers* $dw/du\,(a, \beta)$ *and* $du/dw\,(\beta, a)$ *exist and are* $\neq 0$, *then*

$$dw/du\,(a, \beta) \cdot du/dw\,(\beta, a) = 1.$$

For each $(\xi, \eta) \in \Pi$ such that $u\xi \neq ua$ and $w\eta \neq w\beta$, one has

$$\frac{w\eta - w\beta}{u\xi - ua} \cdot \frac{u\xi - ua}{w\eta - w\beta} = 1.$$

Since (a, β) is Π-binormal, Π contains pairs (ξ, η) arbitrarily close to (a, β) for which both difference quotients mentioned in the preceding formula are numbers $\neq 0$. Since by assumption the two rates of change exist and are $\neq 0$, their product equals 1.

12. The chain rule. Consider three fluents u, v, w and two classes $\Pi \subseteq \mathrm{Dom}\,u \times \mathrm{Dom}\,v$ and $P \subseteq \mathrm{Dom}\,v \times \mathrm{Dom}\,w$. Assume that (a, β) be Π-binormal and that P is quasi-continuous at (β, γ). Clearly, $(a, \gamma) \in$

7*

$\epsilon \, P\Pi \subseteq \mathrm{Dom}\, u \times \mathrm{Dom}\, w$. There exist arbitrarily close ne ghbors $(\xi, \eta) \, \epsilon \, \Pi$ of (a, β) such that $u\xi \neq ua$ and $v\eta \neq v\beta$; and arbitrarily close to γ there exists an element ζ of $\mathrm{Dom}\, w$ such that $(\eta, \zeta) \, \epsilon \, P$. Clearly,

$$\frac{v\eta - v\beta}{u\xi - ua} \cdot \frac{w\zeta - w\gamma}{v\eta - v\beta} = \frac{w\zeta - w\gamma}{u\xi - ua}.$$

If the corresponding rates of change exist, they satisfy the chain rule expressed in the following

THEOREM IV. *If (a, β) is Π-binormal and P is quasi-continuous at (β, γ), then*

$$\underset{\Pi}{dv/du}(a, \beta) \cdot \underset{P}{dw/dv}(\beta, \gamma) = \underset{P\Pi}{dw/du}(a, \gamma)$$

if the three rates of change exist.

13. The rate of change of totally unconnected fluents. Theorems III and IV hold even for fluents that are not functionally connected with one another. Consider in a Cartesian plane, e. g., the ellipse E given by

$$\{x^2 + 4y^2 = 1 \text{ (on } E)\},$$

that is, the class of all points P in the plane such that $x^2P + 4y^2P = 1$. Clearly, neither of the fluents x (on E) and y (on E) is a function of the other relative to $I(E)$, that is, the class of all pairs (P, P) for any P on E. For instance, y assumes unequal values for two points on E whose abscissae are equal. Notwithstanding the lack of any functional connection between x (on E) and y (on E), Theorem III implies

$$\underset{\Pi}{dy/dx}(P, P) \cdot \underset{\Pi^*}{dx/dy}(P, P) = 1, \quad \text{where} \quad \Pi = \Pi^* = I(E),$$

for any point P on E where both rates of change exist and are $\neq 0$.

Of course, in a neighborhood of each such point, x (on E) and y (on E) are functionally connected. For instance, along the open quarter of E in the first quadrant:

$$y = \tfrac{1}{2}\sqrt{1-x^2} = \tfrac{1}{2}\sqrt{1-j^2}\, x \quad \text{and} \quad x = \sqrt{1-4y^2} = \sqrt{1-4j^2}\, y,$$

where $\sqrt{}$ indicates the positive square root, and the juxtaposition of the function and fluent symbols indicates substitution of the fluents into the functions. According to Theorem II,

$$\underset{\Pi}{dy/dx}(P, P) = -\frac{1}{2} \frac{j}{\sqrt{1-j}} \, xP = -\frac{1}{2} \frac{xP}{\sqrt{1-x^2P}}.$$

Similarly, $\underset{\Pi^*}{dx/dy}$ can be connected with y. For each P, the product of the two rates of change is 1, in agreement with Theorem III.

But it is quite different examples that illustrate the true width of the idea of rate of change and the full generality of Theorems III and IV. This is achieved by examples of fluents u and w such that dw/du exists
$$\scriptstyle \Pi$$
without w being a function of u relative to Π *in any open subclass of* Dom u; in other words, in cases where every open subclass of Dom u includes two elements a and a' such that

$$(a, \beta) \in \Pi, \quad (a', \beta') \in \Pi, \quad ua = ua' \quad \text{and} \quad w\beta \neq w\beta'$$

and, nonetheless, dw/du exists. In fact, such examples of u and w exist
$$\scriptstyle \Pi$$
in the realm of functions, where Π is a subclass of j (that is, a class of pairs of equal numbers).

Let g denote Cantor's nowhere decreasing function such that $g0 = 0$, $g1 = 1$, and which is constant on every interval in the complement of Cantor's discontinuum D. Set

$$g_D = g \ (\text{on } D) \quad \text{and} \quad j_D = j \ (\text{on } D).$$

Then it is easy to show that

$$dj_D/dg_D = 0 \ (\text{on } D);$$

that is to say,

$$dj_D/dg_D(\mathrm{x}, \mathrm{x}) = 0 \quad \text{for each} \quad \mathrm{x} \in D.$$

Yet each open subclass of D includes pairs of unequal numbers x, x' such that $g_D\mathrm{x} = g_D\mathrm{x}'$ while, of course, $j_D\mathrm{x} \neq j_D\mathrm{x}'$.

From a remark in Section 8 it follows that, if f is a differentiable function, then

$$d(fg_D)/dg_D = Dfg_D,$$

where Dfg_D is the result of substituting g_D into Df. Clearly, $g\mathrm{x} = g\mathrm{x}'$ implies $fg\mathrm{x} = fg\mathrm{x}'$. Consequently, just like j_D, the function $j_D + fg_D$ is not a function of g_D in any open subclass of D. Yet, in view of the linearity of the rate of change operator, the rate of change of $j_D + fg_D$ with respect to g_D exists, and, more specifically,

$$d(j_D + fg_D)/dg_D = Dfg_D.$$

Any continuous function is the derivative of a differentiable function f. Hence

THEOREM V. *On D, any continuous function connects with g the rate of change with respect to g of some function h even though h is not functionally connected with g in any open subclass of D.*

Karl Menger

References

[1] K. Menger, *Algebra of analysis*, Notre Dame Math. Lectures 1944.

[2] — *The ideas of variable and function*, Proc. Nat. Acad. Sci. U. S. A. 39 (1953), p. 956-961.

[3] — *On variables in mathematics and in natural science*, Br. J. Phil. Sci. 5 (1954), p. 134-142.

[4] — *Variables de diverses natures*, Bull. Sciences Mathématiques 78 (1954), p. 229-234.

[5] — *Random variables and the general theory of variables*, Proc. 3rd Berkeley Symposium Math. Stat. & Prob., vol. II, 1954, p. 215-229.

[6] — *Calculus. A modern approach*, Boston 1955.

ILLINOIS INSTITUTE OF TECHNOLOGY, CHICAGO

Reçu par la Rédaction le 16. 11. 1957

Selected Papers on Probabilistic Geometry

Commentary on Probabilistic Geometry

B. Schweizer

1. Introduction

In our thirty-plus year association with Karl Menger, neither Abe Sklar nor I ever thought of asking him how, in 1942, he came to the idea of a "statistical metric", i.e., of replacing the classical numerical-valued distance $d(p, q)$ between two points p and q by a probability distribution function F_{pq} whose value $F_{pq}(x)$ is to be interpreted as the probability that the "distance" between p and q is less than x. It could have been stimulated by his Rice Institute lecture "Topology without Points" [40] or, more likely, by the lectures on "The Principles of Statistical Inference" which his close friend and former student, Abraham Wald, gave at the University of Notre Dame in 1941 [78]. Or it could have just come to him; after all, he was a man who was constantly getting new ideas.

2. Beginnings

In his original paper on the subject [41], Menger generalizes the identity and symmetry axioms for an ordinary metric in the obvious way and replaces the triangle inequality $d(p, r) \leqq d(p, q) + d(q, r)$ by the inequality

$$F_{pr}(x + y) \geqq T(F_{pq}(x), F_{qr}(y)), \qquad \text{(IIIm)}$$

where T is a 2-place function from $[0, 1] \times [0, 1]$ to $[0, 1]$ which he calls a "triangular norm". He specifies some of the properties that such a function T should satisfy (e.g., the condition $T(1, 1) = 1$ which guarantees that metric spaces are special cases of statistical metric spaces) but, unfortunately, gives no specific examples. He also introduces a notion of betweenness and points out that the number $F_{pq}(0+)$ may be used as a measure of the indistinguishability of p and q. He concludes with some comments regarding possible applications of such statistical metrics, noting in particular that they might be used to resolve the so-called Poincaré Paradox, i.e., the fact that, in contrast to the mathematical continuum where equality is an equivalence relation, in the physical continuum where "equal" means "indistinguishable" one may have A equal to B and B equal to C while A is not equal to C.

Menger's paper was followed almost immediately by a paper by Wald [79]. In his paper Wald criticizes Menger's inequality because it involves an unspecified function and because the notion of betweenness it induces does not have all the properties of metric betweenness – a concept which Menger himself had introduced and studies in [39]. Wald suggests instead the inequality

$$F_{pr} \geqq F_{pq} * F_{qr} \qquad \text{(IIIw)}$$

where $*$ denotes convolution. This inequality has a natural interpretation in terms of independent random variables and it leads to a metric-like notion of betweenness. History has shown, however, that it is much too strong: indeed, aside from metric spaces, it is exceedingly hard to find examples in which (IIIw) is globally satisfied.

Note: In [78], Wald says that "The notion of a random variable and a distribution function are defined as follows: if $F(x)$ is a function expressing the probability that a real variable $X < x$, we say that X is a random variable and that $F(x)$ is the *probability distribution* of X." This conception is different from Kolmogorov's definition of a random variable as a measurable function defined on a probability space. The distinction between a statistician's and a probabilist's use of the term "random variable" is discussed and analyzed by Menger in [47], which is reprinted in this volume.

After 1942 the subject lay dormant until Menger, probably motivated by his remarks in the last section of [42] and the tragic death of Wald, returned to it in three short notes written in 1951.

The first note, "Probabilistic Theories of Relations" [43], begins with a discussion of Poincaré's Paradox. Menger introduces a probabilistic equivalence relation E on some underlying set, with the interpretation that $E(a,b)$ is the probability that the elements a and b are indistinguishable. He then notes that such a relation "in principle" solves Poincaré's Paradox, e.g., if $E(a,b) = E(b,c) = 3/4$, say, then since by hypothesis $E(a,c) \geqq E(a,b)E(b,c)$, we need not have $E(a,c) = 3/4$. Menger also introduces a probabilistic order relation, but this is of limited usefulness since it is defined in terms of the usual order relation on $\mathbb{R}$. The note concludes with remarks indicating that the above probabilities should be replaced by probability distribution functions.

The third note, "Ensembles flous et fonctions aléatoires" [45], is a direct continuation of the first. In it Menger notes that the relation defined via $E(x, y) > 0$ is an equivalence relation and, following the terminology he introduced in [40], calls the resultant equivalence classes "morceaux" (or "lumps"). He then defines an "ensemble flou" as a real-valued function π_F whose value $\pi_F(x)$, for any subset F of some universal set U and any x in U, is interpreted as the probability that x belongs to F. This definition is identical to the definition of a "fuzzy set" given by L. Zadeh in 1965 [80]. Thus Menger anticipated the Theory of Fuzzy Sets by 14 years! However, characteristically, he did not go beyond laying bare the essential idea of the subject. Thus, while he says "La somme et le produit de deux ensembles non rigides ne sont pas déterminés par les ensembles", he does not present any equalities or inequalities relating $\pi_{F \cup G}$ and $\pi_{F \cap G}$ to π_F and π_G. It should also be noted that Menger's probabilistic equivalence relation is a fuzzy relation.

In the second note, "Probabilistic Geometry" [44], Menger returns to the consideration of statistical metrics. He adopts Wald's triangle inequality with its attendant difficulties. The new element here is the introduction of several distinguishability relations that are based on the behavior of the distance distribution functions in a neighborhood of the origin. These are illustrated by a number of examples which, in the end, lead Menger to say that "we may have to weaken, if not give up, the triangle inequality". The note concludes with Menger pointing out and illustrating "two or three main types of applications of a probabilistic geometry", viz., to the psychological continuum, to the physical macrocosm and to the physical microcosm.

After 1951 Menger returned to the subject on two further occasions. In 1954, in the last chapter of the booklet "Géométrie générale" [46], which is entitled "Esquisse d'une Géométrie métrique aléatoire", he reviews the subject and gives several further examples. He states explicitly that "Une trés grande variété d'hypothèses correspondantes s'offre pour l'inégalité triangu-laire" and that "Il est clair que cette hypothèse de Wald est trés forte"; but in spite of these misgivings, he retains Wald's inequality. In 1970, in his contribution to the collection of essays "Ernst Mach, Physicist and Philosopher" [48], which is reprinted in Volume 1 of these Selecta (pp. 575–593), he considers the foundations of geometry from a positivist's point of view. He discusses the shortcomings of classical and differential geometry as models for both the macrocosm and the microcosm. He again returns to Poincaré's Paradox and, again using the values $F_{pq}(0 +)$, gives a more detailed analysis of the indistinguishability relations. Finally combining the ideas of topology without points, fuzzy sets and probabilistic geometry, he expresses the belief that "the ultimate solution of problems of micro-geometry may well lie in a *probabilistic theory of hazy lumps*".

Thus, in typical Menger-style, virtually all the basic ideas are clearly spelled out in these papers. What is missing is the proper framework in which to place them, the mathematical machinery to deal with them, and bona fide examples. Today these all exist; and the remainder of this essay will be devoted to showing how Menger's Gedankengang has grown and flourished in the intervening half-century.

3. Foundations

Abe Sklar and I began our collaboration on probabilistic metric spaces in 1956. At first we too were hampered by the abovementioned difficulties with Wald's inequality. Our "break-through" came with a return to Menger's original inequality, not at first in its general form (IIIm), but via particular examples. Over the years the subject has grown slowly but steadily; its history and status as of 1982 are detailed in our book [66]. It has also branched out in a number of unexpected directions.

In our first papers, Abe Sklar and I defined a statistical metric space as a pair $(S, \mathcal{F})$, where S is a set and $\mathcal{F}$ a mapping from $S \times S$ into the set of one-dimensional probability distribution functions such that, denoting $\mathcal{F}(p, q)$ by F_{pq}, for all p, q, r in S,

(I) $F_{pq}(0) = 0$, and $F_{pq}(x) = 1$ for all $x > 0$ iff $p = q$;

(II) $F_{pq} = F_{qp}$;

(IIIs) If $F_{pq}(x) = 1$ and $F_{qr}(y) = 1$, then $F_{pr}(x + y) = 1$.

We further defined a *Menger space* as a triple $(S, \mathcal{F}, T)$ with $\mathcal{F}$ satisfying (I), (II), (IIIm) and the *triangular norm* (briefly, *t-norm*) T satisfying the conditions

(a) $T(a, 1) = a$,

(b) $T(a, b) = T(b, a)$,

(c) $T(a, b) \leq T(c, d)$ whenever $a \leq c$ and $b \leq d$,

(d) $T(T(a, b), c) = T(a, T(b, c))$,

for all a, b, c, d in $[0, 1]$. And we defined a *Wald space* as a pair $(S, \mathcal{F})$ satisfying (I), (II) and (IIIw).

The conditions (a)–(c) are a strengthening of the conditions (b)–(e) given by Menger in [41]; and the addition of the associativity condition (d) permits the extension of the triangle inequality (IIIm) to a polygonal inequality.

Particular t-norms are the functions Z, W, Prod and Min given, respectively, by

$$Z(x,y) = \begin{cases} x, & y = 1, \\ y, & x = 1, \\ 0, & \text{otherwise}; \end{cases}$$

$$W(x,y) = Max(x+y-1, 0);$$

$$Prod(x,y) = xy;$$

$$Min(x,y) = \text{Minimum}(x,y).$$

Clearly $Z \leq W \leq Prod \leq Min$ and $Z \leq T \leq Min$ for any t-norm T.

Menger's triangle inequality involves a binary operation T on the unit interval whereas Wald's involves a binary operation – convolution – on the space of probability distribution functions. In 1963, A. N. Šerstnev [70] put these two inequalities under the same roof by first noting, as Menger had already done [41], that (IIIm) holds for all $x, y \geq 0$ if and only if

$$F_{pr}(z) \geq \sup_{x+y=z} T(F_{pq}(x), F_{qr}(y))$$

for all $z \geq 0$, and then by going further and showing that if T is a left-continuous t-norm then the function τ_T defined for any pair of distribution functions F, G via

$$\tau_T(F, G)(z) = \sup_{x+y=z} T(F(x), G(y)) \qquad (3.1)$$

is, like convolution, a binary operation on distribution functions that satisfies conditions analogous to (a)–(d). (In convex analysis such operations are called "supremal convolutions" [60].) These observations at once led Šerstnev to the definitive definition of a probabilistic metric space. It goes as follows: Let $\mathbb{R}^+ = [0, +\infty]$, $I = [0, 1]$ and

$$\Delta^+ = \{F \mid Dom\, F = \mathbb{R}^+,\ Ran\, F \subseteq I,\ F(0) = 0,\ F(+\infty) = 1,$$

$$F \text{ is non-decreasing and left-continuous on } (0, +\infty)\}. \qquad (3.2)$$

Let Δ^+ be partially ordered by the usual pointwise ordering of functions, i.e., $F \leq G$ iff $F(x) \leq G(x)$ for all x in $\mathbb{R}^+$; and, for any $a \geq 0$, let ε_a be the unit step-function given by

$$\varepsilon_a(x) = \begin{cases} 0, & x \leq a, \\ 1, & x > a. \end{cases} \qquad (3.3)$$

Definition 3.1. A *triangle function* is a mapping τ from $\Delta^+ \times \Delta^+$ into Δ^+ such that, for all F, G, H, K in Δ^+,

(α) $\tau(F, \varepsilon_0) = F$,
(β) $\tau(F, G) = \tau(G, F)$,
(γ) $\tau(F, G) \leq \tau(H, K)$ whenever $F \leq H$ and $G \leq K$,
(δ) $\tau(\tau(F, G), H) = \tau(F, \tau(G, H))$.

Particular triangle functions are convolution and the functions τ_T defined in (3.1).

Definition 3.2. A *probabilistic metric space* (briefly, a *PM space*) is a triple $(S, \mathcal{F}, \tau)$, where S is a set, $\mathcal{F}$ is a mapping from $S \times S$ into Δ^+, τ is a triangle function, and the following hold:

(I) $\mathcal{F}(p, q) = \varepsilon_0$ iff $p = q$;
(II) $\mathcal{F}(p, q) = \mathcal{F}(q, p)$;
(III) $\mathcal{F}(p, r) \geq \tau(\mathcal{F}(p, q), \mathcal{F}(q, r))$.

If $\tau = \tau_T$, then $(S, \mathcal{F}, \tau)$ is a Menger space (under T); and if $\tau = $ convolution, then $(S, \mathcal{F}, \tau)$ is a Wald space.

Comparing Definition 3.2 to the usual definition of a metric space $(S, d, +)$, we see that the generalization from a metric to a probabilistic metric space consists of: (1) Replacing the range space $\mathbb{R}^+$ of the metric by Δ^+ and: (2) Replacing the operation of addition in the ordinary triangle inequality by a triangle function τ.

4. Examples

A. The simplest non-trivial PM spaces are those obtained by starting with a metric space (S, d) and a distribution function $G (\neq \varepsilon_0$ or $\varepsilon_\infty)$ in Δ^+ and using G to make the distance in S "fuzzy" by defining $\mathcal{F}$ via $\mathcal{F}(p, q)(x) = G(x/d(p, q))$. The resulting space $(S, \mathcal{F})$ is the *simple space* generated by (S, d) and G. It is a Menger space under Min. Moreover, if for any F in Δ^+ we define the inverse function $F^{(-1)}$ on I via

$$F^{(-1)}(y) = \sup\{x | F(x) < y\}$$

then $F_{pq}^{(-1)} = d(p, q)G^{(-1)}$. From this it follows that for any c in $(0, 1)$ the function d_c given by $d_c(p, q) = F_{pq}^{(-1)}(c)$ is a metric on S. Thus any simple space may be viewed as a linearly ordered collection of metrics on S.

Furthermore, if $\lim_{x \to \infty} G(x) = 1$, then $F_{pq}(x)$ is the Lebesgue measure of the set $\{c \,|\, d_c(p, q) < x\}$.

Simple spaces have been used to construct a model for a phenomenological theory of hysteresis (see Section 8.5 of [66] and the references cited therein).

B. Another class of PM spaces is obtained via the following construction, due to H. Sherwood [71]. Let S be a set of functions from a probability space $(\Omega, \mathcal{A}, P)$ into a metric space (M, d). Suppose that for all p, q in S and all $x \geq 0$, the set $A = \{t \in \Omega \,|\, d(p(t), q(t)) < x\}$ is P-measurable. Then $d(p, q)$ is a random variable whose distribution function is determined by $\mathcal{F}(p, q)(x) = P(A)$. The pair $(S, \mathcal{F})$ is an *E-space* with base $(\Omega, \mathcal{A}, P)$ and target (M, d). Every E-space is a Menger space under the t-norm W. Note that this construction generalizes in the sense that if the metric space (M, d) is replaced by any space with a deterministic structure (e.g., a normed space or an inner product space) then one obtains a space in which this structure is probabilistic (e.g., a probabilistic normed or probabilistic inner product space).

For any t in Ω, the function d_t defined on $S \times S$ by $d_t(p, q) = d(p(t), q(t))$ is a pseudometric on S (but not necessarily a metric since $d_t(p, q) = 0$ need not imply $p = q$). Moreover, if $\mathbb{D}$ is the set of all pseudometrics d_t then the function μ defined on all subsets B of $\mathbb{D}$ of the form $B = \{d_t \in \mathbb{D} \,|\, t \in A \text{ and } A \in \mathcal{A}\}$ by $\mu(B) = P(A)$ is a measure on $\mathbb{D}$ and $\mathcal{F}(p, q)(x) = \mu(B)$. Thus if $(S, \mathcal{F})$ is an E-space then $(S, \mathcal{F})$ is *pseudometrically generated* in the sense that there exists a probability space $(\mathbb{D}, B, \mu)$ such that: (a) $\mathbb{D}$ is a collection of pseudometrics on S; (b) all sets of the form $\{d \in \mathbb{D} \,|\, d(p, q) < x\}$ are μ-measurable; (c) $\mathcal{F}(p, q)(x) = \mu\{d \in \mathbb{D} \,|\, d(p, q) < x\}$. Conversely, every pseudometrically generated PM space is (isometric to) an E-space and hence a Menger space under W [71]. With this characterization, the probabilistic extensions of such metric properties as betweenness, convexity, etc., can be defined by simply considering the measure of the set of all pseudometrics for which the given property holds.

C. If, in an E-space, the functions $p, q, \ldots$, in S from $(\Omega, \mathcal{A}, P)$ to (M, d) are themselves random variables, if $d(p, q)$ is also measurable, and if $\mathcal{F}(p, q)$ is the distribution function of $d(p, q)$, then we obtain a *random variable generated E-space*. Here the case when (M, d) is the n-dimensional Euclidean space $\mathbb{R}^n$ and the distribution functions of the elements of S are non-singular, independent, spherically symmetric Gaussian vectors has been studied in detail (see Chapter 10 of [66]). If $N(\bar{c}_p, \sigma_p I)$ and $N(\bar{c}_q, \sigma_q I)$ are the probability densities of p and q, respectively, then $\mathcal{F}(p, q)(x) = G(x/\sigma_{pq}^2)$, where G is the distribution function of the non-central chi-square distribution with n degrees of freedom and non-centrality parameter $d^2(\bar{c}_p, \bar{c}_q)/\sigma_{pq}^2$ and $\sigma_{pq}^2 = \sigma_p^2 + \sigma_q^2$. We think of the elements of S as "particles" and, for each particle p, view $N(\bar{c}_p, \sigma_p I)$ as a "cloud" in $\mathbb{R}^n$ whose "density" at any point of $\mathbb{R}^n$ measures the relative likelihood of finding the particle p, which should be at $\bar{c}_p$, in a vicinity of that point. Accordingly, for any Borel subset A of $\mathbb{R}^n$, the integral $\int_A N(\bar{c}_p, \sigma_p I) dV$ is naturally interpreted as the probability that the particle p is in the set A, and $\mathcal{F}(p, q)(x)$ as the probability that the distance between the particles p and q is less than x.

The random variable $d^2(p, q)$ is known as Mahalanobis' D_1^2-statistic; it is closely related to Hotelling's T^2, and both of these quantities play an important role in discriminatory analysis. Indeed, the "statistical field" introduced by Mahalanobis in [38] is succinctly described as an E-space over $\mathbb{R}^n$ whose elements are (not necessarily spherically symmetric) Gaussian random vectors. Other considerations which also lead to Gaussian E-spaces may be found in the writings of L. de Broglie [11], A. S. Eddington [21], and N. Rosen [61], as well as in the books by D. I. Blokhintsev [9] and E. Prugovečki [57], where further references to the literature may be found.

For any $k \geq 0$ and any p, q in the Gaussian E-space described above, the k'th moment $M_k(p, q)$ of $\mathcal{F}(p, q)$ exists and for $k \geq 1$ the function $M_k^{\frac{1}{k}}$ is a metric on S. Furthermore, as $d(\bar{c}_p, \bar{c}_q) \to \infty$, $M_k^{\frac{1}{k}}(p, q) \to \sqrt{2}\left[\Gamma\left(\frac{n+1}{2}\right)/\Gamma\left(\frac{n}{2}\right)\right]\sigma_{pq}$, a postive number. It follows that, in the large, these *Fréchet-Minkowski Metrics* [25] are asymptotic to the metric d of the Euclidean space $\mathbb{R}^n$ over which the Gaussian E-space was constructed while, in the small, they are discrete. Looking at the E-space as a set of particles and clouds, we see on the one hand, that as the Euclidean distance $d(\bar{c}_p, \bar{c}_q)$ between the centers of the clouds of p and q approaches 0, any Fréchet-Minkowski distance between p and q remains greater than a positive number; and on the other hand, that when it is large, $d(\bar{c}_p, \bar{c}_q)$ is a good estimate of the distance between the particles themselves. If, furthermore, there is a number $C > 0$ such that $\sigma_{pq} \leq C$ for all p, q, in S, then the set of variances of the distributions $\mathcal{F}(p, q)$ is bounded. It follows that as $d(\bar{c}_p, \bar{c}_q)$ increases, the ratio of the standard deviation of $\mathcal{F}(p, q)$ to the mean of $\mathcal{F}(p, q)$, a quantity that measures the relative uncertainty in the probabilistic determination of the distance between p and q, decreases to 0. This means that the "haziness" of the distance between the particles p and q – which is predominant when their clouds are close together – becomes virtually insignificant when these clouds are sufficiently far apart. In this sense, the probabilistic metric, just as any associated Fréchet-Minkowski metric, is asymptotically Euclidean.

D. Returning to E-spaces in general, let $(S, \mathcal{F})$ be an E-space with base $(\Omega, \mathcal{A}, P)$ and target (M, d), let $(\mathbb{D}, \mathcal{B}, \mu)$ be the associated pseudometrically generated space and, for any pair of points p, q in S, let $X(p, q)$ be the function from Ω into $\mathbb{R}^+$ defined by

$$X(p, q)(t) = d(p(t), q(t)) = d_t(p, q).$$

Then $X(p, q)$ is a non-negative random variable and, for all p, q, r in S,

(i) $X(p, q)(t) = 0$ for all t in Ω iff $p = q$;
(ii) $X(p, q)(t) = X(q, p)(t)$ for all t in Ω;
(iii) $X(p, r)(t) \leqq X(p, q)(t) + X(q, r)(t)$ for all t in Ω.

It follows that the notion of a pseudometrically generated space is equivalent to the notion of a space of non-negative random variables satisfying (i), (ii) and (iii). Such spaces were first considered by A. Špaček [74]; and, as M. Regenwetter and A. A. J. Marley have shown, analogs of E-spaces and random metric spaces play a role in mathematical psychology [58].

Relaxing the condition "for all t" in (i), (ii) and (iii) to "almost all" t yields the definition of a *random metric space* (see Section 9.3 of [66]). Such spaces are more general than E-spaces. However, they are endowed with considerably more structure than PM spaces. This is so because the random variables $X(p, q)$ are all defined on a common probability space. Therefore, all their joint distribution functions exist and are completely determined. In an arbitrary PM space, joint distributions are not even defined. Furthermore, since distinct random variables can be identically distributed, the assertion $X(p, q) = X(q, p)$ a.e. is much stronger than the corresponding assertion that $\mathcal{F}(p, q) = \mathcal{F}(q, p)$. Recently Guo Tie-xin has undertaken an in-depth study of random metric, random normed and random inner product spaces [27].

E. As a last example, we consider a class of spaces that are constructed as follows:

Let (S, d) be a metric space and f a mapping from S into S. For any non-negative integer n, let f^n denote the n-th iterate of f, so that, for every x in S, $f^0(x) = x$ and $f^{n+1}(x) = f(f^n(x))$. The sequence $\{f^n(x)\}$ is the *trajectory* of x under f. Next, for any pair of points x, y in S, any positive integer n, and any $t \geqq 0$, let

$$\chi(x, y, t, n) = \sharp\{0 \leqq m < n \mid d(f^m(x), f^m(y)) < t\},$$

where $\sharp A$ denotes the number of elements in the set A. Thus $\chi(x, y, t, n)$ is the number of times in the first $n-1$ iterations that the distance between the m'th iterates of x and y is less than t. The function $F_{xy}^{(n)}(t)$ defined via

$$F_{xy}^{(n)}(t) = \frac{1}{n}\chi(x, y, t, n), \text{ for any } t \geq 0,$$

is in Δ^+ and $F_{xy}^{(n)}(t)$ may be interpreted as the probability that the distance between the initial segments of length n of the trajectories of x and y is less than t. We are interested in the asymptotic behavior of the sequences $\{F_{xy}^{(n)}(t)\}$. To this end, we let

$$
\begin{aligned}
F_{xy}(t) &= \liminf_{n\to\infty} F_{xy}^{(n)}(t), \\
F_{xy}^*(t) &= \limsup_{n\to\infty} F_{xy}^{(n)}(t).
\end{aligned}
\tag{4.1}
$$

For any x, y in S, F_{xy} and F_{xy}^* are non-decreasing functions from $\mathbb{R}^+$ into I and $F_{xy} \leq F_{xy}^*$. Without loss of generality, we may assume that they are in Δ^+, whence if $F_{xy} < F_{xy}^*$ then $F_{xy}(t) < F_{xy}^*(t)$, not only for some t, but for all t in some interval of positive length. The function F_{xy} is the *lower distribution* of x and y and F_{xy}^* is the *upper distribution* of x and y.

If $\mathcal{F}$ is the mapping from $S \times S$ into Δ^+ defined by $\mathcal{F}(x, y) = F_{xy}$, then $(S, \mathcal{F}, \tau_W)$ is a probabilistic pseudo-metric space, namely, the *transformation generated space* determined by the metric space (S, d) and the transformation f. Note that since $\mathcal{F}(f(x), f(y)) = \mathcal{F}(x, y)$ for any x, y in S, the transformation f preserves probabilistic distances.

In 1973, A. Sklar and I considered transformation generated spaces from the point of view of ergodic theory (see Chapter 11 of [66]). Using the Birkhoff Ergodic Theorem, we showed that if (S, d) is separable, if there is a probability measure P defined on a sigma algebra that contains the Borel sets of S, and if f is measure-preserving with respect to P, then for almost all pairs of points (x, y) in $S \times S$, endowed with the product measure P^2, we have $F_{xy} = F_{xy}^*$; and furthermore, if for any $t \geq 0$, $D(t)$ is the set given by

$$D(t) = \{(x, y) \in S \times S \mid d(x, y) < t\},$$

then

$$P^2(D(t)) = \int_{S \times S} F_{xy}(t) dP^2.$$

In the same year, in a joint paper with T. Erber [22] in which we were primarily concerned with Poincaré recurrence – and non-recurrence – we showed that if, in addition to being measure-preserving, f is mixing, then the limiting distribution is independent of the initial points, i.e., there is a distribution function G_f such that, for almost all pairs of points x, y in S, the sequence $\{F_{xy}^{(n)}\}$ converges weakly to G_f. It thus follow that if f is mixing then $G_f(t) = P^2(D(t))$ for any $t \geq 0$.

Stipulating that the limiting distributions F_{xy} and F_{xy}^* are equal leads to regular behavior; stipulating that they are not leads in the diametrically opposite direction – namely to chaos.

A function f mapping compact interval $[a, b]$ into itself is *chaotic in the sense of Li and Yorke* if there exist at least two points, x and y, in $[a, b]$ and an $\varepsilon > 0$ such that

$$
\limsup_{n \to \infty} |f^n(x) - f^n(y)| \geq \varepsilon,
$$
$$
\liminf_{n \to \infty} |f^n(x) - f^n(y)| = 0 \tag{4.2}
$$

The similarity between (4.1) and (4.2) is striking: and the fact that the distributions of the average distances between trajectories give more information than just their limits inferior and limits superior leads one to conjecture that transformation generated spaces might play a role in the study of chaos. They do: and in the last decade J. Smítal, A. Sklar and I have laid the foundations of a theory of distributional chaos. In its barest outlines, it goes as follows:

Definition 4.1. Let (S, d) be a metric space and f a function from S into S. Then f is *distributionally chaotic* if there are two points, x and y, in S such that $F_{xy} < F_{xy}^*$.

For maps on a compact interval it can be shown that if f is distributionally chaotic then f is chaotic in the sense of Li and Yorke, that the converse is false, that f is distributionally chaotic if and only if f has positive topological entropy, and that for spaces other than compact intervals this equivalence is generally false.

We can go beyond merely providing a new definition of chaos. To this end, recall that if $F_{xy} < F_{xy}^*$ then these functions differ on some interval of positive length. Thus if f is distributionally chaotic then there exist at least two points, x and y, in S for which

$$
\int_0^\infty (F_{xy}^*(t) - F_{xy}(t)) dt > 0
$$

Furthermore, if (S, d) has a finite diameter d_S, then for all x and y in S we have

$$0 \leqq \frac{1}{d_S} \int_0^\infty (F_{xy}^*(t) - F_{xy}(t))dt \leqq 1,$$

and this leads naturally to

Definition 4.2. Let (S, d) and f be as in Definition 4.1 and suppose that S has a finite diameter d_S. Then the *principal measure of chaos* of f is the number

$$\mu_p(f) = \sup_{x,y \in S} \frac{1}{d_S} \int_0^\infty (F_{xy}^*(t) - F_{xy}(t))dt.$$

It follows at once that f is distributionally chaotic if and only if $\mu_p(f) > 0$.

There are other, numerical as well as distribution function-valued measures of chaos; but discussing them, as well as the deeper aspects of the theory of distributional chaos would take us too far afield. Suffice it to say that the basics of the theory are presented in [68], that further developments are given in [6, 23, 37, 81], that a short historical survey is given in [64], and that an elementary exposition, with numerous illustrative examples, is given in [67].

5. Topology and Indistinguishability

The set of distribution functions Δ^+ defined in (3.2) can be metrized in various ways. The one best suited to our purposes is the following: For any F, G in Δ^+ and h in $(0, 1]$ let $[F, G; h]$ denote the condition

$$G(x) \leqq F(x + h) + h \text{ for } x \text{ in} \left(0, \frac{1}{h}\right)$$

and let d_L be the mapping from $\Delta^+ \times \Delta^+$ into I given by

$$d_L(F, G) = \inf\{h| \text{ both } [F, G; h] \text{ and } [G, F; h] \text{ hold}\}.$$

Then d_L is a metric on Δ^+. Convergence in the metric d_L – which is a modification of the classical Lévy metric designed to take into account the fact that $\lim_{x \to \infty} F(x)$ may be less than 1 – corresponds to weak convergence of distribution functions, i.e., $d_L(F_n, F) \to 0$ iff $F_n(x) \to F(x)$ at every point of continuity of the limit function F. In addition, the metric space (Δ^+, d_L) is compact, hence complete, and separable [66, Chapter 4].

Convolution and (when T is a continuous t-norm) the operations τ_T given by (3.1) are continuous functions from $\Delta^+ \times \Delta^+$, endowed with the product topology, into Δ^+.

Let $(S, \mathcal{F}, \tau)$ be a *PM* space with τ continuous. There is a natural topology on S which is determined by the system of neighborhoods

$$N_p(\varepsilon, \lambda) = \{q | F_{pq}(\varepsilon) > 1 - \lambda\}.$$

This *strong topology* is metrizable. The distance function $\mathcal{F}$ is uniformly continuous, $(S, \mathcal{F}, \tau)$ has a completion which is unique up to isometry, product topologies are readily definable, etc. [66, Chapter 12]. Nevertheless, it suffers from a serious defect – namely it contains the tacit assumption that statements about arbitrarily small distances can be made with an arbitrarily high degree of certainty. And this goes counter to the entire philosophy underlying the theory of *PM* spaces. This matter led to the consideration of more general topological structures, specifically, the generalized topologies of M. Fréchet and A. Appert – Ky Fan [5]. Some years later these matters were put into a more appropriate context by R. M. Tardiff [75]. The whole development may be summarized as follows:

Let ϕ be a fixed distribution function in Δ^+. Call ϕ a *threshold function* and for any $x \geq 0$ interpret $\phi(x)$ as the maximum probability with which statements about distances $< x$ can be made. For example, if $\phi = \varepsilon_a$ then nothing can be said about distances $< a$; and if $\phi(x) = b$ for all $x > 0$, where $b \in I$, then statements about distances have at most a probability b of being valid. Given ϕ, the system of neighborhoods

$$N_p(\phi, h) = \{q | [F_{pq}, \phi; h] \text{ holds}\}$$

endows the *PM* space $(S, \mathcal{F}, \tau)$ with the structure of a Čech closure space, i.e., a space in which the closure operator need not be idempotent [13]. If p and q are points in S such that $F_{pq} \geq \phi$, then we say that p and q are *indistinguishable relative to* ϕ and write $p(\text{ind}\,\phi)q$. The relation $\text{ind}\,\phi$ is clearly reflexive and symmetric; but unless $\tau(\phi, \phi) = \phi$ (which is rarely the case) it is not transitive: for if $p(\text{ind}\,\phi)q$ and $q(\text{ind}\,\phi)r$ then by virtue of the triangle inequality all we can say is that $F_{pr} \geq \tau(F_{pq}, F_{qr}) \geq \tau(\phi, \phi)$. And so here we have a constructive resolution of Poincaré's Paradox! An alternate – and related – resolution, based on the Lukasiewicz infinite-valued logic and a notion of [0, 1]-valued equality has recently been given by U. Höhle [28].

The relation $\text{ind}\,\phi$ is readily extended to subsets of S. In addition if ϕ^n denotes the n-th τ-power of ϕ, i.e., $\phi^1 = \phi$, $\phi^2 = \tau(\phi, \phi), \ldots, \phi^{n+1} = \tau(\phi^n, \phi)$, and if ϕ is such that for every F in Δ^+ there is a positive integer n

such that $\phi^n < F$, i.e., if is ϕ is an Archimedean element of the semigroup (Δ^+, τ), then for any p, q such that $F_{pq} < \phi$ we can define a *degree of proximity relative to* ϕ of p and q via

$$\delta_\phi(p, q) = \inf\{n | F_{pq} \geq \phi^n\}.$$

This degree of proximity is a metric on S [66, Chapter 13]. Note that the discussion given by Menger in [48] deals, in essence, with the special case of a constant threshold function.

6. Associative Functions on Intervals

To a large extent the structure of a Menger space $(S, \mathcal{F}, \tau_T)$ is determined by its t-norm T. Thus, in 1957, one of the first things on the agenda was the matter of building an arsenal of t-norms. Here A. Sklar and I were aided by the fact that the problem of finding continuous t-norms is essentially the problem of finding continuous solutions of the functional equation of associativity or, in other words, of finding topological semigroups on a real interval. This problem has a long and distinguished history, dating back to Abel's first paper in Volume 1 of Crelle's Journal. Details may be found in the books [1, 2, 66, 82]. Here it suffices to note C. H. Ling's extension of J. Aczél's basic representation theorem, namely the fact that if T is continuous and Archimedean ($T(x,x) < 1$ for $0 < x < 1$), then T admits the representation

$$T(x, y) = f^{(-1)}(f(x) + f(y)),$$

where f is a continuous and strictly decreasing function from I into $\mathbb{R}^+$, with $f(1) = 0$, and

$$f^{(-1)}(x) = \begin{cases} f^{-1}(x), & 0 \leq x \leq f(0), \\ 0, & x \geq f(0). \end{cases}$$

If T is strictly increasing in each place on $(0, 1] \times (0, 1]$ then $f(0) = +\infty$ and $f^{(-1)} = f^{-1}$. The function f is an *additive generator* of the t-norm T; and it follows that large classes of t-norms are as easy to find as additive generators. For example, if $f(x) = -\log x$, then $f^{-1}(x) = e^{-x}$ and $T = \text{Prod}$; and if $f(x) = 1 - x$, then $f^{(-1)}(x) = \text{Max}(1 - x, 0)$ and $T = W$.

T-norms and related associative functions also arise in other contexts. First and foremost, they play a crucial role in the theory of fuzzy sets and multivalued logic. To Illustrate, if π_A is the membership function of the set A, i.e., if $\pi_A(x)$ is the degree of membership of x in A, then it is often assumed that

$$\pi_{A\cap B}(x) = T(\pi_A(x), \pi_B(x)),$$

and

$$\pi_{A\cup B}(x) = S(\pi_A(x), \pi_B(x)),$$

where T is a t-norm and S is a t-conorm, i.e., a function satisfying (b), (c), (d) and the boundary condition $S(a, 0) = a$; often S is the t-conorm of T, i.e., $S(a, b) = 1 - T(1-a, 1-b)$. These matters are discussed in detail in many (indeed, almost all) books on the subject of fuzzy sets and systems, e.g., in [29] and [55]. In his original paper [80], Zadeh stipulated that $T =$ Min and $S =$ Max. The observation that Min could be replaced by any t-norm and Max by any t-conorm is due to Trillas, Alsina and Valverde [77]. Note that Max is the t-conorm of Min.

Associative functions on $[0, \infty]$ arise in the construction of products of metric [49] and normed [10] spaces; they play a central role in the Kampé de Fériet – Forte approach to information theory, where they go under the name of composition laws [35]; and they also appear in the theory of inequalities, e.g., the classical Minkowski inequality may be viewed as a relation between the associative operation of addition and the associative operation m_p given by $m_p(x, y) = (x^p + y^p)^{\frac{1}{p}}$ for $p > 1$ [2, 66].

7. Binary Operations on Distribution Functions

With the advent of Šerstnev's inequality (III) our attention shifted from t-norms to triangle functions, i.e., from the functional equation of associativity on I to the functional equation of associativity on Δ^+; and, in view of the obvious relations to probability theory, we enlarged our scope to include the space Δ of all distribution functions of, possibly defective, real-valued random variables. As with t-norms, our first aim was to build a repertory of triangle functions. But whereas the first problem was essentially solved for us by the Aczél-Ling representation theorem, the second was – and to a large extent still is – terra incognita. To date a number of families of such functions have been found and studied. Here we present several (see [66, Chapter 7] and [62] for details).

First of all, R. Moynihan has studied the arithmetic and analytical properties of the topological semigroups (Δ, τ_T) and their subsemigroups (Δ^+, τ_T), where τ_T is given by (3.1). Here, as in other cases, the aim was – and is – to do for these semigroups what has been done in great detail for the convolution semigroup. Thus Moynihan [50, 51] determined indecomposable elements, infinitely divisible elements, cancellative and other subsemigroups, limit theorems, zero-one laws, etc. He also introduced and applied a "conjugate transform" which plays the role of a characteristic function. When $T = $ Prod, this transform reduces to the maximum transform of Bellman and Karush [7]; and it is also closely related to the conjugate transform of convex analysis.

The operation of addition in (3.1) can be replaced by a suitable binary operation L on $\mathbb{R}$, yielding a family of binary operations $\tau_{T,L}$ on Δ. When L is a binary operation on $\mathbb{R}^+$ and satisfies certain further natural conditions, $\tau_{T,L}$ is a triangle function.

Next, if S is a continuous t-conorm and if τ_S is defined for any F, G in Δ and z in $\mathbb{R}$ via

$$\tau_S(F, G)(z) = \inf_{x+y=z} S(F(x), G(y)),$$

then (Δ, τ_S) is a topological semigroup. When restricted to $\Delta^+ \times \Delta^+$, τ_S is a triangle function; and again the operation of addition can be replaced by a suitable binary operation L on $\mathbb{R}$, yielding a binary operation $\tau_{S,L}$ on Δ which, under further restrictions on L, is a triangle function.

There are relations – equalities and inequalities – among the triangle functions exhibited above. Thus for any continuous t-norms T_1 and T_2 and any continuous t-conorms S_1 and S_2 such that $T_1 \leqq T_2$ and $S_1 \leqq S_2$, we have

$$\tau_{T_1} \leqq \tau_{T_2} \leqq \tau_{\text{Min}} = \tau_{\text{Max}} \leqq \tau_{S_1} \leqq \tau_{S_2}.$$

With the aid of the mathematical machinery outlined above, we can now describe Wald's and Menger's notions of betweenness in an appropriate manner. Thus, if $(S, \mathcal{F}, \tau)$ is a PM space and p, q, r are three distinct points in S, then q is *Wald-between* p and r if $F_{pr} = \tau(F_{pq}, F_{qr})$; and if $(S, \mathcal{F}, \tau_T)$ is a Menger space then q is *Menger-between* p and r if $F_{pr} \leqq \tau_{T^*}(F_{pq}, F_{qr})$, where T^* is the t-conorm of T [66, Chapter 12]. In addition, we can also give appropriate definitions of probabilistic normed and probabilistic inner product spaces [3, 4].

Another important family of binary operations on Δ are the functions $\sigma_{C,L}$ given by

$$\sigma_{C,L}(F, G)(z) = \iint_{L(x,y) < z} dC(F(x), G(y)).$$

Here $L: \mathbb{R} \times \mathbb{R} \to \mathbb{R}$ is a Borel-measurable function and C is a (two-dimensional) *copula*, i.e., a mapping from $I \times I$ onto I that satisfies

(i) $C(0, a) = C(a, 0) = 0$ and $C(1, a) = C(a, 1) = a$ for all a in I,
(ii) $C(c, d) - C(a, d) - C(b, c) + C(a, b) \geq 0$, whenever $a \leq c$, $b \leq d$.

It readily follows that C is continuous, non-decreasing in each place, and that $W \leq C \leq \text{Min}$. There is a close connection between copulas and t-norms: a copula is a t-norm if and only if it is associative and a t-norm is a copula if and only if it satisfies a Lipschitz condition. Note that if L is addition then $\sigma_{\text{Prod},L}$ is convolution.

Copulas were first defined in full generality by A. Sklar [73] in the course of providing an answer to a query of M. Fréchet. Their importance is due to the following results. The first is the two-dimensional case of what is nowadays known as "Sklar's Theorem"; the second is, in essence, due to Fréchet.

Theorem 7.1. Let H be a two-dimensional probability distribution function with one-dimensional margins F and G. Then there exists a copula C, uniquely determined on $\text{Ran}\, F \times \text{Ran}\, G$, such that

$$H(x, y) = C(F(x), G(y)). \tag{7.1}$$

Theorem 7.2. Let X and Y be random variables with individual distribution functions F_X and F_Y, respectively, joint distribution function H_{XY} and copula C_{XY}, where C_{XY} is determined by (7.1). Then

(i) X and Y are independent iff $C_{XY} = \text{Prod}$ on $\text{Ran}\, F_X \times \text{Ran}\, F_Y$.
(ii) Y is a.s. an increasing function of X iff $C = \text{Min}$ on $\text{Ran}\, F_X \times \text{Ran}\, F_Y$.
(iii) Y is a.s. a decreasing function of X iff $C = W$ on $\text{Ran}\, F_X \times \text{Ran}\, F_Y$.

As M. J. Frank has shown, the operations $\sigma_{C,L}$ are rarely associative [24]. Their significance lies instead in the fact that if X and Y are real random variables with distribution functions F and G, respectively, joint distribution function H and copula C, then $\sigma_{C,L}$ is the distribution function of the random variable $L(X, Y)$.

As is well-known, if F and G are distribution functions, then there exist (independent) random variables X and Y, defined on a common probability space, such that F is the distribution function of X, G is the distribution function of Y, and the convolution $F * G$ is the distribution function of $X + Y$. Thus convolution of distribution functions is *derivable* from addition of random variables. Similarly the operations $\sigma_{C,L}$ are derivable in this fashion. This raises the question: Are the operations τ_T derivable, i.e., are there binary operations on random variables that correpond to the binary operations τ_T on distribution functions? The answer, in general, is "No". Specifically, unless $T = \text{Min}$, there is no Borel-measurable function V such that: For any F, G in Δ there exist random variables X and Y, defined on a common probability space, such that F is the distribution function of X, G is the distribution function of Y and τ_T is the distribution function of $V(X, Y)$. In this sense, the operations τ_T, as well as the operations $\tau_{T,L}$ and $\tau_{S,L}$ are intrinsically different from the derivable operations, i.e., from the operations $\tau_{C,L}$. This result is of philosophical as well as practical importance – random variables on a common probability space cannot tell the whole story. It is also relevant in mathematical statistics since such simple operations as the forming of mixtures – $m_\lambda(F, G) = \lambda F + (1-\lambda)G$ for λ in $(0, 1)$ – are not derivable [54].

The study of binary operations on Δ and Δ^+ is an interesting mathematical undertaking with applications to probability theory and with close relations to other parts of mathematical analysis. It can also be motivated and justified by philosophical considerations. For, as stated in [66, p. 123], "looking at the development of measurement processes during the past century, one soon observes that with increasing frequency the raw data are distribution functions (or frequency functions) rather than real numbers. This is so in the physical sciences; and in the biological and social sciences it is the rule rather than the exception. Thus one may convincingly argue that probability distribution functions are the 'numbers' of the future". This being so, it behooves us to study their arithmetic.

8. Clusters

The idea of operating with distribution functions as data is already explicit in Menger's writings [42, 48]. It was – and is – an essential part of the motivation behind the development of probabilistic metric, normed and inner product spaces. It also led A. Sklar and me to develop the rudiments of a probabilistic theory of information based on a generalization of the axiomatics introduced by Kampé de Fériet and Forte [35, 65]; and it led Kampé de Fériet to consider measures of information given by a set of observers [34]. More recently, M. F. Janowitz and I have applied these ideas to cluster analysis [32] and E. F. Diday and his colleagues are introducing them into their work on symbolic data analysis [19, 20].

A large class of cluster methods start with a set S of elements to be clustered and a dissimilarity coefficient, i.e., a mapping d from $S \times S$ into $\mathbb{R}^+$

that satisfies $d(x, x) = 0$ and $d(x, y) = d(y, x)$. The dissimilarity coefficient is first used to partition S, then to define a new dissimilarity coefficient on the elements of the partition, which is then used to partition these elements, etc. The process yields a nested sequence of partitions culminating in S. There are also more general cluster methods that allow overlap among the clustered subsets, i.e., that work with tolerance relations rather than equivalence relations. In a series of papers beginning in 1978, Janowitz has studied such cluster methods from an abstract, lattice-theoretic point of view and in the process developed his ordinal model for clustering [30, 31].

Dissimilarity coefficients are generally derived from the attributes on which the classification is based. Often these attributes are distributed or stochastic rather than deterministic and the dissimilarity coefficient in question is a mean, median or other typical value of the data. In condensing the data in this fashion much information may be lost. It is therefore desirable to design cluster methods which, instead of summarizing and then classifying, classify first and then summarize. In short, the dissimilarity coefficients should be replaced by probabilistic dissimilarity coefficients. These observations led Janowitz and me to develop a theory of *percentile clustering*. In our paper [32] we showed how many diverse cluster methods could be so generalized and how the whole fits smoothly into a slightly generalized version of Janowitz's abstract theory. We also applied the resultant clustering techniques to a number of bona-fide data sets. But here, while the first stage of the clustering algorithm was probabilistic, subsequent stages employed the older deterministic algorithms. For while one can easily define a probabilistic distance between sets (e.g., a probabilistic generalization of the Hausdorff distance), a form of Menger's "probabilistic theory of hazy lumps" [48] that can be implemented algorithmically still needs to be worked out.

It should also be noted that these investigations have led to a study of Δ and Δ^+ from a lattice theoretic point of view [56, 59].

9. Copulas

We conclude this essay with a brief description of some quite unexpected developments which are without doubt one of the most significant outgrowths of the study of PM spaces.

As already mentioned, A. Sklar introduced (n-dimensional) copulas and proved his basic theorem in 1959. During the years 1959–1974 most results concerning copulas were obtained in the course of the development of the theory of PM spaces [66, Chapter 6]. Then it was discovered that two-dimensional copulas play a critical role in nonparametric statistics [69]. First of all, if X and Y

are continuous random variables with copula C_{XY}, then the quantity

$$\sigma(X, Y) = 12 \int_0^1 \int_0^1 |C_{XY}(u, v) - uv| \, du \, dv$$

is a measure of monotone dependence of X and Y having many pleasant properties. Other known nonparametric measures, e.g., Spearman's rho and Kendall's tau can also be elegantly expressed in terms of copulas. But more important, if X and Y are random variables and f and g are a.s. strictly increasing functions on Ran X and Ran Y, respectively, then $C_{f(X)g(Y)} = C_{XY}$; and from this it follows that under such transformations of X and Y the copula is invariant while the one-dimensional margins may be changed at will. Hence it is precisely the copula that captures those properties of the joint distribution which are invariant under a.s. strictly increasing transformations; and consequently, the study of rank statistics – insofar as it is the study of properties invariant under such transformations – may be characterized as the study of copulas and copula-invariant properties.

In the ensuing years the copula concept was rediscoverd on several occasions and these functions began slowly to attract attention. The relationship between copulas, doubly stochastic measures and Markov operators was investigated [36, 72]; a copula-based approach to stochastic processes was developed [16]; extreme value copulas and copula-based tests of independence were studied [12, 17, 18]; and the special properties of associative copulas were exploited [26, 52]. Then, in 1990, a conference on copulas and related matters was held in Rome (see the conference proceedings [15] and, in particular, my paper [63] for a survey of the state of the subject at that time and for an extensive bibliography). Since then the growth in this area has been explosive. Three additional international conferences have been held and three additional volumes of proceedings have been published [8, 14, 76]. R. Nelsen has written "An Introduction to Copulas" [53] and copulas appear prominently in the book by H. Joe [33]. This is not the place to detail these developments. Suffice it to say that copulas are now established entities and that there is little doubt that within several years they will make their appearance in textbooks on mathematical statistics. There is also no doubt that Menger would have been pleased; his best friend Wald would have been delighted.

References

1. J. Aczél: Lectures on Functional Equations and Their Applications, Academic Press, New York, 1966.
2. C. Alsina, M. J. Frank, B. Schweizer: Associative Functions on Intervals, to appear.

3. C. Alsina, B. Schweizer, A. Sklar: On the definition of a probabilistic normed space. Aeq. Math. 46 (1993) 91–96.
4. C. Alsina, B. Schweizer, C. Sempi, A. Sklar: On the definition of a probabilistic inner product space. Rend. Mat. (7) (1997) 17, 115–127.
5. A. Appert, Ky Fan: Espaces topologiques intermédiaires. Actualitiés Sci. Ind. 1121, Hermann et Cie, Paris, 1951.
6. F. Balibrea, B. Schweizer, A. Sklar, J. Smítal: On the generalized specification property and distributional chaos. Intern. J. of Bifurcation and Chaos, to appear.
7. R. Bellman, W. Karush: On the maximum transform. J. Math. Anal. Appl. 6 (1963) 67–74.
8. V. Beneš, J. Štěpán (eds.): Distributions with Given Marginals and Moment Problems, Kluwer, Dordrecht, 1997.
9. D. I. Blokhintsev: Space and Time in the Microworld, D. Reidel, Dordrecht, 1973.
10. H. F. Bohnenblust: An axiomatic characterization of L_p-spaces. Duke Math. J. 6 (1940) 627–640.
11. L. deBroglie: Une remarque sur l'interaction entre la matière et le champ electro-magnétique. C. R. Acad. Sci. Paris 200 (1935) 361–363.
12. P. Capéraà, A.-L. Fougères, C. Genest: Bivariate distributions with given extreme value attractor. J. Multivariate Anal. 72 (2000) 30–49.
13. E. Čech: Topological Spaces, Wiley, New York, 1966.
14. C. M. Cuadras, J. Fortiana, J. A. Rodriguez-Lallena (eds.): Distributions with Given Marginals and Statistical Modelling, Kluwer, Dordrecht, to appear.
15. G. Dall'Aglio, S. Kotz, G. Salinetti (eds.): Advances in Probability Distribution Functions with Given Marginals: Beyond the Copulas, Mathematics and its Applications, v. 67. Kluwer, Dordrecht, 1991.
16. W. F. Darsow, B. Nguyen, E. T. Olsen: Copulas and Markov processes. Ill. J. Math. 36 (1992) 600–642.
17. P. Deheuvels: A nonparametric test for independence. Pub. Inst. Statist. Univ. Paris 26 (1981) 29–50.
18. P. Deheuvels: Probabilistic aspects of multivariate extremes, in Statistical Extremes and Applications, ed. by J. Tiago de Oliveira, D. Reidel, Dordrecht, 1984, 117–130.
19. E. F. Diday, R. Emilion, Y. Hillali: Symbolic data analysis of probabilistic objects by capacities and credibilities, Atti della XXXVIII Riunione Societa Italiana di Statistica, Rimini, 1996, 5–22.
20. E. F. Diday: Symbolic data analysis and the SODAS project; purpose, history, perspective, in Analysis of Symbolic Data, ed. by H.-H. Bock, E. F. Diday, Springer, New York, 2000, 1–23.
21. A. S. Eddington: Fundamental Theory, Cambridge Univ. Press, London, 1953.
22. T. Erber, B. Schweizer, A. Sklar: Mixing transformations on metric spaces. Comm. Math. Phys. 29 (1973) 311–317.
23. G. L. Forti, L. Paganoni, J. Smital: Dynamics of homeomorphisms on minimal sets generated by triangular mappings. Bull. Austral. Math. Soc. 59 (1999) 1–20.
24. M. J. Frank: Associativity in a class of operations on a space of distribution functions. Aeq. Math. 12 (1975) 121–144.
25. M. Fréchet: Les éléments aléatoires de nature quelconque dans un espace distancié. Ann. Inst. H. Poincare 10 (1948) 215–310.
26. C. Genest, L.-P. Rivest: Statistical inference procedures for bivariate Archimedean copulas. J. Amer. Statist. Assoc. 88 (1993) 1034–1043.

27. Guo Tie-xin: Survey of recent developments of random metric theory and its applications in China. Acta Analysis Functionalis Applicata 3 (2001) 129–158 and 208–229.

28. U. Höhle: The Poincaré Paradox and non-classical logics, in Fuzzy Sets, Logics and Reasoning about Knowledge, ed. by D. Dubois et al., Kluwer, Dordrecht, 1999, 7–16.

29. U. Höhle, S. E. Rodabaugh (eds.): Mathematics of Fuzzy Sets: Logic, Topology and Measure Theory, Kluwer, Dordrecht, 1999.

30. M. F. Janowitz: An order theoretic model for cluster analysis. SIAM J. App. Math. 34 (1978) 55–72.

31. M. F. Janowitz: An ordinal model for cluster analysis – 15 years in retrospect, in From Data to Knowledge: Theoretical and Practical Aspects of Classification, Data Analysis and Knowledge Organization, ed. by W. Gaul, D. Pfeifer, Springer, Berlin, 1995, 58–72.

32. M. F. Janowitz, B. Schweizer: Ordinal and percentile clustering. Math. Social Sciences 18 (1989) 135–186.

33. H. Joe: Multivariate Models and Dependence Concepts, Chapman and Hall, London, 1997.

34. J. Kampé de Fériet: Le théorie généralisée de l'information et la mesure subjective de l'information Lecture Notes in Math. 398 (1974) 1–35.

35. J. Kampé de Fériet, B. Forte: Information et probabilité. C. R. Acad. Sci. Paris 265A (1967) 110–114; 142–146; 350–353.

36. X. Li, P. Mikusinski, H. Sherwood, M. D. Taylor: On approximation of copulas, in Distributions with Given Marginals and Moment Problems, ed. by V. Beneš, J. Štěpán, Kluwer, Dordrecht, 1997, 106–116.

37. Gongfu Liao, Qinjie Fan: Minimal subshifts which display Schweizer-Smital chaos and have zero topological entropy. Science in China (Series A) 41 (1998) 33–41.

38. P. C. Mahalanobis: On the generalized distance in statistics. Proc. Nat. Inst. Sci. India 2 (1936) 49–55.

39. K. Menger: Untersuchungen über allgemeine Metric I. Math. Ann. 100 (1928) 75–113.

40. K. Menger: Topology without points. The Rice Institute Pamphlet 27 (1940) 80–107.

41. K. Menger: Statistical metrics. Proc. Nat. Acad. Sci. USA 28 (1942) 535–537.

42. K. Menger: The theory of relativity and geometry, in Albert Einstein, Philosopher-Scientist, Library of Living Philosophers, ed. by P. S. Schilpp, Evanston, IL, VII (1949) 459–474.

43. K. Menger: Probabilistic theories of relations. Proc. Nat. Acad. Sci. USA 37 (1951) 178–180.

44. K. Menger: Probabilistic geometry. Proc. Nat. Acad. Sci. USA 37 (1951) 226–229.

45. K. Menger: Ensembles flous et fonctions aléatoires. C. R. Acad. Sci. Paris 232 (1951) 2001–2003.

46. K. Menger: Géométrie générale, Mem. Sci. Math. 124. Gauthier-Villars, Paris, 1954.

47. K. Menger: Random variables from the point of view of a general theory of variables, in Proc. Third Berkeley Symposium on Mathematical Statistics and Probability, ed. by L. M. LeCam, J. Neyman, Univ. of California Press, Berkeley and Los Angeles, 2 (1956) 215–229.

48. K. Menger: Mathematical implications of Mach's ideas: Positivistic geometry, the clarification of functional connections, in Ernest Mach, Physicist and Philosopher, Boston Studies in the Philosophy of Science, ed. by R. S. Cohen, R. J. Seeger, Reidel, Dordrecht, 6 (1970) 107–125.

49. T. S. Motzkin: Sur le produit d'espaces métriques. C. R. Congres Int. Mathématiciens, Oslo, 2 (1936) 137–138.
50. R. Moynihan: On τ_T-semigroups of probability distribution functions II. Aeq. Math. 17 (1978) 19–40.
51. R. Moynihan: Conjugate transforms and limit theorems for τ_T-semigroups. Studia Math. 69 (1980) 1–18.
52. R. B. Nelsen: Dependence and order in families of Archimedean copulas. J. Multivariate Anal. 60 (1997) 111–122.
53. R. B. Nelsen: An Introduction to Copulas, Lecture Notes in Statistics, v. 139, Springer, New York, 1999.
54. R. B. Nelsen, J. J. Quesada-Molina, B. Schweizer, C. Sempi: Derivability of some operations on distribution functions, in Distributions with Fixed Marginals and Related Topics, ed. by M. D. Taylor, B. Schweizer, L. Rüschendorf, IMS Lecture Notes – Monograph Series, 28 (1996) 233–243.
55. H. T. Nguyen, E. Walker: A First Course in Fuzzy Logic. CRC Press, Boca Raton, 1997.
56. R. C. Powers: Order automorphisms of spaces of nondecreasing functions. J. Math. Anal. Appl. 136 (1988) 112–123.
57. E. Prugovečki: Stochastic Quantum Mechanics and Quantum Spacetime, D. Reidel, Dordrecht, 1984.
58. M. Regenwetter, A. A. J. Marley: Random relations, random utilities and random functions. J. Math. Psychology 45 (2001) 864–912.
59. T. Riedel: Cauchy's equation on Δ^+. Aeq. Math. 41 (1991) 192–211.
60. R. T. Rockafellar: Convex Analysis, Princeton University Press, 1970.
61. N. Rosen: Quantum geometry. Ann. Physics 19 (1962) 165–172.
62. B. Schweizer: Multiplications on the space of distribution functions. Aeq. Math. 12 (1975) 156–183.
63. B. Schweizer: Thirty years of copulas, in Advances in Probability Distribution Functions with Given Marginals: Beyond the Copulas, ed. by G. Dall-Aglio, S. Kotz, G. Salinetti. Mathematics and its Applications, Kluwer, Dordrecht, 67 (1991) 13–50.
64. B. Schweizer: On the genesis of the notion of distributional chaos. Rendiconti del Seminario Matematico e Fisico di Milano 66 (1996) 159–167.
65. B. Schweizer, A. Sklar: Mesures aléatoires de l'information. C. R. Acad. Sci. Paris 269A (1969) 149–152.
66. B. Schweizer, A. Sklar: Probabilistic Metric Spaces, Elsevier – North Holland, New York, 1983.
67. B. Schweizer, A. Sklar, J. Smítal: Distributional (and other) chaos and its measurement. Real Anal. Exchange 27 (2001/2002) 495–524.
68. B. Schweizer, J. Smítal: Measures of chaos and a spectral decomposition of dynamical systems on the interval. Trans. Amer. Math. Soc. 344 (1994) 737–754.
69. B. Schweizer, E. F. Wolff: On nonparametric measures of dependence of random variables. Ann. Statist. 9 (1981) 879–885.
70. A. N. Šerstnev: On a probabilistic generalization of metric spaces. Kazan Gos. Univ. Učen. Zap. 124 (1964) 3–11.
71. H. Sherwood: On E-spaces and their relation to other classes of probabilistic metric spaces. J. London Math. Soc. 44 (1969) 441–448.
72. H. Sherwood, M. D. Taylor: Doubly stochastic measures with hairpin support. Prob. Theory and Related Fields 78 (1988) 617–626.

73. A. Sklar: Fonctions de répartition à n dimensions et leurs marges. Publ. Inst. Statist. Univ. Paris 8 (1959) 229–231.
74. A. Špaček: Random metric spaces, Trans. Second Prague Conf. Information Theory, Decision Functions and Random Processes, Academic Press, New York, 1960, 627–638.
75. R. M. Tardiff: Topologies for probabilistic metric spaces. Pacific J. Math. 65 (1976) 233–251.
76. M. D. Taylor, B. Schweizer, L. Rüschendorf (eds.): Distributions with Fixed Marginals and Related Topics, IMS Lecture Notes – Monograph Series, v. 28 (1996).
77. E. Trillas, C. Alsina, L. Valverde: Do we need Max, Min and $1-j$ in fuzzy set theory?, in Fuzzy Set and Possibility Theory, Recent Developments, ed. by R. R. Yager, Pergamon Press, New York, 1982, 275–297.
78. A. Wald: On the Principles of Statistical Inference, Notre Dame Mathematical Lectures, No. 1 (1942).
79. A. Wald: On a statistical generalization of metric spaces. Proc. Nat. Acad. Sci. USA 29 (1943) 196–197.
80. L. Zadeh: Fuzzy sets. Information and Control 8 (1965) 338–353.
81. A. Sklar, J. Smítal: Distributional chaos on compact metric spaces via specification properties, J. Math. Anal. Appl. 241 (2000) 181–188.
82. E. P. Klement, R. Mesiar, E. Pap: Triangular Norms, Kluwer, Dordrecht, 2000.

STATISTICAL METRICS

By Karl Menger

DEPARTMENT OF MATHEMATICS, UNIVERSITY OF NOTRE DAME

Communicated October 27, 1942

We shall call *statistical metric* a set S such that with each two elements ("points") p and q of S a probability function $\Pi(x; \ p, q)$ is associated satisfying the following conditions:

1. $\Pi(0; \ p, p) = 1$.
2. If $p \neq q$, then $\Pi(0; \ p, q) < 1$.
3. $\Pi(x; \ p, q) = \Pi(x; \ q, p)$.
4. $T[\Pi(x; \ p, q), \Pi(y; \ q, r)] \leq \Pi(x + y; \ p, r)$.

where $T(\alpha, \beta)$ is a function defined for $0 \leq \alpha \leq 1$ and $0 \leq \beta \leq 1$ such that

   (a) $0 \leq T(\alpha, \beta) \leq 1$.
   (b)   T is non-decreasing in either variable.
   (c) $T(\alpha, \beta) = T(\beta, \alpha)$.
   (d) $T(1, 1) = 1$.
   (e)   If $\alpha > 0$, then $T(\alpha, 1) > 0$.

By a probability function we mean a non-decreasing function defined for all non-negative values of x, continuous to the right, with values between 0 and 1, and converging toward 1 as x increases beyond all bounds.

We call $\Pi(x; \ p, q)$ the *distance function* of p and q and interpret it as the probability that the points p and q have a distance $\leq x$. Condition 4, our "triangular inequality," implies that $\Pi(z;\ p, r) \geq$ Max. $T[\Pi(x;\ p, q)$, $\Pi(z - x;\ q, r)]$ for all points q and all numbers x between 0 and z. We shall call the function T the *triangular norm* of the statistical metric, and more specifically refer to the metric defined above as a T-metric. A triangular norm T will be called *simple* if

(f) $0 < T(\alpha, \beta) < 1$ for $0 < \alpha\cdot\beta < 1$.

An ordinary metric space is a statistical metric such that for each pair of points p, q there exists a number $d(p, q) \geq 0$ with the property that $\Pi(x;\ p, q)$ is $= 0$ if $x < d(p, q)$, and $= 1$ if $x \geq d(p, q)$.

On the basis of our postulates large parts of metric geometry can be developed, in particular, a theory of betweenness. We shall say that q lies *between* p and r (and we write pqr) if

$$T[1 - \Pi(x;\ p, q), 1 - \Pi(y;\ q, r)] \leq 1 - \Pi(x + y;\ p, r).$$

Equivalent is the assumption

$\Pi(z;\ p, r) \leq 1 -$ Max. $T[1 - \Pi(x;\ p, q), 1 - \Pi(z - x;\ q, r)]$ for $0 \leq x \leq z$.

Obviously, if pqr then rqp. In metric spaces if q and r are distinct, then pqr and prq are incompatible. In a statistical metric we can only prove: If q and r are apart, then pqr and prq are incompatible, where q and r are said to be apart if there exists a number $y > 0$ such that $\Pi(y;\ q, r) = 0$.

If for each two points of a statistical metric S the distance function $\Pi(x;\ p, q)$ belongs to a family $\mathfrak{P}$ of probability functions, we call S metrized by means of $\mathfrak{P}$. Let $\mathfrak{P}$ be a 2 parameter family of probability functions $\Pi(x;\ a', a'')$ defined for all real numbers a' and a'' such that $0 \leq a' \leq a''$ and satisfying the conditions

$$\Pi(x;\ a', a'') = 0 \text{ if } x \leq a',$$
$$\Pi(x;\ a', a'') = 1 \text{ if } x \geq a'',$$
$$0 < \Pi(x;\ a', a'') < 1 \text{ for } a' < x < a''.$$

If S is metrized by means of $\mathfrak{P}$ and T is simple, then for each three points p, q, r one of which lies between the two other ones, two of the distance functions $\Pi(x;\ p, q)$, $\Pi(y;\ q, r)$, $\Pi(z;\ p, r)$ determine the third. In particular, if pqr and $\Pi(x;\ p, q) = \Pi(x;\ a', a'')$ and $\Pi(y;\ q, r) = \Pi(y;\ b', b'')$, then $\Pi(z;\ p, r) = \Pi(z;\ a' + b', a'' + b'')$. From this theorem one readily derives the classical law: If pqr and prs, then pqs and qrs.

If

$$\Pi(x;\ p, q) = \Pi(x;\ r, s) = \Pi(x;\ a', a'')$$
$$\Pi(x;\ q, r) = \Pi(x;\ p, s) = \Pi(x;\ b', b'')$$
$$\Pi(x;\ p, r) = \Pi(x;\ q, s) = \Pi(x;\ a' + b';\ a'' + b''),$$

then p, q, r, s form what may be called a *pseudo-linear statistical quadruple*, i.e., a quadruple which cannot be ordered by means of the between-relation though for each three of the four points one lies between the other two.

If a statistical T-metric S metrized by means of 𝔓 contains more than four points, then by virtue of the properties of betweenness this relation can be used to order S. Moreover, the other ideas of metric geometry (convexity, geodesics, etc.) can be applied.

The three principal applications of statistical metrics are to macroscopic, microscopic and physiological spatial measurements. Statistical metrics are designed to provide us (1) with a method removing conceptual difficulties from microscopic physics and transferring them into the underlying geometry, (2) with a treatment of thresholds of spatial sensation eliminating the intrinsic paradoxes of the classical theory. For a given point p_0 the number $\Pi(0;\ p_0, q)$ considered as a function of the point q indicates the probability that q cannot be distinguished from p_0. The study of this function should replace the attempt to determine a definite set of points q which cannot be distinguished from p_0. This function could also be used advantageously instead of a relation of physical identity for which, as Poincaré emphasized on several occasions, we always have triples p, q, r for which

$$p = q, q = r, \text{ and } p \neq r.$$

Experiments indicate that q sometimes can and sometimes cannot be distinguished from p_0. Hence, the adequate description of the situation seems to arise from counting the relative frequency of these occurrences.

PROBABILISTIC THEORIES OF RELATIONS

By Karl Menger

Illinois Institute of Technology

Communicated by Marston Morse, January 29, 1951

Poincaré repeatedly emphasized that only in the mathematical continuum the equalities $A = B$ and $B = C$ imply the equality $A = C$. In the observable physical continuum, "equal" means "indistinguishable," and $A = B$ and $B = C$ by no means imply $A = C$. "The raw result of experience may be expressed by the relation

$$A = B, B = C, A < C$$

which may be regarded as the formula for the physical continuum." According to Poincaré, physical equality is a non-transitive relation.

A closer examination of the physical continuum suggests that in describing our observations we should sacrifice more than the transitivity of equality. We should give up the assumption that equality is a relation. For this assumption implies that for every two elements, A and B, the question as to whether or not A and B can be distinguished, is inalterably settled. But simple experiments show that, for instance, the simultaneous irritation of the same two spots, A and B, on the skin sometimes produces one sensation, sometimes two. Only by reliance on the majority of the impressions, by processes of averaging and the formation of means, equality relations have been artificially created.

We obtain a more realistic theoretical description of the equality of two elements *by associating with A and B a number, namely, the probability of finding A and B indistinguishable.* In applications, this number would be represented by the relative frequency of the cases in which A and B are not distinguished.

In principle, this idea solves Poincaré's paradox. For if it is only very likely that A and B are equal, and very likely that B and C are equal, why should it not be less likely that A and C are equal? In fact, why should the equality of A and C not be less likely than the inequality of A and C?

If $E(a, b)$ denotes the probability that a and b be equal, the following postulates seem to be rather natural:

(1) $E(a, a) = 1$ for every a;
(2) $E(a, b) = E(b, a)$ for every a and b;
(3) $E(a, b) \cdot E(b, c) \leq E(a, c)$ for every a, b, c.

(1) and (2) correspond to the reflexivity and symmetry of the equality relation, (3) expresses a minimum of transitivity.

If we call a and b *certainly-equal* provided that $E(a, b) = 1$, we obtain an equality relation. All the elements which are certainly-equal to a may be united in an "equality-set," A. Any two such sets are disjoint unless they are identical. We may define $E(A, B)$ as the probability that any element of A and any element of B be equal. For this number is independent of the particular choice of the two elements.

If we set $-\log E(A, B) = d(A, B)$, then we readily conclude:

$(1_a')$ $d(A, A) = 0$; $(1')$ $d(A, B) \geq 0$; $(1_c')$ $d(A, B) \neq 0$ if $A \neq B$;
$(2')$ $d(A, B) = d(B, A)$;
$(3')$ $d(A, B) + d(B, C) \geq d(A, C)$.

These are Fréchet's postulates for the distance in a metric space. In particular $(3')$ is the triangle inequality. Conversely, if disjoint sets $A, B, \ldots$ form a metric space with the distance $d(A, B)$ and we set $E(A, B) = E(a, b) = e^{-d(A, B)}$ for each element a of A, and b of B, then $E(a, b)$ satisfies the postulates (1), (2), (3), of a probability of equality. *The systems of probabilities of equality in a set, S, are thus identical with the systems of negative antilogarithms of the distance for the various possible metrizations of S.*

If S is a straight line, $E(a, b) = e^{-|b - a|}$ is the probability of equality corresponding to the euclidean metrization of S. Another example is $E(a, b) = e^{-\sqrt{|b-a|}}$ while the smooth function $e^{-(b - a)^2}$ is ruled out by the fact that the corresponding distance $d(a, b) = (b - a)^2$ fails to satisfy the triangle inequality.

Since $d(a, b) < \infty$, for every two points of a metric space, it follows that $E(a, b) > 0$. We may find it more desirable to assume that every element b differing from a by more than a certain exterior threshold be certainly distinguishable from a. But *then we have to give up even the minimum of transitivity expressed in Postulate (3).*

In the same way as we have just treated equality we may treat the order relation "a precedes b." If we introduce the probability $P(a, b)$ of a preceding b, the following postulates are natural.

Monotony. If $a_1 \leq a_2 \leq b_2 \leq b_1$, then $P(a_2, b_2) \leq P(a_1, b_1)$.
Boundary Conditions. $\lim_{y \to \infty} P(a, y) = \lim_{x \to -\infty} P(x, b) = 1$.
The following assumptions reflect properties of the order relation:
Asymmetry. $P(a, b) + P(b, a) = 1$.
Transitivity. $P(a, b) \cdot P(b, c) \leq P(a, c)$.
An example of a function satisfying the above postulates is

$$P(a, b) = \begin{cases} k^{(b - a)/2} & \text{if } b \leq a \\ 1 - k^{(a - b)/2} & \text{if } a \leq b \end{cases}$$

for any $k > 1$. Besides, this function satisfies the following condition of *Homogeneity.* $P(a, b) = P(a', b')$ if $b' - a' = b - a$.
Moreover, the example illustrates the following general fact. If we

wish to satisfy the transitivity postulate we must admit that every number, even a very large one, has a positive probability of preceding 0.

It is clear that probabilistic theories of dyadic, triadic, and higher relations can be developed which are based on functions of 2, 3, and more variables. *This procedure seems to be a natural extension of the idea of replacing qualities by quantities* (monadic relations by functions of one variable).

But the probabilistic point of view far transcends the domain of relations. It is applicable to functions. Instead of a real function of one, two or n variables, we may consider *the association of a distribution function with every element, pair or n-tuple of elements of a set.* Of particular geometric importance is the probabilistic generalization of the concept of a metric space.[1]

[1] An attempt in this direction was made by the author in the note "Statistical Metrics," Proc. Natl. Acad. Sci., 28, 535–537 (1942).

PROBABILISTIC GEOMETRY

By Karl Menger

ILLINOIS INSTITUTE OF TECHNOLOGY

Communicated by Marston Morse, February 11, 1951

We generalize the concept of a metric space by associating a cumulative distribution function, Δ_{ab}, with every ordered pair (a, b) of elements of a set S. The value $\Delta_{ab}(x)$ may be interpreted as the probability that the distance from a to b be $<x$. But S need not be a metric space in the ordinary sense. The distribution functions and the association of these functions with the pairs of elements of S, are all that is assumed. For instance, a probabilistic metric consisting of five elements is a 5-by-5 matrix of distribution functions—as an ordinary metric space consisting of five points is a 5-by-5 matrix of numbers.

In analogy to the assumptions $1'.$, $2'.$, $3'.$ about a metric space,* the nondecreasing functions Δ_{ab}, which are continuous to the right, are assumed to satisfy the following postulates. (Probabilistic interpretations are mentioned parenthetically.)

$1_a''.$ $\Delta_{aa}(x) = 1$ for every point a and every $x > 0$. (The distance from a point to itself is certainly ≤ 0.)

$1_b''.$ $\Delta_{ab}(x) = 0$ for every two points and every $x \leq 0$. (Negative distances are impossible.)

$2''.$ $\Delta_{ab}(x) = \Delta_{ba}(x)$ for every two points and every x. (The distance from a to b is as likely to be $<x$ as is the distance from b to a.)

The triangle inequality included in our original set of postulates[1] will be replaced by the following more stringent version due to Wald.[2]

$3''.$ $[\Delta_{ab} * \Delta_{bc}](x) \leq \Delta_{ac}(x)$ for every three points and every x. Here, $\Delta' * \Delta''$ denotes the convolution of Δ' and Δ'', that is, the distribution of the sum of the random variables with the distributions Δ' and Δ''. [For every x, it is less likely (or, at any rate, not likelier) that the sum of the distances from a to b and from b to c be $< x$, than that the distance from a to c be $< x$.]

An ordinary metric space with the distance function $d(a, b)$ is a special case of a probabilistic metric, namely, the case that

$$\Delta_{ab}(x) = \begin{cases} 0 \text{ if } x \leq d(a, b) \\ 1 \text{ if } x > d(a, b). \end{cases}$$

Normal distributions are ruled out by Postulate $1_b''$.

We call a and a' *certainly-indistinguishable* if $\Delta_{aa'}(x) = 1$ for each $x > 0$. Uniting all elements which are certainly indistinguishable from each other into identity sets we decompose the space into disjoint sets A, B, $\ldots$. We may define $\Delta_{AB}(x) = \Delta_{ab}(x)$ for any a belonging to A, and any b belonging to B. (The number is independent of the choice of a and b.) The identity sets form a perfect analog of an ordinary metric space since they satisfy the condition

$1_c''$. If $A \neq B$, then there exists a positive x such that $\Delta_{AB}(x) > 1$.

Replacing, if necessary, points by identity sets we shall assume that no two points are certainly-indistinguishable unless they are identical.

Let $\delta(a, b)$ denote the greatest lower bound of all numbers x such that $\Delta_{ab}(x) = 0$. If this certainly-lower-distance is positive, then a and b will be said to be *certainly-distinguishable*. If $\delta(a, b) = 0$, we distinguish two cases: we call a and b *barely-distinguishable* if $\Delta_{ab}(+0) = 0$; and *perhaps-indistinguishable* if $\Delta_{ab}(+0) > 0$. Here $\Delta_{ab}(+0)$ denotes the right-side limit of Δ_{ab} at 0.

The relation of being perhaps-indistinguishable is reflexive, symmetrical and (by virtue of 3''.) transitive. Again the space can be decomposed into disjoint lumps consisting of perhaps-indistinguishable elements. But if a and a' belong to the same lump, and b is any element, the functions Δ_{ab} and $\Delta_{a'b}$ need not be equal. Even if the distances between lumps are defined by distribution functions rather than numbers, the definition of these functions must be based on averaging processes.

Now let S be an ordinary metric space with the distance function $d(a, b)$. We call a probabilistic metric *monotonic* on S if $d(a, b) \leq d(a, c)$ implies $\Delta_{ac}(x) \leq \Delta_{ab}(x)$ for every a, b, c, x. Clearly, in a monotonic metric, $d(a, b) = d(a, c)$ implies $\Delta_{ab} = \Delta_{ac}$.

If the metric space is complete, convex and externally convex[3] (so that every two points are on at least one straight line), then it can be shown that either each lump contains only one point or that there is only one lump, namely, the entire space. More precisely, we obtain the following theorem. *One of three cases is present: either every two points of S are perhaps-indistinguishable; or every two points are barely-distinguishable; or every two points are certainly-distinguishable.*

Each of these three cases can occur on the straight line, even in conjunction with the assumption of

Homogeneity. $\Delta_{bc}(x) = \Delta_{0|c-b|}(x)$ for every b, c, x.

In a homogeneous metric of the line it is sufficient to define $\Delta_{0a}(x)$ for $a > 0$ and $x > 0$.

With the following quasi-exponential distributions, every two points are perhaps-indistinguishable.

$$\Delta_{0a}(x) = 1 - \Phi(a)e^{-x} \quad (a > 0, x > 0).$$

Here, Φ must be a nowhere decreasing function for which

$$\Phi(2a) \leq 2\Phi(a) - \Phi^2(a).$$

A solution satisfying the postulates is $\Phi(a) = 1 - e^{-a}$.

Exponential distributions, $\Delta_{ab}(x) = 1 - e^{-\varphi(a, b)x}$, are impossible on the straight line (and in much more general spaces) even if the assumption of homogeneity is dropped.

With the following rectangular distributions, every two points are barely-distinguishable.

$$\Delta_{0a}(x) = x/\varphi(a) \text{ for } 0 < x \leq \varphi(a),$$

where $\varphi(a) + \varphi(b) - \varphi(a + b) \geq \varphi(a)\varphi(b)/2\varphi(a + b)$. A solution satisfying the postulates is $\varphi(a) = a^k$ if

$$1 < 2^k \leq 1 + 1/\sqrt{2} \quad \text{or} \quad 0 < k \leq 0.7714 \ldots$$

A particularly simple example is $\varphi(a) = \sqrt{a}$.

With the following rectangular distributions, every two points are certainly-distinguishable.

$$\Delta_{0a}(x) = \frac{x - \psi(a)}{\varphi(a) - \psi(a)} \text{ for } \psi(a) \leq x \leq \varphi(a),$$

where

$$\varphi(a) + \varphi(b) - \varphi(a + b) \geq \frac{[\varphi(a) - \psi(a)]\,[\varphi(b) - \psi(b)]}{2[\varphi(a + b) - \psi(a + b)]}.$$

The postulates are satisfied if we set $\varphi(a) = a^k$ $(0 < k \leq 0.7714 \ldots)$ as in the preceding example, and $\psi(a) = c\varphi(a)$ for some constant $c < 1$.

The metrics of the last two examples contract the scale in the sense that the distance from 0 to a is certainly $\leq \varphi(a)$, for instance, $\leq \sqrt{a}$. We are reminded of the psychophysical contractions of the scale, especially of Cramer's suggestion[4] that the subjective value of an amount of money, a, does not exceed $\sqrt{a}$. The assumption that the distance from 0 to a be certainly $\leq \log a$ (that is, the analog of the well-known Bernoulli-Fechner contraction) contradicts the triangle inequality for $a < 2^{\sqrt{2}}$.

An inhomogeneous metric, which does not contract the scale is defined by the following rectangular distributions

$$\Delta_{bc}(x) = \Delta_{cb}(x) = \Delta_{0,\,(c-b)c/(c-b/2)}(x) \text{ for } b \leq c \text{ and every } x,$$
$$\Delta_{0a}(x) = x/a \text{ for } 0 < x \leq a.$$

In this metric, every two points are barely-distinguishable.

For some purposes, none of the three main cases is desirable. For instance, none of them is compatible with the existence of a point a, which lies in a bounded threshold consisting of points that are perhaps-indistinguishable from a, while the points outside are certainly (or at least barely) distinguishable from a. From our theorem it follows that, in order to cover this and similar situations, we have to weaken, if not to give up, the triangle inequality.

It appears that *in the transition from the rigid geometry in the large to a probabilistic geometry in the small, the geometric laws to which we are accustomed, not only cease to be certain but, in some cases, become unlikely and even impossible.*

There seem to exist two or three main types of applications of a probabilistic geometry.

1. To the *psychophysical continuum*. The observer starts with an objective metric space (or what, in a first approximation, he considers as a metric space): the skin of the blindfolded observed person, the sets of weights read on a scale, temperatures, and the like. Upon these objective rigid metrics, the observed person superimposes his subjective probabilistic metric.

2. To the *physical macrocosm*. Our theory provides us with a scheme for dealing with bundles of observations of spatial quantities, weights, temperatures, etc. Such a scheme should be useful where the enormously simplifying method of replacing these bundles by single numbers would prove to be an oversimplification. Underlying the (more or less arbitrary) averaging and selection procedures of the simplifying method is the idea that various phases of the macrocosm have the structure of a numerical continuum (for the elements of which we have to discover the "real" values and distances). But the foundation of this idea seems to lie only in scientific usage.

3. Perhaps to the *physical microcosm*. In this application, the fundamental difficulty seems to lie in the lack of an objective metric comparable to that of the observer of another person's psychophysical continuum. The postulate that each world point can be characterized by four numerical coordinates is not more plausible than the assumption that the distance between two points be a number. Consequently, the theory would have to develop criteria for identifying the elements of a set, along with the distance distributions between the elements.

---

* See my note "Probabilistic Theories of Relations" in the preceding issue of these PROCEEDINGS, **37**, 178, 1951, where the assumptions 1.', 2.', 3.' are quoted on p. 179.

[1] PROC. NATL. ACAD. SCI., **28**, 535 (1942).

[2] *Ibid.*, **29**, 196 (1943).

[3] "Untersuchungen ueber allgemeine Metrik," *Math. Ann.*, **100**, 75–163 (1928).

[4] Cf. the author's paper in *Z. Nationaloekonomie*, **5**, 465 (1934).

ANALYSE PROBABILISTE. — *Ensembles flous et fonctions aléatoires.*
Note (*) de M. **Karl Menger**, présentée par M. Arnaud Denjoy.

Nous étudions des relations probabilistes, surtout la probabilité que deux éléments d'un univers soient équivalents. Les relations probabilistes monaires sont des ensembles flous. Au lieu de fonctions classiques nous étudions l'association de nombres aléatoires aux éléments du domaine.

Dans la théorie classique on entend par relation n-aire définie dans un univers U, un ensemble de systèmes ordonnés de n éléments de U. Dans une théorie probabiliste de relations que nous avons esquissée récemment ([1]), nous associions à tout système ordonné de n éléments de U un nombre réel compris entre o et 1, que nous interprétions comme la probabilité que cette relation subsiste entre les n éléments.

Par exemple, au lieu d'une relation binaire d'équivalence qui est réflexive, symétrique et transitive, nous étudions une fonction E des paires d'éléments de U dont les valeurs $E(x, y)$ sont comprises entre o et 1, et qui jouit des proprïétés suivantes :

1. $E(x, x) = 1$ pour tout élément x de U ;
2. $E(x, y) = E(y, x)$ pour tout couple d'éléments de U ;
3. $E(x, y).E(y, z) \leq E(x, z)$ pour tout triplet d'éléments de U.

Les propriétés 1 et 2 correspondent à la réflexivité et à la symétrie de l'équivalence. La propriété 3 est une transitivité rudimentaire ([2]).

Nous dirons que deux éléments x et y donnés sont *certainement équivalents* (*certainement non-équivalents*) si $E(x, y) = 1 (= o)$. Nous appelons *ensemble d'équivalence certaine* un ensemble X qui, avec tout élément x, contient tout y tel que $E(x, y) = 1$. Deux tels ensembles sont disjoints ou identiques. Nous pouvons définir une probabilité d'équivalence entre ces ensembles en posant $E(X, Y) = E(x, y)$ pour tout $x \in X$ et tout $y \in Y$. En remplaçant, s'il le faut, les éléments par les ensembles d'équivalence certaine, nous pouvons donc admettre l'hypothèse suivante :

4. $E(x, y) < 1$ pourvu que $x \neq y$.

Appelons *morceau* un ensemble qui avec tout x contient tout y tel que $E(x, y) > o$. Deux morceaux sont disjoints ou identiques. Deux éléments appartenant à deux morceaux disjoints sont certainement non équivalents.

(*) Séance du 21 mai 1951.
([1]) Cf. *Proc. Nat. Acad. Sci.*, **37**, 1951, p. 178.
([2]) Comme nous l'avons souligné (*loc. cit.*), cette notion est applicable à l'indiscernabilité des éléments du continu physique et explique la remarque de Poincaré que dans ce continu (par contraste avec le continu mathématique) les égalités $x = y$ et $y = z$ soient compatibles avec l'inégalité $x < z$. Dans le continu physique, il s'agit d'une probabilité d'indiscernabilité.

Un morceau devient un espace distancié au sens de M. Fréchet si nous posons $-\log E(x, y) = d(x, y)$. D'autre part, si dans un espace distancié donné nous posons $E(x, y) = e^{-d(x,y)}$, nous obtenons une probabilité d'équivalence d'après laquelle l'espace ne contient qu'un seul morceau.

Une relation monaire au sens classique est un sous-ensemble F de l'univers. Au sens probabiliste, c'est une fonction Π_F définie pour tout $x \in U$. Nous appellerons cette fonction même un *ensemble flou* et nous interpréterons $\Pi_F(x)$ comme la probabilité que x appartienne à cet ensemble. Si Π_F ne prend que les valeurs 1 et 0, il s'agit essentiellement d'un sous-ensemble de U au sens classique et nous parlerons d'un ensemble *rigide*. Par exemple, l'univers Π_U (le vide Π_V) dont toutes les valeurs sont 1(0), sont des ensembles rigides. La somme et le produit de deux ensembles non rigides ne sont pas déterminés par les ensembles. L'égalité de Π_F et Π_G est une relation probabiliste qui dépend de Π_{F+G} et $\Pi_{F.G}$.

Des relations, le point de vue probabiliste peut être étendu aux fonctions réelles. Au lieu des nombres nous associons des nombres aléatoires ou des fonctions de répartition aux systèmes ordonnés de n éléments de U. Un exemple que nous avons étudié ailleurs (^{3}), est une métrique probabiliste. A toute paire d'éléments on associe une fonction de répartition dont les propriétés correspondent à celle d'une distance.

Dans le cas $n = 1$, une fonction de répartition est associée à tout élément x de U. Sa valeur $f(x, t)$, définie pour t réel, peut être interprétée comme la probabilité qu'un nombre $< t$ soit associé à x. Considérons le cas où U est l'intervalle fermé $a \leq x \leq b$. Pour tout t réel, $y = f(x, t)$ est une courbe C_t. L'ensemble des C_t est une courbe floue. Quelle est l'aire du domaine compris entre l'axe des x et cette courbe? Nous choisissons une division de l'intervalle $a = x_0 < x_1 < \ldots < x_m = b$ et des nombres $\xi_i (x_{i-1} \leq \xi_i \leq x_i)$. La probabilité que l'aire sous le polygone en escalier, défini par la division de l'intervalle, soit $< t$, est la valeur pour t de la fonction $g_1 \star g_2 \star \ldots \star g_m$, où l'astérisque indique la composition des fonctions de répartition et où

$$g_i(t) = f\left(\xi_i, \frac{t}{x_i - x_{i-1}}\right) \qquad (i = 1, 2, \ldots, m).$$

Si toute fonction $f(x, t)$ est normale avec la moyenne $\mu(x)$ et la déviation $\sigma(x)$, alors la répartition de l'aire sous le polygone est normale avec la moyenne $\Sigma \mu(\xi_i)(x_i - x_{i-1})$ et la déviation

$$\sqrt{\Sigma \sigma^2(\xi_i)(x_i - x_{i-1})^2} \leq \sqrt{\Sigma \sigma^2(\xi_i)(x_i - x_{i-1})} \sqrt{\max(x_i - x_{i-1})}.$$

Pour toute suite de divisions telles que $\max(x_i - x_{i-1})$ tende vers zéro, les aires convergent donc vers une fonction de répartition qui prend la valeur 0

(^{3}) *Proc. Nat. Acad. Sc.*, **37**, 1951, p. 226.

ou 1 selon que t est $\leq$ ou $> \int_a^b \mu(x)\,dx$, pourvu que les fonctions $\mu(x)$ et $\sigma^2(x)$ soient intégrables au sens de Riemann. L'aire du domaine sous la courbe floue (l'intégrale de la fonction aléatoire) est donc rigide. Évidemment cette simple remarque reste valable pour des classes assez étendues de fonctions de répartition et pour des domaines très généraux.

Cette méthode a des conséquences importantes pour l'idée d'une métrique *convexe* (une métrique d'après laquelle la distance entre deux points est égale à la longueur d'un chemin qui les joint). Il semble que cette idée soit limitée aux métriques rigides.

Selected Papers on Group Theory and Algebra

Commentary on Menger's Work on Algebra

Hans Lausch

1. Distance in Groups

When Menger's book on curve theory [23], written in co-operation with Georg Nöbeling, appeared in 1932 he already had over 60 publications to his credit. Among them were his group theoretical investigations, an offshoot of his seminal paper *"Untersuchungen über allgemeine Metrik"* [18]. The *Mathematisches Kolloquium* at the University of Vienna discussed the ideas developed in this paper and in this way other young mathematicians, especially Abraham Wald (cf. [36]), Olga Taussky, Franz Alt and Gustav Beer, became actively interested in Menger's distance geometry. At the 13th meeting of the *Kolloquium*, held on 14 March 1930, Menger presented new results under the heading *"Über eine metrische Geometrie in Gruppen"* [19]. They were published in the *Mathematische Zeitschrift* as *"Beiträge zur Gruppentheorie. I. Über eine Gruppenmetrik"*, [21][1]. The second of Menger's three investigations in [18], *"Die euklidische Metrik"*, opens with a proof of the following theorem discovered by M. M. Biedermann: a connected metric space M is homeomorphic with a subspace of $\mathbb{R}_1$ if, for each triple (a, b, c) of distinct points of M, one of the points lies between the other two.[2] However, Menger observed that Biedermann's condition does not characterise metric spaces which are *isometric* (*"abstandsgleich"* in [18], *"kongruent"* in [20]) to some subspace of $\mathbb{R}_1$. Thus he posed the problem of finding necessary and sufficient conditions for a semimetric space[3] to be *isometrically* embeddable (*"abstandstreu einbettbar"*) into – or, more specifically, to be isometric to – the euclidean n-space $\mathbb{R}_n$. His first result was: if each set of $n + 3$ points in a semimetric space M is isometrically embeddable into $\mathbb{R}_n$, then M itself is

---

[1] An extract appeared in *Anz. d. Akad. d. Wiss.* (*Wien*) 1930. Menger never published a follow-up under the title *Beiträge zur Gruppentheorie*.

[2] Menger also refers to a thesis by Lindenbaum (Warsaw 1928).

[3] A *semimetric space* is a pair (M, d) consisting of a set M and a map $d : M \times M \to \mathbb{R}$, the *distance*, satisfying all the requirements for a metric other than the triangle inequality.

isometrically embeddable into $\mathbb{R}_n$ (Satz 1). Furthermore, if M has more than $n+3$ points and any subset of $n+2$ points of M is isometrically embeddable into $\mathbb{R}_n$, then M itself is isometrically embeddable into $\mathbb{R}_n$ (Satz 5). On the other hand, for each $n>0$ there exists a semimetric space M containing a subset P of $n+3$ points which is not isometrically embeddable into $\mathbb{R}_n$, while each subset of P with $n+2$ points is isometrically embeddable into $\mathbb{R}_n$ (Satz 2). Such a subset P is called a *pseudo-euclidean* $(n+3)$-tuple. In a semimetric space, the pseudo-euclidean quadruples are characterised (see [18, p. 127]) as those quadruples of points p_1, p_2, p_3, p_4 which permit a relabelling of indices such that $d(p_1,p_2) = d(p_3,p_4)$, $d(p_1,p_3) = d(p_2,p_4)$, $d(p_1,p_4) = d(p_2,p_3) = d(p_1,p_2) + d(p_1,p_3)$. Indeed, as a starting point for his investigation, Menger looked at pseudo-euclidean quadruples $\{p_1,p_2,p_3,p_4\}$ defined by $d(p_1,p_2) = d(p_3,p_4) = d(p_1,p_3) = d(p_2,p_4) = 1$, $d(p_1,p_4) = d(p_2,p_3) = 2$.

Menger's discussion of isometric embeddings in a group-theoretical framework was motivated by his realisation that his results for $\mathbb{R}_1$ were "merely special cases of a new type of simple group-theoretical theorems" (cf. [20, p. 210]). In [21], Menger considers an (additively written) abelian group G, on which an "absolute value" is introduced for each of its elements a as $|a| = \{a, -a\}$. The *order* of $|a|$ is defined to be the order of a. To define distances in arbitrary sets, the set $|G| = \{|a| : a \in G\}$ is used in the same way as the set of all non-negative reals. Let M be a set and $d : M \times M \rightarrow |G|$ a map; then (M, d) is a *G-metrical* set if, for all p, $q \in M$, (i) $d(p,q) = d(q,p)$ and (ii) $d(p,q) = |0|$ if and only if $p = q$; $d(p,q)$ is called the *distance* between p and q. Two G-metrical sets (M,d) and (M',d') are *congruent* if there is a bijection $K : M \rightarrow M'$ such that $d'(K(p),K(q)) = d(p,q)$, for all p, $q \in M$. In [22], the name *congruent correspondence* is given to any such K. Clearly, every abelian group G can be equipped with a distance function in a "natural" way, viz. $d_G : (a,b) \mapsto |a-b|$, for all a, $b \in G$, so that (G, d_G) is a G-metrical set. This distance function leads to a sort of metric geometry, e.g., equilateral triangles, perpendicular bisectors, translations, reflexions, etc. If $\{a,b\}$ and $\{a',b'\}$ are two congruent pairs of elements of (G,d_G), then there exists *exactly one* congruent correspondence from G to itself mapping a to b and a' to b', unless the order of $d_G(a,b)$ is 2, in which case there are *two* (Satz 6). In 1949, David Ellis [8] showed that every congruent correspondence between two subsets of G extends to a congruent correspondence from G to itself. The problem of characterising those semimetric spaces which are isometric to subsets of $\mathbb{R}_1$ is a special case of the problem of characterising those G-metrical sets which are congruent to subsets of (G, d_G). Indeed, the results for $\mathbb{R}_1$ turn out to be instances of the following theorems:

(1) If each four-element subset of a G-metrical set (M, d) is congruent to some subset of (G, d_G), then (M, d) itself is congruent to some subset of (G, d_G) (Satz 8).

(2) If no element of G has order 2 or 3 and (M, d) is a G-metrical set with more than four elements such that each of its three-element subsets is congruent to a subset of (G, d_G), then (M, d) is congruent to some subset of (G, d_G) (Satz 14). As Menger notes, this is the case even if G has elements of order 3 provided M contains no *equilateral triangles*, i.e., three-element subsets $\{a, b, c\}$ with $d(a, b) = d(b, c) = d(c, a)$. Taussky pointed out that if $G = \{0, a, -a\}$ is the group of three elements and M is a set with at least four elements, then putting $d(p, q) = |a|$, whenever p and q are distinct elements of M, turns M into a G-metrical set in which each triple is congruent to (G, d_G) while M itself is not. On the other hand, as Taussky showed [32] for the statement of Satz 14 to be valid, the absence of elements of order 2 is not required.

(3) Pseudo-euclidean quadruples generalise to *pseudo-G-quadruples*; i.e., G-metrical four-element sets (M, d) which are not congruent to any subset of (G, d_G), while each three-element subset of (M, d) is congruent to some subset of (G, d_G). Distances between any two points of a pseudo-G-quadruple do not have order 2 (Satz 9). If (M, d) is a pseudo-G-quadruple and $M = \{p_1, p_2, p_3, p_4\}$, then $2d(p_1, p_2) = 2d(p_3, p_4)$, $2d(p_1, p_3) = 2d(p_2, p_4)$; $2d(p_1, p_4) = 2d(p_2, p_3)$ (Satz 10). For his Satz 12, Menger assumes that G has no elements of order 2 and (M, d) is a G-metrical four-element set, where $M = \{p_1, p_2, p_3, p_4\}$ and each triple of distinct elements of M is congruent to some subset of (G, d_G). Then (M, d) is a pseudo-G-quadruple if and only if

$$d(p_1, p_2) = d(p_3, p_4), d(p_1, p_3) = d(p_2, p_4), d(p_1, p_4) = d(p_2, p_3) \quad (*)$$

The proof of this theorem relies on Satz 9 and Satz 10 as well as on the result that if a_1, a_2, a_3, a_4 are distinct elements of G and $d_G(a_1, a_2) = d_G(a_3, a_4)$, $d_G(a_1, a_3) = d_G(a_2, a_4)$, $d_G(a_1, a_4) = d_G(a_2, a_3)$ then, for least two indices i, j $(1 \le i < j \le 4)$, the distance $d_G(a_i, a_j)$ has order 2 (Satz 11).[4]

---

[4] There is a misprint in the proof: on p. 411, lines 1 and 2 of [21]: the label $\alpha 2)$ should be replaced by $\beta 2)$ and vice versa.

In the 13th meeting of the *Kolloquium*, Taussky's results on group metrics were presented, next to Menger's, under the title *"Bemerkungen zur Metrik der Gruppen* [19, p. 26]. Accordingly, Menger acknowledges Taussky's share in the origin of his paper [21]. Through her doctoral thesis on class field theory, which had been supervised by Philipp Furtwängler, she had already gained "work experience" in group theory. Taussky had pointed out to Menger the exceptional role distances of order 2 would play in his investigation (p. 401, Footnote 3). The first time that Menger had to distinguish between distances of order 2 and those of order greater than 2 was when he took any four elements $a, b, d, e \in G$ and asked how many elements $c \in G$ exist with $d_G(c, a) = |d|$, $d_G(c, b) = |e|$. If $d_G(a, b) > 2$, then there exists at most one such element c. If, however, $d_G(a, b) = 2$, $|d| > 2$, $|e| > 2$, and one such element c exists, then there exists precisely one more element $c' \in G$, the *mirror image* of c with respect to a, b, such that $d_G(c', a) = |d|$, $d_G(c', b) = |e|$ (Satz 2). An example is provided by the cyclic group of order 4 with a and b being its two distinct elements of order 4. Section 6 of [21] *"Über eine Gruppenmetrik"* consists of Taussky's characterisation of pseudo-G-quadruples (cf. [32]), where G is an arbitrary abelian group. When G is the cyclic group of order 16, Taussky provides a pseudo-G-quadruple $M = \{p_1, p_2, p_3, p_4\}$ which does not satisfy condition $(*)^5$. She then characterises pseudo-G-quadruples as G-metrical sets $\{p_1, p_2, p_3, p_4\}$ where the distances $d(p_i, p_j) = \{p_{ij}, p_{ji}\}$ ($p_{ji} = -p_{ij}$) satisfy:

(i) $2 p_{ij} \neq 0$,
(ii) $\varepsilon_{ik} p_{ik} + \varepsilon_{kj} p_{kj} + \varepsilon_{ji} p_{ji} = 0$ ($i \neq j \neq k \neq i$), $\varepsilon_{ij} = -\varepsilon_{ji}$, $i, j = 1, 2, 3, 4.^6$

At the above-mentioned meeting of the *Kolloquium*, Menger [19, p. 26] also proposed a generalisation of the metric in $\mathbb{R}_n$ to Cartesian powers G^n of (additively written) abelian groups G. To define a metric on G^n, he requires an extra (unary) operation $g \mapsto g^s$ on G, which satisfies: (i) $g^s = 0$ if and only if $g = 0$, (ii) $g^s = \varepsilon(-g)^s$, where ε equals either $+1$ or -1 for all $g \in G$, and (iii) $(g + h)^s + (g - h)^s = 2(g^s + h^s)$ for all $g, h \in G$. He called the resultant

---

[5] The smallest group for which such pseudo-G-quadruples exist is the cyclic group of order 6. If a is one of its generators, then as can be readily verified, the distances given by $d(p_1, p_2) = d(p_2, p_3) = d(p_2, p_4) = \{a, -a\}$, $d(p_1, p_3) = d(p_1, p_4) = d(p_3, p_4) = \{2a, -2a\}$ define a pseudo-G-quadruple.

[6] The following restatement of condition (ii) in Taussky's characterisation of pseudo-G-quadruples highlights even more the significance of elements of order 2: after a suitable permutation of the indices, $p_{34} = p_{12} + c$, $p_{14} = p_{23} + c$, $p_{13} = p_{12} + p_{23}$, $p_{24} = p_{13} + c$ where $2c = 0$.

algebraic structure a quadratically-tiered ("*quadratisch gestufte*") abelian group. If $n \cdot g$ denotes the sum of n copies of $g \in G$, then Condition (iii) implies that $(n \cdot g)^s = n^2 \cdot g^s$. This makes the operation $g \mapsto g^s$ resemble the operation of squaring elements in a ring. By way of polarisation, a kind of dot product on G^n can then be obtained. For a "Pythagorean" distance function $d: G^n \times G^n \mapsto G$, Menger subsequently proposes the function defined by $d((g_1, g_2, \ldots, g_n), (g'_1, g'_2, \ldots, g'_n)) = (g'_1 - g_1)^s + (g'_2 - g_2)^s + \cdots + (g'_n - g_n)^s$ and raises the question whether the Artin-Schreier definition of "real" ([2]), when extended to arbitrary quadratically-tiered abelian groups is linked with orderability.

In 1933, at another session of the Kolloquium Menger [25] (see also [28, p. 30–31]) presented a definition of arc length in an abelian *L*-group *G* (in the sense of Schreier [29]). Arc lengths, being certain subsets of *G*, are generalisations of distances on *G*. They are invariant under congruent correspondences, and as is shown, also have other desirable properties. Furthermore, Menger characterises those subsets of vector spaces, especially those of the plane, which occur as lengths.

In a series of lectures given in a seminar at Harvard University in the winter of 1930/31, Menger [22] presented eight of his research topics "from the point of view of metrical geometry and the topology of point sets". Continuing his investigation into group metrics, his sixth topic was a metric criterion for the existence of isomorphisms between two abelian groups. If *G* and *G'* are abelian groups and if (M, d) is a *G*-metrical and (M', d') a *G'*-metrical set, then a bijection $\kappa: M \to M'$ is a *transformation of similarity* if κ maps each pair of congruent subsets of *M* to pair of congruent subsets of *M'*; the metrical sets (M, d) and (M', d') are then said to be *similar*. Clearly, every isomorphism from *G* to *G'* is a transformation of similarity from (G, d_G) to $(G', d_{G'})$. Moreover, each *translation* on *G*, i.e., a map from *G* to itself which is of the form $x \mapsto x + t$ for some $t \in G$, is a transformation of similarity from (G, d_G) to itself. Menger proves that each transformation of similarity from (G, d_G) to $(G', d_{G'})$ is the composite of an isomorphism from *G* to *G'* and a translation on *G'*; in particular, if (G, d_G) and $(G', d_{G'})$ are similar, then *G* and *G'* are isomorphic.

Taussky continued Menger's group-theoretical investigations "in a nice way",[7] as Edmund Hlawka [13, p. 12] comments. Although Menger's results on group metrics refer mostly to abelian groups, his paper [21, p. 399] contains definitions of distances on not necessarily abelian groups as well. If *G* is an (additively written) group and $a, b \in G$, then $d_{G,r}(a, b) = \{b-a, a-b\}$

---

[7] "in schöner Weise".

is the *right distance* (*"rechtsseitiger Abstand"*) between a and b. Similarly, Taussky defined the *left distance* between these two elements as $d_{G,l}(a,b) = \{-a+b, -b+a\}$. In her paper [33], she showed that $d_{G,r}(a,b) = d_{G,l}(a,b)$ for all $a, b \in G$, i.e., right (left) distances are invariant under the mapping $x \mapsto -x$, if and only if G is either abelian or a hamiltonian 2-group (or, equivalently, any two non-commuting elements of G generate a quaternion group).[8] Calling two groups G and G' *similar* if a bijection $\kappa : G \to G'$ exists such that $d_{G,r}(a,b) = d_{G,r}(c,d)$ implies $d_{G',r}(\kappa(a), \kappa(b)) = d_{G',r}(\kappa(c), \kappa(d))$, Taussky then generalised Menger's similarity result in [22] to not necessarily abelian groups: two groups G and G' are similar if and only if they are isomorphic.

In 1949, Kestelman and Smith [14], motivated by a result of Steinhaus [31], proved, albeit without reference to either Menger or Taussky, the following theorem:

Let G_1 be a subgroup of an arbitrary (multiplicatively written) group G, and let G_2 be a normal subgroup of G_1 with $|G_1/G_2| = k$. Then there exist subsets E and R of G satisfying (i) *$\{rE: r \in R\}$ is a partition of G into k distinct subsets and* (ii) *the set of all right distances between the elements in each set rE has no element in common with the set of all right distances between elements in $G_1 \backslash G_2 = \{g \in G_1: g \notin G_2\}$.*

B. H. Neumann (cf. [14, p. 134]) pointed out a consequence of this theorem: if S is a subset of a group G and $k > 1$ is an integer with the property that $\Pi s_i^{a_i} = 1 (s_i \in S)$ implies $k | \sum a_i$, then G can be partitioned into k subsets $H_1, H_2, \ldots, H_k$, where $H_i = r_i H_1$ for some r_i in G, such that the set of all right distances between the elements in each set H_i (which is independent of i) has no element in common with S.

Menger [28, p. 30–32] himself mentioned various other generalisations of the distance function. In 1933, again at a session of the Kolloquium, Abraham Wald [36] presented results about distances that take on complex values, which lead to a *characterisation* of the n-dimensional space $\mathbb{C}_n$ (see also Menger [24]). Taussky [34] studied the more general case, where distances are elements in abstract fields of characteristic zero. Having contributed to the *Kolloquium* in 1936 ([3], [4]), Leonard Blumenthal [5, Chapter 15] took lattices as codomains of distance functions, while in 1951 David Ellis [9]

---

[8] This result was generalised in her joint paper with John Todd in [35]: a necessary and sufficient condition for a group G to be abelian or a hamiltonian 2-group is that distances in G are invariant under any "general inversion", i.e. any transformation of the form $x \to t + (-x) + t (t \in G)$.

singled out Boolean algebras for this purpose. Ellis also wrote a report [9] on distance geometry that includes a survey of results about group metrics.

2. Determinants

There have been, in principle, two different approaches to the axiomatisation of determinant theory. In one approach, going back to Hensel [12], it is postulated that determinants are multiplicative functions on $n \times n$-matrices ($n = 1, 2, \ldots$). Originating with Weierstrass [37, p. 272] and varied by Carathéodory [6, p. 318], Schreier [30, p. 69] and Artin [2, p. 11], the other approach introduces determinants as multilinear and alternating functions on the row or column vectors of matrices. Menger points out that, while every proposition about rows of determinants is also valid for columns, and vice versa, the second approach tends toward one side by favouring either rows or columns. His aim, a self-dual foundation of determinant theory, was one motivation for the investigation reported in his note [27]. In what follows, some of the notation that Menger used in [17] will be adopted, i.e. if A is a matrix, then A_i, A^k and A_i^k will denote the i-th row, the k-th column, and the (i, k)-entry of A.

In a 1953 master's thesis, written under Menger's direction by Frank Kozin (cf. [17]), the following theorem was proved (cf. [16]):

Let X, Y, Z be (additively written) groups[9] and $f: X \times Y \times Z \to \mathbb{R}$ a function. Then $f(x + x_1, y + y_1, z) \geq f(x, y, z) + f(x_1, y_1, z)$ and $f(x + x_1, y, z + z_1) \leq f(x, y, z) + f(x_1, y_1, z_1)$ for all x, $x_1 \in X$, y, $y_1 \in Y$, z, $z_1 \in Z$ if and only if $f(x + x_1, y + y_1, z) \leq f(x, y, z) + f(x_1, y_1, z)$ and $f(x + x_1, y, z + z_1) \geq f(x, y, z) + f(x_1, y_1, z_1)$ for all x, $x_1 \in X$, y, $y_1 \in Y$, z, $z_1 \in Z$.

In his note [27], Menger provides the following characterisation[10] of determinants:

Let n be a positive integer, and $M_{n,n}(\mathbb{R})$ the set of all $n \times n$-matrices with real entries. A function $f: M_{n,n}(\mathbb{R}) \to \mathbb{R}$ is the determinant if and only if f has the following properties:

(1r) For each i ($1 \leq i \leq n$), f is *subadditive in the i-th row* that is to say, for any three $n \times n$-matrices A, B, C the function f has the property: if $A_i = B_i + C_i$ and $A_{i'} = B_{i'} = C_{i'}$ for each $i' \neq i$, then $f(A) \leq f(B) + f(C)$.

---

[9] Kozin's assumption is that X, Y, Z are vector spaces (not necessarily) of the same dimension, but his proof uses only the fact that X, Y, Z are abelian groups with respect to vector addition.

[10] In [17], the term "self-dual" is used to signify that the theorem remains unaltered under the simultaneous exchange of the terms "rows" for "columns" and "super-" for "sub-" (including "$\geq$" for "$\leq$").

(1c) For each $k\,(1 \leq k \leq n)$, f is *superadditive in the k-th column*, that is to say, for any three $n \times n$-matrices A, B, C, the function f has the property: if $A^k = B^k + C^k$ and $A^{k'} = B^{k'} = C^{k'}$ for each $k' \neq k$, then $f(A) \geq f(B) + f(C)$.

(2r) For each $i\,(1 \leq i \leq n)$, f is *positively subhomogeneous in the i-th row* that is, if $c \geq 0$, then for any two $n \times n$-matrices A, B,

$$A_i = cB_i \text{ and } A_{i'} = B_{i'} \text{ for each } i' \neq i \text{ imply } f(A) \leq cf(B)$$

(2c) For each $k\,(1 \leq k \leq n)$, f is *positively superhomogeneous in the k-th column*, that is, if $c \geq 0$, then for any two $n \times n$-matrices A, B,

$$A^k = cB^k \text{ and } A^{k'} = B^{k'} \text{ for each } k' \neq k \text{ imply } f(A) \geq cf(B).$$

(3r) For each pair h, $i\,(1 \leq h,\, i \leq n, h \neq i)$, f is nonnegative if the h-th and the i-th row are equal.

(3c) For each pair h, $k\,(1 \leq h,\, k \leq n, h \neq k)$, f is nonnegative if the h-th and the k-th column are equal.

(4) $f(D) = 1$ if D is the $n \times n$-identity matrix.

A proof of the theorem is contained in Kozin's master's thesis and also in [17]. In the special case, where X, Y, Z are finite direct sums of copies of the additive group of $\mathbb{R}$, Kozin's theorem is used by Menger for that part of the proof which consists of showing conditions (1r) and (1c) together imply the additivity of f in each row and in each column. For this, Menger requires no property of $\mathbb{R}$ other than it being an abelian group with respect to addition. Thus he arrives at the first part of his *Théoreme* in [27] (cf. [17, Theorem 2]):

Let f be a real function defined on the set of all $n \times n$-matrices with entries from an (additively written) group G. If f is subadditive in each row and superadditive in each column, then f is additive in each row and in each column.

If $\pi = \left(\begin{smallmatrix} 1 & 2 & \cdots & n \\ i_1 & i_2 & \cdots & i_n \end{smallmatrix} \right)$ is a permutation on $\{1, 2, \ldots, n\}$ and A is an $n \times n$-matrix, then $A\pi$ is the $n \times n$-matrix with $(A\pi)_i^k = 0$ if $i \neq i_k$, while $(A\pi)_{i_k}^k = A_{i_k}^k$ for $1 \leq k \leq n$.[11]

With this notation, the second part of Menger's *Théoreme* in [27], which is a corollary of the first part, reads (cf. [17, Corollary 2]):

---

[11] It may be noted that these matrices, called *quasi-diagonal* by Menger, while known to group-theorists (if G is not necessarily abelian) as *monomial* matrices, have also been used to describe wreath products of groups, while their determinants taken modulo the commutator subgroup provide group-theoretical transfer.

If f satisfies (1r) *and* (1c), *then* $f(A) = \sum_\pi f(A\pi)$, *where* π *ranges over the* $n!$ *permutations of* $\{1, 2, \ldots, n\}$.

Menger notes that if, in addition, f satisfies (3r) and (3c) and g is the function from the direct sum of n copies of G to the reals defined by

$$g(a_1, a_2, \ldots, a_n) = f\begin{pmatrix} a_1 & 0 & \cdots & 0 \\ 0 & a_2 & \cdots & 0 \\ \cdot & \cdot\cdot & \cdots & \cdot \\ 0 & 0 & \cdots & a_n \end{pmatrix}$$

(which is a symmetric function), then the above sum can be written as $f(A) = \sum_\pi \varepsilon(\pi) g(A^1_{i_1}, A^2_{i_2}, \ldots, A^n_{i_n})$, where $\varepsilon(\pi) = 1$ or -1, depending on whether π is even or odd. For the latter equation to be valid, conditions (3r) and (3c) must hold only for very special types of matrices. The note also discusses the consequences of assuming conditions (2r), (2c) and (4) as well for the case of G being a commutative ring with identity and closes with the observation that the second part of the *Théoreme* may be extended to functions f with values in an arbitrary abelian group G_1, if one assumes additivity of f in each row (column).[12]

The mathematical literature is rich in axiomatic descriptions of the determinant and various generalisations. While the axiom systems have been usually based on either Weierstrass' or Hensel's (for an interesting "mixture" see [11]), generalisations have concerned both the domains and the codomains of the determinant function to be axiomatised. In addition, abstract fields, determinants of matrices over integral domains and commutative rings (with or without identity) have been studied, and – what has led to a highly abstract theory – determinants of endomorphisms of modules over not necessarily commutative rings. Several references to these generalisations can be found in [38].

Bibliography

1. E. Artin, A. N. Milgram: Galois Theory. Second edition. Notre Dame, Ind.: University of Notre Dame, 1944.
2. E. Artin, O. Schreier: Algebraische Konstruktion reeller Körper. Abh. Math. Sem. Hamburg 5 (1926) 83–115.
3. L. M. Blumenthal: Kurzer Beweis eines Satzes von Menger. Erg. Math. Kolloqu. 7 (1936) 6–7.

---

[12] Other generalisations are possible, e.g. the retention of the original conditions ((1r) and (1c)), while G_1 is assumed to be an abstract ordered abelian group rather than $\mathbb{R}$.

4. L. M. Blumenthal: Remarks Concerning the Euclidean Four-point Property. Erg. Math. Koloqu. 7 (1936) 7–10.

5. L. M. Blumenthal: Distance Geometries: A Study of the Development of Abstract Metrics. Columbia, Missouri: University of Missouri, 1938.

6. C. Caratheodory: Vorlesungen über Relle Funktionen, Leipzig, Berlin BG Teubner, 1918.

7. L. M. Blumenthal: Theory and Applications of Distance Geometry. Oxford: Oxford University Press, 1953.

8. D. Ellis: Superposability Properties of Naturally Metrised Groups. Bull. Amer. Math. Soc. 55 (1949) 639–640.

9. D. Ellis: Geometry in Abstract Distance Spaces. Publ. Math. Debrecen 2 (1951) 1–25.

10. D. Ellis: Autometrized Boolean Algebras. Canad. J. Math. 3 (1951) 87–93.

11. J. Gaspar: Eine neue Definition der Determinanten. Publ. Math. Debrecen 3 (1954) 257–260.

12. K. Hensel: Über den Zusammenhang zwischen den Systemen und ihren Determinanten. J. Reine Angew. Math. 159 (1928) 246–254.

13. E. Hlawka: Karl Menger. Internat. Math. News 149 (1988) 11–13.

14. H. Kestelman, C. A. B. Smith: On the Distances Between the Elements of a Subset of a Group. J. London Math. Soc. 24 (1949) 131–135.

15. F. Kozin: On Functions of Three Vectors [abstract]. Bull. Amer. Math. Soc. 58 (1952) 627.

16. F. Kozin: On Functions of Three Vectors. Publ. Math. Debrecen 10 (1963) 120–122.

17. F. Kozin, K. Menger: A Self-dual Theory of Real Determinants. Publ. Math. Debrecen 10 (1963) 123–126.

18. K. Menger: Untersuchungen über allgemeine Metrik. Math. Ann. 100 (1928) 75–163.

19. K. Menger: Bericht über ein mathematisches Kolloquium 1929/30. Monatsh. Math. 38 (1931) 17–38.

20. K. Menger: Bericht über metrische Geometrie. Jber. Deutsch. Math.-Verein. 40 (1931) 201–219.

21. K. Menger: Beiträge zur Gruppentheorie: Über eine Gruppenmetrik. Math. Zeitschrift 33 (1931) 396–418.

22. K. Menger: Some Applications of Point-set Methods. Ann. of Math. 32 (1931) 739–760.

23. K. Menger: Kurventheorie. Hrsg. unter Mitarbeit von Georg Nöbeling. (Mengentheoretische Geometrie in Einzeldarstellungen Hrsg. von Karl Menger. Bd. 2.) Leipzig and Berlin: B. G. Teubner, 1932.

24. K. Menger: Über den imaginären euklidischen Raum. Tohoku Math. J. 37 (1933) 475–478.

25. K. Menger: Zur Begründung einer Theorie der Bogenlänge in Gruppen. Erg. Math. Kolloqu. H. 5 (1933) 1–6.

26. K. Menger: The Formative Years of Abraham Wald and his Work in Geometry. Ann. Math. Statist. 23 (1952) 14–20.

27. K. Menger: Une theorie axiomatique générale des determinants, C.R. Acad. Sci. Paris 234 (1952) 1941–1943.

28. K. Menger: Géométrie générale. Mémorial d. Sci. Math. 124. Paris: Gauthier-Villars, 1954.

29. O. Schreier: Abstrakte kontinuierliche Gruppen. Abh. Math. Sem. Hamburg 4 (1925) 15–32.

30. O. Schreier, E. Sperner: Einführung in die analytische Geometrie und Algebra I. Leipzig: B. G. Teubner, 1931.
31. H. Steinhaus: Sur les distances des points dans les ensembles de mesure positive. Fund. Math. 1 (1920) 93.
32. O. Taussky: Zur Metrik der Gruppen. Anz. d. Akad. d. Wiss. (Wien), 1930, 140–142.
33. O. Taussky: Über isomorphe Abbildungen von Gruppen. Math. Ann. 108 (1933) 615–620.
34. O. Taussky: Abstrakte Körper und Metrik I. Endliche Mengen und Körperpotenzen. Erg. Math. Kolloqu. 6 (1935) 20–23.
35. O. Taussky, J. Todd: Inversion in Groups. Quart. J. Math. Oxford Ser. 12 (1941) 65–67.
36. A. Wald: Komplexe und undefinite Räume. Erg. Math. Kolloqu. 5 (1933) 32–42.
37. K. Weierstrass: Zur Determinantentheorie. In Mathematische Werke 3 (1903) 271–286.
38. K. Wolffhardt: Über eine Charakterisierung der Determinante. Math. Z. 103 (1968) 259–267.

Beiträge zur Gruppentheorie. I.
Über eine Gruppenmetrik [1]).

Von

Karl Menger in Wien.

Inhalt.

Seite

1. Problemstellung 396
2. Metrische Gruppen und ihre Abstandsverhältnisse 399
3. Über Kongruenzen von Gruppen 403
4. Über G-metrische Mengen und die Teilmengen metrischer
   Gruppen 405
5. Über Pseudo-G-Quadrupel 408
6. Sätze von O. Taussky über Pseudo-G-Quadrupel 412
7. Über G-metrische Mengen mit mehr als vier Elementen . . 415

1. Problemstellung.

In der vorliegenden Abhandlung soll erstens ein *Abstandsbegriff* in die Gruppentheorie eingeführt werden, der eine Art einfacher *Geometrie der Gruppen* ermöglicht, nämlich metrische Untersuchungen jeder Gruppe z. B. betreffend kongruente Figuren (bestehend aus Gruppenelementen), betreffend Symmetralen und Spiegelbilder, betreffend kongruente Selbstabbildungen der Gruppe usw. Es sollen zweitens Elemente beliebiger Gruppen als *Metrisierungsmittel* für beliebige Mengen (d. h. zur Definition des Abstandes von Elementepaaren beliebiger Mengen) verwendet werden in derselben Art, in der die reellen oder komplexen Zahlen als Abstände von Elementepaaren beliebiger Mengen (die dann halbmetrische Räume heißen) verwendet werden. Und es sollen drittens unter den mit Hilfe einer Gruppe G metrisierten Mengen speziell die *Teilmengen der Gruppe selbst* gekennzeichnet werden,

---

[1]) Ein Auszug der folgenden Arbeit erschien im Wiener akad. Anzeiger 1930.

so wie unter den metrischen Räumen mit reellen Abstandszahlen jene gekennzeichnet sind [2]), welche mit Teilmengen der Geraden, d. h. der Menge aller reellen Zahlen, kongruent sind.

Ehe wir diese Untersuchungen darstellen, mögen die Probleme und Methoden zunächst an einem ganz einfachen Beispiel erläutert werden, welches sich nicht auf die Metrik selbst, sondern auf Orientiertheit bezieht. Wir werden erstens eine Orientierung von Gruppen erklären, werden zweitens Elemente beliebiger Gruppen als Orientierungsmittel beliebiger Mengen verwenden und werden drittens unter den mit Hilfe dieser Gruppe orientierten Mengen die Teilmengen einer orientierten Gruppe kennzeichnen.

Es sei G eine Gruppe. Das aus den Elementen a und b komponierte Element bezeichnen wir mit $a + b$, das Einheitselement von G mit 0, das zum Element a inverse Element mit $- a$. Statt $a + (- b)$ schreiben wir kurz $a - b$.

Wir wollen die Gruppe G *orientiert* nennen, wenn jedem geordneten Paar a, b von Elementen von G das Element $b - a$, welches wir auch mit $a\,b$ bezeichnen werden, als *„gerichteter Abstand“* zugeordnet ist. Offenbar genügt dieser gerichtete Abstand für je zwei Elemente a und b von G der Beziehung

$$a\,b = - b\,a \begin{cases} \neq 0, & \text{wenn } a \neq b, \\ = 0, & \text{wenn } a = b. \end{cases}$$

Ist beispielsweise G die Menge der reellen (bzw. der komplexen) Zahlen, welche hinsichtlich der Addition eine Gruppe bildet, so ist diese Gruppe G nach der Orientierung, die jedem geordneten Paar a, b von Zahlen die Zahl $b - a$ als gerichteten Abstand zuordnet, mit dem $\vec{R}_1$ (bzw. mit dem $\vec{K}_1$), d. h. mit dem gerichteten eindimensionalen reellen (bzw. komplexen) euklidischen Raum identisch.

Eine Menge M heiße, wenn G eine vorgelegte Gruppe ist, G-*orientiert,* wenn jedem geordneten Paar p, q von Elementen von M ein Element von G, welches wir mit $p\,q$ bezeichnen wollen, zugeordnet ist gemäß der Bedingung

$$p\,q \begin{cases} \neq 0 & \text{für } p \neq q, \\ = 0 & \text{für } p = q. \end{cases}$$

Die orientierte Gruppe G selbst und alle ihre Teilmengen sind dieser Definition zufolge G-orientierte Mengen. Ist G die additive Gruppe der reellen Zahlen, so ist G-orientiert eine Menge M, wenn jedem geordneten Paar p, q von Elementen von M eine reelle Zahl zugeordnet ist, die dann und nur dann verschwindet, falls die beiden Elemente identisch sind.

---

[2]) Zweite Untersuchung über allgemeine Metrik, Math. Annalen **100** (1928).

Zwei G-orientierte Mengen M und M' mögen *kongruent* heißen, wenn die eine, etwa M', eindeutiges Bild der anderen Menge M ist derart, daß je zwei Elementen p und q von M zwei Elemente p' und q' von M' entsprechen, für welche $p'q' = pq$ und $q'p' = qp$ gilt. Eine dieser Bedingung genügende eindeutige Abbildung nennen wir eine *kongruente Abbildung*. Eine kongruente Abbildung führt je zwei verschiedene Elemente von M, d. h. je zwei Elemente von M, deren Abstand $\neq 0$ ist, in zwei Elemente von M', deren Abstand $\neq 0$ ist, die also verschieden sind, über und ist demnach eo ipso eineindeutig.

Es fragt sich nun, wie die Teilmengen der orientierten Gruppe G unter den G-orientierten Mengen kongruent gekennzeichnet sind, m. a. W. unter welchen Bedingungen eine vorgelegte G-orientierte Menge M mit einer Teilmenge von G kongruent ist. Die Antwort ist in folgendem Satze enthalten:

Damit die G-orientierte Menge M mit einer Teilmenge der orientierten Gruppe G kongruent sei, ist notwendig und hinreichend, daß für je drei Elemente p, q, r von M die Beziehung gilt

$$pq + rp + qr = 0.$$

Diese Bedingung ist notwendig. Ist nämlich M eine Teilmenge von G, so gilt für je drei Elemente a, b, c von M die Beziehung

$$(b - a) + (a - c) + (c - b) = 0, \quad \text{d. h.} \quad ab + ca + bc = 0.$$

Dieselbe Beziehung muß daher für je drei Elemente jeder mit M kongruenten G-orientierten Menge gelten.

Die Bedingung ist hinreichend. Wir wollen die Annahme machen, sie sei für je drei Elemente der G-orientierten Menge M erfüllt. Sind p und q irgend zwei Elemente von M, so ist die Bedingung insbesondere für das Elementetripel p, q, q erfüllt, und ergibt $pq + qp + qq = 0$. Da M als G-orientiert vorausgesetzt ist, gilt $qq = 0$, also $pq + qp = 0$ oder, was gleichbedeutend ist, $pq = -qp$ für je zwei Elemente p und q von M.

Zur Herstellung einer kongruenten Abbildung von M auf eine Teilmenge von G wählen wir irgendein Element p von M und ordnen ihm irgendein Element p' von G zu. Ist dann q irgendein Element von M, so ordnen wir ihm das Element $q' = pq + p'$ von G zu, wo pq den gerichteten Abstand von p nach q, also ein Element von G bezeichnet, zu welchem das Element p' von G addiert werden kann. Auf diese Weise ist M eindeutig auf eine Teilmenge von G abgebildet. Wir behaupten, daß diese Abbildung kongruent ist. Es seien also q und r irgend zwei Elemente von M, und q' bzw. r' ihre Bilder in G. Wir haben nachzuweisen, daß die beiden Beziehungen gelten $q'r' = qr$ und $r'q' = rq$. Da q' und r' Elemente der orientierten Gruppe G sind, gilt $q'r' = -r'q'$. In M gilt für q und r die (oben für je zwei Elemente von M bewiesene) Beziehung $q\,r = -\,r\,q$. Es genügt daher, von den beiden zu beweisenden Beziehungen eine, etwa die erste, nachzuweisen. Falls eines der Elemente q und r, etwa r, mit p identisch ist, so gilt $p'q' = q' - p' = (p\,q + p') - p' = p\,q$. In diesem Falle ist also die Behauptung bewiesen. Sind q und r beide von p verschieden, so verwenden wir die voraussetzungsgemäß gültige Beziehung $p\,q + r\,p + q\,r = 0$, aus der $q\,r = -\,r\,p - p\,q$ folgt. Anderseits gilt, da p', q', r' Elemente von G sind, $p'q' + r'p' + q'r' = 0$, also $q'r' = -\,r'p' - p'q'$. Wegen $r'p' = r\,p$ und $p'q' = p\,q$ folgt daraus $q'r' = q\,r$, d. h. die behauptete Beziehung. Damit ist die Bedingung auch als hinreichend bewiesen.

2. Metrische Gruppen und ihre Abstandsverhältnisse.

Wir fassen in der Gruppe G (so wie dies bei der Definition des absoluten Betrages eines Elementes der additiven Gruppe aller reellen Zahlen geschieht) die zueinander inversen Elemente zu Paaren zusammen und bezeichnen mit $|G|$ die Menge aller dieser Paare. Ist $|a|$ ein Element von $|G|$, so bezeichnen wir die (untereinander gleichen) Ordnungen der beiden in $|a|$ enthaltenen Elemente von G als *Ordnung* von $|a|$. Ist das Element $|a|$ von $|G|$, bestehend aus den Elementen a und $-a$ von G, von der Ordnung 2, so gilt $2\,a = 0$, also $a = -a$, und $|a|$ besteht demnach aus zwei gleichen Elementen von G. Umgekehrt gilt für jedes aus zwei gleichen Elementen a, a von G bestehende Element $|a|$ von $|G|$ die Beziehung $a = -a$, und daher $2\,a = 0$. Ein solches Element ist also von höchstens zweiter Ordnung, d. h. entweder von zweiter Ordnung oder von erster Ordnung, in welch letzterem Falle es aus den Elementen 0, 0 von G besteht und im folgenden auch kurz mit 0 bezeichnet werden soll.

Wir nennen eine Gruppe G *rechtsseitig metrisch*, wenn je zwei Elementen a und b von G das aus den Elementen $b - a$ und $a - b$ bestehende Element von $|G|$ als *Abstand* zugeordnet ist. Diesen Abstand bezeichnen wir auch kurz mit $a\,b$. Es gilt dann offenbar für je zwei Elemente a und b von G

$$a\,b = b\,a \begin{cases} \neq 0 & \text{für } a \neq b, \\ = 0 & \text{für } a = b. \end{cases}$$

Entsprechend heißt die Gruppe G *linksseitig metrisch*, wenn als Abstand der Elemente a und b das aus den Elementen $-a + b$ und $-b + a$ bestehende Element von $|G|$ zugeordnet ist. Ist G eine Abelsche Gruppe, so sind diese beiden Abstandsdefinitionen identisch und wir nennen die Gruppe *metrisch* schlechthin. Die folgenden Untersuchungen beschränken wir auf Abelsche Gruppen.

Ist a ein Element von G und $|d|$ ein Element von $|G|$, bestehend aus den Elementen d und $-d$ von G, so muß, wenn das Element b von G vom Element a den Abstand $|d|$ haben soll, eine der vier folgenden Beziehungen gelten:

$$d = b - a, \quad d = a - b, \quad -d = b - a, \quad -d = a - b.$$

Dieselben reduzieren sich offenbar auf zwei Beziehungen: Damit ein Element von a den Abstand $|d|$ habe, ist notwendig und hinreichend, daß es Lösung von einer der beiden Gleichungen $d = x - a$ und $d = a - x$ sei, d. h. das Element muß entweder mit $a + d$ oder $a - d$ identisch sein. Diese beiden Elemente sind identisch dann und nur dann, wenn $2\,d = 0$ gilt, wenn also d von höchstens zweiter Ordnung ist. Wir haben also, indem wir der kürzeren Ausdrucksweise halber ein Gruppenelement g, das nicht eine Ordnung $\leq n$ (wo n eine natürliche Zahl ist) besitzt (für welches also die Beziehung $n\,g \neq 0$ gilt), als ein Element einer Ordnung $> n$ bezeichnen, folgenden

Satz I. *Ist $|d|$ ein Element von $|G|$ von einer Ordnung > 2, so gibt es zu jedem Element a von G genau zwei Elemente, die von a den Abstand $|d|$ besitzen. Ist $|d|$ von der Ordnung 2, so gibt es zu jedem Element a genau ein von a verschiedenes Element im Abstand $|d|$ von a. Den Abstand 0 von a hat nur das Element a selbst.*

Es seien nun zwei verschiedene Elemente a und b von G und zwei Elemente $|d|$ und $|e|$ von $|G|$ bestehend aus den Elementen d, $-d$ bzw. e, $-e$ von G gegeben. Damit ein Element von G existiere, das von a den Abstand $|d|$ und von b den Abstand $|e|$ hat, ist offenbar notwendig und hinreichend, daß eines der höchstens zwei Elemente, die von a den Abstand $|d|$ haben, mit einem der höchstens zwei Elemente, die von b den Abstand $|e|$ haben, identisch ist; m. a. W., daß mindestens eine der vier folgenden Beziehungen besteht:

$$(1) \quad a + d = b + e, \qquad (2) \quad a - d = b + e,$$
$$(3) \quad a + d = b - e, \qquad (4) \quad a - d = b - e.$$

Damit mehr als ein Element existiere, welches von a den Abstand $|d|$ und von b den Abstand $|e|$ hat, ist offenbar notwendig und hinreichend, daß zwei Elemente existieren, die von a den Abstand $|d|$ haben, und zwei Elemente, die von b den Abstand $|e|$ haben, und daß jedes der beiden ersten mit einem der beiden zweiten Elemente identisch sei, daß m. a. W. mindestens zwei von den vier angeführten Beziehungen zusammen bestehen. Nun folgt aus dem Zusammenbestehen von (1) und (2) oder von (3) und (4) die Beziehung $2\,d = 0$, und aus dem Zusammenbestehen von (1) und (3) oder von (2) und (4) die Beziehung $2\,e = 0$. Wenn aber d oder e von der Ordnung 2 ist, so existiert nach Satz I bloß ein Element, welches von a den Abstand $|d|$, bzw. von b den Abstand $|e|$ hat. Aus dem Zusammenbestehen von (1) und (4) oder von (2) und (3) folgt $2a = 2b$, also $2(a - b) = 0$, d. h. die Elemente a und b haben einen Abstand der Ordnung 2. Haben umgekehrt a und b einen Abstand der Ordnung 2 und existiert ein Element, das von a den Abstand $|d|$ und von b den Abstand $|e|$ hat, wo $|d|$ und $|e|$ von Ordnungen > 2 sind, so gilt eine der vier Beziehungen (1), (2), (3), (4) und aus ihr folgt auf Grund von $2a = 2b$ eine zweite dieser Beziehungen, so zwar, daß entweder (1) und (4) oder (2) und (3) zusammen bestehen. Es existiert dann also auch ein zweites Element, welches von a den Abstand $|d|$ und von b den Abstand $|e|$ hat, und dasselbe ist von dem ersten verschieden, da $|d|$ und $|e|$ Ordnungen > 2 haben. Zusammenfassend können wir also aussprechen folgenden

Satz II. *Zu je zwei Elementen a und b, deren Abstand eine Ordnung > 2 besitzt, und zu je zwei Elementen $|d|$ und $|e|$ von $|G|$ existiert höchstens ein Element, das von a den Abstand $|d|$ und von b den Abstand $|e|$ besitzt. Ist der Abstand von a und b von der Ordnung 2, so existiert zu jedem Element c von G, dessen Abstände von a und b Ordnungen > 2 besitzen, genau ein von c verschiedenes Element c', für welches $ca = c'a$ und $cb = c'b$ gilt*[3]). (Wir werden dasselbe als das *Spiegelbild* von c in bezug auf a und b bezeichnen.)

Um ein Beispiel eines Spiegelbildes in bezug auf zwei Punkte, deren Abstand die Ordnung 2 hat, zu erlangen, betrachten wir die multiplikative Gruppe der vierten Einheitswurzeln. Die vier Elemente dieser Gruppe sind $1, -1, i, -i$. Das Kompositionszeichen $+$ der Gruppe ist die Multiplikation, das Einheitselement 1. Der Abstand von 1 und -1 ist $\left(\dfrac{-1}{1}, \dfrac{1}{-1}\right)$ $= (-1, -1)$, also von der Ordnung 2, ebenso der Abstand von i und $-i$, der $\left(\dfrac{i}{-i}, \dfrac{-i}{i}\right) = (-1, -1)$ ist. Die vier anderen Punktepaare haben den Abstand $(i, -i)$. Die Elemente i und $-i$ sind also Spiegelbilder in bezug auf die Elemente 1 und -1. Die Elemente 1 und -1 sind Spiegelbilder in bezug auf i und $-i$.

Sind a und b zwei Elemente der Gruppe G, so bezeichnen wir jenes Element von G, das von a denselben Abstand hat wie b, als das *Spiegelbild von b an a*. Damit das Element b mit seinem Spiegelbild an a identisch sei, ist dem Vorangehenden zufolge notwendig und hinreichend, daß der Abstand von a und b von höchstens zweiter Ordnung ist. Das Spiegelbild eines Elementes c an zwei Elementen a und b, deren Abstand die

---

[3]) Den Hinweis auf die Ausnahmerolle der Abstände der Ordnung 2 verdanke ich Fräulein O. Taussky.

Ordnung 2 hat, ist nichts anderes als das Spiegelbild von c an a und das Spiegelbild von c an b, welche beiden Spiegelbilder eben deshalb, weil der Abstand von a und b die Ordnung 2 hat, identisch sind.

Sind a und b zwei verschiedene Elemente von G, so bezeichnen wir als *Symmetrale von a und b* die (eventuell leere) Menge aller Elemente von G, die denselben Abstand von a wie von b haben, für welche also $ac = bc$ gilt. Für ein Element c der Symmetrale von a und b muß also entweder $a - c = b - c$ oder $a - c = c - b$ gelten. Der erste Fall kann wegen der Verschiedenheit von a und b nicht eintreten, also muß der zweite vorliegen und daher die Beziehung $2c = a + b$ gelten. Ist c' ein anderes Element der Symmetrale von a und b, so gilt $2c' = 2c = a + b$, also $2(c - c') = 0$. Es haben also je zwei Elemente der Symmetrale von a und b einen Abstand der Ordnung 2. Umgekehrt gehört offenbar neben jedem Element c der Symmetrale von a und b auch jedes Element, dessen Abstand von c die Ordnung 2 hat, der Symmetrale an. Diese Tatsachen können kurz formuliert werden, wenn wir die Menge aller Elemente der Gruppe G, deren Ordnung ≤ 2 ist, mit G_2 bezeichnen und eine Menge, bestehend aus den Elementen $a + g$, wo a ein festes Element von G und g ein beliebiges Element von G_2 ist, eine *mit G_2 kongruente Menge* nennen. G_2 ist offenbar eine Untergruppe von G, welche, wenn die Gruppe keine Elemente der Ordnung 2 enthält, bloß aus dem Einheitselement besteht. Wir haben bewiesen:

Satz III. *Die Symmetrale von zwei verschiedenen Elementen der Gruppe G ist entweder leer oder kongruent mit G_2.*

Wir können dies auch dahin ausdrücken, daß zwei Elemente, die Spiegelbilder in bezug auf ein Element a sind, auch Spiegelbilder in bezug auf die Menge $a + G_2$ sind. Insbesondere bestehen nach Satz III in Gruppen ohne Elemente der Ordnung 2 die Symmetralen aus höchstens einem Element. Enthält eine Gruppe Elemente der Ordnung 2, so können dieselben nur in der Symmetrale solcher Elementepaare liegen, in deren Symmetrale auch das Element O liegt, d. h. in der Symmetrale von zueinander inversen Elementen. Wenn eine Gruppe G nur Elemente von höchstens zweiter Ordnung enthält, wenn sie m. a. W. mit G_2 identisch ist, so gibt es keine verschiedenen zueinander inversen Elemente von G. Wir sehen also: *In einer Gruppe ohne Elemente einer Ordnung > 2 ist die Symmetrale von je zwei verschiedenen Elementen leer.*

Anderseits wollen wir nun zeigen: *In einer zyklischen Gruppe von ungerader Ordnung ist die Symmetrale von keinem Elementenpaar leer.* Seien nämlich ka $(0 \leq k \leq 2n)$ die Elemente der Gruppe. Sind ka und la irgend zwei verschiedene Elemente, so ist die ganze Zahl x so bestimmbar, daß xa gleichen Abstand von ka und la hat. Wir haben zu diesem Zweck nur $xa - ka = la - xa$, also $2x \equiv k + l \ (mod \ 2n + 1)$ zu wählen. Man bestätigt endlich noch unmittelbar folgende oft anwendbare Bemerkung: *Sind c_a und c_b Spiegelbilder von c in bezug auf a und b, so gilt $c_a c_b = 2ab$.*

3. Über Kongruenzen von Gruppen.

Wir bezeichnen eine Selbstabbildung der Gruppe G, also eine Zuordnung von einem Element a' von G zu jedem Element a von G als *Kongruenz*, wenn sie abstandstreu ist, d. h. wenn für je zwei Elemente a und b von G und ihre Bildelemente a' und b' die Beziehung gilt: $a'b' = ab$. Die Kongruenzen einer Gruppe bilden offenbar ihrerseits eine Gruppe, denn hintereinander ausgeführte Kongruenzen lassen sich durch eine einzige ersetzen. Die identische Selbstabbildung ist das Einheitselement der Kongruenzengruppe, und zu jeder Kongruenz existiert eine inverse.

Ist t ein bestimmtes Element von G und wird jedem Element x von G das Element $x' = x + t$ zugeordnet, so entsteht eine Selbstabbildung von G, welche als Translation um t oder kurz als *Translation* bezeichnet werden möge. Dieselbe ist offenbar eine Kongruenz. Ferner ist klar, daß die Translationen von G eine Gruppe, also eine Untergruppe der Kongruenzengruppe bilden, welche mit der Gruppe G isomorph ist. Sind a und a' irgend zwei Elemente von G, so existiert genau eine Translation, welche a in a' überführt, nämlich $x' = x + a' - a$.

Ist s ein bestimmtes Element von G und ordnen wir jedem Element von G sein Spiegelbild in bezug auf s zu, so entsteht offenbar eine kongruente Selbstabbildung von G, welche wir als Spiegelung an s oder kurz als *Spiegelung* bezeichnen wollen. Bezeichnet x' das Bild von x vermöge der Spiegelung an s, so gilt $x' - s = s - x$, also $x' = 2s - x$. Zwei Spiegelungen an den Elementen s und r hintereinander ausgeführt ergeben eine Translation um $2(r - s)$, denn aus $x' = 2s - x$ und $x'' = 2r - x'$ folgt $x'' = 2(r - s) + x$. Für $r = s$ ergibt sich naturgemäß die Identität.

Wir beweisen nun

Satz IV. *Sind a, b und a', b' zwei kongruente Paare von Elementen von G, so gibt es eine Kongruenz von G, welche entweder eine Translation ist oder sich aus einer Translation und einer Spiegelung zusammensetzt und a in a' und b in b' überführt.*

Wir betrachten die Translation $x' = x + a' - a$, welche a in a' überführt. Wenn $b' = b + a' - a$ ist, so führt diese Translation auch b in b' über und wir sind am Ziel. Andernfalls ist b' ein von $b + a' - a$ verschiedenes Element von G, das von a' den Abstand ab hat. Nun hat

26*

$b + a' - a$ von a' den Abstand $a'b' = ab$. Wenn also $b' \neq b + a' - a$ gilt, so ist b' das Spiegelbild von $b + a' - a$ an a', d. h. $2a' - (b + a' - a)$ $= a + a' - b$. In diesem Falle führen wir nach der Translation $x' = x + a' - a$ noch die Spiegelung um das Element a' aus, $x'' = 2a' - x'$, und erhalten so die Kongruenz $x'' = a + a' - x$. Diese Abbildung führt a in a' und b in b' über, womit unser Satz für jeden Fall bewiesen ist. —

Es sei nun G eine Gruppe, deren sämtliche Elemente von höchstens zweiter Ordnung sind. Es seien zwei Kongruenzen von G gegeben, welche beide dem Element a das Element a' zuordnen. Es sei dann b irgendein Element von G. Vermöge der ersten Kongruenz sei ihm das Element b', vermöge der zweiten das Element b'' zugeordnet. Sowohl b' als auch b'' müssen wegen der Kongruenz der beiden Abbildungen von a' den Abstand ab haben. Derselbe hat auf Grund der Voraussetzung über alle Gruppenelemente die Ordnung 2. Daher gibt es nur ein einziges Element, das von a' den Abstand ab hat. Es muß mithin $b' = b''$ gelten, d. h. die beiden Kongruenzen ordnen jedem Element von G dasselbe Bildelement zu und sind mithin identisch. In einer Gruppe ohne Element einer Ordnung > 2 ist also jede Kongruenz durch Angabe eines Elementes und seines Bildes festgelegt.

Enthält G auch ein Element einer Ordnung > 2 und sind zwei Kongruenzen gegeben, welche a in a' überführen, so bestimmen wir ein Element b, das von a einen Abstand hat, dessen Ordnung > 2 ist. Es seien b' und b'' die Bilder von b vermöge der beiden Kongruenzen. Wenn erstens $b' = b''$ gilt und c irgendein Element von G ist, so müssen für seine Bilder c' und c'' die Beziehungen gelten:

$$c'a' = ca, \quad c'b' = cb \quad \text{und} \quad c''a' = ca, \quad c''b' = cb.$$

Da es aber nicht zwei verschiedene Elemente geben kann, die von a' und b', deren Abstand eine Ordnung > 2 besitzt, dieselben Abstände haben, so muß $c' = c''$ gelten, d. h. wenn $b' = b''$ gilt, so sind die beiden Kongruenzen identisch. Ist zweitens $b' \neq b''$, so sind, da sowohl b' als auch b'' von a' den Abstand ab haben müssen, b' und b'' Spiegelbilder in bezug auf a', und es sind dann offenbar für jedes Element c seine Bilder c' und c'' Spiegelbilder in bezug auf a'. In diesem Falle gibt es also zwei zueinander in bezug auf a' spiegelbildliche Kongruenzen, welche a in a' überführen. Zusammenfassend können wir sagen:

Satz V. *Sind a und a' zwei Elemente der Gruppe G, so existieren genau zwei Kongruenzen, welche a in a' überführen. Dieselben sind zueinander spiegelbildlich in bezug auf a' und dann und nur dann miteinander identisch, wenn G bloß Elemente von höchstens zweiter Ordnung enthält.*

Zugleich geht aus dem Beweise offenbar hervor folgender

Satz VI. *Sind a, b und a', b' zwei kongruente Paare von Elementen der Gruppe G, so gibt es genau eine Kongruenz, welche a in a' und b in b' überführt, falls der Abstand ab = a'b' von einer Ordnung > 2 ist oder falls alle Elemente von G die Ordnung 2 haben, und es gibt genau zwei (zueinander spiegelbildliche) Kongruenzen, welche a in a' und b in b' überführen, falls ab die Ordnung 2 hat, aber Elemente einer Ordnung > 2 existieren.*

Als Korollar entnehmen wir diesen Überlegungen, *daß jede Kongruenz einer Gruppe entweder eine Translation ist oder sich aus einer Translation und einer Spiegelung zusammensetzt.*

4. Über G-metrische Mengen und die Teilmengen metrischer Gruppen.

Es sei eine Gruppe G gegeben. Eine Menge M von irgendwelchen Elementen heiße metrisch in bezug auf G oder kurz G-*metrisch*, wenn je zwei Elementen p und q von M ein Element von $|G|$, welches wir mit pq bezeichnen und den *Abstand* von p und q nennen, zugeordnet ist gemäß den Bedingungen:

$$pq = qp \begin{cases} = 0 & \text{für } p = q \\ \neq 0 & \text{für } p \neq q \end{cases}, \quad \text{für je zwei Elemente } p \text{ und } q \text{ von } M.$$

Die in bezug auf die additive Gruppe der reellen Zahlen metrischen Mengen sind offenbar identisch mit den halbmetrischen Räumen. So wird nämlich jede Menge genannt, in der je zwei Elementen p und q eine nichtnegative reelle Zahl $pq = qp$ zugeordnet ist, die für Paare identischer Elemente und nur für solche $= 0$ ist. Der Sinn der Definition der G-metrischen Menge ist also die Heranziehung beliebiger Gruppen statt speziell der Gruppe reeller Zahlen als Metrisierungsmittel beliebiger Mengen.

Kongruent heißen zwei G-metrische Mengen M und M', wenn eine eindeutige Abbildung von M auf M' existiert, so daß für je zwei Elemente p und q von M und ihre Bildelemente p' und q' die Beziehung gilt: $pq = p'q'$.

Die metrische Gruppe G selbst sowie alle Teilmengen von G sind offenbar G-metrische Mengen. Aber nicht jede G-metrische Menge muß eine Teilmenge von G oder mit einer Teilmenge von G kongruent sein. Betrachten wir eine Menge, bestehend aus drei Elementen, die paarweise den Abstand 1 haben, so liegt ein halbmetrischer Raum, d. h. eine in bezug auf die additive Gruppe der reellen Zahlen metrische Gruppe vor, die doch nicht mit einer Teilmenge der Geraden kongruent ist, da es nicht drei reelle Zahlen gibt, die paarweise den Abstand 1 haben.

Es erhebt sich infolgedessen die Frage nach der Kennzeichnung jener G-metrischen Mengen, die mit einer Teilmenge von G kongruent sind. Es ist diesbezüglich vor allem klar, *daß jede aus höchstens zwei Elementen bestehende G-metrische Menge mit einer Teilmenge von G kongruent ist.* Bezüglich der aus drei Elementen p_1, p_2, p_3 bestehenden Mengen gilt, wenn der Abstand $p_i p_j$ gleich dem aus den Elementen r_{ij} und $-r_{ij}$ von G bestehenden Element von $|G|$ ist, folgender

Satz VII. *Damit die aus p_1, p_2, p_3 bestehende G-metrische Menge, in welcher p_i und p_j den Abstand r_{ij}, $-r_{ij}$ haben, mit einer Teilmenge von G kongruent sei, ist notwendig und hinreichend das Bestehen von einer der vier Relationen* $r_{12} + r_{23} + r_{31} = 0$, $r_{12} + r_{23} - r_{31} = 0$, $r_{12} - r_{23} + r_{31} = 0$, $-r_{12} + r_{23} + r_{31} = 0$ *oder m. a. W. das Bestehen einer Relation* $\varepsilon_1 r_{23} + \varepsilon_2 r_{31} + \varepsilon_3 r_{12} = 0$, *wo die $\varepsilon_i = \pm 1$ sind.*

Sind die Elemente p_1, p_2, p_3 mit drei Elementen p_1', p_2', p_3' von G kongruent, so gilt

$$(p_1' - p_2') + (p_2' - p_3') + (p_3' - p_1') = 0.$$

Da der Abstand von p_i und p_j gleich dem aus den Elementen $p_i' - p_j'$, $p_j' - p_i'$ von G bestehenden Element von $|G|$ ist, so ergibt sich hieraus die Notwendigkeit der im Satz VII für Tripel ausgesprochenen Bedingung. Nehmen wir umgekehrt für drei Elemente p_1, p_2, p_3 die Gültigkeit einer Beziehung $\varepsilon_1 r_{23} + \varepsilon_2 r_{13} + \varepsilon_3 r_{12} = 0$ ($\varepsilon_i = \pm 1$) an, so können wir drei kongruente Elemente von G angeben. Wir ordnen p_1 irgendein Element p_1' von G zu und dem Element p_2 das Element $p_2' = p_1' + r_{12}$. Ordnen wir sodann p_3 das Element $p_3' = p_1' - \dfrac{\varepsilon_2}{\varepsilon_3} r_{13}$ zu, so sind p_1', p_2', p_3' mit p_1, p_2, p_3 kongruent. Denn es ist $p_3' p_1'$ gleich dem aus r_{13} und $-r_{13}$ bestehenden Element von $|G|$, so wie $p_3 p_1$. Und es ist $p_3' - p_2' = p_1' - \dfrac{\varepsilon_2}{\varepsilon_3} r_{13} - p_1' - r_{12}$, d. i. aber unter Berücksichtigung der vorausgesetzten Beziehung $= \dfrac{\varepsilon_1}{\varepsilon_3} r_{23}$, womit Satz VII bewiesen ist.

Wir ziehen zunächst eine Folgerung aus Satz VII. Wir wollen sagen, drei Elemente a, b, c einer G-metrischen Menge bilden ein *gleichseitiges Dreieck*, wenn $ab = bc = ca$ gilt, welch gemeinsamen Wert wir als die *Seitenlänge* des Dreieckes bezeichnen. Sind a, b, c drei Elemente von G, die ein gleichseitiges Dreieck bilden, dessen Seitenlänge das aus den Elementen s und $-s$ bestehende Element $\neq 0$ von $|G|$ ist, dann muß Satz VII zufolge eine Relation $\varepsilon_1 s + \varepsilon_2 s + \varepsilon_3 s = 0$ ($\varepsilon_i = \pm 1$) gelten, also eine Beziehung $(\varepsilon_1 + \varepsilon_2 + \varepsilon_3) s = 0$. Es ist unmöglich, daß von den drei Zahlen ε_i zwei denselben und die dritte den andern Wert hat, da hieraus $s = 0$ folgen würde, was der Voraussetzung widerspricht. Es müssen also alle drei Werte ε_i einander gleich sein, d. h. es muß $3s = 0$ gelten und somit s die Ordnung 3 haben. Wir haben also zu Satz VII folgendes

Korollar. *Die Seitenlänge eines gleichseitigen Dreieckes von Gruppenelementen besitzt die Ordnung 3. In einer Gruppe ohne Elemente der Ordnung 3 existieren keine gleichseitigen Dreiecke.*

Wir führen nun das Problem der Kennzeichnung der Teilmengen von G unter den G-metrischen Mengen zurück auf die Lösung dieses Problems für die aus vier Elementen bestehenden Mengen durch den Beweis von folgendem

Satz VIII. *Eine G-metrische Menge, von der je vier Elemente mit vier Elementen von G kongruent sind, ist mit einer Teilmenge von G kongruent.*

Wir zeigen vor allem: *Ist M eine G-metrische Menge, in der je zwei verschiedene Elemente einen Abstand der Ordnung 2 haben und von der je drei Elemente mit drei Elementen von G kongruent sind, so ist M mit einer Teilmenge von G kongruent.* Es sei p irgendein Element von M. Wir ordnen ihm irgendein Element p' von G zu. Da der Abstand pq die Ordnung 2 hat, so gibt es nur ein Element q' von G, das von p' den Abstand pq hat. Dieses Element q' ordnen wir dem Element q zu. Auf diese Weise ist eine eindeutige Abbildung von M auf eine Teilmenge von G definiert. Wir behaupten, daß diese Abbildung kongruent ist. Seien also q und r irgend zwei Elemente von M, q' und r' ihre Bilder in G. Wir haben zu zeigen, daß $p'q' = pq$ gilt. Dies ist auf Grund der Definition der Abbildung klar für den Fall, daß einer der Punkte q und r mit p identisch ist. Andernfalls betrachten wir das Tripel p, q, r und wissen auf Grund der Annahme, daß ein kongruentes Elemententripel $\bar{p}, \bar{q}, \bar{r}$ in G existiert. Wir unterwerfen nun G jener Kongruenz, welche $\bar{q}$ in p' überführt. (Nach Satz V gibt es nur eine solche Kongruenz.) Bei dieser Kongruenz gehen $\bar{q}$ und $\bar{r}$ in zwei Elemente über, die von p' die Abstände $\bar{q}\,\bar{p}$ bzw. $\bar{r}\bar{p}$, d. h. die Abstände qp bzw. rp haben. Nun gibt es aber bloß ein Element, das von p' den Abstand qp hat, nämlich q', und nur ein Element, das von p' den Abstand rp hat, nämlich r'. Es muß also das Tripel $\bar{p}, \bar{q}, \bar{r}$ mit p', q', r' kongruent sein. Also gilt $\bar{q}\bar{r} = q'r'$, und daher wegen $\bar{q}\bar{r} = qr$ die Beziehung $q'r' = qr$, womit die Behauptung bewiesen ist.

Es sei nun M irgendeine G-metrische Menge, von der je vier Elemente mit vier Elementen von G kongruent sind. Wir behaupten, daß M mit einer Teilmenge von G kongruent ist. Haben je zwei verschiedene Elemente von G einen Abstand zweiter Ordnung, so ist die Behauptung eine Folge der eben bewiesenen. Wir können also annehmen, daß zwei Elemente p

und q von M existieren, deren Abstand eine Ordnung > 2 hat. Wir ordnen ihnen zwei kongruente Elemente p' und q' von G zu. Ist r irgendein Element von M, so existiert, da $p'q'$ von einer Ordnung > 2 ist, nach Satz I höchstens ein Element r' von G, das von p' den Abstand rp und von q' den Abstand rq hat. Anderseits muß es mindestens ein Element r' dieser Art geben. Denn nach Annahme existiert ein zu p, q, r kongruentes Elementetripel $\bar{p}, \bar{q}, \bar{r}$ von G. Es gibt nach Satz VI nur eine Kongruenz, welche $\bar{p}$ in p' und $\bar{q}$ in q' überführt, da der Abstand $\bar{p}\bar{q}$ von einer Ordnung > 2 ist. Diese Kongruenz führt $\bar{r}$ in ein Element r' über, das von p' den Abstand $\bar{r}\bar{p} = rp$ und von q' den Abstand $\bar{r}\bar{q} = rq$ hat. Es entspricht also jedem Element r von M genau ein Element r', welches von p' bzw. q' denselben Abstand hat wie r von p bzw. q. Dieses Element r' ordnen wir r zu und haben damit eine eindeutige Abbildung von M auf eine Teilmenge von G definiert.

Wir behaupten, daß diese Abbildung kongruent ist. Es seien nämlich r und s irgend zwei Elemente von M und r' bzw. s' ihre Bilder in G. Wir haben zu zeigen, daß $r's' = rs$ gilt. Dies ist sicher der Fall, wenn mindestens eines der Elemente r und s mit p oder q identisch ist. Andernfalls betrachten wir die vier Elemente p, q, r, s. Nach Annahme entsprechen ihnen vier kongruente Elemente $\bar{p}, \bar{q}, \bar{r}, \bar{s}$ von G. Es gibt eine und, da der Abstand $\bar{p}\bar{q}$ von einer Ordnung > 2 ist, nach Satz VI nur eine Kongruenz, welche $\bar{p}$ in p' und $\bar{q}$ in q' überführt. Dieselbe führt $\bar{r}$ und $\bar{s}$ in zwei Elemente über, die den Abstand $\bar{r}\bar{s} = rs$ haben und die von p' bzw. q' dieselben Abstände haben wie $\bar{r}$ und $\bar{s}$ von $\bar{p}$ bzw. von $\bar{q}$, d. h. wie r und s von p bzw. von q. Nun gibt es aber nur ein Element, das von p' und q' den Abstand rp bzw. rq hat, nämlich r', und es gibt nur ein Element, das von p' und q' den Abstand sp bzw. sq hat, nämlich s'. Es muß also also bei der Kongruenz, die $\bar{p}$ in p' und $\bar{q}$ in q' überführt, $\bar{r}$ in r' und $\bar{s}$ in s' übergehen, d. h. es gilt $\bar{r}\bar{s} = r's'$ und mithin $r's' = rs$. Damit ist Satz VIII bewiesen.

5. Über Pseudo-G-Quadrupel.

Wir haben festgestellt, daß aus der Kongruenz von je vier Elementen einer G-metrischen Menge mit einem Quadrupel von G die Kongruenz der ganzen Menge mit einer Teilmenge von G folgt. Wir wollen nun untersuchen, was aus der Kongruenz von je drei Elementen einer G-metrischen Menge mit einem Elementetripel von G für die Menge folgt, und behandeln dieses Problem zunächst für den Fall, daß die Menge aus vier Elementen besteht. Es sind zwei Fälle möglich: Entweder ist die Menge mit einem Quadrupel von G kongruent oder dies ist nicht der Fall. Im letzteren Falle bezeichnen wir die Menge als ein *Pseudo-G-Quadrupel*. So heißt also *eine aus vier Elementen bestehende G-metrische Menge, die nicht mit vier Elementen von G kongruent ist, von der aber je drei Elemente mit einem Elementetripel von G kongruent sind.*

Wir schicken den die diesbezüglichen Hauptresultate enthaltenden Sätzen XII und XIII zunächst drei vorbereitende Sätze voran.

Satz IX. *Sind je drei von den vier Elementen p_1, p_2, p_3, p_4 einer G-metrischen Menge mit drei Elementen von G kongruent und haben zwei von ihnen einen Abstand der Ordnung 2, so sind die vier Elemente mit vier Elementen von G kongruent.*

Es sei etwa $2 p_1 p_2 = 0$. Dann ordnen wir p_1 und p_2 zwei Elemente p_1' und p_2' von G zu, für welche $p_1' p_2' = p_1 p_2$ gilt. Es existieren höchstens zwei verschiedene Elemente $\bar{p}_3$ und $\bar{\bar{p}}_3$ von G, die von p_1' bzw. p_2' die Abstände $p_1 p_3$ bzw. $p_2 p_3$ haben, und höchstens zwei Elemente $\bar{p}_4$ und $\bar{\bar{p}}_4$, welche von p_1' bzw. p_2' die Abstände $p_1 p_4$ bzw. $p_2 p_4$ haben. Dabei liegen $\bar{p}_3$ und $\bar{\bar{p}}_3$ zueinander spiegelbildlich sowohl in bezug auf p_1' als auch in bezug auf p_2', und dasselbe gilt für $\bar{p}_4$ und $\bar{\bar{p}}_4$. Nun ist das Tripel p_1, p_3, p_4 nach Annahme mit drei Elementen von G kongruent. Wir ordnen p_1 und p_3 die Elemente p_1' bzw. $\bar{p}_3$ zu. Der Bildpunkt von p_4 muß von p_1' den Abstand $p_1 p_4$ haben, daher entweder mit $\bar{p}_4$ oder mit $\bar{\bar{p}}_4$ identisch sein. Der Abstand $p_3 p_4$ ist daher entweder $= \bar{p}_3 \bar{p}_4$ oder $= \bar{p}_3 \bar{\bar{p}}_4$. Im ersten Falle sind p_1, p_2, p_3, p_4 mit p_1', p_2', $\bar{p}_3$, $\bar{p}_4$, im zweiten Falle mit p_1', p_2', $\bar{p}_3$, $\bar{\bar{p}}_4$ kongruent. Jedenfalls sind also die Elemente p_1, p_2, p_3, p_4 mit vier Elementen von G kongruent, womit Satz IX bewiesen ist.

Satz X. *Bilden die Elemente p_1, p_2, p_3, p_4 ein Pseudo-G-Quadrupel, so gelten die Beziehungen*

$$2 p_1 p_2 = 2 p_3 p_4, \quad 2 p_1 p_3 = 2 p_2 p_4, \quad 2 p_1 p_4 = 2 p_2 p_3.$$

Es seien also p_1, p_2, p_3, p_4 vier Elemente, die nicht mit vier Elementen von G kongruent sind, während je drei von ihnen mit drei Elementen von G kongruent sind. Es seien p_1', p_2', p_3' bzw. p_1', p_2', p_4' zwei Tripel von G, die mit p_1, p_2, p_3 bzw. mit p_1, p_2, p_4 kongruent sind. Dann ist $p_3' p_4' \neq p_3 p_4$, da sonst p_1, p_2, p_3, p_4 mit p_1', p_2', p_3', p_4', also mit vier Elementen von G kongruent wären, während sie als ein Pseudo-G-Quadrupel vorausgesetzt sind. Nun bestimmen wir zum Tripel p_1, p_3, p_4 ein kongruentes Tripel p_1', p_3', p_4^1 in G. Es ist $p_4^1 \neq p_4'$, da sonst $p_3 p_4 = p_3' p_4'$ gelten würde. Es hat aber p_4^1 von p_1' den gleichen Abstand wie p_4'. Also sind p_4' und p_4^1 spiegelbildlich in bezug auf p_1', und wir sehen wegen $p_3 p_4 = p_3' p_4^1$: Der Abstand $p_3 p_4$ ist gleich dem Abstand von p_3' und dem zu p_4' in bezug auf p_1' spiegelbildlich gelegenen Element. In analoger Weise entnimmt man aus der Betrachtung eines zu p_2, p_3, p_4 kongruenten Tripels p_2', p_3', p_4^2, daß $p_3 p_4$ gleich ist dem Abstand von p_3' und dem zu p_4' in bezug auf p_2' spiegelbildlich gelegenen Element. Daraus folgt, daß $p_3' p_4^1 = p_3' p_4^2$ gilt. Die Elemente p_4^1 und p_4^2 sind also entweder *identisch* oder in bezug auf p_3' *spiegelbildlich*.

Der erste Fall kann nicht vorliegen, denn sonst müßte einer der beiden folgenden Unterfälle vorliegen: *Entweder* wäre $p_4' = p_4^1 = p_4^2$, d. h. das Element p_4' wäre mit seinen Spiegelbildern in bezug auf p_1' und p_2' identisch und die Abstände $p_1' p_4'$ und $p_2' p_4'$ wären von der Ordnung 2. Nach Satz IX aber kann kein Elementepaar des Pseudo-G-Quadrupels p_1, p_2, p_3, p_4 und daher auch keines der Elementepaare p_1', p_4' und p_2', p_4' einen Abstand der Ordnung 2 besitzen, also kann dieser erste Unterfall nicht vorliegen. *Oder* es wäre $p_4' \neq p_4^1 = p_4^2$. Dann existierten zwei verschiedene Elemente, nämlich p_4' und $p_4^1 = p_4^2$, die von p_1' und p_2' die gleichen Abstände besäßen. Dann müßte der Abstand $p_1' p_2'$, welcher $= p_1 p_2$ ist, die Ordnung 2 besitzen. Da p_1, p_2, p_3, p_4 ein Pseudo-G-Quadrupel bilden, ist dies nach Satz IX unmöglich; also kann auch der zweite Unterfall nicht vorliegen. Es kann also nicht die Beziehung $p_4^1 = p_4^2$ gelten und es müssen demnach die Elemente p_4^1 und p_4^2 spiegelbildlich in bezug auf p_3' sein.

Daraus folgt, daß $p_4^1 p_4^2 = 2 p_3' p_4^1 = 2 p_3' p_4^2$ gilt. Wegen $p_3 p_4 = p_3' p_4^1 = p_3' p_4^2$ ergibt sich hieraus

(†) $$2 p_3 p_4 = p_4^1 p_4^2.$$

Anderseits sind p_4^1 und p_4^2 die Spiegelbilder von p_4' an p_1' bzw. p_2'. Mithin gilt nach der Schlußbemerkung des Abschnittes 2 die Beziehung $p_4^1 p_4^2 = 2 p_1' p_2'$. Wegen $p_1' p_2' = p_1 p_2$ folgt hieraus $p_4^1 p_4^2 = 2 p_1 p_2$ und ein Vergleich dieser Beziehung mit (†) ergibt $2 p_3 p_4 = 2 p_1 p_2$, also eine der drei zu beweisenden Beziehungen. Zu den beiden anderen gelangt man ganz analog, indem man bloß statt des Paares p_1, p_2 die Paare p_1, p_3 bzw. p_1, p_4 in der Argumentation auszeichnet, oder m. a. W. durch eine Umnumerierung der Elemente des Quadrupels. Damit ist Satz X bewiesen.

Satz XI. *Sind a_1, a_2, a_3, a_4 vier Elemente der metrischen Gruppe G, deren Abstände für eine gewisse natürliche Zahl n den Beziehungen genügen*

$$n a_1 a_2 = n a_3 a_4, \qquad n a_1 a_3 = n a_2 a_4, \qquad n a_1 a_4 = n a_2 a_3,$$

dann besitzt der Abstand von mindestens zwei der vier Elemente eine Ordnung $\leq 2 n$.

In der Tat:

$n a_1 a_2 = n a_3 a_4$ bedeutet entweder $\alpha 1)$ $n a_1 - n a_2 = n a_3 - n a_4$

oder $\beta 1)$ $n a_1 - n a_2 = n a_4 - n a_3$

$n\,a_1\,a_3 = n\,a_2\,a_4$ bedeutet entweder $\alpha\,2)$ $n\,a_1 - n\,a_3 = n\,a_2 - n\,a_4$

oder $\beta\,2)$ $n\,a_1 - n\,a_3 = n\,a_4 - n\,a_2$

$n\,a_1\,a_4 = n\,a_2\,a_3$ bedeutet entweder $\alpha\,3)$ $n\,a_1 - n\,a_4 = n\,a_2 - n\,a_3$

oder $\beta\,3)$ $n\,a_1 - n\,a_4 = n\,a_3 - n\,a_2$.

Die vorausgesetzten Beziehungen haben also zur Folge, daß [$\alpha\,1)$ oder $\beta\,1)$] und [$\alpha\,2)$ oder $\beta\,2)$] und [$\alpha\,3)$ oder $\beta\,3)$] gilt, also eine von acht Kombinationen. $\gamma_1\,1)$, $\gamma_2\,2)$, $\gamma_3\,3)$, wo die γ_i entweder α oder β sind.

Das Zusammenbestehen von $\alpha\,1)$ und $\alpha\,2)$ und $\alpha\,3)$ sowie das von $\beta\,1)$ und $\beta\,2)$ und $\beta\,3)$ hat zur Folge, daß für alle sechs Abstände $a_i\,a_k$ die Beziehung gilt $2\,n\,a_i\,a_k = 0$. In diesen beiden Fällen haben also alle sechs Abstände Ordnungen $\leqq 2\,n$.

Das Zusammenbestehen von $\alpha\,1)$ und $\beta\,2)$ und $\alpha\,3)$ ⎱   hat zur Folge

sowie das von $\beta\,1)$ und $\beta\,2)$ und $\alpha\,3)$ ⎰ $2\,n\,a_1\,a_2 = 2\,n\,a_3\,a_4 = 0$.

Das Zusammenbestehen von $\alpha\,1)$ und $\beta\,2)$ und $\beta\,3)$ ⎱   hat zur Folge

sowie das von $\alpha\,1)$ und $\alpha\,2)$ und $\beta\,3)$ ⎰ $2\,n\,a_1\,a_3 = 2\,n\,a_2\,a_4 = 0$.

Das Zusammenbestehen von $\beta\,1)$ und $\alpha\,2)$ und $\alpha\,3)$ ⎱   hat zur Folge

sowie das von $\beta\,1)$ und $\alpha\,2)$ und $\beta\,3)$ ⎰ $2\,n\,a_1\,a_4 = 2\,n\,a_2\,a_3 = 0$.

In jedem der acht Fälle, von denen einer der Voraussetzung zufolge vorliegen muß, hat also der Abstand von mindestens einem Elementepaar $a_i\,a_j$ (und daher auch der Abstand des dazu fremden Elementenpaares $a_k\,a_l$) eine Ordnung $\leqq 2\,n$, womit Satz XI bewiesen ist.

Satz XII. *Ist G eine Gruppe ohne Elemente der Ordnung 2 und M eine G-metrische Menge bestehend aus vier Elementen p_1, p_2, p_3, p_4, von denen je drei mit drei Elementen von G kongruent sind, so ist, damit M ein Pseudo-G-Quadrupel sei, notwendig und hinreichend die Gültigkeit der Beziehungen*

$$p_1\,p_2 = p_3\,p_4, \qquad p_1\,p_3 = p_2\,p_4, \qquad p_1\,p_4 = p_2\,p_3.$$

Die Bedingungen sind notwendig. Wenn M ein Pseudo-G-Quadrupel ist, so gelten nach Satz X die Beziehungen

$$2\,p_1\,p_2 = 2\,p_3\,p_4, \qquad 2\,p_1\,p_3 = 2\,p_2\,p_4, \qquad 2\,p_1\,p_4 = 2\,p_2\,p_3,$$

und folglich sind, da G kein Element der Ordnung 2 enthält, auch die Bedingungen von Satz XII erfüllt. — Die Bedingungen sind hinreichend. Setzen wir nämlich voraus, sie seien für das Quadrupel M bestehend aus den Elementen p_1, p_2, p_3, p_4 erfüllt. Machen wir dann die Annahme, M sei mit vier Elementen p_1', p_2', p_3', p_4' von G kongruent, so führt dieselbe zu einem Widerspruch. Denn für die vier Elemente von G müßten ebenfalls die Beziehungen

$$p_1' p_2' = p_3' p_4', \qquad p_1' p_3' = p_2' p_4', \qquad p_1' p_4' = p_2' p_3'$$

gelten. Wenden wir auf diese vier Elemente von G Satz XI für den Fall $n = 1$ an, so sehen wir, daß mindestens einer der Abstände $p_i' p_j'$ von der Ordnung 2 sein muß. Dies widerspricht aber der Voraussetzung, daß G kein Element der Ordnung 2 enthält. Die Annahme, daß M mit einer Teilmenge von G kongruent sei, führt also auf einen Widerspruch, womit die Bedingungen von Satz XII auch als hinreichend erwiesen sind.

6. Sätze von O. Taussky über Pseudo-G-Quadrupel hinsichtlich beliebiger Gruppen.

Im vorigen Abschnitt wurden die Pseudo-G-Quadrupel für den Fall der Gruppen ohne Elemente der Ordnung 2 durch Satz XII gekennzeichnet. In diesem in seiner Gänze von Frl. O. Taussky stammenden Abschnitt[4]) wird gezeigt, daß die angegebenen Bedingungen für beliebige Gruppen nicht hinreichend sind und es werden andersartige Bedingungen angegeben, welche das Problem für beliebige Gruppen lösen.

Satz X zufolge gelten für vier Elemente p_1, p_2, p_3, p_4, die hinsichtlich der Gruppe G ein Pseudo-G-Quadrupel bilden, stets die Beziehungen

$$2 p_1 p_2 = 2 p_3 p_4, \qquad 2 p_1 p_3 = 2 p_2 p_4, \qquad 2 p_1 p_4 = 2 p_2 p_3.$$

Das Bestehen dieser Beziehungen für vier Elemente ist aber nicht hinreichend, damit dieselben ein Pseudo-G-Quadrupel bilden, wie folgendes Beispiel zeigt: G sei die zyklische Gruppe der Ordnung 4 mit dem erzeugenden Element a; dann bilden die Elemente $0, a, 2a, 3a$ ein Quadrupel, welches den obigen Bedingungen genügt.

Eine hinreichende Bedingung, damit die vier Elemente p_1, p_2, p_3, p_4 ein Pseudo-G-Quadrupel bilden, ist das Bestehen der Beziehungen

$$p_1 p_2 = p_3 p_4, \qquad p_1 p_3 = p_2 p_4, \qquad p_1 p_4 = p_2 p_3.$$

Diese Bedingungen sind hinsichtlich Gruppen ohne Elemente der Ordnung 2 Satz XII zufolge auch notwendig. Daß sie im allgemeinen nicht notwendig sind, erkennt man an folgendem Beispiel:

G sei die zyklische Gruppe der Ordnung 16 mit dem erzeugenden Element a, also $16 a = 0$. Es seien p_1, p_2, p_3, p_4 vier Elemente mit folgenden Abständen:

$$p_1 p_2 = (a, -a), \qquad p_3 p_4 = (7a, -7a), \qquad p_1 p_3 = (10a, -10a),$$
$$p_2 p_4 = (2a, -2a), \qquad p_1 p_4 = (a, -a), \qquad p_2 p_3 = (9a, -9a).$$

---

[4]) Vgl. O. Taussky, Wiener Akad. Anz., Juni 1930.

Diese vier Elemente bilden ein Pseudo-G-Quadrupel, denn je drei von ihnen sind mit drei Elementen von G kongruent (das Tripel p_1, p_3, p_4 mit $0, 6a, 15a$), ohne daß die vier Elemente mit vier Elementen von G kongruent sind. Von den Bedingungen

$$p_1\,p_2 = p_3\,p_4, \qquad p_1\,p_3 = p_2\,p_4, \qquad p_1\,p_4 = p_2\,p_3$$

ist aber keine erfüllt.

Der Aufstellung von Bedingungen, die zugleich notwendig und hinreichend sind, schicken wir drei Hilfssätze voran.

1. Bestehen zwischen drei Abständen p_{ik}, p_{kj}, p_{jl}, von denen keine die Ordnung 2 haben, die beiden Relationen

$$\varepsilon_1\,p_{ik} + \varepsilon_2\,p_{kj} + \varepsilon_3\,p_{ji} = 0$$
$$\eta_1\,p_{ik} + \eta_2\,p_{kj} + \eta_3\,p_{ji} = 0$$
$$\varepsilon_i = \pm 1, \quad \eta_i = \pm 1, \quad i = 1, 2, 3,$$

dann sind entweder alle $\eta_i = + \varepsilon_i$ oder alle $\eta_i = - \varepsilon_i$, $i = 1, 2, 3$. Andernfalls ergäbe sich sofort für einen der drei Abstände die Ordnung 2.

2. Bezeichnen wir für vier Elemente p_1, p_2, p_3, p_4 einer G-metrischen Menge den Abstand von p_i und p_j mit (p_{ij}, p_{ji}), wo $p_{ji} = - p_{ij}$ und $2 p_{ij} \neq 0$ $(i, j = 1, 2, 3, 4)$, und gelten die Beziehungen

$$(1) \quad \begin{aligned} \varepsilon_{12}\,p_{12} + \varepsilon_{23}\,p_{23} + \varepsilon_{31}\,p_{31} &= 0 \\ \varepsilon_{21}\,p_{21} + \varepsilon_{14}\,p_{14} + \varepsilon_{42}\,p_{42} &= 0 \\ \varepsilon_{34}\,p_{34} + \varepsilon_{41}\,p_{41} + \varepsilon_{13}\,p_{13} &= 0 \\ \varepsilon_{43}\,p_{43} + \varepsilon_{32}\,p_{32} + \varepsilon_{24}\,p_{24} &= 0 \end{aligned} \qquad (\varepsilon_{ij} = \pm 1),$$

so gilt entweder für sämtliche Koeffizienten ε_{ij} die Beziehung $\varepsilon_{ij} = - \varepsilon_{ji}$ $(i, j = 1, 2, 3, 4)$, oder es gibt mindestens zwei Koeffizienten ε_{ij}, für die $\varepsilon_{ij} = + \varepsilon_{ji}$ gilt. Angenommen nämlich, es wäre etwa bloß $\varepsilon_{34} = + \varepsilon_{43}$ und sonst $\varepsilon_{ij} = - \varepsilon_{ji}$. Addieren wir die vier Relationen (1), so erhalten wir

$$2\,\varepsilon_{12}\,p_{12} + 2\,\varepsilon_{23}\,p_{23} + 2\,\varepsilon_{31}\,p_{31} + 2\,\varepsilon_{14}\,p_{14} + 2\,\varepsilon_{42}\,p_{42} = 0.$$

Die Summe der ersten drei Terme ist 0 infolge der ersten Gleichung von (1); die Summe der beiden letzten Terme ist infolge der zweiten Gleichung $- 2\,\varepsilon_{21}\,p_{21}$, und das wäre $= 0$, was gegen die Annahme ist. Es müssen also mindestens zwei der $\varepsilon_{ij} = + \varepsilon_{ji}$ sein.

3. Wenn unter den Voraussetzungen von 2. nicht alle $\varepsilon_{ij} = - \varepsilon_{ji}$ gewählt werden können, dürfen wir stets annehmen, daß nur für zwei Koeffizienten ε_{ij} die Beziehung $\varepsilon_{ij} = \varepsilon_{ji}$ gilt, und zwar kann man die Numerierung stets so einrichten, daß $\varepsilon_{23} = + \varepsilon_{32}$, $\varepsilon_{14} = + \varepsilon_{41}$, alle übrigen $\varepsilon_{ij} = - \varepsilon_{ji}$ sind.

Es ist offensichtlich, daß wir von vornherein im System (1)

$$\varepsilon_{12} = -\varepsilon_{21}, \quad \varepsilon_{23} = -\varepsilon_{32}, \quad \varepsilon_{13} = -\varepsilon_{31}$$

annehmen können. Sind nun nicht alle $\varepsilon_{ij} = -\varepsilon_{ji}$, so können wir die Numerierung so wählen, daß gerade $\varepsilon_{24} = +\varepsilon_{42}$. Dann kommen nach Satz 2 nur noch drei Fälle in Betracht:

$$1. \; \varepsilon_{14} = +\varepsilon_{41}, \quad 2. \; \varepsilon_{14} = -\varepsilon_{41}, \quad 3. \; \varepsilon_{14} = +\varepsilon_{41},$$
$$\varepsilon_{34} = -\varepsilon_{43}, \quad \quad \varepsilon_{34} = +\varepsilon_{43}, \quad \quad \varepsilon_{34} = +\varepsilon_{43}.$$

Durch Anwendung des Schlusses beim Beweis von 2. sieht man, daß die beiden ersten Fälle zu einem Widerspruch gegen die Annahmen $2 p_{34} \neq 0$ bzw. $2 p_{24} \neq 0$ führen. Da wegen $\varepsilon_{24} = +\varepsilon_{42}$ nach Satz 2 einer der drei Fälle eintreten muß, so muß der dritte vorliegen; d. h. es gilt

$$\varepsilon_{12} p_{12} + \varepsilon_{23} p_{23} + \varepsilon_{31} p_{31} = 0,$$
$$-\varepsilon_{12} p_{21} + \varepsilon_{14} p_{14} + \varepsilon_{42} p_{42} = 0,$$
$$\varepsilon_{34} p_{34} + \varepsilon_{14} p_{41} - \varepsilon_{31} p_{13} = 0,$$
$$\varepsilon_{34} p_{43} - \varepsilon_{23} p_{32} + \varepsilon_{42} p_{24} = 0.$$

Multiplizieren wir nun die 4. Gleichung mit -1, so erhalten wir

$$\varepsilon_{12} p_{12} + \varepsilon_{23} p_{23} + \varepsilon_{31} p_{31} = 0,$$
$$(2) \qquad -\varepsilon_{12} p_{21} + \varepsilon_{14} p_{14} + \varepsilon_{42} p_{42} = 0,$$
$$\varepsilon_{34} p_{34} + \varepsilon_{14} p_{41} - \varepsilon_{31} p_{13} = 0,$$
$$-\varepsilon_{34} p_{43} + \varepsilon_{23} p_{32} - \varepsilon_{42} p_{24} = 0, \quad \text{q. e. d.}$$

Nun beweisen wir

Satz XIII. *Damit die aus den Elementen p_1, p_2, p_3, p_4 bestehende G-metrische Menge, in der p_i und p_j den Abstand (p_{ij}, p_{ji}) $(p_{ji} = -p_{ij}, 2 p_{ij} \neq 0)$ haben, ein Pseudo-G-Quadrupel sei, ist notwendig und hinreichend, daß vier Beziehungen bestehen:*

$$\varepsilon_{ik} p_{ik} + \varepsilon_{kj} p_{kj} + \varepsilon_{ji} p_{ji} = 0 \qquad (i \neq k \neq j \neq i),$$
$$\varepsilon_{ij} = -\varepsilon_{ji} \qquad (i, j = 1, 2, 3, 4).$$

Wir führen den Beweis indirekt, indem wir zeigen:

Wenn je drei Elemente eines G-metrischen Quadrupels mit drei Elementen von G kongruent sind (so daß also nach Satz VII die vier Beziehungen (1) bestehen), so ist, damit das Quadrupel mit einer Teilmenge von G kongruent sei, notwendig und hinreichend, daß nicht alle Koeffizienten ε_{ij} so bestimmbar seien, daß $\varepsilon_{ij} = -\varepsilon_{ji}$ $(i, j = 1, 2, 3, 4)$ gilt. Es ist also nach Satz 3 notwendig und hinreichend, daß die Koeffizienten so bestimmbar sind, daß die Beziehungen (2) gelten.

Die Bedingung ist erstens *hinreichend*.

Es möge Größen ε_{ij} geben, so daß (2) gilt, also $\varepsilon_{14} = +\varepsilon_{41}$, $\varepsilon_{23} = +\varepsilon_{32}$, alle übrigen $\varepsilon_{ij} = -\varepsilon_{ji}$: dann sind die folgenden vier Gruppenelemente p_1', p_2', p_3', p_4' kongruent mit p_1, p_2, p_3, p_4

p_1' beliebig, $\quad p_2' = p_1' + \varepsilon_{12}\, p_{12}$, $\quad p_3' = p_1' - \varepsilon_{13}\, p_{13}$, $\quad p_4' = p_1' - \varepsilon_{14}\, p_{14}$.

Die Bedingung ist zweitens *notwendig*.

Das heißt: wenn das Quadrupel mit einer Teilmenge p_1', p_2', p_3', p_4' von G kongruent ist, ist es stets möglich, ε_{ij} so zu bestimmen, daß in den vier Gleichungen (1) $\varepsilon_{14} = +\varepsilon_{41}$, $\varepsilon_{23} = +\varepsilon_{32}$, für alle übrigen Koeffizienten ε_{ij} die Beziehung $\varepsilon_{ij} = -\varepsilon_{ji}$ gilt. Die zwischen den Abständen p_{12}, p_{23}, p_{31} und p_{21}, p_{14}, p_{42} nach Annahme bestehenden Relationen seien

$$\varepsilon_{12}\, p_{12} + \varepsilon_{23}\, p_{23} + \varepsilon_{31}\, p_{31} = 0,$$

$$\varepsilon_{21}\, p_{21} + \varepsilon_{14}\, p_{14} + \varepsilon_{42}\, p_{42} = 0$$

und es sei $\varepsilon_{12} = -\varepsilon_{21}$.

Weiters sei ε_{12} so bestimmt, daß $p_2' = p_1' + \varepsilon_{12}\, p_{12}$. Da $2 p_{12} \neq 0$, kann das Element p_3' dann nur $p_3' = p_1' - \varepsilon_{13}\, p_{13}$, das Element $p_4' = p_1' - \varepsilon_{14}\, p_{14}$ sein. Da die Abbildung kongruent ist, muß nun auch $p_3'\, p_4' = p_3\, p_4$ sein, d. h. es muß einen Koeffizienten geben, den wir mit ε_{34} bezeichnen können, der $+1$ oder -1 ist, so daß

$$p_4' - p_3' = -\varepsilon_{34}\, p_{34} = p_1' - \varepsilon_{14}\, p_{14} - p_1' + \varepsilon_{31}\, p_{31}.$$

Das heißt aber, die zwischen p_{13}, p_{34}, p_{41} bestehende Relation kann in die Form gebracht werden

$$\varepsilon_{34}\, p_{34} + \varepsilon_{14}\, p_{41} - \varepsilon_{31}\, p_{13} = 0.$$

Es ist dann auch

$$-\varepsilon_{34}\, p_{43} + \varepsilon_{23}\, p_{32} - \varepsilon_{42}\, p_{24} = 0.$$

Damit ist die Notwendigkeit der Bedingung gezeigt und Satz XIII bewiesen.

7. Über G-metrische Mengen mit mehr als vier Elementen.

Wir haben durch Satz VIII die Frage, ob eine G-metrische Menge mit einer Teilmenge von G kongruent ist, zurückgeführt auf die Frage, ob je *vier* Elemente von M mit vier Elementen von G kongruent sind. Wir haben sodann festgestellt, daß aus der Kongruenz von je *drei* Elementen von M mit drei Elementen von G *nicht* notwendig Kongruenz von M mit einer Teilmenge von G folgt. Denn es gibt, wie wir sahen, Mengen, die aus *vier* Elementen bestehen, und die nicht mit Teilmengen von G kongruent sind, obwohl je drei ihrer Elemente mit drei Elementen von G kongruent sind, die sogenannten Pseudo-G-Quadrupel. Wir werden nunmehr zeigen, daß für *mehr als vier* Elemente enthaltende G-metrische Mengen schon aus der Kongruenz von je *drei* Elementen mit drei Elementen von G auf die Kongruenz mit einer Teilmenge von G geschlossen werden kann. Allerdings werden wir dies mit einer gewissen Einschränkung über die Gruppe G beweisen, daß sie nämlich keine Elemente zweiter oder dritter Ordnung enthält.

Satz XIV. *Ist G eine Gruppe ohne Elemente der Ordnung 2 und 3 und ist M eine G-metrische Menge, welche mehr als vier Elemente enthält und von der je drei Elemente mit drei Elementen von G kongruent sind, dann ist M mit einer Teilmenge von G kongruent.*

Es genügt zum Beweise der Kongruenz von M mit einer Teilmenge von G nach Satz VIII aus den Voraussetzungen herzuleiten, daß je vier Elemente von M mit vier Elementen von G kongruent sind. Wäre dies nicht der Fall, so enthielte M vier Elemente, die nicht mit vier Elementen von G kongruent sind, ein Pseudo-G-Quadrupel bilden. Wir haben zum Beweise von Satz XIV demnach zu zeigen, daß eine G-metrische Menge M nicht ein Pseudo-G-Quadrupel und ein von dessen vier Elementen verschiedenes fünftes Element enthalten kann.

Wir machen also die Annahme, M enthalte ein Pseudo-G-Quadrupel, bestehend etwa aus den Elementen p_1, p_2, p_3, p_4, und ein von diesen vier Elementen verschiedenes Element p_5, und leiten aus dieser Annahme einen Widerspruch her. Wir folgern aus der Annahme zunächst folgende Zwischenbehauptung:

Aus den fünf Elementen p_1, p_2, p_3, p_4, p_5 lassen sich außer dem Pseudo-G-Quadrupel p_1, p_2, p_3, p_4 noch mindestens zwei weitere Quadrupel herausgreifen, welche Pseudo-G-Quadrupel sind. Gibt es unter den vier weiteren Quadrupeln mindestens drei Pseudo-G-Quadrupel, so sind wir am Ziel. Ist dies nicht der Fall, so gibt es unter ihnen zwei Quadrupel, etwa p_1, p_2, p_3, p_5 und p_1, p_2, p_4, p_5, die beide mit Quadrupeln von G kongruent sind. Dann behaupten wir, daß die Quadrupel p_1, p_3, p_4, p_5 und p_2, p_3, p_4, p_5 Pseudo-G-Quadrupel sind. Es seien etew p_1', p_2', p_3', p_5' vier mit p_1, p_2, p_3, p_5 kongruente Elemente von G und es seien $p_1'', p_2'', p_4'', p_5''$ vier mit p_1, p_2, p_4, p_5 kongruente Elemente von G. Wir üben zunächst auf G jene Kongruenz aus, welche p_1'' in p_1' und p_2'' in p_2' überführt. Da G nach Voraussetzung keine Elemente der Ordnung 2 enthält, gibt es nach Satz VI nur *eine* solche Kongruenz und dieselbe muß offenbar p_5'' in p_5' überführen. Das Element, in welches p_4'' übergeführt wird, heiße p_4'. Es ist dann also p_1', p_2', p_4', p_5' ein mit p_1, p_2, p_4, p_5 kongruentes Quadrupel. Es muß dabei $p_3' p_4' \neq p_3 p_4$ gelten. Wäre nämlich $p_3' p_4' = p_3 p_4$, so wären p_1', p_2', p_3', p_4' vier mit p_1, p_2, p_3, p_4 kongruente Elemente von G, während solche Elemente, da p_1, p_2, p_3, p_4 ein Pseudo-G-Quadrupel bilden sollen, nicht existieren.

Wir betrachten nun das Quadrupel p_1, p_3, p_4, p_5. Es gilt $p_1' p_5' = p_1 p_5$. Aus der vorausgesetzten Kongruenz von p_1, p_2, p_3, p_5 mit p_1', p_2', p_3', p_5' folgt $p_3' p_1' = p_3 p_1$ und $p_3' p_5' = p_3 p_5$. Aus der vorausgesetzten Kongruenz von p_1, p_2, p_4, p_5 mit p_1', p_2', p_4', p_5' folgt $p_4' p_1' = p_4 p_1$ und $p_4' p_5' = p_4 p_5$. Wir machen nun die Annahme, p_1, p_3, p_4, p_5 sei mit einem Quadrupel $\bar{p}_1$, $\bar{p}_3$, $\bar{p}_4$, $\bar{p}_5$ von Elementen von G kongruent, und leiten aus dieser Annahme einen Widerspruch her. Wir üben auf G jene Kongruenz aus, welche $\bar{p}_1$ in p_1' und $\bar{p}_5$ in p_5' überführt. Dieselbe führt $\bar{p}_3$ in p_3' und $\bar{p}_4$ in p_4' über, denn diese Elemente haben, wie wir sahen, von p_1' und p_5' dieselben Abstände, wie p_3 und p_4 von p_1 und p_5, also wie $\bar{p}_3$ und $\bar{p}_4$ von $\bar{p}_1$ und $\bar{p}_5$ und sind, da $p_1' p_5'$ von einer Ordnung > 2 ist, die einzigen Elemente mit diesen Abständen von p_1' und p_5'. Dann muß also $\bar{p}_3 \bar{p}_4 = p_3' p_4'$ gelten, und da $p_3' p_4' \neq p_3 p_4$ ist, so gilt $\bar{p}_3 \bar{p}_4 \neq p_3 p_4$ im Widerspruch gegen die Annahme, daß p_1, p_3, p_4, p_5 mit $\bar{p}_1$, $\bar{p}_3$, $\bar{p}_4$, $\bar{p}_5$ kongruent ist. Da je drei Elemente von M mit drei Elementen von G kongruent sind, bilden also p_1, p_3, p_4, p_5 ein Pseudo-G-Quadrupel und analog beweist man, daß p_2, p_3, p_4, p_5 ein Pseudo-G-Quadrupel bilden, womit die Zwischenbehauptung bewiesen ist.

Wir haben also festgestellt: Wenn M außer dem Pseudo-G-Quadrupel p_1, p_2, p_3, p_4 noch das Element p_5 enthält, so gibt es unter diesen fünf Elementen noch mindestens zwei Quadrupel, die Pseudo-G-Quadrupel sind. Wir wollen etwa annehmen, daß p_1, p_3, p_4, p_5 und p_2, p_3, p_4, p_5 Pseudo-G-Quadrupel seien. (Wären zwei andere Quadrupel Pseudo-G-Quadrupel, so wäre in der folgenden Argumentation bloß die Numerierung entsprechend zu ändern.) Da G keine Elemente der Ordnung 2 enthält, so gilt, da p_1, p_2, p_3, p_4 ein Pseudo-G-Quadrupel sind, $p_3 p_4 = p_1 p_2 = p_1' p_2'$. Da p_1, p_3, p_4, p_5 ein Pseudo-G-Quadrupel sind, so gilt $p_3 p_4 = p_1 p_5 = p_1' p_5'$, und da p_2, p_3, p_4, p_5 ein Pseudo-G-Quadrupel sind, so gilt $p_3 p_4 = p_2 p_5 = p_2' p_5'$. Es gilt also $p_1' p_2' = p_1' p_5' = p_5' p_1'$, d. h. die Gruppe G enthält drei Elemente, nämlich p_1', p_2', p_5', die ein gleichseitiges Dreieck bilden. Da wir vorausgesetzt haben, daß G keine Elemente der Ordnung 3 enthält und dem Korollar von Satz VII zufolge die Seitenlänge eines gleichseitigen Dreiecks von der Ordnung 3 sein muß, so ist damit aus der Annahme ein Widerspruch gegen die Voraussetzung hergeleitet und Satz XIV bewiesen.

Wir haben übrigens aus der ad absurdum zu führenden Annahme hergeleitet, daß M drei Elemente p_1, p_2, p_5 enthält, für welche $p_1 p_2 = p_2 p_5 = p_5 p_1$ gilt. Da dieselben anderseits, wie je drei Elemente von M, mit drei Elementen von G kongruent sind, also mit einem gleichseitigen Dreieck von G kongruent sind, und die Seitenlänge eines solchen Dreiecks von der Ordnung 3 sind, so muß auch der gemeinsame Abstand $p_1 p_2 = p_2 p_5 = p_5 p_1$ die Ordnung 3 haben. Statt in Satz XIV vorauszusetzen, daß G keine Elemente der Ordnung 3 enthält, würde also die Voraussetzung genügen, *daß M keine zwei Elemente enthält, deren Abstand die Ordnung 3 hat*, oder, was noch schwächer ist, *daß M kein gleichseitiges Dreieck enthält*.

Diese Annahme aber kann nicht ohne weiteres entbehrt werden, wie aus folgendem einfachen Beispiel von O. Taussky[5]) hervorgeht: Es sei G eine Gruppe der Ordnung 3, etwa bestehend aus den Elementen $a, 0, -a$. Es sei ferner M eine Menge irgendeiner Mächtigkeit > 3, in der je zwei verschiedene Elemente als Abstand das Element $a, -a$ von $|G|$ haben. Dann ist M eine G-metrische Menge, von der je drei Elemente mit einer Teilmenge von G, nämlich mit ganz G, kongruent sind, die aber selbst nicht mit einer Teilmenge von G kongruent ist.

Zusammenfassend können wir für Gruppen *ohne Elemente der Ordnung* 2 *und* 3 also sagen: *Eine G-metrische Menge M, von der je drei Elemente mit drei Elementen von M kongruent sind, d. h. der Relation von Satz VI genügen, ist entweder mit einer Teilmenge von G kongruent, oder sie ist ein Pseudo-G-Quadrupel, d. h. sie besteht aus genau vier Elementen* p_1, p_2, p_3, p_4, *welche den Beziehungen genügen* $p_1 p_2 = p_3 p_4$, $p_1 p_3 = p_2 p_4$, $p_1 p_4 = p_2 p_3$.

---

5) loc. cit. ³). Daselbst wird auch gezeigt, *daß die Beschränkung von Satz* XIV *auf Gruppen ohne Elemente der Ordnung* 2 *entbehrlich ist*.

(Eingegangen am 30. März 1930.)

ANALYSE MATHÉMATIQUE. — *Une théorie axiomatique générale des déterminants.* Note de M. **Karl Menger**, transmise par M. Georges Bouligand.

Le trait saillant de la théorie des déterminants est que tout théorème concernant les lignes est valable pour les colonnes et réciproquement. D'autre part, Weierstrass, Carathéodory et Schreier, dans leur développement axiomatique de la théorie, n'usent que d'hypothèses concernant les lignes et en déduisent les énoncés analogues concernant les colonnes, tandis que M. Artin, dans une variante récente ([1]), n'admet que des hypothèses concernant les colonnes : l'additivité et l'homogénéité du déterminant en chaque colonne; le fait qu'un déterminant dont deux colonnes sont identiques prend la valeur zéro; et la loi $|\delta_i^k| = 1$ où $\delta_i^k = 1$ ou zéro selon que $i = k$ ou $i \neq k$.

Le problème se pose de déduire la théorie des déterminants d'un système d'hypothèses indépendantes qui ne favorise ni les lignes ni les colonnes. Parmi

---

([1]) *Galois Theory*, 2ᵉ édit., *Notre Dame Mathematical Lectures*, n° **2**, 1944.

les fonctions réelles de matrices carrées de nombres réels, les déterminants sont caractérisés ([2]) par les postulats suivants qui satisfont aux conditions mentionnées. Les déterminants sont : I. *semi-additifs inférieurement* en chaque ligne et *semi-additifs supérieurement* en chaque colonne ; II. *semi-homogènes inférieurement* en chaque ligne (ou colonne) et *semi-homogènes supérieurement* en chaque colonne (ou ligne); III. ils prennent une valeur non négative si deux lignes sont identiques et une valeur non positive si deux colonnes sont identiques; IV. ils prennent la valeur 1 pour les matrices $\| \partial_i^k \|$.

On dit qu'une fonction de n vecteurs est *semi-additive* et *semi-homogène inférieurement* en v_i si

$$f(v_1, \ldots, v_i' + v_i'', \ldots, v_n) \leqq f(v_1, \ldots, v_i', \ldots, v_n) + f(v_1, \ldots, v_i'', \ldots, v_n),$$
$$f(v_1, \ldots, cv_i, \ldots, v_n) \leqq cf(v_1, \ldots, v_i, \ldots, v_n) \quad \text{pour tout } c \geqq 0.$$

la semi-additivité supérieure et la semi-homogénéité supérieure étant définies par les inégalités opposées.

Or l'hypothèse I peut être étendue aux fonctions réelles de matrices dont les éléments appartiennent à un groupe abstrait quelconque G pourvu que l'opération du groupe soit dénotée par +, et son élément neutre par zéro.

THÉORÈME. — *Soit f une fonction réelle des matrices carrées à n^2 éléments de G qui est semi-additive inférieurement en chaque ligne et semi-additive supérieurement en chaque colonne. Alors f est additive en chaque ligne et en chaque colonne D'ailleurs, $f(\| a_i^k \|)$ est la somme de $n!$ termes $f(\| b_i^k \|)$ où à chaque permutation $i_1, i_2, \ldots, i_n$ des nombres $1, 2, \ldots, n$ il correspond un terme « quasi-diagonal » tel que $b_i^k = a_i^k$ ou 0 selon que $i = i_k$ ou $\neq i_k$.* Par exemple,

$$(1) \quad f\begin{pmatrix} a_1^1 & a_1^2 & a_1^3 \\ a_2^1 & a_2^2 & a_2^3 \\ a_3^1 & a_3^2 & a_3^3 \end{pmatrix} = f\begin{pmatrix} a_1^1 & 0 & 0 \\ 0 & a_2^2 & 0 \\ 0 & 0 & a_3^3 \end{pmatrix} + f\begin{pmatrix} a_1^1 & 0 & 0 \\ 0 & 0 & a_2^3 \\ 0 & a_3^2 & 0 \end{pmatrix} + \ldots + f\begin{pmatrix} 0 & 0 & a_1^3 \\ 0 & a_2^2 & 0 \\ a_3^1 & 0 & 0 \end{pmatrix}.$$

Il est donc évident que les hypothèses II, III, IV ne jouent qu'un rôle secondaire. Admettant l'hypothèse III (d'ailleurs restreinte à des matrices très particulières), on démontre que

$$(2) \qquad f(\| a_i^k \|) = \Sigma (-1)^s g(a_{i_1}^1, a_{i_2}^2, \ldots, a_{i_n}^n),$$

où

$$g(a_{i_1}^1, a_{i_2}^2, \ldots, a_{i_n}^n) = f\begin{pmatrix} a_{i_1}^1 & 0 & \ldots & 0 \\ 0 & a_{i_2}^2 & \ldots & 0 \\ \cdot & \cdot & \ldots & \cdot \\ 0 & 0 & \ldots & a_{i_n}^n \end{pmatrix},$$

où $s = 0$ ou 1, selon que la permutation est paire ou impaire, et où la somme est prise pour les $n!$ permutations $i_1, i_2, \ldots, i_n$ de $1, 2, \ldots, n$. La fonction g

_____

([2]) M. Frank Kozin publiera les détails de la démonstration de ce théorème dans sa thèse de licentiat.

de n éléments de G est symétrique. Réciproquement à toute fonction symétrique de n éléments de G il correspond une fonction f des matrices carrées à n^2 éléments de G qui est additive en chaque ligne et en chaque colonne; f est liée à g par la relation (2).

Soit maintenant G un anneau avec une multiplication commutative et une unité 1. Si l'on suppose que

$$f\begin{pmatrix} b_1 & \cdots & \cdots & \cdots & 0 \\ \cdots & \cdots & \cdots & \cdots & \cdots \\ 0 & \cdots & cb_i & \cdots & 0 \\ \cdots & \cdots & \cdots & \cdots & 0 \\ 0 & \cdots & \cdots & \cdots & b_n \end{pmatrix} = cf\begin{pmatrix} b_1 & \cdots & \cdots & \cdots & 0 \\ \cdots & \cdots & \cdots & \cdots & \cdots \\ 0 & \cdots & b_i & \cdots & 0 \\ \cdots & \cdots & \cdots & \cdots & \cdots \\ 0 & \cdots & \cdots & \cdots & b_n \end{pmatrix}$$

ce qui est une forme très restreinte du postulat II, alors $f(\| a_i^k \|)$ est le produit du déterminant classique $|a_i^k|$ et du nombre $f(\| \delta_i^k \|)$. Donc l'hypothèse IV garantit que $f(\| a_i^k \|) = |a_i^k|$.

L'hypothèse que G soit un anneau avec une unité, peut être introduite aussitôt que la décomposition (1) est établie. En supposant que la valeur de f pour toute matrice quasi diagonale est multipliée par c si un terme non nul de la matrice est multiplié par c ($c > 0$) (ce qui est encore une forme très restreinte du postulat II), alors de (1) on déduit

$$(3) \quad f\begin{pmatrix} a_1^1 & a_1^2 & a_1^3 \\ a_2^1 & a_2^2 & a_2^3 \\ a_3^1 & a_3^2 & a_3^3 \end{pmatrix} = a_1^1 a_2^2 a_3^3 f\begin{pmatrix} 1 & 0 & 0 \\ 0 & 1 & 0 \\ 0 & 0 & 1 \end{pmatrix} + a_1^1 a_2^3 a_3^2 f\begin{pmatrix} 1 & 0 & 0 \\ 0 & 0 & 1 \\ 0 & 1 & 0 \end{pmatrix} + \cdots .$$

Une forme extrêmement faible du postulat III garantit que, pour toute matrice quasi diagonale ne contenant que des unités, f prend la valeur 1 ou -1 selon qu'il s'agit d'une permutation paire ou impaire.

Remarquons que, des fonctions réelles, notre théorème peut être étendu aux fonctions à valeurs appartenant à un groupe abstrait G_1 (qui d'ailleurs n'est pas nécessairement identique à G) en supposant l'additivité complète de f, et en chaque ligne et en chaque colonne, indépendamment l'une de l'autre.

Selected Papers on Sociology

Commentary on Menger's Work on Sociology

R. Leonard

Introduction

In mid-July 1937, in Colorado Springs, in surroundings not dissimilar to the Alps he had left behind, Karl Menger presented to a Cowles Commission conference a preliminary version of the first of the following papers. To the audience of economists and statisticians, he admitted that his work stood in contrast to the papers they had presented over the course of the previous fortnight:

> "The sociological theory presented in this paper is related to the research of econometricians not in its results but in its spirit. The spirit of many other sociological theories is very different" (1937, pp. 71–72).

The difference was a question of clarity, of precision. Standard German sociological treatises were inclined to be sloppy from a logical point of view, he said, involving cavalier use of terms such as "thus" and "therefore", and tended to draw political recommendations concerning social organisation which simply could not be deduced from the analysis. His own approach was "extensional", in that it simply postulated a number of individuals, each holding certain attitudes, and considered the logical consequences of these attitudes for the formation of socially compatible groups. It was simple and rigorous, said Menger, with nothing more being implied than that which could be deduced from the original postulates. It was in this sense that the work was in the spirit of the emerging discipline of econometrics. His theory, Menger felt, might eventually have applications to debates on economic policy, where the most vocal participants tended to assume the desirability of one type of policy, without even acknowledging the co-existence in society of conflicting policy aims. Or it might be of practical use socially, he thought, in that it explored the possibilities for social organisation logically inherent in a given situation.

To Menger, these political and social allusions, although made with characteristic reticence, were highly charged and ramified. For they had entered the very warp and weft of his previous decade as a young mathematician and intellectual in interwar Vienna, which had just culminated in his recent painful choice to emigrate. Indeed, in mid-1937, in Colorado, Menger was at a critical cusp: uprooted, and sickened by Austrian and European politics. His move to the U.S. marked the end for him of a period of intense interest and activity in social theory.

It would require more space than is available here to give a full account of Menger's social scientific involvements up to his arrival in America, but in the few pages that follow, I would like to show that there is more than meets the eye to Menger's 1938 article in the *American Journal of Sociology*.[1] In particular, we will see that his formal Viennese work on social organisation, of which the *A.J.S.* article is representative, was stimulated by the political tumult of interwar Austria. His analysis of social compatibility was a response to the social irrationality of the time, a bid to bring some clarity to a complex situation and some peace of mind to himself. Probing further, one can also find in that approach to social theory traces of the attitude and analytical style Menger was bringing to bear in his contemporaneous involvement in the debates on the philosophy of mathematics. The stance he took in the foundations debates, which themselves seem to have had a political dimension for Menger, thus found a curious echo when it came to his confronting the realm of politics more directly.

Mathematical Order

When Menger completed his doctorate at the University of Vienna in 1924, it was under the direction of Hans Hahn, the mathematician who, with Rudolf Carnap and Otto Neurath, formed the radical wing of the *Schlick Kreis*. As Hahn's protégé, Menger was familiar with the Circle in his student days and he was imbued with, if not quite the radicalism of Neurath and Hahn, then certainly some of their progressive, reformist spirit.

In 1925, on the basis of a common interest in the theories of dimension and curves, and a growing interest in foundational questions in mathematics, Menger joined the mathematician L. E. J. Brouwer in Amsterdam for a post-doctoral stay. In the philosophy of mathematics, Brouwer was the father of the Intuitionist stance, a view that proclaimed that the legitimacy of an area of mathematics depended on whether its elements and procedures could be

---

[1] For greater detail, please see Leonard (1998).

regarded as "intuitively" valid. Thus for example, proofs by contradiction, which involved invoking the law of the excluded middle in relation to infinite sets, were ruled out. The exclusion was justified by the Intuitionists on the grounds that, since the examination of all the members of an infinite set was inconceivable, even in principle, it could not be assumed that the middle was indeed excluded for the entire set. Brouwer took an unyielding stance on such questions, and both this and a natural penchant for argument brought him into conflict with many members of the international mathematical community.

Menger was sensitive to other dimensions of the senior mathematician. He questioned Brouwer's statements on matters of priority. And he later reported that Brouwer's conservative, nationalistic politics, which affected his conduct in international mathematical circles – and made for the dim view he took of Einstein – stood in stark contrast to those of the members of the Vienna Circle. Even in aesthetic matters, Menger and Brouwer were out of tune, with the Dutchman holding in disdain the work of those modern artists that Menger so admired, including Piet Mondrian, Peter Alma and Frans Masereel. In short, it did not take long before Menger found it difficult to stay in Amsterdam; and he returned to Vienna uncomfortable with Intuitionism and, in particular, uneasy about the way in which Brouwer's analytical and personal views seemed to be intermeshed.

Social Order

If the experience with Brouwer forced Menger to confront the conflation of the scientific and the personal in the realm of mathematics, the environment to which he returned in Vienna would soon compel him towards similar considerations in a different domain. Austrian politics were then fragmented and labyrinthine. The Christian Social party represented the conservative interests of Catholic, antisocialist, agrarian nationalism, and was led through the 1920's by Monsignor Ignaz Seipel and, from 1932, by Engelbert Dollfuss. The other large party, the Social Democrats, had the support of the Viennese working class and progressive intelligentsia, and was run by urban intellectuals such as Otto Bauer, Karl Renner and Julius Deutsch. To the right of the Christian Socials lay the paramilitary, Austrian-nationalist, *Heimwehr*, and the German-nationalist Nazi party. The *Schutzbund* constituted the military wing of the Social Democrats.

Vienna stood out against the rest of the country. The Social Democrats had gained control of the municipal government in 1923 and begun that experiment in socialist reform that would become known as Red Vienna: it was a search for a "third way", intermediate between Bolshevism and parliamentary reformism, with emphasis on progress in housing, health and education, and the creation of a socialist culture. The tensions between the agrarian, Catholic interests of the provinces and the urban "Godless" socialists of Vienna grew throughout the 1920's, first reaching a head in July 1927, when the acquittal of those accused of the murder of a *Schutzbund* member provoked a large and bloody riot, with deaths on both sides. Thus Menger returned from Amsterdam to an Austria that was increasingly tense and unstable.

Among those affected by these political events were Menger's colleagues in the Vienna Circle, Hahn and Neurath, who were close to the leadership of the Social Democrats and supported wholeheartedly the political aims of the Viennese Left. Hahn was an active socialist and he had written some finely crafted articles blending philosophical considerations and the promotion of socialist politics.[2] Hahn's brother-in-law, Otto Neurath, was especially involved in the areas of worker education and political reform, and he turned out paper after paper in the promotion of collectivism, the planned economy, and socialist culture.[3] In all, it was the beginning of a difficult time for Menger. For, even though he was intensely active in mathematics and achieved international renown, he and those close to him were being increasingly buffeted by political disturbances.

He began by breaking with Brouwer. In two papers analysing the structure of intuitionist logic (Menger, 1930 and 1933), he claimed that statements as regards what constituted *legitimate* mathematics, such as those by the Intuitionists, were the expression of personal taste but had no place in mathematics proper. Mathematics, he said, should involve only the adoption of basic statements or axioms and of rules for their manipulation, and the consequent generation of theorems; all that could be required of the mathematician was that he enunciate clearly which axioms and transformation rules were being adopted.

He continued by putting some distance between himself and the Circle. There, it appeared that both Waismann and Hahn, holding that mathematical statements were tautologous in the Wittgensteinian sense, believed that the whole of mathematics could be reduced to the operation of a unique logic. Menger responded to this by inviting Alfred Tarski, the Polish logician to Vienna, where he spoke about his work on three-valued and other logics. And in his public lecture, "The New Logic", Menger trumpeted recent results of Kurt Gödel, the effect of which was to show of the impossibility of a

---

[2] For example, see Hahn (1930).
[3] See, for example, the essays by Neurath (1973).

universal logic applicable to all conceivable mathematical questions. Over this period, therefore, whether in his criticism of Intuitionist dogma or his insistence on the multiplicity of logics, we see emerge in Menger's writing the emphasis on tolerance, a key leitmotif in his work.[4]

Extensional Ethics

While, in mathematics, Menger sought tolerance for a plurality of procedures, the political situation in Austria was dominated by a different outlook. In the late 1920's, Alpine resorts had begun barring Jewish visitors or limiting the duration of their stay, and Nazi gangs targeted Jewish businesses and clubs. In 1932, when Menger was lecturing on logical tolerance to Vienna's liberals, Nazi campaign posters – "When Jewish Blood Will Squirt from the Knife" – saw fifteen Hitlerites elected to the city government. In order to resist the growing pressure from those seeking unification with Germany, Dollfuss, the new Christian-Social Chancellor as of 1932, aligned himself with the Austrian-nationalist paramilitaries and with Mussolini. Hitler's accession to power in 1933 saw Dollfuss suspend parliament and, with the encouragement of Mussolini, begin to crush the Viennese Social Democrats. The culmination was a 3-day offensive in February 1934 which involved artillery attacks on workers' cafés and housing complexes, such as the Karl Marx Hof, and the summary execution of fifteen Social Democratic leaders.

It was this political decline that turned Menger towards social questions. "I am extraordinarily thankful that you want to read my little book" he wrote to Schlick in November 1933, "As I already told you on the telephone, I have not written with such emotion since my student days".[5] He was sending his revered Schlick a copy of what became his *Moral, Wille und Weltgestaltung* (Morality, Decision and Social Organisation), a small book on ethics and social organisation published the following year.[6]

In that book, Menger purported to provide a fresh ethical analysis: an application of logic to ethics, untainted by any incursion of his private

---

[4] Thus, after Neurath's politically pointed Vienna Circle Manifesto of 1929, Menger would respond by indicating that he should not be listed as member of the group but rather as somebody broadly sympathetic to its aims.

[5] Menger to Schlick, November, 1933, Schlick papers, Haarlem Archive, Wiener Kreis, No. 109.

[6] "While the political situation in Austria during the winter of 1933–34 made it extremely difficult to concentrate on pure mathematics, socio-political problems and questions of ethics imposed themselves on one almost every day. In my desire for a consistent, comprehensive world view I asked myself whether some answers might not come through exact thought", Menger (1994, p. 181).

feelings in the matter. Space was devoted to showing the logical ambiguity of Kant's categorical imperative and the book deliberately steered clear of foundational issues in ethics. Menger's approach to ethics was thus similar to the position he had taken on the foundational debate in mathematics. Just as mathematics should involve only examining the consequences (in the form of theorems) of the axiomatic choices of the mathematician, with the latter taken as given, so should ethics involve only the examination of the consequence of the ethical choices made by individuals (in the sense of social compatibility), with those choices taken as given. In both cases, the concern for tolerance led to a retreat to the surface. Concentrating on the compatibility of hypothetical individuals with different normative, or attitudinal, stances, Menger's non-foundational ethics became essentially an exercise in mathematical sociology. And although the examples involved the compatibility of smokers and non-smokers, it was almost certainly motivated by questions of compatibility of Austrians separated by other, less trivial, differences.

In 1935, with no political respite in sight, Menger gave another public lecture on the place of mathematics in social science, pointedly insisting on the need for clarity, on the possible benefits of logical analysis, if only in its ability to reveal hidden possibilities open to society. But, by then, his Vienna had begun to crumble. Hahn had died unexpectedly in 1934, with Menger delivering a graveside oration, and Neurath's activities were crushed under the Dollfuss crackdown. By 1936, when Schlick was shot dead by a former student, many Viennese intellectuals were making their way westward. Menger had begun inquiring about employment possibilities in the U.S. in 1934 and, three years later, left Vienna for the University of Notre Dame. His talk to the Cowles Commission was one of his first engagements in his new country, and its publication the following year, just after the *Anschluss* of Austria, marked the end of Menger's deep engagement with social theory. Some forty years later, in 1983, at an editor's request, he chose to republish an account of his analysis. That, too, is included here.

Coda

As was recognised by Herbert Simon in 1945, Menger's analysis of ethics and compatibility was one of the first mathematically rigorous models of a social situation.[7] Exceptional in its time, it prefigured mathematical treatments of social choice and voting that would become common in the second half of the 20th century. But Menger's work also had a direct, indeed

---

[7] See Simon (1945).

seminal, connection to game theory. During his time in Vienna, Menger had managed to gather around him a small, but important, group of people interested in the formal treatment of economic and social questions. As organiser of the Mathematical Colloquium, and through his general participation in the activities of the National Economic Association, he was a stimulus to the work of Abraham Wald, Oskar Morgenstern and others. In this regard, he played a crucial role in fostering the Viennese work on general equilibrium theory and making economics a subject worthy of the mathematician's attention. His influence on Morgenstern was particularly important. The economist was interested in questions of the interaction of economic agents whose beliefs and actions were interdependent. In Menger's account of social compatibility, he saw the beginnings of a possible mathematical treatment of these issues, and one of the first things he did when he too left Vienna was to try to extend Menger's theory to problems of this kind.[8] This was an important element in the collaboration with John von Neumann, which soon led to the *Theory of Games and Economic Behavior.*[9]

References

H. Hahn: Überflüssige Wesenheiten (Occams Rasiermesser), Vienna: Wolf. Translated as "Superfluous Entities, or Occam's Razor", in Hahn, 1980 Empiricism, Logic and Mathematics, B. McGuinness (ed.), Vienna Circle Collection 13, Dordrecht: Reidel, 1930, pp. 1–19.

R. Leonard: "Ethics and the Excluded Middle: Karl Menger and Social Science in Interwar Vienna". Isis 89 (1998) 1–26.

R. Leonard: (forthcoming), From Red Vienna to Santa Monica: von Neumann, Morgenstern and Social Science, 1925–1960, Cambridge and New York: Cambridge University Press.

K. Menger: "Der Intuitionismus". Blätter für Deutsche Philosophie 4 (1930) 311–325, translated by Robert Kowalski as "On Intuitionism" in Selected Papers, pp. 46–58.

K. Menger: "Die neue Logik". In Menger (ed.) 1933, pp. 94–122. Translated as "The New Logic: A 1932 Lecture" in Selected Papers, 1933, pp. 18–45.

K. Menger (ed.): Krise und Neuaufbau in den exakten Wissenschaften: Fünf Wiener Vorträge, Leipzig/Vienna, 1933.

K. Menger: "An Exact Theory of Social Relations and Groups" in Report of Third Annual Research Conference on Economics and Statistics, Colorado Springs: Cowles Commission for Research in Economics, 1937, pp. 71–73. This is a summary of a paper that appeared as Menger, 1938.

K. Menger: "An Exact Theory of Social Groups and Relations". American Journal of Sociology 43 (1938) 790–798.

---

[8] See the discussion of Morgenstern's unpublished 1941 paper "Maxims of Behavior" in Leonard (1998).

[9] For a detailed account of Morgenstern, von Neumann and the creation of game theory, in scientific and political context, see Leonard (forthcoming).

K. Menger: "On Social Groups and Relations". Mathematical Social Sciences 6 (1983) 13–25.

K. Menger: Selected Papers in Logic, Foundations, Didactics and Economics. Dordrecht: Reidel, 1979.

K. Menger: Reminiscences of the Vienna Circle and the Mathematical Colloquium. Edited by Louise Golland, Brian McGuinness and Abe Sklar. Dordrecht: Kluwer, 1994.

O. Neurath: Empiricism and Sociology. Edited by Marie Neurath and Robert S. Cohen. Vienna Circle Collection, Vol. 1. Dordrecht: Reidel, 1973.

H. Simon: Review of von Neumann and Morgenstern (1944) Theory of Games and Economic Behavior, In American Journal of Sociology, 50 (1945) 558–560.

AN EXACT THEORY OF SOCIAL GROUPS
AND RELATIONS

KARL MENGER

We consider a group of men[1] which we shall denote by G and to which we shall refer as the "total group" of the case under consideration. G may be divided into two subgroups which have no members in common. Each member of the total group G belongs to one and only one of these subgroups, which we shall denote by G_1 and G_2 and call the two "fundamental groups" of the considered case. For instance, these very general assumptions are satisfied if the total group consists of the inhabitants of a country, G_1 of the men, G_2 of the women; or if G consists of the inhabitants of a country, G_1 of the white ones, G_2 of the colored ones; or if G consists of the passengers of a train, G_1 of the smokers, G_2 of the non-smokers.

Suppose now we had to divide the total group G into smaller subgroups. Each member of G will have a certain attitude toward the association with each other member of G. In what follows we shall consider the case that each member of G has a common (either positive or negative) attitude toward all members of the same fundamental group; that is to say, the case that each member of G who is willing to associate with one member of G_1 is willing to associate with every member of G_1; that each member of G who dislikes association with one member of G_1 dislikes association with every member of G_1; and that each member of G has also a common (either positive or negative) attitude toward the members of G_2. For example, a passenger of a train objecting to one smoker in his compartment objects in general to every smoker in his compartment, and a smoker admitting one smoker admits in general every smoker. Though

---

[1] Social groups and relations have been the topic of several sociological schools, and even the central topic of some, as of the von Wiese school. But whereas the work of these sociologists consists in descriptions and classifications of social groups and relations, the present paper sketches a rigorous theory of these entities. A special case of this theory has been developed in the author's book, *Moral, Wille und Weltgestaltung: Grundlegung einer Logik der Sitten* (Vienna: Springer, 1934). A brief sketch of the general theory is contained in the author's contribution to the book, *Neuere Fortschritte in den exakten Wissenschaften* (Vienna: Deuticke, 1936).

790

exceptions are conceivable (a man who dislikes smokers in his compartment may nevertheless be willing to admit one particular smoker because he wants to talk to him), the assumption of a common attitude is in this case at least a good approximation to reality.

By no means do we intend to state that for all divisions of a group into two subgroups a common attitude in the described sense can actually be observed. If G_1 and G_2 are the groups of the men and the women of a country, and if the association considered is marriage, then a member of G_1 will certainly not be willing to associate with every member of G_2 though he may be willing to marry a particular member of G_2. What we intend to do is merely to start with the study of those cases in which the assumption of a common attitude is satisfied, and we do so because these cases are comparatively simple.

In such a case each member M of G has one of four possible attitudes toward the association with other members of G. Either M wishes to associate exclusively with members of G_1, or M wishes to associate exclusively with members of G_2, or M is willing to associate with everybody, or M dislikes association with anybody. The total group G is thus divided into four subgroups which shall be called "groups of attitude": a group that shall be denoted by G^1, consisting of those who wish to associate exclusively with members of the fundamental group G_1; a group G^2 consisting of those who wish to associate exclusively with members of the fundamental group G_2; a group G^{12} consisting of those who are willing to associate with everybody; a group G^0 consisting of those who dislike association with anybody.

This division of G into the four groups of attitude G^1, G^2, G^{12}, G^0 overlaps with the division of G into the two fundamental groups G_1 and G_2. The result is a division of G into eight groups to which we shall refer as the eight "principal groups" of the considered case: (1) the group consisting of those who belong to G_1 and G^1—i.e., of those members of G_1 who wish to associate exclusively with members of G_1—this group shall be denoted by G_1^1; (2) the group consisting of those who belong to G_1 and G^2—i.e., of those members of G who wish to associate exclusively with members of G_2—this group shall be denoted by G_1^2; (3) the group, denoted by G_1^{12}, which consists of those who belong to G_1 and G^{12}—i.e., of those members of G_1 who are willing to associate with everybody; (4) the group, denoted by G_1^0, which consists of those who belong to G_1 and G^0—i.e., of those members of G_1 who dislike association with anybody. In an analogous way we define the four principal groups G_2^1, G_2^2, G_2^{12}, G_2^0. We can summarize the definitions of the eight principal groups in the following scheme:

$$
\begin{array}{ccccc}
 & G^{1} & G^{2} & G^{12} & G^{0} \\
G_{1} & G_{1}^{1} & G_{1}^{2} & G_{1}^{12} & G_{1}^{0} \\
G_{2} & G_{2}^{1} & G_{2}^{2} & G_{2}^{12} & G_{2}^{0}
\end{array}
$$

We shall now consider what properties of the single principal groups and what relations between different ones result from the attitude of their members. We shall call a subgroup G' of G "consistent" if any two members chosen from G' are willing to associate with each other; and we shall call two subgroups G' and G'' of G "mutually associating" if each member of G' is willing to associate with each member of G'' and conversely. These two definitions about general subgroups apply in particular to principal groups.

The assumption of the common attitude implies that any two men M and M' who belong to the same principal group have the same attitude toward any third member of G, and that any third member of G has the same attitude toward M as toward M'. Examining the eight principal groups, one after the other, we see: Each member of the principal group G_{1}^{1} is willing to associate with each member of the fundamental group G_{1}, thus with all the members of the principal groups $G_{1}^{1}, G_{1}^{2}, G_{1}^{12}, G_{1}^{0}$, and dislikes association with each member of the fundamental group G_{2}, thus with all the members of the principal groups $G_{2}^{1}, G_{2}^{2}, G_{2}^{12}, G_{2}^{0}$; we see, in particular, that the group G_{1}^{1} is consistent. Each member of the principal group G_{1}^{2} is willing to associate with each member of the fundamental group G_{2}, thus with all the members of the principal groups $G_{2}^{1}, G_{2}^{2}, G_{2}^{12}, G_{2}^{0}$, and dislikes association with each member of the fundamental group G_{1}, thus with all the members of the principal groups $G_{1}^{1}, G_{1}^{2}, G_{1}^{12}, G_{1}^{0}$; the group G_{1}^{2} is not consistent, for each member of the group dislikes association with each other member of the group. Each member of the principal group G_{1}^{12} is willing to associate with everybody, thus with all the members of each of the eight principal groups; the group G_{1}^{12} is consistent. Each member of the principal group G_{1}^{0} dislikes association with everybody, thus with all the members of each of the eight principal groups; the group G_{1}^{0} is inconsistent. In an analogous way the attitude of the members of the four remaining principal groups can be described.

It is easily verified that the groups $G_{1}^{1}, G_{1}^{12}, G_{2}^{2}, G_{2}^{12}$ are consistent, the groups $G_{1}^{2}, G_{1}^{0}, G_{2}^{1}, G_{2}^{0}$ inconsistent, and that the following pairs of principal groups are the only ones which are mutually associating: G_{1}^{1} and G_{1}^{12}; G_{1}^{12} and G_{2}^{12}; G_{2}^{12} and G_{2}^{2}; G_{1}^{12} and G_{2}^{1}; G_{2}^{12} and G_{1}^{2}; G_{1}^{2} and G_{2}^{1}. We may represent these relations in a diagram, a method that for the sake of a comprehensive view is quite useful in more complicated cases. We associate a small circle with each consistent principal group, a small cross with each inconsistent principal group, and join two of these symbols if and only if the principal groups they represent are mutually associating.

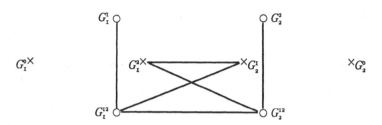

Among the pairs of groups which are not mutually associating we may point out those whose members mutually dislike each other. Mutually antipathetic in this sense are the six following pairs of principal groups: G_1^1 and G_2^2; G_1^1 and G_2^0; G_2^2 and G_1^0; G_1^0 and G_2^2; G_1^0 and G_1^1; G_2^0 and G_1^1. There remain sixteen pairs of principal groups like G_1^1 and G_1^2 in which one group is willing and one group is unwilling to associate with the other.

From what has been stated above one readily deduces that each consistent subgroup of G is necessarily of one of the five following types:

1. A group consisting of an arbitrary number of members either of G_1^1 and G_1^{12} or of G_2^2 and G_2^{12}. A group of this type may be called "pure" because all its members belong to the same fundamental group; all of them belong in the first case to G_1, in the second case to G_2.

2. A group consisting of an arbitrary number (greater than o) of members of G_1^{12} and an arbitrary number (greater than o) of members of G_2^{12}. A group of this type may be called a "mixed" group because it contains members of both the fundamental groups G_1 and G_2.

3. A group consisting of exactly one member of G_1^2 and an arbitrary number of members of G_2^{12}, or of exactly one member of G_2^1 and an arbitrary number of members of G_1^{12}. In the first case it would not be possible to extend the group by adding a second member of G_1^2 or a member of any other group than G_2^{12} without destroying the consistency of the group, because the member of G_1^2 in the group is not willing to associate with any other member of the fundamental group G_1 whereas the members of the fundamental group G_2, except those of G_2^{12} and G_1^1, dislike association with the member of G_1^2. Members of G_2^1 dislike those of G_2^{12}. The only extension of the group with preservation of its consistency is thus the addition of

further members of G_2^{12}. In the second case we meet with an analogous situation. A group of this type with one exceptional member may be called a "centered" group.

4. A group consisting of exactly one member of G_1^2 and exactly one member of G_2^1. We call a group of this type a "singular pair"—singular because the pair differs essentially from other consistent pairs, e.g., from a pair consisting of one member of G_1^{12} and one member of G_2^{12}. While the latter pair can be extended to a larger consistent group by adding other members of G_1^{12} and G_2^{12}, the singular pair cannot be extended at all without losing its consistency; for the member of the pair belonging to G_1^2 dislikes the acceptance of any other member of the fundamental group G_1, the member belonging to G_2^1 dislikes the acceptance of any other member of the fundamental group G_2.

5. A group consisting of exactly one member of G_1^0 or of exactly one member of G_2^0. A group of this type differs essentially from other groups consisting of exactly one member, namely, by the fact that it cannot be extended without losing its consistency. The single member of the group, like a hermit, does not wish any association. We may call a group of this type a "hermit group."

The described scheme, simple as it is, unites a great number of quite different social relations. In special cases some of the principal groups may be vacuous (i.e., contain no members). If G is the population of a country, G_1 and G_2 the inhabitants distinguished according to two different colors or religions or languages, and association is the entering into ordinary social relations, then it is the four consistent principal groups that play a role: G_1^1 and G_2^2, consisting of the intolerant members of G_1 and G_2 who wish to associate merely with members of their own fundamental group, and G_1^{12} and G_2^{12}, consisting of the tolerant members of G_1 and G_2. The diagram in this case reduces to the four circles G_1^1, G_1^{12}, G_2^{12}, G_2^2, each of which is joined with the consecutive circle by a line. The consistent subgroups of G are either of the pure or of the mixed type. We find the same situation if G_1 are the healthy men, G_2 the men with an infectious disease, G_1^1 and G_2^2 consisting of the healthy who are afraid of infection and of the sick who are afraid of infecting the healthy—or if G_1 are the men belonging to a caste, G_2 the outcaste, G_1^{12} and G_2^{12} consisting of the men emancipated from the restrictions of caste.

If we denote by G_1 polite or altruistic men, by G_2 impolite or egotistical men, respectively, then nobody is anxious to associate only with members of G_2; thus the principal groups G_1^2 and G_2^2 are vacuous. There may exist members of G_2^1, i.e., impolite men who merely wish to associate with polite ones, and egotists who wish to associate merely with altruists. The diagram in these cases reduces to the circles G_1^I, G_1^{12}, G_2^{12} and the cross G_1^I, the circle G_1^{12} being joined with each of the three other symbols by a line. The consistent subgroups of G are either of the pure or of the mixed or of the centered type.

If we call G_1 the sane people of a town, G_2 the insane, then the group of attitude G^2 and thus the principal groups G_1^I and G_2^2 are vacuous, for nobody is anxious to abide only with the insane. Furthermore, the principal groups G_2^I, G_1^0, and G_2^0 are of little importance. It is the following principal groups that play a role: G_1^I, consisting of the sane who like association only with the sane; G_1^{12}, consisting of the sane who associate with everybody, like physicians and attendants; and G_2^{12}, consisting of the insane. In this example the diagram reduces to the three circles G_1^I, G_1^{12}, G_2^{12}, the second of which is joined with the two others by straight lines. All consistent subgroups are either of the pure or of the mixed type. If G_1 consists of the nonsmokers, and G_2 of the smokers on a train, especially on a European train whose cars are divided into several compartments, the situation is quite similar, association meaning "sitting in the same compartment."

Singular pairs are important if, for instance, we consider the groups G_1 and G_2 of men and women in a party and the association of dancing. In this case the four consistent principal groups G_1^I, G_1^{12}, G_2^2, G_2^{12} are vacuous. There are no men who wish to dance with men or with both men and women, nor women who wish to dance with women or with both men and women. The principal groups that play a role are the four inconsistent groups G_1^2, G_2^I, G_1^0, G_2^0 of men who wish to dance with women, of women who wish to dance with men, and of men and women who do not wish to dance. The diagram reduces to the four crosses, G_1^2 and G_2^I being joined by a line. The only consistent subgroups of G in this case are singular pairs consisting of a male and a female dancer and groups consisting of exactly one member, a nondancing person or a wallflower.

If we call G_1 the group of despotic men or of political dictators, G_2 the group of the other men, then G_1 reduces to the principal group G_1^2. The members of G_2 are partly nondespotic men who do not wish to be objects of despotism, partly passive natures who do not object to being oppressed. The first ones are united in G_2^2, the latter ones in G_2^{12}. The diagram reduces to the circles G_2^2, G_2^{12} and the cross G_1^2, the circle G_2^{12} being joined with both the other symbols by a line. The consistent groups are of the two following types: pure groups consisting of members of G_2^2 and centered groups consisting of one member of G_1^2 and an arbitrary number of members of

G_2^{12}. In the history of philosophy this case played a role in the discussions of the system of Nietzsche. No matter how we may interpret the terms "rulers" and "slaves," frequently used and seldom explained in this system, it is obvious that Nietzsche overlooked or completely underestimated the class G_2^2 of men who, without the tendency to oppress or to exploit others, refuse to be oppressed or exploited.

Another famous ethical principle, Kant's categorical imperative, is the command to follow motives which are fit to be general rules for all men. Kant was mistaken in his belief that concrete moral rules may be deduced from this principle. First of all, no intelligence provides us with a survey of the consequences which a motive might have, when stated as a general rule. But even with such a survey a decision as to whether or not a principle is fit for a general rule would presuppose some general aims. Furthermore, obviously there are many situations which may be regulated by many different rules. We might add that the imperative does not provide any rule on how to react to intentional violations of the very rule by others.[2] From the point of view of the scheme of associations, the categorical imperative may be interpreted as the command to ask from others what one is one's self. It is the members of the classes G_1^1 and G_2^2 who obey this command. The classes are not mutually associating, whereas a centered group is consistent without obedience to the categorical imperative. Kant's principle is thus neither necessary nor sufficient to guarantee general harmony.

When is it possible to unite all members of G into one consistent group? In other words: Under which conditions is the total group G itself consistent? It is easy to see that in the considered scheme there are only the two following possibilities of general harmony: Either all men belong to the tolerant group of attitude G^{12}, thus each man is a member of G_1^{12} or G_2^{12} in which case G is a consistent mixed group; or one of the fundamental groups, say G_1, contains exactly one member belonging to the inconsistent principal group G_1^2 whereas all other members of G belong to the tolerant principal group G_2^{12}, in which case G is a consistent centered group.

Some of the subsequent problems of the theory that can be solved with methods similar to that applied in the case considered in this paper (in mathematical terminology: with combinatorial methods) consist in the study of the following cases: (1) The total group is divided into three or more fundamental groups, like a population divided into men, women, and children, or into three or more nationalities, religions, or parties, or

[2] A more extensive discussion of the categorical imperative is contained in the author's book, *Moral, Wille und Weltgestaltung*, cited above.

like the state of the bees divided into three sexes. (2) Two or more divisions of a group, each into two or more fundamental groups, superpose, like the divisions of a population, into members of different religions and of different parties. (3) Each man may assume more than two attitudes toward association with other men, e.g., either a positive or a negative or an indifferent attitude. (4) The attitude of a man toward association with another man depends not only on the fundamental group to which the other belongs but also on the attitude of the other toward associations. There are men who not only object to association with criminals but also refuse to associate with men who associate with criminals. Problems of this type seem to have a necessarily circular character; nevertheless some of them can be solved by clearly establishing all possible divisions of the total group into consistent subgroups. (5) The assumptions of a common attitude made in this paper may be replaced by less exigent assumptions. The assumption may be weakened without affecting the results. (6) The properties and the attitudes of the members of G change in time. The study of cases of this type provides us with a scheme of social movements and forces. (7) Supplementary assumptions on the form in which the properties of the fundamental groups and the attitudes toward association are transmitted to following generations imply statements on the roles of environment and of heredity in social movements and provide us with schemes of history.

More complicated are problems involving quantitative questions. There are cases in which the number of members in each group plays a role. The attitude of a man toward a group may depend upon whether the group does not contain less than a certain minimum or more than a certain maximum of members. Still more complicated are problems involving quantities which may assume a continuum of values. In the study of the relations of groups in questions of taxation or in the case of contributions to a common enterprise the amount of money is such a quantity. Some problems of this kind are rather complicated mathematically.

Of the practical applications of a sociological theory one frequently expects advice as to what ought to happen and predictions as to what will happen. Advice, I believe, cannot be obtained from any theory at all. A theory can at the best explain how to attain aims; but the choice between the various and often contradicting ultimate aims is a matter of will. Predictions of more or less probability may result from some inductive sociological research but can hardly be expected from a deductive theory as outlined in this paper. What, then, are its practical applications?

What such a theory does is to provide us with a survey of all forms of organization conceivable in given situations. It exhibits all divisions of a group into consistent harmonious groups. It may decide whether, in cases of incompatible aims of different subgroups, it is necessary to devise totalitarian measures valid for all subgroups but meeting with the resistance of some of them. It may also, in some cases, suggest other possibilities, such as measures to be applied only to certain subgroups.

If we compare systematically the possibilities considered in the theory with social reality we shall certainly find that some of the theoretical possibilities do not at present exist. The reason for their nonexistence in most cases is likely to be that they displease everybody or almost everybody; but in some cases the reason may simply be that they had been overlooked as possibilities actually attainable. The practical applications of the theory lie here in pointing out new possibilities which may have been overlooked. Though naturally leaving to the will of individuals and groups the choice and decision as to which form of organization shall actually be adopted, the theory enlarges the field within which we may choose, and thus may eventually prove of considerable practical importance.

UNIVERSITY OF NOTRE DAME

Mathematical Social Sciences 6 (1983) 13-25
North-Holland

13

ON SOCIAL GROUPS AND RELATIONS

Karl MENGER

Department of Mathematics, Illinois Institute of Technology, Chicago, IL 60640, U.S.A.

Communicated by B.M. Schein
Received 20 May 1983

A group of people is divided (1) into n fundamental groups and (2) into 2^n attitude classes, the members of each class accepting just those belonging to a particular set of fundamental groups. Cohesiveness of the $n \cdot 2^n$ resulting classes, compatibility relations among them, and inextensible (maximal) cohesive sets are studied.

Key words: Group cohesiveness and compatibility; social group.

1. Introduction

The author's application in the 1930s of simple combinatorics to human groups and social relations had a twofold origin. One source was a critique of the famous categorical imperative decreed by the philosopher Kant as the supreme command for human conduct: 'Always act according to that maxim that can become a general law.' In another formulation, often treated interchangeably with the first (even though the two differ from one another about as much as formalistic and intuitionistic mathematics), Kant prescribed 'that maxim of which *one can wish* that it become a general law'.[1] The critique (Menger, 1974) was based on the following observations.

It is difficult, if not impossible, to ascertain what can and what cannot become a general law. Moreover, in most specific situations it is impossible to deduce (in the strict logical sense of deducing – and what other sense is there?) specific precepts for behavior unless the imperative is supplemented by value judgments. Such supplements are often based on authoritative sayings and writings or they concern – sometimes in unexplicit or ill-defined ways – life and liberty, equality of possessions and opportunities, or what the speaker or writer regards as the common good or the general welfare. Furthermore (if less important), evaluations appear in more hidden forms: for example, in the guise of stipulations as to which individuals who differ in age, mental capacity and personal records should be regarded as equal

---

[1] Kant's first-mentioned formulation is contained in his *Kritik der praktischen Vernunft* (*Critique of Practical Reason*, 1788), the other in his *Grundlegung zur Metaphysik der Sitten* (*Foundations of the Metaphysics of Ethics*, 1785).

0165-4896/83/$3.00 © 1983, Elsevier Science Publishers B.V. (North-Holland)

with respect to the maxims. In addition to all that, there remain questions concerning reactions to willful violations of the imperative by others.

But the basic difficulty about the categorical imperative seemed to be the fact, unrecognized or ignored or denied by most students of ethics in the 1930s, that situations can be handled in various mutually incompatible ways. Only individual choices in keeping with individual value judgments can decide between these ways;[2] and various choices are actually being made, whereas the imperative is based on the illusions of unique solutions and uniformity of attitudes.

The author's second inducement to the study of groups and relations was dissatisfaction with a sentiment expressed in the Vienna Circle of the early 1930s in connection with Wittgenstein's *Tractatus* (Wittgenstein, 1922) – the idea that after the complete elimination of value judgments from ethics (a process all of us deemed indispensable) only historical and ethnographical descriptions of moral beliefs and conditions were possible – but certainly no theoretical considerations at all.[3] Just the multiplicity of beliefs and evaluations, however, seemed to suggest a formal (and therefore completely objective) study of the outward manifestations of all those inner judgments and attitudes: the human groups with diverse and often incompatible wishes and conflicting decisions and the resulting relations between individuals and individuals, individuals and groups, and groups and groups.[4] Certainly, before dynamic and quantitative ideas (as are of primary importance, e.g. for game theory) are somehow combined with combinatoric reasoning, one can hope only for scant results of mathematical or sociological significance. But even before such a synthesis is achieved, attempts to externalize inner leanings and decisions in a rigorous theory of their effects may contribute to a clarification of those obscure matters.

2. A simple model of human groups and relations

Long before models became ubiquitous in social sciences, the preceding considerations led to an utterly simple combinatoric model of an elementary, but fairly comprehensive, situation: (1) a group, H, of people is somehow divided into two nonoverlapping (*fundamental*) groups: H_1 and H_2, such as men and women, blacks and whites, smokers and nonsmokers, polite and impolite persons, submissive and autocratic individuals; and (2) each member of H has a certain attitude toward each fundamental group: of *accepting* or *nonaccepting* all of its members,

---

[2] Menger (1974, pp. 31–34). Emphasis on individual decisions was to be of paramount importance in existentialism.

[3] Also expressed in the writings of R. Carnap (see Carnap, 1928).

[4] Menger (1938). Cf. Also both editions of the book (Menger, 1974), in particular the Postscript to the English edition.

e.g. by association as partners, in a club, in a train compartment, in some organization. Mutual acceptance is *compatibility*.

Besides people who accept only the members of H_1 or only the members of H_2, there are also persons who accept both groups or neither. Let H^1, H^2, H^{12} and H^0 be these four *attitude classes*, respectively, and

$$H_1^1, H_1^2, H_1^{12}, H_1^0; \qquad H_2^1, H_2^2, H_2^{12}, H_2^0$$

the *main classes*, i.e. the parts of the attitude classes belonging to H_1 and H_2, respectively. In most situations, some of the eight main classes are empty.

Clearly, $H_1^1, H_1^{12}, H_2^2, H_2^{12}$ are *cohesive* in the sense that, in each of these classes, any two members are compatible; the other four main classes are noncohesive and might even be called *explosive* in the sense that *no* two members of such a class are compatible.

The following examples illustrate a variety of situations and various ways of representing them.

Example 1. On a train, let H_1 and H_2 be the group of the smokers and non-smokers, respectively, and consider admission to one's own compartment as acceptance. Only the main classes $H_2^2, H_1^{12}, H_2^{12}$ and possibly H_1^0 and H_2^0 are nonempty. The last two (explosive) classes break up into *solitaries*, each alone in a compartment, which can of course materialize only if there are sufficiently many compartments on the train. Of the first three (cohesive) classes, H_2^2 is compatible only with H_2^{12}, which itself is also compatible with H_1^{12} (the latter accepts, but is not accepted by, H_2^2). The situation can be described by a *compatibility graph*, whose vertices and sides correspond to the main classes and the compatibilities (Fig. 1). In order to indicate the self-compatibility of the cohesive classes we shall represent the latter

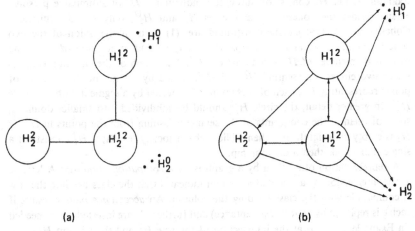

(a)   (b)

Fig. 1.

by little circles, while groups of points indicate explosive classes (each point of such a group of course representing an individual rather than a main class). From a compatibility graph one obtains an *acceptance graph* (in which acceptance is represented by an arrow from the accepting to the accepted class) by replacing each side of the graph by a double arrow and adding single arrows where they are required (in Example 1: an arrow from H_1^{12} to H_2^2, arrows from H_1^{12} and H_2^{12} to H_1^0 and H_2^0, and from H_2^2 to H_2^0).

Example 2. If, in a partly integrated society, H_1 are the whites and H_2 the blacks, and acceptance has to do with clubs or the like, then $H_1^1, H_1^{12}, H_2^2, H_2^{12}$ are significant. The circles representing these cohesive classes have to be joined by three segments in a horseshoe-like arrangement.

Example 3. If, in a social group, H_1 consists of the polite persons, and H_2 of the impolite individuals, then $H_1^1, H_1^{12}, H_2^{12}$ are significant and the graph is L-shaped like that in Example 1. In fact, the graphs in Examples 1 and 3 are identical if in one example H_1 and H_2 are relabelled H_2 and H_1.

Example 4. If in a ballroom, H_1 are the men, H_2 the women and acceptance is for dancing partners, then all four cohesive main classes are empty. Only $H_1^2, H_1^0, H_2^1, H_2^0$ are significant; and at any moment the only cohesive groups that can be formed are *pairs* consisting of one man and one woman. The remaining members of H are of three types: solitaries belonging to H^0 who cannot or will not dance (as in former times, chaperons); members of the sex that is preponderant in the group and who therefore cannot find partners; wallflowers who find no partners because of personal characteristics.

Example 5. Let H_1 consist of autocratic individuals, H_2 of submissive persons. Then, besides the cohesive main groups H_2^2 and H_2^{12}, only H_1^2 is significant. Cohesive groups that can be constituted are: (1) parts of the union of the two cohesive main classes and (2) groups which may be said to be *centered* consisting of a single member of H_1^2 and a part of H_2^{12}. The graph of the situation consists (1) of two circles representing H_2^2 and H_2^{12}, joined by a segment, (2) a group of points representing H_1^2, each of these points connected by a segment with the circle H_2^{12}. In greater detail, the circle H_2^{12} should be subdivided into smaller domains, some of them at the end points of the segments issuing from the points in H_1^2. If H_2^2 is empty or insignificant, then a Nietzschean society results; if H_1^2 consists of a single individual, then a dictatorship.

A representation other than by a graph is by a *compatibility matrix*: A letter c at the intersection of a row and a column indicates that the class heading the row is compatible with the class heading the column. An *acceptance matrix* results if each c is replaced by an α (for acceptance) and further α's are inserted where needed (in Example 5: one α at the intersection of the row H_1^2 and the column H_2^2).

	H_1^2	H_2^2	H_2^{12}
H_1^2			c
H_2^2	c	c	
H_2^{12}	c	c	c

	H_1^2	H_2^2	H_2^{12}
H_1^2		α	α
H_2^2		α	α
H_2^{12}	α	α	α

The formulation of the *categorical imperative* was undoubtedly in part motivated by the desire for universal compatibility and a cohesive society. The preceding simple model shows that the uniformity implied by that law and in fact that law itself are *neither sufficient nor necessary for the cohesiveness of a group*. The group H_1^2 is uniform, yet explosive; and the union of the groups H_1^{12} and H_2^{12} is not at all uniform and yet cohesive; and so is any centred group, even though it represents a flagrant violation of the categorical imperative.

Example 6. Representations by graphs and matrices are of course also possible in the case of eight nonempty main classes. If the fundamental classes are properly arranged, and β, 1 and 0 are written for being accepted, compatible and totally incompatible (i.e. neither $h \, \alpha \, h'$ nor $h' \, \alpha \, h$), the matrix displays antisymmetry with regard to both diagonals:

	H_1^0	H_1^1	H_1^{12}	H_1^2	H_2^2	H_2^{12}	H_2^1	H_2^0
H_1^0	0	β	β	0	β	β	0	0
H_1^1	α	1	1	α	β	β	0	0
H_1^{12}	α	1	1	α	1	1	α	α
H_1^2	0	β	β	0	1	1	α	α
H_2^1	α	α	1	1	0	β	β	0
H_2^{12}	α	α	1	1	α	1	1	α
H_2^2	0	0	β	β	α	1	1	α
H_2^0	0	0	β	β	0	β	β	0

The general case will now be treated by yet another method. In Fig. 2 we represent: each cohesive main group by a circle or an ellipse; each cohesive union of two main groups by a polygon (a rectangle or a triangle); the groups of solitaries as well

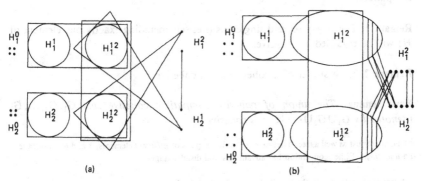

(a) (b)

Fig. 2.

as H_1^2 and H_2^1 by groups of points. Compatibility of two cohesive main classes is expressed by a rectangle including the two representative circles. Each of these overlapping rectangles represents a group that is *maximal* in the sense of not being a part of a larger cohesive group. The (possibly numerous) centered groups are represented by a triangle including a circle and having an apex at a point of an explosive group. A segment between a point of H_1^2 and a point of H_2^1 represents the (possibly numerous) isolated cohesive pairs.

In Fig. 2(b) the ellipse H_1^{12} is broken up into several parts: one is united in a rectangle with the circle H_1^1, another one with a part of H_2^{12}; and into several slices, each of which is united with a point of H_2^1 in a triangle. Each of these polygons represents a cohesive union (the triangles indicate centered groups); and they are nonoverlapping, but not maximal; in fact, each of these unions can be extended at the expense of the others. The ellipse H_2^{12} is broken up in a similar way.

3. The case of n fundamental classes

Turning now to the case of n (≥ 2) fundamental classes we begin with a few general remarks.

Let H be a set with a binary relation, α. If h and h' are elements of H, then $h \, \alpha \, h'$ is to be read: h *accepts* h' or, synonymously, h' *is accepted by* h. The relation will *not* be assumed to be either symmetric or transitive; that is, $h \, \alpha \, h'$ will not be assumed to imply $h' \, \alpha \, h$; nor $h \, \alpha \, h'$ and $h' \, \alpha \, h''$ to imply $h \, \alpha \, h''$. Whether one does or does not say that the individuals accept themselves (i.e. that $h \, \alpha \, h$) is a matter of convention. In the sequel it will be assumed that $h \, \alpha \, h$ for each h, which is expressed by saying that the relation is *reflexive*.[5]

Mutual acceptance, called *compatibility*, is an obviously symmetric and reflexive relation.[6] Two parts (subsets) H' and H'' of H are said to be *compatible* if and only if each member (element) of H' and each member of H'' are compatible. As before, the part H' of H is said to be *cohesive*, if and only if any two members of H' are compatible.

Remark 1. The empty set and singletons (i.e. sets including exactly one element of H) will be regarded as cohesive.

Remark 2. Each subset of a cohesive set is cohesive.

Basic lemma. *The union of pairwise compatible cohesive sets* $G_1, G_2, \ldots, G_k$ *(denoted by $G_1 \cup G_2 \cup \cdots \cup G_k$) is cohesive.*

---

[5] One might just as well agree to define $h \, \alpha \, h'$ only for pairs of *different* elements, h and h', and leave $h \, \alpha \, h$ *undefined*. Either convention has advantages and disadvantages.

[6] A 'tolerance relation' in the terminology of Zeeman (1962).

Any two members of the union are compatible: if they belong to the same set G_i, then because G_i is cohesive; if they belong to different sets, then because the sets are compatible.

Now let H be divided into n (≥ 2) nonoverlapping parts $H_1, H_2, \ldots, H_n$, which will be referred to as the *fundamental classes* and which are uniform with regard to compatibility in the sense of the following

Remark 3. If h does not belong to the fundamental class H_i, then h accepts either all or none of the members of H_i. If h does belong to H_i, then h accepts either all members of H_i or none except h.

If h accepts all members of a set H' (fundamental or not), then we say that h *accepts* H' and write $h \, \alpha \, H'$; similarly, if each member of H' accepts h, we write $H' \, \alpha \, h$. But it will *not* be assumed that if a member of the fundamental class H_j accepts h, then all members of H_j accept h.

Under the uniformity assumption of Remark 3, there are 2^n possible attitudes toward the fundamental classes and hence 2^n *attitude classes*: Let H^0 be the class of the solitaries not accepting anyone else; and $H^1, H^2, \ldots, H^n$ the classes of all those who accept only the members of one fundamental class, $H_1, H_2, \ldots, H_n$, respectively. Let

$$H^{12}(=H^{21}), H^{13}, \ldots, H^{n-1\,n}$$

be the $n(n-1)/2$ classes of those who accept only the members of certain pairs of fundamental classes. Extending this, if $J = \{j_1, j_2, \ldots, j_k\}$ is any nonempty subset of $N = \{1, 2, \ldots, n\}$, then let H^J be the class of all those members who accept the fundamental classes $H_{j_1}, H_{j_2}, \ldots, H_{j_k}$ and no others. When listing the elements of J one is forced to write them in some order. Frequently it is convenient to write the numbers in the order of increasing magnitude. But this (or any other order) is irrelevant for the attitude class; for example, $H^{231} = H^{123}$ and $H^{5472} = H^{2457}$.

The break-up of H into 2^n nonoverlapping attitude classes brings about that each fundamental class H_i is the union of 2^n nonoverlapping parts: the intersections of H_i with the diverse attitude classes. The set of all members of H_i that belong to H^J will be denoted by H_i^J and these $n \cdot 2^n$ sets will be called the *main classes*.

Remark 4. For the main class H_i^J to be cohesive it is necessary and sufficient that the set J includes i.

There are altogether $n \cdot 2^{n-1}$ cohesive and equally many noncohesive main classes. Just as in the case $n = 2$, *each noncohesive main class is explosive.* Some unions of main classes, such as that of H_1^1 and H_2^2, are noncohesive, without being explosive.

The elements of explosive sets are of two kinds: *solitaries* not being compatible with anyone else – belonging to H^0; and elements that will be said to be *heteroclitic* (i.e. leaning on or toward someone of a different kind) – belonging to a main class H_i^J for a nonempty set J not including i, as the elements of H_1^2.

Remark 5. For the main class H_i^J to accept H_k^L it is necessary and sufficient that J include k. The two classes are compatible if and only if J includes k, and L includes i.

If main classes that are parts of the same fundamental class be called *cofundamental*, then Remark 5 implies

Remark 6. Any two cofundamental main classes are compatible if they are cohesive, and incompatible if they are explosive. An explosive main class is incompatible with each cofundamental main class.

A subset of $N = \{1, 2, \ldots, n\}$ that contains a set J as a subset is said to be *between J and N*; and so are, in particular, J and N.

Remark 7. If $H_i^J \alpha H_k^L$ and J_1 lies between J and N, then $H_i^{J_1} \alpha H_k^L$. If H_i^J and H_k^L are compatible then, for each J_1 between J and N, and each L_1 between L and N, also $H_i^{J_1}$ and $H_k^{L_1}$ are compatible.

For each explosive main class H_i^J, let σH_i^J *(the associated cohesive class)* be the class H_i^{J+i}, where $J+i$ denotes the set J enlarged by the element i; and set $\sigma H_i^0 = H_i^i$. A corollary of Remark 7 is

Remark 8. If the explosive main classes H' and H'' are compatible, then so are their cohesive associates, $\sigma H'$ and $\sigma H''$.

For each cohesive main class H_i^J, let τH_i^J *(the associated explosive class)* be H_i^{J-i}, where $J-i$ is the set J after the omission of the element i, and set $\tau H_i^i = H_i^0$. If H' and H'' are cofundamental, then so are $\tau H'$ and $\tau H''$, which are incompatible according to Remark 6. If H' belongs to H_i, and H'' to H_k, where $i \neq k$, then, because of their cohesiveness, there exist sets J and L not including i and k, respectively, such that $H' = H_i^{J+i}$ and $H'' = H_k^{L+k}$. If H' and H'' are compatible, then (because of $i \neq k$) J includes k, and L includes i, whence also the associated explosive sets are compatible. This proves

Remark 9. If H' and H'' are compatible cohesive main classes, then their associated explosive classes, $\tau H'$ and $\tau H'$, are compatible if and only if H' and H'' are not cofundamental.

From Remarks 8 and 9 it follows for each number of fundamental classes that

(1) *the compatibility structure of all explosive sets is isomorphic with that of their cohesive associates*; and (2) *the compatibility structure of all cohesive sets is isomorphic with that of their explosive associates after the compatibilities between cofundamental classes are deleted from the former.* The mutually incompatible classes H_{i}^{i} correspond to the n classes of solitary singletons H_{i}^{0}.

In a rudimentary way, the preceding statement can be verified in the case of only two fundamental classes, where the compatibility structure of the explosive main classes is reduced to the one compatibility between H_{1}^{2} and H_{2}^{1}. Its isomorphic image among the cohesive classes is the compatibility between H_{1}^{12} and H_{2}^{12}. Moreover, these two cohesive classes are respectively compatible with the mutually incompatible classes H_{1}^{1} and H_{2}^{2}.

Of course, as has been shown in Section 2, the compatibility of the two explosive classes means merely that each element of H_{1}^{2} and each element H_{2}^{1} can join in an inextensible cohesive *pair* and not that the classes (unless each happens to be a singleton) can be united in a cohesive *class*.

4. Inextensible cohesive (i.c.) sets

Sets that are not subsets of larger cohesive sets will be said to be *inextensible*. Since by Remark 2 all noncohesive sets are inextensible, inextensibility is significant only in connection with cohesiveness. *Inextensible cohesive (i.c.) sets* are in a sense maximal.

Remark 10. A cohesive set S is inextensible if and only if S includes every element of H compatible with S.

A cohesive set S that is extensible is a subset of a cohesive set T containing an element h not belonging to S and yet (since T is cohesive) compatible with S. Conversely, if the cohesive set S is compatible with an element h not belonging to S, then the union of S and h is a larger cohesive set containing the subset S. This concludes the proof of Remark 10.

Of importance for the study of i.c. sets is the concept of the *compatibility closure* – biefly, the *closure* – of a cohesive main class H_{i}^{J}, that is, the union of the main classes H_{i}^{K} for all sets K between J and N (including J and N). This closure of course depends on N (that is, on n). Therefore, where in the course of an argument N is being changed, the closure must be denoted by $\mathrm{cl}_{n} H_{i}^{J}$, as in

$$\mathrm{cl}_{2} H_{1}^{1} = H_{1}^{1} \cup H_{1}^{12}, \quad \mathrm{cl}_{3} H_{1}^{1} = H_{1}^{1} \cup H_{1}^{12} \cup H_{1}^{13} \cup H_{1}^{123}, \quad \mathrm{cl}_{n} H_{i}^{N} = H_{i}^{N},$$
$$\mathrm{cl}_{4} H_{2}^{23} = H_{2}^{23} \cup H_{2}^{231} \cup H_{2}^{234} \cup H_{2}^{2314} = H_{2}^{23} \cup H_{2}^{123} \cup H_{2}^{234} \cup H_{2}^{1234}.$$

In what follows, however, N will not be changed, whence the closure may be denoted simply by *cl*. From its definition, one obtains:

Remark 11. If L is a subset of J, then cl H_i^{J+i} is a subset of cl H_i^{L+i}. In particular, cl H_i^{J+i} is a subset of cl H_i^i.

Since H_i^J is assumed to be cohesive, each of the main classes H_i^K united in cl H_i^J is cohesive and any two of those classes are compatible. From the Basic Lemma, we thus obtain:

Remark 12. The closure of each cohesive main class is cohesive.

Remark 13. The closure of each of the n classes H_i^i is inextensible; that of any other cohesive main class is extensible.

For let h be an element of H compatible with the set $C = $ cl H_i^i. In order to be accepted by C the element h must be accepted in particular by H_i^i and thus belong to H_i. In order to accept H_i^i the element h must belong to a main class $H_i^{i+J} = H_i^{J+i}$ for some (possibly empty) subset J of $N-i$. But all these classes are united in C, whence h belongs to C, and cl H_i^{i+J} is extensible, one of its extensions being cl H_i^i (by Remark 11). Another cohesive extension is supplied by

Remark 14. The union – to be denoted by $*H^K$ – of the sets cl H_k^K, for all k belonging to a nonempty subset K of N, is an i.c. set. $*H^K$ is the attitude class H^K after the exclusion of all its heteroclitic elements.

Since the sets cl H_k^K are cohesive and mutually compatible, their union is cohesive. If h is any element of H compatible with $*H^K$ then, in order to accept all classes H_k^K, h must belong to an attitude class H^{K+L} for some set L between the empty set and $N-K$; and in order to be accepted by members of H^K, the element h must belong to a fundamental class H_k for some k belonging to K. But H_k^{K+L} is a subset of $*H^K$ which therefore includes h and is an i.c. set.

If K consists of a single number, i, then $*H^K = $ cl H_i^i. On the other end of the compatibility spectrum, $*H^N = H^N$. The $2^n - 1$ sets $*H^J$ for the nonempty subsets J of N are all the i.c. sets that are free of heteroclitic elements or, as we shall say, they are all the i.c. sets that are *solid*.

A second class of i.c. sets consists exclusively of heteroclitic elements. By Remark 2, the simplest are singletons h_i^0 belonging to the attitude class H^0. Next are the pairs of elements of the form $\{h_i^j, h_j^i\}$ which, for $n = 2$, have been mentioned in Section 2. Such pairs are i.c. sets even if $n > 2$. If K and L are any two subsets of $N-i-j$ without a common element, then also $\{h_i^{K+j}, h_j^{L+i}\}$ is an i.c. pair, K and L representing what may be called an *ineffective tolerance* of the members. If $2 < m \leq n$, then

$$\{h_1^{23\cdots m}, h_2^{13\cdots m}, \ldots, h_m^{12\cdots m-1}\}$$

is an i.c. set of m elements. So is, if $n > m$, the m-tuple (with ineffective tolerance) obtained from the preceding by adding to the upper indices of the h's any subsets $L_1, L_2, ..., L_m$ of $\{m+1, m+2, ..., n\}$ that have no element in common.

I.c. sets of a third type will be called *mixed*. Each consists of a solid set extended by a cohesive finite set of heteroclitic elements or (which is equivalent) of such a finite set extended by a solid. The simplest examples are the centered sets

$$H_1^{12} \cup h_2^1 \quad \text{and} \quad H_2^{12} \cup h_1^2,$$

which have been considered in Section 2 for $n = 2$. For $n > 2$, these sets are extensible, the former, for example, to $H_1^{12} \cup H_1^{123} \cup h_2^1$. But cl $H_1^{12} \cup h_2^1$ is i.c. Thus, cl H_1^{12} (besides being extensible to the solid i.c. sets $*H^1$ and $*H^{12}$; see Remarks 13 and 14), can also be extended to a mixed i.c. set by the adjunction of a single element h_2^1. If $n > 2$, then one can, instead, adjoin a single element h_2^{K+1}, where K is any subset of $\{3, 4, ..., n\}$ indicating ineffective tolerance of the adjoined element.

The set H_1^{123} can be extended to an i.c. mixed set by adjoining either one of the singletons h_2^1 or h_3^1 or the cohesive pair $h_2^{13} \cup h_3^{12}$. More generally, cl H_i^J can be extended to an i.c. set by adjoining either one singleton h_j^i for a number j belonging to J, or one of certain cohesive sets consisting of 2 or 3... or $m-1$ heteroclitic elements if the number of elements in J is m. In the extreme case, cl H_i^N becomes an i.c. mixed set upon adjunction of

$$h_2^{N-2} \cup h_3^{N-3} \cup \cdots \cup h_n^{N-n}.$$

5. The case of three fundamental classes

(a) In the case $n = 3$, there are 24 main classes, 12 of which are cohesive. The latter are united in 7 inextensible cohesive sets:

$*H^1 = H_1^1 \cup H_1^{12} \cup H_1^{13} \cup H_1^{123}$ and two analogues;

$*H^{12} = H_1^{12} \cup H_2^{12} \cup H_1^{123} \cup H_2^{123}$ and two analogues;

$*H^{123} = H_1^{123} \cup H_2^{123} \cup H_3^{123}$.

(b) These are all the *solid* i.c. sets. Among the explosive main classes, those associated with the cohesive classes in $*H^{123}$, that is, $H_1^{23}, H_2^{13}, H_3^{12}$, are pairwise compatible. Consequently, any three members selected from them constitute an inextensible cohesive triple of heteroclitic elements,

$$h_1^{23} \cup h_2^{13} \cup h_3^{12}.$$

The entire classes (unless they are singletons) of course cannot be united in a cohesive set. Nor are two triples of heteroclites corresponding to differnt selections of elements compatible.

Among the explosive associates of the main classes in $*H^{12}$ any two consecutive classes in the (cyclical) order

$$H_1^2, H_2^1, H_1^{23}, H_2^{13}, (H_1^2)$$

are compatible, while the pairs H_1^2, H_1^{23} and H_2^1, H_2^{13} consist of associates of co-fundamental cohesive classes and hence are incompatible. Consequently, for any selection of elements from the cycle of explosive associates of $*H^{12}$, there are 4 cohesive pairs

$$h_1^2 \cup h_2^1, \qquad h_2^1 \cup h_1^{23}, \qquad h_1^{23} \cup h_2^{13}, \qquad h_2^{13} \cup h_1^2,$$

and 8 more for the analogues of $*H^{12}$. But there are no cohesive quadruples or further triples of heteroclites. It is easily seen that all but the third of the preceding cohesive pairs are inextensible. (The second and the fourth each include an element with ineffective tolerance.)

The four main classes in $*H^1$ are pairwise cofundamental, whence their explosive associates are pairwise incompatible.

(c) The compatibility structures of the cohesive and the explosive classes (which are in a restricted sense isomorphic) are connected. H_1^2 is compatible with the mutually compatible classes H_2^{12} and H_2^{123}. Hence, for each choice of an element h_1^2, there is an inextensible cohesive *centered* set

$$h_1^2 \cup H_2^{12} \cup H_2^{123} \text{ and 5 analogues.}$$

The cohesive pair $h_1^{23} \cup h_2^{13}$ can be extended to the inextensible *bicentered* set

$$h_1^{23} \cup h_2^{13} \cup H_3^{123},$$

which has two analogues.

Significant tripartitions of human groups are obtained, for example, by refining certain bipartitions (as those in Examples 3 and 5 in Section 2) through the introduction of a middle fundamental group.

6. Concluding remarks

A comparison of the preceding models with the social reality shows two main discrepancies. (1) The model is static and does not take change into account. (2) While the model includes multicentered cohesive sets (with a small number of individuals in each 'center') it does not make allowance for the important phenomenon of larger groups with a separate identity within cohesive sets and even main classes.

These shortcomings of the model can be traced to the basic assumptions in Section 3: (1) the division into unchanging attitude classes (the partition into rigid fundamental classes being less objectionable); (2) the all-or-nothing assumption of Remark 3.

Improvement in the first respect might be achieved by introducing a *time element*, perhaps in the form of a sequence of divisions into attitude classes allowing transitions from one main class to another. An advance in the second direction might result from raising the number of fundamental classes to the point where the attitude classes could be treated by *statistical* methods. More promising would perhaps be the introduction of *degrees of acceptance* (and compatibility). A certain improvement in both directions could probably be achieved through a procedure of *class formation in stages* or by approximation in the following sense. After e.g. the class H_1^1 (consisting of all members of the fundamental class H_1 who accept only members of the same class) has been constituted, there could develop a subclass of members who confine their acceptance to members of H_1^1; and similarly for other cohesive classes – a procedure that obviously could not have been carried out at the initial stage.

In the third place, as indicated in Section 1, the mathematical and the sociological interest of the model might be enhanced by the introduction of quantitative elements. The simplest and, in connection with a combinatoric study of classes, the most natural quantitative idea is that of the numbers of elements in the considered classes. (Minor hints in this direction can be found in Examples 1 and 4 in Section 2.) All such amendments might make the model more fit for applications to the theory of organizations and of politics.

In conclusion it may be mentioned that seminal ideas presented with the original rudimentary model in the 1930s included proposals for *social experiments* under minimal assumptions, that is, in a juridico-constitutional vacuum or semivacuum, perhaps on a remote island – recently one has considered such experiments in space; and the organization of *diverse legal systems* (each confined to its adherents) co-existing in the same territory, and of the same system in different territories – arrangements which in recent years may have become feasible through computers.

References

R. Carnap, Scheinprobleme der Philosophie (Weltkreis Verlag, Berlin, 1928).

K. Menger, An exact theory of social groups and relations, Amer. J. Sociology 43 (1938) 790–798.

K. Menger, Morality, Decision and Social Organization (Reidel, Dordrecht, 1974) 9–12. Translated from Moral Wille und Weltgestaltung (Springer, Wien, 1934).

L. Wittgenstein, Tractatus Logico-Philosophicus (London, 1922).

E.C. Zeeman, The topology of the brain, in: M.K. Fort, Jr., ed., Topology of 3-Manifolds (Prentice-Hall, 1962).

Menger's Work on Economics

Commentary on Menger's 'Austrian Marginalism and Mathematical Economics'

Karl Sigmund

This essay, one of Karl Menger's last papers, written in much the same style and spirit as his 'Reminiscences' (Menger, 1994), may be viewed as a return to his roots. It is a somewhat expanded version of a paper he had published in the 'Zeitschrift für Nationalökonomie' (Menger, 1972), after having spent a sabbatical in Vienna at his friend Oskar Morgenstern's 'Institut für höhere Studien'. The fact of finding himself back in Vienna and together with Morgenstern, some three decades after the political upheaval which had driven them both away, must have evoked poignant memories. In his paper, Menger not only offers a fascinating glimpse at some of the finest achievements of his 'Mathematisches Kolloquium', the group of highly gifted young mathematicians who had met under his guidance, but also analyses some aspects of the work of the school founded by his father Carl Menger. Anyone born into Freud's Vienna cannot fail to be sensitive to the tensions of such a return to the sources, and in fact Menger strikes an appropriate note by quoting, right at the start, Goethe's dictum that 'two souls reside within my breast'. But after that briefest of subjective confidences, the utmost objectivity is summoned forth throughout the paper.

In 1871, Carl Menger, as a young civil servant, published a ground-breaking book, 'Grundsätze der Volkswirtschaftslehre' (Principles of National Economics), in which he derived economic values from individual human wants (rather than from some inherent quality in the goods, or from the work spent in producing them). In the same year, the economists Jevons and Walras independently proposed the same idea, which entailed a complete reformulation of the classical theory. Carl Menger's book earned him, at the age of 33, a position as associate professor at the University of Vienna. From 1876 to 1878, he was tutor to Crown Prince Rudolph, the emperor's only son, and subsequently became full professor at the Unversity of Vienna. Here, he could count on some brilliant supporters, such as Friedrich Wieser and Eugen

Böhm-Bawerk, to carry on with the theoretical work which soon became known world-wide under the heading of 'Austrian Marginalism'. In 1903, one year after his son was born, sixty-three years old Carl Menger retired from the University to concentrate on the second edition of his 'Grundsätze'. But when he died in 1922, it fell upon his twenty-year old son to complete this edition and see it through to publication (Carl Menger, 1922). Karl Menger, the brilliant student at the Mathematical Seminar in Vienna, thus became very well versed in economics. However, his first theoretical work in this field went unrecognised. It was only ten years later when, as associate professor and head of the 'Mathematisches Kolloquium', he was able, through Oskar Morgenstern and Karl Schlesinger, to establish close ties with Austrian economists. His contacts led to a historical breakthrough in mathematical economics; and this tale is the high point of Menger's essay (see also K. Sigmund, 1998).

In a long introduction that cannot fail to strike non-mathematicians as a 'captatio benevolentiae', Karl Menger compares the usual formulations of certain basic economic principles as given by Austrian marginalists and by mathematical economists, and concludes that it is erroneous to assume that the mathematical formulations are necessarily more precise. On the contrary, they often rely on hidden assumptions – as, for instance, the differentiability of the price function. Nevertheless, for more complex economic statements, the verbal argumentation is usually so involved that it becomes impracticable. Menger's analysis is, as usual with him, of superb clarity and precision, and reminds one of his early, devastating criticism of the sloppy methods in theoretical economics (1936a, b) where, in Schumpeter's term, 'he read us the logician's riot act.' (Schumpeter, 1954, p. 587).

Menger next goes on to ask 'why men of such high intelligence as the members of the old Austrian school did not have a better understanding of calculus'. The training in mathematical analysis that they received at school was practically nil. Some (like Carl Menger) felt this and tried to teach themselves calculus. But in this they failed; and Karl Menger pinpoints the confusing notation, especially concerning variables, as the cause of their failure. This leads him to embark on a topic that had long fascinated him (see the Commentary on Didactics, Variables, Fluents in this volume).

Menger had been instrumental in teaching calculus to Viennese economists: in the early 'thirties, he gave a set of courses in Morgenstern's 'Institut für Konjunkturforschung' and, more importantly still, he provided Oskar Morgenstern and Karl Schlesinger with tutors. Oskar Morgenstern, who was about the same age as Menger, was at the time head of the Institut für Konjunkturforschung (Institute for Business Cycle Research) and editor in chief of the Zeitschrift für Nationalökonomie, where he had hastened to repair his predecessor's blunder and publish Menger's previously rejected paper on the St. Petersburg paradox (K. Menger, 1934b), a paper to which Menger returned in (K. Menger, 1934a, c) as well as in the last section of the present essay. Karl Schlesinger was a banker with a strong bent for theory. And the tutors were the geometers Franz Alt and Abraham Wald, two former students of Menger and mainstays of the Colloquium who, in spite of their brilliance, had failed to find jobs in an Austria beset with an economic recession and a pervasive anti-semitism. The impact on Morgenstern is discussed in Robert Leonard's commentary.

Menger describes in detail how Schlesinger discovered that in the usual formulation of Walras' equations relating the prices and quantities of factors and products, a system of equations due to Cassel, it is assumed that there are no surplus factors. Schlesinger proposed that this systems of equations should be turned into a system of equations and inequalities expressing the fact that, for each factor, either the surplus or the price is zero. This modified system was presented in a Kolloquium lecture (K. Schlesinger, 1934). In the very next lecture, Wald proved the existence of a unique equilibrium solution for that system (Wald, 1934). This led to further ground-breaking work by Wald (Wald, 1935, 1936) culminating in his proof of the existence of a unique equilibrium of prices in a pure exchange economy. (The corresponding manuscript for the Ergebnisse eines Mathematischen Kolloquiums was lost in the turmoil of Austria's annexation to the Third Reich, and has since become something of a legend.) Wald's result led, again through Menger's mediation, to the publication (in the Ergebnisse) of John von Neumann's dynamical model of a growth economy, based on a closed production loop (J. von Neumann, 1937). After the war, the results of both Wald and von Neumann gave rise to a veritable explosion of mathematical methods in economics, mostly based on fixed point arguments. The 'general equilibrium theory' became for decades the leading paradigm in economical theory, leading to many Nobel Prizes (see the articles by Arrow, Samuelson and Punzo in Dore et al., 1989).

In his subtle analysis of the crucial turning point of this story – Schlesinger's model and Wald's solution – Menger stresses that a mathematical economist (Walras) had made an unwarranted assumption which was brought to light by an Austrian marginalist whose verbal, rather than mathematical, mode of reasoning may actually have helped him psychologically. That insight led in turn to enormous progress in the mathematical theory. But the story did not end there. The assumptions behind Wald's proof are still fueling critical debates in economic circles (W. Hildenbrand, 1998).

By bringing together economists and mathematicians, Menger had played a catalytic role for which he was uniquely predisposed by his upbringing and education.

References

W. Hildenbrand: An exposition of Wald's existence proof. In: Karl Menger: Ergebnisse eines Mathematischen Kolloquiums (eds.: E. Dierker, K. Sigmund), Springer-Verlag, 1998, pp. 51–61.

K. Sigmund: Menger's Ergebnisse – a biographical introduction. In: Karl Menger: Ergebnisse eines Mathematischen Kolloquiums (eds.: E. Dierker, K. Sigmund), Springer-Verlag, 1998, pp. 51–61.

C. Menger: Grundsätze der Volkswirtschaftslehre, second edition, Springer-Verlag Wien, 1922.

K. Menger: Bernoullische Wertlehre und Petersburger Spiel. Ergebnisse eines mathematischen Kolloquiums 6 (1934a) 26–7.

K. Menger: Das Unsicherheitsmoment in der Wertlehre. Zeitschrift für Nationalökonomie 5 (1934b) 459–485.

K. Menger: Bemerkungen dazu. Zeitschrift für Nationalökonomie 6 (1935) 283–5.

M. Dore, P. Chakravarty, R. Goodwin (eds.): John von Neumann and Modern Economics, Oxford University Press, 1989.

J. Schumpeter: History of Economic Analysis, Oxford University Press, 1954.

J. von Neumann: Über ein ökonomisches Gleichgewichtssystem und eine Verallgemeinerung des Brouwerschen Fixpunktsatzes. Ergebnisse eines mathematischen Kolloquiums 8 (1937) 73–83.

K. Schlesinger: Über die Produktionsgleichungen der ökonomischen Wertlehre. Ergebnisse eines mathematischen Kolloquiums 6 (1934) 10.

A. Wald: Über die eindeutige positive Lösbarkeit der neuen Produktionsgleichungen. Ergebnisse eines mathematischen Kolloquiums 6 (1934) 12–18.

A. Wald: Über die Produktionsgleichungen der ökonomischen Wertelehre (II). Ergebnisse eines mathematischen Kolloquiums 7 (1935) 1–6.

A. Wald: Über einige Gleichungssysteme der mathematischen Ökonomie. Zeitschrift für Nationalökonomie 7 (1936) 637–70.

K. Menger: Österreichischer Marginalismus und mathematische Ökonomie. Zeitschrift für Nationalökonomie 32 (1972) 19–28.

K. Menger: Reminiscences of the Vienna Circle and the Mathematical Colloquium, Vienna Circle Collection Vol. 20, Kluwer, Dordrecht, 1994.

K. Menger: Selected Papers in Logic and Foundations, Didactics, Economics, Vienna Circle Collection Vol. 13, Kluwer, Dordrecht, 1979.

K. Menger: Bemerkungen zu den Ertragsgesetzen. Zeitschrift für Nationalökonomie 7 (1936a) 25–26.

K. Menger: Weitere Bemerkungen zu den Ertragsgesetzen. Zeitschrift für Nationalökonomie 7 (1936b) 388–397.

3

Austrian Marginalism and Mathematical Economics

KARL MENGER

THE Austrian marginalists and the mathematical economists agree in most of their fundamental economic views. Yet the relations between the two schools have not always been marked by sympathy and mutual understanding. Since I am the son of the author of the *Grundsätze* as well as a mathematician, 'two souls reside within my breast'. I hope that this fact will help me in discussing various points of the controversy objectively.

The difference between the two methods that first meets the eye is the fact that the Austrians formulate their assertions in sentences of the common language and connect them by logical reasoning whereas the mathematical economists express theirs in formulae which they transform and combine mathematically.

In order to illustrate this difference in a neutral field, Irving Fisher[1] pointed out that even the laws of physics lend themselves to verbal as well as formal formulations. Actually, one can take a further step into mathematics itself. Even algebra can be presented with or without formulae although countless elementary textbooks claim that the (phylogenetic as well as ontogenetic) beginning of algebra is marked by the introduction of letters as symbols. But this view is both historically and logically untenable. *Historically* the claim is false since the branch of mathematics called algebra had its origin in the Persian book *al jebr'* in which some time before A.D. 1000 Alkhowarizmi developed the theory of quadratic equations in sentences of the common language without introducing a single symbol; and in this purely verbal form

---

[1] *Mathematical Investigations in the Theory of Value and Prices* (1892).

algebra remained for almost 500 years. *Logically* the claim is unjustified since each algebraic assertion expressed in a formula can, in principle, also be formulated purely verbally. The sentence

(1) Any real number increased by 1 is equal to 1 increased by that number

is perfectly equivalent with the assertion[2]

(1′) $x+1 = 1+x$ for any real number x.

But only in principle can formulae be avoided throughout algebra; practically, they are inevitable in assertions of greater complexity, especially in statements concerning several arbitrary numbers. In verbal form, even a relatively simple proposition such as the general binomial theorem is much too involved for a listener to grasp or for readers to retain in their memory. This is clearly what Fisher had in mind when he compared expressions in formulae to a trip across America by railroad, and verbal formulations to a transcontinental walk.

But to what extent is Fisher's comparison valid for the Austrian marginalists and the mathematical economists? A study of this question may appropriately begin with a comparison of the two methods in the case of the Principle of Marginal Utility, even though this case is mainly of historical interest. But since the relations described in that principle are similar to those that obtain for return, demand, supply, etc., the scope of the study transcends that principle and includes concepts of great current interest; and since the measurability of return, etc., is unquestioned, even that of utility need not be discussed at this point where the interest is concentrated on the form rather than the contents of the principle.

AUSTRIAN FORMULATION. *For each good, the utility of a larger quantity is greater (or at any rate not less) than that of a smaller quantity, whereas the marginal utility of the larger quantity is less (or at any rate not greater) than that of the smaller.*

---

[2] The formula (1′) includes what in logic and algebra is called a *variable*, i.e. a letter that stands for any element of a certain class, viz. the letter x which, according to the legend in (1′), may be replaced in the formula by the designation of any element of the class of all real numbers, e.g. by the numeral 5 for the number five, thus yielding the particular assertion $5+1 = 1+5$.

828181   D

MATHEMATICAL FORMULATION. *If q denotes the quantity of a good, and u its utility, then*

$$u = f(q),^3 \quad \frac{du}{dq} = f'(q) \geqslant 0, \quad \text{and} \quad \frac{d^2u}{dq^2} = f''(q) \leqslant 0.$$

To this day, some mathematical economists believe that these formulae express *more* than the simple words of the Austrians and, furthermore, that they describe the situation *more precisely*. But neither of these claims seems to be justified.

Far from saying more, those formulae actually say *less* than the Austrian formulation since they express *the same assertion under an additional, if tacit, hypothesis*, viz. the assumption that the function connecting utility with quantity admits a second derivative and its graph, therefore, has a curvature at each of its points—an additional hypothesis that clearly is not anchored in economical facts. The Austrian formulation of decreasing marginal utility is *more general* since it is valid even if there are places where the function does not admit a second derivative and its graph has no curvature,[4] whereas at such places the mathematical formulation fails to assert anything. Thus if one wants to compare the two methods to crossings of a continent, then, as far as convexity principles (in particular, the Principle of Marginal Utility) are concerned, one must not say that the first is by foot and the second by train, but rather that the first is unencumbered while the second is a crossing with unnecessary and obnoxious baggage.

Many mathematical economists since Cournot have used tacit assumptions in the theories of return, supply, demand, and so on, by assuming continuity and differentiability of functions as though these properties were matters of course, whereas they are nothing but prerequisites for the application of classical analysis and not based on facts. This point deserves being stressed since mathematical economists consider it as one of the advantages of their

---

[3] Here f denotes some unidentified function with non-negative first, and non-positive second, derivative. The first to realize the decrease of marginal utility, namely in the case of monetary gains, the mathematicians Cramer (1728) and D. Bernoulli (1738), identified f with well-known simple functions. They equated $f(q)$ with $C\sqrt{q}$ and $C \log(1+q/Q)$, respectively, for some number C, where Q is the fortune to which q is added.

[4] The function assuming for x the value $3x-x^2$ if $0 \leqslant x \leqslant 1$, and the value $\frac{1}{2}+2x-\frac{1}{2}x^2$ if $1 \leqslant x \leqslant 2$ has for $x = 1$ the value 2, the derivative 1 but no second derivative. Its graph has everywhere a tangent, in particular at the point with the abscissa 1, but no curvature at that point. Cf. p. 42 n. 10 below.

method that it brings tacit assumptions of verbal formulations to the surface. The Principle of Marginal Utility is a case—and as we shall see, not the only one—illustrating the opposite situation. As far as *precision* is concerned, consider, for example, the statements[5]

(2) *To a higher price of a good, there corresponds a lower (or at any rate not a higher) demand.*

(2') *If p denotes the price of, and q the demand for, a good, then*

$$q = f(p) \quad and \quad \frac{dq}{dp} = f'(p) \leqslant 0.$$

Those who regard the formula (2') as more precise or 'more mathematical' than the sentence (2) are under a complete misapprehension just as are the textbook writers who consider the formula (1') as more precise or 'more algebraic' than the verbal sentence (1). The only difference between (2) and (2') is this: since (2') is limited to functions that are differentiable and whose graphs, therefore, have tangents (which from an economic point of view are not more plausible than curvature), the sentence (2) is *more general*, but it is by no means less precise; it is *of the same mathematical precision* as (2').

The point at which the precision of some Austrian economists does indeed leave something to be desired is, as Schumpeter[6] noticed as early as 1908, their definition of *Grenznutzen (marginal utility)*—a concept which, by the way, is not expounded in Menger's *Grundsätze*. Jevons[7] introduced the *final degree of utility* as 'the degree of utility of the last addition or the next possible addition of a very small, or infinitely small, quantity to the existing stock'. Since by *degree of utility* Jevons means the ratio of utility to quantity, in modern terms his final degree thus is either $du/dq = f'(q)$, i.e. the limit of the difference quotients

$$\frac{f(q+h)-f(q)}{h} \tag{3}$$

as *h* gets arbitrarily small, or the quotient (3) itself for the smallest usable increment *h*. Only if (a) a good comes in equal pieces

---

[5] The well-known exceptions to these statements due to snobbism are irrelevant for our present purpose.

[6] *Wesen und Hauptinhalt der theoretischen Nationalökonomie* (Leipzig, 1908), p. 105 f.   [7] *Theory of Political Economy* (1871).

any one of which serves as unit and (b) such a piece is the smallest usable increment (so that $h = 1$) is that final degree of utility equal to *the utility of one unit* as which Wieser[8] originally defined *Grenznutzen*. If, however, one can utilize, for example, one-half or one-tenth of a unit (in other words, if the good is measured in pairs or tens of usable amounts), then the utility of such a part is of course much smaller than Wieser's marginal utility and one must divide it by $\frac{1}{2}$ and $\frac{1}{10}$, i.e. multiply it by 2 and 10, respectively, in order to obtain the marginal utility and an approximation to the final degree.[9]

The Austrians, as has just been shown, not only steer clear of the concepts of calculus (limit, derivative, rate of change); in keeping with their ordinal conception of utility, they even eschew arithmetical operations such as dividing or multiplying a utility by a number, even though dividing utility by the quantity of a good would be indispensable if they considered difference quotients such as (3) for $h \neq 1$, i.e. at least average rates of change. The problem thus arises whether the Principle of Marginal Utility can at all be stated in conformity with those Austrian restrictions. This is indeed possible and even in a form that is valid both for Wieser's marginal utility and for Jevons's final degree.

Stripped of superfluous differentiability assumptions, which creep even into the very definitions of marginal utility by Gossen, Jevons, and Walras, the principle asserts that the function f connecting utility with quantity be (non-decreasing and) *convex*, from above, or, briefly, convex, i.e. satisfy the condition

If b is between a and c, then

$$f(b) \geqslant b \cdot \frac{f(c)-f(a)}{c-a} - \frac{af(c)-cf(a)}{c-a}; \qquad (4)$$

in geometric terms, that the graph of f nowhere dip below any of its chords.[10] While this definition of convexity is altogether inde-

[8] *Ursprung und Hauptgesetze des wirthschaftlichen Werthes* (1884), p. 128.

[9] This is why Böhm-Bawerk's and Wieser's later definition of marginal utility as 'the smallest obtainable utility' (cf. Wieser's *Der natürliche Werth* (1889), p. 12) is (as far as Böhm-Bawerk is concerned, admittedly) imprecise.

[10] A non-decreasing convex function has everywhere a left-side and a right-side derivative—at the initial point of course only one on the right side, which may be infinite. The assumption that the function f connecting utility with quantity be convex thus guarantees the existence of finite degrees of utility with regard to decrements and with regard to increments of the quantity, the latter

pendent of differentiability, (4) does require division of utilities by numbers and thus still fails to meet one of the Austrian requirements.

In the two most important cases, however, viz. where the quantity varies continuously or where the good comes in equal pieces, the convexity of f can be expressed not only without an explicit or tacit assumption of differentiability but even without arithmetical operations on utilities beyond the comparison of differences between two utilities. In these cases, the Principle of Marginal Utility thus can be generalized *without assuming the existence of a marginal utility* to what may be called *ordinal* principles. The claim that these be generalizations of the traditional principle means that in cases where the function happens to be differentiable (or twice differentiable) one can deduce from them the Austrian (or the mathematical) form of the principle. The two following generalizations must be taken in conjunction with the (ordinal) first half of the Austrian formulation: For each good, the utility of a larger quantity is greater (or at any rate not less) than that of a smaller quantity.

FIRST ORDINAL GENERALIZATION. *If two quantities of a good are both increased or both decreased by the same amount, then the utility of the greater changes less (or at any rate not more) than that of the smaller.* In symbols,

$$\text{If } x_1 < x_2, \quad \text{then } |f(x_2+h)-f(x_2)| \leqslant |f(x_1+h)-f(x_1)|.$$

The preceding inequality implies the one in the following generalization if $x_1+h = x_2 = x$.

SECOND ORDINAL GENERALIZATION. *For any quantity of a good, the increase in utility due to the addition of an increment to the quantity is less (or at any rate not greater) than the loss of utility due to the subtraction of the same amount.* In symbols,

$$f(x+h)-f(x) \leqslant f(x)-f(x-h) \text{ for each } x \text{ and } h \text{ (even for } h < 0)$$

These generalizations[10a] can easily be shown to imply the convexity of f in the two principal cases, i.e. for continuously varying

degree at no point greater, but at each corner of the graph of f actually smaller, that the former (e.g. at each corner of a polygonal graph). The derivative f', however, may at some places not even possess derivatives at either side (which it does possess in the simple example on p. 40 n. 4 above). At such places, the traditional formula $d^2u/dq^2 \leqslant 0$ is quite meaningless.

[10a] The second in conjunction with the fact that the utility never decreases as the quantity increases.

quantity and for goods that come in equal pieces (even if one-half or one-tenth of a unit can be utilized; that is to say, if 2 or 10 pieces serve as unit). On the other hand, *they do not necessarily imply convexity if the good comes in unequal pieces.* For example, if a good comes only in two equal pieces that serve as units and in one piece of measure 1·4, then the possible quantities are

$$0, 1, 1·4, 2, 2·4, 3·4.$$

If the corresponding utilities are

$$0, 1, 1·5, 2, 2·4, 3·3,$$

then all inequalities postulated in the two preceding generalizations are easily seen to be satisfied and yet the graph of f is not convex, as is shown by the triples of points above 0, 1, 1·4, and 1·4, 2, 2·4.

As has just been shown, one can formulate in the Austrian fashion (i.e. in ordinal and verbal form) general principles without using, or referring to, any ideas of calculus or even the very concept of marginal utility in any sense. Yet a question of an altogether different nature remains open—a psychological problem which has always puzzled me. How is it to be conceived that men of such high intelligence and such specific logical-analytic talent as the members of the old Austrian School did not have a better know-ledge and understanding of mathematics, especially of calculus (Schumpeter being the only exception), even if, for whatever reasons, they decided to refrain from using it in economic theory? It is known that in contrast to Walras and Pareto who had a sound training in analysis as engineering students, and to Jevons, who was an accomplished mathematical logician, the Austrians came to economic theory from jurisprudence, the government, and economic activities; and in the old Austrian *Gymnasien*, they had not received any instruction in mathematical analysis. Still, they might have resorted to a self-study from textbooks in later life; and in the 1890s my father indeed started such a self-study, as is clear from a three-page introduction into the elements of differential calculus in his handwriting, which he had bound into his copy of the second edition of Walras's *Éléments d'économie politique pure.*[11]

[11] Also in my possession is a series of about twenty notebooks, dated 1867–8, which contain excerpts from the classics of economy interspersed with the first sketches of his own ideas on the theory of value. One of these notebooks includes a couple of pages with proportions and graphs, in which utility is

But I am afraid that he did not acquire an operative knowledge, let alone a critical insight into calculus.

The psychological problem is thus reduced to explaining why such eminent minds as the founders (and perhaps also several younger members) of the Austrian School were, as mature men, unsuccessful in their self-study of analysis. My explanation of this phenomenon is this: the textbooks, while adequate, and in a few cases excellent, *when accompanied by oral explanations*, were (and, I am afraid, in most cases still are) such that without oral help from an experienced teacher the reader can hardly overcome the difficulties created by numerous equivocations, symbolic shortcomings, and conceptual confusions. Of these difficulties I mention here only the following:[12] about a dozen of totally different uses of the letter x; the lack of symbols for the identity, the power functions, and the constant functions with the resulting equivocal use of the same symbol for a function ('the functions x, $f(x)$, log x,...') and for its value for x ('the numbers x, $f(x)$, log x,...'); the utter confusion of the variables in the logico-algebraic sense explained in footnote 2, p. 39 above, with the totally different scientific concept of what, reviving a term introduced by Newton, I call *fluents*.[13] Even today successful beginners usually overcome

represented by the area under an (approximately linear) graph, the latter thus representing marginal utility. But these pages are crossed out and none of the later entries in those notebooks return to that 'mathematical' exposition of his ideas.

[12] For details see the author's papers: 'What are x and y?', *Mathematical Gazette*, 40 (1956). 'On Variables in Mathematics and in Natural Science', *British Journal of the Philosophy of Science*, 5 (1954). 'The Ideas of Variable and Function', *Proceedings of the National Academy of Science of the U.S.A.*, 39 (1953). 'An Axiomatic Theory of Functions and Fluents' in Henkin *et al.* (eds.) *The Axiomatic Method* (Amsterdam, 1959). 'A Counterpart of Occam's Razor', *Synthese*, 12 (1960), and 13 (1961). In the field of economics, H. Dickson in Göteborg was the first to discuss equivocal uses of the term *variable*, especially in connection with R. G. D. Allen's books. Cf. Dickson's book *Variable and Function. A semantic study in mathematics and economics* (Göteborg, 1967).

[13] An economic fluent is the result of pairing a number with each element of a class of economic objects or elements relevant for economics, e.g. connected with time. That class is called the *domain*, the associated numbers are referred to as the *values*, of the fluent. For example, consider a good G, a buyer B, and a market visited by B where G is being sold. With every instant τ one can pair a number $q_B(\tau)$ of units of G which, in view of the market situation (as far as it is known to him) B is willing to buy. The result of this pairing, i.e. the class of the pairs $(\tau, q_B(\tau))$ for all instants τ, is a fluent q_B, called B's demand for G. By aggregation, one obtains the demand q for G, a fluent consisting of the pairs $(\tau, q(\tau))$ for all τ. Similarly, one can pair with each τ the price $p(\tau)$ of a unit of G at the instant τ (usually concocted from various prices arrived at in more or less

those difficulties only if and when a teacher shows in numerous *examples* how certain symbols are used in a certain context and in what other ways those same symbols are used in certain other contexts. They do not find *articulate rules* for the use of those symbols in elementary textbooks. They learn the elements of calculus somewhat as a child learns his native tongue rather than the way an adult learns a foreign language (beyond a smattering)— by explicit rules of grammar. While a logical and analytic mind is perfectly sufficient for the self-study of any of the numerous good books on the theory of real functions or on abstract algebra (fields as yet of little use for economics) such a disposition hinders more than helps a mature man, with limited time at his disposal, in a self-study of calculus.[14] Austrian economists seem to have been victims of this situation.

I am now turning to problems that can be translated into systems of equations. These problems also lend themselves to

simultaneous deals) and thus arrive at a fluent p, the price of G, consisting of the pairs $(\tau, p(\tau))$ for all τ. It is evident that fluents in this sense and variables in the logico-algebraic sense defined in footnote 2 p. 39 above belong to completely different ontological categories, even though in applied mathematics, science, and economics both are indiscriminately referred to as *variables*. The situation is complicated by the fact that general assertions about fluents utilize symbols that stand for any element of a class of fluents, thus *fluent variables*. Mathematical economists elaborate classifications into endogenous, exogenous, autonomous, etc., 'variables', thereby always thinking of fluents without, however, explicitly and in full generality defining those objects which they classify. This is in keeping with the somewhat neo-Pythagorean attitude of the applied mathematicians, who consider only the numerical values and neglect the objects belonging to the domain of fluents although a completely articulate application of mathematics to fluents must, of course, take those objects into account (cf. p. 53 n. 21).

[14] A corroboration of this point of view seems to come from some of the so-called tests for logical thinking, which include questions such as 'What is the next number in the (!) sequence 1, 3, 6, 10 ... ?' and (in slightly simplified form) 'Which of the following four figures differs from the other three: a square; a cross; a circle; a triangle?' Questions of this kind do indeed test a valuable ability, viz. the ability to guess what the man who formulated them had in mind, e.g. the circle, because it is the only figure that is round (although, of course, the square is the only one with four corners; the cross the only one with a ramification point; the triangle the only one with exactly three corners). What in this way certainly is *not* tested is logical thinking. Yet the marks youngsters make in such tests highly correlate with their achievements in elementary mathematics! Far from demonstrating that those test questions have more to do with logic than with guessing and empathy this correlation rather seems to indicate that the presentation of elementary mathematics has more to do with guessing and empathy than with logical thinking.

perfect verbal *formulations*. But he who looks for specific *solutions* is practically forced to switch, in Fisher's simile, from pedestrian verbalism to the vehicle of mathematics.

For various reasons, it seems appropriate to choose as the first example the famous question that Wieser has called *Zurechnungs-problem* (problem of imputation). To begin with, there is complete agreement between the Austrians and the mathematical economists about the economic background of the problem. Whereas the classical economists had taught that the prices of products were determined by the prices of the means of production—briefly, of the factors—both schools founded a century ago recognized that, on the contrary, the prices of the factors were determined by the expected prices of the products; and naturally both schools asked how the prices of the latter are to be imputed to the former. Another reason for choosing this example is the fact that, prices and quantities being the only objects referred to in the problem, all objects are clearly measurable; and that, in connection with the *Zurechnungsproblem*, even Wieser mentioned equations, if only in very primitive examples. Thirdly, the (in a way final) solution of the problem was achieved in Vienna through the co-operation of a mathematician, Abraham Wald, and an economist connected with the Austrian School, Karl Schlesinger.

I had frequently discussed the problem with members of the Mathematical Colloquium that I conducted in Vienna in the years 1928–36, especially with my student and friend Wald. But I must confess that I was on the wrong track. I believed imputing the price of a product to its factors to be somewhat analogous to finding the distribution of the weight of a horizontal plate over its various points of support. In statics this problem is insoluble if the plate is supported at more than three points, since statics supplies only three linear equations. A table with more than three rigid legs on a rigid ground always rests, in a more or less unstable way, on three legs and shakes. In constructing bridges and the like, where it is imperative to rule out any shaking, the support is not rigid and by supplementing the static considerations with the theory of elasticity one indeed obtains a unique solution of the (non-linear) problem. I asked myself whether there was perhaps an economic analogue of the elasticity considerations.

At that time Schlesinger, who complained about difficulties in his self-study of mathematics, on my recommendation asked Wald to tutor him. What made Schlesinger particularly eager to study more mathematics was his idea that Walras's method of imputation was based on an unjustified tacit assumption; and Schlesinger suspected that that very assumption might be responsible for the fact that Walras's equations have in general no solution.

Let us consider m factors $R_1,..., R_i,..., R_m$ and n products $S_1,..., S_j,..., S_n$ and assume that the production of each unit of S_j require exactly a_{ij} units of R_i. These mn numbers a_{ij} are called *technical coefficients*. (They are of course non-negative; and since each S_j requires at least one factor, at least one of the m numbers $a_{1j},..., a_{ij},..., a_{mj}$ is positive for each j.)

With each process π of production of the n products S_j from the m factors R_i there are associated $2(m+n)$ numbers:

(1) the m quantities $r_i(\pi)$ of R_i available for the process π, and the n quantities $s_j(\pi)$ of the S_j resulting from it;

(2) the m prices $\rho_i(\pi)$ of a unit of R_i at the beginning and the n expected prices $\sigma_j(\pi)$ of S_j at the end of π.

By pairing those numbers with π one defines $2(m+n)$ fluents, $r_i, s_j, \rho_i, \sigma_j$, viz. the classes of all pairs (cf. p. 45, n. 13).

$$(\pi, r_1(\pi)), ..., (\pi, \sigma_n(\pi)).$$

These fluents are interconnected by economic laws the first of which can be expressed in verbal form as follows. *The price of each product is equal to the sum of the prices of the factors used in its production.* This is a conjunction of n assertions, one for each product $S_1,..., S_j,..., S_n$. The assertion about S_j can be translated into the following formula about *the values for π of the fluents* $\rho_1,..., \rho_i,..., \rho_m$ and σ_j

$$a_{1j}\rho_1(\pi)+...+a_{ij}\rho_i(\pi)+...+a_{mj}\rho_m(\pi) = \sigma_j(\pi)$$

for any process π, as well as into the following connection of *the fluents themselves*

$$a_{1j}\rho_1+...+a_{ij}\rho_i+...+a_{mj}\rho_m = \sigma_j.$$

Altogether Walras thus obtained

$$n \text{ FORMULAE ABOUT PRICES}$$

$$a_{11}\rho_1+...+a_{ij}\rho_i+...+a_{mn}\rho_m = \sigma_1,$$
$$\cdots \cdots \cdots \cdots$$
$$a_{1n}\rho_1+...+a_{in}\rho_i+...+a_{mn}\rho_m = \sigma_n.$$

Laws connecting fluents can be used in attempts to express some of them in terms of the other ones and thereby become *equations* in the sense in which this word is used in algebra: problems of 'finding' certain 'unknowns', by which is meant; expressing them in terms of certain other entities said to be 'known'.

From the premarginalistic point of view, the m factor prices $\rho_1, ..., \rho_m$ are known and the prices of the n products are to be determined. Since the formulae (I)—the formulae as they stand, without any transformation!—express the σ_j in terms of the ρ_i the problem is solved without the application of any mathematical machinery. From the marginalistic point of view, however, the n expected prices $\sigma_1, ..., \sigma_n$ are known and the prices of the m factors are to be determined. But the formulae (I), even after all kinds of transformations, do not in general permit us to express the ρ_i in terms of the σ_j.

Walras had the splendid idea to introduce also the quantities r_i and s_j into the problem of imputation. Of the $r_i(\pi)$ units of the factor R_i that are available for the process π, the number of units used in the production of $s_j(\pi)$ units of S_j is $a_{ij} s_j(\pi)$, by definition of the technical coefficients. Walras postulated

$$a_{i1} s_1(\pi) + ... + a_{ij} s_j(\pi) + ... + a_{in} s_n(\pi) = r_i(\pi)$$

for each process π and thus, in terms of fluents

$$a_{i1} s_1 + ... + a_{ij} s_j + ... + a_{in} s_n = r_i.$$

Altogether he obtained in this way another system of linear connections, viz. the following

m Formulae about Quantities

$$
\begin{aligned}
a_{11} s_1 + ... + a_{1j} s_j + ... + a_{1n} s_n &= r_1, \\
\cdots \cdots \cdots \cdots \cdots \cdots & \qquad \text{(II)} \\
a_{m1} s_1 + ... + a_{mj} s_j + ... + a_{mn} s_n &= r_m.
\end{aligned}
$$

In verbal form, these postulates can be expressed—although it has not often been done—in the sentence: *For each factor, the quantity available for a production is used in the process.*

Apart from the inverted matrix of the technical coefficients, the formulae (I) and (II) have nothing in common: they connect prices and connect quantities, respectively. Walras bridges this gap by adjoining the connections of the expected prices of the products with the quantities produced. Let F_j be the (in general non-linear) function expected to connect the price σ_j of S_j with the quantities $s_1, ..., s_j, ..., s_n$ of the various products. (While σ_j mainly depends on the quantity s_j of S_j one cannot neglect its dependence on the other products because of possible substitutes for S_j.) Altogether, Walras adjoined the following

n Formulae about the Products

$$\sigma_1 = F_1(s_1, ..., s_j, ..., s_n),$$
$$\cdots \cdots \cdots \qquad \qquad \text{(III)}$$
$$\sigma_n = F_n(s_1, ..., s_j, ..., s_n).$$

Clearly (I) and (III) can be combined into the n formulae

$$a_{11}\rho_1 + ... + a_{i1}\rho_i + ... + a_{m1}\rho_m = F_1(s_1, ..., s_j, ..., s_n),$$
$$\cdots \cdots \cdots \cdots \cdots \cdots \qquad \text{(A)}$$
$$a_{1n}\rho_1 + ... + a_{in}\rho_i + ... + a_{mn}\rho_n = F_n(s_1, ..., s_j, ..., s_n).$$

Thereby the fluents σ_j are eliminated and all that has to be considered are the quantities of the factors and products and the prices of the former, these $2m+n$ fluents being interconnected by the $m+n$ relations (A) and (II).

The quantities of the m factors being known, Walras conceived the ingenious plan to use those $m+n$ formulae as a system of $m+n$ equations for the $m+n$ 'unknowns' ρ_i and s_j; that is to say, to determine, in terms of the m fluents r_i and the technical coefficients, not only the prices of the m factors but also the quantities of the n products. The Austrians assumed the expected prices of the products to be known. Walras replaced this by the mere assumption that the s functions F_j expected to connect the prices with the produced quantities be known, whence also the expected prices σ_j can be ascertained after the quantities have been determined. These n quantities s_j together with the m prices ρ_i of the factors are the unknowns in the $m+n$ equations.

For the next sixty years, the coincidence of the number of equations and the number of unknowns was sufficient to satisfy all mathematical economists (unbothered by the fact that even two linear equations in two unknowns may have *no* solution, as $x+y = 1$ and $2x+2y = 3$, or *infinitely many* solutions, as $x+y = 1$ and $2x+2y = 2$) of the existence of a unique solution of Walras's equations. Actually, these equations need not have any solution; and the reason lies in the fact pointed out by Zeuthen[15] and

15 *Zeitschrift für Nationalökonomie,* 4 (1933).

Schlesinger[16] that equations (II) are based on the tacit assumption that the entire available quantities of all factors, including those whose price will turn out to be 0, are used up in the production. Evidently this assumption is false, e.g. for water—an indispensable factor in many productions. Of course, obviously worthless goods might be struck from the list of the factors and be ignored in the equations. But there may be factors R_i about which at the outset it is doubtful whether a positive surplus (an *Überschuss*, $\ddot{u}_i$, as Schlesinger said) will be left and $\rho_i = 0$, or nothing will be left over (that is, $\ddot{u}_i = 0$) and $\rho_i > 0$; and such factors cannot be ignored. This suggests that the equations (II) should be replaced either by inequalities or by m equations with m new unknowns $\ddot{u}_1,..., \ddot{u}_i,..., \ddot{u}_m$, viz.

$$a_{i1} s_1 + ... + a_{ij} s_j + ... + a_{in} s_n + \ddot{u}_i = r_i \qquad \text{(for } i = 1, ..., m) \qquad \text{(B)}$$

and m additional non-linear equations

$$\ddot{u}_i \rho_i = 0 \qquad \text{(for } i = 1, ..., m) \qquad \text{(C)}$$

expressing that for each factor R_i either the surplus or the price is 0 (or both are 0).

The great achievement of Wald[17] was the proof that the $2m+n$ equations (A), (B), (C) do indeed have a unique solution consisting of $2m+n$ fluents ρ_i, $\ddot{u}_i$, s_j assuming only non-negative values, provided that the functions F_j connecting the prices of the products with the quantities produced satisfy certain conditions implied by the Principle of Marginal Utility. Wald's proof is far from simple, in fact it involves deeper mathematical considerations than any study in mathematical economics published before 1934. As I predicted[18] in the discussion following his presentation of his results in the Colloquium, Wald's paper brought to an end the epoch when mathematical economics was confined to setting up equations without solving them and to merely counting equations and unknowns.

What is of particular interest in connection with this present paper is (1) the fact that, with regard to the problem of imputation, again a tacit assumption made by a mathematical economist (in

---

[16] *Ergebnisse eines mathematischen Kolloquiums*, 6 (March 1934).
[17] Ibid.; also 7 (November 1934). For a summary of Wald's results concerning Walras's equations (both about production and about exchange), but without proofs, cf. *Zeitschrift für Nationalökonomie*, 7 (1936), 637.
[18] *Ergebnisse eines mathematischen Kolloquiums*, 6 (March 1934).

this case, by the greatest of them) was brought to the surface by a member of the Austrian School, and (2) that *psychologically* the erroneous character of that assumption seems to become more obvious in its (seldom mentioned) verbal formulation than in the formulae (II), although, of course, *logically* the verbal statement and the formulae are equivalent.

There are, as Wald pointed out, several other tacit assumptions underlying the equations (A), (B), (C): that products are not factors; that no product can be produced in different ways; and the like. And there is another difficulty. Some factor, says R_1, may be superabundant in the production of the n goods S_j so that $\ddot{u}_1 > 0$ and $\rho_1 = 0$. Yet when other uses of R_1 are taken into account there may be a shortage of R_1 and ρ_1 may turn out to be positive. In order to be consistent one therefore must not only take into account all factors of the products considered but also all products of the factors considered. But this leads to a system of equations embracing almost all goods. In view of this unmanageable situation, some of the Austrian attempts to impute by trial and error and successive approximations[19] seem to me still worth reading although they do not precisely determine the prices of the factors. But since physicists in statics, the oldest chapter of their science, must resign themselves to rickety tables it is not clear why mathematical economists expect unique solutions in infinitely more complex situations.

The history of the problem of imputation is also of interest in connection with one of three *methodological* differences between the Austrian and the mathematical economists that ultimately can be traced to the correspondence between Walras and Menger in the early 1880s.[20] While Walras claimed that mathematics was a means of research Menger only conceded that it might be a method of presentation; and some later Austrians denied even this.

In the 1880s, even the best mathematicians and philosophers lacked our present insight into the role of mathematics with regard to extra-mathematical objects. As we now know, mathematics cannot create any statements about the physical or social universe. It is confined to *transformations* of assertions (formulae, verbal assumptions, etc.). Walras's famous systems of equations

[19] Cf. e.g. Wieser, op. cit., p. 88.   [20] *Économie appliquée*, 6 (1953).

concerning production as well as exchange certainly are examples of mathematical *presentations* of economic ideas—formulations without which one could not aspire at numerical solutions!—but they cannot be said to be the result of mathematical research. Such research with regard to imputation—and deep research at that—began with Wald's proof that Walras's corrected equations have a unique solution. This proof consists of transformations of those equations and uses ready-made mathematical machinery; that is to say, it applies mathematical theorems earlier discovered and proved for their own sake or for other purposes. Such a proof certainly is not a matter of mere presentation of economical ideas.

Skilful transformations may, of course, reveal most important consequences which had not even been suspected. Every wave that we are using in radio and television ultimately stems from Hertz's transformation of Maxwell's equations describing electro-magnetic fields. In economic theory, however, be it mathematical or Austrian, neither mathematical transformations of formulae nor logical transformations of verbal assumptions seem to have been carried out so far that would be of comparable importance.

The reasons for the difference between economics and classical physics are fairly clear. The functions connecting some of the most fundamental fluents of mechanics and the theory of electro-magnetic fields are (1) simple, i.e. identifiable with well-known functions that had been previously studied, especially powers, exponential, trigonometric, and logarithmic functions; (2) not changing in time. On the other hand, of the basic economic fluents only few are related by functions (especially by non-stochastic functions) at all, still fewer by identifiable functions, and of these few (usually econometrically determined) functions most are reminiscent of the rather complicated ('empirically' determined) functions familiar from the theory of frictions and not of the functions governing mechanics and field theory.[21] Besides there

---

[21] The attempts of Cramer and D. Bernoulli (cf. p. 40 n. 3) and later of Gossen to identify economic connections with simple functions were oversimplifications and failures; and even the simple production functions of the Cobb–Douglas type do not seem to be as significant as similar functions occurring in thermodynamics.

If u and w are fluents, say, the classes of pairs $(\alpha, u(\alpha))$ and $(\beta, w(\beta))$ respectively, then a functional connection between u and w presupposes a correlation of their domains, i.e. a class Γ of pairs (α, β). Relative to Γ, the fluent w is (1) a function of u, or (2) connected with u by the particular function f, or (3) functionally related with u by $F = 0$ if and only if for all pairs (α, β) belonging to Γ the

are important connections between economical fluents that are not permanent, as is especially demonstrated in Morgenstern's very interesting discussion of the connection between demand for a good and its price.[22]

A second methodological difference between some Austrian and some mathematical economists stems from the fact that the former look for *causal explanations* of some economic phenomena, whereas the latter wish to confine themselves to the study of *functional relations*, thereby, consciously or unconsciously, following the methodological programme of Mach. The Austrians consider the restriction of economics to the study of ratios of quantities (*Grössenverhältnisse*) as unwarranted while mathematical economists, quoting a somewhat bantering remark of Schumpeter's, call the Austrian aim illusory. But even physical science offers certain causal explanations which Mach himself might have slightly restated but certainly never ridiculed. Take, for example, the phenomenon of thunder. The road from the blow of the hammer of an irate god to the disturbance of the air, caused (*sit venia verbo!*) by an immense electric spark, is not paved with the discoveries of functional relations; neither is Wells's beautiful way of explaining dew, as presented by Herschel and J. S. Mill. Nor do ratios of quantities play a great role in those and similar studies which, though not world-shaking, certainly have produced legitimate insights. What makes mathematical economists expect that all of economics should be on a higher level than those scientific studies?

A third difference lies in the fact that the mathematical economists pay less attention than the Austrians do—perhaps sometimes

class of the pairs of numbers $(u(\alpha), w(\beta))$ satisfies the respective condition (1) it is a function, i.e. a class of pairs of numbers not including two pairs with equal first, and unequal second, members, or (2) it belongs to the function f, that is to say, $w(\beta) = f(u(\alpha))$, or (3) it satisfies the condition $F(u(\alpha), w(\beta)) = 0$ (cf. 'An axiomatic theory of functions and fluents', loc. cit. footnote 12). In the case of the equations (A), (B), (C) all $2m+n$ fluents have the same domain consisting of all processes of production and their correlation is relative to the set Γ of all $(2m+n)$-tuples $(\pi, \pi, ..., \pi)$.

'If u changes, then so does w', which countless books propose as a definition of w being a function of u, actually is neither necessary nor sufficient. 'If u does not change, neither does w' would be better but still superficial since it does not relativise the concept to a definite pairing of the domains. Between variables in the logico-algebraic sense, if one takes their definition in footnote 2 (p. 39, above) seriously, a functional connection is, of course, out of the question.

[22] *Quarterly Journal of Economics*, 62 (1948).

not enough attention!—to the definition of concepts represented by symbols whereas the Austrians, following the opposite extreme, are looking for the *essence* (*das Wesen*) of economic phenomena, thereby moving on dangerous ground, surrounded by swamps of pseudo-problems. What in the last analysis such studies may achieve seems to be the formulation of definitions of terms that are loosely used in everyday language. Somewhat similar problems arise, not so much in physical science as in geometry. Geometers in a way do look for the essence of connectedness, of curves, and the like, viz. in the sense of defining the geometric objects that deserve those names. A definition, for example, of curves circumscribes a class that (*a*) includes all objects that are always called curves, e.g. circles; (*b*) excludes all objects that are never called curves, e.g. cubes; and (*c*) extends the use of the term into the wide realm of objects between those two categories. The particular (and inevitably arbitrary) choice of the extension (*c*) must be justified by its mathematical fruitfulness; i.e. one must derive from, or by means of, the definition an extensive satisfactory theory, preferably one that is applicable, i.e. connects the subject with other domains of research. It was Schumpeter who in 1930, in a conversation at Harvard, pointed out the similarity of this problem and its solution to work of the Austrian School on value, capital, national product, etc. At any rate, such procedures of definition avoid pseudo-problems, provided, of course, that the ultimately *un*defined concepts in terms of which the others are defined are never lost sight of (and preferably are listed at the outset).

Wald's paper on the equations concerning production greatly interested von Neumann, as he told me when passing through Vienna soon after its publication. It reminded him of equations he had formulated and solved in 1932 and now offered to present to our Colloquium.[23]

von Neumann drops the distinction between factors and products. He studies n goods $G_1, ..., G_n$ whose prices are called $y_1, ..., y_n$. The goods can be transformed into each other in m processes $P_1, ..., P_m$. Thus a product in one process may be a factor in another and can be produced in various ways—assumptions that are much more realistic than Walras's. Let a_{ij} and b_{ij} be the numbers of units of G_j which in the process P_i are

[23] *Ergebnisse eines mathematischen Kolloquiums*, 8 (1937).

consumed and produced, respectively; and let x_i be the intensity with which P_i is carried out, $x_i = 0$ indicating that P_i is not being carried out at all. Since, altogether, not more of a good G_j can be consumed than is being produced, we have for $j = 1,..., n$ the relations

$$\alpha(a_{1j} x_1 + ... + a_{ij} x_i + ... + a_{mj} x_m)$$
$$\leqslant b_{1j} x_1 + ... + b_{ij} x_i + ... + b_{mj} x_m, \quad \text{(D)}$$

where α is a number indicating the expansion of the total economy. If less of G_j is consumed than produced, then G_j is a free good of price o. Hence

if in (D) the *in*equality sign holds, then $y_j = 0$. \hfill (D')

Clearly (D) and (D') are similar to the formulae (B) and (C), respectively. Indeed the latter might be written, without explicit introduction of the $\ddot{u}_i$, in the following form:

$$a_{i1} s_1 + ... + a_{in} s_n \geqslant r_i; \quad \text{and} \quad \rho_i = 0 \text{ if the *in*equality holds.}$$

Looser than the relation between (B), (C) and (D), (D') concerning quantities is the relation between (A) and von Neumann's following formulae (E) concerning prices. In money, the process P_i costs $a_{i1} y_1 + ... + a_{in} y_n$ and brings in $b_{i1} y_1 + ... + b_{in} y_n$. Since, in equilibrium, P_i cannot bring in more than it costs, we have for $i = 1,..., m$ the relations

$$\beta(a_{i1} y_1 + ... + a_{ij} y_j + ... + a_{in} y_n)$$
$$\geqslant b_{i1} y_1 + ... + b_{ij} y_j + ... + b_{in} y_n, \quad \text{(E)}$$

where β is a number indicating interest. If the cost of P_i is higher than its proceeds, then the process P_i is unprofitable and will not be carried out.

If in (E) the *in*equality sign holds, then $x_i = 0$. \hfill (E')

In contrast to Walras's formulae (III), no direct marginalistic connection between prices and quantities is assumed. But a remarkable duality exists between (D), (D') and (E), (E') and hence between uneconomical goods and unprofitable processes. Deep mathematical methods enabled von Neumann to prove the existence of a unique solution of the equations (D), (D'), (E), (E').

Even more fundamental than Walras's equations concerning production are his formulae concerning exchange and equilibrium prices. Here again Wald succeeded in proving the existence of a unique solution using mathematical methods that were not yet available when Walras published an inadequate proof.[24]

The Austrians in their theory of exchange, instead of looking for an exact equilibrium price, are satisfied, more realistically, with ascertaining an interval within which any price is advantageous to both bargaining parties. In this way, they only set up limits within which the price will be agreed upon as a result of bargaining; and the exact outcome depends on the skill and the knowledge of the parties. In Menger's *Grundsätze* we read:

The efforts of either party will be directed toward securing for himself as much of the economic advantage [in the afore-mentioned interval] as possible. This gives rise to the phenomenon called haggling. . . . The demands of either party will be the higher the less he knows of the economic situation of the other or the limits to which the other can go.

This passage—and a similar remark can be found in Jevons's work—may be homely, but it is a bridge to Morgenstern's economic applications of von Neumann's game theory, whereas the mathematical work of the Lausanne School, for all its ingenuity, does not lend itself to such a connection.

'Misers and mathematicians estimate money proportional to its amount' said Buffon in support of subjective, marginalistic estimation, reflecting normal behaviour, as proposed by Cramer and D. Bernoulli (cf. n. 3, p. 40). These two mathematicians developed their ideas in connection with *expectations*, i.e. chances of gains.[25]

[24] The publication of Wald's proof, which was to appear in Issue 8 of the *Ergebnisse*, had to be deferred to Issue 9. But not only did Issue 8 in 1937 mark the end of the Vienna Mathematical Colloquium but in the upheavals of 1938 Wald's manuscript most unfortunately got lost and Wald never found the time to rewrite that paper. For an outline of Wald's ideas including an interesting example for the non-existence of solutions under slightly weakened conditions, cf. the summary, loc. cit. above, p. 51 n. 17.

[25] With his subjective evaluation Bernoulli attempted to solve the so-called Petersburg Paradox—a game in which Peter, if he is in his senses, will not risk any appreciable part of his wealth even though his mathematical expectation is infinite (i.e. greater than any finite amount). What Bernoulli and his successors took for a solution of this paradox is the fact that Peter's expectation turns out to be finite if the amounts of his possible gains are replaced by their subjective values. But I showed (cf. *Zeitschrift für Nationalökonomie*, 5 (1934); English translation in *Essays in Mathematical Economics in Honor of Oskar Morgenstern* (Princeton, 1967)) that Bernoulli's is only a solution *ad hoc*. For I proved that no evaluation without a finite upper bound for the values of gains can solve all

The traditional evaluation of gaining the amount w with the probability p is the product pw, called the *mathematical expectation*. D. Bernoulli replaced this product by $pF(w, W)$, where $F(w, W)$ is the subjective estimate of the addition w to the wealth W of the evaluating person. He thus modified only the factor w in the traditional expression leaving the factor p unchanged; and for two centuries this factor remained a *noli me tangere*.

But, to paraphrase Buffon, only statisticians and mathematicians estimate chances proportional to their probability. The estimates of the man in the street, not to speak of the gambler in Monte Carlo or Las Vegas, are quite different. If $\Phi(w, p, W)$ is the amount a man will pay, out of his wealth W, for the chance of a gain w with the probability p, then, as I pointed out in the paper cited above (p. 57 note 25), $\Phi(w, p, W)$ is not proportional to either w or p. In a fair game of dice a man may be willing to risk a dollar while refusing to risk 100 dollars unless he is promised much more than 600 dollars should he win, as well as refusing to risk even one dollar for a chance of one in a billion to win a billion.

Just as both Austrians and mathematical economists replaced Bernoulli's logarithmic evaluation by an unidentified function (cf. n. 3 above, p. 40) subject only to the conditions of being non-decreasing and convex, varying from person to person, I left Φ unidentified and stressed its dependence upon the character of the evaluating person. Just as approximately $F(w, W) = w$ for Buffon's miser, it is characteristic of a gambler that $\Phi(w, p, W) \geqslant pw$ even if p is small and pw is an appreciable part of W.

Writing $\ll$ for *much less than* one has not only

$$\Phi(w, p, W) \ll pw \qquad \text{if } p \text{ is very close to o,}$$

but also

$$\Phi(w, p, W) \ll pw \qquad \text{if } p \text{ is very close to 1.}$$

Indeed, who would risk a dollar for the chance of getting \$1·01 with a probability of ·99 (that is, a net gain of a cent against an unlikely but possible loss of a dollar)? On the other hand, there is a middle range of probabilities in which chances are overestimated.

$$\Phi(w, p, W) > pw \quad \text{for certain } p \text{ and } w/W. \tag{*}$$

the analogous paradoxes arising from similar games. More recently this result was restated by P. A. Samuelson (*International Economic Review*, 1 (1960)). It is odd that just in the case that led Bernoulli to his ingenious anticipation of the Principle of Marginal Utility its application is inadequate.

Millions are spent in casinos by people who pay the amount a for the chance of getting $35a$ with the probability $1/36$. Thus

$$\Phi(35a, \tfrac{1}{36}, W) = a \; (> \tfrac{1}{36} \cdot 35a).$$

Actually the gambler's subjective estimation of his chance *exceeds a* or he would not pay that amount for the chance. Similarly, if M and M' play a fair game of dice, then

$$\Phi(w, \tfrac{1}{6}, W) \geqslant \tfrac{1}{6}w \quad \text{and} \quad \Phi'(w, \tfrac{5}{6}, W') \geqslant \tfrac{5}{6}w, \qquad (**)$$

which illustrates the subjective productivity of the exchange.

For normal people, (*) is valid only if $pw \ll W$. The ratio pw/W up to which (*) is valid measures the evaluator's *propensity to gamble*. More specific parameters describing his attitude are the limits for p and w/W within which (*) holds for him. Except under pathological conditions or in certain desperate situations (e.g. when a man absolutely needs six times what he owns, say, for a medical treatment, in order to continue to live) no one will risk his entire wealth on a single throw of a dice; that is to say, the first formula (*) holds only if $w \ll 6W$.

The von Neumann–Morgenstern axioms for measuring utility are, as their authors of course realize,[26] incompatible with formulae such as (*) or (**); that is to say, their axiomatic makes no allowance for any propensity to gamble. Yet, except for strict puritans, this propensity seems to be a widespread human trait (and even puritans confine their idiosyncrasy to games of chance while in ordinary life and in business transactions they cannot help taking risks as everyone else has to). On the other hand, it may be possible to question people about their preferences in the realm of combinations of chances such as

the quantity q of the good G with the probability p combined with the quantity q' of the good G' with the probability p'

bracketing out, so to speak, their pleasure in gambling as such, which then would not affect the axioms concerning the measurement of utility.

[26] Cf. the Appendix to the second and later editions of *The Theory of Games and Economic Behavior*, where probability is introduced into the measuring of utility, after the important paper 'Truth and Probability' in this direction by F. P. Ramsey, reprinted in his book *The Foundations of Mathematics and Other Logical Essays* (1931). Professor Morgenstern reports (cf. *Bulletin of the American Mathematical Society*, 64 (1958), no. 3, part 2, p. 108) that a primary role in persuading von Neumann to undertake a formal treatment of utility was played by the paper quoted in footnote 25.

I have tried to present the relations between Austrian marginalism and mathematical economics objectively—probably with the result that both sides will be dissatisfied with the presentation, as is so often the fate of objectivity. The important fact is one that Professor Hicks once pointed out: that each of the three founders of marginalism, Jevons, Menger, and Walras, apart from their common ideas, made valuable contributions of his own to the development of economic theory; and this tradition has been continued by their schools. This is a fact which, rather than playing off the schools against one another, we should welcome and utilize.

Miscellaneous

Editor's Comments

In addition to his extraordinary mathematical gifts, Menger had a great talent for exposition. This is clearly in evidence in his mathematical writings which, as the reader of these Selecta will no doubt have discovered, are still a pleasure to read.

Throughout his life, Menger also wrote a large number of surveys, expository articles, etc. Some of these, e.g., the papers entitled "On the Origin of the n-arc Theorem" and "The New Foundations of Hyperbolic Geometry" are included in Volume 1. Many others are to be found in his "Selected Papers in Logic and Foundations, Didactics, Economics".[1] In this section we present several more.

One of the items that we include is somewhat unusual: In 1950, the directors of the Museum of Science and Industry in Chicago asked Menger to design an exhibit on Geometry. The result was one of the first, if not the first, interactive exhibit on this topic. To accompany it, Menger produced a little booklet entitled "You Will Like Geometry". It is written for the layman; but the professional mathematician will notice that one could easily use it as the basis for a one-semester (or longer) course at the university level. It exhibits Menger's clear vision of the subject and is a wonderful example of his didactic skills.

---

[1] The Table of Contents of this collection is reprinted at the end of this section.

Otto Schreier.

Nachruf, gehalten in der Wiener mathematischen Gesellschaft am 8. November 1929.

Von Karl Menger in Wien.

In einer der letzten Sitzungen des vorigen Studienjahres hat der Vorsitzende unserer mathematischen Gesellschaft die erschütternde Nachricht von dem schweren Verlust gemacht, den die Gesellschaft durch den Tod ihres Mitgliedes Otto Schreier erlitten hat. Meine traurige Aufgabe ist es heute, den Lebenslauf meines lieben Freundes zu schildern und seine wissenschaftlichen Leistungen sowie seine akademische Tätigkeit in wenigen Worten zu würdigen.

Schreier wurde am 3. März 1901 in Wien geboren. Nach Absolvierung des Gymnasiums ging er im Herbst 1919 an die Wiener Universität. Er hörte hier bei Furtwängler Algebra und Zahlentheorie, bei Wirtinger Funktionentheorie und Differentialgleichungen und seit 1921 bei Hahn Mengenlehre, Theorie der reellen Funktionen und mengentheoretische Geometrie. Die alle Seiten der Mathematik umfassende Ausbildung, welche er bei diesen drei Lehrern erwarb, legte den Grund für sein späteres immenses mathematisches Wissen, auf welches ich noch zurückkomme. Schreier besuchte überdies mathematische Spezialvorlesungen, so bei Rella und Lense, und (im letzten Jahre seiner Studien) insbesondere ein Seminar über kombinatorische Topologie bei Reidemeister. Außerdem studierte er Physik und bei Schlick Erkenntnistheorie.

Während Schreiers Interessen gleichmäßig über alle Gebiete der Mathematik verteilt waren, gehörte seine besondere Vorliebe von Anfang an der Gruppentheorie. So übernahm er denn 1922 bei Furtwängler ein gruppentheoretisches Dissertationsthema, das Problem der Erweiterung von Gruppen. 1923 besuchte er die Marburger Mathematikertagung und trug auf derselben über seine im Anschluß an das Reidemeistersche Seminar entstandene Arbeit betreffend die Gruppen $A^a B^b = 1$ vor. Von diesem Jahre an blieb Schreier ein regelmäßiger Besucher der deutschen Mathematikertagungen. Im Herbst 1923 promovierte er in Wien auf Grund seiner

Monatsh. für Mathematik und Physik. XXXVII. Band.   1

Dissertation „Über die Erweiterung von Gruppen" und ging gleich
darauf nach Hamburg, wo er am mathematischen Seminar im Laufe
der folgenden Jahre Assistent wurde, einen Lehrauftrag erhielt und
sich habilitierte. Schreier, der im Frühling 1928 in Hamburg hei-
ratete, kam aus seinem neuen Heim häufig nach Wien zu Besuch und
unsere mathematische Gesellschaft, der Schreier seit 1921 als Mitglied
angehörte, konnte ihn bei diesen Gelegenheiten wiederholt als Vor-
tragenden begrüßen. Im Herbst 1928 wurde Schreier zum außer-
ordentlichen Professor der Universität Rostock ernannt. Trotz seiner
Schonungsbedürftigkeit — er war von Kind auf herzleidend —
dachte Schreier, während des ersten Semesters seiner Rostocker Vorlesungen auch seine Hamburger Seminartätigkeit durchzuführen. Zu
Beginn des Jahres 1929 erkrankte er an einer Grippe, die bald in
eine unheilbare Form allgemeiner Blutvergiftung ausartete, der
Schreier nach fünfmonatigem Leiden, glücklicherweise ohne sich
selbst seines Zustandes bewußt zu werden, am 2. Juni 1929 erlag.

*   *   *

Durch Schreiers Tod verliert die Wissenschaft einen ausgezeich-
neten Forscher, einen der bedeutsamsten Förderer aller Teile der
Gruppentheorie in letzter Zeit.

Einem Problem der Theorie der *endlichen* Gruppen ist Schreiers
Dissertation „Über die Erweiterung von Gruppen"[1]) gewidmet: zu
zwei gegebenen abstrakten Gruppen $\mathfrak{A}$ und $\mathfrak{B}$ alle Erweiterungen
von $\mathfrak{A}$ mit Hilfe von $\mathfrak{B}$, d. h. alle Gruppen $\overline{\mathfrak{B}}$ zu bestimmen, welche $\mathfrak{A}$
als Normalteiler enthalten, während die zugehörige Faktorgruppe
mit $\mathfrak{B}$ einstufig isomorph ist. In der letzten Zeit haben Spezialfälle
der das Erweiterungsproblem betreffenden Sätze zahlentheoretische
Anwendungen gefunden, so bei Hasse und insbesondere in Furt-
wänglers Beweis des Hauptidealsatzes. Im zweiten Teil der Arbeit gelangt Schreier auf Grund seiner Entwicklungen zu einer sehr
durchsichtigen Klassifikation aller Gruppen von den Ordnungen p^3,
p^4, p^5, wo p eine Primzahl bedeutet.

Der Theorie der endlichen Gruppen gehört auch Schreiers
Untersuchung „Über den Jordan-Hölderschen Satz"[2]) an. Bezeichnet
man als „Normalkette" der Gruppe G eine mit G beginnende und
dem Einheitselement endende Kette K von Untergruppen von G,
von welcher jede ein Normalteiler der vorangehenden ist, — nennt
man ferner eine Normalkette K' eine „Verfeinerung" von K, wenn
sie mindestens alle Glieder von K enthält, — und bezeichnet man
zwei Normalketten von G als „isomorph", wenn für beide Ketten
die Systeme der Faktorgruppen aufeinanderfolgender Gruppen (ab-

---

[1]) I. Teil, Monatshefte f. Math. u. Phys., 34, S. 165, II. Teil, Abh. d. Math.
Sem. Hamburg, 4, S. 321.
[2]) Abh. d. Math. Sem. Hamburg, 6, S. 300.

gesehen von der Reihenfolge) isomorph sind, — dann gilt nach Schreier der Satz, daß je zwei Normalketten einer Gruppe isomorphe Verfeinerungen besitzen. Da eine Kompositionsreihe von G eine Normalkette ist, in der die Faktorgruppen aufeinanderfolgender Glieder einfach sind und eine solche Kette nur durch Wiederholung von gewissen ihrer Glieder verfeinert werden kann, so ist im Schreierschen Theorem der Jordan-Höldersche Satz enthalten. Eine andere Teilaussage findet Verwendung in der von Schreier gemeinsam mit van der Waerden konzipierten Abhandlung „Die Automorphismen der projektiven Gruppe" [3]), deren Hauptresultat der Satz ist, daß jeder Automorphismus der projektiven Gruppe eines n-dimensionalen projektiven Raumes, dessen Koordinaten irgend einem Körper entnommen sind, durch Transformation mit einem Element der Gruppe der Kollineationen und Korrelationen erhalten werden kann.

Zur Theorie der *diskontinuierlichen* Gruppen schrieb Schreier erstens eine Arbeit betreffend die Gruppen $A^a B^b = 1$ [4]). Da unter diesen Gruppen die Fundamentalgruppen aller auf dem Torus gelegenen Knoten vorkommen, ergeben Schreiers Sätze eine äußerst durchsichtige Begründung Dehnscher Sätze über die Torusknoten, welche bis dahin bloß auf komplizierterem Wege bewiesen worden waren. — Der Theorie der diskontinuierlichen Gruppen gehört zweitens Schreiers Habilitationsschrift „Über die Untergruppen der freien Gruppen" [5]) an. Schreier beweist in dieser — wohl seiner originellsten — Arbeit allgemein den (bis dahin nur für einige spezielle Untergruppen der freien Gruppe bekannten) Satz, daß jede Untergruppe einer freien Gruppe bei geeigneter Wahl von erzeugenden Elementen selbst eine freie Gruppe ist. Für die freie Gruppe von n Erzeugenden, wo n eine endliche Zahl ist, ist jede Untergruppe von endlichem Index j eine freie Gruppe von $1 + j \cdot (n - 1)$ Erzeugenden und jeder Normalteiler von unendlichem Index eine freie Gruppe von abzählbar vielen Erzeugenden. Für die freie Gruppe von n Erzeugenden, wo n eine unendliche Mächtigkeit ist, ist jede Untergruppe, deren Index $< n$ ist, mit der Gesamtgruppe einstufig isomorph.

In der Theorie der *kontinuierlichen Gruppen* untersuchte Schreier nicht, gleich Lie, Transformationsgruppen, d. h. Gruppen, deren Elemente Transformationen eines Cartesischen Raumes sind, sondern er abstrahierte völlig von der Natur der Elemente der betrachteten kontinuierlichen Gruppe, bezeichnete demgemäß den Gegenstand seiner Untersuchungen als *abstrakte* kontinuierliche Gruppen [6]) und faßte die Gesamtheit der Gruppenelemente als Raum auf. Eine Menge irgend welcher Elemente, für welche erstens eine Limesdefinition vorliegt (so daß also die Menge in einem gewissen allgemeinen Sinn

---

[3]) Abh. d. Math. Sem. Hamburg, S. 303.
[4]) Abh. d. math. Sem. Hamburg, 3, S. 167.
[5]) Abh. d. math. Sem. Hamburg, 5, S. 161.
[6]) „Abstrakte kontinuierliche Gruppen", Abh. d. math. Sem. Hamburg, 4, S. 15, und „Die Verwandtschaft stetiger Gruppen im großen", ebenda, 5, S. 233.

1*

als „Raum" bezeichnet werden kann) und für welche zweitens eine Kompositionsvorschrift erklärt ist, derzufolge die Menge eine Gruppe ist, wobei drittens diese Kompositionsvorschrift mit der Limesdefinition so zusammenhängt, daß das Kompositum zweier Elemente und das inverse Element im Sinne der Limesdefinition stetige Funktionen sind — eine solche Menge bezeichnet Schreier als L-Gruppe. Einen wichtigen Vorteil dieser allgemeinen Begriffsbildung, nämlich die Möglichkeit, durch sukzessive Spezialisierungen zu den gewöhnlichen kontinuierlichen Gruppen zu gelangen und durch Anwendung der Methoden der mengentheoretischen Geometrie, insbesondere der Dimensionstheorie, Zwischenresultate herzuleiten, hat Schreier unausgenützt gelassen. Er machte bloß den letzten Schritt dieses sicher aussichtsreichen Weges, indem er nämlich sogleich jene Gruppen studierte, welche, als Raum betrachtet, Mannigfaltigkeiten im Sinne der kombinatorischen Topologie sind. Er bewies, daß die Fundamentalgruppe jeder kombinatorischen Mannigfaltigkeit, die als Gruppe auftreten kann, Abelsch ist. Die Frage nach notwendigen und hinreichenden Bedingungen, damit eine Mannigfaltigkeit als Gruppe auftreten könne, ist noch offen. Von Wichtigkeit ist Schreiers Begriff der stetigen Isomorphie zweier Umgebungen der Einheitselemente zweier kontinuierlicher Gruppen, welcher zum Begriff der im kleinen isomorphen Gruppen und zu einer entsprechenden Klasseneinteilung der kontinuierlichen Gruppen führt. Jede Klasse, bestehend aus allen untereinander im kleinen isomorphen Gruppen, enthält eine „Überlagerungsgruppe", welche durch Zusammenziehbarkeit aller geschlossenen Wege unter den Gruppen der Klasse gekennzeichnet ist und die zu jeder gegebenen Gruppe G der Klasse einen diskreten Normalteiler besitzt, mit dessen Faktorgruppe G stetig isomorph ist. — Auf der Düsseldorfer Mathematikertagung 1926 berichtete Schreier zusammenfassend „Über neuere Untersuchungen in der Theorie der kontinuierlichen Gruppen" [7]).

Außer Schreiers gruppentheoretischen Leistungen liegen *zahlentheoretische* Arbeiten von ihm vor: eine einen Beweis von Tschebotareff vereinfachende Note [8]) und insbesondere die gemeinsam mit Artin abgefaßte Abhandlung „Algebraische Konstruktion reeller Körper" [9]), welche wohl als eines der schönsten Ergebnisse des Steinitzschen Ideenkreises bezeichnet werden muß. Während unter den abstrakten Körpern bis dahin die reellen Körper lediglich durch den der Körpertheorie fremden Begriff der Anordnung gekennzeichnet werden konnten, bezeichnen Artin und Schreier einen Körper als reell, wenn in ihm — 1 nicht als Quadratsumme darstellbar ist. Die beiden Autoren beweisen sodann, daß jeder reell abgeschlossene Körper, d. h. jeder reelle Körper, von dem keine algebraische Er-

---

[7]) Jahresber. d. D. m. V., 37, S. 113.
[8]) „Über eine Arbeit von Herrn Tschebotareff", Abh. d. math. Sem. Hamburg, 5, S. 1.
[9]) Abh. d. math. Sem. Hamburg, 5, S. 85.

weiterung reell ist, auf eine und nur eine Weise geordnet werden kann. Für diese reell abgeschlossenen Körper werden die Sätze der reellen Algebra bewiesen, während für beliebige reelle Körper die Existenz von reell abgeschlossenen algebraischen Erweiterungen nachgewiesen wird. In einer späteren Note[10]) beweisen die beiden Autoren, daß die reell abgeschlossenen Körper mit jenen Körpern identisch sind, welche durch endliche Erweiterung algebraisch abgeschlossen werden können, ohne selbst algebraisch abgeschlossen zu sein. Artin hat, an die mit Schreier gemeinsame Untersuchung anknüpfend, Hilberts Problem betreffend die Zerlegbarkeit definiter Funktionen mit rationalen Koeffizienten in eine Summe von Quadraten rationaler Funktionen mit rationalen Koeffizienten in positivem Sinne gelöst. —

Die *Dimensionstheorie* verdankt Schreier eine Konstruktion, welche zu jedem Punkt einer eindimensionalen ebenen Menge M (dieselbe mag abgeschlossen sein oder nicht) beliebig kleine konvexe Minkowskische Strahlumgebungen ergibt, deren Begrenzungen mit M diskontinuierliche Durchschnitte haben.[11]) Die Verallgemeinerungen und Analoga dieses Satzes habe ich in einer Note „Eine dimensionstheoretische Bemerkung von O. Schreier"[12]) zusammengestellt, in welcher eine weitere diesem Problemenkreis der Dimensionstheorie angehörige Bemerkung, welche Schreier mir mündlich mitteilte, bewiesen wird.

In Buchform werden, von Artin und Schreiers Schüler E. Sperner herausgegeben, die Vorlesungen erscheinen, welche Schreier an der Hamburger Universität über elementare Algebra und analytische Geometrie gehalten hat. Das Buch soll den Titel „Einführung in die Algebra und analytische Geometrie" erhalten und demnächst im Verlage von J. Springer, Berlin, erscheinen. — In der von Blaschke besorgten Bearbeitung und Herausgabe von F. Klein's „Vorlesungen über höhere Geometrie" (3. Aufl., 1926) stammt von Schreier das die Elementarteilertheorie (im Anschluß an Weyl) behandelnde 5. Kapitel des dritten Hauptteiles (S. 379).

* * *

Schreiers Dozententätigkeit erstreckte sich nur auf wenige Jahre, war aber infolge seines die gesamte Mathematik umspannenden Wissens von außerordentlichem Umfange: Die Theorie des Klassenkörpers und der Reziprozitätsgesetze, die Siegelschen Untersuchungen über die Darstellung von Körperzahlen als Summe von Quadraten, die Hardy-Littlewoodschen Methoden der analytischen Zahlentheorie, die Weylsche Darstellungstheorie der kontinuierlichen

---

[10]) „Eine Kennzeichnung der reell abgeschlossenen Körper", Abh. d. math. Sem. Hamburg, 5, S. 225.

[11]) Vgl. Monatshefte f. Math. u. Phys., 33, S. 153.

[12]) Monatshefte f. Math. u. Phys., 37.

Gruppen, die Theorie der Θ-Funktionen, die neuen Untersuchungen
über Fourier sche Reihen, die Bohrsche Theorie der fastperiodischen
Funktionen, die Dimensionstheorie, die Kurventheorie, — alle diese
so verschiedenartigen Ergebnisse der modernen Mathematik waren
Schreier nicht nur völlig geläufig, sondern waren Gegenstand seiner
Vorlesungen, Seminare und Referate.

Und ein Referat von Schreier war nicht nur vortragstechnisch
meisterhaft, sondern stets auch eine wissenschaftliche Leistung.
Zumindest wurde die Eleganz des Originals erhöht und eine Ab-
kürzung seiner Beweise geboten, meistens aber bei diesem Anlasse
die Formulierung der vorgetragenen Theoreme verallgemeinert, wenn
nicht neue Sätze hinzugefügt. Daß Schreier sein außerordentliches
Wissen nicht durch eine lange Lehrtätigkeit zur Auswirkung bringen
konnte, ist für den mathematischen Vorlesungsbetrieb der deutschen
Universitäten ein unersetzlicher Verlust. Und derselbe ist um so
tragischer, als die ganze Ökonomie von Schreiers Leben darauf
gerichtet war, oft unter Hintansetzung seiner eigenen Produktion
Wissen zu erwerben, um dasselbe zum Vorteile Einzelner oder im
Interesse seiner Vorlesungstätigkeit zu verwerten. Wie viel er dabei
an fördernder und abkürzender Arbeit geleistet hat, das läßt sich
leider wohl nicht zusammenstellen. Mir selbst hat Schreier in den
Jahren 1921—1923, als wir gemeinsam in Wien studierten, bei
meinen ersten dimensions- und kurventheoretischen Arbeiten in
ebenso unermüdlicher wie uneigennütziger Weise seine Hilfe an-
gedeihen lassen und war mir ein Berater, dem ich mich zu tiefem
Danke verpflichtet fühle.

*   *   *

Persönlich hatte Schreier keine Gegner, geschweige denn Feinde.
Sein durch besondere Liebenswürdigkeit ausgezeichnetes Wesen ver-
bunden mit außerordentlicher Hilfsbereitschaft, seine stete Rücksicht-
nahme auf andere und seine Anspruchslosigkeit für die eigene Person
schufen ihm bei allen, zu denen er in irgend welche Beziehungen
trat, Sympathien. Für die wenigen Menschen, denen er nähertrat
und die die ganze Liebenswertheit seines Wesens kennen lernten,
ist im Leben durch sein frühes Scheiden eine unausfüllbare Lücke
entstanden.

Vorträge von Otto Schreier in der Wiener mathematischen Gesellschaft:

11. IV. 1923: „Neueres über Fouriersche Reihen."
 4. IV. 1924: „Über abstrakte Gruppen."
 6. III. 1926: „Kontinuierliche Gruppen im Großen."
19. III. 1926: „Weyls Beiträge zur Gruppentheorie."
18. III. 1927: „Reelle Algebra in abstrakten Körpern."

Hans Hahn †

Von

Karl Menger (Wien).

Mit Hans Hahn, der am 24. VII. 1934 unerwartet in seinem
55. Lebensjahre starb *), ist der Mathematik ein in vielen Richtungen
erfolgreicher Forscher entrissen worden. Seine Jugendarbeiten enthalten bedeutsame Beiträge zur Variationsrechnung; eine andere
Arbeit bezieht sich auf Funktionen zweier komplexer Veränderlicher;
weiters hat er in zahlreichen Abhandlungen die Theorie der Reihen-
und Integraldarstellungen bereichert und eine besonders bemerkens-
werte und schöne Anwendung dieser Methoden auf das Interpola-
tionsproblem gegeben (Math. Zeitschr. 1); wichtig sind seine Beiträge
zum allgemeinen Funktionalkalkül; in der Elementargeometrie führte
er den ersten auf Verknüpfungs- und Anordnungsaxiomen be-
ruhenden lückenlosen Beweis des Jordan'schen Satzes für Polygone
(Monatshefte f. Math. u. Phys. 19). Schon in allen den erwähnten
Arbeiten (vgl. meinen Nachruf auf Hahn in den Ergebnissen eines
math. Kolloquiums 6) zeigt sich Hahn als scharfer Logiker mit außer-
ordentlich klarer Darstellungsgabe, Eigenschaften, welche auch seinen
Bericht über die Theorie der linearen Integralgleichungen aus dem
Jahre 1911 trotz der inzwischen erfolgten Fortschritte heute noch
lesenswert machen.

Was aber die Redaktion der *Fundamenta Mathematicae* zweifellos besonders zum Wunsche veranlaßt hat, an dieser Stelle eine
Würdigung des Verstorbenen erscheinen zu lassen, ist der Umstand.

---

*) 1879 in Wien geboren, studierte Hahn in seiner Vaterstadt, in Strassburg,
München und Göttingen, habilitierte sich 1905 in Wien, war 1909—16 Professor
in Czernowitz, dann bis 1921 in Bonn und seither an der Universität Wien.

daß Hahn einer der hervorragendsten Vertreter der in dieser Zeitschrift kultivierten Forschungsrichtung war: einer der besten Kenner
und Förderer der Theorie der reellen Funktionen und der mengentheoretischen Geometrie, sowie einer der ersten Gelehrten im deutschen Sprachgebiet, welche die Bedeutung dieser Teile der Mathematik erkannten und durch glänzende Vorlesungen der nächsten
Mathematikergeneration bekannt machten.

Sein Buch „*Theorie der reellen Funktionen*" ist eines der Stan-
dardwerke dieses Gebietes und viele von Hahn's Leistungen, vor
allem die Entdeckung des Zusammenhanges im Kleinen zur Kennzeichnung der stetig durchlaufbaren Mengen, die gleichzeitig und
unabhängig von Mazurkiewicz erfolgte, sind so allgemein bekannt, daß hier jede Besprechung derselben überflüssig ist. Es sei
deshalb im Folgenden bloß auf einige Leistungen des Verstorbenen
aus den Publikationsgebieten der *Fundamenta Mathematicae* ver-
wiesen, die aus verschiedenen Gründen bisher minder bekannt ge-
worden sind.

Da erwähne ich die von Hahn 1929 gegebene Einführung des
Lebesgue-Stieltjes-Integrals (Wiener Ak. Anz. 66): Ist E irgend
eine Menge, $\varphi(M)$ eine für alle Mengen eines σ-Körpers $\Re$ von
Teilmengen von E definierte total-additive Mengenfunktion; E liege
in $\Re$ und $\varphi(E)$ sei endlich; ist $\varphi(M) = 0$ für eine Menge M aus $\Re$,
so soll auch jede Teilmenge M' von M in $\Re$ liegen. Als φ-inte-
grierbar bezeichnet Hahn eine φ-meßbare Funktion f dann, wenn
eine für alle M aus $\Re$ definierte Mengenfunktion $\lambda(M)$, auch $\int\limits_{M} f\, d\varphi$

genannt, existiert, welche total additiv in $\Re$ ist und für jede Menge M
aus $\Re$, auf welcher $c' \leqslant f \leqslant c''$ gilt, der Bedingung

$$c'\, \varphi(M) \leqslant \lambda(M) \leqslant c''\, \varphi(M)$$

genügt (wobei, wenn $\varphi(M) = 0$ ist, $c\,\varphi(M) = 0$ zu setzen ist, auch
wenn $c = \pm\,\infty$ ist).

Hinzuweisen ist ferner auf Hahns Heranziehung einfacher neuer
Sätze über unendliche Reihen zum Beweis von Sätzen über Mengen-
funktionen (Wiener Ak. Anz. 65 u. Bull. Calcutta Math. Soc. 20).
Dabei wird durch Hahns Untersuchungen nahegelegt, wenn eine
unendliche Reihe $\overset{\infty}{\underset{n=1}{\Sigma}} a_n$ gegeben ist, ganz allgemein die Menge aller
Zahlen $\overset{\infty}{\underset{i=1}{\Sigma}} a_{n_i}$ für alle Teilfolgen $\{a_{n_i}\}$ der Folge $\{a_n\}$ zu untersuchen.

Es wäre wohl sehr wünschenswert, wenn dieses Problem gleich allgemein für Reihen von Vektoren des R_n behandelt würde und mit den bekannten Steinitz'schen Untersuchungen über die Umordnungen von Vektorreihen, sowie mit den Untersuchungen über Längenmengen von Bögen des Vektorraumes (vgl. Ergebn. eines math. Kolloquiums 5) in Zusammenhang gebracht würde, wodurch zweifellos neue wichtige Kapitel der Theorie der unendlichen Reihen entstehen würden.

Hahn war einer der ersten, welche die Wichtigkeit von Fréchet's abstrakten Raumbegriffen erkannten. Er bewies (Monatsh. für Math. u. Phys. 19), daß in jeder Klasse V nicht konstante stetige Funktionen existieren (bekanntlich der Kernsatz des Metrisationsproblems), indem er die später von Urysohn zum Beweis des Hauptlemmas der Metrisationstheorie benützte Methode der Umgebungsringe entwickelte.

Auf seine bekannten Untersuchungen über die stetigen Streckenbilder wurde Hahn durch das Studium der stetigen Abbildungen der Strecken auf das Quadrat geführt, wobei er fand, daß bei jeder solchen Abbildung im Quadrat Punkte mit mindestens je drei Urbildpunkten dicht liegen und die Menge der Quadratpunkte mit mindestens zwei Urbildpunkten die Mächtigkeit des Kontinuums besitzt. Dieses Ergebnis ist der erste Spezialfall der allgemeinen dimensionstheoretischen Sätze von Hurewicz über die Multiplizität der Bildpunkte bei dimensionserhöhenden stetigen Abbildungen.

Hahn's Darstellung der irreduziblen Kontinua als Summe von Primteilen löst die irreduziblen Kontinua nicht so stark auf, wie die von Kuratowski und Vietoris eingeführten Teilmengen, ermöglicht aber dafür die Formulierung des abgerundeten Satzes, daß der Raum der Primteile eines irreduziblen Kontinuums einpunktig oder ein Bogen ist, ein Ergebnis, das durch den Satz von R. L. Moore, daß der Raum der Primteile eines beliebigen Kontinuums stetig durchlaufbar ist, eine schöne Ergänzung erfuhr (kurze Beweise beider Sätze in Hausdorff's *Mengenlehre*, 2. Aufl.).

Schließlich sei noch auf die in der abstrakten Algebra m. E. zu wenig beachtete wichtige Arbeit Hahn's über nichtarchimedische Größensysteme (Wien. Ak. Ber. 116) hingewiesen. In der heutigen Terminologie lassen sich Hahn's Resultate folgendermaßen aussprechen. Es sei G eine geordnete Abelsche Gruppe, deren Kompositionsoperation wir Addition nennen wollen, während wir eine

Summe von n Elementen a mit na bezeichnen. Wir fassen je zwei Elemente a und b von G dann und nur dann in eine „Klasse" zusammen, wenn zu jeder ganzen Zahl m eine ganze Zahl n existiert, so daß $ma < nb$ ist, und zu jedem n' ein m' existiert, so daß $m'a > n'b$. Die Menge dieser Klassen ist eine geordnete Menge Γ, welche der Klassentypus von G heißt. Zu jeder geordneten Menge Γ existieren geordnete Abelsche Gruppen G mit dem Klassentypus Γ, nämlich Systeme von Vektoren mit weniger als $\aleph$ nichtverschwindenden Komponenten eines Γ-dimensionalen reellen Raumes, womit folgendes gemeint ist: wir bilden alle absteigend wohlgeordneten Teilmengen einer Mächtigkeit $< \aleph$ von Γ und belegen die Elemente dieser Mengen auf verschiedene Weisen mit reellen Zahlen; dabei nennen wir absteigend wohlgeordnet eine Menge N, von der jede Teilmenge, N inbegriffen, ein Element höchsten Ranges besitzt. Es ist klar, wie Gleichheit, Anordnung und Addition für die Belegungen der absteigend wohlgeordneten Teilmengen von Γ mit Mengen reeller Zahlen zu definieren sind. Ein Beispiel einer geordneten Abelschen Gruppe liefert ein System von Vektoren der geschilderten Art dann, wenn es gegenüber der Addition abgeschlossen ist. Umgekehrt beweist H a h n, dass zu jeder geordneten Abelschen Gruppe ein $\aleph$ und eine geordnete Menge Γ existiert, so dass G isomorph ist mit einem System von Vektoren mit weniger als $\aleph$ nichtverschwindenden Komponenten eines Γ-dimensionalen Raumes. Archimedisch heißt G, wenn Γ nur ein Element enthält, vollständig wird G genannt, wenn jede geordnete Gruppe, die umfassender ist als G, Klassen enthält, die in G nicht auftreten. Ist Γ selbst eine geordnete Abelsche Gruppe, so kann in G eine Multiplikation definiert werden, derzufolge G ein geordneter Körper ist.

Mit H a h n ist nicht nur ein erfolgreicher Forscher und ein vortrefflicher Lehrer dahingegangen — aus der großen Zahl derer, die sich dankbar als seine Schüler bezeichnen, darf ich wohl K. G ö d e l, W. H u r e w i c z und mich selbst nennen — sondern auch ein gültiger und stets für seine Überzeugung eintretender Mensch. Insbesondere hat sich H a h n jederzeit für internationale wissenschaftliche Zusammenarbeit eingesetzt. Dem Kreise der Herausgeber und Mitarbeiter der *Fundamenta* und ihrer, seiner eigenen in vielen Punkten verwandten Arbeitsrichtung brachte er stets besondere Sympathien entgegen.

Memories of Moritz Schlick

By Karl Menger*

1

My first encounter with Moritz Schlick came in 1923 when, as a student, I listened to some of his lectures on *Erkenntnistheorie* (Epistemology) at the University of Vienna. Having read older epistemological writings I appreciated his undogmatic way of speaking in simple terms. My impression at the time was that of an extremely refined and very sincere man, unassuming almost to the point of diffidence.

In later years, as I came to know him more closely, I admired his sincerity even more. Empty phrases from his lips or the slightest trace of pompousness were unthinkable. The apparent diffidence in his dealings with students actually was an effect of his sometimes exaggerated politeness, which was especially directed toward the younger and the weaker. In discussions with peers he was perfectly self-assured, and even not quite free of good-natured sarcasm. If someone lengthily elaborated an opinion that Schlick did not deem to be quite up to his standards, he would almost imperceptibly shake his head with a faint mocking smile, and then try to exchange glances with those whom he thought to be sharing his feelings. Most speakers so castigated would shorten their speech.

If a slight doubt in Schlick's self-assuredness occasionally recurred, it was only because of the intensity and the exclusiveness of his admiration for a single person. Such an attachment of Schlick —

* I am grateful to Dr. Phyllis Kittel-Light and Dr. Abe Sklar for valuable suggestions I have utilized in writing this paper.

6*

always to a figure of the first order — apparently occurred at each
stage of his development. In the early years of this century, he
studied physics in Berlin under Max Planck, the great originator
of quantum theory. In the mid 1910's, he became profoundly interested
in relativity theory to which he devoted an excellent monograph,
Space and Time in Contemporary Physics, and venerated Albert
Einstein. Toward the end of the second decade, there followed
the period of his *General Theory of Knowledge* and deep admiration
for David Hilbert. During this period Schlick promulgated the
axiomatic method, with particular stress on implicit definitions,
that is, on the simultaneous introduction of several concepts of a
theory by its postulates. Upon coming to Vienna in 1922, Schlick
for a while became interested in mathematical logic and greatly
admired Bertrand Russell. A few years later he fell under the ever
increasing spell of Ludwig Wittgenstein. This last phase I had
opportunity to observe closely; and I was sometimes rather unhappy
to see Schlick extol his idol to the point of self-effacement: he ascribed to Wittgenstein ideas that he himself had uttered before
he had seen the *Tractatus*.

2

While Schlick's lectures were always very clear he delivered them
in a monotone and habitually spoke in a low voice.[1] But when I
attended his seminar in 1923/24 I soon became aware of his talent
for conducting stimulating discussions. In that seminar he brought up,
among other ideas, the elusive concept of simplicity of theories —
a notion of great importance in connection with Mach's, Avenarius',
Duhem's and especially Kirchoff's conception of a desirable physical
theory as the simplest description of nature. For my part, I was
particularly interested in that notion because of Henri Poincaré's
claim that Euclidean geometry was simpler than its non-Euclidean

---

1 The first time I heard Schlick lecture I had a slight difficulty understanding him because of his most unusual habit of breaking up some words
into syllables — and strangely, syllables in the English sense of *rid-er*, rather
than the German *Rei-ter*. But within an hour I got completely used to this
peculiarity of his at that time. While Schlick's speech was definitely not
of the Austrian variety, neither was it characteristic of someone coming
from Northern Germany, so that I was surprised to learn that he was born
in Berlin.

sister geometries "in the sense in which a linear polynomial is simpler than a quadratic polynomial." Schlick assigned the seminar topic of simplicity to me. One of the points to which I called attention, without arriving at general criteria, was the fact that theories can be only partially (not totally or linearly) ordered according to simplicity: there are pairs of theories neither of which can be said to be simpler than the other. F. Hausdorff's concept of partial order was little known at that time and Schlick expressed great interest in my talk.[2]

Soon thereafter, in the spring of 1924, my *Philosophicum* came up — the examination which every student of mathematics, of the physical or biological sciences, and of history or philology in an Austrian university had to pass before receiving a PhD degree. Schlick made the occasion appear to be a relaxed exchange of ideas. We spoke mainly about Kant's and Poincaré's views on mathematics and physics. For the first time I experienced the delight that in the course of the years I felt in every conversation with this highly cultivated man.

3

Early in 1925 I left for Amsterdam. When I returned to the University of Vienna in the fall of 1927, in order to teach in the chair for geometry, the mathematician Hans Hahn asked me whether I would like to join the *Schlick-Kreis*, the group that abroad became famous under the name of Vienna Circle. "I attend regularly," he said, "and so do Carnap, Neurath and a few younger people; and Philipp Frank visits us whenever he comes to Vienna [from Prague]. We meet every other Thursday evening on the ground floor of the wing of this building on the Boltzmanngasse." We were talking in the L-shaped university building that housed the mathematics and physics institutes. "I have not seen Schlick since my Philoso-

---

2 Schlick was not to hear an epilogue. For only two years after his untimely death in 1936 did I find that Poincaré's statement about the relative simplicity of Euclidean geometry was disputable. In a deductive development, the geometry of Euclid requires much more complicated basic terms and assumptions than the geometry of Bolyai and Lobatchevsky. [Cf. K. Menger, "New Foundation of Non-Euclidian, Affine, Real Projective and Euclidian Geometry", *Proc. National Academy of Science USA* 24 (1938), pp. 486—490; Chapter 5 of L. M. Blumenthal and K. Menger, *Studies in Geometry* (San Francisco: Freeman, 1970)].

phicum", I said. "I don't think that he remembers me". "He remembers you well and wants you to come", Hahn answered. "He and Carnap are interested in your work on dimension theory." I thanked and accepted the invitation with anticipation.

The room on the ground floor in which the Circle met — rarely more than 20 persons — was rather drab. We would stand in little groups talking until Schlick would clap his hands and we would be seated. Sometimes Schlick would begin by reading to us a letter that he had received dealing with problems that we had discussed or planned to consider. His correspondents included Einstein and Russell. He would open other sessions by reading announcements of new publications (especially British ones) and would promise to report about some or ask for volunteers to review them. Occasionally, Schlick would introduce a guest passing through Vienna. Then there would begin a discussion on the topic proposed in the preceding meeting or someone's report about work in progress. But in none of the many meetings that I attended in the course of the years did the debates ever touch political or economic problems. Even men with strong political convictions never expressed them in the discussions of the Circle. It was in these discussions that Schlick particularly excelled both as a stimulating participant and as a moderator of ideal poise.

4

Only in 1926/27 had the program of the Circle been different, Hahn told me. Various disagreements about Wittgenstein's *Tractatus* had surfaced in the group.[3] Thus, at Carnap's suggestion, Schlick had decided to devote as many consecutive meetings as necessary to read the book aloud. The joint reading, sentence by sentence, filled that entire academic year before I joined. In the fall of 1927, this common experience was still fresh in all members' minds — a fact that made it difficult for a newcomer to join the group. I never quite overcame this handicap.

A new kind of jargon had developed. In particular, two terms, new in the 1920's, had been completely integrated in the vocabulary of Schlick and all other members of the Circle and were freely and

---

3 Wittgenstein himself never attended a meeting of the Circle either before or after 1927.

perpetually used by the Circle: *elementary propositions* and *tautologies*. Schlick seemed very impressed by the twin ideas that the world could be described by the conjunction of all true elementary propositions — a rather grandiose conception reminiscent of Democritus' atomistic cosmology — and that the true propositions not dealing with the world (that is, all logico-mathematical propositions) were tautologies.

In trying to understand the new terms fully I asked for definitions. When I did not receive answers that fully satisfied me — in the case of elementary propositions I got no definition at all, in the case of tautologies more than one — I asked for more examples.

As to elementary propositions, some members mentioned as an instance: *this is red*. Others referred me to *Tractatus* 4.221, according to which it is obvious that in the analysis of propositions we must come to elementary propositions. But all that has ever been obvious to me is that every analysis must break off somewhere. If one so desires he may call the proposition at his break-off point *elementary in that particular analysis*. This, however, would make elementariness relative — a relation between propositions and analysts — rather than a predicate.[4]

---

4 In *Tractatus* 4.221, elementary propositions are said to consist of names in immediate connection. But I did not find in the book examples of immediate connections any more than of elementary propositions. In 1928, the second edition of Whitehead's and Russell's *Principia* reached Vienna. There, at the very beginning of the Introduction I found two definitions of atomic propositions (a concept jointly introduced by Russell and Wittgenstein just before the first World War): one is *negative* — propositions containing neither parts that are propositions nor the quantifiers *all* or *some*; the other is *positive* — propositions of the form *x has the predicate P* or $x_1, x_2, \ldots, x_n$ *are in the n-ary relation R*. Besides the ineluctable *this is red*, the authors mention a second example: *this is earlier than that*. But I saw difficulties. If in a statistical study of color in a population of roses I pick an example and say "this is red", then, according to Russell, I have uttered an atomic sentence. The statement prompts a mark on the tally-sheet thereby representing one indivisible element of the study. To a botanist interested in the color of the petals of roses, however, the same sentence about the same rose means *all the petals are red*, which according to Russell is nonelementary. In the botanist's study, *this is red* is elementary only in regard to a single petal. Whether or not *this is red* is elementary thus seems to depend on the context in which the sentence is uttered. Further problems arise in connection with the word *red*. If we say about the same object *this is red* and *this is light red*, which sentence (if either) is elementary? And what about *this is pink* and *this is colored*?

As to the term *tautology*, I found that it was used in various ways, but mainly in the following two:

a) in the sense of the *Tractatus*,[5] that is, for any proposition compounded of (component) propositions by particles such as *not, and, or, if . . . then,* provided that the components are free of quantifying particles such as *all, some, any* and that the compound is true regardless of the truth or falsehood of the components. (Example: it is raining or it is not raining.)

b) in the sense of propositions that are consequences of mere axioms of logic (including also the example: all men are mortal or some men are not mortal).

Despite the obvious differences between the two concepts they were often used interchangeably.

In the fall of 1927, the *Tractatus* as such was no longer on the agenda. But it loomed over the discussions and especially over all that Schlick said and thought.[6] Moreover, Schlick was at that time preoccupied with his first encounters with Wittgenstein, whose personality fascinated him as much as his book. He often deplored the philosophical inactivity of the author of the *Tractatus*, which I once heard him ascribe to *ressentiment*.[7]

5

Since I had just returned from Amsterdam and an association with L. E. J. Brouwer, Schlick asked me to report about Intuitionism. I presented the intuitionistic-formalistic dictionary of set theory that I had devised as well as what just at that time had begun to take a firmer shape in my mind — the epistemological consequences of my critique of intuitionism: the plurality of logics and languages entailing some kind of logical conventionalism.[8] Schlick expressed

---

5 Under a less suggestive name, this concept of tautology has also been independently introduced by the great American logician, Emil L. Post (*Amer. J. of Math.* 43, 1921).

6 Only a few years later did I realize that there was an important exception (see Section 12).

7 Cf. K. Menger, "Wittgenstein betreffende Seiten aus einem Buch über den Wiener Kreis", *Proc. 2nd Internat. Wittgenstein Symposium,* Kirchberg/Wechsel (Wien: Hölder – Pichler – Tempsky, 1979), p. 27.

8 Cf. my *Selected Papers in Logic and Foundations, Didactics, Economics* (Dordrecht: D. Reidel, 1978), Chapter 5.

some gratification about understanding for the first time the meaning of Brouwer's page-long definition of a set, as did Hahn and Carnap. The epistemological considerations did not at that time elicit any response.

As I became better aquainted with the Circle I began to realize that those epistemological considerations were incompatible with views then held by all other members. Their views on these matters were at that time expressed in four ways:

1) constant references to *the* logic and *the* language;

2) the ever more prevailing use of the term *tautology* in the (absolute) sense b) with reference to *the* logic

3) the claim (advanced by Schlick and Waismann, Hahn and Carnap, and, I believe, Feigl) that all mathematical propositions are tautologies;

4) an employment of the term *meaningless* that appeared to me arbitary and loose, since it was not supported by any rules as to what was regarded as *meaningful*.

Sometimes during the year 1928/29, I presented a second paper, criticizing the four points just mentioned and developing, in contrast, a thesis of what Carnap[9] later so aptly called *logical tolerance*: the possibility of choosing from a variety of logics and languages. This second paper was received very unfavorably. Schlick soon began to shake his head ever more perceptibly and I realized that for the first time (and, it turned out, the only time) I was in for the mild punishment he meted out to babblers. And indeed he soon tried to exchange glances with others. Only Waismann returned Schlick's glance with a smile and both then attacked my arguments vigorously. But even Hahn, while averse to the indiscriminate use of *meaningless*, kept referring to the tautological character of mathematics. Carnap, who had always spoken about *the* logic and *the* language,[10] seemed to become rapt in deep thought early in the talk and said nothing. Neurath obviously was not very interested.

---

9 R. Carnap, *Logical Syntax of Language* (London: Kegan Paul, Trench, Trubner & Co. Ltd., 1937), p. 51f. This is the English version of Carnap's *Die logische Syntax der Sprache* (Prag 1934).

10 As he reports in *Phil. of Science* 4 (1937), 20. Cf. also my book *loc. cit.* 8, pp. 12 ff.

Victor Kraft listened attentively but remained silent.[11] Certainly
I had no support from anyone — except for assenting nods from a
young man whom at that time I only knew as one of the students
in my course on dimension theory and whose name was Kurt Gödel.

6

During 1928/29 I became aware of, and ever more disturbed by, the
imprecision in the epistemological ideas and formulations of the
Circle. During an evening walk home from a meeting I tried to
formulate to my satisfaction the claims and counterclaims that
had been raised in the debate, but I did not succeed. Since I merely
aimed at clarifying (not at deciding) the issues I began to suspect
that the roots of my difficulties might be general deficiencies of
the language used in the debates, in particular, the absence of explicit
rules. I remembered that it was this shortcoming on which I had
also blamed the anarchy in the use of the term *meaningless*. In the
Circle, a few remarks voicing my suspicions fell on deaf ears; and
lack of time forced me to postpone writing an epistemological
paper on the difficulties. For in addition to my mathematical research,
I directed a group of excellent students, soon joined by guests from
abroad, which was called my *Mathematical Colloquium*. I tried
to model the Colloquium a little after Schlick's Philosophical Circle,
though in some ways this ideal was of course unattainable.[12]

7

During 1928, Neurath advocated ever more emphatically the need
for a public forum, where the insights gained in the privacy of the
Circle could be disseminated. "We have the Viennese Philosophical
Society", Schlick responded; and indeed that organization arranged
lectures and published writings of the type that few German-speaking

---

11 He later described the development of the idea in his book *Der
Wiener Kreis* (Wien—New York: Springer-Verlag, 1950), p. 55, n. 2.

12 There was one aspect of conducting the group in which I deliberately
deviated from Schlick's procedure. From the beginning I kept short, but precise records of our meetings. They were published in eight issues (1928—1936)
under the title *Ergebnisse eines Mathematischen Kolloquiums* and included
contributions by Gödel, Nöbeling, von Neumann, Wald and others. I have
always regretted that, as far as I know, no complete record was kept of
the Schlick Circle.

groups (most of which were metaphysically oriented) would have considered. But Neurath insisted emphasizing the need for greater concentration on Mach's kind of philosophy than the Viennese Philosophical Society offered, and stressing the fact that Schlick occupied Mach's chair at the University. Eventually — late in 1928 — he prevailed on Schlick to accept the presidency of the newly founded *Ernst Mach Society*, which Neurath conceived as a means of propagandizing ideas conceived in the Circle.

Early in 1929, Schlick received an attractive offer of a professorship from the University of Bonn; and soon thereafter he visited the United States. (The meetings of the Circle during his absence were conducted by Hahn and Carnap.) Great was the joy of all of us when we learned that Schlick had decided to return to Vienna for good. "This must be celebrated", Neurath said and we all agreed. "We must write a book outlining our views — a manifesto of the Circle — and dedicate it to Schlick when he comes home in the fall", Neurath added; and with his habitual expeditiousness and energy he went to work.

Most members of the Circle attributed Schlick's decision to stay in Vienna to his devotion to the Circle and undoubtably this feeling was one of his principal motives. But there were also other reasons. Vienna in the mid-twenties had a general intellectual atmosphere Schlick would have hardly found elsewhere.[13] I observed on several occasions that he greatly enjoyed social relations with those well-read physicians, lawyers and businessmen, who were seriously interested in philosophy; and this extra-academic intelligentsia highly appreciated Schlick.

I did not hear about the progress of Neurath's project until late in the summer when Hahn asked me about some bibliographical details in the last proofsheets of a pamphlet that was ready for publication. It was entitled *Wissenschaftliche Weltauffassung: Der Wiener Kreis* (The Scientific Conception of the World: The Vienna Circle).[14] Neurath had completed the manuscript after several conferences with Carnap. In the process, he gave the name *Vienna Circle* to what had been called (and in Vienna of course

---

13 More about the Viennese intelligentsia between the two world wars in my book, *loc. cit.* 8, p. 17f.

14 An English translation of the pamphlet (without the bibliography) is included in M. Neurath and R. S. Cohen (eds.), *O. Neurath: Empiricism and Sociology* (Dordrecht: D. Reidel, 1973), Chapter 9.

continued to be called) the *Schlick-Kreis*. Hahn had consented to
sign as the principal author (with Neurath and Carnap as co-authors)
following, I suspect, his irrepressible penchant for arranging meetings
of minds. For I doubt that he liked the booklet very much. While
informative, it was not of the quality of his own writings on the
subject.[15] Under the circumstances, however, the main question
seemed to me to be whether *Schlick* would like it; and this I doubted
even more. He certainly would notice (though in his overly modest
way probably not complain about) the fact that his own views
were rather inadequately presented. Moreover, Schlick believed
in strict separation of cognitive material and evaluations of any
kind. Neurath, on the other hand, seemed to me to have one over-
powering motive in whatever he said or did — a quite unselfish,
but certainly extracognitive desire for a reform of society. And so
the book blended ideals with insights — perhaps not quite as freely
as Neurath would have desired, but certainly too freely from Schlick's
point of view.[16]

8

In September of 1929 I visited Warsaw at the invitation of the Polish
mathematicians. There, I became personally aquainted with members
of the Polish school of logic. At that time the Polish logicians, whose
philosophical outlook was partly derived from the philosopher
K. Twardowski (a student of Franz Brentano's in Vienna), were
little known outside of Poland. I was greatly impressed not only
by their technical achievements in logic and mathematics, but also
by the precision of their approach to epistemological problems.

My first thoughts were of Schlick who, I was sure, would also
be very interested. Beyond that, I felt that it would be a real service
to philosophy to establish a rapproachement between the schools

---

15 Cf. in particular, H. Hahn, "Die Bedeutung der wissenschaftlichen
Weltauffassung insbesondere für Mathematik und Physik", *Erkenntnis* 1
(1924), pp. 9—106. English edition: "Empiricism, Logic, and Mathematics",
in B. F. McGuinness (ed.), *H. Hahn: Philosophical Papers,* Vienna Circle
Collection 13 (Dordrecht: D. Reidel, 1980).

16 For all these reasons, the pamphlet estranged me from the Circle to
the point where I asked Neurath to list me henceforth merely as *close* to,
rather than as a *member* of, the Circle (Cf. *Erkenntnis* 1, p. 335). The
booklet produced an even greater disaffection in Gödel.

of Vienna and Warsaw. I invited A. Tarski to visit my Mathematical Colloquium and give us three talks: one on mathematics and two on logic. This was Tarski's first opportunity to make Polish logic known in the West and he accepted with pleasure.

Upon my return to Vienna, I reported my impressions of Poland to the Circle and announced Tarski's Colloquium lectures. In mid-February, I extended a cordial invitation to all members to join the Colloquium for two talks: "Some Fundamental Concepts of Mathematics" and "Studies in the Calculus of Propositions".[17] However, Hahn and Carnap were the only members of the Circle who joined the Colloquium for the first of these lectures.[18] The absence of Schlick and the others greatly disappointed me. So, late in the evening, I rang up each member, reporting briefly about the first lecture and urgently reiterating the invitation to attend the second. The next evening, all members of the Circle including Schlick appeared.

Tarski presented, among other discoveries, Łukasiewicz's three-valued and many-valued logics, which were quite new to the Circle.[19] I had also asked him to make a few remarks about Łukasiewicz's parentheses-free notation.[20] Today this device is being used even in some mass-produced electronic calculators; but when I learned about it in Warsaw in 1929, it probably was not known to any mathematician outside of Poland. The Circle, I felt, should be particularly interested in the fact that parentheses could be dispensed with, since Wittgenstein believed in their indispensability, and emphasized the philosophical significance of this supposed fact (*Tractatus*, 5.161). In the discussion that followed, Waismann advanced a rather inane argument in defense of Wittgenstein. But

---

17 Cf. *Ergebnisse e. Math. Koll.*, 2, p. 13 f. (11. and 12. Colloquium, February 20 and 21, 1930).

18 They of course were interested in the talk and, at my suggestion, Hahn offered to publish Tarski's paper on metamathematics in the Viennese mathematical journal, *Monatshefte für Mathematik*, where it appeared in vol. 38 (1932).

19 It was unknown to the Circle that E. L. Post had simultaneously with, and independently of, Lukasiewicz introduced multi-valued logics (*loc. cit.*, 5). Relativization to multi-valued logics brings about a narrowing of the class of tautologies in sense a) of Section 4.

20 Cf. *Ergebnisse e. Math. Koll.*, 2, p. 14. See also Chapter 3 of my book *loc. cit.* 8.

Hahn said somewhat impatiently, "Why not admit that on this point Wittgenstein was obviously mistaken?" And I distinctly remember that looking at Schlick I saw him nodding assent to Hahn.

I then asked Tarski to summarize briefly the main ideas of his first talk on metamathematics for the benefit of those who had been unable to attend. In the ensuing discussion, Hahn, Carnap, and I spoke strongly in favor of a precise metalanguage; Waismann equally strongly against it basing his objections essentially on Wittgenstein's relegation of language to the domain about which one cannot speak. Schlick was unusually reserved that evening, but what little he said was in support of Waismann.

In the first meeting of the Circle after Tarski's visit, I was delighted to hear Carnap emphatically propounding what I had been suggesting to deaf ears for almost a year: that a more exact philosophical meta-language would be of great value to our discussions. Carnap's description of this meeting in his Autobiography[21] ends with the sentences: "I pointed out that most of the puzzles, disagreements and misunderstandings in our discussions arose from the inexactness of our metalanguage", and "My [i. e. Carnap's] talks with Tarski were fruitful for my further studies of the problems of speaking about language".

Sooner or later a rapprochement between the schools of Vienna and Warsaw would of course have occured even without Tarski's visit — but certainly not as early as 1930. My purpose in inviting him to Vienna would thus have been fully achieved had it not been for Schlick. I had always found him to be a man unusually open to new ideas and eager to learn as well as prone to respond; but in the case of Polish logic, he proved to be more than reserved — Carnap (loc. cit.[21]) says that Schlick "remained sceptical".

As I soon could observe in Circle discussions, however, Schlick's reserved attitude was not confined to logic coming from abroad. His interest in all of logic and mathematics, which for a couple of years

---

21 P. A. Schilpp (ed.), *The Philosophy of Rudolf Carnap* (La Salle, Ill.: Open Court, 1962), p. 30. Several minor details in Carnap's description are inaccurate (e. g., the statement that Tarski came to Vienna at the invitation of the Department of Mathematics, which actually had nothing whatsoever to do with my guest. Fortunately, in arranging the meeting of the Circle with Tarski, I did not expect acknowledgement from either side.)

had been quite strong, was fading just at that time. Only later did the reason become clear to me. In the years after their first personal contacts in 1927, Schlick had a number of philosophical conversations with Wittgenstein in many of which Waismann took part and which had an immense influence on both men.[22] In the light of what I have read and heard since then, there is no doubt in my mind that it was Wittgenstein who created in them, perhaps unwittingly, a bias against formal logic and mathematics.

9

During the academic year 1930/31, I lectured in various parts of the United States. Feigl, who likewise spent that year in America, energetically publicized the Vienna Circle applying himself especially to prepare the way for Carnap and deciding to stay on himself.

When I returned to Vienna in the fall of 1931 Carnap had left to join Philipp Frank in Prague, but he visited the Circle from time to time. He no longer talked about "the" language and in fact, to the displeasure of Waismann, *used* several languages. Yet I continued belaboring the point until (I believe after his second visit) he also gave up references to "the" logic despite strong protests not only from Waismann, but also from Schlick. And two new topics came to prominence in the discussions of the Circle at that time: one epistemological, the other methodological.

Like most epistemologists, Schlick wanted to give a description of the foundation of empirical knowledge *(das Fundament der Erkenntnis)*.[23] He saw this foundation in utterly simple *Konstatierungen*, by which he meant the establishment of comment-free primitive observations such as

The pointer of this instrument, here and now, coincides with such and such a mark.

At about the same time, Neurath and Carnap advanced the doctrine that (empirical) knowledge be based on what they called *protocol sentences*, by which they meant propositions of the form

---

22 Conversations held after December 1929 are recorded from Waismann's notes in the book B. F. McGuinness (ed.), *Ludwig Wittgenstein und der Wiener Kreis* (Oxford 1967). But the meetings that decisively influenced Schlick's philosophical attitude (held between the end of 1927 and the end of 1929) do not seem to have been recorded.

23 Cf. *Erkenntnis* 4 (1934), pp. 79—99.

Observer O at the place p and the time t has read mark m on the instrument I.

Many discussions in the Circle centered on the differences between the two views, which obviously exist, but seemed to me to be outweighed by similarities. In fact, under certain assumptions (such as that the person uttering the protocol sentence does not lie) one might well establish a correspondence (perhaps even a one-to-one correspondence) between *Konstatierungen* and protocol sentences. (Incidentally, Philipp Frank had always emphasized the basic role of meter readings for our knowledge in physics.)

The second new topic was Carnap's and Neurath's doctrine of *unity of science*,[24] originally called *panphysicalism*. It was directed against the thesis that there are fundamental differences between human and physical sciences. The idea of unity was rooted in Carnap's attempt to develop in a systematic way our empirical knowledge of the entire physical and social universe in his book *The Logical Structure of the World*.[25] Neurath eagerly seized this view in his fight against social scientists claiming the possession of insights unavailable to physical science. On this point he had the full support of Schlick, who had denied the existence of intuitive knowledge as early as 1913.[26] Schlick now was in favor of the unity idea, if less fervently than Neurath. For my part, I have always taken a pragmatic view on questions of method and disliked all monistic schemes which may a priori limit the objects and/or the methods of research. But there was a point on which we all agreed: that in the 1930's many social scientists used their alleged insights for the justification of value judgments and especially of their (mostly fascist) political views in a logically quite unacceptable way. Earlier in the century such arguments had played a role in the promotion of a fascist mentality in the universities of Germany

---

24 *Unity of Science* is the perfect translation of the German *Einheit der Wissenschaft*. Actually, Neurath and Carnap spoke of *Einheitswissenschaft*, a sobriquet which, because of various connotations, cannot be quite faithfully translated into English.

25 University of California Press, 1969 [Translation of *Der Logische Aufbau der Welt* (Berlin, 1928).]

26 "Is there Intuitive Knowledge?", Chapter 6 in H. L. Mulder and Barbara F. B. van de Velde-Schlick (eds.), *Moritz Schlick: Philosophical Papers, Vol. I* (1909—1922) (Dordrecht: D. Reidel, 1979), (Translation of a 1913 paper).

and Austria. But when Neurath seemed to me to dream of halting that development by a unity-of-science movement the situation in Central Europe had gone far beyond the point of intellectual arguments.

10

In 1933 the year of Hitler's coming to power in Germany there were periods when life in Vienna was almost intolerable. The newspapers published extras around the clock and vendors ran shouting through the streets offering the latest editions. Groups of young people, many wearing swastikas, marched on the side walks singing Nazi songs. Now and then, members of one of the rivaling paramilitary groups paraded through the wider avenues. I found it almost impossible to concentrate and rushed out hourly to buy the latest extra. On one of these days, I met Dr. and Mrs. Schlick in a street car. "It is impossible to concentrate", the professor said. "I read extras from morning to night".

Schlick's position at the University became precarious. He was not and, as far as I knew, never had been politically active. His political views can probably best be described as those of a British-style liberal. But this was far from satisfactory to the nationalistic professors and students. Moreover, in all adminstrative and academic matters, especially with regard to proposals for new appointments, Schlick was always objective, unprejudiced and impartial — qualities that failed to ingratiate him with the majority of his colleagues at that time. And that majority had more specific objections to his activity: his insistence on keeping Waismann (who was Jewish) as his assistant; his supposed friendship with the radical Neurath; his *Kreis* (the Vienna Circle!) which, notwithstanding the strictly apolitical way in which Schlick conducted it, was regarded as a sort of secret conspiracy; and finally his sponsorship of the Ernst Mach Society, which was almost treated as an open rebellion (it would be dissolved by the Dollfuss regime in the wake of the first civil war in 1934).[27]

---

27 Nationalistic students tried to discredit Schlick even because of his given name, Moritz, which many Austrians (perhaps because of its alliteration with Moses) regarded as Jewish even though the name clearly derived from the Latin *Mauritius*. Moreover, Schlick was named after a

To some extent, all of us were of course affected by these conditions making it difficult even to the most faithful friends of Austria to feel about the country as they had before. It was sad also to see Schlick's quiet serenity slowly disappear. In one of my conversations with him during that terrible period he said that in his opinion Hitler meant the *Untergang des deutschen Volkes* — the decline and fall (more precisely, the perdition) of the German people; and that all should join in a supreme effort to avert this imminent calamity. He also mentioned to me his intention to write a letter to Cardinal Innitzer, the archbishop of Vienna, whom many of us had met when he was professor of theology at the University. We had admired him for restoring that institution for the year of his rectorship to a place of teaching and learning: he energetically suppressed riots of Nazi students that constantly plagued the university before and after that year and forced most other rectors to keep the institution closed for long periods. In his letter to the cardinal, while mentioning his unrelatedness to Catholicism, Schlick planned to offer his support of any action that might stem the flood.[28] (Whether Schlick ever dispatched that letter I don't know).

In 1933/34, the university was closed for extended periods of time. Both Schlick's Circle and my Colloquium, however, met regularly though Schlick, Hahn and I, being the only members with keys to the deserted building, had to let the others in. Upon entering one had the feeling of having reached a quiet oasis.

Austria's chancellor Dollfuss, who ruled without Parliament, not only wanted to rid the country of Nazis but also tried, probably abetted by Mussolini, to destroy their only absolutely implacable Austrian opponents, the Social Democrats. He subjected the latter to provocations culminating in the suppression of their widely read daily newspaper. At the beginning of 1934 an explosion was

---

close relative of his mother, the once famous Ernst Moritz Arndt, one of the greatest heroes of Germany in the Napoleonic period. His books and poems inspired the liberation of the country from French occupation. In his writings, Arndt developed ideas about German national aims which some National Socialist theoreticians, knowingly or unknowingly, paralleled. To the end of his long life in 1861, he was called "the most German of all Germans" (*der teutscheste aller Teutschen, teutsch* being a chauvinist version of *deutsch*, related to *teutonic*).

28 Because of his activity as rector, the cardinal appeared to Schlick as a permanent pillar against Nazism. But those who hoped that he would stand up to the Nazis after Hitler's occupation of Austria were disappointed.

clearly imminent. In February, Dollfuss under some pretext attacked the main city-owned buildings (chiefly the *Karl Marx Hof*) of the organized Viennese laborers with heavy artillery. The ensuing civil war ended with the destruction of the Social Democrat forces. Hitler's Austrian followers, greatly strengthened by these events, made ever increasing demands on the government. In July 1934, in a surprise attack on the chancellery, a group of Nazis assassinated Dollfuss thereby starting a short second civil war. But the new chancellor, K. von Schuschnigg, succeeded in suppressing that uprising; and he continued his own brand of fascist regime until Hitler occupied Austria in 1938.

The day before Dollfuss was murdered, Schlick and the few left in his Circle were deeply shocked by the totally unexpected death of Hans Hahn after a short illness. Neurath, who happened to be abroad in February, never returned to Austria.

11

In the months before Hahn's death, the Circle had reached an impasse. In a perennial debate, Waismann denied that one could talk about language; Hahn asked why one could not make statements about the ordinary or "object" language, in a language of higher order. Waismann motivated his position essentially by saying that Hahn's suggestion did not fit into Wittgenstein's scheme of ideas — an argument that impressed Hahn less and less. Later, most discussions centered on Waismann's so-called *Theses*,[29] supposedly written around 1930, but distributed in the Circle (in mimeographed form) only after Hahn's death. They were a new version of the "ontology" at the beginning of Wittgenstein's *Tractatus*. I felt that in order to bring out clearly the ideas behind those theses they ought to be subjected to the same analysis that the beginning sections of the *Tractatus* seemed to me to call for.[30] Actually the discussions were a relapse into the imprecise metalanguage that had so much disturbed me in 1929. In fact, the debates of the *Theses* made me so impatient and restless that I sometimes

---

29 See *loc. cit.* p. 22.

30 See my paper "Language and Mathematics", *Proc. 4th Intern. Wittgenstein Symposium* Kirchberg/Wechsel (Wien: Hölder – Pichler – Tempsky, 1980), pp. 21—26.

7*

slipped out of the room soon after they began. It was another example of Schlick's kindness toward me that he tolerated this behavior.

12

While the atmosphere during 1933/34 was anything but conducive to the study of epistemological subtleties, one encountered political and ethical problems on every turn. For my part, I decided to reread the writings on ethics by Kant, Nietzsche and Brentano, and to study for the first time Schlick's *Problems of Ethics*. I had not gotten round to reading the book when it appeared in 1930.

An aspect of that work that struck me at once was the apparent lack of any influence of Wittgenstein. In 1927, when Schlick was already strongly affected by the *Tractatus*, he had published a booklet *Der Sinn des Lebens* (The Meaning of Life), whose very title had been, as it were, foreclosed in the *Tractatus*. There (6.521), Wittgenstein wrote that people could not say in what the meaning of life consisted even if, after long pondering, that meaning became clear to them, because "the solution of the problem of life is seen by the disappearance of that problem". But I had imagined that Schlick's more poetical than scientific booklet on the meaning of life with its dithyrambic praise of youth and play had perhaps been written before the author was familiar with the *Tractatus*. When Schlick brought out his *Problems of Ethics*, however, all his epistemological thinking was along Wittgenstein's lines and yet also this major work of his on ethics lacked any refence to the *Tractatus*. In fact, some statements in the book, such as "Ethics seeks nothing but knowledge" were contrary to Wittgenstein's basic ideas on this subject.

In the *Problems of Ethics*, Schlick proposed to develop ethics in a strictly objective way unencumbered by wishes, hopes or fears and incompatible with the positing or the absolute justification of commands. He criticized Kant's categorical imperative as an absolute command without a commander, "comparable to an absolute uncle without nephews and nieces". And he attempted to replace Kant's ethics of duty with the ideal that individuals desire what society demands — an ideal that seemed to him to presuppose gradual changes both in the attitudes of individuals and in the claims of society. According to Schlick, individuals are determined by the motive of greatest pleasure and least pain, but altruistic behavior and even sacrificial suffering are among the possible sources of pleasure. The book breathes its author's spirit of gentleness and all-embracing benevolence. But while free from the idle talk that fills so many traditional writings on ethics neither does Schlick's book include logical analyses or an explicit critique of language.

The program of strict objectivity and of refraining from positing or justifying commands greatly appealed to me (as did, incidentally, some critical remarks about ethics at the end of the *Tractatus*). But I wanted to carry out that program more consistently than Schlick had done (and without any mysticism). After one strictly excludes all evaluations and subjective elements from consideration, what is left is nothing but individuals evaluating, choosing, deciding and acting. The results are relations between individuals and individuals, between individuals and groups of individuals, and between groups and groups — relations amenable to treatment by logic and mathematics. In the absence of quantitative parameters (which play a dominant role in game theory), the considerations remained unfortunately on the most elementary level. Yet, long before models became ubiquitous in the social sciences, I developed a model for the constitution of a population in groups of mutually compatible individuals. Adherence of all members of the population, or of a group, to Kant's categorical imperative turned out to be neither necessary nor sufficient, for their mutual compatibility.

I did not speak about my ideas in the Circle. But I certainly wanted Schlick to know about them and hear his opinion. In the spring of 1934, I brought him the manuscript of my booklet *Moral, Wille und Weltgestaltung (Morality, Decisions and Social Organization)*. He read and returned it within ten days. His comments were very favorable; and Schlick was the only reader — then or later — who read between the lines of the book: my hope that models such as the one dealing with the categorical imperative might eventually suggest schemes for an organized coexistence of antagonistic (especially democratic and dictatorship tolerating) individuals and groups. He then spontaneously offered to recommend the book to the Springer-Verlag in Vienna, where it appeared early in the summer of 1934. During the summer vacation, Schlick wrote me that he had asked the editors of the journal *Erkenntnis* to let him review my book. But his review never appeared and *Erkenntnis* never even mentioned the publication.

13

There was in Vienna in the mid 1930's one respected politician who kept out of all the conflicts — the *Bundespräsident*, president Miklas of Austria. I had never seen him in person, but in the spring of 1936 I received from his office an invitation to attend the opening of some exhibition in a building of the former imperial court. I arrived a little late and found the room so packed with guests that it was almost impossible to move. The first familiar person whom I saw was Schlick standing not far from the entrance. We were talking when a passage opened in the crowd near us. Through it, the President, having completed the opening ceremony, was leaving with his party. To my surprise, one of the men behind the President waved rather intimately to Schlick, and the latter responded in the same way. "You have friends in the government?" I asked teasingly. But Schlick's expression changed to one of utmost seriousness and he said in a grave tone, "This is not a friend. This is a security man who used to be my body-guard". I must have looked totally bewildered so that Schlick explained further. "For some time now I have been threatened by an insane person who is in and out of mental institutions; and the man behind the president used to be assigned to my protection". "So you are no longer threatened", I said. Schlick sighed. "Until quite recently the fellow had been interned", he said. "But just three days ago he was released again; and yesterday I had another threatening telephone call from him. Yet, for all his threats, he has never actually harmed me. So I don't dare to complain to the police again". And, as though it had happened only yesterday, I remember how Schlick added with a forced smile, "I fear they begin to think it is I who is mad"; and he changed the topic. But I felt that he actually lived in great fear and must have done so for a number of years.

A few weeks later, Vienna was stunned by the news that, on the steps to the auditorium in the university building, a paranoic former student had shot and killed Professor Schlick.

The grief of the philosopher's numerous friends and students was deep. Since Schlick had held Austria's most prestigious chair in philosophy all newspapers published long, if uncordial obituaries. They also mentioned facts about the assassin. He was a disgruntled former student of philosophy who had the paranoic idea that Schlick frustrated his attempts to find employment. But as soon as the first impact of the tragedy had subsided some newspapers close to the government and paramilitary organizations changed their tone to open hostility. They now wrote that while murder was always condemnable it was not altogether surprising that students of Schlick's corrosive philosophy *(zersetzende Philosophie)* would etc., etc... that the instruction in philosophy at Austrian universities had to be reformed and so forth.

It would be futile to speculate about what might have happened had the bullets missed the victim or had the madman been permanently interned and Schlick allowed to live out his life without fear. But there is little doubt that he would have deservedly attained an even higher stature as a philosopher in the English-speaking world and eventually in Germany than he has, since being cut off from his work and further philosophical development. And many more would have enjoyed his wisdom and his kindness.

INTRODUCTION TO THE
SIXTH AMERICAN EDITION, 1960

Ernst Mach's book *Die Mechanik in Ihrer Entwicklung Historisch-Kritisch Dargestellt,* one of the great scientific achievements of the last century, remains a model for the presentation of the development of ideas in any field. In its own domain, the work is still full of vitality. It is an inspiration to philosophers of science, a valuable source of information for historians of physics, and a splendid help to teachers of mechanics. Its first half is a most stimulating introduction of unsurpassed clarity and depth for beginners.

The book follows the development of mechanics up to the turn of the century.[1] As the title indicates, the work is historical and critical.

That the *historical* presentation of a branch of science is the most penetrating approach to the subject matter and leads to the deepest insight was one of Mach's general methodological ideas. Nor is anything more conducive to creative thinking than an exposition of ideas as they have developed, of notions abandoned long since, and of the role that historical accidents have played in the genesis of current concepts—techniques that Mach introduced and superbly developed in his *Science of Mechanics.* Mach also applied the historical method of presentation to the theory of heat and to

[1] Nine German editions of this book have been published. Seven of them appeared during Mach's lifetime (1838-1916), in 1883, 1888, 1897, 1901, 1904, 1908, and 1912. The eighth and ninth German editions appeared in 1921 and 1933. English translations were published by the Open Court Publishing Company in 1893, 1902, 1915, 1919, 1942, and 1960.

parts of optics and the theory of electricity.[2] But the method has further potentialities. Other parts of mathematics, especially algebra, would profit from a similar treatment; and, most of all, the science of mechanics itself might gain if its development since 1900, including the theory of relativity, wave mechanics, and quantum mechanics, were presented *à la* Mach.

The *critical* parts of Mach's book culminate in his analysis of the concept of mass and his examination of Newton's ideas of absolute space and absolute time. The latter critique is quoted in almost every presentation of the theory of relativity. "This book," Einstein wrote in his autobiography[3] about Mach's *Science of Mechanics*, "exercised a profound influence upon me . . . while I was a student."

*   *   *

In this century, the analysis of the mass concept and, even more, physicists' views on space and time, have advanced beyond Mach. Yet his original discussions remain classics not only of physics but also of *philosophy of science*.

Against Newton's definition of mass as a quantity of matter, Mach raised the objection that it was of no help in actual operations with masses; and he formulated a new definition, based on Newton's Third Law— a definition that made it possible to measure masses. Mach's treatment initiated a method that was later greatly elaborated and applied to the philosophy of

---

[2]E. Mach, *Prinzipien der Wärmelehre* (Leipzig: 1896);
———, *Principles of Physical Optics* (Leipzig: J. A. Barth, 1921; London: Methuen Co., 1926; New York: Dover Press, 1953.)
———, "On the Fundamental Concepts of Electrostatics," in *Popular Scientific Lectures* (5th ed., La Salle, Illinois: Open Court, 1943), pp. 107-136.

[3]"Albert Einstein: Philosopher-Scientist," in *Library of Living Philosophers*, ed. P. A. Schilpp (Evanston: 1949), pp. 20-21.

physics by P. W. Bridgman[4] under the name of *operationalism.*

Mach rejected absolute space and time because they are unobservable. More generally, he proposed to eliminate from science notions which lack counterparts that are actually or at least potentially observable. He thereby became one of the initiators of *antimetaphysical positivism.*

A third point that Mach stressed over and over again was his view that science had the purpose of saving mental effort. General laws are shorter and easier to grasp than enumerations of specific instances. Simpler theories are preferable to more complicated ones. His theory of "economy of thought" is Mach's main point of contact with R. Avenarius who, in his *empirio-criticism,*[5] regarded philosophy as thinking about the world with minimum effort.

A fourth point of philosophical importance in Mach's program was the replacement of causal explanations by functional connections. In this respect, Newton had been a shining model. Without entering into the question that was uppermost in the minds of his contemporaries—the question as to *why* bodies attract each other—Newton was satisfied with formulating the specific connection of the attractive force between two bodies with their masses and with the distance between them. In the 18th and 19th centuries, countless attempts, now all but forgotten, were made to *explain* gravitation. Physicists hypostasized vortices, or tensions in media, or bombardments of the bodies by particles traversing space at random and driving, for instance, a stone toward the earth because the latter shields the stone against the particles coming from

[4]P. W. Bridgman, *The Nature of Physical Theories* (Princeton: 1936; New York: Dover Press, 1936).

[5]R. Avenarius, *Philosophie als Denken der Welt gemäss dem Princip des kleinsten Kraftmasses* (Leipzig, 1876).

below. But the real triumphs of human insight into gravitation—Newton's deduction of Kepler's Laws, the prediction of new planets which were subsequently observed, and, in the present age of experimental astronomy, the control of the motion of artificial satellites—these triumphs are independent of any explanation of gravitation and are entirely anchored in Newton's law that the attractive force between two bodies is proportional to their masses and inversely proportional to the square of the distance between them. Similarly, Hertz' prediction of electromagnetic waves is based on Maxwell's equations connecting the fundamental electric and magnetic quantities with each other, without aiming at an explanation of phenomena.

Mathematics has the tremendous creative power of evolving hidden consequences from assumptions about the observable universe and thus prompting predictions of previously unobserved phenomena. Not that the universe is under any obligation to conform to those predictions! But if these consequences are verified, then mathematics has led to new discoveries; and if they are not borne out by observations, then mathematics has necessitated a revision of the underlying general assumptions.

Mach's emphasis on functional connection raises two questions: What are functions? And what is it that functions connect?

Once the logarithm of any positive number has been explained as the exponent to which 10 must be raised to give that number, mathematicians define the logarithmic function by pairing, to every positive number, its logarithm. So, e.g., to 10, they pair 1; to 100, 2; to .1, -1. The *logarithmic function* (the result of this definition) is the class of all pairs of numbers thus obtained—a class including in particular the pairs (10,

1), (100, 2), (.1, —1). More generally, mathematicians say that a function has been defined if, to every number or to every number of a certain kind, a number somehow has been paired. The *function* (the result of this definition) is the class of all pairs of numbers thus obtained.

The traditional answer to the second question is: Functions connect variable quantities or, briefly, *variables*. A thorough analysis has revealed that the term variable is used in several totally discrepant meanings.[6] For instance, it is applied to the letters x and y in the mathematical statement:

(1) $x + \log y = \log y + x$ for any number x and any positive number y. Variables in this sense are used according to the following rules: (a) In the formula, according to the accompanying legend, the letters x and y may be replaced with numerals, say, —3 and 10, each such replacement yielding a specific formula, e.g., $-3 + \log 10 = \log 10 - 3$. (b) Without any change in the meaning of the statement (1), any two unlike letters may be used as variables; e.g., one may write:

$x + \log b = \log b + x$ for any number x and any positive number b; or x and y may even be interchanged as in:

$y + \log x = \log x + y$ for any positive number x and any number y.

The so-called variables that are functionally connected in physical laws, however, are of a completely different nature. Suppose, e.g., that, in the course of a process of a certain type, the work w in joules done by changing the pressure p in atmospheres of a gas

[6] K. Menger, "On Variables in Mathematics and in Natural Science," *British Journal for Philosophy of Science,* V., 1954, pp. 134-142. An elementary presentation of the concepts of variable, function, and fluent is contained in K. Menger, *Calculus. A Modern Approach:* (Boston: Ginn & Co., 1955). See particularly chapters IV and VII.

from its initial value p_0 is connected with p/p_0 by the logarithmic function—in a formula:

(2) $w = \bar{w} \log p/p_0$.

The contrast between p and w in (2) on the one hand, and x and y in (1), on the other, could hardly be greater than it is. (a) The letters p and w must not be replaced with just any two values of pressure and work. For instance, the work done by compressing the helium in one tank on Monday is not in general the logarithm of the pressure of oxygen in another tank on Wednesday. Nor is the formula (2) accompanied by any legend authorizing any replacements. (b) The meaning of (2) changes completely if, instead of w, say, the designation v of the gas volume is introduced in the formula or if p and w are interchanged. (In fact, the formulae thus obtained are, in general, false.) For, whereas x and y in (1) stand for *any* number and do not designate anything specific, p and w do; they designate pressure and work.

But exactly what is gas pressure in atmospheres? One way of defining it is by pairing, to each state S of a gas, the pressure in atmospheres $p(S)$ of the gas in the state S. *Gas pressure* (the result of this definition) is the class of all pairs $(S, p(S))$ thus obtained. Similarly, work in joules is a class of pairs $(S, w(S))$. The traditional formula (2) is nothing but an abbreviation of the following law:

(2_s) $w(S) = \bar{w} \log p(S)/p_0$ for any state S of any gas undergoing a process of the type under consideration.

In a strictly positivistic and operationalist spirit, one would take a step beyond the preceding definition of pressure in atmospheres and define *observed pressure in atmospheres* by pairing, to each act A of reading a pressure gauge calibrated in atmospheres, the number $p^*(A)$ that is read as the result of the act. The result of this definition is the class of all pairs

(A, p^*(A)) thus obtained.[7] Similarly, one can define w^*, the *observed work in joules* as a class of pairs (B, w^*(B)) for any act B of measurement of a certain other kind. The formula (2) then is an abbreviation of the following more articulate formulation of the physical law: (2*) w^*(B) $=$ log p^*(A)$/p_0$ for any two acts of reading pressure gauges and work meters, respectively—acts simultaneously directed to the same gas sample undergoing a process of the type under consideration.

According to their definitions, p^*, w^*, p and w are classes of pairs, yet not only, as such, fundamentally different from the variables x and y in (1) but also basically different from functions. For while the latter e.g., the logarithmic functions, are purely logico-mathematical objects, the definitions of pressure, observed pressure, and the like include references to physical states or observations. The work, which is the logarithm of the pressure, bears to the logarithmic function a relation similar to the relation that a yard, which is equal to three feet, bears to the number three.[8] In order to distinguish objects of scientific studies such as p and p^* both from variables and from functions, the writings quoted in [6] and [8] have revived the term coined by Newton for time, distance traveled, gas pressure, work and the like—the term *fluents*. During the 18th century, not only did this term fall into almost

[7]An extreme positivist might question whether, beyond the *observed* pressure p^*, there is any (so to speak *objective*) pressure p. How, indeed, is the pressure p(S) of a gas in the state S ascertained? It is derived (by more or less arbitrary averaging processes) from values of the observed pressure, namely, by somehow averaging numbers p^*(A₁), p^*(A₂), ... that result from acts A₁, A₂, ... of reading various guages—acts that various observers direct to the gas in the state S. In the case of helium in the corona of the sun or the terrestial atmosphere a million years ago, the pressure is ascertained by what Bridgman calls pencil-and-paper-operations.

[8]Detailed discussions of these and related points are contained

complete oblivion, being replaced by the word *variable*, but the underlying concept was confused with that of a variable in the logico-mathematical sense of a letter meant to be replaced with any numeral or, more generally, with the designation of any element of a certain class of objects.

The questions raised by Mach's emphasis on functional connections thus can be answered as follows: Functions are certain classes of pairs of numbers. The objects that, in physical laws, are connected by functions are fluents—classes of pairs, each of which results from pairing a number to a physical state or an act of observation.

The fifth and last philosophical point to be mentioned here (which is only briefly touched on at the end of the present book) is Mach's emphasis on the immediate sense data. He called them *elements* and used them as building blocks in constituting *complexes* such as the idea of the various things surrounding us as well as the idea of ourselves. He refused to look for objective causes of the phenomena or data.

"As several ideas imprinted on the senses are observed to accompany each other, they become marked by one name and so to be reputed as one thing. Thus, for example, a certain color, taste, smell, figure and consistence having been observed to go together, are accounted one distinct thing, signified by the name apple. Other collections of ideas constitute a stone, a tree, a book and the like sensible things." This passage, which precisely renders the first part of the contention,

in the book quoted in footnote 6, and in the following articles by the same author: K. Menger, "Mensuration and Other Mathematical Connections of Observable Material," in *Measurement: Definitions and Theories*, ed. C. W. Churchman and P. Ratoosh (New York: Wiley & Sons, 1959), pp. 97-128; and
———, "An Axiomatic Theory of Functions and Fluents," in *The Axiomatic Method* ed. Henkin, Suppes, and Tarski (Amsterdam: North-Holland Publishing Co., 1959), pp. 454-473.

might well have been written by Mach; actually, however, it is a quotation from the very first section of the *Treatise concerning the Principles of Human Knowledge* written by Bishop Berkeley in 1710. In developing these thoughts further, however, Mach diverged altogether from Berkeley, who assumed extra-physical, spiritual causes of sense data in which, because he was a theologian, he was greatly interested. Mach shunned a search for causes of data, in particular, for extra-physical and spiritual causes, and confined himself to the phenomena. His fear of being identified with Berkeley's spiritualism probably explains why Mach did not mention the development of what he called the theory of elements and complexes in the first part of Berkeley's classical *Treatise*—a book that can hardly have escaped Mach's attention.[9] In contrast to Berkeley's idealistic metaphysics, Mach's philosophy of science has often been called *phenomenalism*.

*   *   *

Mach's operationalist, antimetaphysical, anticausal views and his ideas on economy of thought pervade his presentation of the science of mechanics. Yet an introduction to the present book seems to call less for a discussion of some moot points of Mach's philosophy[10] than for an appraisal of its author as a scientist.

Mach repeatedly emphasized that admiration for a great physicist of the past should not keep historians from discussing the master's limitations. In keeping

[9]On the occasion of the 200th anniversary of Berkeley's death, K. R. Popper published "A Note on Berkeley as a Precursor of Mach," *British Journal for Philosophy of Science*, IV, 1953, including an amazing list of quotations from lesser known writings by Berkeley wherein also some of Mach's scientific ideas (connected with the concept of force, the critique of absolute space, time, and motion and with the economy of thought) are clearly anticipated.

[10]Debates have centered on the question whether references to basically unobservable entities can be, and should be, eliminated

with this admonition, our respect for Mach should not prevent us from observing that, in the course of this century, three limitations of Mach himself as a scientist have become apparent.

Even though Mach is generally recognized as one of the principal precursors of the theory of relativity, he himself not only ignored that theory in the editions of the present book[1] that he published after the appearance of Einstein's first paper in 1905, but actually underlined his aloofness. Remarks to that effect are included in the Preface to his book *Principles of Physical Optics;*[2] and his son, Ludwig Mach, quoted the following passage from papers left by his father: "I do not consider the Newtonian principles as completed and perfect; yet, in my old age, I can accept the theory of relativity just as little as I can accept the existence of atoms and other such dogma."

This leads to the second point where the actual development of physics has completely diverged from Mach's views. Mach seems to have been unimpressed by Boltzmann's triumphs in the kinetic theory of gases as well as by Perrin's experiments on Brownian motion; at least, it appears from the quoted passage, he was not sufficiently impressed to attribute physical significance to the assumption of atoms. We can only speculate as to how he would react to science of the mid-twentieth century, which is completely dominated by atomic physics. Would he admit that the phenomena in a Wilson cloud chamber make a granular structure

only from the final statements of a theory or also from all intermediate steps and from the basic assumptions. Other discussions have dealt with the concept of the simplicity of a theory. On questions of this type there is not only disagreement between various scientists, but some of them, in the course of their lives, have changed their own opinions. For instance, Einstein mentions in his autobiography[*] that, while always admiring Mach as a physicist, at a later age he abandoned many of Mach's philosophical views which had impressed and influenced him greatly in his youth.

of matter and electricity almost visible? Would he perhaps, while admitting granular structure, question the
precise equality of all grains of like type and point to
galaxies, which also have a granular structure without
all grains of like type (e.g., all red giants) having
precisely equal masses or sizes? How would he react to
the discovery of ever more types of elementary particles?

Strangely, Mach, who had such a sharp eye for
the difficulties of atomism, did not seem to appreciate
that the idea of matter continuously filling space leads
to other conceptual problems and perhaps to even
greater difficulties. One can imagine advances in the
technique of experimentation and observation that will
make statements about hitherto unobservable particles
verifiable. Certain statements about the behavior of
matter continuously filling space, however, are fundamentally unverifiable.

The writer of this Introduction strongly believes
that some of the difficulties common to *all* theories of
microphysical phenomena have a common cause—the
lack of an adequate microgeometry. The current views
on geometry are still essentially Euclid's. The various
non-Euclidean geometries developed in the 19th century differ from Euclid's geometry only with regard
to assumptions such as the parallel postulate which,
when applied to nature, are reflected in properties of
space *in the large,* wherefore the main domain of application of those geometries is astronomy or cosmology. *In the small,* all the 19th century non-Euclidean
geometries are indistinguishable from Euclid's geometry. Any two points are assumed to have an exact
distance from one another and (at least if they are
not too far apart) to be joined by exactly one straight-
est line. And even the sum of the angles in any small
non-Euclidean triangle is, according to the assumptions,
indistinguishable from two right angles. Only rather

recent work has abandoned the conception that the same laws are valid or even the same general notions are applicable in the very small that everyone knows from the geometry of larger regions. In fact, the very ideas of numerical distance and of points have been challenged.[11] And this is what microphysics probably needs: a microgeometry built on assumptions that are completely different from Euclid's; a theory of lumps rather than of points, and of distance distributions rather than of exact numerical distances; that is to say, given any two lumps and any interval of numbers, all that can be determined is the probability that the distance between the two lumps belongs to the said interval.

Mach's third limitation lies in his neglect of logic and of the critique of language—even the prelogical critique developed by his contemporary F. Mauthner[12] who, unfortunately, has been sadly neglected not only in his own day but by his successors as well. The way in which a man combines immediate sense data or elements in constituting complexes is profoundly influenced by others: mainly by those who, long since, taught him to speak; then by his teachers and educators; and finally by people with whom he exchanges information and views. He is, in other words, strongly influenced by language and all the wisdom and all the folly which, since time immemorial, his ancestors have stored in that means of communication.

Connected with Mach's alogical orientation are two

[11]See K. Menger, "Theory of Relativity and Geometry," in *Albert Einstein: Philosopher-Scientist,*[8] especially pp. 472-74; ———, "Statistical Metric," *Proceedings National Academy of Science,* XXVIII, 1942, p. 535 et seq. ———, "Probabilistic Geometry," *ibid,* XXXVII, 1951, p. 226. B. Schweizer and A. Sklar, "Statistical Metric Spaces," *Pacific Journal of Mathematics,* X, 1960, p. 313 et seq.

[12]F. Mauthner, *Beiträge zu einer Kritik der Sprache* (Stuttgart: 1901-2), I, II, III.

other aspects of his work: a certain lack of precision in the formulation of some philosophical ideas and strictly empiristic views on mathematics. Mach considers even arithmetic as entirely based on experience. While in recent times empiristic views in mathematics have been decidedly underemphasized and, therefore, still seem to present unexplored potentialities, they unquestionably are one-sided and in definite need of a complementation by logic, by the logical analysis of language and, perhaps, by some of the ideas advanced by H. Poincaré.[13]

*   *   *

Of the many scientists who have been influenced by Mach, only Einstein and Bridgman have been mentioned on the preceding pages; of the kindred philosophers, only Avenarius and Poincaré. The 1920's witnessed the constitution of a group of philosophers of science who may be considered as direct successors and continuators of Mach—even in a geographical sense: they taught at the two schools with which Mach had been connected: the Universities of Vienna and Prague.[14] This group has become widely known under the name of *Wiener Kreis or Vienna Circle.**

In its beginnings, the Vienna Circle was altogether Mach-oriented. The philosopher M. Schlick emphasized[15] that the postulates of a theory are implicit definitions of its basic concepts—an idea which, in the

[13]H. Poincaré, *Science and Hypothesis* (New York: Dover Press, 1952). First French edition, 1912.

[14]Mach, who was born in Turas, Moravia, was professor at the University of Prague from 1867 to 1895, and at the University of Vienna from 1895 until his retirement in 1901. He died near Munich in 1916.

*The writer of this Introduction was a member of the Vienna Circle (remark by the editor of the 6th American Edition, 1960).

[15]M. Schlick, *Allgemeine Erkenntnislehre* (2d ed.; Berlin: Springer, 1923).

field of mechanics, goes back to Mach's use of Newton's Third Law as a definition of mass. In 1927, O. Neurath founded an Ernst Mach Society in Vienna; and the first meeting of the philosophers of science in Prague stood, as P. Frank emphasized, in the sign of Mach.

It was the mathematician H. Hahn who first directed the interest of the Vienna Circle to logic by his detailed presentation of the ideas of B. Russell and the *Principia Mathematica*.[16] In this way, L. Wittgenstein's *Tractatus*[17] became a topic—and in the years 1925-27 the dominant topic—of the discussions in the Circle. The interest shifted from Mach's elements and complexes to the ways of talking about observations and of formulating laws; from the analysis of sensations to the analysis of language.[18]

Even in this respect, and in spite of his limited interest in logic, Mach was a precursor. His antimetaphysical attitude, exemplified in his views on absolute space and time, anticipated the statement that only verifiable propositions are meaningful or, to put it somewhat less dogmatically, the positivistic postulate that extra-logical propositions should be verifiable. Moreover, long before Wittgenstein and Carnap, Mach had used, if only on the base of common sense and unsystematically, the terms *Scheinprobleme* (pseudo— or apparent problems) and *meaningless questions*. "Refraining from answering questions that have been recognized as meaningless," Mach wrote in the *Anal-*

[16]B. Russell, *Introduction to Mathematical Philosophy* (London: 1919); and A. N. Whitehead and B. Russell, *Principia Mathematica* (2d ed.; Cambridge: Cambridge University Press, 1925-27), I, II, III.

[17]L. Wittgenstein, *Tractatus Logico-Philosophicus* (London: 1922).

[18]A more detailed account of this transition is given in R. von Mises, *Positivism. A Study in Human Understanding* (Cambridge: Harvard University Press, 1951).

ysis of Sensations, "is by no means resignation. It is, in the presence of the enormous material that may be meaningfully investigated, the only reasonable attitude of scientific investigators." Later, on the basis of linguistic analysis, R. Carnap tried to eradicate pseudo-problems systematically. The philosophy of the Vienna Circle developed into *logical positivism.*

A description of Mach's influence would be incomplete without a mention of the impact of his ideas in the first decade of this century on philosophers in Russia. That development seems to have culminated in 1908, when an outline of philosophy with empirio-critical contributions by A. Bodgdanov and A. Lunacharsky was published in St. Petersburg. The man who was to shape the future of Russia, however, was opposed to Mach. Even in letters from his exile in Siberia in the first years of this century, Lenin had been critical of "Machism." In 1908, he went to London for extensive studies of the philosophical literature and, as their result, in the following year, he published a violent attack on Mach, Avenarius, Poincaré, and related thinkers in a book *Materialism and Empirio-Criticism. Critical Notes on a Reactionary Philosophy.*[19] Lenin begins by emphasizing the strong similarity of Mach's constitution of the idea of objects with Berkeley's. He then quotes several passages from Mach (e.g., references to the world of which we form pictures) which actually are at variance with Mach's own general views, whose presentation here and there lacks precision. Lenin further criticizes some statements that were also abandoned by logical positivists—two decades later, but on the basis of an analysis of language, whereas Lenin's critique is based on dialectical ma-

[19]An English translation with a Foreword by A. Deborin describing the book's background was published as volume 13 of Lenin's *Collected Works* (New York: International Publishers, 1927).

terialism. Lenin asserts, for instance, that "there is nothing in the world but matter in motion; and matter can not move save in space and time"; and "on the basis of relative conceptions we arrive at the absolute truth"—statements that logical positivists analyzing language find just about as unacceptable as the (it goes without saying, un-Machian) opposite statements that one used to hear from idealistic and relativistic philosophers: "there is nothing but ideas and minds wherein they exist; and ideas cannot exist without having a cause"; and "all truth is relative; there is no absolute truth." Lenin then proceeds to identify Mach's philosophy with Berkeley's idealistic and theological views, from which Mach definitely kept aloof; and he finally condemns the author of the *Science of Mechanics* because of the pietistic utterances of some of Mach's minor followers. In the Soviet Union, Lenin's views on Mach's philosophy became authoritative.

*   *   *

Mach's life, his son Ludwig once wrote, was dominated by a fundamental impulse toward personal clarity. He was a champion of mass education and progress and always a fearless advocate of truth as he saw it. "I see Mach's greatness," Einstein wrote,[3] "in his incorruptible skepticism and independence." In the prosperous but nationalistic and militaristic atmosphere of Central Europe in the late Victorian and Edwardian era, Mach seems to have felt a strong affinity to the English speaking world. Special ties connected him with Paul Carus and the Edward Hegeler family, who founded the Open Court Publishing Company in La Salle, Illinois. To them Mach dedicated his last book, *The Principles of Physical Optics,* in gratitude for their help in disseminating his ideas. In that dedication Mach expressed the wish that in discussions of his work mention should be made of their names.

The first English translation of Mach's *Science of Mechanics* was made and published by Open Court in 1893, and it is indeed appropriate that the first book to appear in a new series of quality paperbound books now being published by Open Court is this new edition of Mach's great work on *The Science of Mechanics*.

Karl Menger

Illinois Institute of Technology
Chicago, March, 1960

INTRODUCTION

The role Hans Hahn played in the Vienna Circle has not always been sufficiently appreciated. It was important in several ways.

In the first place, Hahn belonged to the trio of the original planners of the Circle. As students at the University of Vienna and throughout the first decade of this century, he and his friends, Philipp Frank and Otto Neurath, met more or less regularly to discuss philosophical questions. When Hahn accepted his first professorial position, at the University of Czernowitz in the northeast of the Austrian empire, and the paths of the three friends parted, they decided to continue such informal discussions at some future time — perhaps in a somewhat larger group and with the cooperation of a philosopher from the university. Various events delayed the execution of the project. Drafted into the Austrian army during the first world war, Hahn was wounded on the Italian front. Toward the end of the war he accepted an offer from the University of Bonn extended in recognition of his remarkable mathematical achievements.[1] He remained in Bonn until the spring of 1921 when he returned to Vienna and a chair of mathematics at his alma mater. There, in 1922, the Mach–Boltzmann professorship for the philosophy of the inductive sciences became vacant by the death of Adolf Stöhr; and Hahn saw a chance to realize his and his friends' old plan. It was mainly through Hahn's influence that the chair was offered to Moritz Schlick, then in Kiel. Soon after his arrival in Vienna, Schlick began to arrange discussions for a small invited group, with Hahn and Neurath as

ix

the first principal participants. Frank was in Prague where he had
gone before the war as the successor of Einstein; however, he visited
Vienna at least twice a year. Then there was Victor Kraft and, at
Hahn's suggestion, the geometer, Kurt Reidemeister, who in 1923
came to Vienna for about two years. Soon after he left, Rudolf
Carnap arrived and joined the group to which Schlick had also
brought promising students of philosophy working in his seminars,
among them Herbert Feigl and Friedrich Waismann. Thus, ultimately through Hahn's efforts, his, Neurath's and Frank's old
plan had become a reality. *"Man kann,"* Frank wrote in his
obituary of Hahn in *Erkenntnis*, Vol. 4, *"Hahn als den eigentlichen
Begründer des Wiener Kreises ansehen."* ("Hahn may be regarded
as the real founder of the Vienna Circle.")

Secondly, it was Hahn who directed the interest of the Circle
toward logic. Schlick, a student of Planck's and admirer of Einstein,
had until then been mainly interested in the philosophy of nature
and in epistemology up to (but not beyond) studies of the axiomatic
method. Otto Neurath and Olga Hahn (the blind sister of Hans, later
Otto's wife) had written — individually and jointly — papers on
Boolean algebra[2] in the years 1909 and 1910; but thereafter and
especially during the war Neurath's interests turned again to economics, sociology and history, while Frank was engrossed in the
philosophy of physics and the study of causality. Carnap, having
been a student of Frege's, was well-versed in logic, but was mainly
concerned with the philosophy of science at the time he moved to
Vienna. Hahn, however, right after his return, began an intense study
of symbolic logic with an eye to related philosophical problems. In
1922 he offered a course on Boolean algebra.[3] During the year 1924/
25 he conducted a memorable seminar on the *Principia Mathematica*
of Whitehead—Russell in which, after some introductory lectures,
he let advanced students, young Ph.D.s and lecturers report on
the contents of the book, chapter by chapter.[4] This seminar had
a very large audience and was of great influence not only on the

development of many Viennese students of mathematics and philosophy but also on the trend of the discussions in the Circle.

Thirdly, until his untimely death in 1934, Hahn greatly contributed to the Circle as a prominent participant in the discussions. Since he was carrying out very interesting mathematical research[5] in addition to his extensive activity at the University he unfortunately found little time to publish many of the ideas he proposed in the meetings. But his penetrating criticism, the clarity of his ideas and his skill in presenting them greatly impressed everyone, and often influenced Neurath and Carnap as well as Schlick and Waismann. "One can say," Frank wrote in the obituary already quoted, "that in a certain sense Hahn was always a center of the group. He always represented its central ideas without entering into differences of opinion on side issues. No one knew as well as he how to present those leading ideas in such a simple as well as thorough way, in such a logical as well as suggestive form."

While Hahn's mathematical knowledge was unusually extensive,[6] his familiarity with traditional philosophy was more limited. His favorite author was Hume, whose works, as he once told me, he found not only intellectually delightful but also morally uplifting. Furthermore he greatly admired Leibniz to whose *identitas indiscernibilium* he gave much thought, and Bolzano, whose *Paradoxes of Infinity* he edited. Kant, on the other hand, he strongly disliked because of the changes – from chapter to chapter and often from sentence to sentence – of the meaning of the terms used in his writings. Of the more modern philosophers Hahn favored Ernst Mach; and during the early 1920's he developed a great admiration for the works of Bertrand Russell. He reviewed some of them in the *Monatshefte für Mathematik und Physik*. In one of these reviews Hahn suggested that one day Russell might well be regarded as the most important philosopher of his time – a statement remarkable at a period when few philosophers in Central Europe knew or even cared to know Russell's writings.[7]

The subsequent development of his views on Wittgenstein was described to me by Hahn himself upon my return to Vienna in the fall of 1927, when he and Schlick invited me to join the Circle. He asked me whether I had heard of the *Tractatus*. I said that some time ago I had started reading the book but had not continued beyond the first pages. "This was my original experience also," Hahn said, "and I did not have the impression that the book was to be taken seriously. Only after hearing Reidemeister give an excellent report about it in the Circle three years ago and then carefully reading the *entire* work myself did I realize that it probably represented the most important contribution to philosophy since the publication of Russell's basic writings. Yet we had controversies in the Circle about the book, and there were so many differences of opinion about details that, a year ago, Carnap suggested we should, in order to clear up the confusion, devote as many consecutive meetings of the Circle as necessary to a reading of the work paragraph by paragraph; and we have indeed devoted the entire past academic year (1926/27) to this task. To *me*," Hahn continued, "the *Tractatus* has explained the role of logic." In later writings, which are included in this volume, Hahn expounded his own (somewhat oversimplifying) view by describing logic as "a prescription for saying the same thing in various ways, and for extracting from what is said all that is (in a strict sense) connoted."

In the early 1930's, after Carnap had gone to Prague, a controversy about a related topic arose in the Circle when Waismann proclaimed that one could not speak about language. Hahn took strong exception to this view. Why should one not – if perhaps in a higher-level language – speak about language? To which Waismann replied essentially that this would not fit into the texture of Wittgenstein's latest ideas. The debate recurred several times up to Hahn's death: Hahn asking, "Why?" and Waismann answering, in the last analysis, "Because". Schlick sided with Waismann,

Neurath with Hahn. Kurt Gödel and I, though reticent in most debates in the Circle, strongly supported Hahn in this one.

Another topic that Hahn raised in the Circle was a view on history that he had developed – his contribution to the Neurath–Carnap program of Unity of Science. Despite great reservations that I had with regard to the general program I felt that Hahn's interesting and, as far as I knew, original idea was quite independent of those generalities. Its starting point was Poincaré's remark that physicists are not interested in historical propositions such as *Caesar crossed the Rubicon* because they yield no predictions: Caesar will not cross the Rubicon again. Hahn took issue with a part of Poincaré's remark. He pointed out that assertions about Caesar are based on countless old manuscripts, paintings, documents, excavations, coins and the like; and that the core of the proposition, which is capable of further verification or of falsification, actually is a *prediction*, namely, the prediction that all old manuscripts, paintings, documents etc. to be found in the future will be consistent with the present assertions.

On several occasions, the mathematicians in the Circle helped the philosophers by providing them with technical information – often concerning their own results. I remember, for instance, a discussion about inductive processes in physics. In order to obtain formulae describing actual as well as potential observations, Carnap and Schlick suggested interpolation – the passing of a polynomial through the finite number of observed data. Hahn convinced them of the impracticability of this idea by reporting results of a paper of his.[8] If one starts with a given continuous curve, C, the following phenomenon may occur: One begins with a finite set, S_1, of points on C and interpolates, thereby obtaining a curve, C_1; then one adds a few points to S_1 obtaining a set S_2 and by interpolation a curve C_2; and so on. However, the curves C_1, C_2, ... need not approach C. For between the points of the set S_n, where C_n and C agree, the curve C_n may swing far away from C; and with increasing n, more and more oscillations with appreciable amplitudes may accumulate and prevent the curves C_n, C_{n+1}, ... from approaching *any* continuous curve. *A fortiori*, the same may occur in the more realistic situation where no curve C is given at the outset, but only a (potentially infinite) set of points representing observed data.

Carnap and Neurath as well as Schlick and Waismann often gratefully acknowledged Hahn's role. So, during Schlick's absence on a visit to the United States in 1929, Frank asked Hahn to deliver the inaugural address at the First Congress of the Epistemology of Science (in Prague), while Neurath and Carnap asked him to be the principal signer of the pamphlet *Wissenschaftliche Weltauffassung*.

Hahn's address (Chapter II in this volume) formulated with masterful succinctness and precision a kind of creed for logical positivism and the scientific *Weltanschauung*. It contrasted them with mysticism, metaphysics and various other tendencies then prevailing in philosophy (especially in German philosophy), and also delimited them *vis-à-vis* pure empiricism and rationalism.

The pamphlet, whose genesis I had occasion to observe, was mainly written by Neurath[9]; Carnap cooperated to some extent; Hahn received the final draft.[10] It was well written and informative in various ways. Yet Schlick, to whom the booklet was dedicated, was not altogether pleased with it when he received it upon his return to Vienna. It lacked the depth and the precision of Hahn's address to the Prague congress. Also, it introduced political views and tendencies, which Neurath never completely separated from epistemological insights, whereas Schlick always insisted on the strict separation of the latter from value judgments of any kind. Moreover, some of those views were, if only in passing, presented as common to all members of the highly individualistic group, while some members, including Schlick himself, did not fully share them.[11]

In his later years, Hahn spent a minor but appreciable part of his free time on parapsychological studies. Many friends of his and admirers of his intellect found this interest of his very odd and wondered how the topic of parapsychology could even be broached in a group as strictly scientific in its orientation as the Vienna Circle. A partial explanation lies in the way Hahn's interest originated. In the first years after World War I a new influx of mediums had appeared in Vienna. They were viewed by the intelligentsia with the utmost skepticism. Finally, one day in the early 1920s, the newspapers claimed that two professors of physics at the University of Vienna, Stephan Meyer and Karl Przibram, had exposed the entire spiritualistic swindle. What had happened, the newspapers elaborated, was that Professor Meyer had invited many people to his house for a séance. The guests sitting in a circle and holding hands in a dark room, had clearly observed the phenomenon of levitation – more specifically, a figure in white rising several feet into the air. But at that moment the host unexpectedly turned on the lights and everyone could see that the apparition was none other than the tall professor Przibram who, in the dark, had managed to cover himself with a bed sheet and to climb on a chair. In the midst of general laughter the two physicists claimed that they had produced a levitation and exposed the mediums.

It goes without saying that the parapsychological groups were outraged; and for once, in a reversal of the ordinary situation, the mediums called all scientists swindlers. But there was also great indignation in the intellectual community; and a group including besides Schlick and Hahn the eminent neurophysiologist and physician Julius von Wagner-Jauregg, the physicist Hans Thirring and a number of others (most of them scientists) formed a committee for the serious investigation of mediums. Very soon, however, members began to drop out: first Wagner-Jauregg; soon after him Schlick. By 1927 apart from non-scientists only Hahn and Thirring were left in the group. They were, as they told me, not convinced that any of the phenomena produced by the mediums were genuine; but they were even less sure that all of them were not. They believed, rather, that some parapsychological claims might well be justified; and that certainly the matter warranted further serious investigation.

Hahn devoted a few interesting public lectures to the subject. I remember especially two points from these lectures. One was taken from the physiologist Charles Richet, an early French Nobel laureate, who suggested that one should imagine a world in which all men with the exception of a few scattered individuals had completely lost the sense of smell. Walking between two high stone walls one of those few might say, "There are roses behind these walls." And to everyone's amazement, his assertion would be verified. Upon beginning to open some drawer he might say, "There is lavender in this drawer," and if none should be found he would insist, "Then there *was* lavender in this drawer." And sure enough, it would be established that two years earlier there was indeed some lavender in that drawer. Are mediums with extra-ordinary perceptions and exceptional abilities in our world what the few people with a sense of smell are in Richet's?

The second point goes back to Hahn himself. Some mediums claim that great thinkers and poets speak through them while they are in trance, but actually those mediums utter only lines that are far below the level of their supposed authors. This well-known fact is usually construed as proving that the uneducated mediums simply say what *they* think their sources would say. Hahn, however, pointed out that many mediumistic revelations are *so* trivial and incoherent that even a medium with little education would not consider them as utterances of their supposed sources, and that in fact they are definitely below the *medium*'s own level. To Hahn this indicated that such chatter was not the product of the medium's conscious mind, but was generated subconsciously. Its very triviality combined with the tormented stammering in which the babble is frequently uttered, suggested to Hahn that in many cases one is dealing with a *genuine* phenomenon *of some kind*.

In his public lectures Hahn was of supreme clarity; but he also prepared his daily courses meticulously. He applied a technique that I have never seen anyone else carry to such extremes: he proceeded by almost imperceptible steps and at the end of each hour left his audience amazed at the mass of material covered. How stimulating his lectures were I have tried to describe – in Chapter 21 of my book *Selected Papers in Logic and Foundations, Didactics, Economics* (Vol. 10 of the Vienna Circle Collection) – by the example of the effect on myself of the first lecture he gave after his return to Vienna in 1921.

Politically, Hahn was a socialist of deep conviction and took part in several phases of the social-democratic movement. He never spared any pains to serve the public. In particular, he furthered adult education and did whatever he could for competent under-privileged students. Where he saw injustice or oppression, he tried to help the injured. Once on the street, when a coachman mal-treated his horse and Hahn's protest was ignored, he dragged the ruffian to the police. Hahn was respected even by his opponents.

KARL MENGER

NOTES

[1] Hahn's first results were contributions to the classical calculus of variations. He then turned to the study of real functions and set functions, especially integrals. He further published a fundamental paper on non-Archimedean systems, and early recognized the significance of Fréchet's abstract spaces. In a paper introducing local connectedness he characterized the sets which a point can traverse in a continuous motion; that is, the continuous images of a time interval or a segment (now often called Peano continua). The paper is a classic of the early set-theoretical geometry. About Hahn's work after World War I, see note [5] below.

[2] Two of these papers, one of Olga's and one jointly written, are marked with an asterisk by C. L. Lewis in his book *A Survey of Symbolic Logic*, Berkeley, 1918, indicating that at that time Lewis ranked these studies among those "that are considered the most important contributions to symbolic logic."

[3] For administrative reasons, this lecture course was listed as a "seminar."

[4] I remember having reported about one chapter of the Principia myself before leaving Vienna in the spring of 1925. Contrary to what has been sometimes written about Hahn's seminars, Wittgenstein's name had, certainly up to that time, never been mentioned in them.

[5] After World War I, Hahn published volume I of his monumental *Theorie der Reellen Funktionen* (volume II came out posthumously). He then returned to the calculus of variations and to the theory of integrals from modern points of view. He applied some of his results to problems of interpolation, which later turned out to be of interest to the Circle (see Note [8] below). Perhaps most importantly, he developed the concept and parts of the theory of general normed linear spaces, simultaneously with and independently of Stefan Banach, after whom they are now called 'Banach spaces.'

[6] At one of the large yearly meetings of the German-speaking mathematicians a group including the most prominent members in Hahn's age bracket decided to test their general mathematical knowledge. According to a system agreed on in advance, they asked each other questions of the level and the type of those in doctoral examinations. Anyone missing an answer was excluded from further competition. One by one the contestents were eliminated leaving Hahn the ultimate winner, with the well-known analyst and number-theoretician Edmund Landau as runner-up.

[7] Apart from this remark, Hahn's reviews of Russell's, Meyerson's, and a few other philosophers' books are merely brief summaries of their contents, and have therefore not been included in this volume.

[8] *Mathematische Zeitschrift*, vol. 1. This is the paper referred to above in Note [5]

[9] It is only fair that a translation of the pamphlet has been included in Neurath's collected papers, *Empiricism and Sociology*, (Volume 1 of the Vienna Circle Collection, Chapter IX).

[10] Hahn signed even though he would have written the pamphlet somewhat differently and was not in complete agreement with all details — one of the concessions he was occasionally prepared to make for the sake of peace. He had an irrepressible penchant for mediating between conflicting views and between quarreling people.

[11] For the same reasons and because of what I regarded as superficial views on social sciences I felt myself somewhat estranged by the pamphlet — in fact to the point where I asked Neurath to list me henceforth only among those *close* to the Circle (Cf. *Erkenntnis*, 1, p. 312). And the pamphlet alienated Gödel even more.

YOU WILL LIKE GEOMETRY

"Impossible" you say. "Geometry is a bore. It has been dead and petrified for centuries."

But you are wrong. Geometry is amazing and ingenious and beautiful and profound; and, most important, it is alive and growing.

Just follow the growth of the geometric world of plane figures through the ages.

THE GROWTH OF THE GEOMETRIC WORLD

In **ANTIQUITY,** Greek geometers studied simple forms such as the circle and the square.

In the **MIDDLE AGES,** designers combined simple forms into patterns.

We find such arabesques on the floors and the walls of cathedrals and mosques.

The **RENAISSANCE** of geometry started about 1640. At that time Descartes conceived the idea of identifying every point of the plane by its distances from two perpendicular lines — just as we identify any block in Chicago by its distances (in blocks) from Madison and State Streets. This idea was helpful in studying the simple forms then known and it led to the discovery of innumerable curves hitherto unknown. For instance,

if x and y are the distances of a point from two perpendicular lines, then the (dotted) circle consists of the points for which $x^2 + y^2 = 1$. In the same way, $x^4 + y^4 = 1$ describes the outer curve, $\sqrt[3]{x^2} + \sqrt[3]{y^2} = 1$ the inner curve.

These flow lines of water around a circular object correspond to other equations.

So does this curious curve which is shaped like some 17th century picture frames.

In **MODERN TIMES** (about 1875) Cantor took the final step. He included in geometry all possible forms and shapes. Some of these figures correspond to equations, others can only be defined by more complicated processes. Although this unlimited field contains figures with the most amazing properties, Cantor and his successors discovered general laws governing all of them.

is a so-called Simple Closed Curve—the shape of a piece of string after an arbitrary deformation. Each such curve divides the plane into two parts, the interior and the exterior (Jordan's Theorem).

is an approximation to a zigzag line which divides the plane into three parts (a black, a gray, and a white part). So does the symbol 8 but only one point in this symbol belongs to the boundaries of all three parts. On the zigzag line, every point belongs to the boundaries of all three parts. (Such zigzag lines were discovered by Brouwer and Knaster.)

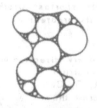

is the area inside a simple closed curve covered by circles no two of which overlap (a so-called Vitali covering).

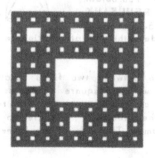

is an approximation to a Plane Universal Curve. Every plane curve can be so deformed that it becomes a part of that curve (discovered by Sierpinski).

GEOMETRY IS AMAZING

CURVES OF CONSTANT WIDTH.

A circle revolves between two fixed parallel lines.

A revolving ellipse alters the distance between two parallel lines between which it revolves.

Which of the curves to the right turn like a circle and which like an ellipse?

All except the first turn like a circle.

Here is a curve lying between two parallel tangent lines. The perpendicular distance between the two lines is called the **WIDTH** of the curve in this perpendicular direction. A circle has the same width in every direction. An ellipse has its shortest width in the direction of its short axis; its greatest width in the direction of its long axis.

Curves which, like a circle, have the same width in every direction are called **CURVES OF CONSTANT WIDTH.** You can easily construct such a curve. First draw an equilateral triangle A, B, C. Then, with each of the three corners as center, draw the circular arc passing through the other two corners. You obtain a curve with corners at A, B, and C. It is named **REULEAUX TRIANGLE** after the man who discovered it in 1875. The width of this curve in every direction is equal to the side of the equilateral triangle ABC.

Any curve of constant width can be revolved between two fixed parallel lines. In fact, it can be revolved within a square each side of which is touched by the curve at every stage of its revolution. To the left you see the Reuleaux Triangle in an oblique position within a square frame. The side of the frame is equal to the width of the curve.

Some curves of constant width are smooth, others have corners; some are highly symmetric, others quite irregular; some consist of circular arcs, others are made up of more complicated curves. But they all have a property in common: If a curve is of the constant width d, then it has the length πd, the same length as the circle of diameter d (Theorem of Minkowski).

CAN A CIRCLE BE A SQUARE?

You may not have noticed the fact but you can find square circles in every modern city where all streets intersect at right angles.

Which points lie on a circle with a radius of half a mile? The points in the plane half a mile from the center.

If you measure the distance **AS THE CROW FLIES**, then the points half a mile from the center form an ordinary circle, and the laws of **EUCLIDEAN GEOMETRY** are valid.

But **IN A TAXI** we measure distances in a different way. Suppose the city is laid out in square blocks, 8 blocks to a mile. Which points can you reach by riding in a taxi half a mile?

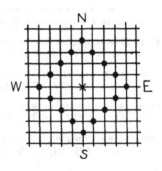

The driver may go four blocks East, or three blocks East and one block North, or two blocks East and two blocks North, or one block East and three blocks North, etc. etc.

ALL THE POINTS WHERE YOU MIGHT END UP, LIE ON A SQUARE.

Measuring the distances in a taxi we obtain a so-called **NON-EUCLIDEAN GEOMETRY.** The circles are square and many laws of Euclid are not valid.

In Euclid's Geometry, between any two points, C and D, there is a shortest path, the straight path; every other path is longer.

In the Taxicab Geometry, between two intersections, C and D, (D two blocks East and two blocks North of C) each path is at least half a mile long. But there are several shortest paths joining C and D, each precisely half a mile long: you may first drive East and then North, or first North and then East, etc.

In Euclid's Geometry, two circles have at most two points in common. Is this true in the Taxicab Geometry?

In our century, Einstein has made extensive use of various Non-Euclidean spaces in describing the universe.

A FINITE PLANE

is the name given to a strange-looking colored model in space. "Finite" because you can count the objects: twelve triangles and nine rods. "Plane" – although the model lies in space – because its triangles are related to its rods just as in an ordinary plane the lines are related to the points.

As candles on a birthday-cake, the nine rods are attached (evenly spaced) to the edge of a circle. Call the rods

I, II, III, IV, V, VI, VII, VIII, IX,

in the order they come.

The three vertices of each colored triangle are attached to three of the rods. No two of the twelve triangles intersect or touch one another since they are attached to the rods at different levels. There are

three **blue** triangles whose vertices touch the rods
I, IV, VII;   II, V, VIII;   III, VI, IX;

three **red** triangles whose vertices touch the rods
I, V, VI;   IV, VIII, IX;   VII, II, III;

three **yellow** triangles whose vertices touch the rods
I, III, VIII;   II, IV, VI;   V, VII, IX;

three **green** triangles whose vertices touch the rods
I, II, IX;   III, IV, V;   VI, VII, VIII;

The main features about points and straight lines in the plane are:

1) For any two points, there is one and only one line joining both of them.

2) For any two lines, there is either one point or none common to both of them. In the later case, the lines are parallel.

3) If L is any line, then each point which does not lie on L, lies on exactly one line parallel to L.

4) No line passes through every point.

The main features about rods and triangles in the model are:

1) For any two rods, there is one and only one triangle touching both of them.

2) For any two triangles, there is either one rod or none touched by both of them. In the latter case, the triangles are of the same color.

3) If T is any triangle, then each rod which does not touch T, touches exactly one triangle of the same color as T.

4) No triangle is touched by every rod.

A comparison of these statements bears out the contention that the triangles are related to the rods just as in the plane the lines are related to the points.

Until a few years ago, finite planes were studied only because of their mathematical interest.

Today, the U. S. Department of Agriculture uses such planes in testing relative efficiency of fertilizers.

CAN A MOVING POINT SWEEP OVER
THE ENTIRE SURFACE OF A SQUARE?

YES.

A point may move from one corner of a square to the opposite corner occupying on the way every possible position within the square.

Mathematically, the course of the point may be so defined that it can be completed within a finite time, say in 9 seconds (although a physical particle, moving at a finite speed, could not complete such a course within a finite time).

The first figure shows roughly how the point moves from the start to the end: during the first second, it moves to the opposite corner of the left lower square; during the next second, it moves to the opposite corner of the next ninth of the square; and so on.

The second figure shows roughly how the point moves during each of the nine seconds: during the first ninth of the first second, it moves from the start to the opposite corner of the left lower ninth of the square; during the next ninth of the first second, it moves to the opposite corner of the next ninth of the first ninth of the square; and so on.

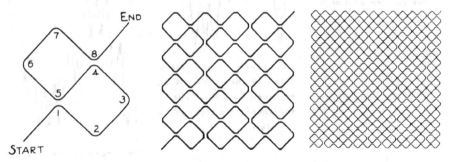

The third figure gives a still more detailed picture and shows where the moving point travels during the first 1/81 part of the first second, the next 1/81 part of the first second, and so on.

Proceeding in this way, Peano in 1890 precisely defined the position of the moving point at any moment.

Not only will the point occupy every position within the square but some of them as often as four times. Look at the position at the end of the first second. At the end of five seconds, the point occupies the same position again.

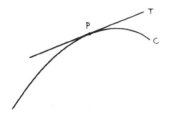

This curve C has a tangent at the point P, namely, the line T.

In contrast, the curve D has so many ups and downs near the point Q that it has no tangent at Q.

Here you see a letter N.

Here each of the three lines of the above N is replaced by 3 Ns. These together form a zigzag line consisting of 9 Ns.

Here each of the above 9 Ns has been replaced by 9 Ns. Altogether there are 81 Ns.

In this way you might go on. At each stage you would obtain a more jittery line. The lines would become more and more like a curve without a tangent at any point.

The first curve without any tangent was discovered by Weierstrass about 1870. It was then considered very exceptional. Today it is known that there exist more curves without any tangents than there are smooth curves with tangents.

- 8 -

A UNIVERSAL CURVE.

On page 3 you have seen a Plane Universal Curve obtained from a square by omitting smaller squares. A similar construction can be performed in space. Divide a cube into 27 small cubes by trisecting each of its sides. Of the small cubes, omit the innermost and the six adjacent ones. The remaining 20 small cubes taken together look like a cube cut through by three perpendicular channels. If this construction is repeated over and over indefinitely, a Universal Curve is obtained. Its front face looks like the Plane Universal Curve which has the property that every plane curve can be deformed into one of its parts. The Universal Curve (discovered by Menger) has the property that every curve of our space can be deformed into one of its parts. Even more can be asserted. Every curve of four-dimensional and higher-dimensional spaces (of which you may have heard) can be so deformed that it becomes a part of the Universal Curve.

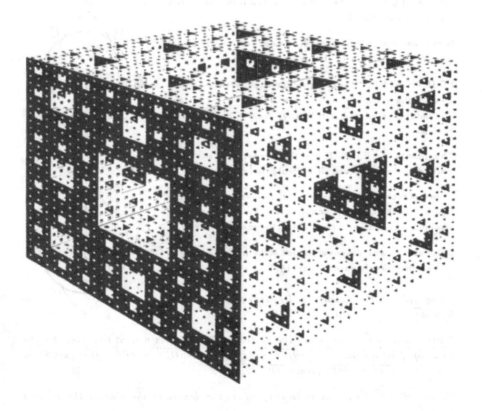

GEOMETRY IS INGENIOUS

Almost everyone knows how to mark on a circle the CORNERS OF A REGULAR HEXAGON if a compass is available. The fact that no straightedge is needed is one of the oldest geometrical discoveries.

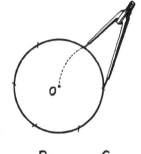

But it took Emperor Napoleon to raise the following question. Can
THE CORNERS OF A SQUARE
be marked on a circle without the use of a straightedge?

The Abbe Mascheroni found the answer.

FIRST STEP. On the circle with the center O mark four vertices A, B, C, D, of a regular hexagon.

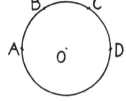

SECOND STEP. Draw the circle with the center A which passes through C and the circle with the center D which passes through B.

THIRD STEP. Denote either point where these two circle intersect by P.

It can be shown* that the distance between P and O is equal to the side of the desired square.

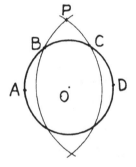

FOURTH STEP. Measure the distance between P and O with the compass and lay it off four times on the circumference of the circle say, from A to E to D to F to A.

The points A, E, D, F are the corners of a square marked without the use of a straightedge.

Later Mascheroni found that the corners of every figure which can be constructed by compass and straightedge, can be constructed by compass alone.

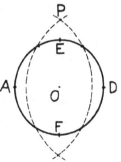

Still later it was discovered that Mohr in Denmark had proved this result as early as 1672.

*The side of the square $AEDF$ is equal to $\sqrt{2}\,\overline{OA}$. Now $\overline{AP} = \overline{AC} = \sqrt{3}\,\overline{OA}$.

Furthermore, $\overline{OP}$ is perpendicular to $\overline{OA}$. Thus AOP is a right triangle. By the Law of Pythagoras, $\overline{OA}^2 + \overline{OP}^2 = \overline{AP}^2$. Thus $\overline{OP}^2 = \overline{AP}^2 - \overline{OA}^2$ which is $= 3\,\overline{OA}^2 - \overline{OA}^2 = 2\,\overline{OA}^2$ (since $\overline{AP} = \sqrt{3}\,\overline{OA}$).

Hence $OP = \sqrt{2}\,\overline{OA}$; that is to say, $\overline{OP}$ is the length of the side of the desired square.

EUCLID knew how to inscribe in a circle **A REGULAR POL-YGON WITH 3 SIDES** (an equilateral triangle)

and

A REGULAR POLYGON WITH 5 SIDES (a regular pentagon). Compass and straight edge was all he needed for these constructions.

350 B.C.

Now 3 and 5 are the smallest odd prime numbers. 7, 11, 13, 17, 19, 23, . . . follow.
(A number is prime if it has no factors except itself and 1) IS IT POSSIBLE BY MEANS OF COMPASS AND STRAIGHT EDGE TO INSCRIBE IN A CIRCLE A REGULAR POLYGON WITH 7 OR 13 OR 17 OR 19 OR 23. . . SIDES?

1796 A.D.

This question waited 2000 years for an answer. In 1796, the 19 year old GAUSS, by means of the theory of algebraic equations, proved: Of all the regular polygons with an odd prime number of sides,

ONLY THOSE WITH 3, 5, 17, 257, 65537 SIDES
CAN BE CONSTRUCTED BY COMPASS AND STRAIGHT EDGE.

These five prime numbers may be expressed as follows:

$$
\begin{aligned}
3 &= 2+1 = 2^1 + 1 \\
5 &= 4+1 = 2^2 + 1 \\
17 &= 16+1 = 2^4 + 1 \\
257 &= 256+1 = 2^8 + 1 \\
65537 &= 65536+1 = 2^{16} + 1.
\end{aligned}
\left.\vphantom{\begin{aligned}3\\5\\17\\257\\65537\end{aligned}}\right\}
$$

Each of these numbers is of the form

$$2^{2^n} + 1.$$

Now the next number of this form, $2^{32} + 1 = 4,294,967,927$, is 641 times 6,700,417 and thus not a prime number. Similarly, $2^{64} + 1$, $2^{128} + 1$ . . . . are not prime numbers. Should there exist further prime numbers of the form $2^{2^n} + 1$, then the corresponding regular polygons could be constructed by compass and straightedge. But no other prime numbers of this form have yet been found.

- 11 -

BISECTING AND TRISECTING ANGLES

Here are two well-known ways of
BISECTING AN ANGLE
by compass and straightedge.

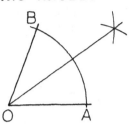

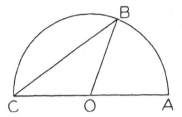

Triangle BOC shows that
∠ OBC = ∠ OCB since OB = OC.
Since ∠ AOB is an exterior angle
∠ AOB = ∠ OBC + ∠ OCB.
Thus ∠ AOB = 2 ∠ OCB and
∠ OCB = 1/2 ∠ AOB.

A compass is not the only machine to
draw a circle.

Move a right angle in such a way that its
sides pass through two fixed points, A and B.
Then the vertex, R, describes a semi-circle
with the segment AB as diameter.

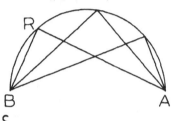

If one side of the right
angle DRO is extended to
S, where RS = 1/2 DO, then,
as R describes the semi-
circle, S moves along a so-
called SNAIL – CURVE (the
limacon of Pascal) .

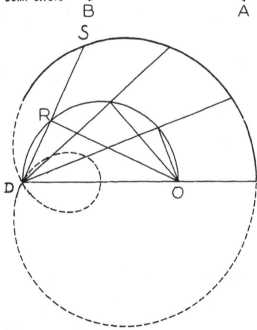

In the figure, the rest of
the snail-curve is dotted.

IS IT POSSIBLE TO TRISECT AN ANGLE OF 60 DEGREES?

NO, if compass and straight-edge are the only tools available.
YES, if a machine for drawing snail-curves is used.

By means of this machine, every angle can be trisected.

IN ORDER TO TRISECT THE ANGLE ∠ AOB, PROCEED AS FOLLOWS.

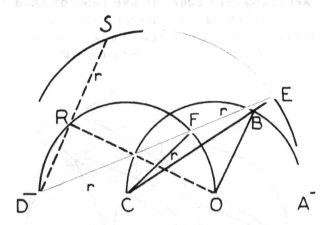

1) Draw the semi-circle $\widehat{AC}$ of radius OA = r with the center O.
2) Draw the semi-circle $\widehat{OD}$ of radius r with the center C.
3) Form a right angle DRO with the sides passing through D and O.
    Then extend the side DR to S, the point a distance r =1/2 DO from R.
4) By moving R along the semi-circle $\widehat{OD}$, let S describe the snail-cuve.
5) Extend the line CB to E where it intersects the snail-curve.
        Then ∠ CDE = 1/3 ∠ AOB.
Thus by finding the angle ∠ CDE you have trisected the angle ∠ AOB.
        Proof.
    Let F be the point where the line DE intersects the semicircle $\widehat{OD}$.
    In triangle CFE
        FC = FE = r, thus ∠ FEC = ∠ FCE.
    In triangle DCF
        CD = CF = r, thus ∠ CFD = ∠ CDF.
    Now since ∠ CFD is an exterior angle of the traingle CFE
            ∠ CFD = ∠ FEC + ∠ FCE.
        Thus ∠ FEC = 1/2 ∠ CFD
        and ∠ FEC = 1/2 ∠ CDF.
    Since ∠ OCE is an exterior angle of the triangle CDE,
        ∠ OCE = ∠ CDF + ∠ FEC.
    In view of ∠ FEC = 1/2 ∠ CDF,
        it follows that ∠ OCE = 3/2 ∠ CDF.
    Now ∠ OCE is the same angle as ∠ OCB
        and obviously ∠ OCB = 1/2 ∠ AOB.
    From the last two equalities
        It follows that 3/2 ∠ CDF = 1/2 ∠ AOB.
        Thus 3 ∠ CDF = ∠ AOB
        and ∠ CDF = 1/3 ∠ AOB.
    Since ∠CDF is the same angle as ∠ CDE this is what was to be proved.

· 13 ·

APPROXIMATELY EQUAL TO ONE THIRD OF ∠AOB

can be found by compass and straightedge.
One of the most accurate constructions was discovered by
the German master-tailor Kopf in 1933.

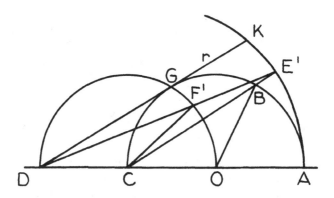

1) As before, draw the semi-circles $\overarc{AC}$ and $\overarc{OD}$.

2) If G is the point where the semi-circles intersect, extend DG to K where GK = r.

3) Instead of the snail-curve, draw the circle passing through K and A with the center on the line AD. (This can be done by compass and straightedge.)

4) Using this circle instead of the snail-curve repeat the procedure of the exact trisection; that is to say, extend the line CB to E′ where it intersects the circle.

The angle ∠ CDE′ at which you arrive can be shown to differ from 1/3 ∠ AOB by less than one quarter of a minute.

If ∠AOB is an angle of less than 20 degrees, the difference is less than 1 second (1/60 of a minute).

SQUARING A FIGURE

means

Constructing a Square of the same Area.
In order to square the RECTANGLE ABCD

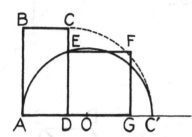

1) Extend AD.
2) Lay off DC' = DC.
3) Find the midpoint O of AC'.
4) Draw the semi-circle with the center O and the diameter AC'.
5) At E, the intersection of the semi-circle with the line DC, construct the square DEFG.

The area of this square is equal * to the area of ABCD.
Only compass and straightedge were used in this construction.

*Proof

$\angle AEC'$ is a right angle since it is inscribed in the semi-circle.
It follows that

$$\angle DAE = \angle DEC'.$$

Therefore

$$\overline{DA} : \overline{DE} = \overline{DE} : \overline{DC'}$$

and

$$\overline{DA}.\overline{DC'} = \overline{DE}^2$$

Since

$$DC' = DC$$

it follows that

$$\overline{DA}.\overline{DC} = \overline{DE}^2$$

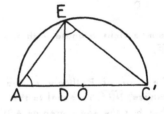

that is to say, the area of ABCD is equal to the area of DEFG.

In order to square
the RIGHT TRIANGLE ADC

1) Draw the rectangle ABCD whose area is twice that of the triangle.
2) Draw the square DEFG with the same area as the rectangle.
3) Form the square DHGI made up of two quarters of the square DEFG.

Clearly, the square has the same area as the triangle.

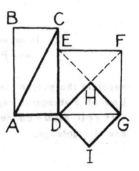

ANY TRIANGLE ADZ

can be transformed into a right
triangle ACD having the same area as ADZ.

In order to square
the QUADRANGLE ABCD

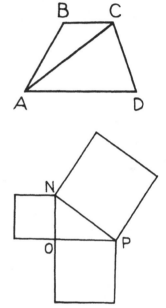

1) Separate the quadrangle into two triangles ABC and ACD.

2) Construct a square with the side $\overline{ON}$ whose area is equal to that of ABC and a square with the side $\overline{OP}$ whose area is equal to that of ACD.

3) Form a right triangle with the sides $\overline{ON}$ and $\overline{OP}$.

By the Law of Pythagoras, the square of the hypotenuse, $\overline{NP}^2$, is equal to the sum $\overline{ON}^2 + \overline{OP}^2$ and thus has the same area as ABCD.

In order to square
any POLYGON

separate the polygon into triangles and apply the Law of Pythagoras repeatedly. But also some **CURVED FIGURES** can be squared by compass and straightedge.

SQUARING A CRESCENT

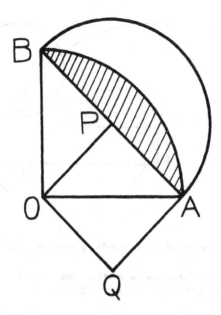

The crescent between the semi-circle with the center P and the quarter-circle with the center O both passing through A and B, has the same area as the square OPAQ.

For the semi-circle and the quarter-circle which have the shaded segment in common, can be shown* to have the same area.-

Hence the non-shaded part of the semi-circle (that is, the crescent) has the same area as the non-shaded part of the quarter-circle (that is, the triangle AOB).

But this triangle has the same area as the square OPAQ.

* The entire circle of radius PA has the area $\pi \overline{PA}^2$.

The entire circle of radius OA has the area $\pi \overline{OA}^2$.

Now $1/2 \, \overline{OA}^2 = \overline{PA}^2$. For the right triangle AOB has the same area as the square OPAQ.

It follows that a quarter-circle of the radius $\overline{OA}$ has the same area as a semi-circle of the radius $\overline{PA}$.

IS IT POSSIBLE TO SQUARE THE CIRCLE?

No, if only compass and straight edge are available.
Yes, if the machine below is used.

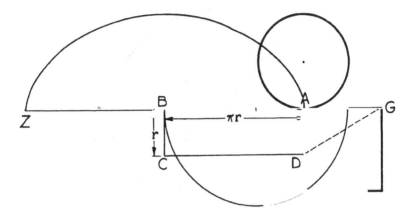

The machine consists of a circle rolling along a straight line.

As the circle completes one revolution, the point A on the circumference, moving from A to Z describes a curve called CYCLOID.

The length of the straight line from A to Z is equal to that of the circle -- that is, 2 π r.

Thus if B is the midpoint of AZ, then AB = πr.

Hence, if AD = r, the area of the rectangle ABCD is
$$\pi \, r \times r = \pi \, r^2$$
which is also the area of the rolling circle.

Squaring this rectangle yields the square with the side AG. Thus the circle too is squared.

In order to square the circle with the center O

1) Lay off the radius OP on the vertical line from P to R.
2) Draw the tangent at P.
3) Through the point R draw the line RQ parallel to the line DG in the square obtained by using the rolling circle.

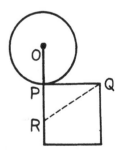

The square with the side PQ has exactly the same area as the circle with the center O.

- 18 -

In 1882, Lindemann showed that the number π has a property which makes it impossible to square the circle by compass and straightedge alone.

But compass and straightedge are sufficient to construct a square with an area very nearly that of the circle with the center O.

One of the easiest ways to do this was found by C. O. Johnston in Chicago.

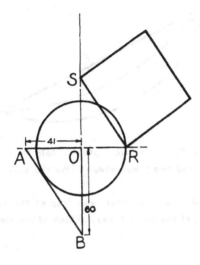

1) Choose any unit of length.
2) From O, the center of the circle, lay off 41 units to A and, in the perpendicular direction, 60 units to B.
3) Draw the hypotenuse AB of the right triangle.
4) Extend AO to the point R on the circumference of the circle.
5) Through R draw the line RS parallel to AB (where S is on the extension of OB).

The area of the square with the side RS differs from the area of the circle with radius OR by less than one part is 300,000.

For the area of the circle is

$$\pi r^2 = (3.14159265...)r^2$$

and the area of the square is

$$r^2 + \left(\frac{60}{41}r\right)^2 = \frac{5281}{1681}r^2 = (3.14158239...)r^2$$

- 19 -

GEOMETRY IS USEFUL

Its applications range from Einstein's description of the universe in terms of Non-Euclidean geometries to the testing of fertilizers by means of Finite Planes. Here is another example.

Take two points, A and B, in a vertical plane. Join the two points by three thin tubes: a straight tube and two tubes shaped like the curves in the figure.

Now suppose that, at the same instant, you drop at the point A three identical marbles, one into each of the three tubes. Which of the three marbles will reach the point B first?

You may be inclined to think that the marble dropped into the straight tube will arrive first because it moves along the shortest path from A to B.

However, the marble descending the upper curve reaches a higher speed in the early part of its descent and thus reaches B earlier than the marble in the straight tube.

Neither of these two tubes, however, has the shape of fastest descent which is a CYCLOID, the curve used in squaring a circle. The longest of the three tubes has the shape of a cycloid. Although it dips below the horizontal level of B and then rises again, the marble moving in this tube reaches B first.

GEOMETRY IS BEAUTIFUL

A solid is called **REGULAR** if it has the following four properties.

1) All its faces are polygons with the same number of vertices.

2) All its vertices are common to the same number of faces.

3) All its edges are of equal length. (The length of an edge E is the distance between the two vertices which lie on E.)

4) All its dihedral angles are equal. (The dihedral angle with the edge E is the angle between the two faces which meet in E.)

Notice that corresponding statements have been made about vertices and faces.

The Greeks proved that there are only five regular solids.

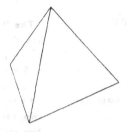

TETRAHEDRON
(tetra = Greek four)
4 faces
(equilateral triangles)
3 vertices each.
4 vertices
each common to 3 faces.
6 edges.

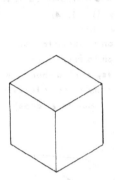

CUBE or HEXADEDRON
(hex = Greek six)
6 faces
(squares)
4 vertices each.
8 vertices
each common to 3 faces.
12 edges.

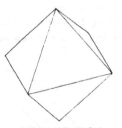

OCTAHEDRON
(octo = Greek eight)
8 faces
(equilateral triangles)
3 vertices each.
6 vertices
each common to 4 faces.
12 edges.

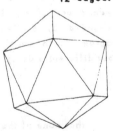

DODECAHEDRON
(dodeca = Greek twelve)
12 faces
(regular pentagons)
5 vertices each.
20 vertices
each common to 3 faces.
30 edges.

ICOSAHEDRON
(icosi = Greek 20)
20 faces
(equilateral triangles)
3 vertices each.
12 vertices.)
each common to 5 faces.
30 edges.

- 21 -

Here are four solids
each of which has only three of the four properties
of a regular solid.

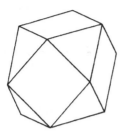

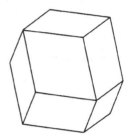

The CUBOCTAHEDRON

has Properties 2), 3), 4)
but of its 14 faces
   8 are triangles,
   6 are squares.
(It has 14 vertices, each common to
two triangles and two squares.)

The RHOMBE-DODECAHEDRON

has Properties 1), 3), 4)
but of its 14 vertices
   8 are common to three faces,
   6 are common to four faces.
(It has 14 faces, each a rhombe with
two vertices common to 3 faces and
two vertices common to 4 faces).

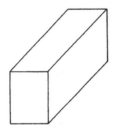

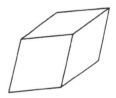

This PRISM

has Properties 1), 2), 4) but not all
of its edges are of equal length.
(It is bounded by 6 rectangles,
parallelograms with equal (right)
angles but different sides.)

This RHOMBOHEDRON

has Properties 1), 2), 3) but not all
of its dihedral angles are equal.
(It is bounded by 6 rhombes,
parallelograms with equal sides
but different angles.)

These examples show

that none of the four properties defining a regular solid

is a consequence of the others.
Such properties are said to be <u>independent</u> of one another.

The Greeks in proving that only five regular solids exist
tacitly assumed that the surfaces of such solids must have
no reentrant angles.

In 1610, Kepler dropped this restriction and
found two more regular solids
with stars for faces.

In 1809, Poinsot found another two
with stars for corners.

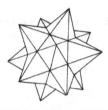

SMALL STAR-FACE DODECAHEDRON

12 faces (star-shaped pentagons)
5 vertices each.
12 vertices (ordinary corners)
each common to 5 faces.
30 edges.

STAR-CORNER DODECAHEDRON

12 faces (ordinary pentagons)
5 vertices each.
12 vertices (star-shaped corners)
each common to 5 face.
30 edges.

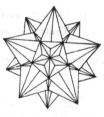

GREAT STAR-FACE DODECAHEDRON

12 faces (star-shaped pentagons)
5 vertices each.
20 vertices (ordinary corners)
each common to 3 faces.
30 edges.

STAR-CORNER ICOSAHEDRON

20 faces (equilateral triangles)
3 vertices each.
12 vertices (star-shaped corners)
each common to 5 faces.
30 edges.

In 1813, Cauchy proved that no further regular stars exist.

- 23 -

A REGULAR COMPOUND

consists of congruent regular solids arranged as follows

1) The combined vertices of the constituent solids
are the vertices of one regular solid.
2) The combined faces of the constituent solids
contain the faces of one regular solid.

There are only four regular compounds.

TWO TETRAHEDRA

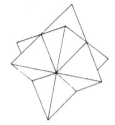

1) Their 8 vertices are the
vertices of a cube.

2) Their 8 faces contain the
faces of an octahedron.

FIVE TETRAHEDRA

1) Their 20 vertices are the vertices of a dodecahedron.
2) Their 20 faces contain the faces of an icosahedron.

Either of the above two compounds is a mirror image of the other.

TEN TETRAHEDRA

This compound is composed of
the above two compounds of five
tetrahedra.

- 24 -

GEOMETRY IS PROFOUND

Here are Polygons (broken lines) with eight sides.

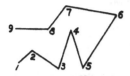

The polygon above is not closed.

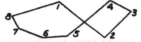

This polygon is called **CLOSED** because the last side ends where the first side begins.

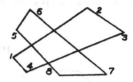

This is a **COMPOUND** consisting of two quadrangles.

The closed polygon to the right is called **SIMPLE** because no two sides cross one another as do the sides 1, 2, and 4, 5, of the polygon above.

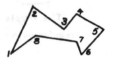

A simple polygon divides the plane into two parts, an interior and an exterior. Moving along such a polygon until you return to your starting point you perform one revolution about each point of the interior.

The polygon below is not simple. It divides the plane into more than two parts. Starting at 1 and moving along the polygon until you return to 1, you perform

3 revolutions about the point a,
2 revolutions about the point b,
1 revolution about the point c.

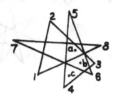

A polygon without reentrant angles is called **CONVEX**. Any line segment joining two points of a convex polygon lies within the polygon.

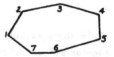

This convex octagon has equal sides but unequal angles.

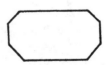

This convex octagon has equal angles but unequal sides.

A polygon is called REGULAR if it has the following two properties:
1) all its sides are equal 2) all its angles are equal. These two properties are
said to be INDEPENDENT because either one can occur without the other as is
demonstrated by the last two polygons on the preceding page.

Here are three regular octagons.

| The simple regular octagon. | The regular octagonal star. | A regular compound of two squares. |

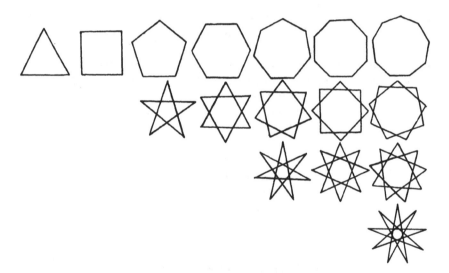

While there is only a small number of regular solids, there is an infinitude of
regular polygons. How many regular polygons with 17 and with 18 sides are
there? How many of them are star-shaped, how many are compounds?

DUALITY

A cube and an octahedron are in a peculiar mutual relation. The relation is brought out clearly if the two solids are labelled as in the two figures:

The faces of the cube and the vertices of the octahedron by
A, B, C, A ', B ', C ',

The vertices of the cube and the faces of the octahedron by
P, Q, R, S, P', Q', R', S'.

Primes indicate opposites such as the top, A, and the bottom, A '.

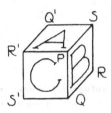

In the cube:	In the octahedron:
The faces A, B, C, meet in the vertex P.	The vertices A, B, C, are the corners of the face P.

The vertices P, Q, S', R', are the corners of the face C.

The faces P, Q, S', R', meet in the vertex C.

The edge in which the faces A and C intersect joins the vertices P and R'.

The edge joining the vertices A and C is the intersection of the faces P and R'.

Every statement relating faces, vertices, and edges which is true for either one of the two solids, becomes true for the other one if the words "face" and "vertex" are interchanged.

This fact is often expressed by saying
the cube and the octahedron are dual solids.

The centers of the six faces of a cube are the vertices of an octahedron. The centers of the eight faces of an octahedron are the vertices of a cube.

In the same way,
the dodecahedron and the icosahedron are dual solids.

The tetrahedron is called self-dual since the dual of a tetrahedron is a tetrahedron.

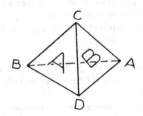

Here is an example of duality in the plane.

The plane is covered with domains of two kinds: squares and triangles.

The plane is covered with congruent domains. Each domain is a pentagon with two kinds of angles: 90° and 120°.

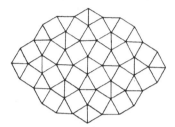

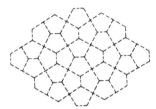

Each vertex is the common corner

of 5 domains of which 2 are squares, 4 vertices each, and 3 are triangles, 3 vertices each.

Each (pentagonal) domain is bounded by a polygon with 5 vertices of which

2 are the common corners of 4 domains, and 3 are the common corners of 3 domains.

If in either figure you join the centers of every two domains having a common side you obtain the other figure.

Below, the two figures are shown together.

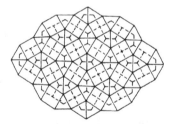

Everyone knows the covering of the plane by regular hexagons, each vertex the common corner of three domains. The pattern resembles that of a honeycomb. If you join the centers of adjacent hexagons you obtain the dual covering by equilateral triangles, each vertex the common corner of six domains.

What is the dual of the covering of the plane by squares, each vertex the common corner of four domains?

Duality is a special case of a more general relation which is important in geomtry as well as in other fields. The relation is called Isomorphism.

ISOMORPHISM

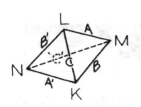

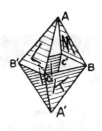

In the figures, the edges of the tetrahedron and the vertices of the octahedron are marked, A, B, C, A', B', C'. (Primes indicate opposites.) K, L, M, N, are the labels of the tetrahedron and of four faces of the octahedron no two of which have a common edge. (In the figure, the faces K, L, M, N, are shaded.)

In the tetrahedron:

A is the edge joining the vertices L and M. The edges A', B, C, meet in K.

In the octahedron:

A is the vertex common to the faces L and M. The vertices A', B, C, are the corners of K.

Every statement relating the edges and vertices of a tetrahedron is true for the vertices and shaded faces of the octahedron - a fact expressed by saying:

The edges and vertices of the tetrahedron   **are isomorphic with**   The vertices and shaded faces of the octahedron

You find further isomorphic objects outside the realm of geometry.

Suppose, for instance, that six men wish to form clubs of three in such a way that:

    1) each man belongs to at least two clubs,
    2) no two clubs have more than one member in common.

It is easy to show that the six men have to form four clubs and that:

the six men and the four clubs   **are isomorphic with**   the six edges and the four vertices of a tetrahedron.

This means that the four clubs can be named after the vertices: K-club, L-club, M-club, N-club, and the six men can be nick-named after the edges: Mr. A, Mr. B, Mr. C, Mr. A', Mr. B', Mr. C', in such a way that whatever can be said about relations between edges and vertices can also be said about relations between the six men and their four clubs.

For instance:

Each edge joins precisely two vertices. The edge A joins the vertices L and M.

Each man belongs to precisely two clubs. Mr. A is a member of the L-club and the M-club.

The vertex K is common to the edges A', B, C.

The K-Club consists of Messrs. A', B, C.

The edge A meets every other edge, except A', in some vertex.

Mr. A and every other man, except Mr. A', are fellow members of some club.

29

TRANSLATIONS AND REFLECTIONS

Here is a Mexican pattern which may be indefinitely continued to the right as well as to the left.

Each rectangle contains a part of the pattern which mathematicians call a **PERIOD** while designers refer to it as a **REPEAT** of the pattern.

If you trace the pattern on a strip of paper and slide the strip any number of periods to the right or to the left, the two patterns will again coincide. Each such operation is called a **TRANSLATION** of the pattern. If you first translate the pattern 2 periods to the right, then further translate it 3 periods to the right, the effect is the same as if you had translated the pattern 5 periods.

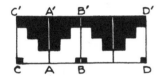

Consider the rectangle AA'B'B. Place a mirror perpendicular to the paper along AA' and draw the reflection of the rectangle in the mirror. You obtain the rectangle AA'C'C. Now place the mirror along BB' and draw the reflection of the rectangle BB'C'C (consisting of the two rectangles). You obtain the rectangle BB'D'D. You would obtain this same rectangle by sliding the rectangle CC'B'B to the right.

Now return to the border pattern and the strip that you traced. Apart from translations, you can make the two patterns coincide by turning the strip about a vertical side or middle line of one of the rectangles. This operation brings about a reflection of the pattern in such a vertical line.

If you reflect the pattern first in one vertical line and then in another, the effect is the same as that of one translation.

Perform two reflections on a row of numbers:

$$\ldots -2, -1, 0, 1, 2, 3, \quad 4, \quad 5, \quad 6, \quad 7, \quad 8, \quad 9, 10, \ldots$$

Reflect this row in 4

$$\ldots 10, 9, 8, 7, 6, 5, \quad 4, \quad 3, \quad 2, \quad 1, \quad 0, \quad -1, -2, \ldots$$

Then reflect the new row in the original place of 1 (now occupied by 7)

$$\ldots 4, 5, 6, 7, 8, 9, 10, 11, 12, 13, 14, 15, 16, \ldots$$

The two reflections have the same result as a translation of the original row 6 units to the right. What is the effect of reflecting the original row first in 1, then in 4? What translation has the same effect on the original row as a reflection in the number m, followed by a reflection in the number n?

30

GROUPS OF OPERATIONS

In the triangle TUV two sides are equal, UT = UV.

Cut out the triangle along its sides. There are two possible positions in which the cutout can be fitted into the triangular hole left in the paper.

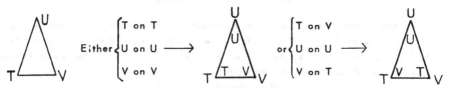

Either
$\begin{cases} \text{T on T} \\ \text{U on U} \\ \text{V on V} \end{cases} \longrightarrow$
or
$\begin{cases} \text{T on V} \\ \text{U on U} \\ \text{V on T} \end{cases} \longrightarrow$

Either we put the triangle into its original position or we first turn over the triangle and then put it back into the hole.

KLM is a triangle with three different sides.

If you cut this triangle out, there is only one way of fitting it into the hole:

$\begin{cases} \text{K on K} \\ \text{L on L; that is, you have to restore the original position.} \\ \text{M on M} \end{cases}$

ABC is an equilateral triangle.

There are six possible positions in which the cut-out can be placed into the hole. Each position of the cutout can be attained as the result of a simple operation on the triangle.

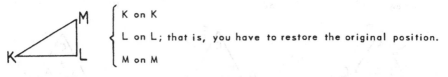

I. $\begin{cases} \text{A on A} \\ \text{B on B} \\ \text{C on C} \end{cases}$

Operation I: Restore the triangle to its original position.

II. $\begin{cases} \text{C on A} \\ \text{A on B} \\ \text{B on C} \end{cases}$

Operation II: Turn the triangle 120° clockwise about the center O.

III. $\begin{cases} \text{B on A} \\ \text{C on B} \\ \text{A on C} \end{cases}$

Operation III: Turn the triangle 120° counter-clockwise.

IV. $\begin{cases} \text{A on A} \\ \text{C on B} \\ \text{B on C} \end{cases}$

Operation IV: Turn the triangle about the axis AO.

V. $\begin{cases} \text{C on A} \\ \text{B on B} \\ \text{A on C} \end{cases}$

Operation V: Turn the triangle about the axis BO.

VI. $\begin{cases} \text{B on A} \\ \text{A on B} \\ \text{C on C} \end{cases}$

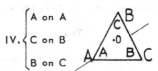

Operation VI: Turn the triangle about the axis CO.

What happens if you perform the operation VI twice in succession? If you turn the triangle about the axis CO, and then, in this new position, turn it another time about the axis CO, you restore its original position, that is, you perform Operation I. We express this fact by writing

first VI, then VI equals I.

Next perform operation II twice in succession. If you turn the triangle 120° clockwise about O, and then in this new position turn it another time 120° clockwise, altogether you put: C on A, A on B, B on C. The result is the same as that of Operation III. We write

first II, then II equals III.

Now first perform the operation VI, then the operation IV.

that is, first bring the triangle into the position	then turn the triangle about the axis AO.	The result is the same as that of operation III.

We write

first VI, then IV equals III.

Proceeding in this way you will find that if you perform any two operations in succession, the result is the same as that of a single one of the six operations.

The six operations are said to form a GROUP.

The above results can be incorporated in a table, the so-called group-table of the equilateral triangle ABC. You may try to complete this table.

The group-table may remind you of a multiplication-table.

first
then→

	I	II	III	IV	V	VI
I	.	.	.	.	.	.
II	.	III	.	.	.	.
III	.	.	.	.	.	.
IV	.	.	.	.	.	.
V	.	.	.	.	.	.
VI	.	.	.	III	.	I

	1	2	3	4	5	6
1	1	2	3	4	5	6
2	2	4	6	8	10	12
3	3	6	9	12	15	18
4	4	8	12	16	20	24
5	5	10	15	20	25	30
6	6	12	18	24	30	36

From the group-table you find
first VI, then IV equals III

in the same way as

from the multiplication-table
6 times 4 equals 24.

If you complete the group-table, you will see that Operation I plays a role similar to that of Number I. For instance.

first I, then II equals II   1 times 2 equals 2

comparable to

first III, then I equals III   3 times 1 equals 3.

For numbers you find

$$(2 \times 3) \times 4 = 2 \times (3 \times 4) \text{ because } 6 \times 4 = 2 \times 12.$$

You can show that the corresponding equality holds for operations. In other respects, however, operations and numbers follow different laws.

While $4 \times 6 = 24 = 6 \times 4$; you can show that

first IV, then VI equals II   while   first VI, then IV equals III.

There are 8 different ways of fitting a square ABCD into a square hole. You may:

    I. Place it into the original position.
    II. Turn it 90° clockwise about its center.
    III. Turn it 90° counter-clockwise.
    IV. Turn it 180°.
    V. Turn it about the diagonal AC.
    VI. Turn it about the diagonal BD.
    VII. Turn it about the vertical middle-line.
    VIII. Turn it about the horizontal middle-line.

You may try to set up a group-table for these eight operations.

If you cut out a tetrahedron of a wooden block, there are 12 ways of putting it back into the cavity. What is the corresponding number for the cube? Is there a plane figure of which the cut-out can be fit into the hole in precisely 3 ways? And are there plane figures for which the number is 4, 5, and 10?

Why, you ask, should one waste time on such idle games and notions?

But these questions are not idle at all. The simple concepts of which you have seen examples are very profound, and lend themselves to numerous applications. Here are a few examples.

1. **Algebraic equations.** The key to a full understanding of a cubic equation, such as $8x^3 - 6x - 1 = 0$, is the group-table of the equilateral triangle. And a penetrating study of the above cubic equation results in the proof than an angle of 60° cannot be trisected by compass and straightedge.

2. **Geometric Patterns.** The study of patterns is not only attractive for its own sake but important for another reason. Crystals owe their special physical properties to the fact that their molecules are arranged in patterns.

3. **The Structure of Atoms** has been clarified by applications of the concept of symmetry.

THE EXHIBIT FOR WHICH THIS BOOKLET WAS PRE-
PARED is on permanent display at Chicago's Museum of Science
and Industry.

It was built under the direction of Dr. Karl Menger, professor
of mathematics at Illinois Institute of Technology, author of
this booklet. Design and construction of the exhibit is the work
of Otto Kolb, professor of visual design at Illinois Tech.
Mechanical devises were built by Samuel Rusinoff, professor of
mechanical processes. Frederick Rest, physicist at Armour
Research Foundation of Illinois Institute of Technology, was a
consultant on the necessary electrical work. James E. Logan,
graduate student, built the solid models.

Dr. Menger first conceived the idea of the exhibit several
years ago while studying a conventional mathematics display.
He began developing ideas for an exhibit along more modern
lines. He suggested this project to Illinois Tech authorities
but funds were not available to build it until 1950 when an
anonymous donor made a gift to the Institute for the purpose
of creating a permanent exhibit for the Museum. Dr. Menger has
since presented ideas for further exhibits.

PUBLICATIONS BY THE SAME AUTHOR

CALCULUS. A MODERN APPROACH
    Ginn and Co., Boston 1955.   xviii and 354 pp. $7.00.

It is unusual for this department to review textbooks, but this
is an unusual textbook. . .   The renovations are likely to
benefit all concerned, sophisticates as well as innocents,
for it is doubtful that a sounder and more sensible entry into
the subject has appeared in many years.
                                    Scientific American, July 1956.

WHY JOHNNY HATES MATH. In The Mathematics Teacher,
vol. 49, December 1958, pp. 578 – 584.

WHAT MATHEMATICS IS REALLY LIKE. A Round Table
Discussion by Karl Menger, Gordon Pall,   Haim Reingold,
S. S. Shu, and L. R. Wilcox, Professors of Mathematics,
Illinois Institute of Technology.   This pamphlet can be
obtained by writing to the department of Mathematics, Illinois
Institute of Technology, Chicago 16, Ill., and enclosing 12
cents in coins or stamps.

Table of Contents – Vol. I

Schweizer, B.: Introduction 1

Sigmund, K.: Karl Menger and Vienna's Golden Autumn 7

Johnson, D. M.: Commentary on Menger's Work
on Dimension Theory 23

Selected Papers on Dimension Theory 33

Zur Dimensions- und Kurventheorie. Unveröffentlichte
Aufsätze aus den Jahren 1921–1923, Monatshefte
für Mathematik und Physik 36 (1929), 411–432 35

Das Hauptproblem über die dimensionelle Struktur der
Räume, Proceedings Amsterdam 30 (1927), 138–144 57

(With W. Hurewicz). Dimension und Zusammenhangsstufe,
Mathematische Annalen 100 (1928), 618–633 65

Allgemeine Räume und Cartesische Räume. I., Proceedings
Amsterdam 29 (1926), 476–482 81

Allgemeine Räume und Cartesische Räume. II.: Über
umfassendste *n*-dimensionale Mengen, Proceedings
Amsterdam 29 (1926), 1125–1128 89

Über die Dimension von Punktmengen III. Zur
Begründung einer axiomatischen Theorie der Dimension,
Monatshefte für Mathematik und Physik 36 (1929), 193–218 93

Axiomatische Einführung des Dimensionsbegriffes,
Comptes Rendus du Premier Congrès des Mathématiciens
des Pays Slaves. Warszawa. 1929, 57–65 119

(With C. Kuratowski). Remarques sur la théorie
axiomatique de la dimension, Monatshefte für
Mathematik und Physik 37 (1930), 169–174 129

Über die Hinweise auf Brouwer in Urysohns Mémoire,
Selbstverlag, Wien 1932 135

Crilly, T., Moran, A.: Commentary on Menger's Work
on Curve Theory and Topology 141

Selected Papers on Curve Theory and Topology 153

Einige Überdeckungssätze der Punktmengenlehre,
Akademie der Wissenschaften zu Wien,
Sitzungsberichte 133 (1924), 421–444 155

Grundzüge einer Theorie der Kurven, Mathematische
Annalen 95 (1925), 277–306 179

Grundzüge einer Theorie der Kurven, Proceedings
Amsterdam 28 (1925), 67–71 . . . . : . . 209

Une forme abstraite du théorème de Borel-Lebesgue
généralisé, Comptes Rendus Acad. Paris 206 (1938), 563–565 . . . 215

On the Origin of the *n*-Arc Theorem, Journal
of Graph Theory 5 (1981), 341–350 219

Plaumann, P., Strambach, K.: Commentary on Menger's
"Untersuchungen über allgemeine Metrik" 229

"Untersuchungen über allgemeine Metrik" 235

Untersuchungen über allgemeine Metrik (Erste Untersuchung:
Theorie der Konvexität; Zweite Untersuchung: Die
euklidische Metrik; Dritte Untersuchung: Entwurf einer
Theorie der *n*-dimensionalen Metrik), Mathematische
Annalen 100 (1928), 75–163 237

Bemerkungen zur zweiten Untersuchung über allgemeine
Metrik, Proceedings Amsterdam 30 (1927), 710–714 327

Untersuchungen über allgemeine Metrik. Teil IV: Zur Metrik
der Kurven, Mathematische Annalen 103 (1930), 466–501 333

Sagan, H.: Commentary on Menger's Work on the Calculus
of Variation and Metric Geometry 369

Selected Papers on Calculus of Variations 377

Sur un théoréme général du calcul des variations,
Comptes Rendus Acad. Paris 201 (1935), 705–707 379

Calcul des variations dans les espaces distanciés généraux,
Comptes Rendus Acad. Paris 202 (1936), 1007–1009 383

Courbes minimisantes non rectifiables et champs généraux
de courbes admissibles dans le calcul des variations,
Comptes Rendus Acad. Paris 202 (1936), 1648–1650 387

Metric Methods in Calculus of Variations,
Proc. Nat. Acad. Sci. 23 (1937), 244–250 391

A Theory of Length and its Applications to the Calculus
of Variations, Proc. Nat. Acad. Sci. 25 (1939), 474–478 399

Benz, W.: Commentary on Menger's Work on the Algebra
of Geometry 405

Selected Papers on Algebra of Geometry 417

Bemerkungen zu Grundlagenfragen. IV: Axiomatik
der endlichen Mengen und der elementargeometrischen
Verknüpfungsbeziehungen, Jahresbericht der Deutschen
Mathematikervereinigung 37 (1928), 309–325 419

(With F. Alt and O. Schreiber). New Foundations of Projective
and Affine Geometry. Algebra of Geometry, Annals
of Mathematics, II. Ser. 37 (1936), 456–482 437

(With F. Alt). A Note on a Previous Paper
"New Foundations of Projective and Affine Geometry",
Annals of Mathematics, II. Ser. 38 (1937), 450 465

A New Foundation of Non-Euclidean, Affine,
Real Projective and Euclidean Geometry,
Proc. Nat. Acad. Sci. 24 (1938), 486–490 467

Self-Dual Fragments of the Ordinary Plane,
The American Mathematical Monthly 56 (1949), 545–546 473

The Projective Space, Duke Math. Journal 17 (1950), 1–14 475

Frammenti piani autoduali e relative sostituzioni,
Rendiconti Accad. Nazionale di Lincei 30 (1961), 713–717 489

The New Foundation of Hyperbolic Geometry, in:
A Spectrum of Mathematics (Essays Presented
to H. G. Forder), ed. by J. C. Butcher,
Auckland University Press 1971, 86–97 495

Benz, W.: Commentary on Menger's Expository
Papers on Geometry 507

Selected Expository Papers on Geometry 515

Some Applications of Point-Set Methods,
Annals of Mathematics, II. Ser. 32 (1931), 739–760 517

Generalized Vector Spaces. I. The Structure
of Finite-Dimensional Spaces, Canadian Journal
of Mathematics 1 (1949), 94–104 539

The Theory of Relativity and Geometry, in:
Albert Einstein, Philosopher-Scientist, The Library
of Living Philosophers, vol. VII, ed. by P. A. Schilpp,
Evanston, Illinois 1949, 459–474 551

The Formative Years of Abraham Wald and His Work
in Geometry, The Annals of Mathematical
Statistics 23 (1952), 14–20 567

Mathematical Implications of Mach's Ideas: Positivistic Geometry,
The Clarification of Functional Connections, in: Ernst Mach,
Physicist and Philosopher, Boston Studies in the Philosophy
of Science VI, eds. R. S. Cohen and R. J. Seeger, D. Reidel,
Dordrecht 1970, 107–125 575

List of Publications – Karl Menger 595

Table of Contents of Karl Menger: "Selected Papers in Logic and Foundations, Didactics, Economics"

Acknowledgments xi

Introduction 1

Part I. Papers Introducing Logical Tolerance

             Logical Tolerance in the Vienna Circle 11
Chapter 1   The New Logic (1933) 17
             Appendix (1937) 42
Chapter 2   On Intuitionism (1930) 46

Part II. Opuscula Logica

Chapter 3   Meaningfulness and Structure (1930) 61
             Appendix (1978) 63
Chapter 4   A New Point of View on the Logical
             Connectives (1978) 68
Chapter 5   An Intuitionistic-Formalistic Dictionary
             of Set Theory (1928) 79
Chapter 6   Ultrasets and the Paradoxes of Set Theory (1928) 88
Chapter 7   A Logic of the Doubtful. On Optative
             and Imperative Logic (1939) 91

Part III. Fundamental Concepts in Pure and Applied Mathematics

Chapter 8   A Counterpart of Occam's Razor (1960, 1961) 105
Chapter 9   A Theory of the Application of the Function
             Concept to Science (1970) 136

Chapter 10   Variables, Constants, Fluents (1961) 144
Chapter 11   Wittgenstein on Formulae and Variables (1978) 153

Part IV. Didactics of Mathematics

Introduction 161
Chapter 12   A New Approach to Teaching Intermediate
Mathematics (1958) 163
Chapter 13   Why Johnny Hates Math (1956) 174
Chapter 14   On the Formulation of Certain Questions
in Arithmetic (1956) 185
Chapter 15   On the Design of Grouping Problems
and Related Intelligence Tests (1953) 189
Chapter 16   The Geometry Relevant to Modern
Education (1971) 199

Part V. Philosophical Ramifications of Some Geometric Ideas

Chapter 17   On Definition, Especially of Dimension
(1921–1923, 1982) 207
Chapter 18   Square Circles (The Taxicab Geometry)
(1952, 1978) 217
Chapter 19   The Algebra of Geometry (1978) 220
Chapter 20   Geometry and Positivism. A Probabilistic
Microgeometry (1970) 225

Part VI.

Chapter 21   My Memories of L. E. J. Brouwer (1978) 237

Part VII. Economics. Meta-Economics

Chapter 22   The Role of Uncertainty in Economics (1934) 259
Chapter 23   Remarks on the Law of Diminishing Returns.
A Study in Meta-Economics (1936) 279

Part VIII. Gulliver's Interest in Mathematics

Chapter 24   Gulliver in the Land without One, Two,
Three (1959) 305

Chapter 25 Gulliver's Return to the Land without One,
            Two, Three (1960) 315
Chapter 26 Gulliver in Applyland (1960) 320

Karl Menger: Principal Dates 324

Fields of Research 325

Bibliography of Works by Karl Menger 327

Index of Names 339

List of Publications – Karl Menger

Über die Dimensionalität von Punktmengen I, Monatshefte für Mathematik und Physik 33 (1923), 148–160.

Introduction to: Carl Menger; Grundsätze der Volkswirtschaftslehre, Wien 1923, Second Edition.

Über die Dimensionalität von Punktmengen II, Monatshefte für Mathematik und Physik 34 (1924), 137–161.

Einige Überdeckungssätze der Punktmengenlehre, Akademie der Wissenschaften zu Wien, Sitzungsberichte 133 (1924), 421–444.

Über die Dimension von Punktmengen, Proceedings Amsterdam 27 (1924), 639–643.

Grundzüge einer Theorie der Kurven, Proceedings Amsterdam 28 (1925), 67–71.

Grundzüge einer Theorie der Kurven, Mathematische Annalen 95 (1925), 277–306.

Über geodätische Linien in allgemeinen metrischen Räumen, Proceedings Amsterdam 29 (1926), 166–169.

Allgemeine Räume und Cartesische Räume. I., Proceedings Amsterdam 29 (1926), 476–482.

Dimensionstheoretische Konsequenzen des Verhältnisses von allgemeinen Räumen und Zahlenräumen, Proceedings Amsterdam 29 (1926), 648–649.

Zur Entstehung meiner Arbeiten über Dimensions- und Kurventheorie, Proceedings Amsterdam 29 (1926), 1122–1124.

Allgemeine Räume und Cartesische Räume. II: Über umfassendste n-dimensionale Mengen, Proceedings Amsterdam 29 (1926), 1125–1128.

Bericht über die Dimensionstheorie, Jahresbericht der Deutschen Mathematiker-Vereinigung 35 (1926), 113–150.

Über reguläre Baumkurven, Mathematische Annalen 96 (1926), 572–582.

Das Hauptproblem über die dimensionelle Struktur der Räume, Proceedings Amsterdam 30 (1927), 138–144.

Zur allgemeinen Kurventheorie, Fundamenta Mathematicae 10 (1927), 96–115.

Zusammenhangsstufen und Cantorsche Mannigfaltigkeiten, Proceedings Amsterdam 30 (1927), 705–709.

Bemerkungen zur zweiten Untersuchung über allgemeine Metrik, Proceedings Amsterdam 30 (1927), 710–714.

Die Haupttheoreme der Dimensionstheorie, Jahresbericht der Deutschen Mathematiker-Vereinigung 36 (1927), 8–12.

Untersuchungen über allgemeine Metrik (Erste Untersuchung. Theorie der Konvexität; Zweite Untersuchung. Die euklidische Metrik; Dritte Untersuchung. Entwurf einer Theorie der n-dimensionalen Metrik), Mathematische Annalen 100 (1928), 75–163.

(With W. Hurewicz) Dimension und Zusammenhangsstufe, Mathematische Annalen 100 (1928), 618–633.

Bemerkungen zu Grundlagenfragen. I: Über Verzweigungsmengen, Jahresbericht der Deutschen Mathematiker-Vereinigung 37 (1928), 213–226.

Bemerkungen zu Grundlagenfragen. II: Die mengentheoretischen Paradoxien, Jahresbericht der Deutschen Mathematiker-Vereinigung 37 (1928), 298–302.

Bemerkungen zu Grundlagenfragen. III: Über Potenzmengen, Jahresbericht der Deutschen Mathematiker-Vereinigung 37 (1928), 303–308.

Bemerkungen zu Grundlagenfragen. IV: Axiomatik der endlichen Mengen und der elementargeometrischen Verknüpfungsbeziehungen, Jahresbericht der Deutschen Mathematiker-Vereinigung 37 (1928), 309–325.

Ein Theorem der Topologie, Akademie der Wissenschaften zu Wien, Anzeiger 65 (1928), 4 and Ergebnisse eines mathematischen Kolloquiums 1 (1931), 2–3.

Dimensionstheoretische Notizen, Akademie der Wissenschaften zu Wien, Anzeiger 65 (1928), 5–8, 11–12 and Ergebnisse eines mathematischen Kolloquiums 1 (1931), 6–8.

Konvexitätstheoretische Notizen: Über konvexe Hüllen; Über Vollkonvexität; Über Verallgemeinerungen des Zwischenbegriffes; Über den Begriff der Konkavität, Akademie der Wissenschaften zu Wien, Anzeiger 65 (1928), 154–156 and Ergebnisse eines mathematischen Kolloquiums 1 (1931), 23–26.

Die Metrik des Hilbertschen Raumes, Akademie der Wissenschaften zu Wien, Anzeiger 65 (1928), 159–160 and Ergebnisse eines mathematischen Kolloquiums 1 (1931), 26–27.

Ein Theorem über die Bogenlänge, Akademie der Wissenschaften zu Wien, Anzeiger 65 (1928), 264–266.

Der allgemeine Trennungssatz, Akademie der Wissenschaften zu Wien, Anzeiger 65 (1928), 266–268 and Ergebnisse eines mathematischen Kolloquiums 1 (1931), 103–104.

Die Halbstetigkeit der Bogenlänge, Akademie der Wissenschaften zu Wien, Anzeiger 65 (1928), 278–281.

Dimensionstheorie, B. G. Teubner, Leipzig-Berlin 1928.

Review of A. Fraenkel, Einleitung in die Mengenlehre, Monatshefte für Mathematik und Physik 36 (1928), 5–7.

Review of M. Fréchet, Les espaces abstraits, Monatshefte für Mathematik und Physik 36 (1928), 35–37.

Zur Frage nach der Herleitung des Dimensionsbegriffes aus Forderungen, Akademie der Wissenschaften zu Wien, Anzeiger 66 (1929), 87–90.

Allgemeine Räume und Cartesische Räume III. Beweis des Fundamentalsatzes, Proceedings Amsterdam 32 (1929), 330–340.

Dimensionstheoretische Notizen: Ein Zerlegungssatz für rational- und irrational-dimensionale Mengen; Über die nirgends dichten Teilmengen des R_n; Über die Summe regulärer Kurven, Akademie der Wissenschaften zu Wien, Anzeiger 66 (1929), 5–8 and Ergebnisse eines mathematischen Kolloquiums 1 (1931), 6–8.

Über eine neue Definition der Bogenlänge. Eine weitere Verallgemeinerung des Längenbegriffes, Akademie der Wissenschaften zu Wien, Anzeiger 66 (1929), 23–25.

Die allgemeine Kurventheorie, Forschung und Fortschritt 5 (1929 Nr. 20), 232–233.

Zur Dimensions- und Kurventheorie. Unveröffentlichte Aufsätze aus den Jahren 1921–1923, Monatshefte für Mathematik und Physik 36 (1929), 411–432.

Über die Dimension von Punktmengen III. Zur Begründung einer axiomatischen Theorie der Dimension, Monatshefte für Mathematik und Physik 36 (1929), 193–218. Abstract in: Akademie der Wissenschaften zu Wien, Anzeiger 66 (1929), 148–149.

Axiomatische Einführung des Dimensionsbegriffes, Comptes Rendus du Premier Congrès des Mathématiciens des Pays Slaves, Warszawa 1929, 57–65.

(With C. Kuratowski) Remarques sur la théorie axiomatique de la dimension, Monatshefte für Mathematik und Physik 37 (1930), 169–174.

Untersuchungen über allgemeine Metrik Teil IV: Zur Metrik der Kurven, Mathematische Annalen 103 (1930), 466–501.

Zum Entwurf einer neuen Theorie des Maßes, Akademie der Wissenschaften zu Wien, Anzeiger 67 (1930), 12–16 and Ergebnisse eines mathematischen Kolloquiums 2 (1932), 6–10.

Bemerkungen über dimensionelle Feinstruktur und Produktsatz, Prace matematyczno-fizyczne. 37 (1930), 77–90.

Otto Schreier, Nachruf, Monatshefte für Mathematik und Physik 37 (1930), 1–6.

Eine dimensionstheoretische Bemerkung von O. Schreier, Monatshefte für Mathematik und Physik 37 (1930), 7–12. Abstract in: Ergebnisse eines mathematischen Kolloquiums 2 (1932), 1.

Antwort auf eine Note von Brouwer, Monatshefte für Mathematik und Physik 37 (1930), 175–182.

Über einen Abstandsbegriff in Gruppen, Akademie der Wissenschaften zu Wien, Anzeiger 67 (1930), 70–74.

(With G. Nöbeling) Über den n-Beinsatz in lokal-zusammenhängenden Kontinua, Akademie der Wissenschaften zu Wien, Anzeiger 67 (1930), 86–88.

Einführung des Komplexes in die allgemeine Metrik und metrische Untersuchungen in abstrakten Gruppen, Akademie der Wissenschaften zu Wien, Anzeiger 67 (1930), 39–44 and Ergebnisse eines mathematischen Kolloquiums 2 (1932), 34–38.

Über plättbare Dreiergraphen und Potenzen nichtplättbarer Graphen, Akademie der Wissenschaften zu Wien, Anzeiger 67 (1930), 85–86 and Ergebnisse eines mathematischen Kolloquiums 2 (1932), 30–31.

Über die sogenannte Konstruktivität bei arithmetischen Definitionen, Akademie der Wissenschaften zu Wien, Anzeiger 67 (1930), 257–258.

Der Intuitionismus, Blätter für Deutsche Philosophie 4 (1930), 311–325.

Beiträge zur Gruppentheorie I. Über eine Gruppenmetrik, Mathematische Zeitschrift 33 (1931), 396–418.

Some Applications of Point Set Methods, Annals of Mathematics, II. Ser. 32 (1931), 739–760.

New Foundations of Euclidian Geometry, American Journal of Mathematics 53 (1931), 721–745.

Bericht über metrische Geometrie, Jahresbericht der Deutschen Mathematiker-Vereinigung 40 (1931), 201–219.

Remarks Concerning the Paper of W. L. Ayres on the Regular Points of a Continuum, Transactions of the American Mathematical Society 33 (1931), 663–667.

Über den Konstruktivitätsbegriff. Zweite Mitteilung, Akademie der Wissenschaften zu Wien, Anzeiger 68 (1931), 7–9.

Ein Problem von Blaschke, Akademie der Wissenschaften zu Wien, Anzeiger 68 (1931), 75–76.

Eine neue Kennzeichnung der Geraden, Akademie der Wissenschaften zu Wien, Anzeiger 68 (1931), 103–105.

Bericht über ein mathematisches Kolloquium 1929/30, Monatshefte für Mathematik und Physik 38 (1931), 17–38.

Eine neue Definition der Bogenlänge, Ergebnisse eines mathematischen Kolloquiums 2 (1932), 11–12.

Das Botenproblem, Ergebnisse eines mathematischen Kolloquiums 2 (1932), 12.

Bemerkungen von Mazurkiewicz und Tarski über die Axiomatik des Dimensionsbegriffes, Ergebnisse eines mathematischen Kolloquiums 2 (1932), 16.

Über Mengensysteme, Ergebnisse eines mathematischen Kolloquiums 2 (1932), 16–17.

Probleme der allgemeinen metrischen Geometrie, Ergebnisse eines mathematischen Kolloquiums 2 (1932), 20–22.

Bericht über die mengentheoretischen Überdeckungssätze, Ergebnisse eines mathematischen Kolloquiums 2 (1932), 23–27.

Eine elementare Bemerkung über die Struktur logischer Formeln, Ergebnisse eines mathematischen Kolloquiums 3 (1932), 22–23.

(With G. Nöbeling) Kurventheorie, Leipzig 1932, reprint, Bronx N. Y. 1967.

Neuere Methoden und Probleme der Geometrie, Verhandlungen des Internationalen Mathematiker-Kongresses Zürich 1932, 1, 310–323.

Über die Hinweise auf Brouwer in Urysohns Mémoire, Selbstverlag, Wien 1932.

Review of Jahrbuch über die Fortschritte der Mathematik, Sonderhefte "Geschichte, Philosophie, Pädagogik, Mengenlehre", "Analysis", "Arithmetik und Algebra" der Jahrgänge 1925, 1928, 1929, Monatshefte für Mathematik und Physik 39 (1932), 6–8.

Eine Zuschrift von Karl Menger an Hans Hahn betreffend Mengers Antwort auf eine Note von Brouwer in Bd. 37 der Monatshefte, Monatshefte für Mathematik und Physik 40 (1933), 233.

Über die lokale Dimension von Mengensummen, Ergebnisse eines mathematischen Kolloquiums 4 (1933), 1–2.

Eine NZ-Kurve, Ergebnisse eines mathematischen Kolloquiums 4 (1933), 7.

Über eine Limesklasse, Ergebnisse eines mathematischen Kolloquiums 4 (1933), 9.

Diskussion über die Verzweigungsordnung von Flächenpunkten, Ergebnisse eines mathematischen Kolloquiums 4 (1933), 11–12.

Zur mengentheoretischen Behandlung des Tangentenbegriffes und verwandter Begriffe, Ergebnisse eines mathematischen Kolloquiums 4 (1933), 23.

Ein Überdeckungssatz für F_σ, Ergebnisse eines mathematischen Kolloquiums 4 (1933), 40–41.

Eine neue Kennzeichnung der Geraden, Ergebnisse eines mathematischen Kolloquiums 4 (1933), 41–43.

Foreword to: Krise und Neuaufbau in den exakten Wissenschaften, 5 Wiener Vorträge, Leipzig und Wien 1933, i–ii.

Die neue Logik, in: Krise und Neuaufbau in den exakten Wissenschaften, 5 Wiener Vorträge, Leipzig und Wien 1933, 93–122.

Über den imaginären euklidischen Raum, The Tôhoku Mathematical Journal 37 (1933), 475–478.

Zur Begründung einer Theorie der Bogenlänge in Gruppen, Ergebnisse eines mathematischen Kolloquiums 5 (1933), 1–6.

Eine Bemerkung über die Potenzen schwach eindimensionaler Mengen, Ergebnisse eines mathematischen Kolloquiums 5 (1933), 9–10.

Rein imaginäre Räume, indefinite Metriken und verwandte Probleme, Ergebnisse eines mathematischen Kolloquiums 5 (1933), 16–17.

(With Kurt Gödel and Abraham Wald) Diskussion über koordinatenlose Differentialgeometrie, Ergebnisse eines mathematischen Kolloquiums 5 (1933), 25–26.

Neuer Aufbau der Vektoralgebra, Ergebnisse eines mathematischen Kolloquiums 5 (1933), 27–29.

Ist die Quadratur des Kreises lösbar?, in: Alte Probleme – neue Lösungen in den exakten Wissenschaften, 5 Wiener Vorträge, 2. Zyklus, Leipzig und Wien 1934, 1–28.

Postscript to: Alte Probleme – neue Lösungen in den exakten Wissenschaften, 117–122.

Bericht über neueste Ergebnisse der metrischen Geometrie, Comptes Rendus du Deuxième Congrès des Mathématiciens des Pays Slaves, Praha 1934, 116–117.

Das Unsicherheitsmoment in der Wertlehre. Betrachtungen in Anschluß an das sogenannte Petersburger Spiel, Zeitschrift für Nationalökonomie 5 (1934), 459–485.

Bemerkungen zu: Das Unsicherheitsmoment in der Wertlehre. Betrachtungen in Anschluß an das sogenannte Petersburger Spiel, Zeitschrift für Nationalökonomie 6 (1934), 283–285.

Moral, Wille und Weltgestaltung, Grundlegung zur Logik der Sitten, Wien 1934.

Eine Bemerkung über Längenmengen, Ergebnisse eines mathematischen Kolloquiums 6 (1935), 1–2.

Remarks in the Discussion of K. Schlesinger's Modification of the Walras-Cassel Production Equations and A. Wald's Solution of the Modified Equations, Ergebnisse eines mathematischen Kolloquiums 6 (1935), 18–20.

Ein Satz über endliche Mengen mit Anwendungen auf die formale Ethik, Ergebnisse eines mathematischen Kolloquiums 6 (1935), 23–26.

Bernoullische Wertlehre und Petersburger Spiel, Ergebnisse eines mathematischen Kolloquiums 6 (1935), 26–27.

Hans Hahn, Ergebnisse eines mathematischen Kolloquiums 6 (1935), 40–44.

Hans Hahn, Fundamenta Mathematicae 24 (1935), 317–320.

Metrische Geometrie und Variationsrechnung, Fundamenta Mathematicae 25 (1935), 441–458.

Sull Indirizzo di Idee e sulle Tendenze Principali del Colloquio Matematico di Vienna, Annali di Pisa 4 (1935), 1–13. An English translation is contained in the reissue of the "Ergebnisse eines mathematischen Kolloquiums", ed. by E. Dierker and K. Sigmund, Springer, Wien 1998.

Sur un théorème général du calcul des variations, Comptes Rendus Acad. Paris 201 (1935), 705–707.

Bericht über neueste Ergebnisse der metrischen Geometrie, Časopis 64 (1935), 116–117.

Algebra der Geometrie. Zur Axiomatik der projektiven Verknüpfungsbeziehungen, Ergebnisse eines mathematischen Kolloquiums 7 (1936), 11–12.

Über die φ-Metrik und φ-Bogenlänge, Ergebnisse eines mathematischen Kolloquiums 7 (1936), 13–14.

Programmatisches zur Anwendung der metrischen Geometrie auf die Variationsrechnung, Ergebnisse eines mathematischen Kolloquiums 7 (1936), 40–51.

Calcul des variations dans les espaces distanciés généraux, Comptes Rendus Acad. Paris 202 (1936), 1007–1009.

Courbes minimisantes non-rectifiables et champs généraux des courbes admissibles dans le calcul des variations, Comptes Rendus Acad. Paris 202 (1936), 1648–1650.

Metric Methods in Calculus of Variations, Congrés International des Mathématiciens, Oslo 1936, II, 45–46.

New Ways in Differential Geometry, Congrès International des Mathématiciens, Oslo 1936, II, 171–173.

Bemerkungen zu den Ertragsgesetzen, Zeitschrift für Nationalökonomie 7 (1936), 25–56 and 388–397.

Einige neuere Fortschritte in der exakten Behandlung sozialwissenschaftlicher Probleme, in: Neuere Fortschritte in den exakten Wissenschaften, 5 Wiener Vorträge, 3. Zyklus, Leipzig und Wien 1936, 103–132.

Foreword to: Friedrich Waismann; Einführung in das mathematische Denken. Die Begriffsbildung der modernen Mathematik, Wien 1936, iii–iv (English Translation: Introduction to Mathematical Thinking by T. J. Benac, New York 1951).

(With F. Alt and O. Schreiber) New Foundations of Projective and Affine Geometry. Algebra of Geometry, Annals of Mathematics, II. Ser. 37 (1936), 456–482.

Metric Methods in Calculus of Variations, Proc. Nat. Acad. Sci. 23 (1937), 244–250.

What Is the Calculus of Variations and What Are Its Applications?, Scientific Monthly 44 (1937), 250–253. Reprinted in: The World of Mathematics 2, ed. by J. R. Newman, New York 1956, 886–890.

La géométrie des distances et ses relations avec les autres branches des mathématiques, L'Enseignement mathématique 35 (1937), 348–372.

An Exact Theory of Social Relations and Groups, in: Reports of the Third Annual Research Conference on Economics and Statistics, Cowles Commission for Research in Economics, Colorado Springs 1937, 71–73.

The New Logic, Philosophy of Science 4 (1937), 299–336.

Die metrische Methode in der Variationsrechnung, Ergebnisse eines mathematischen Kolloquiums 8 (1937), 1–32.

(With F. Alt) A Note on a Previous Paper "New Foundations of Projective and Affine Geometry", Annals of Mathematics, II. Ser. 38 (1937), 450.

Non-Euclidean Geometry of Joining and Intersecting, Bulletin of the American Mathematical Society 44 (1938), 821–824.

Axiomatique simplifiée de l'algèbre de la géométrie projective, Comptes Rendus Acad. Paris 206 (1938), 306–308.

A Foundation of Projective Geometry, Proceedings Indiana Academy of Science 47 (1938), 189–191.

A Symposium on the Algebra of Geometry and Related Subjects, Science 87 (1938), 324.

An Abstract Form of the Covering Theorems of Topology, Annals of Mathematics (2) 39 (1938), 794–803.

A New Foundation of Non-Euclidean, Affine, Real Projective and Euclidean Geometry, Proc. Nat. Acad. Sci. 24 (1938), 486–490.

Nouvelle base pour le développement de la géométrie de Bolyai et Lobatchefski, Comptes Rendus Acad. Paris 206 (1938), 458–460.

Une forme abstraite du théorème de Borel-Lebesgue généralisé, Comptes Rendus Acad. Paris 206 (1938), 563–565.

Introduction to: L. M. Blumenthal; Distance Geometries, University of Missouri Studies 13 (1938).

An Exact Theory of Social Groups and Relations, American Journal of Sociology 43 (1938), 790–798.

(With A. Milgram) On Linear Sets in Metric Spaces, Reports of Math. Colloq. Notre Dame, Indiana 1 (1939), 16–17.

A Logic of the Doubtful. On Optative and Imperative Logic, Reports of Math. Colloq. Notre Dame, Indiana 1 (1939), 53–64.

On Necessary and on Sufficient Conditions in Elementary Mathematics, School, Science and Mathematics (1939), 631–642.

A Theory of Length and Its Applications to the Calculus of Variations, Proc. Nat. Acad. Sci. 25 (1939), 474–478.

On Cauchy's Integral Theorem in the Real Plane, Proc. Nat. Acad. Sci. 25 (1939), 621–625.

Analysis and Metric Geometry. Line Integrals, Their Semicontinuity Properties and Their Independence of the Path, The Rice Institute Pamphlet 27 (1940), 1–40.

On Algebra of Geometry and Recent Progress in Non-Euclidean Geometry, The Rice Institute Pamphlet 27 (1940), 41–79.

Topology Without Points, The Rice Institute Pamphlet 27 (1940), 79–107.

On Shortest Polygonal Approximations to a Curve, Reports of Math. Colloq. Notre Dame, Indiana 2 (1940), 33–38.

Redundancies in the Classical Treatment of the Cauchy-Riemann Conditions, Reports of Math. Colloq. Notre Dame, Indiana 2 (1940), 45–48.

On Green's Formula, Proc. Nat. Acad. Sci. 26 (1940), 660–664.

Statistical Metrics, Proc. Nat. Acad. Sci. 28 (1942), 535–537.

Projective Generalization of Metric Geometry (abstract), Bulletin of the American Mathematical Society 48 (1942), 833.

Review of W. Hurewicz and H. Wallman, Dimension Theory, Science 95 (1942), 554–556.

What is Dimension?, The American Mathematical Monthly 50 (1943), 2–7.

Differential Equations, in: Practical Mathematics, ed. by R. S. Kimball, New York 1943, Issue 9, vol. 8, 513–540.

Algebra of Analysis, Notre Dame Mathematical Lectures, vol. 3, 1944.

Tri-Operational Algebra, Reports of Math. Colloq. Notre Dame, Indiana 5/6 (1944), 3–10.

Projective Generalizations of Metric Geometry, Reports of Math. Colloq. Notre Dame, Indiana 5/6 (1944), 60–76.

On the Teaching of Differential Equations, The American Mathematical Monthly 51 (1944), 392–395.

Why Study Mathematics?, Al-Geo-Trig (Central Catholic High School, Toledo, Ohio) 7 (1944), 5–6.

On the Relation Between Calculus of Probability and Statistics, Notre Dame Mathematical Lectures, vol. 4, 1944, 44–53.

Définition intrinsèque de la notion de chemin, Comptes Rendus Acad. Paris 221 (1945), 739–741.

Methods of Presenting e and π, The American Mathematical Monthly 52 (1945), 28–33.

New Projective Definitions of the Concepts of Hyperbolic Geometry, Reports of Math. Colloq. Notre Dame, Indiana 7 (1946), 20–28.

General Algebra of Analysis, Reports of Math. Colloq. Notre Dame, Indiana 7 (1946), 46–60.

Analysis without Variables, Journal of Symbolic Logic 11 (1946), 30–31.

The Topology of the Triangle Inequality, Revista de Ciencias, Lima 50 (1948), 155–165.

Stieltjes Integrals Considered as Lengths, Annales Société Polonaise Math. 21 (1948), 173–175.

Fundamental and Applied Research in Geometry, Illinois Tech Engineer 13 (1948), 13–14 and 38–46.

Independent Self-Dual Postulates in Projective Geometry, Reports of Math. Colloq. Notre Dame, Indiana 8 (1948), 81–87.

(With Y. R. Simon) Aristotelian Demonstration and Postulational Method, The Modern Schoolman 25 (1948), 183–192.

What Paths Have Length?, Fundamenta Mathematicae 36 (1949), 109–118.

Self-Dual Fragments of the Ordinary Plane, The American Mathematical Monthly 56 (1949), 545–546.

Modern Geometry and the Theory of Relativity, in: Albert Einstein, Philosopher–Scientist, The Library of Living Philosophers, vol. VII, ed. by P. A. Schilpp, Evanston, Illinois 1949, 459–474.

Generalized Vector Spaces. I. The Structure of Finite-Dimensional Spaces, Canadian Journal of Mathematics 1 (1949), 94–104.

La géométrie axiomatique de l'espace projectif, Comptes Rendus Acad. Paris 228 (1949), 1273–1274.

Are Variables Necessary in Calculus?, The American Mathematical Monthly 56 (1949), 609–620.

The Projective Space, Duke Mathematical Journal 17 (1950), 1–14.

The Mathematics of Elementary Thermodynamics, American Journal of Physics 18 (1950), 89–103.

Probabilistic Theories of Relations, Proc. Nat. Acad. Sci. 37 (1951), 178–180.

Probabilistic Geometry, Proc. Nat. Acad. Sci. 37 (1951), 226–229.

Ensembles flous et fonctions aléatoires, Comptes Rendus Acad. Paris 232 (1951), 2001–2003.

Espaces vectoriels généraux, topologies triangulaires, transformations linéaires généralisées, Comptes Rendus Acad. Paris 232 (1951), 2176–2178.

A Topological Characterization of the Length of Paths, Proc. Nat. Acad. Sci. 38 (1952), 66–69.

Une théorie axiomatique générale des déterminants, Comptes Rendus Acad. Paris 234 (1952), 1941–1943.

The Formative Years of Abraham Wald and His Work in Geometry, The Annals of Mathematical Statistics 23 (1952), 14–20.

You Will Like Geometry, A Guide Book for the Illinois Institute of Technology Geometry Exhibition at the Museum of Science and Industry, Chicago 1952.

Calculus. A Modern Approach, The Bookstore, Illinois Institute of Technology, Chicago 1953.

On the Design of Grouping Problems and Related Intelligence Tests, Journal of Educational Psychology 44 (1953), 275–287.

The Ideas of Variable and Function, Proc. Nat. Acad. Sci. 39 (1953), 956–961.

Variables de diverses natures, Bulletin des Sciences Mathématiques (2) 78 (1954), 229–234.

Is Calculus a Perfect Tool?, Journal of Engineering Education 45 (1954), 261–264.

The Logic of the Laws of Return. A Study in Meta-Economics, in: Economic Activity Analysis, ed. by O. Morgenstern, New York 1954, 419–482.

Tossing a Coin, The American Mathematical Monthly 61 (1954), 634–636.

Géométrie générale, Mémorial des Sciences Mathématiques 124 (1954).

On Variables in Mathematics and in Natural Science, British Journal for the Philosophy of Science 5 (1954), 134–142.

A Simple Definition of Analytic Functions and General Multi-Functions, Proc. Nat. Acad. Sci. 40 (1954), 819–821.

Remarks to L. Sommer's Translation of Bernoulli's "Exposition of a New Theory on the Measurement of Risk", Econometrica 22 (1954), 28, 31–32 and 34.

The Behavior of a Complex Function at Infinity, Proc. Nat. Acad. Sci. 41 (1955), 512–513.

(With S. S. Shü) Generalized Derivates and Expansions, Proc. Nat. Acad. Sci. 41 (1955), 591–595.

Calculus. A Modern Approach, Ginn and Company, Boston 1955.

What Are x and y?, The Mathematical Gazette 40 (1956), 246–255.

Random Variables from the Point of View of a General Theory of Variables, Proceedings of the Third Berkeley Symposium on Mathematical Statistics and Probability (1954–1955), eds. L. M. Le Cam and J. Neyman, University of California Press, Berkeley 1956 vol. 2, 215–229.

(With H. J. Curtis) On the Formulation of Certain Arithmetical Questions, The Mathematics Teacher 49 (1956), 528–530.

Calculus 1950 – Geometry 1880, I, Scripta mathematica 22 (1956), 89–96.

Calculus 1950 – Geometry 1880, II, Scripta mathematica 22 (1956), 203–206.

What Are Variables and Constants?, Science 123 (1956), 547–548.

Why Johnny Hates Math, The Mathematics Teacher 49 (1956), 578–584.

The Basic Concepts of Mathematics. A Companion to Current Textbooks on Algebra and Analytic Geometry. Part I. Algebra, Illinois Institute of Technology, Chicago 1957.

Multiderivatives and Multi-Integrals, The American Mathematical Monthly 64, No 8, Part II (1957), 58–70.

Rates of Change and Derivatives, Fundamenta Mathematicae 46 (1958), 89–102.

New Approach to Teaching Intermediate Mathematics, Science 127 (1958), 1320–1323.

Is w a Function of u?, Colloquium Mathematicum 6 (1958), 41–47.

Optimal Differences in Computing Probable Derivatives, Journal de Mathématiques pures et appliquées, Paris (9) 38 (1959), 245–252.

An Axiomatic Theory of Functions and Fluents. The Axiomatic Method. With Special Reference to Geometry and Physics, Proceedings of an International Symposium held at the Univ. of California, Berkeley 1959, eds. L. Henkins, P. Suppes and A. Tarski, Studies in Logic and the Foundations of Mathematics, North-Holland Publishing Co., Amsterdam 1959, 454–473.

Gulliver in the Land Without One, Two, Three, The Mathematical Gazette 43 (1959), 241–250.

(With B. Schweizer und A. Sklar) On Probabilistic Metrics and Numerical Metrics with Probability 1 (Russian summary), Czechoslovak Mathematical Journal 9 (84) (1959), 459–466.

Mensuration and Other Mathematical Connections of Observable Material, in: Measurement, Definitions and Theories, eds. C. W. Churchman and P. Ratoosh, New York 1959, 97–128.

Gulliver's Return to the Land Without One, Two, Three, The American Mathematical Monthly 67 (1960), 641–648.

Introduction to: Ernst Mach; The Science of Mechanics, 6th American Edition, Open Court, La Salle, Illinois 1960, iii–xxi.

Gulliver in Applyland, Eureka 23 (1960), 5–8.

A Counterpart of Occam's Razor in Pure and Applied Mathematics; Ontological Uses, Synthese 12 (1960), 415–428.

A Counterpart of Occam's Razor in Pure and Applied Mathematics; Semantic Uses, Synthese 13 (1961), 331–349.

Frammenti piani autoduali e relative sostituzioni, Rendiconti Accad. Nazionale di Lincei 30 (1961), 713–717.

The Algebra of Functions: Past, Present, Future, Rendiconti di Matematica 20 (1961), 409–430.

(With F. Henmüller) What is Length?, Philosophy of Science 28 (1961), 172–177.

Variables, Constants, Fluents, in: Current Issues in the Philosophy of Science, eds. H. Feigl and G. Maxwell, New York 1961, 304–313 and 316–318.

Eine Algebra der Funktionen, Nachrichten der Österreichischen Mathematischen Gesellschaft, Wien 68/69 (1961), 99–100.

A Group in the Substitutive Algebra of the Calculus of Propositions, Archiv der Mathematik 13 (1962), 471–478.

On Compositive Functions of Matrices, Annali di Matematica pura et applicata (4) 58 (1962), 69–84.

Function Algebra and Propositional Calculus, In: Self-Organizing Systems, eds. M. C. Yovits, G. T. Jakobi and G. D. Goldstein, Washington D. C. (Spartan books) 1962, 525–532.

Review of L. Félix, The Modern Aspect of Mathematics, Philosophy of Science 29 (1962), 95–96.

(With F. Kozin) A Self-Dual Theory of Real Determinants, Publ. Math. Debrecen 10 (1963), 123–127.

(With M. Schulz) Postulates for the Substitutive Algebra of the 2-Place Functors in the 2-Valued Calculus of Propositions, Journal of Formal Logic 4 (1963), 188–192.

Calculus of Variations, in: Harper Encyclopedia of Science I, New York and Evanston 1963, 195–196.

On Substitutive Algebra and Its Syntax, Zeitschrift für mathematische Logik und Grundlagen der Mathematik 10 (1964), 81–104.

Superassociative Systems and Logical Functors, Mathematische Annalen 157 (1964), 278–295.

A Characterization of Weierstrass Analytic Functions, Proc. Nat. Acad. Sci. 54 (1965), 1025–1026.

Analytische Funktionen, Wissenschaftliche Abhandlungen der Arbeitsgemeinschaft für Forschung des Landes Nordrhein-Westfalen vol. 33, Köln 1965, 609–612.

Une caractérisation des fonctions analytiques, Comptes Rendus Acad. Paris 261 (1965), 4968–4969.

Weierstrass Analytic Functions, Mathematische Annalen 167 (1966), 177–194.

(With H. I. Whitlock) Two Theorems on the Generation of System of Functions, Fundamenta Mathematicae 58 (1966), 229–240.

Calculus. The Elements. Introduction into Pre-Limit Calculus for Everyone, Chapter I, Illinois Institute of Technology, Chicago 1966.

Calculus. The Elements. Introduction into Pre-Limit Calculus for Everyone, Chapter II, Illinois Institute of Technology, Chicago 1967.

The Role of Uncertainty in Economics, in: Essays in Mathematical Economics in Honor of O. Morgenstern, ed. by M. Shubik, Princeton 1967, 211–231.

General Algebraic Equations and Functional Equations, Aequationes Mathematicae 1 (1968), 281.

(With L. M. Blumenthal) Introduction; Projective and Related Structures (Part 2) and Curve Theory (Part 4), in: Studies in Geometry, San Francisco 1970 iv–viii, 135–223 and 391–506.

Mathematical Implications of Mach's Ideas: Positivistic Geometry, The Clarification of Functional Connections, in: Ernst Mach, Physicist and Philosopher, Boston Studies in the Philosophy of Science VI, eds. R. S. Cohen and R. J. Seeger, D. Reidel, Dordrecht 1970, 107–125.

The New Foundation of Hyperbolic Geometry, in: A Spectrum of Mathematics (Essays presented to H. G. Forder), ed. by J. C. Butcher, Auckland University Press 1971, 86–97.

The Geometry Relevant to Modern Education, in: Educational Studies in Mathematics IV, ed. by H. Freudenthal, Dordrecht 1971, 1–17; reprinted in: The Teaching of Geometry at the Pre-College Level, ed. by H. G. Steiner, Dordrecht 1971, 225–241.

Bericht über metrische Geometrie in: Geometrie, Wege der Forschung 177, ed. by K. Strubecker, Darmstadt 1972, 371–395. (Reprint of the 1931 article, with an appendix written in 1970.)

Österreichischer Marginalismus und mathematische Ökonomie, Zeitschrift für Nationalökonomie 32 (1972), 19–28.

Austrian Marginalism and Mathematical Economics, in: Carl Menger and the Austrian School of Economics, eds. R. J. Hicks and W. Weber, Oxford 1973, 38–60.

Morality, Decision and Social Organization. Toward a Logic of Ethics, Vienna Circle Collection vol. 6, D. Reidel, Dordrecht-Boston 1974.

Wittgenstein betreffende Seiten aus einem Buch über den Wiener Kreis, in Wittgenstein, der Wiener Kreis und der kritische Rationalismus, Akten des 3. Internationalen Wittgenstein Symposiums, Kirchberg am Wechsel (Österreich) 1978.

Selected Papers in Logic and Foundations, Didactics, Economics, Vienna Circle Collection vol. 10, D. Reidel Publishing Co., Dordrecht-Boston-London 1979.

Language and Mathematics. Language, Logic and Philosophy, Proc. Fourth International Wittgenstein-Symposium, Kirchberg am Wechsel 1979, Wien 1980, 21–26.

Introduction to: Hans Hahn; Empiricism, Logic and Mathematics. Vienna Circle Collection vol. 13, D. Reidel Publishing Co., Dordrecht-Boston-London 1980.

Some Thoughts About the History and the Teaching of Mathematics, in: Proceedings of the Symposium on the Use of History in the Teaching of Mathematics, Valparasio University, Valparaiso, Indiana, 1980.

On the Origin of the n-Arc Theorem, Journal of Graph Theory 5 (1981), 341–350.

Memories of Moritz Schlick, in: Rationality and Science, ed. by E. T. Gadol, Springer, Wien 1982, 83–103.

On Social Groups and Relations, Mathematical Social Sciences 6 (1983), 13–25.

Reminiscences of the Vienna Circle and the Mathematical Colloquium, Vienna Circle Collection, vol. 20, eds. L. Golland, B. McGuinness and A. Sklar, Kluwer 1994.

Ergebnisse eines mathematischen Kolloquiums, reissued and ed. by E. Dierker and K. Sigmund, Springer, Wien 1998.

SpringerMathematik

Karl Menger

Ergebnisse eines
Mathematischen Kolloquiums

Herausgegeben von / Edited by Egbert Dierker, Karl Sigmund

Mit Beiträgen von / With contributions by
J. W. Dawson jr., R. Engelking, W. Hildenbrand.
Geleitwort von / Foreword by G. Debreu.
Nachwort von / Afterword by F. Alt

1998. IX, 470 Seiten.
Text: deutsch/englisch (großteils deutsch)
Gebunden **EUR 110,–**, sFr 166,50
ISBN 3-211-83104-5

Die von Karl Menger und seinen Mitarbeitern (darunter Kurt Gödel)
herausgegebenen „Ergebnisse eines Mathematischen Kolloquiums"
zählen zu den wichtigsten Quellenwerken der Wissenschafts- und
Geistesgeschichte der Zwischenkriegszeit, mit bahnbrechenden Beiträgen von Menger, Gödel, Tarski, Wald, John von Neumann und vielen anderen. In diesem Band liegt der Inhalt erstmals gesammelt vor.
Der Nobelpreisträger Gerard Debreu schrieb die Einleitung, die
Kommentare wurden vom Logiker und Gödel-Biographen John
Dawson jr., dem Topologen Ryszard Engelking und dem Wirtschafts-
theoretiker Werner Hildenbrand verfasst.
Außerdem enthält der Band einen biographischen Aufsatz über Karl
Menger sowie einen von Menger verfassten Überblick über die
wichtigsten topologischen und geometrischen Arbeiten des Kollo-
quiums.

Springer Wien New York

A-1201 Wien, Sachsenplatz 4–6, P.O. Box 89, Fax +43.1.330 24 26, e-mail: books@springer.at, Internet: **www.springer.at**
D-69126 Heidelberg, Haberstraße 7, Fax +49.6221.345-229, e-mail: orders@springer.de
USA, Secaucus, NJ 07096-2485, P.O. Box 2485, Fax +1.201.348-4505, e-mail: orders@springer-ny.com
Eastern Book Service, Japan, Tokyo 113, 3–13, Hongo 3-chome, Bunkyo-ku, Fax +81.3.38 18 06 64, e-mail: orders@svt-ebs.co.jp

SpringerMathematik

Hans Hahn

Gesammelte Abhandlungen / Collected Works

Herausgegeben von Leopold Schmetterer, Karl Sigmund

Band 1 / Volume 1
Mit einem Geleitwort von / With a Foreword by Karl Popper.
1995. XII, 511 Seiten.
ISBN 3-211-82682-3

Band 2 / Volume 2	**Band 3 / Volume 3**
1996. XIII, 545 Seiten.	1997. XIII, 581 Seiten.
ISBN 3-211-82750-1	ISBN 3-211-82781-1

Text: deutsch/englisch
Jeder Band: Gebunden **EUR 110,–**, sFr 166,50
Vorzugspreis bei Abnahme aller 3 Bände:
EUR 88,–, sFr 136,50 (ca. 20 % reduziert)

Hans Hahn (1879–1934) war einer der bedeutendsten Mathematiker des 20. Jahrhunderts. Er hat zahlreiche Gebiete der Analysis entscheidend geprägt und zählt zu den Vätern der Funktionalanalysis. Aber auch die Maßtheorie und die harmonische Analyse wurden von ihm nachhaltig angeregt; die allgemeine Topologie verdankt ihm wesentliche Impulse.

„... Den Herausgebern der gesammelten Werke des bedeutenden Mathematikers und Gründers des Wiener Kreises, Hans Hahn, ist man zu großem Dank verpflichtet für die Realisierung eines in jeder Hinsicht hervorragend (auch drucktechnisch) gelungenen, dreibändigen Werkes mit interessanten Kommentaren zu den einzelnen Teilgebieten ... Ein wichtiges, zeitloses Werk ..."

Monatshefte für Mathematik

SpringerWienNewYork

A-1201 Wien, Sachsenplatz 4–6, P.O. Box 89, Fax +43.1.330 24 26, e-mail: books@springer.at, Internet: **www.springer.at**
D-69126 Heidelberg, Haberstraße 7, Fax +49.6221.345-229, e-mail: orders@springer.de
USA, Secaucus, NJ 07096-2485, P.O. Box 2485, Fax +1.201.348-4505, e-mail: orders@springer-ny.com
Eastern Book Service, Japan, Tokyo 113, 3–13, Hongo 3-chome, Bunkyo-ku, Fax +81.3.38 18 08 64, e-mail: orders@svt-ebs.co.jp

*Springer-Verlag
and the Environment*

Printed in the United States
By Bookmasters